T0296106

LONDON MATHEMATICAL SOCIETY LECTURE NOTE SERIES

Managing Editor: Professor M. Reid, Mathematics Institute,
University of Warwick, Coventry CV4 7AL, United Kingdom

The titles below are available from booksellers, or from Cambridge University Press at
http://www.cambridge.org/mathematics

299 Kleinian groups and hyperbolic 3-manifolds, Y. KOMORI, V. MARKOVIC & C. SERIES (eds)
300 Introduction to Möbius differential geometry, U. HERTRICH-JEROMIN
301 Stable modules and the D(2)-problem, F.E.A. JOHNSON
302 Discrete and continuous nonlinear Schrödinger systems, M.J. ABLOWITZ, B. PRINARI & A.D. TRUBATCH
303 Number theory and algebraic geometry, M. REID & A. SKOROBOGATOV (eds)
304 Groups St Andrews 2001 in Oxford I, C.M. CAMPBELL, E.F. ROBERTSON & G.C. SMITH (eds)
305 Groups St Andrews 2001 in Oxford II, C.M. CAMPBELL, E.F. ROBERTSON & G.C. SMITH (eds)
306 Geometric mechanics and symmetry, J. MONTALDI & T. RATIU (eds)
307 Surveys in combinatorics 2003, C.D. WENSLEY (ed.)
308 Topology, geometry and quantum field theory, U.L. TILLMANN (ed)
309 Corings and comodules, T. BRZEZINSKI & R. WISBAUER
310 Topics in dynamics and ergodic theory, S. BEZUGLYI & S. KOLYADA (eds)
311 Groups: topological, combinatorial and arithmetic aspects, T.W. MÜLLER (ed)
312 Foundations of computational mathematics, Minneapolis 2002, F. CUCKER *et al* (eds)
313 Transcendental aspects of algebraic cycles, S. MÜLLER-STACH & C. PETERS (eds)
314 Spectral generalizations of line graphs, D. CVETKOVIC, P. ROWLINSON & S. SIMIC
315 Structured ring spectra, A. BAKER & B. RICHTER (eds)
316 Linear logic in computer science, T. EHRHARD, P. RUET, J.-Y. GIRARD & P. SCOTT (eds)
317 Advances in elliptic curve cryptography, I.F. BLAKE, G. SEROUSSI & N.P. SMART (eds)
318 Perturbation of the boundary in boundary-value problems of partial differential equations, D. HENRY
319 Double affine Hecke algebras, I. CHEREDNIK
320 L-functions and Galois representations, D. BURNS, K. BUZZARD & J. NEKOVÁR (eds)
321 Surveys in modern mathematics, V. PRASOLOV & Y. ILYASHENKO (eds)
322 Recent perspectives in random matrix theory and number theory, F. MEZZADRI & N.C. SNAITH (eds)
323 Poisson geometry, deformation quantisation and group representations, S. GUTT *et al* (eds)
324 Singularities and computer algebra, C. LOSSEN & G. PFISTER (eds)
325 Lectures on the Ricci flow, P. TOPPING
326 Modular representations of finite groups of Lie type, J.E. HUMPHREYS
327 Surveys in combinatorics 2005, B.S. WEBB (ed)
328 Fundamentals of hyperbolic manifolds, R. CANARY, D. EPSTEIN & A. MARDEN (eds)
329 Spaces of Kleinian groups, Y. MINSKY, M. SAKUMA & C. SERIES (eds)
330 Noncommutative localization in algebra and topology, A. RANICKI (ed)
331 Foundations of computational mathematics, Santander 2005, L.M PARDO, A. PINKUS, E. SÜLI & M.J. TODD (eds)
332 Handbook of tilting theory, L. ANGELERI HÜGEL, D. HAPPEL & H. KRAUSE (eds)
333 Synthetic differential geometry (2nd Edition), A. KOCK
334 The Navier–Stokes equations, N. RILEY & P. DRAZIN
335 Lectures on the combinatorics of free probability, A. NICA & R. SPEICHER
336 Integral closure of ideals, rings, and modules, I. SWANSON & C. HUNEKE
337 Methods in Banach space theory , J.M.F. CASTILLO & W.B. JOHNSON (eds)
338 Surveys in geometry and number theory, N. YOUNG (ed)
339 Groups St Andrews 2005 I, C.M. CAMPBELL, M.R. QUICK, E.F. ROBERTSON & G.C. SMITH (eds)
340 Groups St Andrews 2005 II, C.M. CAMPBELL, M.R. QUICK, E.F. ROBERTSON & G.C. SMITH (eds)
341 Ranks of elliptic curves and random matrix theory, J.B. CONREY, D.W. FARMER, F. MEZZADRI & N.C. SNAITH (eds)
342 Elliptic cohomology, H.R. MILLER & D.C. RAVENEL (eds)
343 Algebraic cycles and motives I, J. NAGEL & C. PETERS (eds)
344 Algebraic cycles and motives II, J. NAGEL & C. PETERS (eds)
345 Algebraic and analytic geometry, A. NEEMAN
346 Surveys in combinatorics 2007, A. HILTON & J. TALBOT (eds)
347 Surveys in contemporary mathematics, N. YOUNG & Y. CHOI (eds)
348 Transcendental dynamics and complex analysis, P.J. RIPPON & G.M. STALLARD (eds)
349 Model theory with applications to algebra and analysis I, Z. CHATZIDAKIS, D. MACPHERSON, A. PILLAY & A. WILKIE (eds)
350 Model theory with applications to algebra and analysis II, Z. CHATZIDAKIS, D. MACPHERSON, A. PILLAY & A. WILKIE (eds)
351 Finite von Neumann algebras and masas, A.M. SINCLAIR & R.R. SMITH
352 Number theory and polynomials, J. MCKEE & C. SMYTH (eds)
353 Trends in stochastic analysis, J. BLATH, P. MÖRTERS & M. SCHEUTZOW (eds)
354 Groups and analysis, K. TENT (ed)
355 Non-equilibrium statistical mechanics and turbulence, J. CARDY, G. FALKOVICH & K. GAWEDZKI
356 Elliptic curves and big Galois representations, D. DELBOURGO
357 Algebraic theory of differential equations, M.A.H. MACCALLUM & A.V. MIKHAILOV (eds)
358 Geometric and cohomological methods in group theory, M.R. BRIDSON, P.H. KROPHOLLER & I.J. LEARY (eds)
359 Moduli spaces and vector bundles, L. BRAMBILA-PAZ, S.B. BRADLOW, O. GARCÍA-PRADA & S. RAMANAN (eds)
360 Zariski geometries, B. ZILBER

361 Words: Notes on verbal width in groups, D. SEGAL
362 Differential tensor algebras and their module categories, R. BAUTISTA, L. SALMERÓN & R. ZUAZUA
363 Foundations of computational mathematics, Hong Kong 2008, F. CUCKER, A. PINKUS & M.J. TODD (eds)
364 Partial differential equations and fluid mechanics, J.C. ROBINSON & J.L. RODRIGO (eds)
365 Surveys in combinatorics 2009, S. HUCZYNSKA, J.D. MITCHELL & C.M. RONEY-DOUGAL (eds)
366 Highly oscillatory problems, B. ENGQUIST, A. FOKAS, E. HAIRER & A. ISERLES (eds)
367 Random matrices: High dimensional phenomena, G. BLOWER
368 Geometry of Riemann surfaces, F.P. GARDINER, G. GONZÁLEZ-DIEZ & C. KOUROUNIOTIS (eds)
369 Epidemics and rumours in complex networks, M. DRAIEF & L. MASSOULIÉ
370 Theory of p-adic distributions, S. ALBEVERIO, A.YU. KHRENNIKOV & V.M. SHELKOVICH
371 Conformal fractals, F. PRZYTYCKI & M. URBANSKI
372 Moonshine: The first quarter century and beyond, J. LEPOWSKY, J. MCKAY & M.P. TUITE (eds)
373 Smoothness, regularity and complete intersection, J. MAJADAS & A. G. RODICIO
374 Geometric analysis of hyperbolic differential equations: An introduction, S. ALINHAC
375 Triangulated categories, T. HOLM, P. JØRGENSEN & R. ROUQUIER (eds)
376 Permutation patterns, S. LINTON, N. RUŠKUC & V. VATTER (eds)
377 An introduction to Galois cohomology and its applications, G. BERHUY
378 Probability and mathematical genetics, N. H. BINGHAM & C. M. GOLDIE (eds)
379 Finite and algorithmic model theory, J. ESPARZA, C. MICHAUX & C. STEINHORN (eds)
380 Real and complex singularities, M. MANOEL, M.C. ROMERO FUSTER & C.T.C WALL (eds)
381 Symmetries and integrability of difference equations, D. LEVI, P. OLVER, Z. THOMOVA & P. WINTERNITZ (eds)
382 Forcing with random variables and proof complexity, J. KRAJÍCEK
383 Motivic integration and its interactions with model theory and non-Archimedean geometry I, R. CLUCKERS, J. NICAISE & J. SEBAG (eds)
384 Motivic integration and its interactions with model theory and non-Archimedean geometry II, R. CLUCKERS, J. NICAISE & J. SEBAG (eds)
385 Entropy of hidden Markov processes and connections to dynamical systems, B. MARCUS, K. PETERSEN & T. WEISSMAN (eds)
386 Independence-friendly logic, A.L. MANN, G. SANDU & M. SEVENSTER
387 Groups St Andrews 2009 in Bath I, C.M. CAMPBELL *et al* (eds)
388 Groups St Andrews 2009 in Bath II, C.M. CAMPBELL *et al* (eds)
389 Random fields on the sphere, D. MARINUCCI & G. PECCATI
390 Localization in periodic potentials, D.E. PELINOVSKY
391 Fusion systems in algebra and topology, M. ASCHBACHER, R. KESSAR & B. OLIVER
392 Surveys in combinatorics 2011, R. CHAPMAN (ed)
393 Non-abelian fundamental groups and Iwasawa theory, J. COATES *et al* (eds)
394 Variational problems in differential geometry, R. BIELAWSKI, K. HOUSTON & M. SPEIGHT (eds)
395 How groups grow, A. MANN
396 Arithmetic differential operators over the p-adic integers, C.C. RALPH & S.R. SIMANCA
397 Hyperbolic geometry and applications in quantum chaos and cosmology, J. BOLTE & F. STEINER (eds)
398 Mathematical models in contact mechanics, M. SOFONEA & A. MATEI
399 Circuit double cover of graphs, C.-Q. ZHANG
400 Dense sphere packings: a blueprint for formal proofs, T. HALES
401 A double Hall algebra approach to affine quantum Schur–Weyl theory, B. DENG, J. DU & Q. FU
402 Mathematical aspects of fluid mechanics, J.C. ROBINSON, J.L. RODRIGO & W. SADOWSKI (eds)
403 Foundations of computational mathematics, Budapest 2011, F. CUCKER, T. KRICK, A. PINKUS & A. SZANTO (eds)
404 Operator methods for boundary value problems, S. HASSI, H.S.V. DE SNOO & F.H. SZAFRANIEC (eds)
405 Torsors, étale homotopy and applications to rational points, A.N. SKOROBOGATOV (ed)
406 Appalachian set theory, J. CUMMINGS & E. SCHIMMERLING (eds)
407 The maximal subgroups of the low-dimensional finite classical groups, J.N. BRAY, D.F. HOLT & C.M. RONEY-DOUGAL
408 Complexity science: the Warwick master's course, R. BALL, V. KOLOKOLTSOV & R.S. MACKAY (eds)
409 Surveys in combinatorics 2013, S.R. BLACKBURN, S. GERKE & M. WILDON (eds)
410 Representation theory and harmonic analysis of wreath products of finite groups, T. CECCHERINI-SILBERSTEIN, F. SCARABOTTI & F. TOLLI
411 Moduli spaces, L. BRAMBILA-PAZ, O. GARCÍA-PRADA, P. NEWSTEAD & R.P. THOMAS (eds)
412 Automorphisms and equivalence relations in topological dynamics, D.B. ELLIS & R. ELLIS
413 Optimal transportation, Y. OLLIVIER, H. PAJOT & C. VILLANI (eds)
414 Automorphic forms and Galois representations I, F. DIAMOND, P.L. KASSAEI & M. KIM (eds)
415 Automorphic forms and Galois representations II, F. DIAMOND, P.L. KASSAEI & M. KIM (eds)
416 Reversibility in dynamics and group theory, A.G. O'FARRELL & I. SHORT
417 Recent advances in algebraic geometry, C.D. HACON, M. MUSTATA & M. POPA (eds)
418 The Bloch–Kato conjecture for the Riemann zeta function, J. COATES, A. RAGHURAM, A. SAIKIA & R. SUJATHA (eds)
419 The Cauchy problem for non-Lipschitz semi-linear parabolic partial differential equations, J.C. MEYER & D.J. NEEDHAM
420 Arithmetic and geometry, L. DIEULEFAIT *et al* (eds)
421 O-minimality and Diophantine geometry, G.O. JONES & A.J. WILKIE (eds)
422 Groups St Andrews 2013, C.M. CAMPBELL *et al* (eds)
423 Inequalities for graph eigenvalues, Z. STANIĆ
424 Surveys in combinatorics 2015, A. CZUMAJ *et al* (eds)

London Mathematical Society Lecture Note Series: 420

Arithmetic and Geometry

Edited by

LUIS DIEULEFAIT
Universitat de Barcelona

GERD FALTINGS
Max-Planck-Institut für Mathematik (Bonn)

D. R. HEATH-BROWN
University of Oxford

YU. V. MANIN
Max-Planck-Institut für Mathematik (Bonn)

B. Z. MOROZ
Max-Planck-Institut für Mathematik (Bonn)

JEAN-PIERRE WINTENBERGER
Université de Strasbourg

CAMBRIDGE UNIVERSITY PRESS

CAMBRIDGE
UNIVERSITY PRESS

University Printing House, Cambridge CB2 8BS, United Kingdom

Cambridge University Press is part of the University of Cambridge.

It furthers the University's mission by disseminating knowledge in the pursuit of education, learning and research at the highest international levels of excellence.

www.cambridge.org
Information on this title: www.cambridge.org/9781107462540

First published 2015

A catalogue record for this publication is available from the British Library

Library of Congress Cataloguing in Publication data
Arithmetic and geometry / edited by Luis Dieulefait, Universitat de Barcelona [and five others].
pages cm. – (London Mathematical Society lecture note series ; 420)
Papers presented at the trimester on "Arithmetic and Geometry" at the Hausdorff Research Institute for Mathematics (University of Bonn), January–April 2013.
ISBN 978-1-107-46254-0
1. Number theory – Congresses. 2. Algebraic number theory – Congresses. 3. Geometry of numbers – Congresses. I. Dieulefait, Luis, editor.
QA241.A695 2015
510–dc23
2015001829

ISBN 978-1-107-46254-0 Paperback

Contents

Preface

The trimester on "Arithmetic and Geometry" at the Hausdorff Research Institute for Mathematics (University of Bonn) took place in January – April 2013. In the next few pages the reader will find a list of the participants of the trimester, the descriptions of the session on Serre's conjecture, conducted by L.V. Dieulefait and J.-P. Wintenberger, and of the session on counting rational points on algebraic varieties, conducted by D.R. Heath-Brown, the programmes of the workshop on Serre's conjecture and of the final research conference, and a list of the talks given at the HIM during the trimester. The participants were invited to submit their papers for publication in this volume. The papers appearing in the volume have been carefully refereed.

Acknowledgements. We wish to record our gratitude to the Hausdorff Research Institute, and in particular to its director, Professor Dr W. Lück, for the hospitality and financial support we received.

L.V. Dieulefait, G. Faltings, D.R. Heath-Brown, Yu.I. Manin,
B.Z. Moroz, and J.-P. Wintenberger (editors).

Introduction

The main theme of the trimester was the interplay of different methods used in modern number theory. We wish to emphasize the new results and conjectures in arithmetic geometry, having direct bearing on the classical number theoretic problems. Two sessions, on the recently proved Serre's conjecture from 15 January to 14 February (organizers: L. Dieulefait and J.-P. Wintenberger) and on counting rational points on algebraic varieties from 15 March to 14 April (organizer: D.R. Heath-Brown), as well as a couple of shorter workshops, several seminars, and mini-courses were organized. The trimester culminated in a research conference from 15 to 19 April.

The aim of the session "Serre's conjecture" was to report on recent works linked to that conjecture, in particular about Galois representations and automorphic representations. During the weeks starting on 14 January and 21 January, Henri Carayol lectured on his work on the algebraic properties of Griffiths-Schmid varieties. The Griffiths-Schmid varieties are analytic varieties classifying Hodge structures. Studying their algebraic properties might be a step towards constructing Galois representations associated to automorphic representations appearing in the cohomology of these varieties. Our second theme related to the recent work of Michael Harris, Kai-Wen Lan, Richard Taylor and Jack Thorne, who have constructed Galois representations associated to not necessarily self-dual automorphic representations. The proof heavily relies on p-adic properties of automorphic representations.

The aim of the session "counting rational points on algebraic varieties" was to report on recent works on the existence, frequency and distribution of rational points on algebraic varieties. Thus the main themes were local to global principles, Manin's conjecture, developments of the Hardy-Littlewood method and the determinant method.

List of participants

Victor Abrashkin (University of Durham)
Rajender Adibhatla (Universität Regensburg)
Shabnam Akhtari (University of Oregon)
Paloma Bengoechea (College de France)
Denis Benois (Université Bordeaux 1)
Tobias Berger (University of Sheffield)
Yuri Bilu (IMB Université Bordeaux I)
Marco Boggi (University of Los Andes)
Pierre Le Boudec (Institute of Advanced Study)
Régis de la Bretéche (Institut de Mathématiques de Jussieu – Paris Rive Gauche, UMR 7586 Université Paris-Diderot)
Christophe Breuil (Université Paris-Sud)
Tim Browning (University of Bristol)
Jörg Brüdern (Universität Göttingen)
Roman Budylin (Steklov Mathematical Institute)
Alexandru Buium (University of New Mexico)
Alberto Cámara (University of Nottingham)
Henri Carayol (L'Université de Strasbourg et du CNRS)
Magnus Carlson (University of Stockholm)
Tommaso Centeleghe (Universität Heidelberg)
Huan Chen (Ecole normale superieure ENS)
Narasimha Kumar Cheraku (Ruprecht Karls Universität Heidelberg)
Przemyslaw Chojecki (Institut Mathématique de Jussieu)
Laurent Clozel (Université de Paris Sud 11)
Jean-Louis Colliot-Thélène (Université Paris-Sud)
David Mendes da Costa (University of Bristol)
Tuan Ngo Dac (Université de Paris 13)

Ulrich Derenthal (Universität München)
Fred Diamond (King's College London)
Rainer Dietmann (Royal Holloway, University of London)
Luis Dieulefait (Universitat de Barcelona)
Gerd Faltings (Max-Planck-Institut für Mathematik)
Ivan Fesenko (University of Nottingham)
Nuno Freitas (Universitat de Barcelona)
Wojciech Jerzy Gajda (The Adam Mickiewicz University)
David Geraghty (Institute for Advanced Study)
Sergey Gorchinskiy (Steklov Mathematical Institute)
Frank Gounelas (Humboldt University)
Xavier Guitart (Universität Duisburg-Essen)
Shuvra Gupta (University of Iowa)
Shai Haran (Technion – Israel Institute of Technology)
Michael Harris (Institut de Mathématiques de Jussieu)
Roger Heath-Brown (University of Oxford)
Florian Herzig (University of Toronto)
Alexander Ivanov (Universität Heidelberg)
Mikhail Ivanov (Saint-Petersburg State University)
Andrew Kresch (Universität Zürich)
Lars Kühne (SNS Pisa)
Boris Kunyavskii (Bar-Ilan University)
Leonid Kuzmin (National Research Center Kurchatov Institute)
Kai-Wen Lan (University of Minnesota)
Dmitry Logachev (Universidad Simon Bolivar)
Oliver Lorscheid (IMPA)
Daniel Loughran (Leibniz Universität Hannover)
Yuri Manin (Max-Planck-Institut für Mathematik)
Oscar Marmon (Georg-August-Universität Göttingen)
Lilian Matthiesen (University of Bristol)
David McKinnon (University of Waterloo)
Boris Moroz (Universität Bonn)
Jeanine Van Order (The Hebrew University of Jerusalem)
Denis Osipov (Steklov Mathematical Institute)
Ambrus Pál (Imperial College London)
Aftab Pande (Cidade Universitária Ilha do Fundao)
Alexej Parshin (Steklov Mathematical Institute)
Florian Pop (University of Pennsylvania)
Dinakar Ramakrishnan (Caltech)
Giovanni Rosso (Université Paris 13)

Mohamed Saidi (Exeter University)
Per Salberger (Chalmers University of Technology)
Damaris Schindler (University of Bristol)
Mehmet Haluk Sengun (University of Warwick)
Evgeny Shinder (Max-Planck-Institut für Mathematik)
Ceclia Salgado Guimaraes da Silva (UFRJ)
Alexei Skorobogatov (Imperial College London)
Arne Smeets (Université Paris-Sud 11)
Aleksander Smirnov (Steklov Institute of Mathematics in St. Petersburg)
Efthymios Sofos (University of Bristol)
Cesar Alejandro Soto Posada (Universität Tübingen)
Mike Swarbrick Jones (University of Bristol)
Jack Thorne (Harvard University)
Jacques Tilouine (Université Paris 13)
Yuri Tschinkel (New York University)
Pankaj Hemant Vishe (University of York)
Sergei Vostokov (University of St. Petersburg)
Yosuhiro Wakabayashi (Kyoto University)
Gabor Wiese (Université du Luxembourg)
Nicholas Williams (University of Exeter)
Jean-Pierre Wintenberger (Université Strasbourg)
Trevor Wooley (University of Bristol)
Yanhong Yang (Universität Mainz)
Adrin Zenteno (Universidad Nacional Autónoma de México)

Trimester Seminar

January 8, Jeanine Van Order, Iwasawa main conjectures for $GL(2)$ via Howard's criterion (abstract). In this talk, I will present the Iwasawa main conjectures for Hilbert modular eigenforms of parallel weight two in dihedral or anticyclotomic extensions of CM fields. The first part will include an overview of known results, as well as some discussion of open problems and applications (e.g. to bounding Mordell-Weil ranks), and should be accessible to the non-specialist. The second part will describe the p-adic L-functions in more detail, as well as the non-vanishing criterion of Howard (and its implications for the main conjectures).

January 15, Oliver Lorscheid, A blueprinted view on F_1-geometry (abstract). A blueprint is an algebraic structure that "interpolates" between multiplicative monoids and semirings. The associated scheme theory applies to several problems in F_1-geometry: Tits's idea of Chevalley groups and buildings over F_1, Euler characteristics as the number of F_1-rational points, total positivity, K-theory, Arakelov compactifications of arithmetic curves; and it has multiple connections to other branches of algebraic geometry: Lambda-schemes (after Borger), log schemes (after Kato), relative schemes (after Toen and Vaquie), congruence schemes (after Berkovich and Deitmar), idempotent analysis, analytic spaces and tropical geometry. After a brief overview and an introduction to the basic definitions of this theory, we focus on the combinatorial aspects of blue schemes. In particular, we explain how to realize Jacques Tits's idea of Weyl groups as Chevalley groups over F_1 and Coxeter complexes as buildings over F_1. The central concepts are the rank space of a blue scheme and the Tits category, which make the idea of "F_1-rational points" rigorous.

January 16, Jean-Pierre Wintenberger, Introduction to Serre's modularity conjecture (abstract). This lecture is intended for non-specialists. We state Serre's modularity conjecture and give some consequences and hints on its proof.

January 17, Henri Carayol, Realization of some automorphic forms and rationality questions (Part I) (abstract). In this first (and mostly introductory) talk I shall recall some (well-known) facts on the realization of automorphic forms in the cohomology groups of some geometric objects, and the relation with the arithmetic properties of such forms. I shall introduce locally symmetric varieties, Shimura varieties and the more exotic Griffiths-Schmid varieties. I shall discuss the case of automorphic forms whose archimedean component is a limit of discrete series. In the case of degenerate limits, the only known realization uses the coherent cohomology of Griffiths-Schmid varieties.

January 21, Günter Harder, Modular construction of mixed motives and congruences (Part I) (abstract). Starting from a Shimura variety S and its compactification S^v we construct certain objects, which can be thought of as being mixed motives. These mixed motives give rise to certain elements of Ext_1 groups. We can use the theory of Eisenstein cohomology to compute the Hodge-de-Rham extension classes of these extensions. We also have some conjectural formulas for these extensions as Galois modules. Assuming the correctness of these formulas for the Galois extension class we can derive congruences between eigenvalues of Hecke operators acting on the cohomology of different arithmetic groups, these congruences are congruences modulo primes l dividing certain special values of L-functions. These congruences have been verified experimentally in many cases. They imply the reduciblity of certain Galois-representations mod l.

January 22, Yuri Manin, Non-commutative generalized Dedekind symbols (abstract). Classical Dedekind symbol was introduced and studied in connection with functional equation of Dedekind eta-function. Later it was generalized and had multiple applications, in particular to topological invariants. I will define and study generalized Dedekind symbols with values in non-necessarily commutative groups, extending constructions of Sh. Fukuhara done in the commutative context. Basic examples of such symbols are obtained by replacing period integrals of modular forms by iterated period integrals. I will also explain the interpretation of such symbols in terms of non-commutative 1-cocycles.

January 23, Henri Carayol, Realization of some automorphic forms and rationality questions (Part II) (abstract). This talk is a continuation of part I.

January 24, Michael Harris, Eisenstein cohomology and construction of Galois representations (Part I) (abstract). I will report on some aspects of the joint work with Lan, Taylor, and Thorne, which attaches compatible families of l-adic Galois representations to a cuspidal cohomological automorphic representation of $GL(n)$ of a CM field. Earlier work by many authors had treated the case where the automorphic representation is dual to its image under complex conjugation; under this hypothesis, the Galois representations in question, or closely related representations, can be obtained directly in the cohomology with twisted coefficients of Shimura varieties attached to unitary groups. Without the duality hypothesis, this is no longer possible; instead, the representations are constructed by p-adic approximation of Eisenstein cohomology classes by cuspidal classes in an appropriate (infinite-dimensional) space of p-adic modular forms. The lectures will concentrate on the construction of Eisenstein classes, the relation to p-adic modular forms and the definition of Galois representations by p-adic approximation.

January 25, Michael Harris, Eisenstein cohomology and construction of Galois representations (Part II) (abstract). This talk is a continuation of part I.

January 28, Günter Harder, Modular construction of mixed motives and congruences (Part II) (abstract). This talk is a continuation of part I.

January 28, Fred Diamond, The weight part of Serre's conjecture for $GL(2)$ over totally real fields (abstract). I will review the statement of the weight part of Serre's conjecture for $GL(2)$ over totally real fields. I will describe what has been proved by Gee and his coauthors, and give a brief overview of the methods.

January 30, Luis Dieulefait, Non-solvable base change for $GL(2)$ (abstract). We will show that any classical cuspidal modular form can be lifted to any totally real number field. The proof uses a recent Modularity Lifting Theorem proved by Barnet-Lamb, Gee, Geraghty and Taylor (plus a variant of it proved by Gee and the speaker) and another one by Kisin that is used in the "killing ramification" step. The core of the proof is the construction of a "safe" chain of congruences linking to each other any given pair of cuspforms. The safe chain that we will construct is also a key input in the proof of other cases of Langlands functoriality, but this will be explained in another talk (see the abstracts for the conference week).

January 31, Kai-Wen Lan, Galois representations for regular algebraic cuspidal automorphic representations over CM fields (part I) (abstract). I will report on my joint work with Michael Harris, Richard Taylor and Jack Thorne on the construction of p-adic Galois representations for regular algebraic cuspidal automorphic representations of $GL(n)$ over CM (or totally real) fields, without hypotheses on self-duality or ramification. (This should be considered as part III of a series of four talks, the first two being given by Michael Harris in the previous week.)

February 1, Kai-Wen Lan, Galois representations for regular algebraic cuspidal automorphic representations over CM fields (part II) (abstract). This is a continuation of part I.

February 12, David Geraghty, The Breuil-Mezard conjecture for quaternion algebras (abstract). The Breuil-Mezard conjecture relates the complexity of certain deformation rings for mod p representations of the Galois group of Q_p with the representation theory of $GL_2(F_p)$. Most cases of the conjecture were proved by Kisin who established a link between the conjecture and modularity lifting theorems. In this talk I will discuss a generalization of the conjecture to quaternion algebras (over an arbitrary finite extension of Q_p) and show how it follows from the original conjecture for $GL(2)$. This is a joint work with Toby Gee.

February 18, Boris Kunyavskii, Geometry and arithmetic of word maps in simple matrix groups (abstract). We wil discuss various geometric and arithmetic properties of matrix equations of the form

$$P(X_1, \ldots, X_d) = A,$$

where the left-hand side is an associative non-commutative monomial in X_i's and their inverses, and the right-hand side is a fixed matrix. Solutions are sought in some group $G \subset GL(n, R)$. We will focus on the case where the group G is simple, or close to such. We will give a survey of classical and recent results and open problems concerning this equation, concentrating around the following questions (posed for geometrically and/or arithmetically interesting rings and fields R): is it solvable for any A?, is it solvable for a "typical" A?, does it have "many" solutions?, does the set of solutions possess "good" local–global properties?, to what extent does the set of solutions depend on A?. The last question will be discussed in some detail for the case $G = SL(2, q)$ and $d = 2$, where criteria for equidistribution were obtained in our recent joint work with T. Bandman.

February 19, Shai Haran, Non-additive geometry (abstract). We give a language for algebraic geometry based on non-additive generalized rings. In this language, number fields look more like curves over a finite field. The initial object of generalized rings is the "field with one element". This language "sees" the real and complex primes of a number field, and there is a compactificaton $\widetilde{\operatorname{Spec} \mathfrak{O}_K}$ of $\operatorname{Spec} \mathfrak{O}_K$, $\mathfrak{O}_K$ being the ring of integers of a number field K. The arithmetic surface $\widetilde{\operatorname{Spec} \mathfrak{O}_K} \sqcap \widetilde{\operatorname{Spec} \mathfrak{O}_K}$ exists and is not reduced to its diagonal. And yet most of the Grothendieck algebraic geometry works with generalized rings replacing commutative rings.

February 20, Aleksander Smirnov, The internal and external problems of algebraic geometry over F_1 (abstract). An introduction to Durov's approach will be given. The theory will be illustrated with several explicit examples. Besides, we plan to discuss some problems caused by both the development of the theory and the demands of its applications.

February 21, Marina Viazovska, CM values of higher Green's functions and regularized Petersson products (abstract). Higher Green functions are real-valued functions of two variables on the upper half-plane, which are bi-invariant under the action of a congruence subgroup, have a logarithmic singularity along the diagonal, and satisfy the equation $\Delta f = k(1-k)f$; here Δ is a hyperbolic Laplace operator and k is a positive integer. The significant arithmetic properties of these functions were disclosed in the paper of B. Gross and D. Zagier "Heegner points and derivatives of L-series" (1986). In the particular case when $k = 2$ and one of the CM points is equal to $\sqrt{-1}$, the conjecture has been proved by A. Mellit in his Ph.D. thesis. In this lecture we prove that conjecture for arbitrary k, assuming that all the pairs of CM points lie in the same quadratic field. The two main parts of the proof are as follows. We first show that the regularized Petersson scalar product of a binary theta-series and a weight one weakly holomorphic cusp form is equal to the logarithm of the absolute value of an algebraic integer and then prove that the special values of weight k Green's function, occurring in the conjecture of Gross and Zagier, can be written as the Petersson product of that type, where the form of weight one is the $k-1$st Rankin-Cohen bracket of an explicitly given holomorphic modular form of weight $2-2k$ and a binary theta-series. Algebraicity of regularized Petersson products was also proved at about the same time by W. Duke and Y. Li by a different method; however, our result is stronger since we also give a formula for the factorization of the algebraic number in question.

February 26, David Mendes da Costa, Integral points on elliptic curves and the Bombieri-Pila bounds (abstract). In 1989, Bombieri and Pila found upper bounds for the number of integer points of (naive exponential) height at most B lying on a degree d affine plane curve C. In particular, these bounds are both uniform with respect to the curve C and the best possible with this constraint. It is conjectured though that if we restrict to curves with positive genus then the bounds can be broken. In this talk we shall discuss progress towards this conjecture in the case of elliptic curves and an application to counting rational points on degree 1 del Pezzo surfaces.

February 27, Lars Kühne, Effective and uniform results of André-Oort type (abstract). The André-Oort Conjecture (AOC) states that the irreducible components of the Zariski closure of a set of special points in a Shimura variety are special subvarieties. Here, a special variety means an irreducible component of the image of a sub-Shimura variety by Hecke correspondence. The AOC is an analogue of the classical Manin-Mumford conjecture on the distribution of torsion points in abelian varieties. In fact, both conjectures are considered as special instances of the far-reaching Zilber-Pink conjecture(s). I will present a rarely known approach to the AOC that goes back to Yves André himself: Before the model-theoretic proofs of the AOC in certain cases by the Pila-Wilkie-Zannier approach, André presented in 1998 the first proof of the AOC in a non-trivial case, namely, a product of two modular curves. In my talk, I discuss several results in the style of André's method, allowing to compute all special points in a non-special curve of a product of two modular curves. These results are effective – as opposed to the results that could be obtained by the Pila-Wilkie-Zannier approach – and have also the further advantage of being uniform in the degrees of the curve and its field of definition. For example, this allows to show that, in fact, there are no two singular moduli x and y satisfying $x + y = 1$.

February 28, Nuno Freitas, Fermat-type equations of signature (r, r, p) (abstract). In this talk I plan to discuss how a modular approach via the Hilbert cusp-forms can be used to attack equations of the form $x^r + y^r = Cz^p$, where r is a fixed prime and p varies. We first relate a possible solution of that equation to solutions of several related Diophantine equations over certain totally real fields F. Then we attach Frey curves E over F to the solutions of the latter equations. After proving modularity of E and irreducibility of certain Galois representations attached to E we can use the modular approach. We apply the method to solve equations in the particular case of signature $(13, 13, p)$.

March 4, Leonid Kuzmin, l-adic regulator of an algebraic number field and Iwasawa theory (abstract). We give a new definition of the l-adic regulator, which makes sense for any (not necessarily totally real) algebraic number field, present a few results and conjectures, relating to that notion, and discuss the behaviour of the l-adic regulator in a $\mathbb{Z}_l$-cyclotomic extension of the field.

March 4, Alexander Ivanov, Arithmetic and anabelian geometry of stable sets of primes in number fields (abstract). We define a new class of sets – stable sets – of primes in number fields. For example, Chebotarev sets are very often stable. Those sets have positive (but arbitrary small) Dirichlet density and generalize the notion of a set of density 1, in the sense that arithmetic theorems like certain Hasse principles, the Grünwald-Wang theorem, the Riemann existence theorem, etc. hold for them. Geometrically, this allows to give examples of infinite sets with arbitrarily small positive density such that the corresponding arithmetic curves are algebraic $K(\pi, 1)$ and, using some further ideas, to generalize (a part of) the Neukirch-Uchida birational anabelian theorem to stable sets.

March 5, Tuan Ngo Dac, On the problem of counting shtukas (abstract). I will introduce the stacks of shtukas, explain its role in the Langlands program, and then report on my work on the problem of counting shtukas.

March 7, Rajender Adibhatla, Modularity of certain two-dimensional mod p^n representations of $G_{\mathbb{Q}}$ (abstract). For an odd rational prime p and integer $n > 1$, we consider certain continuous representations

$$\rho_n : G_{\mathbb{Q}} \to GL_2(\mathbb{Z}/p^n\mathbb{Z})$$

with fixed determinant, whose local restrictions "look like arising" from modular Galois representations, and whose mod p reductions are odd and irreducible. Under suitable hypotheses on the size of their images, we use deformation theory to lift ρ_n to ρ in characteristic 0. We then invoke a modularity lifting theorem of Skinner-Wiles to show that ρ is modular.

March 11, Frank Gounelas, Rationally connected varieties and free curves (abstract). The first part of this talk will be a general introduction to rationally connected varieties. I will then discuss various ways in which a variety can be "connected by curves of a fixed genus, mimicking the notion of rational connectedness". At least in characteristic zero, in the specific case of the existence of a single curve with a large deformation space of morphisms to a variety implies that the variety is in fact rationally connected. Time permitting I will discuss attempts to show this result in positive characteristic.

March 12, Tommaso Centeleghe, On the decomposition of primes in torsion fields of an elliptic curve (abstract). Let E be an elliptic curve over a number field K and N be a positive integer. In this talk we consider the problem of describing how primes P of K of good reduction for E and away from N decompose in the extension $K(E[N])|K$. As it turns out, the class Frob P in Gal$(K(E[N]|K))$ can be completely described, apart of finitely many primes P, in terms of the error term $a_P(E)$ and the j-invariant of E. The Hilbert class polynomials, associated to imaginary quadratic orders, play a role in the description. The main result relies on a theorem on elliptic curves over finite fields.

March 13, Shuvra Gupta, Noether's problem and rationality of invariant spaces (abstract). In the early 1900s Emmy Noether asked the following question: If a group G acts faithfully on a vector space V (over a field k), is the field of invariants $k(V)^G$ rational, i.e. purely transcendental over k? The answer (for $k = \mathbb{Q}$ or a number field) in general is no, and we will discuss some consequences and variants of Noether's problem. We will also discuss the problem when the field k is algebraically closed, and techniques of testing rationality of the fields of invariants using unramified cohomology groups.

March 14, Mohamed Saidi, Some problems/results related to the Grothendieck anabelian section conjecture (abstract). I will discuss some (major) problems related to the Grothendieck anabelian section conjecture. I will discuss two new results related to these problems and the section conjecture. First result: there exists a local–global principle for torsors under the geometric prosolvable fundamental group of a proper hyperbolic curve over a number field. Second result: the passage in the section conjecture from number fields to finitely generated fields is possible under the assumption of finiteness of suitable Shafarevich-Tate groups.

March 18, Sergey Gorchinskiy, Parameterized differential Galois theory (abstract). Classical Galois theory studies symmetry groups of solutions of algebraic equations. Differential Galois theory studies symmetry groups of solutions of linear differential equations. We discuss the so-called parameterized differential Galois theory which studies symmetry groups of solutions of linear differential equations with parameters. The groups that arise are linear differential groups given by differential equations (not necessarily linear) on functions in parameters. We also discuss, in this connection, derivations on Abelian categories and differential Tannakian categories.

March 19, Alexandru Buium, The concept of linearity for an arithmetic differential equation (abstract). The concept of an ordinary differential equation has an arithmetic analogue in which the derivation operator is replaced by a Fermat quotient operator. We would like to understand which arithmetic differential equations should be considered as being "linear".

Classical linear differential equations arise from differential cocycles of linear algebraic groups into their Lie algebras and their differential Galois groups are algebraic groups with coefficients in the field of constants. On the other hand one can prove that there are no such cocycles in the arithmetic context. This leads one to introduce, in the arithmetic context, a new concept of "Lie algebra", "cocycles", "linear" equations, and "differential Galois groups"; the latter can be viewed as subgroups of the general linear group with coefficients in the algebraic closure of the "field with one element".

March 20, Roman Budylin, Adelic Bloch formula (abstract). Chern class $c_2(X)$ is involved in the functional equation for two-dimensional schemes. To get functional equation by the Tate method we need a local decomposition of the Chern class, satisfying some properties. Bloch proves that the second Chern class of a vector bundle with trivial determinant can be obtained by the boundary homomorphism for the universal central extension of the sheaf $\mathrm{SL}(\mathcal{O}_X)$. In the talk, this construction will be used to get an adelic formula for the second Chern class in terms of trivializations in scheme points. We will also discuss a generalization of this formula for c_n of vector bundles with $c_i = 0$ for $i < n$.

March 21, Alexej Parshin, A generalization of the Langlands correspondence and zeta-functions of a two-dimensional scheme (abstract). We introduce Abelian Langlands correspondence for algebraic surfaces defined over a finite field. When the surface is a semi-stable fibration over an algebraic curve, we define two operations, automorphic induction and base change, which connect this correspondence with the classical Langlands correspondence on the curve. Some conjectural properties of these operations imply the standard theorems for zeta- and L-functions on the surface (analytic continuation and functional equation). In this approach we do not need to use the étale cohomology theory.

March 22, Denis Osipov, Unramified two-dimensional Langlands correspondence (abstract). We will describe the local unramified Langlands correspondence for two-dimensional local fields (following an approach of M. Kapranov). For this goal, we will construct a categorical analogue of principal series representations of general linear groups of even degrees over two-dimensional local fields and describe their properties. The main ingredient of this construction is a central extension of a general linear group defined over a two-dimensional local field or over an adelic ring of a two-dimensional arithmetic scheme. We will prove reciprocity laws for such central extensions, i.e. splittings of the central extensions over some subgroups defined over rings constructed by means of points or by integral one-dimensional subschemes of a two-dimensional arithmetic scheme.

March 25, Rainer Dietmann, On quantitative versions of Hilbert's irreducibility theorem (abstract). If $f(X, Y)$ is an irreducible rational polynomial, then by Hilbert's irreducibility theorem for infinitely many rational specialisations of X the resulting polynomial in Y is still irreducible over the rationals. In this talk we want to discuss quantitative versions of this result, using recent advances from the determinant method on bounding the number of points on curves.

March 27, Roger Heath-Brown, Pairs of quadratic forms in 8 variables (abstract). We show that a smooth intersection of two quadrics in P^7, defined over a number field, satisfies the Hasse principle and weak approximation. The proof is based on the work of Colliot-Thélène, Sansuc and Swinnerton-Dyer on Chatelet surfaces, which enables one to reduce the problem to a purely local problem. The first part of the talk will discuss the background and the overall strategy of the proof, and the second part will look in a little more detail at some of the methods involved.

April 2, Pankaj Vishe, Cubic hypersurfaces and a version of the circle method over number fields (abstract). A version of the Hardy-Littlewood circle method is developed for number fields K and is used to show that any non-singular projective cubic hypersurface over K of dimension ≥ 8 always has a K-rational point. This is a joint work with T. Browning.

April 3, Efthymios Sofos, Counting rational points on the Fermat surface (abstract). In this talk we shall discuss progress towards finding lower bound for the number of rational points of bounded height on the Fermat cubic surface. The argument is based on a uniform asymptotic estimate for the associated counting function on conics.

April 5, Jörg Brüdern and Trevor Wooley, Systems of cubic forms at the convexity barrier (abstract). We describe recent joint work concerning the validity of the Hasse principle for systems of diagonal cubic forms. The number of variables required meets the convexity barrier. Certain features of our methods are motivated by work of Gowers on Szemeredi's theorem.

April 8, Mike Swarbrick Jones, Weak approximation on cubic hypersurfaces of large dimension (abstract). A natural question in arithmetic geometry is to investigate weak approximation on varieties. If the dimension of the variety is large compared to the degree, our most successful tool is the circle method, however there are cases where using this is not feasible given our current state of knowledge. In this talk I will sketch a proof that weak approximation holds for generic cubic hypersurfaces of dimension at least 17, in particular discussing a fibration method argument that applies to the cases where the usual application of the circle method is not possible.

April 10, Arne Smeets, Local–global principles for fibrations in torsors under tori (abstract). This talk is a report on work in progress about local–global principles for varieties fibred over the projective line. In particular, we will study the Brauer-Manin obstruction to the Hasse principle and weak approximation for certain fibrations in torsors under tori, e.g. (multi-)norm form equations. Our results are conditional on Schinzel's hypothesis.

April 12, David Mendes da Costa, On uniform bounds for integral points on elliptic curves (abstract). In 1989, Bombieri and Pila proved that given a plane algebraic affine curve of degree d there are no more than $O(N^{1/d+\epsilon})$ integral points on the curve within a box of size $N \times N$. Moreover, the implied constant in their bound depended only on the degree of the curve and not on the equation. Such bounds are, in general, the best possible, however, it is believed that by restricting to curves which have positive genus one can do much better. In this talk we consider the problem of improving these uniform bounds for integral points on elliptic curves. An application of this work to degree one del Pezzo surfaces will be presented.

Workshop on Serre's conjecture

Monday, February 4

10:00 – 11:00 Fred Diamond (London King's college): Explicit Serre weights for two-dimensional Galois representations.

11:30 – 12:30 Denis Benois (Université Bordeaux 1): Trivial zeros of p-adic L-functions and Iwasawa theory.

14:30 – 15:30 Tommaso Centeleghe (Universität Heidelberg): Computing the number of certain mod p Galois representation.

16:00 – 17:00 Wojciech Gajda (UAM Poznań): Abelian varieties and l-adic representations.

Tuesday, February 5

10:00 – 11:00 Tobias Berger (University of Sheffield): Eisenstein congruences and modularity of Galois representations.

11:30 – 12:30 Christophe Breuil (Université de Paris-Sud): Ordinary representations of $\mathrm{GL}_n(\mathbb{Q}_p)$ and fundamental algebraic representations I.

14:30 – 15:30 Jeanine Van Order (EPFL Lausanne): Critical values of GL(2) Rankin-Selberg L-functions.

16:00 – 17:00 Luis Dieulefait (Universitat de Barcelona): Some new cases of Langlands functoriality solved.

Wednesday, February 6

10:00 – 11:00 Aftab Pande (Cidade Universitária Ilha do Fundao): Deformations of Galois representations and the theorems of Sato-Tate, Lang-Trotter and others.

11:30 – 12:30 Florian Herzig (University of Toronto): Ordinary representations of $\mathrm{GL}_n(\mathbb{Q}_p)$ and fundamental algebraic representations II.

14:30 – 15:30 Mehmet Sengun (University of Warwick): Mod p Cohomology of Bianchi groups and Mod p Galois representations.

16:00 – 17:00 Jack Thorne (Harvard University): Symmetric power functoriality for GL(2).

Thursday, February 7

10:00 – 11:00 Dinakar Ramakrishnan (California Institute of Technology): Picard modular surfaces, residual Albanese quotients, and rational points.

11:30 – 12:30 Jacques Tilouine (Université Paris 13): Image of Galois and congruence ideals, a program of a joint work with H. Hida.

Friday, February 8

10:00 – 11:00 David Geraghty (IAS Princeton): Modularity lifting beyond the "numerical coincidence" of the Taylor-Wiles method.

11:30 – 12:30 Florian Pop (University of Pennsylvania): Faithful representations of absolute Galois groups.

Abstracts

Denis Benois. Trivial zeros of p-adic L-functions and Iwasawa theory. We prove that the expected properties of Euler systems imply quite general Mazur-Tate-Teitelbaum type formulas for derivatives of p-adic L-functions. We also discuss the Iwasawa theory of p-adic representations in the trivial zero case.

Tobias Berger. Eisenstein congruences and modularity of Galois representations. I will report on joint work with Kris Klosin (CUNY) on congruences of Eisenstein series and cuspforms modulo prime powers and its application in proving the modularity of residually reducible Galois representations.

Tommaso Centeleghe. Computing the number of certain mod p Galois representations. We report on computations aimed to obtain, for a given prime p, the number $R(p)$ of two-dimensional odd mod p Galois representations of $G_{\mathbb{Q}}$ which are irreducible and unramified outside p. Thanks to Serre's conjecture, this amounts to a computation of the number of non-Eisenstein systems of Hecke eigenvalues arising from mod p modular forms of level one. Using well-known dimension formulas for modular forms (and the fact that any mod p eigensystem can be Tate-twisted to one arising from a weight $\leq p+1$), an explicit upper bound $U(p)$ of $R(p)$ can be drawn. While discussing the reasons which might make the "error term" $U(p) - R(p)$ large, we stress how, in practice, one can control it from above using only one of the first Hecke operators. This gives a lower bound for $R(p)$, which coincides in many cases with $R(p)$ itself.

Fred Diamond. Explicit Serre weights for two-dimensional Galois representations. I will discuss joint work with Savitt making the set of Serre weights more explicit for indecomposable two-dimensional mod p representations of Galois groups over ramified extensions of $\mathbb{Q}_p$. In particular the results indicate a structure on the set of such weights.

Luis Dieulefait. Some new cases of Langlands functoriality solved. We combine the method of Propagation of Automorphy with recent Automorphy Lifting Theorems (A.L.T.) of Barnet-Lamb, Gee, Geraghty, and Taylor to prove some new cases of Langlands functoriality (tensor products and symmetric powers). In particular, we establish automorphy for lots of Galois representations of $G_{\mathbb{Q}}$ of arbitrarily large dimension (and their base changed counterparts). We also prove some variants of the available A.L.T., which are needed at some steps of our proof. Remark: Some technical improvements required to extend some A.L.T. to the case of "small primes" were accomplished with the kind cooperation of R. Guralnick and T. Gee.

Wojciech Gajda. Abelian varieties and l-adic representations. We will discuss monodromies for abelian varieties, and independence (in the sense of Serre) for families of some geometric l-adic representations over finitely generated fields.

David Geraghty. Modularity lifting beyond the "numerical coincidence" of the Taylor-Wiles method. Modularity lifting theorems have proven very useful since their invention by Taylor and Wiles. However, as explained in the introduction to Clozel-Harris-Taylor, they only apply in situations where a certain numerical coincidence holds. In this talk, I will describe a method to overcome this restriction. The method is conditional on the existence of Galois representations associated to integral cohomology classes (which can be established in certain cases). This is joint work with Frank Calegari.

Aftab Pande. Deformations of Galois Representations and the Theorems of Sato-Tate, Lang-Trotter and others. We construct infinitely ramified Galois representations ρ such that the sequences $a_l(\rho)$ have distributions in contrast to the statements of Sato-Tate, Lang-Trotter and others. Using similar methods, we deform a residual Galois representation for number fields and obtain an infinitely ramified representation with very large image, generalizing a result of Ramakrishna.

Florian Pop. Faithful representations of absolute Galois groups. In his "Esquisse d'un Programme" Grothendieck suggested to study the absolute Galois group of the rationals via its representations on the algebraic fundamental group of natural categories of varieties, e.g. the Teichmueller modular tower. This leads to the intensive study of the so-called Grothendieck-Teichmueller group and its variants, and the I/OM (Ihara/Oda-Matsumoto conjecture). I plan to explain variants of I/OM, and discuss its state of the art.

Dinakar Ramakrishnan. Picard modular surfaces, residual Albanese quotients, and rational points. The Picard modular surfaces X are at the crossroads of rich interplay between geometry, Galois representations, and automorphic forms on $G = U(2, 1)$ associated to an imaginary quadratic field K. The talk will introduce an ongoing project with M. Dimitrov on the quotients of the albanese variety $Alb(X)$ coming from residual automorphic forms on G, give examples with finite Mordell-Weil group, and investigate possible consequences, inspired by classical arguments of Mazur, for the K-rational points on X.

Mehmet Sengun. Mod p Cohomology of Bianchi Groups and Mod p Galois Representations. Given an imaginary quadratic field K with ring of integers R, consider the Bianchi group GL(2, R). It is suspected since the numerical investigations of Fritz Grunewald in the late 1970s that there is a connection between the Hecke eigenclasses in the mod p cohomology of (congruence subgroups of) Bianchi groups and the two-dimensional continuous mod p representations of the absolute Galois group of K. Most of the basic tools used for establishing this connection (and its surrounding problems) in the classical setting fail to work in the setting of Bianchi groups. The situation has an extra layer of complication by the fact that there are "genuinely mod p" Hecke eigenvalue systems, resulting from the existence of torsion in the integral cohomology. In this talk I will elaborate on the above, presenting numerical examples for illustration. Towards the end, I will also talk about how the "even" two-dimensional continuous mod p representations of the absolute Galois group of $\mathbb{Q}$ come into the picture.

Jack Thorne. Symmetric power functoriality for GL(2). We discuss some new automorphy lifting theorems, and their applications to the existence of new cases of Langlands' functoriality for GL(2). This is joint work with L. Clozel.

Jacques Tilouine. Image of Galois and congruence ideals, a programme of a joint work with H. Hida. In a recent preprint, H. Hida showed that the image of the Galois representation associated to a non-CM Hida family contains a congruence subgroup of GL(2) over Λ, whose level is given in terms of p-adic L-functions. We try to generalize this to Hida families for bigger groups, replacing p-adic L-functions by congruence ideals.

Jeanine Van Order. Critical values of GL(2) Rankin-Selberg L-functions. The aim of this talk is to explain the subtle but powerful link between the algebraicity of critical values of automorphic L-functions, the existence of associated p-adic L-functions, and the generic non-vanishing of these values, particularly in the setting of Rankin-Selberg L-functions of GL(2) over a totally real number field. More precisely, the aim is to explain how to extend the conjectures of Mazur to the non self-dual setting, thereby extending the works of Vatsal, Cornut and Cornut-Vatsal, via a combination of techniques from Iwasawa theory, analytic number theory and the theory of automorphic forms. If time permits, then some open problems will also be introduced.

The research conference

Monday, April 15

09:30 – 10:30 Trevor Wooley (University of Bristol): Applications of efficient congruencing to rational points.

11:00 – 12:00 Per Salberger (Chalmers University of Technology): Heath-Brown's determinant method and Mumford's geometric invariant theory.

13:30 – 14:30 Jeanine Van Order (EPFL): Stable Galois averages of Rankin-Selberg L-values and non-triviality of p-adic L-functions.

15:00 – 16:00 Przemyslaw Chojecki (Institut Mathématique de Jussieu): On mod p non-abelian Lubin-Tate theory for GL(2).

Tuesday, April 16

09:30 – 10:30 Alexei Skorobogatov (Imperial College London): Applications of additive combinatorics to rational points.

11:00 – 12:00 Yuri Bilu (IMB Université Bordeaux I): Integral points on modular curves.

13:30 – 14:30 Ulrich Derenthal (Universität München): Counting points over imaginary quadratic number fields.

15:00 – 16:00 Oscar Marmon (Universität Göttingen): The density of twins of k-free numbers.

Wednesday, April 17

09:30 – 10:30 Jörg Brüdern (Universität Göttingen): Random Diophantine equations.

11:00 – 12:00 Florian Pop (University of Pennsylvania): Local–global principles for rational points.

13:30 – 14:30 Mohamed Saidi (Exeter University): On the anabelian section conjecture over finitely generated fields.

15:00 – 16:00 Victor Abrashkin (University of Durham): p-extensions of local fields with Galois groups of nilpotence class less than p.

Thursday, April 18

09:30 – 10:30 Ambrus Pál (Imperial College London): New two-dimensional counter-examples to the local–global principle.

11:00 – 12:00 David McKinnon (University of Waterloo): Approximating points on varieties.

13:30 – 14:30 Tim Browning (University of Bristol): Norm forms as products of linear polynomials, I.

15:00 – 16:00 Lilian Matthiesen (University of Bristol): Norm forms as products of linear polynomials, II.

Friday, April 19

09:30 – 10:30 Gabor Wiese (Université du Luxembourg): Symplectic Galois representations and applications to the inverse Galois problem.

11:00 – 12:00 Damaris Schindler (University of Bristol): Manin's conjecture for certain smooth hypersurfaces in biprojective space.

13:30 – 14:30 Roger Heath-Brown (University of Oxford): Simultaneous representation of pairs of integers by quadratic forms.

15:00 – 16:00 Jean-Louis Colliot-Thélène (Université Paris-Sud): Strong approximation in a family.

Abstracts

Victor Abrashkin. p-extensions of local fields with Galois groups of nilpotence class less than p. Nilpotent analogue of the Artin-Schreier theory was developed by the author about twenty years ago. It has found already applications in an explicit description of the ramification filtration modulo p-th commutators and the proof of an analogue of the Grothendieck Conjecture for local fields. We remind basic constructions of this theory and indicate further progress in the study of local fields of mixed characteristic, especially, higher dimensional local fields.

Yuri Bilu. Integral points on modular curves. The problem of determination of rational points on modular curves reduces grosso mode to three types of curves of prime level, corresponding to three types of maximal subgroups of the linear group $GL_2(\mathbb{F}_p)$: the curve $X_0(p)$, corresponding to the Borel subgroup; the curve $X_{sp}^+(p)$, corresponding to the normalizer of a split Cartan subgroup; the curve $X_{ns}^+(p)$, corresponding to the normalizer of a non-split Cartan subgroup. The rational points on the curves of the first two types are determined (almost) completely: Mazur (1978), B.-Parent-Rebolledo (2012). In particular, it is proved that for $p > 13$ the rational points are either cusps or the CM-points. Little is known, however, about the rational points on $X_{ns}^+(p)$. I will speak about recent progress in a simpler problem: classification of integral points on $X_{ns}^+(p)$ (i.e. rational points P such that $j(P) \in \mathbb{Z}$). My students Bajolet and Sha obtained a rather sharp upper bound for the size of integral points. Also, in a joint work with Bajolet we proved that for $7 < p < 71$ there are no integral points on $X_{ns}^+(p)$ other than the CM-points; this improves on a recent work of Schoof and Tzanakis, who proved that for $p = 11$.

Tim Browning, Lilian Matthiesen. Norm forms as products of linear polynomials. We report on recent progress using additive combinatorics to prove the Hasse principle and weak approximation for certain varieties defined by systems of equations involving norm forms. This is used to show that the Brauer-Manin obstruction controls weak approximation on normic bundles of the shape $N_K(x_1, \ldots, x_n) = P(t)$, where $P(t)$ is a product of linear polynomials all defined over the rationals and K is an arbitrary degree n extension of the rationals.

Jörg Brüdern. Random Diophantine equations. We address the classical questions, concerning diagonal forms with integer coefficients. Does the Hasse principle hold? If there are solutions, how many? If there are solutions, what is the size of the smallest solutions? In a joint work with Dietmann, nearly optimal answers to such questions were obtained for almost all forms (in the sense typically attributed to "almost all" in the analytic theory of numbers) provided that the number of variables exceeds three times the degree of the forms under consideration.

Przemyslaw Chojecki. On mod p non-abelian Lubin-Tate theory for GL(2). Non-abelian Lubin-Tate theory for GL(2) describes the l-adic cohomology of the Lubin-Tate tower for $GL(2, \mathbb{Q}_p)$ in terms of the Langlands program. Until recently, all results were stated under the assumption that p is different from l. We discuss the case when $l = p$ and give a partial description of the mod p étale cohomology of the Lubin-Tate tower in terms of the mod p local Langlands correspondence.

Jean-Louis Colliot-Thélène. Strong approximation in a family. Let $a_i(t)$, $i = 1, 2, 3$ and $p(t)$ be polynomials in $\mathbb{Z}[t]$. Assume the product $p(t) \prod a_i(t)$ is not constant and has no square factor. Consider the equation $\sum a_i(t)x_i^2 = p(t)$. Assume that for all $t \in \mathbb{R}$ the real conic $\sum a_i(t)x_i^2 = 0$ has point over $\mathbb{R}$. Then strong approximation holds for the integral solutions of this equation. The set of solutions with coordinates in $\mathbb{Z}$ is dense in the product of integral local solutions at all finite primes. In particular, there is a local–global principle for integral points. The case where all the $a_i(t)$ are constant was dealt with in an earlier paper with F. Xu (Beijing). The result above is itself a special case of a general theorem on families of homogenous spaces of semisimple groups, obtained jointly with D. Harari (Paris-Sud).

Ulrich Derenthal. Counting points over imaginary quadratic number fields. For Fano varieties over number fields, the distribution of rational points is predicted by Manin's conjecture. One approach uses universal torsors and Cox rings; over the field Q of rational numbers, this was started by Salberger for toric varieties and was extensively studied for many examples of del Pezzo surfaces. In this talk, I present an extension of this approach to imaginary quadratic number fields, in particular for some singular quartic del Pezzo surfaces (joint work with C. Frei).

Roger Heath-Brown. Simultaneous representation of pairs of integers by quadratic forms. This is joint work with Lillian Pierce. We examine the simultaneous integer equations $Q_1(x_1, \ldots, x_n) = m_1$ and $Q_2(x_1, \ldots, x_n) = m_2$, and show, under a smoothness condition, that "almost all" pairs m_1, m_2 which have local representations also have a global solution, as soon as n is at least 5. One can use this to attack $Q_1(x_1, \ldots, x_n) = Q_2(x_1, \ldots, x_n) = 0$ when n is at least 10. The proof uses a two-dimensional circle method with a Kloostermann refinement.

Oscar Marmon. The density of twins of k-free numbers. For $k \geq 2$, we consider the number $A_k(Z)$ of positive integers $n \leq Z$ such that both n and $n+1$ are k-free. In joint work with Dietmann, we prove an asymptotic formula $A_k(Z) = ckZ + O(Z^{14/9k+\varepsilon})$, where the error term improves upon previously known estimates. The main tool used is the approximative determinant method of Heath-Brown.

David McKinnon. Approximating points on varieties. Famous theorems of Roth and Liouville give bounds on how well algebraic numbers can be approximated by rational numbers. In my talk, I will describe not-yet-famous generalizations of these theorems to arbitrary algebraic varieties, giving bounds on how well algebraic points can be approximated by rational points. This is joint work with Mike Roth of Queen's University, who is not yet related to famous-theorem Roth.

Ambrus Pál (Imperial College London). New two-dimensional counterexamples to the local–global principle. In my talk I will describe some new two-dimensional counter-examples to the local–global principle.

Florian Pop. Local–global principles for rational points. One of the most exciting developments originating from capacity theory (Fekete-Szego, Robinson, Cantor, Rumely) is Rumely's local–global principle, and its applications to the solvability of incomplete global Skolem problems (Cantor-Roquette, Roquette, and Moret-Bailly). My talk will be about complete Skolem problems and how they relate to some recent "complete capacity theory" results for curves developed by Rumely.

Mohamed Saidi. On the anabelian section conjecture over finitely generated fields. After introducing the section conjecture and some basic facts I will discuss in the first part of the talk a conditional result on the section conjecture over number fields. In the second part of the talk I will discuss the following result: the section conjecture holds over all finitely generated fields if it holds over all number fields, under the condition of finiteness of suitable Tate-Shafarevich groups.

Per Salberger. Heath-Brown's determinant method and Mumford's geometric invariant theory. Heath-Brown's p-adic determinant method is used to count rational points on hypersurfaces and was extended to subvarieties of codimension > 1 in projective space by Broberg and the author. The determinants that occur give rise to embeddings of Hilbert schemes in Grassmannians, which enables the use of techniques from geometric invariant theory. We describe some Diophantine applications and an unexpected link to the theory of Donaldson and Tian on Kähler metrics of constant scalar curvature.

Damaris Schindler. Manin's conjecture for certain smooth hypersurfaces in biprojective space. So far, the circle method has been a very useful tool to prove many cases of Manin's conjecture. Work of B. Birch back in 1961 establishes this for smooth complete intersections in projective space as soon as the number of variables is large enough depending on the degree and number of equations. In this talk we are interested in subvarieties of biprojective space. There is not much known so far, unless the underlying polynomials are of bidegree (1, 1). In this talk we present recent work which combines the circle method with the generalized hyperbola method developed by V. Blomer and J. Brüdern. This allows us to verify Manin's conjecture for certain smooth hypersurfaces in biprojective space of general bidegree.

Alexei Skorobogatov. Applications of additive combinatorics to rational points. This is a joint work with Y. Harpaz and O. Wittenberg. In 1982 Colliot-Thélène and Sansuc noticed that Schinzel's Hypothesis (H), used in a fibration method going back to Hasse, has strong implications for local–to-global principles for rational points. In the case of the ground field $\mathbb{Q}$ when the degenerate fibres are all defined over $\mathbb{Q}$, we show that in their method Hypothesis (H) can be replaced by the finite complexity case of the generalised Hardy-Littlewood conjecture, which is a recent theorem of Greeen, Tao and Ziegler. We sketch some of the applications of this observation.

Jeanine Van Order. Stable Galois averages of Rankin-Selberg L-values and non-triviality of p-adic L-functions. I will present the notion of a stable Galois average of central values of GL(2) Rankin-Selberg L-functions, and then explain how to derive some consequences for the nonvanishing of certain families of p-adic L-functions (with interesting applications).

Gabor Wiese (Université du Luxembourg). Symplectic Galois representations and applications to the inverse Galois problem. We give an account of joint work with Sara Arias-de-Reyna and Luis Dieulefait about compatible systems of symplectic Galois representations and how they can possibly be applied to the inverse Galois problem. In the beginning of the talk the overall strategy will be outlined, starting from a previous joint work with Dieulefait on the two-dimensional case. We will then explain the existence of a minimal global field such that almost all the residual representations (of the compatible system) can be defined projectively over its residue fields. Moreover, we shall report on a very simple classification of the symplectic representations containing a non-trivial transvection in their image. Finally, we shall combine the two points in an application to the inverse Galois problem.

Trevor Wooley. Applications of efficient congruencing to rational points. We provide an overview of the implications of the new "efficient congruencing" method, first developed for Vinogradov's mean value theorem, so far as applications to problems involving rational points are concerned. Some of these conclusions are close to the convexity barriers in the circle method.

1

Galois groups of local fields, Lie algebras and ramification

Victor Abrashkin

Department of Mathematical Sciences, Durham University, Science Laboratories, South Rd, Durham DH1 3LE, United Kingdom & Steklov Institute, Gubkina str. 8, 119991, Moscow, Russia

E-mail address: victor.abrashkin@durham.ac.uk

ABSTRACT. Suppose K is a local field with finite residue field of characteristic $p \neq 2$ and $K_{<p}(M)$ is its maximal p-extension such that $\mathrm{Gal}(K_{<p}(M)/K)$ has period p^M and nilpotent class $< p$. If char $K = 0$ we assume that K contains a primitive p^M-th root of unity. The paper contains an overview of methods and results describing the structure of this Galois group together with its filtration by ramification subgroups.

Introduction

Everywhere in the paper p is a prime number. For any profinite group Γ and $s \in \mathbb{N}$, $C_s(\Gamma)$ denotes the closure of the subgroup of commutators of order s.

Let K be a complete discrete valuation field with a finite residue field $k \simeq \mathbb{F}_{p^{N_0}}$, $N_0 \in \mathbb{N}$. Let K_{sep} be a separable closure of K and $\Gamma_K = \mathrm{Gal}(K_{sep}/K)$. Denote by $K(p)$ the maximal p-extension of K in K_{sep}. Then $\Gamma_K(p) = \mathrm{Gal}(K(p)/K)$ is a profinite p-group. As a matter of fact, the major information about Γ_K comes from the knowledge of the structure of $\Gamma_K(p)$. This structure is very well known and is related to the following three cases (ζ_p is a primitive p-th root of unity) [17]:

— char $K = p$;
— char $K = 0$, $\zeta_p \notin K$;
— char $K = 0$, $\zeta_p \in K$.

Date: Nov 8, 2013
Key words and phrases: local field, Galois group, ramification filtration.

Arithmetic and Geometry, ed. Luis Dieulefait *et al.* Published by Cambridge University Press.

In all these cases the maximal abelian quotient of period p of $\Gamma_K(p)$ is isomorphic to K^*/K^{*p}. Therefore, $\Gamma_K(p)$ has infinitely many generators in the first case, has $[K:\mathbb{Q}_p]+1$ generators in the second case and $[K:\mathbb{Q}_p]+2$ generators in the third case. In the first two cases $\Gamma_K(p)$ is free and in the last case it has one relation of a very special form, cf. [17, 23, 24].

The above results can't be considered as completely satisfactory because they do not essentially reflect the appearance of $\Gamma_K(p)$ as a Galois group of an algebraic extension of a local field. In other words, let LF be the category of couples (K, K_{sep}) where the morphisms are compatible continuous morphisms of local fields and let PGr be the category of profinite groups. Then the functor $(K, K_{sep}) \mapsto \Gamma_K(p)$ (as well as the functor $(K, K_{sep}) \mapsto \Gamma_K$) is not fully faithful.

The situation can be cardinally improved by taking into account a natural additional structure on $\Gamma_K(p)$ and Γ_K given by the decreasing filtration of ramification subgroups. The ramification filtration $\{\Gamma_K(p)^{(v)}\}_{v\geqslant 0}$ of $\Gamma_K(p)$ (as well as the appropriate filtration $\{\Gamma_K^{(v)}\}_{v\geqslant 0}$ of Γ_K) has many non-trivial properties. For example, it is left-continuous at any $v_0 \in \mathbb{Q}$, $v_0 > 1$, i.e. $\bigcap_{v<v_0}\Gamma_K(p)^{(v)} = \Gamma_K(p)^{(v_0)}$, but is not right-continuous, i.e. the closure of $\bigcup_{v>v_0}\Gamma_K(p)^{(v)}$ is not equal to $\Gamma_K(p)^{(v_0)}$. Another example [14, 15], for any $v_1, v_2 < v_0$, $(\Gamma_K(p)^{(v_1)}, \Gamma_K(p)^{(v_2)}) \not\subset \Gamma_K(p)^{(v_0)}$ and $(\Gamma_K(p)^{(v_1)})^p \not\subset \Gamma_K(p)^{(v_0)}$ and in some sense the groups $\Gamma_K(p)/\Gamma_K(p)^{(v_0)}$ have no "simple" relations [5].

The significance of study of ramification filtration was very well understood long ago, e.g. cf. Shafarevich's Introduction to [17]. (The author also had interesting discussions on this subject in the IAS with A.Weil, P.Deligne and F.Pop.) As a matter of fact, the knowledge of ramification filtration is equivalent to the knowledge of the original field K due to the following local analogue of the Grothendieck conjecture.

Theorem 1.1. *The functor* $(K, K_{sep}) \mapsto (\Gamma_K(p), \{\Gamma_K(p)^{(v)}\}_{v\geqslant 0})$ *from* LF *to the category of profinite* p*-groups with filtration is fully faithful.*

This result was first proved in the mixed characteristic case in the context of the whole Galois group Γ_K by Mochizuki [21] as a spectacular application of p-adic Hodge-Tate theory. The case of arbitrary characteristic was established by the author by a different method in [7] under the assumption $p \neq 2$. Note that the characteristic p case was obtained via the explicit description of ramification filtration modulo the subgroup of third commutators from [2]. Then the mixed characteristic case was deduced from it via the Fontaine-Wintenberger field-of-norms functor. In paper [11] we removed the restriction $p \neq 2$ and reproved the statement in the context of the pro-p-group $\Gamma_K(p)$.

The study of ramification filtration in full generality seems not to be a realistically stated problem: it is not clear how to specify subgroups of a given profinite p-group. If we replace $\Gamma_K(p)$ by its maximal abelian quotient $\Gamma_K(p)^{ab}$ then the appropriate ramification filtration is very well known but reflects very weak information about the original filtration of $\Gamma_K(p)$. This can be seen from class field theory where we have the reciprocity map $K^* \longrightarrow \Gamma_K^{ab}$ and the ramification subgroups appear as the images of the subgroups of principal units of K^*. In particular, we can observe only integral breaks of our filtration.

As a matter of fact, the ramification subgroups can be described on the abelian level without class field theory. The reason is that cyclic extensions of K can be studied via much more elementary tools: we can use the Witt-Artin-Schreier theory in the characteristic p case and the Kummer theory in the mixed characteristic case. Trying to develop this approach to the case of nilpotent Galois groups we developed in [1, 2] a nilpotent analogue of the Witt-Artin-Schreier theory. This theory allows us to describe quite efficiently p-extensions of fields of characteristic p with Galois p-groups of nilpotent class $< p$. Such groups arise from Lie algebras due to the classical equivalence of the categories of p-groups and Lie $\mathbb{F}_p$-algebras of nilpotent class $< p$, [20]. In [1, 2, 4] we applied our theory to local fields $\mathcal{K} = k((t_0))$, where $k \simeq \mathbb{F}_{p^{N_0}}$, and constructed explicitly the sets of generators of the appropriate ramification subgroups. This result demonstrates the advantage of our techniques: it is stated in terms of extensions of scalars of involved Lie algebras but this operation does not exist in group theory.

A generalization of our approach to local fields K of mixed characteristic was sketched earlier by the author in [6]. This approach allowed us to work with the groups $\Gamma_K / \Gamma_K^{p^M} C_p(\Gamma_K)$ under the assumption that a primitive p^M-th root of unity $\zeta_{p^M} \in K$. At that time we obtained explicit constructions of our theory only modulo subgroup of third commutators. Recently, we can treat the general case. First results are related to the case $M = 1$ and can be found in [11] (we discuss them also in Subsection 1.3.6 of this paper). The case of arbitrary M as well as the case of higher dimensional local fields will be considered in upcoming papers. In the case of local fields it would be very interesting to relate our theory to constructions of "nilpotent class field theory" from [19].

Note that the main constructions of the nilpotent Artin-Schreier theory do not suggest that the basic field is local. They can be applied also to global fields but it is not clear what sort of applications we can expect in this direction.

On the other hand, we can't expect the existence of an easy "nilpotent Kummer theory" for global fields. According to anabelian philosophy, for global fields E, the quotient of $\Gamma_E(p)$ by the subgroup of third commutators should already reflect all basic properties of the field E.

1.1 Nilpotent Artin-Schreier theory

In this section we discuss basic constructions of nilpotent Artin-Schreier theory. The main reference for this theory is [2]. We shall call this version contravariant and introduce also its covariant analogue, cf. Subsection 1.1.2 below. Everywhere M is a fixed natural number.

1.1.1 Lifts modulo p^M, $M \in \mathbb{N}$.

Suppose K is a field of characteristic p and K_{sep} is a separable closure of K. Let $\{x_i\}_{i\in I}$ be a p-basis for K. This means that the elements $x_i \bmod K^{*p}$, $i \in I$, form a basis of the $\mathbb{F}_p$-module K^*/K^{*p}. Note that if E is any subfield of K_{sep} containing K then $\{x_i\}_{i\in I}$ can be taken also as a p-basis for E.

Let W_M be the functor of Witt vectors of length M. For a field $K \subset E \subset K_{sep}$, define $O_M(E)$ as the subalgebra in $W_M(E)$ generated over $W_M(\sigma^{M-1}E)$ by the Teichmuller representatives $[x_i] \in W_M(K) \subset W_M(E)$ of all x_i. Then $O_M(E)$ is a lift of E modulo p^M: it is a flat $W_M(\mathbb{F}_p)$-algebra such that $O_M(E)/pO_M(E) = E$. The system of lifts $O_M(E)$ essentially depends on the original choice of a p-basis in K. If σ is the absolute Frobenius (i.e. the morphism of p-th powers) then $W_M(\sigma)$ induces a σ-linear morphism on $O_M(E)$ and we usually denote it again by σ. Note that $O_M(E)|_{\sigma=\mathrm{id}} = W_M(\mathbb{F}_p)$, if E is normal over K then the Galois group $\mathrm{Gal}(E/K)$ acts on $O_M(E)$ and the invariants of this action coincide with $O_M(K)$.

A (continuous) automorphism $\psi \in \mathrm{Aut}(E)$ generally can't be extended to $\mathrm{Aut}O_M(E)$ if ψ changes the original p-basis. But the morphism $\sigma^{M-1}\psi$ admits "almost a lift" $\sigma^{M-1}O_M(E) \longrightarrow O_M(E)$ given by the following composition

$$\sigma^{M-1}O_M(E) \subset W_M(\sigma^{M-1}E) \xrightarrow{W_M(\sigma^{M-1}\psi)} W_M(\sigma^{M-1}E) \subset O_M(E).$$

The existence of such lift allowed us to extend the modulo p methods from [1] to the modulo p^M situation in [2, 4].

1.1.2 Covariant and contravariant nilpotent Artin-Schreier theories

Suppose L is a Lie algebra over $W_M(\mathbb{F}_p)$. For $s \in \mathbb{N}$, let $C_s(L)$ be an ideal of s-th commutators in L, e.g. $C_2(L)$, resp., $C_3(L)$, is generated by the comutators $[l_1, l_2]$, resp. $[[l_1, l_2].l_3]$, where all $l_i \in L$. The algebra L has nilpotent class $< p$ if $C_p(L) = 0$.

The basic ingredient of our theory is the equivalence of the categories of p-groups of nilpotent class $< p$ and the category of Lie $\mathbb{Z}_p$-algebras of the same nilpotent class. This equivalence can be described on the level of objects killed by p^M as follows.

Suppose L is a Lie $W_M(\mathbb{F}_p)$-algebra of nilpotent class $< p$. If A is enveloping algebra for L and J is the augmentation ideal in A then there is a natural embedding of L into A/J^p (and L can be recovered as a submodule of the module of primitive elements modulo J^p in A, cf. [1] Section 1.1). The Campbell-Hausdorff formula is the map $L \times L \longrightarrow L$,

$$(l_1, l_2) \mapsto l_1 \circ l_2 = l_1 + l_2 + \frac{1}{2}[l_1, l_2] + \dots$$

such that in $A \bmod J^p$ we have $\widetilde{\exp}(l_1)\widetilde{\exp}(l_2) = \widetilde{\exp}(l_1 \circ l_2)$, where $\widetilde{\exp}(x) = \sum_{0 \leqslant i < p} x^i/i!$ is the truncated exponential. The set L can be provided with the composition law $(l_1, l_2) \mapsto l_1 \circ l_2$ which gives a group structure on L. We denote this group by $G(L)$. Clearly, this group has period p^M. Then the correspondence $L \mapsto G(L)$ is the above mentioned equivalence of the categories of p-groups of period p^M and Lie $W_M(\mathbb{F}_p)$-algebras.

Here and below we shall use the notation $L_K := L \otimes_{W_M(\mathbb{F}_p)} O_M(K)$ and $L_{K_{sep}} = L \otimes_{W_M(\mathbb{F}_p)} O_M(K_{sep})$. Then Γ_K and the absolute Frobenius σ act through the second factor on $L_{K_{sep}}$, $L_{K_{sep}}|_{\sigma=\mathrm{id}} = L$ and $(L_{K_{sep}})^{\Gamma_K} = L_K$. The covariant nilpotent Artin-Schreier theory states that for any $e \in G(L_K)$, the set $F(e) = \{f \in G(L_{K_{sep}}) \mid \sigma(f) = e \circ f\}$ is not empty and the map $g \mapsto (-f) \circ g(f)$ is a group homomorphism $\pi_f(e) : \Gamma_K \longrightarrow G(L)$. The correspondence $e \mapsto \pi_f(e)$ has the following properties:

a) if $f' \in F(e)$ then $f' = f \circ c$, where $c \in G(L)$, and $\pi_f(e)$ and $\pi_{f'}(e)$ are conjugated via c;
b) for any $\pi \in \mathrm{Hom}(\Gamma_K, G(L))$, there are $e \in G(L_K)$ and $f \in F(e)$ such that $\pi_f(e) = \pi$;
c) for appropriate elements $e, e' \in G(L_K)$ and $f, f' \in G(L_{K_{sep}})$, we have $\pi_f(e) = \pi_{f'}(e')$ iff there is an $x \in G(\mathcal{L}_K)$ such that $f' = x \circ f$ and (therefore) $e' = \sigma(x) \circ e \circ (-x)$; e and e' are called R-equivalent via $x \in G(L_K)$.

According to above properties a)–c), the correspondence $e \mapsto \pi_f(e)$ establishes an identification of the set of all R-equivalent elements in $G(L_K)$ and the set of all conjugacy classes of $\mathrm{Hom}(\Gamma_K, G(L))$.

The above theory can be proved in a similar way to its contravariant version established in [2]. In the contravariant theory for any $e \in G(L_K)$, the set $\{f \in G(L_{K_{sep}}) \mid \sigma(f) = f \circ e\}$ is not empty, the correspondence $g \mapsto g(f) \circ (-f)$

establishes a group homomorphism from Γ_K^0 to $G(L_K)$, where Γ_K^0 coincides with Γ_K as a set but has the opposite group law $(g_1g_2)^0 = g_2g_1$. (Equivalently, if $a \in K_{sep}$ then $(g_1g_2)a = g_2(g_1a)$.) We have also the properties similar to above properties a)–c) but in c) there should be $f' = f \circ x$ and $e' = x \circ e \circ (-\sigma x)$.

The both (covariant and contravariant) theories admit a pro-finite version where L becomes a profinite $W_M(\mathbb{F}_p)$-Lie algebra and the set $\mathrm{Hom}(\Gamma_K, G(L))$ is the set of all continuous group morphisms.

1.1.3 Identification η_0

Suppose $\mathcal{K} = k((t_0))$ where t_0 is a fixed uniformiser in $\mathcal{K}$ and $k \simeq \mathbb{F}_{p^{N_0}}$ with $N_0 \in \mathbb{N}$. Then $\{t_0\}$ is a p-basis for $\mathcal{K}$, and we have the appropriate system of lifts $O_M(\mathcal{E})$ modulo p^M for all subfields $\mathcal{K} \subset \mathcal{E} \subset \mathcal{K}_{sep}$. In addition, fix an element $\alpha_0 \in W(k)$ such that $\mathrm{Tr}(\alpha_0) = 1$, where Tr is the trace map for the field extension $W(k) \otimes_{\mathbb{Z}_p} \mathbb{Q}_p \supset \mathbb{Q}_p$.

Let $\mathbb{Z}^+(p) = \{a \in \mathbb{N} \mid (a, p) = 1\}$ and $\mathbb{Z}^0(p) = \mathbb{Z}^+(p) \cup \{0\}$.

For $M \in \mathbb{N}$, let $\widetilde{\mathcal{L}}_M$ be a profinite free Lie $\mathbb{Z}/p^M$-algebra with the (topological) module of generators $\mathcal{K}^*/\mathcal{K}^{*p^M}$ and $\mathcal{L}_M = \widetilde{\mathcal{L}}_M/C_p(\widetilde{\mathcal{L}}_M)$. From time to time we drop the subscript M off to simplify the notation.

Let $\mathcal{L} = \mathcal{L}_M$. Then $\mathcal{L}_k := \mathcal{L} \otimes W_M(k)$ has the generators

$$\{D_0\} \cup \{D_{an} \mid a \in \mathbb{Z}^+(p), n \in \mathbb{Z}/N_0\}$$

due to the following identifications (where $t = [t_0]$ is the Teichmuller representative of t_0):

$$\begin{gathered}
\mathcal{K}^*/\mathcal{K}^{*p^M} \otimes_{W_M(\mathbb{F}_p)} W_M(k) = \\
\mathrm{Hom}_{W_M(\mathbb{F}_p)}(O_M(\mathcal{K})/(\sigma - \mathrm{id})O_M(\mathcal{K}), W_M(k)) = \\
\mathrm{Hom}_{W_M(\mathbb{F}_p)}((W_M(\mathbb{F}_p)\alpha_0) \oplus_{a \in \mathbb{Z}^+(p)} (W_M(k)t^{-a}), W_M(k)) = \\
W_M(k)D_0 \times \prod_{\substack{a \in \mathbb{Z}^+(p) \\ n \in \mathbb{Z} \bmod N_0}} W_M(k)D_{an}
\end{gathered}$$

Note that the first identification uses the Witt pairing, D_0 appears from $t_0 \otimes 1 \in \mathcal{K}^*/\mathcal{K}^{*p^M} \otimes W_M(k)$ and for all $a \in \mathbb{Z}^+(p)$ and $w \in W_M(k)$, $D_{an}(wt^{-a}) = \sigma^n w$.

For any $n \in \mathbb{Z}/N_0$, set $D_{0n} = t \otimes (\sigma^n \alpha_0) = (\sigma^n \alpha_0) D_0$.

Let $e_0 = \sum_{a \in \mathbb{Z}^0(p)} t^{-a} D_{a0} \in \mathcal{L}_K$, choose $f_0 \in F(e_0)$ and set $\eta_0 = \pi_{f_0}(e_0)$. Then η_0 is a surjective homomorphism from $\Gamma_{\mathcal{K}}$ to $G(\mathcal{L})$ and it induces a group isomorphism $\Gamma_{\mathcal{K}}/\Gamma_{\mathcal{K}}^{p^M} C_p(\Gamma_{\mathcal{K}}) \simeq G(\mathcal{L})$. Note that the construction of

η_0 depends up to conjugacy only on the original choice of the uniformizer t_0 and the element $\alpha_0 \bmod p^M \in W_M(k)$. On the level of maximal abelian quotients of period p^M, η_0 induces the isomorphism of local class field theory $\Gamma_{\mathcal{K}}^{ab} \otimes_{\mathbb{Z}_p} W_M(\mathbb{F}_p) \simeq \mathcal{K}^*/\mathcal{K}^{*p^M}$.

1.1.4 Why Campbell-Hausdorff?

In this subsection it will be explained that in our theory, we are, essentially, forced to use the Campbell-Hausdorff composition law.

Assume for simplicity, that $M = 1$ and $\mathcal{K} = \mathbb{F}_p((t_0))$. Let $\mathcal{K}(p)$ be the maximal p-extension of $\mathcal{K}$ and $\Gamma_{\mathcal{K}}(p) = \mathrm{Gal}(\mathcal{K}(p)/\mathcal{K})$. For $s \in N$ and $a_1, a_2, \dots, a_s, \dots \in \mathbb{Z}^0(p)$, consider the elements $T_{a_1\dots a_s} \in \mathcal{K}(p)$ such that:

$$T_{a_1}^p - T_{a_1} = t_0^{-a_1},$$
$$T_{a_1a_2}^p - T_{a_1a_2} = t_0^{-a_1}T_{a_2}$$
$$\dots\dots\dots$$
$$T_{a_1\dots a_s}^p - T_{a_1\dots a_s} = t_0^{-a_1}T_{a_2\dots a_s}$$
$$\dots\dots\dots$$

Then the system $\{T_{a_1\dots a_s} \mid s \geqslant 0, a_i \in \mathbb{Z}^0(p)\}$ is linearly independent over $\mathcal{K}$ and if $\mathcal{M} = \bigoplus_{\substack{a_1,\dots,a_s\\ s\geqslant 0}} \mathbb{F}_p T_{a_1\dots a_s}$ then $\mathcal{K}(p) = \mathcal{M} \otimes_{\mathbb{F}_p} \mathcal{K}$ and $\Gamma_{\mathcal{K}}(p)$ acts on $\mathcal{M}$ via a natural embedding $\Gamma_{\mathcal{K}}(p) \hookrightarrow \mathrm{GL}_{\mathbb{F}_p}(\mathcal{M})$. This construction would have given us an efficient approach to an explicit construction of the maximal p-extension $\mathcal{K}(p)$ if we could describe explicitly the image of $\Gamma_{\mathcal{K}}(p)$ in $\mathrm{GL}_{\mathbb{F}_p}(\mathcal{M})$.

Analyze the situation at different levels $s \geqslant 1$.

- **1st level.** Here all equations are independent and we can introduce a minimal system of generators τ_a, $a \in \mathbb{Z}^0(p)$, of $\Gamma_{\mathcal{K}}(p)$ with their explicit action via $\tau_a : T_{a_1} \mapsto T_{a_1} + \delta(a, a_1)$ at this level. (Here and below δ is the Kronecker symbol.)
- **2nd level.** Here the roots $T_{a_1a_2}$ are not (algebraically) independent. For example, the following identity

  $$(T_{a_1}T_{a_2})^p = (T_{a_1}+t_0^{-a_1})(T_{a_2}+t_0^{-a_2}) = T_{a_1}T_{a_2}+t_0^{-a_1}T_{a_2}+t_0^{-a_2}T_{a_1}+t_0^{-(a_1+a_2)}$$

  implies under the assumption $(a_1 + a_2, p) = 1$ (and after a suitable choice of involved roots of Artin-Schreier equations) that

  $$T_{a_1}T_{a_2} = T_{a_1a_2} + T_{a_2a_1} + T_{a_1+a_2}.$$

The presence of the term $T_{a_1+a_2}$ creates a problem: $\tau_{a_1+a_2}$ should act non-trivially on either $T_{a_1a_2}$ or $T_{a_2a_1}$ but they both do not depend on the index a_1+a_2. The situation can be resolved by a slight correction of involved equations. Namely, let $T_{a_1a_2}$ be such that

$$T^p_{a_1a_2} - T_{a_1a_2} = t_0^{-a_1} T_{a_2} + \eta(a_1, a_2) t_0^{-(a_1+a_2)}$$

where the constants $\eta(a_1, a_2) \in k$, $a_1, a_2 \in \mathbb{Z}^0(p)$, satisfy the relations

$$\eta(a_1, a_2) + \eta(a_2, a_1) = 1. \tag{1.1}$$

With the above correction, the elements $T_{a_1a_2}$, $a_1, a_2 \in \mathbb{Z}^0(p)$, can be chosen in such a way that we have the following:

— relations: $T_{a_1}T_{a_2} = T_{a_1a_2} + T_{a_2a_1}$;
— Galois action: $\tau_a(T_{a_1a_2}) = T_{a_1a_2} + T_{a_1}\delta(a_2, a) + \eta(a_1, a_2)\delta(a_1, a_2, a)$.

Relation (1.1) will look more natural if we introduce the constants on the first level via $\eta(a) = 1$, $a \in \mathbb{Z}^0(p)$. Then (1.1) can be rewritten as $\eta(a_1)\eta(a_2) = \eta(a_1, a_2) + \eta(a_2, a_1)$. These relations can be satisfied only if $p \neq 2$ and the simplest choice is $\eta(a_1, a_2) = 1/2$ for all a_1, a_2.

The above picture can be generalized to higher levels as follows.

- **s-th level, $s < p$.** Here we have:
  - the equations: $T^p_{a_1\dots a_s} = T_{a_1\dots a_s} + \eta(a_1)t^{-a_1}T_{a_2\dots a_s} + \dots$
    $+\eta(a_1, \dots, a_{s-1})t_0^{-(a_1+\dots+a_{s-1})}T_{a_s} + \eta(a_1, \dots, a_s)t_0^{-(a_1+\dots+a_s)}$
  - the relations: $T_{a_1\dots a_k}T_{b_1\dots b_l} = \sum T_{\text{insertions of } a\text{'s into } b\text{'s}}$, where $k + l < p$;
  - the Galois action: $\tau_a(T_{a_1\dots a_s}) = T_{a_1\dots a_s} + T_{a_1\dots a_{s-1}}\delta(a, a_s)\eta(a_s) +$
    $\dots + T_{a_1}\delta(a, a_2, \dots, a_s)\eta(a_2, \dots, a_s) + \delta(a, a_1, \dots, a_s)\eta(a_1, \dots, a_s)$
  - the constants: if $k + l < p$ then
    $\eta(a_1, \dots, a_k)\eta(b_1, \dots, b_l) = \sum \eta(\text{insertions of } a\text{'s into } b\text{'s})$
  with their simplest choice $\eta(a_1, \dots, a_s) = 1/s!$

Remark. An insertion of the ordered collection $a_1, \dots, a_k$ into the ordered collection $b_1, \dots, b_l$ is the ordered collection $c_1, \dots, c_{k+l}$ such that

— $\{1, \dots, k+l\} = \{i_1, \dots, i_k\} \coprod \{j_1, \dots, j_l\}$;
— $i_1 < \dots < i_k$ and $j_1 < \dots < j_l$;
— $a_1 = c_{i_1}, \dots, a_k = c_{i_k}$ and $b_1 = c_{j_1}, \dots, b_l = c_{j_l}$.

The following formalism allows us to present the above information on all levels $1 \leqslant s < p$ in the following compact way.

Let $\widetilde{\mathcal{A}}$ be a pro-finite associative $\mathbb{F}_p$-algebra with the set of free generators $\{D_a \mid a \in \mathbb{Z}^0(p)\}$. Introduce the elements of the appropriate extensions of scalars of $\mathcal{A}$

$$E = 1 + \sum_{\substack{1 \leqslant s < p \\ a_i \in \mathbb{Z}^0(p)}} \eta(a_1, \ldots, a_s) t_0^{-(a_1 + \cdots + a_s)} D_{a_1} \ldots D_{a_s}$$

$$= \widetilde{\exp}\Big(\sum_{a \in \mathbb{Z}^0(p)} t_0^{-a} D_a \Big) \in \mathcal{A}_{\mathcal{K}},$$

$$\mathcal{F} = 1 + \sum_{\substack{1 \leqslant s < p \\ a_i \in \mathbb{Z}^0(p)}} \eta(a_1, \ldots, a_s) T_{a_1 \ldots a_s} D_{a_1} \ldots D_{a_s} \in \mathcal{A}_{\mathcal{K}_{sep}}$$

Define the diagonal map as the morphism of $\mathbb{F}_p$-algebras

$$\Delta : \mathcal{A} \operatorname{mod} \deg p \longrightarrow \mathcal{A} \otimes \mathcal{A} \operatorname{mod} \deg p$$

such that for any $a \in \mathbb{Z}^0(p)$, $D_a \mapsto D_a \otimes 1 + 1 \otimes D_a$. Then we have the following properties:

— $\Delta(E) \equiv E \otimes E \operatorname{mod} \deg p$; $\Delta(\mathcal{F}) \equiv \mathcal{F} \otimes \mathcal{F} \operatorname{mod} \deg p$;
— $\sigma(\mathcal{F}) \equiv E\mathcal{F} \operatorname{mod} \deg p$; $\tau_a(\mathcal{F}) \equiv \mathcal{F}\, \widetilde{\exp}(D_a) \operatorname{mod} \deg p$.

Now we can verify the existence of $f \in \mathcal{L}_{\mathcal{K}_{sep}}$ such that $\mathcal{F} = \widetilde{\exp}(f)$ modulo deg p, and recover the basic relations $\sigma(f) = (\sum_a t_0^{-a} D_a) \circ f$ and $\tau_a(f) = f \circ D_a$, $a \in \mathbb{Z}^0(p)$, of our nilpotent Artin-Schreier theory.

1.2 Ramification filtration in $\mathcal{L} = \mathcal{L}_{M+1}$

In this section we describe and illustrate the main trick used in papers [1, 2, 4]. This trick allowed us to find explicit generators of ramification subgroups under the identification η_0 from Subsection 1.1.3. Remind that we work over $\mathcal{K} = k((t_0))$, where $k \simeq \mathbb{F}_{p^{N_0}}$, $N_0 \in \mathbb{N}$.

1.2.1 Auxiliary field $\mathcal{K}' = \mathcal{K}(r^*, N)$, [1, 2, 4]

The field $\mathcal{K}'$ is a totally ramified extension of $\mathcal{K}$ in $\mathcal{K}_{sep}$. It depends on two parameters: $r^* \in \mathbb{Q}$ such that $r^* > 0$ and $v_p(r^*) = 0$, and $N \in \mathbb{N}$ such that if $q = p^N$ then $b^* := r^*(q-1) \in \mathbb{N}$. Note that for a given r^*, there are infinitely many ways to choose N, in particular, we can always assume that N is sufficiently large.

By definition, $[\mathcal{K}' : \mathcal{K}] = q$ and the Herbrand function $\varphi_{\mathcal{K}'/\mathcal{K}}$ has only one edge point (r^*, r^*). It can be proved that $\mathcal{K}' = k((t'_0))$, where $t_0 = t_0'^{q} E(-1, t_0'^{b^*})$. Here for $w \in W(k)$,

$$E(w, X) = \exp(wX + \sigma(w)X^p/p + \dots + \sigma^n(w)X^{p^n}/p^n + \dots) \in \mathbb{Z}_p[[X]]$$

is the Shafarevich version of the Artin-Hasse exponential.

Note that if $r^* \notin \mathbb{N}$, $\mathcal{K}'/\mathcal{K}$ is neither Galois nor a p-extension.

1.2.2 The criterion

Consider the following lifts modulo p^{M+1} with respect to the p-basis $\{t_0\}$ of $\mathcal{K}$

$$O_{M+1}(\mathcal{K}) = W_{M+1}(k)((t)) = W_{M+1}(\sigma^M \mathcal{K})[t]$$
$$O_{M+1}(\mathcal{K}_{sep}) = W_{M+1}(\sigma^M \mathcal{K}_{sep})[t] \subset W_{M+1}(\mathcal{K}_{sep}).$$

Remember that $t = [t_0] \in O_{M+1}(\mathcal{K})$ is the Teichmüller representative of t_0 in $W_{M+1}(\mathcal{K})$.

For $\mathcal{K}' = \mathcal{K}(r^*, N)$ and its uniformiser t'_0 from Subsection 1.2.1 consider the appropriate lifts $O'_{M+1}(\mathcal{K}')$ and $O'_{M+1}(\mathcal{K}'_{sep})$. If $t' = [t'_0]$ then t and t' can be related one-to-another in $W_{M+1}(\mathcal{K}')$ via

$$t^{p^M} = t'^{p^M q} \exp(-p^M t'^{b^*} - \dots - p t'^{p^{M-1} b^*}) E(-1, t'^{p^M b^*}).$$

This implies the following relations between the lifts for $\mathcal{K}$ and $\mathcal{K}'$

$$\sigma^M O_{M+1}(\mathcal{K}) \subset W_{M+1}(\sigma^M \mathcal{K}) \subset O'_{M+1}(\mathcal{K}')$$
$$\sigma^M O_{M+1}(\mathcal{K}_{sep}) \subset W_{M+1}(\sigma^M \mathcal{K}_{sep}) \subset O'_{M+1}(\mathcal{K}'_{sep})$$

As earlier, take $e_0 = \sum_{a \in \mathbb{Z}^0(p)} t^{-a} D_{a0} \in \mathcal{L}_{\mathcal{K}}$, $f_0 \in \mathcal{L}_{\mathcal{K}_{sep}}$ such that $\sigma f_0 = e_0 \circ f_0$ and consider $\pi_{f_0}(e_0) : \Gamma_{\mathcal{K}} \longrightarrow G(\mathcal{L})$. Similarly, let $e'_0 = \sum_{a \in \mathbb{Z}^0(p)} t'^{-a} D_{a,-N}$, choose $f'_0 \in \mathcal{L}_{\mathcal{K}_{sep}}$ such that $\sigma f'_0 = e'_0 \circ f'_0$ and consider $\pi_{f'_0}(e'_0) : \Gamma_{\mathcal{K}'} \longrightarrow G(\mathcal{L})$.

For $Y \in \mathcal{L}_{\mathcal{K}_{sep}}$ and an ideal $\mathcal{I}$ in $\mathcal{L}$, define the field of definition of $Y \bmod \mathcal{I}_{\mathcal{K}_{sep}}$ over $\mathcal{K}$ as $\mathcal{K}(Y \bmod \mathcal{I}_{\mathcal{K}_{sep}}) := \mathcal{K}_{sep}^{\mathcal{H}}$, where $\mathcal{H} = \{g \in \Gamma_{\mathcal{K}} \mid g(Y) \equiv Y \bmod \mathcal{I}_{\mathcal{K}_{sep}}\}$.

For any finite field extension $\mathcal{E}/\mathcal{K}$ in $\mathcal{K}_{sep}$ define its biggest ramification number $v(\mathcal{E}/\mathcal{K}) = \max\{v \mid \Gamma_{\mathcal{K}}^{(v)} \text{ acts non-trivially on } \mathcal{E}\}$.

For $v_0 \in \mathbb{Q}_{>0}$, let the ideal $\mathcal{L}^{(v_0)}$ of $\mathcal{L}$ be such that $G(\mathcal{L}^{(v_0)}) = \eta_0(\Gamma_{\mathcal{K}}^{(v_0)})$. Let $f_M = \sigma^M f_0$ and $f'_M = \sigma^M f'_0$. Our method from [1, 2, 4] is based on the following criterion.

Proposition 1.2. *Let $X \in \mathcal{L}_{\mathcal{K}_{sep}}$ be such that $f_M = X \circ \sigma^N(f'_M)$. Suppose $v_0, r^* \in \mathbb{Q}_{>0}$, $v_p(r^*) = 0$ and $r^* < v_0$. Then $\mathcal{L}^{(v_0)}$ is the minimal ideal in the family of all ideals $\mathcal{I}$ of $\mathcal{L}$ such that*

$$v(\mathcal{K}'(X \bmod \mathcal{I}_{\mathcal{K}_{sep}})/\mathcal{K}') \leqslant v_0 q - b^*.$$

The proof is quite formal and is based on the following properties of upper ramification numbers. If $v = v(\mathcal{K}(f \bmod \mathcal{I}_{\mathcal{K}_{sep}})/\mathcal{K})$ then:

a) $v(\mathcal{K}'(f' \bmod \mathcal{I}_{\mathcal{K}'_{sep}})/\mathcal{K}') = v$;
b) $v(\mathcal{K}'(f' \bmod \mathcal{I}_{\mathcal{K}'_{sep}})/\mathcal{K}) = \varphi_{\mathcal{K}'/\mathcal{K}}(v)$;
c) if $v > r^*$ then $\varphi_{\mathcal{K}'/\mathcal{K}}(v) < v$.

1.2.3 Illustration of the criterion

The criterion from Proposition 1.2 was applied in [1, 2, 4] to describe the structure of $\eta_0(\Gamma_{\mathcal{K}}^{(v)})$ by induction by proceeding from the situation modulo p^M to the situation modulo p^{M+1} and from the situation modulo $C_s(\mathcal{L})$ to the situation modulo $C_{s+1}(\mathcal{L})$, where $2 \leqslant s < p$.

Typically, for an ideal $I \subset \mathcal{L}$, we used the knowledge of the structure of $\mathcal{L}^{(v_0)} \bmod I$ to prove (after choosing r^* sufficiently close to v_0) that $X \in \mathcal{L}_{\mathcal{K}'} \bmod (\mathcal{L}_{\mathcal{K}'}^{(v_0)} + I_{\mathcal{K}'})$. Then we could apply our Criterion to $\mathcal{L}^{(v_0)} \bmod J$ for an appropriate (slightly smaller than I) ideal J because $X \bmod J_{\mathcal{K}_{sep}}$ satisfied over $\mathcal{L}_{\mathcal{K}'}$ just an Artin-Schreier equation of degree p. Notice that $f_0 \bmod J_{\mathcal{K}_{sep}}$ and $f'_0 \bmod J_{\mathcal{K}_{sep}}$ satisfy very complicated relations over $\mathcal{L}_{\mathcal{K}'}$. We give below two examples to illustrate how our method works in more explicit but similar situations.

1.2.3.1 First example

Suppose $M \geqslant 0$ and $F \in O_{M+1}(\mathcal{K}_{sep})$ is such that $F - \sigma^{N_0} F = t^{-a}$, $a \in \mathbb{Z}^0(p)$.

If $M = 0$ then F is a root of the Artin-Schreier equation $F - F^{q_0} = t^{-a}$ with $q_0 = p^{N_0}$ and directly from the definition of ramification subgroups it follows that $v(\mathcal{K}(F)/\mathcal{K}) = a$. The case of arbitrary M corresponds to the Witt theory. Here the left-hand side $F - \sigma^{N_0} F$ is already a Witt vector of length $M + 1$, and careful calculations with components of Witt vectors give that $v(\mathcal{K}(F)/\mathcal{K}) = p^M a$. Our criterion allows us to obtain this result in a much easier way.

Take the field $\mathcal{K}' = \mathcal{K}(r^*, N)$ where r^* and N (recall that $q = p^N$) are such that

$$ap^M > r^* > ap^{M-1} q/(q-1). \tag{1.2}$$

Consider the appropriate lift $O'_{M+1}(\mathcal{K}'_{sep})$ and let $F' \in O'_{M+1}(\mathcal{K}'_{sep})$ be such that $F' - \sigma^{N_0} F' = t'^{-a}$.

Set $F_M = \sigma^M F$ and $F'_M = \sigma^M F'$.

Clearly, $\mathcal{K}(F_M) = \mathcal{K}(F)$ and $\mathcal{K}'(F'_M) = \mathcal{K}'(F')$.

According to restrictions (1.2) we have

$$\begin{aligned} t^{-ap^M} &= t'^{-aqp^M} \exp(ap^M t'^{b^*} + \cdots + apt'^{p^{M-1}b^*}) E(a, t'^{p^M b^*}) \\ &= t'^{-aqp^M} + ap^M t'^{-ap^M q + b^*} + f_0, \end{aligned}$$

where $f_0 \in t' W_M(k)[[t']] \subset O'_{M+1}(\mathcal{K}')$. Therefore,

$$F_M = \sigma^N F'_M + p^M X + \sum_{i \geqslant 0} \sigma^{iN_0}(f_0),$$

where $X - \sigma^{N_0} X = at'^{-ap^M q + b^*}$.

In this situation an analogue of our criterion states that

$$v(\mathcal{K}(F_M)/\mathcal{K}) = ap^M \quad \Leftrightarrow v(\mathcal{K}'(X)/\mathcal{K}') = ap^M q - b^*$$

But the right-hand side of this assertion corresponds to the case $M = 0$ and was explained in the beginning of this section.

1.2.3.2 Second example

Consider the following modulo p situation, i.e. the situation where $M = 0$. (The appropriate case of arbitrary M can be considered similarly.)

Let $F, G \in \mathcal{K}_{sep}$ be such that

$$F - \sigma^{N_0} F = t_0^{-a}, \quad G - \sigma^{N_0} G = t_0^{-b} \sigma^{n_0}(F) \tag{1.3}$$

with $a, b \in \mathbb{Z}^0(p)$ and $0 \leqslant n_0 < N_0$. Let

$$A = \max\left\{a + bp^{-n_0}, ap^{-N_0+n_0} + b\right\}$$

Prove that $v(\mathcal{K}(F, G)/\mathcal{K}) = A$ if either $a \neq b$ or $2n_0 \neq N_0$.

Take $\mathcal{K}' = \mathcal{K}'(r^*, N)$, where r^* and $q = p^N$ satisfy the following restrictions

$$\frac{qA}{2(q-1)} < r^* < \frac{qA}{q-1}, \quad r^* > \frac{q}{q-1} \max\{a, ap^{-N_0} + b\}$$

(We can take $r^* \in (A/2, A)$ such that $r^* > \max\{a, b\}$ and then choose sufficiently large N to satisfy these conditions.) Consider $F', G' \in \mathcal{K}'_{sep}$ such that

$$F' - \sigma^{N_0} F' = t_0'^{-a}, \quad G' - \sigma^{N_0} G' = t_0'^{-b} \sigma^{n_0}(F')$$

Notice that

$$t_0^{-a} = t_0'^{-aq} + at_0'^{-aq+b^*} + o_1$$

where $o_1 \in t_0'^{-aq+2b^*}k[[t_0']]$. Therefore,

$$F = F'^q + T_F + \sum_{i\geqslant 0} \sigma^{iN_0} o_1, \tag{1.4}$$

where $T_F - \sigma^{N_0} T_F = at_0'^{-aq+b^*}$. We can choose F' in such a way that (use that $-aq + b^* > 0$) $T_F = at_0'^{-aq+b^*} + o_2$, where $o_2 \in t_0'^{p^{N_0}(-aq+b^*)}k[[t_0']]$.

As earlier,

$$t_0^{-b} = t_0'^{-bq} + bt_0'^{-bq+b^*} + o_3 \tag{1.5}$$

with $o_3 \in t_0'^{-bq+2b^*}k[[t_0']]$. Then (1.4) and (1.5) imply that

$$\begin{aligned} t_0^{-b}\sigma^{n_0} F = (t_0'^{-bq}\sigma^{n_0} F')^q + at_0'^{-bq+p^{n_0}(-aq+b^*)} + \\ bt_0'^{-bq+b^*}(\sigma^{n_0} F')^q + o_3(\sigma^{n_0} F')^q + o_4 \end{aligned}$$

where $o_4 \in t_0' k[[t_0']]$. Therefore, $G = G'^q + T_G + \sum_{i\geqslant 0} \sigma^{iN_0} o_4$, where

$$\begin{aligned} T_G - \sigma^{N_0} T_G = at'^{-bq+p^{n_0}(-aq+b^*)} \\ +bt'^{-bq+b^*}(\sigma^{n_0} F')^q + o_3(\sigma^{n_0} F')^q \end{aligned}$$

An appropriate analogue of our criterion gives that

$$v(\mathcal{K}(F, G)/\mathcal{K}) = A \iff v(\mathcal{K}'(T_F, T_G)/\mathcal{K}') = qA - b^*$$

First of all, $T_F \in k[[t_0']]$ and, therefore, we should prove that $v(\mathcal{K}'(T_G)/\mathcal{K}') = qA - b^*$. Notice that

$$F'^{qp^{n_0}} = (\sigma^{N_0} F')^{qp^{n_0-N_0}} = F'^{qp^{n_0-N_0}} - t_0'^{-aqp^{n_0-N_0}}.$$

Let $O_{\mathcal{K}'(F')}$ and $\mathfrak{m}_{\mathcal{K}'(F')}$ be the valuation ring and, resp., the maximal ideal for $\mathcal{K}'(F')$. Clearly, $t_0'^a \sigma^{N_0} F' \in O_{\mathcal{K}'(F')}$ and, therefore,

$$t_0'^{-bq+b^*} F'^{qp^{n_0-N_0}} \in \mathfrak{m}_{\mathcal{K}'(F')}$$

(use that $b^* > q(ap^{n_0-2N_0} + b)$) and $o_3(\sigma^{n_0} F')^q \in \mathfrak{m}_{\mathcal{K}'(F')}$ (use that $2b^* - (ap^{n_0-N_0} + b) > 0$).

This implies that $\mathcal{K}'(T_G) \subset \mathcal{K}'(T_G, F') = \mathcal{K}'(T, F')$, where

$$T - \sigma^{N_0} T = a\sigma^{n_0}\left(t'^{-(a+bp^{-n_0})q+b^*}\right) - bt_0'^{-(ap^{n_0-N_0}+b)q+b^*},$$

and $v(\mathcal{K}'(T_G)/\mathcal{K}') = v(\mathcal{K}'(T, F')/\mathcal{K}')$.

If either $a \neq b$ or $2n_0 \neq N_0$ the right-hand side of this equation is not trivial and (use that $v(\mathcal{K}'(F')/\mathcal{K}') = a < A$) $v(\mathcal{K}'(T, F')/\mathcal{K}') = v(\mathcal{K}'(T)/\mathcal{K}') = qA - b^*$.

1.2.4 Ramification subgroups modulo $\Gamma_{\mathcal{K}}^{p^M} C_p(\Gamma_{\mathcal{K}})$

As earlier, let $\mathcal{L} = \mathcal{L}_M$ and for $v \geqslant 0$, let $\mathcal{L}^{(v)} = \eta_0(\Gamma_{\mathcal{K}}^{(v)}) \subset \mathcal{L}$. The ideal $\mathcal{L}^{(v)}$ was described in [4] as follows.

For $\gamma \geqslant 0$ and $N \in \mathbb{Z}$, introduce the elements $\mathcal{F}^0_{\gamma,-N} \in \mathcal{L}_k$ via

$$\mathcal{F}^0_{\gamma,-N} = \sum_{\substack{1\leqslant s<p \\ a_i,n_i}} a_1 p^{n_1} \eta(n_1, \ldots, n_s)[\ldots[D_{a_1\bar{n}_1}, D_{a_2\bar{n}_2}], \ldots, D_{a_s\bar{n}_s}]$$

Here:

- all $a_i \in \mathbb{Z}^0(p)$, $n_i \in \mathbb{Z}$, $n_1 \geqslant 0$, $n_1 \geqslant n_2 \geqslant \cdots \geqslant n_s \geqslant -N$, $\bar{n}_s = n_s \bmod N_0$;
- $a_1 p^{n_1} + a_2 p^{n_2} + \cdots + a_s p^{n_s} = \gamma$;
- if $n_1 = \cdots = n_{s_1} > \cdots > n_{s_{r-1}+1} = \cdots = n_{s_r}$ then $\eta(n_1, \ldots, n_s) = (s_1! \ldots (s_r - s_{r-1})!)^{-1}$.

Let $\mathcal{L}(v)_N$ be the minimal ideal of $\mathcal{L}$ such that $\mathcal{L}(v)_N \otimes W_M(k)$ contains all $\mathcal{F}^0_{\gamma,-N}$ with $\gamma \geqslant v$. Then there is an $N^*_M(v) \in \mathbb{N}$ and an ideal $\mathcal{L}(v)$ of $\mathcal{L}$ such that for all $N \geqslant N^*_M(v)$, $\mathcal{L}(v)_N = \mathcal{L}(v)$.

Theorem 1.3. *For any $v \geqslant 0$, $\mathcal{L}^{(v)} = \mathcal{L}(v)$.*

This statement was obtained in the contravariant setting in [4] and uses the elements $\mathcal{F}_{\gamma,-N}$ given by the same formula (as for $\mathcal{F}^0_{\gamma,-N}$) but with the factor $(-1)^{s-1}$. Indeed, the contravariant version of Theorem 1.3 appears by replacing the Lie bracket $[l_1, l_2]$ in $\mathcal{L}$ by the bracket $[l_1, l_2]^0 = [l_2, l_1]$. Therefore, $[\ldots[D_1, D_2], \ldots, D_s]$ should be replaced by $[D_s, \ldots, [D_2, D_1] \ldots] = (-1)^{s-1}[\ldots[D_1, D_2], \ldots, D_s]$.

Remark. For the ideal $\mathcal{L}^{(v)} \bmod C_2(\mathcal{L})$ we have the generators coming from $\mathcal{F}^0_{\gamma,-N}$ taken modulo the ideal of second commutators. Such generators are non-zero only if γ is integral. Therefore, $\mathcal{L}^{(v)} \otimes W(k)$ is generated on the abelian level by the images of $p^n D_{am}$, where $m \in \mathbb{Z}/N_0$ and $p^n a \geqslant v$.

The proof of Theorem 1.3 is quite technical and it would be nice to put it into a more substantial context. This could be done on the basis of the following interpretation.

Assume for simplicity that $M = 0$. Choose r^*, N as in Subsection 1.2.1. We can assume that $N \equiv 0 \bmod N_0$.

If we replace t_0 by t_0^q then the identification η_0 will be not changed. Indeed, $e_0 = \sum_{a\in\mathbb{Z}^0(p)} t_0^{-a} D_{a0}$ is replaced by $\iota(e_0) = \sum_{a\in\mathbb{Z}^0(p)} t_0^{-aq} D_{a0} = (\sigma c) \circ e_0 \circ (-c)$, where $c = (\sigma^{N-1} e_0) \circ \cdots \circ (\sigma e_0) \circ e_0$.

When proving above Theorem 1.3 in [4] we actually established that $\mathcal{L}^{(v_0)}$ appears as the minimal ideal $\mathcal{I}$ of $\mathcal{L}$ such that the replacement (deformation) $d(1) : t_0 \mapsto t_0^q E(-1, t_0^{b^*})$ does not affect the identification $\eta_0 \bmod \mathcal{I}$. The same holds also for the one-parameter deformation $\mathcal{D}(u) : t_0 \mapsto t_0^q E(-u, t_0^{b^*})$ with parameter u. In terms of the nilpotent Artin-Schreier theory this means the existence of $c = c(u) \in \mathcal{L}_{\mathcal{K}[u]}$ such that

$$\mathcal{D}(u)(e_0) \equiv (\sigma c) \circ e_0 \circ (-c) \bmod \mathcal{I}_{\mathcal{K}[u]}$$

(we assume that $\sigma(u) = u$). This condition is satisfied on the linear level (i.e. for the coefficients of u) if and only if the above elements $\mathcal{F}_{\gamma,-N}$, where $\gamma \geqslant v$ and $N \geqslant N_M^*(v)$, belong to $\mathcal{I}_k$. As a matter of fact, the main difficulty we resolved in [1, 4] was that on the level of higher powers u^i, $i > 1$, we do not obtain new conditions. We do expect to obtain this fact in an easier way by more substantial use of ideas of deformation theory.

1.3 The mixed characteristic case

In this section we sketch main ideas which allowed us to apply the above characteristic p results to the mixed characteristic case.

Suppose K is a finite etension of $\mathbb{Q}_p$ with the residue field k. We fix a choice of uniformising element π_0 in K and assume that K contains a primitive p^M-th root of unity ζ_M.

1.3.1 The field of norms functor [26]

Let $\widetilde{K}$ be a composite of the field extensions $K_n = K(\pi_n)$, where $n \geqslant 0$, $K_0 = K$, and $\pi_{n+1}^p = \pi_n$. The field of norms functor $\mathcal{X}$ provides us with:

1) a complete discrete valuation field $\mathcal{K} = \mathcal{X}(\widetilde{K})$ of characteristic p. The residue field of $\mathcal{K}$ can be canonically identified with k, and $\mathcal{K}$ has a fixed uniformizer t_0: by definition, $\mathcal{K}^* = \varprojlim K_n^*$, where the connecting morphisms are induced by the norm maps, and $t_0 = \varprojlim \pi_n$;
2) if E is an algebraic extension of $\widetilde{K}$, then $\mathcal{X}(E)$ is separable over $\mathcal{K}$, and the correspondence $E \mapsto \mathcal{X}(E)$ gives equivalence of the category of algebraic

extensions of $\widetilde{K}$ and the category of separable extensions of $\mathcal{K}$. In particular, $\mathcal{X}$ gives the identification of $\Gamma_{\mathcal{K}} = \mathrm{Gal}(\mathcal{K}_{sep}/\mathcal{K})$ with $\Gamma_{\widetilde{K}} \subset \Gamma_K$;

3) the above identification $\Gamma_{\widetilde{K}} = \Gamma_{\mathcal{K}}$ is compatible with the ramification filtrations in Γ_K and $\Gamma_{\mathcal{K}}$; this means that if $\varphi_{\widetilde{K}/K} = \lim_{n\to\infty}\varphi_{K_n/K}$ then for any $x \geqslant 0$, $\Gamma_{\mathcal{K}}^{(x)} = \Gamma_K^{(y)} \cap \Gamma_{\widetilde{K}}$ with $y = \varphi_{\widetilde{K}/K}(x)$.

1.3.2 Three questions

Let

$$K_{<p}(M) = \bar{K}^{\Gamma_K^{p^M} C_p(\Gamma_K)}\;, \quad \mathcal{K}_{<p}(M) = \mathcal{K}_{sep}^{\Gamma_{\mathcal{K}}^{p^M} C_p(\Gamma_{\mathcal{K}})}.$$

Then $\mathcal{K}_{<p}(M) \supset \mathcal{X}(K_{<p}(M)\widetilde{K}) \supset \mathcal{K}$, and there is a subgroup $\mathcal{H}_M$ of $\Gamma_{\mathcal{K}}(M) := \mathrm{Gal}(\mathcal{K}_{<p}(M)/\mathcal{K})$, such that

$$\mathrm{Gal}(\mathcal{X}(K_{<p}(M)\widetilde{K})/\mathcal{K}) = \Gamma_{\mathcal{K}}(M)/\mathcal{H}_M.$$

Under the identification η_0 from Subsection 1.1.3 we have $\mathcal{H}_M \simeq G(\mathcal{J}_M)$, where $\mathcal{J}_M$ is an ideal of the Lie algebra $\mathcal{L}_M$.

Question A. *What is the ideal* $\mathcal{J}_M$?

Remember that the Lie algebra $\mathcal{L}_M \otimes W(k)$ has a system of generators $\{D_{an} \mid a \in \mathbb{Z}^+(p), n \in \mathbb{Z}/N_0\} \cup \{D_0\}$. So, more precisely,

What are explicit generators of the ideal $\mathcal{J}_M$?

It is easy to see that $K_{<p}(M) \cap \widetilde{K} := K_M = K(\pi_M)$. Therefore, for $\Gamma_K(M) := \mathrm{Gal}(K_{<p}(M)/K)$, we have the following exact sequence of p-groups

$$1 \longrightarrow \Gamma_{\mathcal{K}}(M)/\mathcal{H}_M \longrightarrow \Gamma_K(M) \longrightarrow \langle\tau_0\rangle^{\mathbb{Z}/p^M} \longrightarrow 1,$$

where $\tau_0 \in \mathrm{Gal}(K_M/K)$ is defined by the relation $\tau_0(\pi_M) = \zeta_M\pi_M$. Using the equivalence of the category of Lie $\mathbb{Z}/p^M$-algebras of nilpotent class $< p$ and of the category of p-groups of the same nilpotent class, cf. Subsection 1.1.2, we can rewrite the above exact sequence of p-groups as the following exact sequence of Lie $\mathbb{Z}/p^M$-algebras

$$0 \longrightarrow \mathcal{L}_M/\mathcal{J}_M \longrightarrow L_M \longrightarrow (\mathbb{Z}/p^M)\tau_0 \longrightarrow 0$$

Here L_M is the Lie $\mathbb{Z}/p^M$-algebra such that $G(L_M) = \Gamma_K(M)$. This sequence of Lie algebras splits in the category of $\mathbb{Z}/p^M$-modules and, therefore, can be given by a class of differentiations $\mathrm{ad}\hat{\tau}_0$ of $\mathcal{L}_M$, where $\hat{\tau}_0$ is a lift of τ_0 to an automorphism of L_M.

Question B. *What are the differentiations* $\mathrm{ad}\hat{\tau}_0$?

More precisely,

Find the elements $\mathrm{ad}(\hat{\tau}_0)(D_{a0})$, $a \in \mathbb{Z}^0(p)$.

As we have mentioned in Subsection 1.3.1 the ramification filtrations in $\Gamma_{\mathcal{K}}$ and Γ_K are compatible.

The ramification filtration of $\mathrm{Gal}(K_M/K) = \mathbb{Z}/p^M\tau_0$ has a very simple structure. Let e_K be the absolute ramification index of K and for $s \in \mathbb{Z}$, $s \geqslant 0$, $v_s = e_K p/(p-1) + se_K$. Then

if $0 \leqslant v \leqslant v_0$, then $(\mathbb{Z}/p^M\tau_0)^{(v)}$ is generated by τ_0;
if $v_s < v \leqslant v_{s+1}, 0 \leq s < M$, then $(\mathbb{Z}/p^M\tau_0)^{(v)}$ is generated by $p^s\tau_0$;
if $v > v_M$, then $(\mathbb{Z}/p^M\tau_0)^{(v)} = 0$.

Therefore, we shall obtain a description of the ramification filtration $\Gamma_K(M)^{(v)}$ of $\Gamma_K(M)$ by answering the following question.

Question C. *How to construct "good" lifts* $\widehat{p^s\tau_0} \in L_M^{(v_s)}$, $0 \leq s \leq M$?

Below we announce partial results related to the above questions.

1.3.3 The ideal $\mathcal{J}_M$

Consider the decreasing central filtration by the commutator subgroups $\{C_s(\Gamma_K(M))\}_{s\geq 2}$. This filtration corresponds to the following decreasing central filtration of ideals of $\mathcal{L}_M$

$$J_1 := \mathcal{L}_M \supset J_2 \supset \cdots \supset J_p = \mathcal{J}_M$$

We can treat $\mathcal{L}_M$ as a free pro-finite object in the category of Lie $\mathbb{Z}/p^M$-algebras of nilpotent class $< p$. Its module of generators is $\mathcal{K}^*/\mathcal{K}^{*p^M}$. In this subsection we announce an explicit description of the filtration $\{J_s\}_{2\leqslant s\leqslant p}$. Particularly, in the case $s = p$ we obtain an answer to the above question A.

Let $\mathcal{U}$ be the submodule of $\mathcal{K}^*/\mathcal{K}^{*p^M}$ generated by the images of principal units. $\mathcal{U}$ can be identified with a submodule in the power series ring $W_M(k_0)[[t]]$ via the correspondences

$$E(w, t_0^a) \bmod \mathcal{K}^{*p^M} \mapsto wt^a \bmod p^M W(k)[[t]],$$

where $w \in W(k)$, $a \in \mathbb{Z}^+(p) = \{a \in \mathbb{N} \mid (a, p) = 1\}$ and $E(w, X)$ is the Shafarevich function from Subsection 1.2.1.

Let $H_0 \in W(k)[[t]]$ be a power series such that $\zeta_M = H_0(\pi_0)$. Let $\widetilde{H}_0 = H_0 \bmod p^M \in W_M(k)[[t]]$. Then there is a unique $S \in W_M(k)[[t]]$, such that $\widetilde{H}_0^{p^M} = E(1, S)$. Note that the differential $dS = 0$, in particular, $S \in W_M(k_0)[[t^p]]$, and therefore, $S\mathcal{U} \subset \mathcal{U}$ under the above identification of $\mathcal{U}$ with the submodule $\bigoplus_{a\in\mathbb{Z}^+(p)} W_M(k)t^a$ in $W_M(k)[[t]]$.

For $s \geq 1$ define a decreasing filtration of $\mathcal{K}^*/\mathcal{K}^{*p^M}$ as follows:
$\left(\mathcal{K}^*/\mathcal{K}^{*p^M}\right)^{(1)} = \mathcal{K}^*/\mathcal{K}^{*p^M}$ and $\left(\mathcal{K}^*/\mathcal{K}^{*p^M}\right)^{(s)} = S^{s-1}\mathcal{U}$, if $s \geq 2$.
This filtration determines a decreasing filtration of ideals $\{\mathcal{L}_M(s)\}_{s\geqslant 1}$ in $\mathcal{L}_M$ which can be characterized as the minimal central filtration of $\mathcal{L}_M$ such that for all s, $(\mathcal{K}^*/\mathcal{K}^{*p^M})^{(s)} \subset \mathcal{L}_M(s)$.

Theorem 1.4. *For* $1 \leqslant s \leqslant p$, $J_s = \mathcal{L}_M(s)$.

Remark. a) The element $\widetilde{H}_0^{p^M} - 1$ appears as the denominator in the Brückner-Vostokov explicit formula for the Hilbert symbol and can be replaced in that formula by S, cf. [3]; this element S can be considered naturally as an element of Fontaine's crystalline ring of p-adic periods and coincides with the p-adic period of the multiplicative p-divisible group.

b) A first non-abelian case of the above theorem corresponds to $s = 2$ and is equivalent to the formula

$$S\mathcal{U} = \mathrm{Ker}(\Gamma_{\mathcal{K}}(M)^{ab} \longrightarrow \Gamma_K(M)^{ab}).$$

By Laubie's theorem [18] the functor $\mathcal{X}$ is compatible with the reciprocity maps of class field theories of $\mathcal{K}$ and K cf. also [12]. Therefore,

$$S\mathcal{U} = \mathrm{Ker}(\mathcal{N} : \mathcal{K}^*/\mathcal{K}^{*p^M} \longrightarrow K^*/K^{*p^M}), \tag{1.6}$$

where $\mathcal{N}$ is induced by the projection $\mathcal{K}^* = \varprojlim K_n^* \longrightarrow K_0^* = K^*$. The right-hand side of (1.6) can be interpreted as the set of all $(w_a)_{a\in\mathbb{Z}^+(p)} \in W(k)^{\mathbb{Z}^+(p)} \bmod p^M$ such that $\prod_{a\in\mathbb{Z}^+(p)} E(w_a, \pi_0^a) \in K^{*p^M}$. So, formula (1.6) can be deduced from the Brückner-Vostokov explicit reciprocity law.

1.3.4 Differentiation $\mathrm{ad}\hat{\tau}_0 \in \mathrm{Diff}(\mathcal{L}_M)$

It can be proved that there are only finitely many different ideals $\mathcal{L}_M^{(v)} \bmod \mathcal{J}_M$. Therefore, we can fix sufficiently large natural number N_1 (which depends only on N_0, e_K and M), set $\mathcal{F}_\gamma^0 := \mathcal{F}_{\gamma,-N_1}^0$ and use these elements to describe the ramification filtration $\{\mathcal{L}_M^{(v)} \mathrm{mod} \mathcal{J}_M\}_{v\geq 0}$.

The elements $\mathcal{F}_\gamma^0$, $\gamma > 0$, can be given modulo the ideal of third commutators $C_3(\mathcal{L}_{M\,k})$ as follows :

if $\gamma = ap^l \in \mathbb{N}$, where $a \in \mathbb{Z}^+(p)$ and $l \in \mathbb{Z}_{\geqslant 0}$, then

$$\mathcal{F}_\gamma^0 = ap^l D_{a\bar{l}} + \sum_{s,n,a_1,a_2} \eta(n)a_1 p^s [D_{a_1\bar{s}}, D_{a_2\bar{s}-\bar{n}}];$$

if $\gamma \notin \mathbb{N}$, then

$$\mathcal{F}(\gamma) = \sum_{s,n,a_1,a_2} \eta(n) a_1 p^s [D_{a_1,\bar{s}}, D_{a_2,\bar{s}-\bar{n}}].$$

In the above sums $0 \leqslant s < M$, $0 \leqslant n < N_1$, $a_1, a_2 \in \mathbb{Z}^0(p)$, $p^s(a_1 + a_2 p^{-n}) = \gamma$, $\eta(n) = 1$ if $n \neq 0$ and $\eta(0) = 1/2$.

Let $S \in W_M(k)[[t]]$ be the element introduced in Subsection 1.3.3. Remember that $S = \sigma S'$, where $S' \in W_M(k_0)[[t]]$. For $l \geqslant 1$, let $\alpha_l \in W_M(k)$ be such that

$$S - pS' = \sum_{l \geqslant 1} \alpha_l t^l.$$

Note that

a) if $l < e_K p/(p-1)$ then $\alpha_l = 0$;
b) for any $l \geqslant 1$, we have $l\alpha_l = 0$.

Theorem 1.5. *There is a lift $\hat{\tau}_0 \in L_M$ of τ_0 such that*

$$\mathrm{ad}\hat{\tau}_0(D_0) = \sum_{\substack{l \geqslant 1 \\ 0 \leqslant n < N_0}} \sigma^n(\alpha_l \mathcal{F}_l^0) \mathrm{mod} C_3(\mathcal{L}_M)$$

and, for $a \in \mathbb{Z}^+(p)$,

$$\mathrm{ad}\hat{\tau}_0(D_{a0}) = \sum_{\substack{f \in \mathbb{Z} \\ l \geqslant 1}} \sigma^{-f}(\alpha_l \mathcal{F}^0_{l+ap^f}) \mathrm{mod} C_3(L_{Mk}).$$

1.3.5 Good lifts $\widehat{p^s\tau_0}$, $0 \leq s < M$

If $s = 0$, let $\hat{\tau}_0 \in L_M$ be a lift from Theorem 1.5. Define the lifts $\widehat{p^s\tau_0}$ by induction on s as follows

$$\widehat{p^s\tau_0} = p(\widehat{p^{s-1}\tau_0}) + \frac{1}{2} \sum_{\substack{l \geqslant 1 \\ 0 \leqslant n < N_0}} \alpha_l \sum_{\substack{a_1, a_2 \in \mathbb{Z}^+(p) \\ a_1 + a_2 = \frac{p^s e_K}{p-1}}} a_1 [D_{a_1 n}, D_{a_2 n}].$$

Theorem 1.6. *A lift $\hat{\tau}_0$ from Theorem 1.5 can be chosen in such a way that all $\widehat{p^s\tau_0}$, $0 \leq s < M$, are "good" modulo $C_3(L_{Mk})$, i.e. $\widehat{p^s\tau_0} \in L_M^{\varphi(v_s)} \mathrm{mod}\, C_3(L_{Mk})$, where $v_s = e_K p^s/(p-1)$ and φ is the Herbrand function of the extension $K_{<p}(M)/K$.*

Theorem 1.6 gives a complete description of the ramification filtration of $\Gamma_K(M)$ modulo $C_3(\Gamma_K(M))$. This result can be compared with the description of the filtration $\Gamma_K(1)^{(v)} \cap C_2(\Gamma_K(1))$ modulo $C_3(\Gamma_K(1))$ in [27].

1.3.6 The modulo p case.

In this subsection we give an overview of the results from [11] related to the modulo p aspect of problems discussed in this paper.

Let $M = 1$ and $c_0 := e_K p/(p-1)(= v_1)$. We can simply drop off M from all above notation instead of substituting $M = 1$.

The ideals $\mathcal{L}(s)$, $1 \leqslant s \leqslant p$, can be described now quite explicitly as follows. Define a weight filtration on $\mathcal{L}$ by setting for $a \in \mathbb{Z}^0(p)$ and $n \in \mathbb{Z}/N_0$, $\mathrm{wt}(D_{an}) = s \in \mathbb{N}$ if $(s-1)c_0 \leqslant a < sc_0$. Then for all s, $\mathcal{L}(s) = \{l \in \mathcal{L} \mid \mathrm{wt}(l) \geqslant s\}$, cf. Section 17.3 of [11].

The field-of-norms $\mathcal{K}$ admits a standard embedding into $R_0 = \mathrm{Frac}\, R$, where R is Fontaine's ring [26]. This also identifies R_0 with the completion of $\mathcal{K}_{sep}$ and, therefore, we have a natural embedding of $\mathcal{K}_{<p}$ into R_0. In particular, f_0 can be considered as an element of $\mathcal{L}_{\mathcal{R}_0}$.

Choose a continuous automorphism h_0 of $\mathcal{K}$ such that

$$h_0(t) \equiv \tau_0(t) \bmod t^{c_0(p-1)} \mathrm{m}_R,$$

where m_R is the maximal ideal in R. The formalism of nilpotent Artin-Schreier theory allows us to describe efficiently the lifts $\hat{h}_0 \in \mathrm{Aut}\mathcal{K}_{<p}$ of h_0, [2]. This can be done by specifying the image $\hat{h}_0(f_0)$ in the form

$$\hat{h}_0(f_0) = c(\hat{h}_0) \circ (\mathrm{Ad}\hat{h}_0 \otimes 1) f_0$$

where $c(\hat{h}_0) \in \mathcal{L}_{\mathcal{K}}$ and $\mathrm{Ad}\hat{h}_0$ is the conjugation of $G(\mathcal{L})$ via $\hat{h}_0$.

It can be proved then that the lifts $\hat{\tau}_0 \in \mathrm{Aut}K_{<p}$ satisfy

$$\hat{\tau}_0(f_0) \equiv \hat{h}_0(f_0) \bmod t^{c_0(p-1)} \mathcal{M}_{R_0},$$

where $\mathcal{M}_{R_0} = \sum_{1 \leqslant s < p} t^{-c_0 s} \mathcal{L}(s)_{\mathrm{m}_R} + \mathcal{L}(p)_{R_0}$, and are uniquely determined by this conditions. Therefore, the lifts $\hat{\tau}_0$ can be uniquely described via the morphisms $\mathrm{Ad}\hat{\tau}_0$ and the elements $c(\hat{h}_0) \bmod t^{c_0(p-1)} \mathcal{M}_{\mathcal{K}}$, where $\mathcal{M}_{\mathcal{K}} = \sum_{1 \leqslant s < p} t^{-c_0 s} \mathcal{L}(s)_{\mathrm{m}_{\mathcal{K}}} + \mathcal{L}(p)_{\mathcal{K}}$.

The above elements $c(\hat{h}_0)$ satisfy complicated relations but the whole situation can be linearized as follows.

In [11] we proved that the action of the group $\langle \hat{h}_0 \rangle^{\mathbb{Z}/p}$ on f_0 comes from the action of the additive group scheme $\mathbb{G}_{a,\mathbb{F}_p} = \mathrm{Spec}\mathbb{F}_p[u]$ on $\mathcal{M}_{\mathcal{K}_{<p}} / t^{c_0(p-1)} \mathcal{M}_{\mathcal{K}_{<p}}$, where $\mathcal{M}_{\mathcal{K}_{<p}}$ is defined similarly to $\mathcal{M}_{R_0}$. This implies that if $c[u] = c_0 + c_1 u + \cdots + c_{p-1} u^{p-1} \in \mathcal{L}_{\mathcal{K}}[u]$ is the polynomial with coefficients in $\mathcal{L}_{\mathcal{K}}$ such that for $0 \leqslant k < p$, $c[u]|_{u=k} = c(\hat{h}_0^k)$ then its residue modulo $t^{c_0(p-1)} \mathcal{M}_{\mathcal{K}}$ is well defined and can be uniquely recovered from its first coefficient $c_1 \bmod t^{c_0(p-1)} \mathcal{M}_{\mathcal{K}}$.

Now we can state the main results from [11].

1.3.6.1

There is a bijection

$$\hat{\tau}_0 \mapsto (c_1, \{\mathrm{ad}\hat{\tau}_0(D_{a0}) \mid a \in \mathbb{Z}^0(p)\})$$

of the set of lifts $\hat{\tau}_0$ and solutions $(c_1, \{V_a \mid a \in \mathbb{Z}^0(p)\})$, where $c_1 \in \mathcal{L}_{\mathcal{K}}$, $V_0 \in \alpha_0\mathcal{L}$, $V_a \in \mathcal{L}_k$ with $a \in \mathbb{Z}^+(p)$, of the following equation

$$\sigma c_1 - c_1 + \sum_{a\in\mathbb{Z}^0(p)} t^{-a} V_a =$$

$$\sum_{k,l\geqslant 1} \frac{1}{k!} t^{l-(a_1+\cdots+a_k)} \alpha_l [\ldots [a_1 D_{a_1 0}, D_{a_2 0}], \ldots, D_{a_k 0}]$$

$$-\sum_{k\geqslant 2} \frac{1}{k!} t^{-(a_1+\cdots+a_k)} [\ldots [V_{a_1}, D_{a_2 0}], \ldots, D_{a_k 0}]$$

$$-\sum_{k\geqslant 1} \frac{1}{k!} t^{-(a_2+\cdots+a_k)} [\ldots [\sigma c_1, D_{a_2 0}], \ldots, D_{a_k 0}]$$

(Remember that $\alpha_l \in k$ were defined in Subsection 1.3.4 and they equal 0 if $l \neq c_0, c_0 + p, c_0 + 2p, \ldots$.)

1.3.6.2

A lift $\hat{\tau}_0$ is "good", i.e. $\hat{\tau}_0 \in L^{(c_0)}$ iff all $V_a \in \mathcal{L}_k^{(c_0)}$ and

$$c_1 \equiv -\sum_{\substack{\gamma>0\\ l\geqslant 1}} \sum_{0\leqslant i<N^*} \sigma^i (\alpha_l \mathcal{F}^0_{\gamma,-i} t^{-\gamma+l}) \bmod (\mathcal{L}_{\mathcal{K}}^{(c_0)} + t^{c_0(p-1)} \mathcal{M}_{\mathcal{K}})$$

1.3.6.3

Let the operators F_0 and G_0 on $\mathcal{L}_k$ be such that for any $l \in \mathcal{L}_k$,

$$F_0(l) = \sum_{k\geqslant 1} \frac{1}{k!} [\ldots [l, \underbrace{D_{00}], \ldots, D_{00}}_{k-1 \text{ times}}], \quad G_0(l) = \sum_{k\geqslant 0} \frac{1}{k!} [\ldots [l, \underbrace{D_{00}], \ldots, D_{00}}_{k \text{ times}}]$$

Then the correspondence

$$(c_1^0, V_0) \mapsto (c_1^0 - \sum_{\substack{0\leqslant i<N^*\\ l\geqslant 1}} \sigma^i (\alpha_l \mathcal{F}^0_{l,-i}), V_0)$$

establishes a bijection between the set of all good lifts $\hat{\tau}_0$ and the set of all (x, y) such that $x \in \mathcal{L}_k^{(c_0)}$, $y \in \alpha_0 \mathcal{L}^{(c_0)}$, and

$$(G_0\sigma - \mathrm{id})(x) + F_0(y) = \sum_{l\geqslant 1} \sigma^{N^*} \left(\alpha_l \mathcal{F}^0_{l,-N^*}\right) \tag{1.7}$$

1.3.6.4

For any above solution (x, y) of (1.7) we have

$$y \equiv \alpha_0 \sum_{l\geqslant 1} \mathrm{Tr}_{k/\mathbb{F}_p}(\alpha_l \mathcal{F}^0_{l,-N^*}) \mathrm{mod} \alpha_0 [D_0, \mathcal{L}^{(c_0)}]$$

This implies the existence of a good lift $\hat{\tau}_0$ such that the (only) relation in the Lie algebra L_k (recall that $G(L) = \mathrm{Gal}(K_{<p}/K)$) appears in the form

$$\mathrm{ad}\hat{\tau}_0(D_0) = \sum_{l\geqslant 1} \mathrm{Tr}_{k/\mathbb{F}_p}(\alpha_l \mathcal{F}^0_{l,-N^*}) .$$

References

[1] V.A.ABRASHKIN, *Ramification filtration of the Galois group of a local field*, Proceedings of the St. Petersburg Mathematical Society III, Amer. Math. Soc. Transl. Ser. 2, (1995) **166**, Amer. Math. Soc., Providence, RI

[2] V.A. ABRASHKIN, *Ramification filtration of the Galois group of a local field. II*, Proceedings of Steklov Math. Inst. (1995) **208**

[3] V.ABRASHKIN, *The field of norms functor and the Brückner-Vostokov formula*, Math. Ann., (1997) **308**, no.1, 5-19

[4] V.ABRASHKIN, *Ramification filtration of the Galois group of a local field. III*, Izvestiya RAN: Ser. Mat., **62**, no.5 (1998), 3-48; English transl. Izvestiya: Mathematics **62**, no.5, 857-900

[5] V.A. ABRASHKIN, *A group-theoretical property of the ramification filtration*, Izvestiya RAN: Ser. Mat., (1998) **62**, no.6 3-26; English transl. Izvestiya: Mathematics (1998), **62**, no.6 1073-1094

[6] V.ABRASHKIN, *Report on the ramification filtration of the Galois group of a local field*, Proceedings of the Research Conference on "Number Theory and Arithmetical Geometry: Arithmetical applications of Modular Forms" (San Feliu de Guixols, Spain, 24-29 October, 1997) Inst. fur Exp. Math. Universitat Essen, 1998, 47-53

[7] V. ABRASHKIN, *On a local analogue of the Grothendieck Conjecture*, Int. J. Math. (2000) **11**, no.1, 3-43

[8] V. ABRASHKIN, *Ramification theory for higher dimensional fields*, Contemp. Math. (2002) **300**, 1-16

[9] V. ABRASHKIN, *Characteristic p case of the Grothendieck conjecture for 2-dimensional local fields*, Proceedings of Steklov Institute (2003) **241**, 1-35

[10] V. ABRASHKIN, *Characteristic p analogue of modules with finite crystalline height*, Pure Appl. Math. Q., (2009) **5**, 469-494

[11] V.ABRASHKIN, *Automorphisms of local fields of period p and nilpotent class* $< p$ (in preparation)

[12] V.ABRASHKIN, R.JENNI, *The field-of-norms functor and the Hilbert symbol for higher local fields*, J.Théor. Nombres Bordeaux, (2012) **24**, no.1, 1-39

[13] P.DELIGNE *Les corps locaux de caractéristique p, limites de corps locaux de caractéristique* 0, Representations of reductive groups over a local field, Travaux en cours, Hermann, Paris, 1973, 119-157

[14] N.L.GORDEEV, *Ramification groups of infinite p-extensions of a local field*, Soviet Math. (1982) **20**, no.6 2566-2595

[15] N.L.GORDEEV, Infinity of the number of relations in the Galois group of the maximal p-extension of a local field with bounded ramification, Izvestia AN SSSR, Ser. Matem. (1981) **45**, 592-607

[16] U.JANNSEN, K.WINGBERG, *Die Struktur der absoluten Galoisgruppe p-adischer Zahlkörper*, Invent. math. (1982) **70**, 71-98

[17] H.KOCH, Galoissche Theorie der p-Erweiterungen, Mathematische Monographien, Band 10, VEB Deutsche Verlag der Wissenschaften, Berlin 1970

[18] F. LAUBIE, *Extensions de Lie et groupes d'automorphismes de corps locaux*, Comp. Math., (1988) **67**, 165-189

[19] F.LAUBIE, *Une théorie du corps de classes local non abélien*, Compos. Math., (2007) **143**, no. 2, 339-362.

[20] M. LAZARD, *Sur les groupes nilpotentes et les anneaux de Lie*, Ann. Ecole Norm. Sup. (1954) **71**, 101-190

[21] SH.MOCHIZUKI, *A version of the Grothendieck conjecture for p-adic local fields*, Int. J. Math., (1997) **8**, no.4, 499-506

[22] J.-P.SERRE, *Local Fields*, Berlin, New York: Springer-Verlag, 1980

[23] J.-P.SERRE, *Cohomologie Galoisienne*, Springer Verlag, Berlin-Gottingen-Heidelberg-New York, 1964

[24] J.-P.SERRE, *Structure de certains pro-p-groupes (d'aprés Demuŝkin)* Séminaire Bourbaki, **8**, Exp. No. 252, 145-155, Soc. Math. France, Paris, 1995.

[25] I.R. SHAFAREVICH, *On p-extensions (In Russian)*, Mat. Sbornik (1947) **20**, 351-363

[26] J.-P. WINTENBERGER, *Le corps des normes de certaines extensions infinies des corps locaux; application*, Ann. Sci. Ec. Norm. Super., IV. Ser, (1983) **16** 59-89

[27] W. ZINK, *Ramification in local Galois groups; the second central step*, Pure Appl. Math. Q. (2009) **5**, no. 1, 295-338.

2

A characterisation of ordinary modular eigenforms with CM

Rajender Adibhatla[1] *and Panagiotis Tsaknias*[2]

[1]Mathematics Research Unit, University of Luxembourg, 6 rue Coudenhove-Kalergi, L-1359 Luxembourg

E-mail address: rajender.adibhatla@uni.lu

[2]Mathematics Research Unit, University of Luxembourg, 6 rue Coudenhove-Kalergi, L-1359 Luxembourg

E-mail address: panagiotis.tsaknias@uni.lu

ABSTRACT. For a rational prime $p \geq 3$ we show that a p-ordinary modular eigenform f of weight $k \geq 2$, with p-adic Galois representation ρ_f, mod p^m reductions $\rho_{f,m}$, and with complex multiplication (CM), is characterised by the existence of p-ordinary CM companion forms h_m modulo p^m for all integers $m \geq 1$ in the sense that $\rho_{f,m} \sim \rho_{h_m,m} \otimes \chi^{k-1}$, where χ is the p-adic cyclotomic character.

2.1 Introduction

For a rational prime $p \geq 3$ let f be a primitive, cusp form with q-expansion $\sum a_n(f)q^n$ and associated p-adic Galois representation ρ_f. In this paper, we prove some interesting arithmetic properties of such a form which, in addition, has complex multiplication (CM). One of the reasons that CM forms have historically been an important subclass of modular forms is the simplicity with which they can be expressed. They arise from algebraic Hecke characters of imaginary quadratic fields and this makes them ideal initial candidates on which one can check deep conjectures in the theory of modular forms. The specific arithmetic property we establish involves higher congruence companion forms which were introduced in [1]. The precise definition and properties of these forms are given in Section 2.2, but for now we will only remark that

Date: June 2014
2010 *Mathematics Subject Classification*, 11F33, 11F80

Arithmetic and Geometry, ed. Luis Dieulefait *et al.* Published by Cambridge University Press.

companion forms mod p^m are defined as natural analogues of the classical (mod p) companion forms of Serre and Gross [6].

The main theorem of this work (Theorem 2.6, stated and proved in Section 2.4) establishes that given a p-ordinary CM form f, one can always find a CM companion form mod p^m for any integer $m \geq 1$. The proof explicitly finds the desired companion forms in a Hida family of CM forms, thereby circumventing the deformation theory and modularity lifting approach of the companion form theorem in [1]. In addition, we show that the converse is true as well: Any p-ordinary form f which has CM companions mod p^m for each $m \geq 1$ must necessarily have CM. We therefore have a complete arithmetical characterisation of p-ordinary CM forms. In Section 2.5 we present a different proof of the main theorem. It is elementary in the sense that we work directly with the Hecke characters and avoid the heavy machinery of Hida families and overconvergent modular forms. Moreover, the result in this section is, in fact, a strengthening of the main theorem because it shows the existence of companion forms for odd composite moduli M and not just modulo p^m. On the other hand, the argument appears to work only under the hypothesis that the class number of the imaginary quadratic extension from which f arises is coprime to M.

2.2 Higher companion forms

Let $p \geq 3$ be a rational prime and let f be a primitive cusp form of weight $k \geq 2$, level N prime to p, Nebentypus Ψ and q-expansion $\sum a_n(f)q^n$. Here, f is primitive in the sense that it is a normalised newform that is a common eigenform for all the Hecke operators. For a place $\mathfrak{p}|p$ in K_f, the number field generated by the $a_n(f)$'s, let $\rho_f : \mathrm{Gal}(\overline{\mathbb{Q}}/\mathbb{Q}) \longrightarrow \mathrm{GL}_2(K_{f,\mathfrak{p}})$ be a continuous, odd, irreducible Galois representation that can be attached to f. We may, after conjugation, assume that ρ_f takes values in the ring of integers of some finite extension of $\mathbb{Q}_p$. This allows us to consider reductions of ρ_f modulo p^m (for integers $m \geq 1$) which, with the exception of the mod p reduction $\overline{\rho}$, we will denote by $\rho_{f,m}$. For the rest of this paper we assume that $\overline{\rho}$ is absolutely irreducible. We can, and will, define congruences and reduction mod p^m even when the elements do not lie in $\mathbb{Z}_p$ by using the following notion of congruence due to Ventossa and Wiese [12].

Fix an embedding $\overline{\mathbb{Q}} \hookrightarrow \overline{\mathbb{Q}}_p$ and let $\overline{\mathbb{Z}}_p$ denote the integers of $\overline{\mathbb{Q}}_p$. Let K be a number field or a local field with valuation v_p, normalised such that $v_p(p) = 1$. Let $\alpha, \beta \in \overline{\mathbb{Z}}_p$ and $n \in \mathbb{N}$. Then, $\alpha \equiv \beta \bmod p^n$, if and only if $v_p(\alpha - \beta) > n - 1$.

With f an eigenform as above, p-adic cyclotomic character χ and $2 \leq k \leq p^{m-1}(p-1)+1$, we define a companion form of f to be as follows.

Definition 2.1. Let $k_m \geq 2$ be the smallest integer such that $k_m+k-2 \equiv 0 \bmod p^{m-1}(p-1)$. Then, a **companion form** g of f, modulo p^m, is a p-ordinary normalised eigenform of weight k_m such that $\rho_{f,m} \simeq \rho_{g,m} \otimes \chi^{k-1} \bmod p^m$. An equivalent formulation of the above criterion in terms of the Fourier expansions is: $a_n(f) \equiv n^{k-1} a_n(g) \bmod p^m$ for $(n, Np) = 1$.

Remark 2.2. The equivalence between the Galois side and the coefficient side in the above definition perhaps needs further justification. One direction is immediate if we take the traces of Frobenii of the Galois representations. The other direction follows from the absolute irreducibility of $\overline{\rho}$, Chebotarev and Théorème 1 [3], which is essentially a generalisation of the Brauer-Nesbitt theorem to arbitrary local rings.

Following Wiles [14], we say that f is **ordinary** at p (or simply p-**ordinary**) if $a_p(f) \not\equiv 0 \bmod \mathfrak{p}$ for each prime $\mathfrak{p}|p$ (in K_f). Then, by Wiles [14] and Mazur-Wiles [9], we have

$$\rho_f|_{G_p} \sim \begin{pmatrix} \chi^{k-1}\lambda(\Psi(p)/a_p) & * \\ 0 & \lambda(a_p) \end{pmatrix}$$

where G_p is a decomposition group at p and, for any α in $(\mathcal{O}_{K_f})^\times$, $\lambda(\alpha)$ is the unramified character taking Frob_p to α.

If ρ_f mod p is absolutely irreducible, then we note that the reductions $\rho_{f,m}|_{G_p} = \rho_f|_{G_p} \bmod p^m$ for $m \geq 1$ are independent of the choice of lattice used to define ρ_f. A natural question is to ask when the restriction $\rho_{f,m}|_{G_p}$ actually splits. It is an easy check that a sufficient condition for $\rho_{f,m}$ to split at p is that f has a p-ordinary companion form modulo p^m.

That the splitting at p of $\rho_{f,m}$ implies the existence of a companion form mod p^m is considerably more difficult to prove. This was shown in [1] but only under the hypothesis that $\overline{\rho}_f$ has full image. However, when f has CM, $\mathrm{Im}(\overline{\rho}_f)$ is necessarily projectively dihedral. This renders that method of *loc. cit.* inapplicable even though, *a priori*, we know that $\rho_f|_{G_p}$ splits and therefore expect that f should have a companion mod p^m for all integers m prime to N.

In the sequel we avoid the use of lifting theorems altogether and exhibit the companion forms by working directly with the Hecke character associated to the CM form and visualising them as part of a Hida family.

2.3 Hecke characters and CM forms

In this section we briefly describe the well-known connection between forms with complex multiplication (CM forms) and Hecke characters of imaginary quadratic fields. The interested reader may consult [11, §3] and [10, Chapter VII, §6].

Let K be a number field, $\mathcal{O} = \mathcal{O}_K$ its ring of integers and $\mathfrak{m}$ an ideal of K. We denote by $J^{\mathfrak{m}}$ the group of fractional ideals of $\mathcal{O}_K$ that are coprime to $\mathfrak{m}$.

Definition 2.3. A Hecke character ψ of K, of modulus $\mathfrak{m}$, is a group homomorphism $\psi : J^{\mathfrak{m}} \to \mathbb{C}^*$ such that there exists a character $\psi^{\infty} : (\mathcal{O}_K/\mathfrak{m})^* \to \mathbb{C}^*$ and a group homomorphism $\psi_{\infty} : K_{\mathbb{R}}^* \to \mathbb{C}^*$, where $K_{\mathbb{R}} := K \otimes_{\mathbb{Q}} \mathbb{R}$, such that:

$$\psi((\alpha)) = \psi^{\infty}(\alpha)\psi_{\infty}(\alpha) \text{ for all } \alpha \in \mathcal{O}_K \text{ prime to } \mathfrak{m}. \tag{2.1}$$

We refer to ψ^{∞} and ψ_{∞} as the finite and the infinity type of ψ respectively. The conductor of ψ is the conductor of ψ^{∞} and we will call ψ primitive if its conductor and modulus are equal. We now specialise to the imaginary quadratic field case. Let $K = \mathbb{Q}(\sqrt{-d})$, where d is a squarefree positive integer. We will denote by D its discriminant and by $(D|.)$ the Kronecker symbol associated to it. The only possible group homomorphisms ψ_{∞} are of the form σ^u, where σ is one of the two conjugate complex embeddings of K and $u \geq 0$ is an integer. The following well-known theorem [8, Theorem 4.8.2] due to Hecke and Shimura then associates a modular form to the Hecke character ψ of K.

Theorem 2.4. *Given a Hecke character ψ of infinity type σ^u and finite type ψ^{∞} with modulus $\mathfrak{m}$, assume $u > 0$ and let $\mathbf{N}\mathfrak{m}$ be the norm of $\mathfrak{m}$. Then, $f = \sum_n (\sum_{\mathbf{N}\mathfrak{a}=n} \psi(\mathfrak{a}))q^n$ is a cuspidal eigenform in $S_{u+1}(N, \epsilon)$, where $N = |D|\mathbf{N}\mathfrak{m}$ and $\epsilon(m) = (D|m)\psi^{\infty}(m)$ for all integers m prime to N.*

The eigenform associated with a Hecke character ψ is new if and only if ψ is primitive. (See Remark 3.5 in [11].) An important feature of forms which arise in this way is that they coincide with CM forms whose definition we recall.

Definition 2.5. A newform f is said to have **complex multiplication**, or just CM, by a quadratic character $\phi : G_{\mathbb{Q}} \longrightarrow \{\pm 1\}$ if $a_q(f) = \phi(q)a_q(f)$ for almost all primes q. We will also refer to CM by the corresponding quadratic extension.

It is clear that the cusp form f in the above theorem has CM by $(D|.)$. Indeed, the coefficient $a_q(f)$ in the Fourier expansion of f is 0 if no ideal of K has norm equal to q. Since $(D|q) = -1$ exactly when this holds for q, $a_q(f) = (D|q)a_q(f)$, and f has CM. On the other hand, Ribet [11] shows that if f has CM by an imaginary quadratic field K then it is induced from a Hecke character on K.

2.4 The main theorem

We first describe the setting in which we prove our main result. Let $M_k^\dagger(N)$ denote the space of overconvergent modular forms of level N and weight k in the sense of Coleman [4] and let $M^\dagger(N)$ be the graded ring of overconvergent modular forms of level N. The θ operator, $\theta := q\frac{d}{dq}$, then maps $\sum a_n q^n$ $to \sum na_n q^n$. Note in particular that if f has (finite) slope α then $\theta(f)$ has slope $\alpha + 1$. Furthermore, by [4, Proposition 4.3],

$$\theta^{k-1} : M_{2-k}^\dagger(N) \to M_k^\dagger(N),$$

and classical CM forms of slope $k-1$ lie in this image ([4, Proposition 7.1]). We exploit this property of CM forms to prove our main theorem which is the following.

Theorem 2.6. *Let $p \geq 3$ be a rational prime and $f = \sum a_n(f)q^n$ be a p-ordinary primitive CM cuspidal eigenform of weight $k \geq 2$ and level N prime to p. Then for every integer $m \geq 1$, there exists a p-ordinary CM eigenform $h_m = \sum a_n(h_m)q^n$ of weight k_m that is a companion form for f mod p^m.*

Proof. Let $K = \mathbb{Q}(\sqrt{-d})$ be the imaginary quadratic field by which f has CM, D be the discriminant of K and $\sigma : K \hookrightarrow \mathbb{C}$ one of its two complex embeddings which we fix for the rest of this section. Let g be the (p-old) eigenform of level Np and weight k, whose p-th coefficient has p-adic valuation equal to $k-1$ and whose q-expansion agrees with that of f at all primes n coprime to p. Clearly g has CM by K as well. Let ψ be the Hecke character over K of modulus $\mathfrak{m}$ and infinity type σ^{k-1} associated with the CM form g. The p-ordinariness of f implies that $p = \mathfrak{p}\bar{\mathfrak{p}}$ is split in K. If r is a rational prime coprime to $\mathfrak{m}$ that splits in K, say $r = \mathfrak{r}\bar{\mathfrak{r}}$, then we have:

$$a_r(g) = \psi(\mathfrak{r}) + \psi(\bar{\mathfrak{r}})$$

In the proof of Proposition 7.1 in [4] it is shown how to obtain a p-ordinary (CM) p-adic eigenform h_{2-k} of weight $2-k$ such that

$$\theta^{k-1}(h_{2-k}) = g.$$

It can be easily seen that θ^{k-1} has the following effect on q-expansions:

$$\theta^{k-1}\left(\sum c_n q^n\right) = \sum n^{k-1} c_n q^n.$$

Therefore,

$$a_n(f) = n^{k-1} a_n(h_{2-k}) \text{ for } (n, p) = 1. \tag{2.2}$$

We will denote by L the extension of $\mathbb{Q}_p$ generated by the coefficients of ψ and λ and by $\mathfrak{P}$ its prime ideal above p. We will also denote by E the extension of $\mathbb{Q}_p$ in which λ takes its values. It is clear from (2.2) that it is enough to find classical forms h_{k_m} of weight k_m that are congruent to h_{2-k} mod $\mathfrak{P}^m$, where $\mathfrak{P}$ is the ideal above p in L. By Proposition 7.1 in [4],

$$h_{2-k} = \sum_{\mathfrak{a}} \bar{\psi}^{-1}(\mathfrak{a}) q^{\mathrm{N}\mathfrak{a}}$$

where the sum runs over all the integral ideals of $\mathfrak{a}$ of K away from $\mathfrak{m}$. Ghate in [5, pp 234-236], following Hida, shows how to construct a p-adic CM family admitting a specific CM form as a specialisation. We outline the construction. Let λ be a Hecke character of conductor $\mathfrak{p}$ and infinity type σ. We have that $\mathcal{O}_E^\times \cong \mu_E \times W_E$, where W_E is the pro-p part of $\mathcal{O}_E^\times$. Let $\langle\rangle$ denote the projection from $\mathcal{O}_E^\times$ to W_E. One then gets (part of) the family mentioned above by

$$G(w) := \sum_{\mathfrak{a}} \bar{\psi}^{-1}(\mathfrak{a}) \langle\lambda(\mathfrak{a})\rangle^{w-(2-k)} q^{\mathrm{N}\mathfrak{a}}.$$

For any integer $w \geq 2$, $\psi_w(\mathfrak{a}) = \bar{\psi}^{-1}(\mathfrak{a})\langle\lambda(\mathfrak{a})\rangle^{w-(2-k)}$ defines a Hecke character of infinity type $w-1$, so that by Theorem 3.1, $G(w)$ is a p-adic CM eigenform of weight w. Moreover all of them are p-ordinary ([5, pp 236]) and therefore classical for weight $w \geq 2$ ([7, Theorem I]). Clearly $G(2-k) = h_{2-k}$. Notice also that all the ψ_w have coefficients in L. Let k_E, k_L be the residue fields of E and L, respectively, and consider the composition $\mathcal{O}_E^\times \to k_E^\times \to k_L^\times$, where the first map is the obvious surjection and the second one is the obvious injection. The image has prime-to-p order so the kernel contains W_E. In particular $\langle\lambda\rangle \equiv 1$ mod $\mathfrak{P}$. This implies:

$$\langle\lambda(\mathfrak{a})\rangle^{w-w'} \equiv 1 \bmod \mathfrak{P} \text{ for all } w, w' \in \mathbb{Z}_p.$$

It then follows easily that if $w \equiv w'$ mod $p^{m-1}(p-1)$ then,

$$\langle\lambda(\mathfrak{a})\rangle^{w-w'} \equiv 1 \bmod \mathfrak{P}^{m'},$$

where $m' = (m-1)e + 1$, with e being the ramification degree of L over $\mathbb{Q}_p$. Consider the members of the family with weight $k_m \geq 2$ which is the smallest integer such that $k + k_m \equiv 2 \bmod p^{m-1}(p-1)$. The previous identity then gives:

$$\langle\lambda(\mathfrak{a})\rangle^{k_m-(2-k)} \equiv 1 \bmod \mathfrak{P}^{m'}. \tag{2.3}$$

Since $G(w) = \sum_n \left(\sum_{\mathfrak{a}=n} \bar{\psi}^{-1}(\mathfrak{a})\langle\lambda(\mathfrak{a})\rangle^{w-(2-k)} \right) q^n$, the r-th coefficient of $G(w)$ (for r a rational prime is):

$$a_r(G(w)) = \bar{\psi}^{-1}(\mathfrak{r})\langle\lambda(\mathfrak{r})\rangle^{w-(2-k)} + \bar{\psi}^{-1}(\bar{\mathfrak{r}})\langle\lambda(\bar{\mathfrak{r}})\rangle^{w-(2-k)}.$$

The identity $\mathfrak{r}\bar{\mathfrak{r}} = r$ along with (2.3) then shows that

$$a_q(G(k_m)) \equiv a_q(h_{2-k}) \bmod \mathfrak{P}^{m'}.$$

For the primes r that are inert in K the above equivalence is trivially true since in this case $a_r(G(k_m)) = 0 = a_r(h_{2-k})$. We thus get that for all primes r away from Np the following holds:

$$a_r(G(k_m)) \equiv a_r(h_{2-k}) \bmod \mathfrak{P}^{m'}.$$

As we mentioned before, all the members of G with weight $w \geq 2$ are classical forms so every $h_m := G(k_m)$ is classical. Finally the last identity implies that $a_n(h_m)$ is congruent to $a_n(h_{2-k})$ modulo $\mathfrak{P}^{m'}$ for all $(nN, p) + 1$, as required. □

Note that the Fourier coefficient version of Definition 2.1 was used to show companionship in the above proof. As noted in Remark 2.2 following the definition, this formulation can be reconciled with the Galois representation formulation if one knows that $\overline{\rho}_f$ is absolutely irreducible. For f as in the theorem above, $\overline{\rho}_f$ is irreducible because it has projectively dihedral image. Absolute irreducibility then follows because $\overline{\rho}_f$ is odd and $p \geq 3$.

We conclude this section with by observing that the converse to Theorem 2.6 is also true; therefore giving a complete arithmetic characterisation of p-ordinary CM forms. Assume that f has CM companions h_m for all $m \geq 1$ and assume that h_m has CM with respect to a non-trivial quadratic character $\epsilon_m : (\mathbb{Z}/D_m\mathbb{Z})^\times \longrightarrow \{\pm 1\} \subset \mathbb{C}^\times$. The companionship property between f and each of the h_ms enforces the following compatibility congruences:

$$h_{m_1} \equiv' h_{m_2} \bmod p^{m_1} \text{ for all } m_2 \geq m_1$$
$$\epsilon_{m_1} \equiv' \epsilon_{m_2} \bmod p^{m_1} \text{ for all } m_2 \geq m_1$$

where $\equiv'$ means "away from p". The second compatibility congruence, combined with the fact that the characters ϵ_m are valued in ± 1 and $p \geq 3$, implies that there exists a non-trivial quadratic character ϵ such that $\epsilon_m = \epsilon$ for all

$m \geq 1$. In particular ϵ has conductor $D = D_m$. Since h_m is a companion of f mod p^m and has CM by ϵ,

$$a_\ell(f) \equiv l^{k-1} a_\ell(h_m) \equiv \epsilon(\ell) l^{k-1} a_\ell(h_m) \equiv \epsilon(\ell) a_\ell(f) \bmod p^m$$

for all $m \geq 1$ and primes $l \nmid DNp$. Thus $a_l(f) = \epsilon(l) a_l(f)$ for $l \nmid DNp$, and f has CM by definition.

2.5 An elementary approach

In the section we present a proof of Theorem 2.6 which relies purely on carefully manipulating the Hecke character from which the CM form f arises and then appealing to Theorem 2.4. The weakness of this approach, which we have been unable to overcome, is a condition on the class number of the imaginary quadratic field K from which f arises. Nevertheless, this method allows us to prove that f, in fact, has companion forms modulo an odd integer $M \geq 3$ – not necessarily a prime power. Let $k' \geq 2$ be the smallest integer such that $k + k' \equiv 2 \bmod \phi(M)$, with ϕ being the Euler totient function. We say that f has a companion h modulo M of weight k' if, keeping the terminology of Definition 2.1, $a_n(f) \equiv n^{k-1} a_n(h) \bmod M$ for $(n, NM) = 1$.

Theorem 2.7. *Let $M \geq 3$ be an odd integer and $f = \sum a_n(f) q^n$ be a CM eigenform of weight $k \geq 2$, level N coprime to M and p-ordinary for each $p|M$. Assume that the class number of the imaginary quadratic field K from which f arises is coprime to M. Let $k' \geq 2$ be the smallest integer such that $k + k' \equiv 2 \bmod \phi(M)$, with ϕ being the Euler totient function. Then, there exists a CM eigenform $h = \sum a_n(h) q^n$ of weight k' that is p-ordinary for every prime p dividing M and such that it is a companion form for f mod M.*

Proof. Let $K = \mathbb{Q}(\sqrt{-D})$ be the quadratic imaginary field by which f has CM and $M' = \prod_{i=1}^r p_i$ for primes p_i dividing M. Let g be the eigenform of level NM' and weight k, whose p_i-th coefficient has p_i-adic valuation equal to $k - 1$ for each $p_i | M'$ and whose q-expansion agrees with that of f at all primes away from NM'. Such a g exists and one can construct it iteratively as follows. Let g_1 be the p_1-twin oldform of weight k and level Np_1. Now g_1 is still p_2-ordinary so that one can associate to it the p_2-twin eigenform of weight k and level $Np_1 p_2$. Proceeding iteratively in this manner we have the desired eigenform $g = g_r$ of weight k, and level NM'. Clearly g has CM by K as well. Let $\psi_g : J^{\mathrm{m}} \to \mathbb{C}^*$ be the Hecke character over K associated with g. We will denote by ψ_g^∞ its finite type and by $\psi_{g,\infty}$ its infinity type. Let $\sigma : K \to \mathbb{C}$ be the identity embedding of K in $\mathbb{C}$. Then $\psi_{g,\infty} = \sigma^{k-1}$. Theorem 2.4 implies that $\mathbf{N}\mathfrak{m}|NM'$.

Consider the Hecke character $\varphi = \bar{\psi_g}^{-1}$. We have that $\varphi^\infty = \psi_g^\infty$ and $\varphi_\infty = \sigma^{1-k} = \sigma^{(2-k)-1}$. The reason we are interested in this character is the identity

$$s^{k-1}(\varphi(\mathfrak{s}) + \varphi(\bar{\mathfrak{s}})) = \psi_g(\mathfrak{s}) + \psi_g(\bar{\mathfrak{s}}) = a_s(g), \tag{2.4}$$

for every rational prime s that splits in K as $\mathfrak{s}\bar{\mathfrak{s}}$. In view of this identity it is enough to find a Hecke character ψ' congruent to φ and with infinity type $\sigma^{k'-1}$.

Assume for simplicity that $D \neq 1, 3$. It is then easy to see (for instance the discussion in Section 2.4 of [13]) that the compatibility of a finite and an infinity type is guaranteed by the identity $\psi_g^\infty(-1) = (-1)^{k-1}$. One can therefore find a Hecke character with finite type equal to ψ_g^∞ and infinity type $\sigma^{k'-1}$, for any $k' \equiv 2 - k \bmod 2$. This Hecke character is not unique. In fact it turns out (see for example Lemma 4.1 in [13]) that there are h_K choices, where h_K is the class number of K. We explain this in more detail. Consider the group $J^{\mathfrak{m}}/P^{\mathfrak{m}}$ where $J^{\mathfrak{m}}$ is as in Section 2.2 and $P^{\mathfrak{m}}$ is the subgroup of principal fractional ideals coprime to $\mathfrak{m}$. It is a finite abelian group of order h_K. Let $\mathfrak{a}_i$ denote a fixed choice of representatives that generate $J^{\mathfrak{m}}/P^{\mathfrak{m}}$, c_i be their respective orders and α_i be the generator of the ideal $\mathfrak{a}_i^{c_i}$. Then, given a pair $(\psi^\infty, \sigma^{k'-1})$ consisting of a finite and an infinity type, any Hecke character ψ' corresponding to that pair is completely determined by the values $\psi'(\mathfrak{a}_i)$. Let $d_i(\psi')$ be a non-negative integer and ζ_{c_i} denote a primitive c_i-th root of unity. The only possible values for $\psi'(\mathfrak{a}_i)$ are of the form

$$\sqrt[c_i]{\psi^\infty(\alpha_i)}\,\sqrt[c_i]{\alpha_i^{k'-1}}\,\zeta_{c_i}^{d_i(\psi')}$$

where $\sqrt[c_i]{\psi^\infty(\alpha_i)}$ and $\sqrt[c_i]{\alpha_i^{k'-1}}$ are fixed choices of c_i-th roots of $\psi^\infty(\alpha_i)$ and $\alpha_i^{k'-1}$ respectively.

Let k' be the integer that we defined previously. Since k' has the same parity as $2-k$ we can define a Hecke character ψ' of infinity type $\sigma^{k'-1}$ and finite type ψ_g^∞. Let L be the extension of K with $\sqrt[c_i]{\psi_g^\infty(\alpha_i)}$, $\sqrt[c_i]{\alpha_i^{k'-1}}$ and ζ_{c_i} adjoined for all i (and for the fixed k'). Let $M = \prod_p p^{t_p}$. Furthermore, for every $p|M$, let $\mathfrak{p}$ be a prime in K such that $p = \mathfrak{p}\bar{\mathfrak{p}}$ and pick a prime $\mathfrak{P}$ above $\mathfrak{p}$ in L. Our goal is to show that the $d_i(\psi')$'s can always be chosen so that

$$\varphi(\mathfrak{a}_i) \equiv \psi'(\mathfrak{a}_i) \pmod{\mathfrak{P}^{t'_{\mathfrak{P}}}}$$

for all i, where $t'_{\mathfrak{P}} = e(\mathfrak{P}/\mathfrak{p})(t_p - 1) + 1$. First, notice that if $\alpha_i \equiv 0 \pmod{\mathfrak{p}}$ then the desired congruence is immediate. In what follows we can therefore freely assume that α_i is a unit in $\mathcal{O}_K/\mathfrak{p}^{t_p} \subseteq \mathcal{O}_L/\mathfrak{P}^{t'_{\mathfrak{P}}}$. Moreover, $\mathcal{O}_K/\mathfrak{p}^{t_p}$ is isomorphic to $\mathbb{Z}/p^{t_p}\mathbb{Z}$ because of the splitting condition on the primes dividing M. Since the order of $(\mathbb{Z}/p^{t_p}\mathbb{Z})^*$ is $\phi(p^{t_p})$, we get that $\alpha^{k'-(2-k)} \equiv 1 \pmod{\mathfrak{p}^{t_p}}$. This in turn implies that $\sqrt[c_i]{\alpha^{k'-(2-k)}}$ is a c_i-th root of unity in $\mathcal{O}_L/\mathfrak{P}^{t'_{\mathfrak{P}}}$. Since h_K is coprime to M this c_i-th root of unity will lift to one in $\mathcal{O}_L$ so all one has to do is to choose $d_i(\psi')$ so that $\zeta_{c_i}^{d_i(\varphi)-d_i(\psi')}$ is this root. In conclusion, the CM form h that one may associate to the Hecke character ψ' by Theorem 2.4 is the desired companion of f mod M. Equation (5.1) ensures that h is p-ordinary for each p dividing M. □

Example 2.8. Let f be the CM newform of weight 3 and level 8 with the following Fourier expansion,

$$q - 2q^2 - 2q^3 + 4q^4 + 4q^6 - 8q^8 - 5q^9 + 14q^{11} - 8q^{12} + 16q^{16} + 2q^{17} \\ +10q^{18} - 34q^{19} - 28q^{22} + 16q^{24} + 25q^{25} + \cdots$$

It is ordinary at 3, 11 and 17. Using MAGMA [2] we find companions h_{19}, h_{31} and h_{59} modulo 33, 51 and 99 respectively. The indices denote the weights of the companions; each has level 8 and CM by the quadratic Dirichlet character of conductor 8. Their Fourier expansions are:

$$h_{19} = q - 512q^2 - 3266q^3 + 262144q^4 + 1672192q^6 - 134217728q^8 - \\ 376753733q^9 - 354349618q^{11} - 856162304q^{12} + 68719476736q^{16} + \\ 119842447106q^{17} + 192897911296q^{18} + 335013705758q^{19} + \\ 181427004416q^{22} + 438355099648q^{24} + 3814697265625q^{25} + \cdots$$

$$h_{31} = q - 32768q^2 - 26595314q^3 + 1073741824q^4 + 871475249152q^6 - \\ 35184372088832q^8 + 501419594663947q^9 + 6656187998706302q^{11} - \\ 28556500964212736q^{12} + 1152921504606846976q^{16} - \\ 4422784932886529086q^{17} - 16430517277948215296q^{18} - \\ 23964789267887608402q^{19} - 218109968341608103936q^{22} + \\ 935739423595322933248q^{24} + 931322574615478515625q^{25} - \cdots$$

$$h_{59} = q - 536870912q^2 + 57281430144478q^3 + 288230376151711744q^4 - \\ 30752733642330195623936q^6 - 154742504910672534362390528q^8 - \\ 142896645784953192696771 1205q^9 + \\ 2900908653579886134108518505134q^{11} + \\ 16510248157050893929760929349632q^{12} +$$

$$
\begin{aligned}
&830767497365572420564879412675215 36q^{16} + \\
&955027058519269179716584293727217282q^{17} + \\
&767170525443087764404272509180968960q^{18} - \\
&1584046322102856179371815160177917 4594q^{19} - \\
&155741347447612553367499463682016726 2208q^{22} - \\
&886387198542223265448601408190449870 4384q^{24} + \\
&346944695195361418882384896278381347 65625q^{25} - \cdots
\end{aligned}
$$

Acknowledgments

The first author thanks DFG GRK 1692 at Regensburg, the Hausdorff Research Institute for Mathematics (HIM) and the AFR Grant Scheme of the Fonds National de la Recherche Luxembourg (FNR) for postdoctoral support during the course of this work. The second author acknowledges the support of the DFG Priority Program SPP 1489, the FNR and the invitation extended by the HIM during its Arithmetic and Geometry Trimester Program.

References

[1] R. Adibhatla and J. Manoharmayum. Higher congruence companion forms. *Acta Arith.*, 156: 159–175, 2012.

[2] W. Bosma, J. Cannon and C. Playoust. The Magma algebra system. I. The user language *J. Symbolic Comput.*, 24(3-4):235–265, 1997.

[3] H. Carayol Formes modulaires et représentations galoisiennes à valeurs dans un anneau local complet *Contemp. Math.*, 165:213–237, 1994.

[4] Robert F. Coleman. Classical and overconvergent modular forms. *Invent. Math.*, 124(1-3):215–241, 1996.

[5] Eknath Ghate. Ordinary forms and their local Galois representations. In *Algebra and number theory*, pages 226–242. Hindustan Book Agency, Delhi, 2005.

[6] Benedict H. Gross. A tameness criterion for Galois representations associated to modular forms (mod p). *Duke Math. J.*, 61(2):445–517, 1990.

[7] H. Hida. Galois representations into $\mathrm{GL}_2(\mathbf{Z}_p[[X]])$ attached to ordinary cusp forms. *Invent. Math.*, 85(3):545–613, 1986.

[8] Toshitsune Miyake. *Modular forms*. Springer Monographs in Mathematics. Springer-Verlag, Berlin, english edition, 2006. Translated from the 1976 Japanese original by Yoshitaka Maeda.

[9] B. Mazur and A. Wiles. On p-adic analytic families of Galois representations. *Compositio Math.*, 59(2):231–264, 1986.

[10] Jürgen Neukirch. *Algebraic number theory*, volume 322 of *Grundlehren der Mathematischen Wissenschaften [Fundamental Principles of Mathematical Sciences]*. Springer-Verlag, Berlin, 1999. Translated from the 1992 German original and with a note by Norbert Schappacher, With a foreword by G. Harder.

[11] Kenneth A. Ribet. Galois representations attached to eigenforms with Nebentypus. In *Modular functions of one variable, V (Proc. Second Internat. Conf., Univ. Bonn, Bonn, 1976)*, pages 17–51. Lecture Notes in Math., Vol. 601. Springer, Berlin, 1977.

[12] Xavier Taixési Ventosa and Gabor Wiese. Computing congruences of modular forms and Galois representations modulo prime powers. In *Arithmetic, geometry, cryptography and coding theory 2009*, volume 521 of *Contemp. Math.*, pages 145–166. Amer. Math. Soc., Providence, RI, 2010.

[13] P. Tsaknias. A possible generalization of Maeda's conjecture. *Computational aspects of modular forms, Conference Proceedings Heidelberg*, pages 1–21. Springer-Verlag 2013.

[14] A. Wiles. On ordinary λ-adic representations associated to modular forms. *Invent. Math.*, 94(3):529–573, 1988.

3

Selmer complexes and p-adic Hodge theory

Denis Benois

Institut de Mathématiques, Université de Bordeaux, 351, cours de la Libération 33405 Talence, France

E-mail address: denis.benois@math.u-bordeaux1.fr

Contents

Arithmetic and Geometry, ed. Luis Dieulefait *et al.* Published by Cambridge University Press.

Introduction

3.0.1 Selmer groups

In this paper we discuss some applications of p-adic Hodge theory to the algebraic formalism of Iwasawa theory. Fix an odd prime number p. For every finite set S of primes containing p, we denote by $\mathbf{Q}^{(S)}/\mathbf{Q}$ the maximal algebraic extension of $\mathbf{Q}$ unramified outside $S\cup\{\infty\}$ and set $G_{\mathbf{Q},S}=\mathrm{Gal}(\mathbf{Q}^{(S)}/\mathbf{Q})$. Let $G_{\mathbf{Q}_v}=\mathrm{Gal}(\overline{\mathbf{Q}}_v/\mathbf{Q}_v)$ denote the decomposition group at v, and let I_v be its inertia subgroup.

Let M be a pure motive over $\mathbf{Q}$. The complex L-function $L(M,s)$ is a Dirichlet series which converges for $s\gg 0$ and is expected to admit a meromorphic continuation on the whole $\mathbf{C}$ with a functional equation of the form

$$\Gamma(M,s)L(M,s)=\varepsilon(M,s)\Gamma(M^*(1),-s)L(M^*(1),-s).$$

Here $\Gamma(M,s)$ is a product of Γ-factors explicitly defined in terms of the Hodge structure of M, and $\varepsilon(V,s)$ is a factor of the form $\varepsilon(M,s)=ab^s$, where $a\in\mathbf{C}^*$ and b is a positive integer (see [55], [27]).

Let V denote the p-adic realization of M. We consider V as a p-adic representation of the Galois group $G_{\mathbf{Q},S}$, where S contains p, ∞ and the places where M has bad reduction. We will write $H^*_S(\mathbf{Q},V)$ and $H^*(\mathbf{Q}_v,V)$ for the continuous cohomology of $G_{\mathbf{Q},S}$ and $G_{\mathbf{Q}_v}$ respectively. The Bloch–Kato Selmer group $H^1_f(\mathbf{Q},V)$ is defined as

$$H^1_f(\mathbf{Q},V)=\ker\left(H^1_S(\mathbf{Q},V)\to\bigoplus_{v\in S}\frac{H^1(\mathbf{Q}_v,V)}{H^1_f(\mathbf{Q}_v,V)}\right),\tag{3.1}$$

where the "local conditions" $H^1_f(\mathbf{Q}_v,V)$ are given by

$$H^1_f(\mathbf{Q}_v,V)=\begin{cases}\ker(H^1(\mathbf{Q}_v,V)\to H^1(I_v,V)) & \text{if } v\neq p,\\ \ker(H^1(\mathbf{Q}_p,V)\to H^1(\mathbf{Q}_p,V\otimes\mathbf{B}_{\mathrm{cris}})) & \text{if } v=p\end{cases}\tag{3.2}$$

(see [13]). Here $\mathbf{B}_{\mathrm{cris}}$ is Fontaine's ring of crystalline periods [25]. Note that, using the inflation–restriction sequence, it is easy to see that the definition (3.1) does not depend on the choice of S (see [27], Chapter II). The Beilinson conjecture (in the formulation of Bloch and Kato [13]) relates the rank of $H^1_f(\mathbf{Q}, V^*(1))$ to the order of vanishing of $L(M,s)$, namely one expects that

$$\mathrm{ord}_{s=0}L(M,s) = \dim_{\mathbf{Q}_p} H^1_f(\mathbf{Q}, V^*(1)) - \dim_{\mathbf{Q}_p} H^0(\mathbf{Q}, V^*(1)).$$

3.0.2 p-adic L-functions

Assume that V satisfies the following three conditions (see Subsection 3.2.1 below):

C1) $H^0_S(\mathbf{Q}, V) = H^0_S(\mathbf{Q}, V^*(1)) = 0.$

C2) The restriction of V on the decomposition group at p is semistable in the sense of Fontaine [26].

We denote by $\mathbf{D}_{\mathrm{st}}(V)$ the semistable module associated to V. Recall that $\mathbf{D}_{\mathrm{st}}(V)$ is a filtered $\mathbf{Q}_p$-vector space equipped with a linear Frobenius $\varphi : \mathbf{D}_{\mathrm{st}}(V) \to \mathbf{D}_{\mathrm{st}}(V)$ and a monodromy operator $N : \mathbf{D}_{\mathrm{st}}(V) \to \mathbf{D}_{\mathrm{st}}(V)$ such that $N\varphi = p\varphi N$. Let $\mathbf{D}_{\mathrm{cris}}(V) = \mathbf{D}_{\mathrm{st}}(V)^{N=0}$.

C3) $\mathbf{D}_{\mathrm{cris}}(V)^{\varphi=1} = 0.$

Recall the notion of a regular submodule of $\mathbf{D}_{\mathrm{st}}(V)$, which plays the key role in this paper. It was first introduced by Perrin-Riou [51] in the crystalline case. See also [5] and [7]. Replacing, if necessary, M by $M^*(1)$, we will assume that M is of weight $wt(M) \leqslant -1$. We consider two cases:

The weight $\leqslant -2$ case. Let $wt(M) \leqslant -2$. From Jannsen's conjecture [34] (see Subsection 3.2.1 for more detail) it follows that, in this case, the localization map

$$H^1_f(\mathbf{Q}, V) \to H^1_f(\mathbf{Q}_p, V)$$

should be injective. The condition **C3)** implies that the Bloch–Kato exponential map

$$\exp_V : \mathbf{D}_{\mathrm{st}}(V)/\mathrm{Fil}^0\mathbf{D}_{\mathrm{st}}(V) \to H^1_f(\mathbf{Q}_p, V)$$

is an isomorphism, and we denote by

$$\log_V : H^1_f(\mathbf{Q}_p, V) \to \mathbf{D}_{\mathrm{st}}(V)/\mathrm{Fil}^0\mathbf{D}_{\mathrm{st}}(V)$$

its inverse. A (φ, N)-submodule D of $\mathbf{D}_{\mathrm{st}}(V)$ is regular if $D \cap \mathrm{Fil}^0\mathbf{D}_{\mathrm{st}}(V) = \{0\}$ and the composition of the localisation map with $\log_V$ induces an isomorphism

$$r_{V,D} : H^1_f(\mathbf{Q}, V) \to \mathbf{D}_{\mathrm{st}}(V)/(\mathrm{Fil}^0\mathbf{D}_{\mathrm{st}}(V) + D).$$

Note that the map $r_{V,D}$ is closely related to the syntomic regulator.

The weight -1 case. Let $wt(M) = -1$. In this case, we say that a (φ, N)-submodule D of $\mathbf{D}_{\mathrm{st}}(V)$ is regular if

$$\mathbf{D}_{\mathrm{st}}(V) = D \oplus \mathrm{Fil}^0\mathbf{D}_{\mathrm{st}}(V)$$

as vector spaces. Each regular D gives a splitting of the Hodge filtration on $\mathbf{D}_{\mathrm{st}}(V)$ and therefore defines a p-adic height pairing

$$\langle\, ,\, \rangle_{V,D} \,:\, H^1_f(\mathbf{Q}, V) \times H^1_f(\mathbf{Q}, V^*(1)) \to \mathbf{Q}_p$$

(see [45], [46], [49]), which is expected to be non-degenerate[1].

The special value $L^*(M,n)$ of $L(M,s)$ at $n \in \mathbf{Z}$ is defined as the first non-zero coefficient in the Taylor expansion of $L(M,s)$ at $s = n$. The theory of Perrin-Riou [51] suggests that to each regular D one can associate a p-adic L-function $L_p(M, D, s)$ interpolating the rational parts of $L^*(M,n)$. Set $r(M) = \mathrm{ord}_{s=0}L(M,s)$. The interpolation formula relating special values of the complex and the p-adic L-functions at $s = 0$ should have the form

$$\lim_{s\to 0} \frac{L_p(M, D, s)}{s^{r(M)}} = \mathcal{E}(V, D)\, \Omega_p(M, D)\, \frac{L^*(M,0)}{\Omega_\infty(M)}.$$

Here $\Omega_\infty(M)$ is the irrational part of $L^*(M,0)$ predicted by Beilinson–Deligne's conjectures, and $\Omega_p(M, D)$ is essentially the determinant of $r_{V,D}$ (in the weight $\leqslant -2$ case) or the determinant of $\langle\, ,\, \rangle_{V,D}$ (in the weight -1 case). Finally, $\mathcal{E}(V, D)$ is an Euler-like factor. Its explicit form in the crystalline case was conjectured by Perrin-Riou [51]. Namely, let $E_p(M,X) = \det(1 - \varphi X | \mathbf{D}_{\mathrm{cris}}(V))$. Note that $E_p(M, p^{-s})$ is the Euler factor of $L(M,s)$ at p. One expects that

$$\mathcal{E}(V, D) = E_p(M, 1) \det\left(\frac{1 - p^{-1}\varphi^{-1}}{1 - \varphi} \mid D\right) \qquad \text{if } V \text{ is crystalline at } p$$

(see [51], Chapter 4 and [17], Conjecture 2.7).

3.0.3 Selmer complexes

We introduce basic notation of the Iwasawa theory. Let μ_{p^n} denote the group of p^n-th roots of unity and $\mathbf{Q}^{\mathrm{cyc}} = \bigcup_{n=1}^{\infty} \mathbf{Q}(\mu_{p^n})$. We fix a system $(\zeta_{p^n})_{n\geqslant 0}$ of primitive p^nth roots of unity such that $\zeta_{p^{n+1}}^p = \zeta_{p^n}$ for all n. The Galois

[1] The non-degeneracy of the p-adic height pairing was first conjectured by Schneider [56] for abelian varieties having good ordinary reduction at p.

group $\Gamma = \mathrm{Gal}(\mathbf{Q}^{\mathrm{cyc}}/\mathbf{Q})$ is canonically isomorphic to $\mathbf{Z}_p^*$ *via* the cyclotomic character

$$\chi : \Gamma \to \mathbf{Z}_p^*, \qquad g(\zeta_{p^n}) = \zeta_{p^n}^{\chi(g)} \qquad \text{for all } g \in \Gamma.$$

Let $\Delta = \mathrm{Gal}(\mathbf{Q}(\mu_p)/\mathbf{Q})$ and $\Gamma_0 = \mathrm{Gal}(\mathbf{Q}^{\mathrm{cyc}}/\mathbf{Q}(\mu_p))$. We denote by $F_\infty = \mathbf{Q}(\mu_{p^\infty})^\Delta$ the cyclotomic $\mathbf{Z}_p$-extension of $\mathbf{Q}$ and set $F_n = \mathbf{Q}(\mu_{p^{n+1}})^\Delta$, $n \in \mathbf{N}$. Fix a generator $\gamma \in \Gamma$ and set $\gamma_0 = \gamma^{p-1} \in \Gamma_0$. Let Λ denote the Iwasawa algebra $\mathbf{Z}_p[[\Gamma_0]]$. The choice of γ fixes an isomorphism $\Lambda \simeq \mathbf{Z}_p[[X]]$ such that $\gamma_0 \mapsto 1+X$. Let $\mathscr{H}$ denote the ring of power series, which converge on the open unit disk. We consider Λ as a subring of $\mathscr{H}$ *via* the above chosen isomorphism $\Lambda \simeq \mathbf{Z}_p[[X]]$. Note that $\mathscr{H}$ is the large Iwasawa algebra introduced in [50]. The map $\tau \mapsto \tau^{-1}$, $\tau \in \Gamma_0$ defines a canonical involution on Λ, which we denote by $\iota : \Lambda \to \Lambda$.

Let T be a $\mathbf{Z}_p$-lattice of V stable under the action of $G_{\mathbf{Q},S}$. We denote by $T \otimes \Lambda^\iota$ the tensor product $T \otimes_{\mathbf{Z}_p} \Lambda$ equipped with the following structures:

a) The continuous action of $G_{\mathbf{Q},S}$ defined by

$$g(a \otimes \lambda) = g(a) \otimes \bar{g}\lambda, \qquad g \in G_{\mathbf{Q},S},$$

where $\bar{g} \in \Gamma_0$ is the image of g under the canonical projection $G_{\mathbf{Q},S} \to \Gamma_0$;

b) The structure of a Λ-module defined by

$$\tau(a \otimes \lambda) = a \otimes \iota(\tau)(\lambda) = a \otimes \tau^{-1}\lambda, \qquad \tau \in \Gamma_0.$$

Note that the action of $G_{\mathbf{Q},S}$ on $T \otimes \Lambda^\iota$ is Λ-linear.

Let $C^\bullet(G_{\mathbf{Q},S}, T \otimes \Lambda^\iota)$ (respectively $C^\bullet(G_{\mathbf{Q}_v}, T \otimes \Lambda^\iota)$) be the complex of continuous cochains of $G_{\mathbf{Q},S}$ (respectively $G_{\mathbf{Q}_v}$) with coefficients in $T \otimes \Lambda^\iota$. We denote by $\mathbf{R}\Gamma_{\mathrm{Iw},S}(\mathbf{Q},T)$ and $\mathbf{R}\Gamma_{\mathrm{Iw},S}(\mathbf{Q}_v,T)$ these complexes viewed as objects of the bounded derived category $D^{\mathrm{b}}(\Lambda)$ of Λ-modules. Set $H^i_{\mathrm{Iw},S}(\mathbf{Q},T) = \mathbf{R}^i\Gamma_{\mathrm{Iw},S}(\mathbf{Q},T)$ and $H^i_{\mathrm{Iw}}(\mathbf{Q}_v,T) = \mathbf{R}^i\Gamma_{\mathrm{Iw}}(\mathbf{Q}_v,T)$. From Shapiro's lemma it follows that

$$H^i_{\mathrm{Iw},S}(\mathbf{Q},T) = \varprojlim_n H^i_S(F_n,T), \qquad H^i_{\mathrm{Iw}}(\mathbf{Q}_v,T) = \varprojlim_n H^i_S(F_{n,v},T)$$

as Λ-modules. In [46], Nekovář considers the diagrams of the form

$$\begin{array}{ccc} \mathbf{R}\Gamma_{\mathrm{Iw},S}(\mathbf{Q},T) & \longrightarrow & \bigoplus_{v\in S}\mathbf{R}\Gamma_{\mathrm{Iw}}(\mathbf{Q}_v,T) \\ & & \uparrow \\ & & \bigoplus_{v\in S} U_v^\bullet(T). \end{array} \tag{3.3}$$

The cone of this diagram can be viewed as the derived version of Selmer groups, where the complexes $U_v^\bullet(T)$ play the role of local conditions. For $v \neq p$, the natural choice of $U_v^\bullet(T)$ is to put $U_v^\bullet(T) = \mathbf{R}\Gamma_{\mathrm{Iw},f}(\mathbf{Q}_v, T)$, where

$$\mathbf{R}\Gamma_{\mathrm{Iw},f}(\mathbf{Q}_v, T) = \left[T^{I_v} \otimes \Lambda^\iota \xrightarrow{\mathrm{Fr}_v - 1} T^{I_v} \otimes \Lambda^\iota \right].$$

Here Fr_v denotes the geometric Frobenius, and the terms are placed in degrees 0 and 1. Note that the cohomology groups of

$$\mathbf{R}\Gamma_{\mathrm{Iw},f}(\mathbf{Q}_v, T) \otimes_\Lambda^{\mathbf{L}} \mathbf{Q}_p = \left[V^{I_v} \xrightarrow{\mathrm{Fr}_v - 1} V^{I_v} \right]$$

are $H^0(\mathbf{Q}_v, V)$ and $H^1_f(\mathbf{Q}_v, V)$. The local conditions at p are more delicate to define. First assume that V satisfies the Panchishkin condition, i.e. the restriction of V on the decomposition group at p has a subrepresentation $F^+V \subset V$ such that

$$\mathbf{D}_{\mathrm{dR}}(V) = \mathbf{D}_{\mathrm{dR}}(F^+V) \oplus \mathrm{Fil}^0\mathbf{D}_{\mathrm{dR}}(V)$$

as vector spaces. Set $U_p^\bullet(T) = \mathbf{R}\Gamma_{\mathrm{Iw}}(\mathbf{Q}_p, F^+T)$. Then $U_p^\bullet(T)$ is the derived version of Greenberg's local conditions [28], and the cohomology of Nekovář's Selmer complex is closely related to the Pontryagin dual of Greenberg's Selmer group [46]. If we assume, in addition, that $\mathbf{D}_{\mathrm{cris}}(V^*(1))^{\varphi=1} = 0$, then

$$H^1_f(\mathbf{Q}_p, V) = \ker(H^1(\mathbf{Q}_p, V) \to H^1(I_p, V/F^+V)),$$

and we can compare Greenberg and Bloch–Kato Selmer groups. Roughly speaking, in this case different definitions lead to pseudo-isomorphic Λ-modules, which have therefore the same characteristic series in the case they are Λ-torsion (see [20], [46], Chapter 9 and [48]). If $\mathbf{D}_{\mathrm{cris}}(V^*(1))^{\varphi=1} \neq 0$, the situation is more complicated. The analytic counterpart of this problem is the phenomenon of extra-zeros of p-adic L-functions studied in [29], [5], [7].

We no longer assume that V satisfies the Panchishkin condition. The theory of (φ, Γ)-modules (see [23], [15], [9]) associates to V a finitely generated free module $\mathbf{D}^\dagger_{\mathrm{rig}}(V)$ over the Robba ring $\mathscr{R}$ equipped with a semilinear actions of the group Γ and a Frobenius φ, which commute with each other (see Subsection 3.1.1 below). The category of (φ, Γ)-modules has a nice cohomology theory, whose formal properties are very close to properties of local Galois cohomology [32], [33], [42]. In particular, $H^*(\mathbf{D}^\dagger_{\mathrm{rig}}(V))$ is canonically isomorphic[2] to the continuous Galois cohomology $H^*(\mathbf{Q}_p, V)$. Moreover, to each (φ, Γ)-module $\mathbf{D}$ we can associate the complex

2 Up to the choice of a generator of Γ_0.

$$\mathbf{R}\Gamma_{\mathrm{Iw}}(\mathbf{D}) = \left[\mathbf{D} \xrightarrow{\psi-1} \mathbf{D}\right]^{\Delta},$$

where ψ is the left inverse to φ (see [14]), and the first term is placed in degree 1. We will write $H^*_{\mathrm{Iw}}(\mathbf{D})$ for the cohomology of $\mathbf{R}\Gamma_{\mathrm{Iw}}(\mathbf{D})$. The action of Γ_0 induces a natural structure of $\mathscr{H}$-module on $\mathbf{D}^{\Delta}$. From a general result of Pottharst ([54], Theorem 2.8) it follows that

$$H^i_{\mathrm{Iw}}(\mathbf{Q}_p, T) \otimes_{\Lambda} \mathscr{H} \xrightarrow{\sim} H^i_{\mathrm{Iw}}(\mathbf{D}^{\dagger}_{\mathrm{rig}}(V))$$

as $\mathscr{H}$-modules.

The approach to Iwasawa theory discussed in this paper is based on the observation that the (φ, Γ)-module of a semistable representation looks like an ordinary Galois representation, in particular, it is a successive extension of (φ, Γ)-modules of rank 1. This was first pointed out by Colmez in [19], where the structure of trianguline (φ, Γ)-modules of rank 2 over $\mathbf{Q}_p$ was studied in detail. Therefore in the non-ordinary setting we can adopt Greenberg's approach working with (φ, Γ)-modules instead of p-adic representations. This idea was used in [2] and [3] to study families of Selmer groups, and in [5], [6], [7] to study extra-zeros of p-adic L-functions. Pottharst [52] developed a general theory of Selmer complexes in this setting and related it to the Perrin-Riou's theory from [53].

3.0.4 The Main Conjecture

Fix a regular submodule D of $\mathbf{D}_{\mathrm{st}}(V)$. By [11], we can associate to D a canonical (φ, Γ)-submodule $\mathbf{D}$ of $\mathbf{D}^{\dagger}_{\mathrm{rig}}(V)$. Consider the diagram

$$\begin{array}{ccc} \mathbf{R}\Gamma_{\mathrm{Iw},S}(\mathbf{Q},T) \otimes^{\mathbf{L}}_{\Lambda} \mathscr{H} & \longrightarrow & \bigoplus\limits_{v\in S} \mathbf{R}\Gamma_{\mathrm{Iw}}(\mathbf{Q}_v,T) \otimes^{\mathbf{L}}_{\Lambda} \mathscr{H} \\ & & \uparrow \\ & & \bigoplus\limits_{v\in S} U^{\bullet}_v(V,D) \end{array}$$

in the derived category of $\mathscr{H}$-modules, where the local conditions $U^{\bullet}_v(V, D)$ are

$$U^{\bullet}_v(V, D) = \begin{cases} \mathbf{R}\Gamma_{\mathrm{Iw},f}(\mathbf{Q}_v, T) \otimes^{\mathbf{L}}_{\Lambda} \mathscr{H} & \text{if } v \neq p \\ \mathbf{R}\Gamma_{\mathrm{Iw}}(\mathbf{D}) & \text{if } v = p. \end{cases}$$

Consider the Selmer complex associated to these data

$$\mathbf{R}\Gamma_{\mathrm{Iw}}(V,D) = \mathrm{cone}\left(\left(\mathbf{R}\Gamma_{\mathrm{Iw},f}(\mathbf{Q}_v,T)\otimes^{\mathbf{L}}_{\Lambda}\mathscr{H}\right)\bigoplus\left(\bigoplus_{v\in S}U_v^{\bullet}(V,D)\right)\to \bigoplus_{v\in S}\mathbf{R}\Gamma_{\mathrm{Iw}}(\mathbf{Q}_v,T)\otimes^{\mathbf{L}}_{\Lambda}\mathscr{H}\right)[-1]. \quad (3.4)$$

Set

$$H^i_{\mathrm{Iw}}(V,D) = \mathbf{R}^i\Gamma_{\mathrm{Iw}}(V,D).$$

We propose the following conjecture.

Main Conjecture. *Let $M/\mathbf{Q}$ be a pure motive of weight $\leqslant -1$ which does not contain submotives of the form $\mathbf{Q}(m)$. Assume that the p-adic realization V of M satisfies the conditions* **C1–3)** *from Subsection 3.0.2. Let D be a regular submodule of $\mathbf{D}_{\mathrm{st}}(V)$. Then*

i) $H^i_{\mathrm{Iw}}(V,D) = 0$ *for* $i\neq 2$.
ii) $H^2_{\mathrm{Iw}}(V,D)$ *is a coadmissible*[3] *torsion $\mathscr{H}$-module, and*

$$\mathrm{char}_{\mathscr{H}}\left(H^2_{\mathrm{Iw}}(V,D)\right) = (f_D),$$

where $L_p(M,D,s) = f_D(\chi(\gamma_0)^s-1)$.

Remarks. 1) Since $\mathscr{H}^* = \Lambda[1/p]^*$, our Main Conjecture determines f_D up to multiplication by a unit in $\Lambda[1/p]$.

2) Assume that M is critical and that V is ordinary at p. Then the restriction of V to the decomposition group at p is equipped with an increasing filtration $(F_iV)_{i\in\mathbf{Z}}$ such that for all $i\in\mathbf{Z}$

$$\mathrm{gr}_i(V) = W_i(i),$$

where W_i is an unramified representation of $G_{\mathbf{Q}_p}$. From the definition of a regular submodule it follows immediately that $D = \mathbf{D}_{\mathrm{st}}(F^1V)$ is regular. Coates and Perrin-Riou [16] conjectured that in this situation $f_D\in\Lambda$. In [28], Greenberg defined a cofinitely generated Λ-module (Greenberg's Selmer group) $S(F_\infty, V^*(1)/T^*(1))$ and conjectured that its Pontryagin dual $S(F_\infty, V^*(1)/T^*(1))^\wedge$ is related to the p-adic function $L_p(V,D,s)$ as follows

$$\mathrm{char}_\Lambda S(F_\infty, V^*(1)/T^*(1))^\wedge = f_D\Lambda.$$

Using the results of [46], one can check that this agrees with our Main Conjecture, but Greenberg's conjecture is more precise because it determines f_D up to multiplication by a unit in Λ. See [53] and Section 3.2.4 for more detail.

[3] See Section 3.1.6 for the definition of a coadmissible module and its characteristic ideal.

3) Assume that V is crystalline at p. In [51], for any regular D, Perrin-Riou defined a free Λ-module $\mathbf{L}_{\mathrm{Iw}}(D, V)$ together with a canonical trivialization

$$i_{V,D} \,:\, \mathbf{L}_{\mathrm{Iw}}(D, V) \tilde{\to} \mathscr{H}.$$

See also [7] for the interpretation of Perrin-Riou's theory in terms of Selmer complexes. Perrin-Riou's main conjecture says that

$$i_{V,D}(\mathbf{L}_{\mathrm{Iw}}(D, V)) = \Lambda f_D.$$

In [53], Pottharst proved that

$$i_{V,D}(\mathbf{L}_{\mathrm{Iw}}(D, V)) \otimes_{\mathbf{Z}_p} \mathbf{Q}_p = \mathrm{char}_{\mathscr{H}} H^2_{\mathrm{Iw}}(V, D)$$

i.e. for crystalline representations our conjecture is compatible with Perrin-Riou's theory.

4) Let $f = \sum\limits_{n=1}^{\infty} a_n q^n$ be a newform of weight k on $\Gamma_0(N)$, and let W_f be the p-adic representation associated to f. Then W_f can be seen as the p-adic realization of a pure motive M_f of weight $k-1$. The representation W_f is crystalline if $(p, N) = 1$ and semistable non-crystalline if $p \parallel N$. (As usual, we write $p \parallel N$ if $N = pN'$ with $(p, N') = 1$.) Fix $1 \leqslant m \leqslant k-1$. Then the motive $M_f(m)$ is critical, and we denote by $V = W_f(m)$ its p-adic realization. The subspace $\mathrm{Fil}^0\mathbf{D}_{\mathrm{st}}(V)$ is one dimensional. Using the extension of scalars to a finite field extension $E/\mathbf{Q}_p$, we can assume that $\mathbf{D}_{\mathrm{st}}(V) = Ee_\alpha + Ee_\beta$, where $\varphi(e_\alpha) = \alpha e_\alpha$ and $\varphi(e_\beta) = \beta e_\beta$ for some $\alpha, \beta \in E$. If $(p, N) = 1$, the eigenspaces $D_\alpha = Ee_\alpha$ and $D_\beta = Ee_\alpha$ are regular and

$$L(M_f(m), D_\alpha, s) = L_{p,\alpha}(f, s+m), \quad L(M_f(m), D_\beta, s) = L_{p,\beta}(f, s+m),$$

where $L_{p,\alpha}(f, s)$ and $L_{p,\beta}(f, s)$ are the usual p-adic L-functions associated to α and β [43], [61], [44]. If $p \parallel N$, we have $Ne_\beta = e_\alpha$ and $Ne_\alpha = 0$ and D_α is the unique regular subspace of $\mathbf{D}_{\mathrm{st}}(V)$. The results of Kato [36] have the following interpretation in terms of our theory. Assume that $(p, N) = 1$ and set $L(V, D_\alpha, s) = f_{D_\alpha}(\chi(\gamma_0)^s - 1)$. Then $H^2_{\mathrm{Iw}}(V, D_\alpha)$ is $\mathscr{H}$-torsion and

$$\mathrm{char}_{\mathscr{H}} H^2_{\mathrm{Iw}}(V, D_\alpha) \subset (f_{D_\alpha})$$

(see [53], Theorem 5.4). In the ordinary case, the opposite inclusion was recently proved under some technical conditions by Skinner and Urban [58]. It would be interesting to understand whether their methods can be generalized to the non-ordinary case.

5) It is certainly possible to formulate the Main Conjecture for families of p-adic representations in the spirit of [30].

3.0.5 Organization of the paper

The first part of the paper is written as a survey article. In Section 3.1, we review the theory of (φ, Γ)-modules. In particular, we discuss recent results of Kedlaya, Pottharst and Xiao [40] about the cohomology of families of (φ, Γ)-modules and its applications to the Iwasawa cohomology. In Section 3.2, we apply this technique in the context of global Iwasawa theory. The notion of regular submodule is discussed in Section 3.2.1. In Subsections 3.2.2 and 3.2.3, we construct the complex $\mathbf{R}\Gamma_{\mathrm{Iw}}(V, D)$ and review its basic properties following Pottharst [53].

The Main Conjecture is formulated in Subsection 3.2.4. In the rest of the paper, we discuss the relationship between the nullity of $H^1_{\mathrm{Iw}}(V, D)$ and the structure of the semistable module $\mathbf{D}_{\mathrm{st}}(V)$. In Section 3.3, we consider the weight $\leqslant -2$ case and prove that $H^1_{\mathrm{Iw}}(V, D) = 0$ (and therefore $H^2_{\mathrm{Iw}}(V, D)$ is $\mathscr{H}$-torsion) if the ℓ-invariant $\ell(V, D)$, constructed in [5], [7], does not vanish. In Section 3.4, we consider the weight -1 case and prove that the nullity of $H^1_{\mathrm{Iw}}(V, D)$ follows from the non-degeneracy of the p-adic height pairing. The proof is based on a generalization of Nekovář's construction of the p-adic height pairing [46] to the non-ordinary case. A systematic study of p-adic heights via (φ, Γ)-modules will be done in [8].

Acknowledgement

I would like to thank the referee for numerous suggestions in improving the paper and for a very patient correction of many minor points.

3.1 Overview of the theory of (φ, Γ)-modules

3.1.1 Galois representations and (φ, Γ)-modules

In this section we summarize the results about (φ, Γ)-modules, which will be used in subsequent sections. The notion of a (φ, Γ)-module was introduced by Fontaine in his fundamental paper [23]. We consider only (φ, Γ)-modules over $\mathbf{Q}_p$ and their families, because this is sufficient for applications we have in mind.

Let p be a prime number. Fix an algebraic closure $\overline{\mathbf{Q}}_p$ of $\mathbf{Q}_p$ and set $G_{\mathbf{Q}_p} = \mathrm{Gal}(\overline{\mathbf{Q}}_p/\mathbf{Q}_p)$. Let $\mathbf{Q}_p^{\mathrm{cyc}} = \mathbf{Q}_p(\mu_{p^\infty})$ and $\Gamma = \mathrm{Gal}(\mathbf{Q}_p^{\mathrm{cyc}}/\mathbf{Q}_p)$. The cyclotomic character $\chi : G_{\mathbf{Q}_p} \to \mathbf{Z}_p^*$ induces an isomorphism of Γ onto $\mathbf{Z}_p^*$, which we denote again by $\chi : \Gamma \to \mathbf{Z}_p^*$. Let $\mathbf{C}_p$ be the p-adic completion of $\overline{\mathbf{Q}}_p$. We denote by $|\cdot|_p$ the p-adic absolute value on $\mathbf{C}_p$ normalized by $|p|_p = 1/p$.

We fix a coefficient field E, which is a finite extension of $\mathbf{Q}_p$, and consider the following objects:

- The field $\mathscr{E}_E$ of power series $f(X) = \sum\limits_{k\in\mathbf{Z}} a_k X^k$, $\quad a_k \in E$ such that a_k are p-adically bounded and $a_k \to 0$ when $k \to -\infty$. Thus, $\mathscr{E}_E$ is a complete discrete valuation field with residue field $k_E((X))$, where k_E is the residue field of E.
- For each $0 \leqslant r < 1$ the ring $\mathscr{R}_E^{(r)}$ of p-adic functions

$$f(X) = \sum_{k\in\mathbf{Z}} a_k X^k, \qquad a_k \in E, \tag{3.5}$$

  which are holomorphic on the p-adic annulus

$$A(r,1) = \{z \in \mathbf{C}_p \mid p^{-1/r} \leqslant |z|_p < 1\}.$$

  The Robba ring of power series with coefficients in E is defined as

$$\mathscr{R}_E = \bigcup_r \mathscr{R}_E^{(r)}.$$

  Note that $\mathscr{R}_E$ is a Bézout ring (every finitely generated ideal in $\mathscr{R}_E$ is principal) [41] but it is not noetherian. Its group of units coincides with the group of units of $\mathscr{O}_E[[X]] \otimes \mathbf{Q}_p$, where $\mathscr{O}_E$ is the ring of integers of E.
- For each $0 \leqslant r < 1$ the ring $\mathscr{E}_E^{\dagger,r}$ of p-adic functions (3.5) that are bounded on $A(r,1)$. Then $\mathscr{E}_E^{\dagger} = \bigcup\limits_r \mathscr{E}_E^{\dagger,r}$ is a field, which is contained both in $\mathscr{E}_E$ and in $\mathscr{R}_E$. Its elements are called overconvergent power series.

The rings $\mathscr{E}_E$, $\mathscr{E}_E^{\dagger}$ and $\mathscr{R}_E$ are equipped with an E-linear continuous action of Γ defined by

$$g(f(X)) = f((1+X)^{\chi(g)} - 1), \qquad g \in \Gamma$$

and a linear operator φ called the Frobenius and given by

$$\varphi(f(X)) = f((1+X)^p - 1).$$

Note that the actions of Γ and φ commute with each other. Set $t = \log(1 + X) = \sum\limits_{n=1}^{\infty}(-1)^{n-1}\dfrac{X^n}{n}$. Then $t \in \mathscr{R}_{\mathbf{Q}_p}$ and $\gamma(t) = \chi(\gamma)\,t$, $\varphi(t) = pt$.

Definition. i) A (φ, Γ)-module over $R = \mathscr{E}_E$, $\mathscr{E}_E^{\dagger}$ or $\mathscr{R}_E$ is a finitely generated free R-module $\mathbf{D}$ equipped with commuting semilinear actions of Γ and φ and such that $A \otimes_{\varphi(A)} \varphi(\mathbf{D}) = \mathbf{D}$. In other words, if $e_1, \dots, e_d$ is an R-basis of $\mathbf{D}$, then $\varphi(e_1), \dots, \varphi(e_d)$ is also a basis of $\mathbf{D}$ over R.

We denote by $\mathbf{M}_R^{\varphi,\Gamma}$ the category of (φ,Γ)-modules over R. The category $\mathbf{M}_{\mathscr{E}_E}^{\varphi,\Gamma}$ contains the important subcategory $\mathbf{M}_{\mathscr{E}_E}^{\varphi,\Gamma,\text{ét}}$ of étale modules. A (φ,Γ)-module $\mathbf{D}$ over $\mathscr{E}_E$ is étale if it is isoclinic of slope 0 in the sense of Dieudonné–Manin's theory. More explicitly, $\mathbf{D}$ is étale if there exists a (φ,Γ)-stable lattice $\mathscr{O}_{\mathscr{E}_E}e_1+\cdots+\mathscr{O}_{\mathscr{E}_E}e_d$ of $\mathbf{D}$ over the ring of integers $\mathscr{O}_{\mathscr{E}_E}$ of $\mathscr{E}_E$ such that the matrix of φ in the basis $\{e_1,\dots,e_d\}$ is invertible over $\mathscr{O}_{\mathscr{E}_E}$. The category $\mathbf{M}_{\mathscr{E}_E^\dagger}^{\varphi,\Gamma,\text{ét}}$ of étale modules over $\mathscr{E}_E^\dagger$ can be defined by the same manner.

Let $\mathbf{Rep}_E(G_{\mathbf{Q}_p})$ denote the $\otimes$-category of p-adic representations of the Galois group $G_{\mathbf{Q}_p}=\mathrm{Gal}(\overline{\mathbf{Q}}_p/\mathbf{Q}_p)$ on finite dimensional E-vector spaces. Using the field-of-norms functor [62], Fontaine constructed in [23] an equivalence of $\otimes$-categories

$$\mathbf{D}\ :\ \mathbf{Rep}_E(G_{\mathbf{Q}_p})\to\mathbf{M}_{\mathscr{E}_E}^{\varphi,\Gamma,\text{ét}}$$

and conjectured that each p-adic representation V is overconvergent, i.e. $\mathbf{D}(V)$ has a canonical $\mathscr{E}_E^\dagger$-lattice $\mathbf{D}^\dagger(V)$ stable under the actions of φ and Γ. This was proved later by Cherbonnier and Colmez in [14].

The relationship between p-adic representations and (φ,Γ)-modules over the Robba ring can be summarized in the following diagram

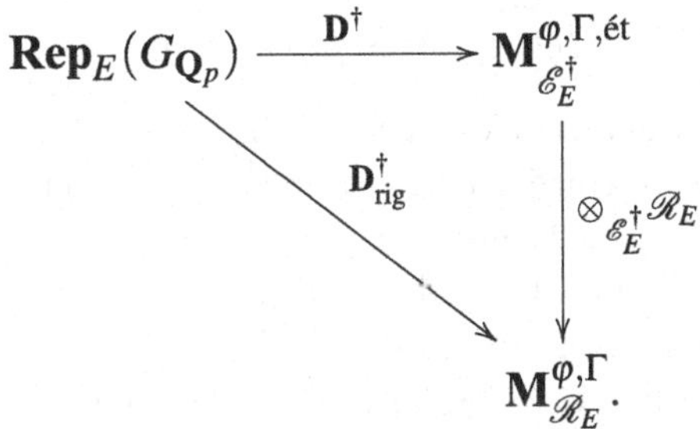

A striking fact is that the vertical arrow is a fully faithful functor. This follows from Kedlaya's generalization of Dieudonné–Manin theory [38]. More precisely, the functor $\mathbf{D}\mapsto\mathbf{D}\otimes_{\mathscr{E}_E^\dagger}\mathscr{R}_E$ establishes an equivalence between $\mathbf{M}_{\mathscr{E}_E^\dagger}^{\varphi,\Gamma,\text{ét}}$ and the category of (φ,Γ)-modules over $\mathscr{R}_E$ of slope 0 in the sense of Kedlaya. See [19], Proposition 1.7 for details.

3.1.2 Cohomology of (φ,Γ)-modules

On the categories of (φ,Γ)-modules one can define cohomology theories whose formal properties are very similar to the properties of continuous Galois cohomology of local fields. The main idea is particulary clear if we work with the category of étale modules. Since $\mathbf{M}_{\mathscr{E}_E}^{\varphi,\Gamma,\text{ét}}$ is equivalent to $\mathbf{Rep}_E(G_{\mathbf{Q}_p})$ it is possible to compute the continuous Galois cohomology $H^*(\mathbf{Q}_p, V)$ in terms of $\mathbf{D}(V)$. In practice, since the category of p-adic representations has no enough of injective objects, one should first work in the context of p^m-torsion modules and consider the category of inductive limits of (φ, Γ)-modules over $\mathscr{O}_{\mathscr{E}_E}/p^m$. This leads to the following results [32], [33]. Let $\Delta = \mathrm{Gal}(\mathbf{Q}_p(\mu_p)/\mathbf{Q}_p)$. Then $\Gamma \simeq \Delta \times \Gamma_0$ where Γ_0 is a procyclic p-group. Fix a topological generator γ_0 of Γ_0 and set $\gamma_n = \gamma_0^{p^n}$, $n \geqslant 0$. Let $K_n = \mathbf{Q}_p(\mu_{p^n})^{\Delta}$, and $K_\infty = \underset{n\geqslant 0}{\cup} K_n$. Then $\Gamma_0 = \mathrm{Gal}(K_\infty/\mathbf{Q}_p)$. If M is a topological module equipped with a continuous action of Γ and an operator φ which commute with each other, we denote by $C^\bullet_{\varphi,\gamma_n}(M)$ the complex

$$C^\bullet_{\varphi,\gamma_n}(M) \; : \; 0 \to M^\Delta \xrightarrow{d_0} M^\Delta \oplus M^\Delta \xrightarrow{d_1} M^\Delta \to 0, \qquad n \geqslant 0, \tag{3.6}$$

where $d_0(x) = ((\varphi - 1)x, (\gamma_n - 1)x)$ and $d_1(y, z) = (\gamma_n - 1)y - (\varphi - 1)z$ and define

$$H^*(K_n, M) = H^*(C^\bullet_{\varphi,\gamma_n}(M)).$$

Let $\mathbf{D}$ be an étale (φ, Γ)-module over $\mathscr{E}_E$, and let V be a p-adic representation of $G_{\mathbf{Q}_p}$ such that $\mathbf{D} = \mathbf{D}(V)$. Then there are isomorphisms

$$H^*(K_n, \mathbf{D}) \simeq H^*(K_n, V),$$

which are functorial and canonical up to the choice of the generator $\gamma \in \Gamma$ (see [32]). This gives an alternative approach to the Euler-Poincaré characteristic formula and the Poincaré duality for Galois cohomology of local fields [32], [33]. If $\mathbf{D}$ is an an étale (φ, Γ)-module over $\mathscr{E}_E^\dagger$, the theorem of Cherbonnier-Colmez implies that again $H^i(K_n, \mathbf{D}) \simeq H^i(K_n, V)$, where V is a p-adic representation such that $\mathbf{D} = \mathbf{D}^\dagger_{\mathrm{rig}}(V)$. The cohomology of (φ, Γ)-modules over $\mathscr{R}_E$ was studied in detail in [42] using a non-trivial reduction to the slope 0 case. For any (φ, Γ)-module $\mathbf{D}$ over $\mathscr{R}_E$ we consider $\mathbf{D}(\chi) = \mathbf{D} \otimes_E E(\chi)$ equipped with diagonal actions of Γ and φ (here φ acts trivially on $E(\chi)$). The main properties of the cohomology groups $H^i(K_n, \mathbf{D})$ of $\mathbf{D}$ are:

1) Long cohomology sequence. A short exact sequence of (φ, Γ)-modules over $\mathscr{R}_E$

$$0 \to \mathbf{D}' \to \mathbf{D} \to \mathbf{D}'' \to 0$$

gives rise to an exact sequence

$$\begin{aligned} 0 \to H^0(K_n, \mathbf{D}') \to H^0(K_n, \mathbf{D}) \to H^0(K_n, \mathbf{D}) \xrightarrow{\delta^0} H^1(K_n, \mathbf{D}') \to \\ \to H^1(K_n, \mathbf{D}') \to H^1(K_n, \mathbf{D}) \to H^1(K_n, \mathbf{D}'') \xrightarrow{\delta^1} H^2(K_n, \mathbf{D}') \to \\ \to H^2(K_n, \mathbf{D}) \to H^2(K_n, \mathbf{D}'') \to 0. \end{aligned} \tag{3.7}$$

2) Euler–Poincaré characteristic. Let $\mathbf{D}$ be a (φ, Γ)-module over $\mathscr{R}_E$. Then for all i, $H^i(\mathbf{D})$ are finite dimensional E-vector spaces, and

$$\chi(K_n, \mathbf{D}) = \sum_{i=0}^{2} (-1)^i \dim_E H^i(K_n, \mathbf{D}) = -[K_n : \mathbf{Q}_p] \operatorname{rank}_{\mathscr{R}_E}(\mathbf{D}). \quad (3.8)$$

3) Computation of the Brauer group. The map

$$\operatorname{cl}(f) \mapsto -\frac{p^n}{\log \chi(\gamma_n)} \operatorname{res}(f\,dt), \quad (3.9)$$

where $\operatorname{res}(f\,dt)$ denotes the residue of the differential form $f\,dt = f\frac{dX}{1+X}$, is well defined and induces an isomorphism $\operatorname{inv}_{K_n} : H^2(K_n, \mathscr{R}_E(\chi)) \xrightarrow{\sim} E$.

4) Cup-products. Let $\mathbf{D}'$ and $\mathbf{D}''$ be two (φ, Γ)-modules over $\mathscr{R}_E$. For all i and j such that $i + j \leqslant 2$ define a bilinear map

$$\cup : H^i(K_n, \mathbf{D}') \times H^j(K_n, \mathbf{D}'') \to H^{i+j}(K_n, \mathbf{D}' \otimes \mathbf{D}'')$$

by

$$\begin{aligned} \operatorname{cl}(x) \cup \operatorname{cl}(y) &= \operatorname{cl}(x \otimes y) && \text{if } i = j = 0, \\ \operatorname{cl}(x) \cup \operatorname{cl}(y_1, y_2) &= \operatorname{cl}(x \otimes y_1, x \otimes y_2) && \text{if } i = 0, j = 1, \\ \operatorname{cl}(x_1, x_2) \cup \operatorname{cl}(y_1, y_2) &= \operatorname{cl}(x_2 \otimes \gamma_n(y_1) - x_1 \otimes \varphi(y_2)) && \text{if } i = 1, j = 1, \\ \operatorname{cl}(x) \cup \operatorname{cl}(y) &= \operatorname{cl}(x \otimes y) && \text{if } i = 0, j = 2. \end{aligned} \quad (3.10)$$

These maps commute with connecting homomorphisms in the usual sense.

5) Duality. Let $\mathbf{D}^* = \operatorname{Hom}_{\mathscr{R}_E}(\mathbf{D}, \mathscr{R}_E)$. For $i = 0, 1, 2$ the cup product

$$H^i(K_n, \mathbf{D}) \times H^{2-i}(K_n, \mathbf{D}^*(\chi)) \xrightarrow{\cup} H^2(K_n, \mathscr{R}_E(\chi)) \simeq E \quad (3.11)$$

is a perfect pairing.

6) Comparison with Galois cohomology. Let $\mathbf{D}$ be an etale (φ, Γ)- module over $\mathscr{E}_E^\dagger$. Then the map

$$C^\bullet_{\varphi,\gamma_n}(\mathbf{D}) \to C^\bullet_{\varphi,\gamma_n}(\mathbf{D} \otimes \mathscr{R}_E),$$

induced by the natural map $\mathbf{D} \to \mathbf{D} \otimes_{\mathscr{E}_E^\dagger} \mathscr{R}_E$, is a quasi-isomorphism. In particular, if V is a p-adic representation of $G_{\mathbf{Q}_p}$, then

$$H^*(K_n, V) \xrightarrow{\sim} H^*(K_n, \mathbf{D}^\dagger_{\mathrm{rig}}(V)).$$

3.1.3 Relation to the p-adic Hodge theory

In [23], Fontaine proposed to classify the p-adic representations arising in the p-adic Hodge theory in terms of (φ, Γ)-modules (Fontaine's program). More precisely, the problem is to recover the classical Fontaine's functors $\mathbf{D}_{\mathrm{dR}}(V)$, $\mathbf{D}_{\mathrm{st}}(V)$ and $\mathbf{D}_{\mathrm{cris}}(V)$ (see for example [26]) from $\mathbf{D}^{\dagger}_{\mathrm{rig}}(V)$. The complete solution was obtained by Berger in [9], [11]. His theory also allowed him to prove that each de Rham representation is potentially semistable. See also [18] for introduction and relation to the theory of p-adic differential equations. In this section, we review some of results of Berger in the case where the ground field is $\mathbf{Q}_p$. Consider the following categories:

- The category $\mathbf{MF}_E$ of finite dimensional E-vector spaces M equipped with an exhaustive decreasing filtration $(\mathrm{Fil}^i M)_{i\in\mathbf{Z}}$.
- The category $\mathbf{MF}_E^{\varphi,N}$ of finite dimensional E-vector spaces M equipped with an exhaustive decreasing filtration $(\mathrm{Fil}^i M)_{i\in\mathbf{Z}}$, a linear bijective Frobenius map $\varphi : M \to M$ and a nilpotent operator (monodromy) $N : M \to M$ such that $\varphi N = p\,\varphi N$.
- The subcategory $\mathbf{MF}_E^{\varphi}$ of $\mathbf{MF}_E^{\varphi,N}$ formed by filtered (φ, N)-modules M such that $N = 0$ on M.

Let $\mathbf{Q}_p^{\mathrm{cyc}}((t))$ be the ring of formal Laurent power series equipped with the filtration $\mathrm{Fil}^i\mathbf{Q}_p^{\mathrm{cyc}}((t)) = t^i\mathbf{Q}_p^{\mathrm{cyc}}[[t]]$ and the action of Γ given by

$$g\left(\sum_{k\in\mathbf{Z}} a_k t^k\right) = \sum_{k\in\mathbf{Z}} g(a_k)\chi(g)^k t^k.$$

The ring $\mathscr{R}_E$ can not be naturally embedded in $E\otimes\mathbf{Q}_p^{\mathrm{cyc}}((t))$ but for any $r>0$ small enough and $n \gg 0$ there exists a Γ-equivariant embedding $i_n : \mathscr{R}_E^{(r)} \to E\otimes\mathbf{Q}_p^{\mathrm{cyc}}((t))$ which sends X to $\zeta_{p^n}e^{t/p^n} - 1$. Let $\mathbf{D}$ be a (φ, Γ)-module over $\mathscr{R}_E$. One can construct for such r a natural Γ-invariant $\mathscr{R}_E^{(r)}$-lattice $\mathbf{D}^{(r)}$. Then

$$\mathscr{D}_{\mathrm{dR}}(\mathbf{D}) = \left(E\otimes\mathbf{Q}_p^{\mathrm{cyc}}((t))\otimes_{i_n}\mathbf{D}^{(r)}\right)^{\Gamma}$$

is a finite dimensional E-vector space equipped with a decreasing filtration

$$\mathrm{Fil}^i\mathscr{D}_{\mathrm{dR}}(\mathbf{D}) = \left(E\otimes\mathrm{Fil}^i\mathbf{Q}_p^{\mathrm{cyc}}((t))\otimes_{i_n}\mathbf{D}^{(r)}\right)^{\Gamma},$$

which does not depend on the choice of r and n.

Let $\mathscr{R}_E[\ell_X]$ denote the ring of power series in variable ℓ_X with coefficients in $\mathscr{R}_E$. Extend the actions of φ and Γ to $\mathscr{R}_E[\ell_X]$ by setting

$$\varphi(\ell_X) = p\ell_X + \log\left(\frac{\varphi(X)}{X^p}\right), \qquad \tau(\ell_X) = \ell_X + \log\left(\frac{\tau(X)}{X}\right), \qquad \tau\in\Gamma.$$

(Note that $\log(\varphi(X)/X^p)$ and $\log(g(X)/X)$ converge in $\mathscr{R}_E$.) Define a monodromy operator $N : \mathscr{R}_E[\ell_X] \to \mathscr{R}_E[\ell_X]$ by $N = -\left(1-\frac{1}{p}\right)^{-1} \frac{d}{d\ell_X}$. For any (φ, Γ)-module $\mathbf{D}$ define

$$\mathcal{D}_{\mathrm{st}}(\mathbf{D}) = \left(\mathbf{D} \otimes_{\mathscr{R}_E} \mathscr{R}_E[\ell_X, 1/t]\right)^{\Gamma}, \quad t = \log(1+X), \tag{3.12}$$

$$\mathscr{D}_{\mathrm{cris}}(\mathbf{D}) = \mathcal{D}_{\mathrm{st}}(\mathbf{D})^{N=0} = (\mathbf{D}[1/t])^{\Gamma}. \tag{3.13}$$

Then $\mathscr{D}_{\mathrm{st}}(\mathbf{D})$ is a finite dimensional E-vector space equipped with natural actions of φ and N such that $N\varphi = p\varphi N$. Moreover, it is equipped with a canonical exhaustive decreasing filtration induced by the embeddings i_n. We have therefore three functors

$$\mathscr{D}_{\mathrm{dR}} : \mathbf{M}^{\varphi,\Gamma}_{\mathscr{R}_E} \to \mathbf{MF}_E, \quad \mathscr{D}_{\mathrm{st}} : \mathbf{M}^{\varphi,\Gamma}_{\mathscr{R}_E} \to \mathbf{MF}^{\varphi,N}_E, \quad \mathscr{D}_{\mathrm{cris}} : \mathbf{M}^{\varphi,\Gamma}_{\mathscr{R}_E} \to \mathbf{MF}^{\varphi}_E.$$

Theorem 3.1 (BERGER). *Let V be a p-adic representation of $G_{\mathbf{Q}_p}$. Then*

$$\mathbf{D}_*(V) \simeq \mathscr{D}_*(V), \qquad * \in \{\mathrm{dR}, \mathrm{st}, \mathrm{cris}\}.$$

Proof. See [9]. □

For any (φ, Γ)-module over $\mathscr{R}_E$ one has

$$\dim_E \mathscr{D}_{\mathrm{cris}}(\mathbf{D}) \leqslant \dim_E \mathcal{D}_{\mathrm{st}}(\mathbf{D}) \leqslant \dim_E \mathscr{D}_{\mathrm{dR}}(\mathbf{D}) \leqslant \mathrm{rk}_{\mathscr{R}_E}(\mathbf{D}).$$

One says that $\mathbf{D}$ is de Rham (resp. semistable, resp. crystalline) if $\dim_E \mathscr{D}_{\mathrm{dR}}(\mathbf{D}) = \mathrm{rk}_{\mathscr{R}_E}(\mathbf{D})$ (resp. $\dim_E \mathcal{D}_{\mathrm{st}}(\mathbf{D}) = \mathrm{rk}_{\mathscr{R}_E}(\mathbf{D})$, resp. $\dim_E \mathscr{D}_{\mathrm{cris}}(\mathbf{D}) = \mathrm{rk}_{\mathscr{R}_E}(\mathbf{D})$). Let $\mathbf{M}^{\varphi,\Gamma}_{\mathscr{R}_E,\mathrm{st}}$ and $\mathbf{M}^{\varphi,\Gamma}_{\mathscr{R}_E,\mathrm{cris}}$ denote the categories of semistable and crystalline (φ, Γ)-modules, respectively. Berger proved (see [11]) that the functors

$$\mathcal{D}_{\mathrm{st}} : \mathbf{M}^{\varphi,\Gamma}_{\mathscr{R}_E,\mathrm{st}} \to \mathbf{MF}^{\varphi,N}_E, \qquad \mathscr{D}_{\mathrm{cris}} : \mathbf{M}^{\varphi,\Gamma}_{\mathscr{R}_E,\mathrm{cris}} \to \mathbf{MF}^{\varphi}_E \tag{3.14}$$

are equivalences of $\otimes$-categories. If $\mathbf{D}$ is de Rham, the jumps of the filtration $\mathrm{Fil}^i \mathscr{D}_{\mathrm{dR}}(\mathbf{D})$ will be called the Hodge–Tate weights of $\mathbf{D}$.

3.1.4 Families of (φ, Γ)-modules

In this section, we review the theory of families of (φ, Γ)-modules and its relation to the families of p-adic representations following [12], [39], [52] and [40]. Let A be an affinoid algebra over $\mathbf{Q}_p$. For each $0 \leqslant r < 1$ the ring $\mathscr{R}^{(r)}_{\mathbf{Q}_p}$ is equipped with a canonical Fréchet topology (see [9]) and we define $\mathscr{R}^{(r)}_A = A \widehat{\otimes}_{\mathbf{Q}_p} \mathscr{R}^{(r)}_{\mathbf{Q}_p}$. Set $\mathscr{R}_A = \underset{0 \leqslant r<1}{\cup} \mathscr{R}^{(r)}_A$. The actions of φ and Γ on $\mathscr{R}_{\mathbf{Q}_p}$

extend by linearity to $\mathscr{R}_A$. Note that $\mathscr{R}_A^{(r)}$ is stable under the action of Γ, and that $\varphi(\mathscr{R}_A^{(r)}) \subset \mathscr{R}_A^{(r^{1/p})}$ (but $\varphi(\mathscr{R}_A^{(r)}) \not\subset \mathscr{R}_A^{(r)}$).

Definition. *i)* Let $0 \leqslant r < 1$. A (φ, Γ)-module over $\mathscr{R}_A^{(r)}$ is a finitely generated projective $\mathscr{R}_A^{(r)}$ -module $\mathbf{D}^{(r)}$ equipped with the following structures:

a) A φ-semilinear injective map

$$\varphi \,:\, \mathbf{D}^{(r)} \to \mathscr{R}_A^{(r^{1/p})} \otimes_{\mathscr{R}_A^{(r)}} \mathbf{D}^{(r)}$$

such that

$$\mathscr{R}_A^{(r^{1/p})} \otimes_{\varphi\left(\mathscr{R}_A^{(r)}\right)} \varphi\left(\mathbf{D}^{(r)}\right) = \mathscr{R}_A^{(r^{1/p})} \otimes_{\mathscr{R}_A^{(r)}} \mathbf{D}^{(r)}.$$

b) A semilinear continuous action of Γ on $\mathbf{D}^{(r)}$ which commutes with φ.

ii) $\mathbf{D}$ is a (φ, Γ)-module over $\mathscr{R}_A$ if $\mathbf{D} = \mathbf{D}^{(r)} \otimes_{\mathscr{R}_A^{(r)}} \mathscr{R}_A$ for some (φ, Γ)-module $\mathbf{D}^{(r)}$ over $\mathscr{R}_A^{(r)}$.

The following proposition shows that, if $A = E$, this definition is compatible with the definition given in Subsection 3.1.1.

Proposition 3.2. *Let $\mathbf{D}$ be a (φ, Γ)-module over $\mathscr{R}_E$. Then there is $0 \leqslant r(\mathbf{D}) < 1$ such that for any $r(\mathbf{D}) \leqslant r < 1$ there is a unique free $\mathscr{R}_E^{(r)}$-submodule $\mathbf{D}^{(r)}$ of $\mathbf{D}$ with the following properties:*

a) $\mathbf{D}^{(s)} = \mathbf{D}^{(r)} \otimes_{\mathscr{R}_E^{(r)}} \mathscr{R}_E^{(s)}$ *for* $r \leqslant s < 1$.

b) $\mathbf{D} = \mathbf{D}^{(r)} \otimes_{\mathscr{R}_E^{(r)}} \mathscr{R}_E$.

c) The Frobenius φ induces isomorphisms

$$\varphi^* \,:\, \mathbf{D}^{(r)} \otimes_{\mathscr{R}_E^{(r)}, \varphi} \mathscr{R}_E^{(r^{1/p})} \xrightarrow{\sim} \mathbf{D}^{(r^{1/p})}, \quad r(\mathbf{D}) \leqslant r < 1.$$

Proof. This is Theorem 1.3.3 of [11]. □

Let $\mathbf{Rep}_A(G_{\mathbf{Q}_p})$ be the category of projective A-modules of finite rank equipped with an A-linear continuous action of $G_{\mathbf{Q}_p}$. The construction of the functor $\mathbf{D}_{\mathrm{rig}}^{\dagger}$ can be directly generalized to this case, cf. [12], [39]. This means the existence of a fully faithful exact functor

$$\mathbf{D}_{\mathrm{rig},A}^{\dagger} \,:\, \mathbf{Rep}_A(G_{\mathbf{Q}_p}) \to \mathbf{M}_{\mathscr{R}_A}^{\varphi,\Gamma},$$

which commutes with base change. More precisely, let $\mathscr{X} = \mathrm{Spm}(A)$. For each $x \in \mathscr{X}$ we denote by $\mathfrak{m}_x$ the maximal ideal of A associated to x and set

$E_x = A/\mathfrak{m}_x$. If V (resp. $\mathbf{D}$) is an object of $\mathbf{Rep}_A(G_{\mathbf{Q}_p})$ (resp. of $\mathbf{M}^{\varphi,\Gamma}_{\mathscr{R}_A}$), we set $V_x = V \otimes_A E_x$ (resp. $\mathbf{D}_x = \mathbf{D} \otimes_A E_x$). Then the diagram

$$\begin{array}{ccc} \mathbf{Rep}_A(G_{\mathbf{Q}_p}) & \xrightarrow{\mathbf{D}^{\dagger}_{\mathrm{rig},A}} & \mathbf{M}^{\varphi,\Gamma}_{\mathscr{R}_A} \\ \downarrow{\scriptstyle \otimes E_x} & & \downarrow{\scriptstyle \otimes E_x} \\ \mathbf{Rep}_{E_x}(G_{\mathbf{Q}_p}) & \xrightarrow{\mathbf{D}^{\dagger}_{\mathrm{rig}}} & \mathbf{M}^{\varphi,\Gamma}_{\mathscr{R}_{E_x}} \end{array}$$

commutes, i.e. $\mathbf{D}^{\dagger}_{\mathrm{rig},A}(V)_x \simeq \mathbf{D}^{\dagger}_{\mathrm{rig}}(V_x)$. Note that in general the essential image of $\mathbf{D}^{\dagger}_{\mathrm{rig},A}$ does not coincide with the subcategory of étale modules. See [12] [40], [31] for further discussion.

Let $\mathbf{D}$ be a (φ, Γ)-module over $\mathscr{R}_A$. As in the case $A = E$, we attach to $\mathbf{D}$ the Fontaine–Herr complexes $C^{\bullet}_{\varphi,\gamma_n}(\mathbf{D})$ and consider the associated cohomology groups $H^*(K_n, \mathbf{D})$. We summarize the main properties of these cohomology in the theorem below. The key result here is the finiteness of the rank of $H^*(K_n, \mathbf{D})$.

Theorem 3.3. *Let A be an affinoid algebra over $\mathbf{Q}_p$ and let $\mathbf{D}$ be a (φ, Γ)-module over $\mathscr{R}_A$. Then*

i) *Finiteness. The cohomology groups $H^i(K_n, \mathbf{D})$ are finitely generated A-modules. More precisely, for each $n \geqslant 0$ the complex $C^{\bullet}_{\varphi,\gamma_n}(\mathbf{D})$ is quasi-isomorphic to the complex of projective A-modules of finite rank concentrated in degrees 0, 1 and 2.*

ii) *Base change. If $f : A \to B$ is a morphism of affinoid algebras, then*

$$C^{\bullet}_{\varphi,\gamma_n}(\mathbf{D}) \otimes^{\mathbf{L}}_{\mathscr{R}_A} \mathscr{R}_B \xrightarrow{\sim} C^{\bullet}_{\varphi,\gamma_n}(\mathbf{D} \widehat{\otimes}_{\mathscr{R}_A} \mathscr{R}_B).$$

In particular, if $x \in \mathscr{X}$, then

$$C^{\bullet}_{\varphi,\gamma_n}(\mathbf{D}) \otimes^{\mathbf{L}}_{\mathscr{R}_A} E_x \xrightarrow{\sim} C^{\bullet}_{\varphi,\gamma_n}(\mathbf{D}_x).$$

iii) *Euler-Poincaré formula. One has*

$$\chi(K_n, \mathbf{D}) = \sum_{i=0}^{2} (-1)^i \mathrm{rk}_A \left(H^i(K_n, \mathbf{D}) \right) = -[K_n : \mathbf{Q}_p] \, \mathrm{rk}_{\mathscr{R}_A}(\mathbf{D})$$

where the rank is considered as a locally constant function $\mathrm{rk}_A : \mathrm{Spm}(A) \to \mathbf{N}$.

iv) *Duality. Formulas* (3.10) *define a duality*

$$C^{\bullet}_{\varphi,\gamma_n}(\mathbf{D}) \xrightarrow{\sim} \mathbf{R}\mathrm{Hom}_A(C^{\bullet}_{\varphi,\gamma_n}(\mathbf{D}^*(\chi)), A)[-2].$$

v) Comparision with Galois cohomology. Let V be a p-adic representation with coefficients in A. Then there are functorial isomorphisms

$$H^*(K_n, V) \xrightarrow{\sim} H^*(K_n, \mathbf{D}^{\dagger}_{\mathrm{rig},A}(V))$$

Proof. See [40], Theorems 4.4.1, 4.4.2, 4.4.5 and [52], Theorem 2.8. □

3.1.5 Derived categories

There is a derived version of the comparision isomorphisms v) of Theorem 3.3, which is important for the formalism of Iwasawa theory [7], [52]. Let A be an affinoid algebra over $\mathbf{Q}_p$. For any (φ, Γ)-module $\mathbf{D}$ over $\mathscr{R}_A$ define

$$C^{\bullet}_{\gamma_n}(\mathbf{D}) = \left[\mathbf{D} \xrightarrow{\gamma_n - 1} \mathbf{D}\right],$$

where the first term is placed in degree 0. Then $C^{\bullet}_{\varphi,\gamma_n}(\mathbf{D})$ can be defined as the total complex

$$C^{\bullet}_{\varphi,\gamma_n}(\mathbf{D}) = \mathrm{Tot}^{\bullet}\left(C^{\bullet}_{\gamma_n}(\mathbf{D}) \xrightarrow{\varphi-1} C^{\bullet}_{\gamma_n}(\mathbf{D})\right).$$

If $\mathbf{D}'$ and $\mathbf{D}''$ are two (φ, Γ)-modules over $\mathscr{R}_A$, the cup product

$$\cup_{\gamma} \; : \; C^{\bullet}_{\gamma_n}(\mathbf{D}') \otimes C^{\bullet}_{\gamma_n}(\mathbf{D}'') \to C^{\bullet}_{\gamma_n}(\mathbf{D}' \otimes \mathbf{D}'')$$

defined by

$$\cup_{\gamma}(x_i \otimes y_j) = \begin{cases} x_i \otimes \gamma^n(y_j), & \text{if } x_i \in C^i_{\gamma_n}(\mathbf{D}'),\ y_j \in C^j_{\gamma_n}(\mathbf{D}'') \text{ and } i+j = 0 \text{ or } 1, \\ 0, & \text{if } i+j \geqslant 2 \end{cases}$$

gives rise to a map of complexes

$$\cup : C^{\bullet}_{\varphi,\gamma_n}(\mathbf{D}') \otimes C^{\bullet}_{\varphi,\gamma_n}(\mathbf{D}'') \to C^{\bullet}_{\varphi,\gamma_n}(\mathbf{D}' \otimes \mathbf{D}'').$$

Explicitly

$$\cup\left((x_{i-1}, x_i) \otimes (y_{j-1}, y_j)\right) = \left(x_i \cup_{\gamma} y_{j-1} + (-1)^j x_{i-1} \cup_{\gamma} \varphi(y_j),\ x_i \cup_{\gamma} y_j\right) \tag{3.15}$$

if $(x_{i-1}, x_i) \in C^i_{\varphi,\gamma_n}(\mathbf{D}') = C^{i-1}_{\gamma_n}(\mathbf{D}') \oplus C^i_{\gamma_n}(\mathbf{D}')$ and $(y_{j-1}, y_j) \in C^j_{\varphi,\gamma_n}(\mathbf{D}'') = C^{j-1}_{\gamma_n}(\mathbf{D}'') \oplus C^j_{\gamma_n}(\mathbf{D}'')$. It is easy to see that this is a bilinear map, which induces the cup-product (3.11) on cohomology groups.

Let $H_{\mathbf{Q}_p} = \mathrm{Gal}(\overline{\mathbf{Q}}_p/\mathbf{Q}_p^{\mathrm{cyc}})$. In [9], Berger constructed a topological ring $\widetilde{\mathbf{B}}^{\dagger,r}_{\mathrm{rig},A}$ equipped with commuting actions of φ and $G_{\mathbf{Q}_p}$ such that $\mathbf{D}^{\dagger,r}_{\mathrm{rig},A}(V) \subset$

$V \otimes_A \widetilde{\mathbf{B}}^{\dagger,r}_{\mathrm{rig},A}$ for each Galois representation V. Moreover, one has an exact sequence

$$0 \to A \to \widetilde{\mathbf{B}}^{\dagger,r}_{\mathrm{rig},A} \xrightarrow{\varphi-1} \widetilde{\mathbf{B}}^{\dagger,r^{1/p}}_{\mathrm{rig},A} \to 0 \tag{3.16}$$

(see [10], Lemma 1.7). Set $\widetilde{\mathbf{B}}^{\dagger}_{\mathrm{rig},A} = \varinjlim_{0<r} \widetilde{\mathbf{B}}^{\dagger,r}_{\mathrm{rig},A}$. Passing to limits in (3.16), we obtain the following exact sequence

$$0 \to A \to \widetilde{\mathbf{B}}^{\dagger}_{\mathrm{rig},A} \xrightarrow{\varphi-1} \widetilde{\mathbf{B}}^{\dagger}_{\mathrm{rig},A} \to 0. \tag{3.17}$$

Set $G_{K_n} = \mathrm{Gal}(\overline{\mathbf{Q}}_p/\mathbf{Q}_p(\mu_{p^n}))$. Tensoring (3.17) with a p-adic representation V, and taking continuous cochains, one obtains an exact sequence

$$\begin{aligned} 0 \to C^\bullet(G_{K_n}, V) \to C^\bullet\left(G_{K_n}, V \otimes \widetilde{\mathbf{B}}^{\dagger}_{\mathrm{rig},A}\right) \\ \xrightarrow{\varphi-1} C^\bullet\left(G_{K_n}, V \otimes \widetilde{\mathbf{B}}^{\dagger}_{\mathrm{rig},A}\right) \to 0. \end{aligned}$$

Define

$$K^\bullet(K_n, V) = \mathrm{Tot}^\bullet\left(C^\bullet\left(G_{K_n}, V \otimes \widetilde{\mathbf{B}}^{\dagger}_{\mathrm{rig},A}\right) \xrightarrow{\varphi-1} C^\bullet\left(G_{K_n}, V \otimes \widetilde{\mathbf{B}}^{\dagger}_{\mathrm{rig},A}\right)\right). \tag{3.18}$$

Consider the diagram

$$\begin{array}{ccc} C^\bullet(G_{K_n},V) & \xrightarrow{\beta_V} & K^\bullet(K_n,V) \\ & & \uparrow{\scriptstyle \alpha_V} \\ & & C^\bullet_{\varphi,\gamma_n}(V), \end{array} \tag{3.19}$$

where the maps α_V and β_V are constructed as follows. Let $\alpha_{V,\gamma} : C^\bullet_{\gamma_n}(V) \to C^\bullet(G_{\mathbf{Q}_p}, V \otimes \widetilde{\mathbf{B}}^{\dagger}_{\mathrm{rig},A})$ be the morphism defined by

$$\alpha_{V,\gamma}(x_0) = x_0, \qquad \text{if } x_0 \in C^0_{\gamma_n}(V),$$
$$\alpha_{V,\gamma}(x_1)(g) = \frac{g-1}{\gamma_n - 1}(x_1), \qquad \text{if } x_1 \in C^1_{\gamma_n}(V).$$

Then α_V is the map induced by $\alpha_{V,\gamma}$ by passing to total complexes. The map β_V is defined by

$$\beta_V : C^\bullet(G_{K_n}, V) \to K^\bullet(K_n, V),$$
$$x_n \mapsto (0, x_n), \qquad x_n \in C^n(G_{K_n}, V).$$

Let $\mathbf{R}\Gamma(K_n, V)$ and $\mathbf{R}\Gamma(K_n, \mathbf{D}^{\dagger}_{\mathrm{rig}}(V))$ denote the images of complexes $C^\bullet(G_{K_n}, V)$ and $C^\bullet_{\varphi,\gamma_n}(\mathbf{D}^{\dagger}_{\mathrm{rig}}(V))$ in the derived category $\mathrm{D}^{\mathrm{b}}(A)$ of A-modules.

Proposition 3.4. *i) In the diagram* (3.19)*, the maps* α_V *and* β_V *are quasi-isomorphisms and therefore*

$$\mathbf{R}\Gamma(K_n, V) \xrightarrow{\sim} \mathbf{R}\Gamma(K_n, \mathbf{D}^{\dagger}_{\mathrm{rig}}(V)) \tag{3.20}$$

in $\mathrm{D}^{\mathrm{b}}(A)$.

Proof. See [7], Proposition A.1. □

3.1.6 Iwasawa cohomology

Let $\mathcal{O}_E$ denote the ring of integers of the field of coefficients E. Recall that Γ_0 denote the Galois group of $\mathbf{Q}_p^{\mathrm{cyc}}/\mathbf{Q}_p(\mu_p)$. We equip the Iwasawa algebra $\Lambda = \mathcal{O}_E[[\Gamma_0]]$ with the involution $\iota : \Lambda \to \Lambda$ defined by $\iota(\tau) = \tau^{-1}$, $\tau \in \Gamma_0$. Let V be a p-adic representation of $G_{\mathbf{Q}_p}$ with coefficients in E. Fix a $\mathcal{O}_E$-lattice T of V stable under the action of $G_{\mathbf{Q}_p}$. The induced module $\mathrm{Ind}_{K_\infty/\mathbf{Q}_p}(T) = \Lambda \otimes_{\mathcal{O}_E} T$ is equipped with the diagonal action of $G_{\mathbf{Q}_p}$ and the natural structure of a Λ-module given by $\lambda * m = \iota(\lambda)m$ (see for example [46], Chapter 8). To fix these structures we will write $\mathrm{Ind}_{K_\infty/\mathbf{Q}_p}(T) = (\Lambda \otimes_{\mathbf{Z}_p} T)^{\iota}$. Let $\mathbf{R}\Gamma_{\mathrm{Iw}}(\mathbf{Q}_p, T)$ denote the class of $C^{\bullet}(G_{\mathbf{Q}_p}, \mathrm{Ind}_{K_\infty/\mathbf{Q}_p}(T))$ in the derived category $D^{\mathrm{b}}(\Lambda)$. Define

$$H^i_{\mathrm{Iw}}(\mathbf{Q}_p, T) = \mathbf{R}^i\Gamma_{\mathrm{Iw}}(\mathbf{Q}_p, T), \qquad i \in \mathbf{N}.$$

From Shapiro's lemma it follows that there are canonical and functorial isomorphisms in $\mathrm{D}^{\mathrm{b}}(\mathbf{Z}_p[G_n])$, where $G_n = \mathrm{Gal}(K_n/\mathbf{Q}_p)$,

$$H^i_{\mathrm{Iw}}(\mathbf{Q}_p, T) \simeq \varprojlim_{\mathrm{cores}_{K_n/K_{n-1}}} H^i(K_n, T),$$

$$\mathbf{R}\Gamma_{\mathrm{Iw}}(\mathbf{Q}_p, T) \otimes^{\mathbf{L}}_{\Lambda} \mathbf{Z}_p[G_n] \simeq \mathbf{R}\Gamma(K_n, T).$$

We review the computation of Iwasawa cohomology in terms of (φ, Γ)-modules. It was found by Fontaine (unpublished, but see [15]). Let

$$\psi : \mathscr{E}^{\dagger}_E \to \mathscr{E}^{\dagger}_E$$

denote the operator

$$\psi(f(X)) = \frac{1}{p}\mathrm{Tr}_{\mathscr{E}^{\dagger}_E/\varphi(\mathscr{E}^{\dagger}_E)}(f(X)).$$

More explicitly, the polynomials $1, (1+X), \dots, (1+X)^{p-1}$ form a basis of $\mathscr{E}_E^\dagger$ over $\varphi(\mathscr{E}_E^\dagger)$, and one has

$$\psi\left(\sum_{i=0}^{p-1}\varphi(f_i)(1+X)^i\right) = f_0.$$

In particular, $\psi \circ \varphi = \mathrm{id}$. Let $e_1, \dots, e_d$ be a base of $\mathbf{D}^\dagger(V)$ over $\mathscr{E}_E^\dagger$. Then $\varphi(e_1), \dots, \varphi(e_d)$ is again a base of $\mathbf{D}^\dagger(V)$, and we define

$$\psi : \mathbf{D}^\dagger(V) \to \mathbf{D}^\dagger(V)$$

by $\psi\left(\sum_{i=1}^{d} a_i\varphi(e_i)\right) = \sum_{i=1}^{d}\psi(a_i)e_i$. Note that this definition extends to (φ, Γ)-modules over the Robba ring.

Consider the complex

$$C^\bullet_{\mathrm{Iw}}(\mathbf{D}^\dagger(T)) : \mathbf{D}^\dagger(T)^\Delta \xrightarrow{\psi-1} \mathbf{D}^\dagger(T)^\Delta,$$

where the first term is placed in degree 0, and denote by $\mathbf{R}\Gamma_{\mathrm{Iw}}(\mathbf{D}^\dagger(T))$ the corresponding object in $\mathrm{D^b}(\Lambda)$.

Theorem 3.5. *The complexes* $\mathbf{R}\Gamma_{\mathrm{Iw}}(\mathbf{Q}_p, T)$, $\mathbf{R}\Gamma_{\mathrm{Iw}}(\mathbf{D}^\dagger(T))$ *and* $\mathbf{R}\Gamma(\mathbf{Q}_p, \mathbf{D}^\dagger(\mathrm{Ind}_{K_\infty/\mathbf{Q}_p} T))$ *are isomorphic in* $\mathrm{D^b}(\Lambda)$.

Proof. See [15] and [7] for a derived version. □

An analog of previous results for (φ, Γ)-modules over the Robba ring was obtained by Pottharst [53]. We start with some preliminary results about coadmissible $\mathscr{H}$-modules and the Grothedieck duality. Let $B(0,1) = \{z \in \mathbf{C}_p \mid |z|_p < 1\}$ denote the open unit disk. Define

$$\mathscr{H} = \{f(X) \in E[[X]] \mid f(X) \text{ converges on } B(0,1)\}.$$

In the context of Iwasawa theory, the ring $\mathscr{H}$ appeared in [50]. A natural description of $\mathscr{H}$ is as follows. Write

$$B(0,1) = \bigcup_n W_n,$$

where, for every n, $W_n = \{z \in \mathbb{C}_p \mid |z|_p \leqslant p^{-1/n}\}$ is the closed disk of radius $p^{-1/n} < 1$, which we consider as an affinoid space. Then

$$\Gamma(W_n, \mathscr{O}_{W_n}) = \left\{ f(X) = \sum_{k=0}^{\infty} a_k X^k \mid |a_k|_p p^{-k/n} \to 0 \text{ when } k \to +\infty \right\}$$

and $\mathscr{H} = \varprojlim_n \Gamma(W_n, \mathscr{O}_{W_n})$. We consider Λ and $\Lambda \otimes_{O_E} E$ as subalgebras of $\mathscr{H}$. Set $\mathscr{H}_n = \Gamma(W_n, \mathscr{O}_{W_n})$. It is easy to see that all $\mathscr{H}_n$ are euclidian rings. A coadmissible $\mathscr{H}$-module M is the inverse limit of a system $(M_n)_{n\geqslant 1}$ where each M_n is a finitely generated $\mathscr{H}_n$-module and the natural maps $M_n \otimes_{\mathscr{H}_n} \mathscr{H}_{n-1} \to M_{n-1}$ are isomorphisms [57]. The structure of admissible modules is given by the following proposition ([53], Proposition 1.1).

Proposition 3.6. *i) A coadmissible $\mathscr{H}$-module is torsion free if and only if it is a finitely generated $\mathscr{H}$-module.*
ii) Let M be a coadmissible torsion $\mathscr{H}$-module. Then

$$M \simeq \prod_{i\in I} \mathscr{H}/\mathfrak{p}_i^{n_i}, \tag{3.21}$$

where $(\mathfrak{p}_i)_{i\in I} \subset \bigcup_{n\geqslant 1} \mathrm{Spec}(\mathscr{H}_n)$ is a system of maximal ideals such that for each n there are only finitely many i with $\mathfrak{p}_i \in \mathrm{Spec}(\mathscr{H}_n)$.

Proof. The proof follows easily from Lazard's theory [41]. □

By Proposition 3.6, any coadmissible $\mathscr{H}$-torsion module M can be decomposed into direct product $M \simeq \prod_{i\in I} \mathscr{H}/\mathfrak{p}_i^{n_i}$ and from [41], Théorème 1, it follows that there exists a unique, up to multiplication by a unit of $\Lambda[1/p]$, element $f \in \mathscr{H}$ such that $\mathrm{div}(f) = \sum_{i\in I} n_i \mathfrak{p}_i$.

Definition. The characteristic ideal $\mathrm{char}_{\mathscr{H}} M$ is defined to be the principal ideal generated by f.

Let $D^{\mathrm{b}}_{\mathrm{coad}}(\mathscr{H})$ denote the category of bounded complexes of $\mathscr{H}$-modules with coadmissible cohomology. Let $\mathscr{K}$ denote the field of fractions of $\mathscr{H}$. Consider the complex

$$\omega = \mathrm{cone}\left[\mathscr{K} \to \mathscr{K}/\mathscr{H}\right][-1],$$

and for any object $C^\bullet$ of $D^{\mathrm{b}}_{\mathrm{coad}}(\mathscr{H})$ define

$$\mathscr{D}(C^\bullet) = \mathrm{Hom}_{\mathscr{H}}(C^\bullet, \omega).$$

Then $\mathscr{D} : D^{\mathrm{b}}_{\mathrm{coad}}(\mathscr{H}) \to D^{\mathrm{b}}_{\mathrm{coad}}(\mathscr{H})$ is an anti-involution, which can be seen as the "limit" of the Grothendieck dualizing functors $\mathbf{R}\mathrm{Hom}(-, \mathscr{H}_n)$. For any coadmissible module M, let $\mathscr{D}^k(M)$ denote the k-th cohomology group of $\mathscr{D}([M])$. Then

$$\mathscr{D}^0(M) = \mathrm{Hom}_{\mathscr{H}}(M/M_{\mathrm{tor}}, \mathscr{H}), \qquad \mathscr{D}^1\left(\prod_{i\in I} \mathscr{H}/\mathfrak{p}_i^{n_i}\right) = \prod_{i\in I} \mathfrak{p}_i^{-n_i}/\mathscr{H},$$

and $\mathscr{D}^k(M) = 0$ for all $k \geqslant 2$ (see [53], Section 3.1).

Let $\mathbf{D}$ be a (φ, Γ)-module over $\mathscr{R}_E$. Consider the complexes

$$C^\bullet_{\mathrm{Iw}}(\mathbf{D}) \,:\, \mathbf{D}^\Delta \xrightarrow{\psi-1} \mathbf{D}^\Delta$$

and $C^\bullet_{\varphi,\gamma}(\overline{\mathbf{D}})$, where $\overline{\mathbf{D}} = \mathbf{D}\otimes_{\mathbf{Q}_p}\mathscr{H}^\iota$, and denote by $\mathbf{R}\Gamma_{\mathrm{Iw}}(\mathbf{D})$ and $\mathbf{R}\Gamma(\mathbf{Q}_p, \overline{\mathbf{D}})$ the corresponding objects of the derived category $D^{\mathrm{b}}_{\mathrm{coad}}(\mathscr{H})$. For each $\mathscr{H}$-module M we set $M_{\Gamma_0} = M \otimes_{\mathscr{H}} E$.

Theorem 3.7. *Let $\mathbf{D}$ be a (φ, Γ)-module over $\mathscr{R}_E$. Then*

i) $H^i_{\mathrm{Iw}}(\mathbf{D})$ $(i = 1, 2)$ are coadmissible $\mathscr{H}$-modules. Moreover, $H^1_{\mathrm{Iw}}(\mathbf{D})_{\mathrm{tor}}$ and $H^2_{\mathrm{Iw}}(\mathbf{D})$ are finite dimensional E-vector spaces.

ii) The complexes $C^\bullet_{\mathrm{Iw}}(\mathbf{D})$ and $C^\bullet_{\varphi,\gamma}(\overline{\mathbf{D}})$ are quasi-isomorphic, and therefore

$$\mathbf{R}\Gamma_{\mathrm{Iw}}(\mathbf{D}) \simeq \mathbf{R}\Gamma(\mathbf{Q}_p, \overline{\mathbf{D}}).$$

iii) One has an isomorphism

$$C^\bullet_{\varphi,\gamma}(\overline{\mathbf{D}}) \otimes_{\mathscr{H}} E \xrightarrow{\sim} C^\bullet_{\varphi,\gamma}(\mathbf{D}),$$

which induces the Hochschild–Serre exact sequences

$$0 \to H^i_{\mathrm{Iw}}(\mathbf{D})_\Gamma \to H^i(\mathbf{D}) \to H^{i+1}_{\mathrm{Iw}}(\mathbf{D})^\Gamma \to 0.$$

iv) One has a canonical duality in $D^{\mathrm{b}}_{\mathrm{coad}}(\mathscr{H})$

$$\mathscr{D}\mathbf{R}\Gamma_{\mathrm{Iw}}(\mathbf{D}) \simeq \mathbf{R}\Gamma_{\mathrm{Iw}}(\mathbf{D}^*(\chi))^\iota[2].$$

v) Let V be a p-adic representation of $G_{\mathbf{Q}_p}$. There are canonical and functorial isomorphisms

$$\mathbf{R}\Gamma_{\mathrm{Iw}}(\mathbf{Q}_p, T) \otimes^{\mathbf{L}}_\Lambda \mathscr{H} \simeq \mathbf{R}\Gamma(\mathbf{Q}_p, T \otimes_{\mathbf{Z}_p} \mathscr{H}^\iota) \simeq \mathbf{R}\Gamma(\mathbf{Q}_p, \overline{\mathbf{D}^\dagger_{\mathrm{rig}}(V)}).$$

Proof. See [53], Theorem 2.6. □

3.1.7 Crystalline extensions

In this subsection we consider (φ, Γ)-modules over the Robba ring $\mathscr{R}_E$. As usual, the first cohomology group $H^1(\mathbf{Q}_p, \mathbf{D})$ can be interpreted in terms of extensions. Namely, to any cocycle $\alpha = (a, b) \in Z^1(C_{\varphi,\gamma}(\mathbf{D}))$ we associate the extension

$$0 \to \mathbf{D} \to \mathbf{D}_\alpha \to \mathscr{R}_E \to 0$$

such that $\mathbf{D}_\alpha = \mathbf{D} \oplus \mathscr{R}_E e$ with $\varphi(e) = e + a$ and $\gamma(e) = e + b$. This defines a canonical isomorphism

$$H^1(\mathbf{Q}_p, \mathbf{D}) \simeq \mathrm{Ext}^1(\mathscr{R}_E, \mathbf{D}).$$

We say that $\mathrm{cl}(\alpha) \in H^1(\mathbf{Q}_p, \mathbf{D})$ is crystalline if

$$\dim_E \mathscr{D}_{\mathrm{cris}}(\mathbf{D}_\alpha) = \dim_E \mathscr{D}_{\mathrm{cris}}(\mathbf{D}) + 1,$$

and define

$$H^1_f(\mathbf{Q}_p, \mathbf{D}) = \{\mathrm{cl}(\alpha) \in H^1(\mathbf{Q}_p, \mathbf{D}) \mid \mathrm{cl}(\alpha) \text{ is crystalline}\}.$$

It is easy to see that $H^1_f(\mathbf{Q}_p, \mathbf{D})$ is a subspace of $H^1(\mathbf{Q}_p, \mathbf{D})$. If $\mathbf{D}$ is semistable (even potentially semistable), the equivalence (3.14) between the category of semistable (φ, Γ)-modules and filtered (φ, N)-modules allows us to compute $H^1_f(\mathbf{Q}_p, \mathbf{D})$ in terms of $\mathcal{D}_{\mathrm{st}}(\mathbf{D})$. This gives a canonical exact sequence

$$0 \to H^0(\mathbf{Q}_p, \mathbf{D}) \to \mathscr{D}_{\mathrm{cris}}(\mathbf{D}) \to \mathscr{D}_{\mathrm{cris}}(\mathbf{D}) \oplus t_{\mathbf{D}}(\mathbf{Q}_p) \to H^1_f(\mathbf{Q}_p, \mathbf{D}) \to 0, \tag{3.22}$$

where $t_{\mathbf{D}}(\mathbf{Q}_p) = \mathcal{D}_{\mathrm{st}}(\mathbf{D})/\mathrm{Fil}^0\mathcal{D}_{\mathrm{st}}(\mathbf{D})$ is the tangent space of $\mathbf{D}$ ([5], Proposition 1.4.4 and [45], Sections 1.19–1.21). In particular, one has

$$\begin{aligned} &H^0(\mathbf{Q}_p, \mathbf{D}) = \mathrm{Fil}^0\mathcal{D}_{\mathrm{st}}(\mathbf{D})^{\varphi=1, N=0}, \\ &\dim_E H^1_f(\mathbf{Q}_p, \mathbf{D}) = \dim_E t_{\mathbf{D}}(\mathbf{Q}_p) + \dim_L H^0(\mathbf{Q}_p, \mathbf{D}) \end{aligned} \tag{3.23}$$

(see [5], Proposition 1.4.4 and Corollary 1.4.5). Moreover, $H^1_f(\mathbf{Q}_p, \mathbf{D})$ and $H^1_f(\mathbf{Q}_p, \mathbf{D}^*(\chi))$ are orthogonal complements to each other under duality (3.11) ([5], Corollary 1.4.10).

We will call exponential map the connecting map in (3.22)

$$\exp_{\mathbf{D}} \;:\; t_{\mathbf{D}}(\mathbf{Q}_p) \to H^1_f(\mathbf{Q}_p, \mathbf{D}). \tag{3.24}$$

Let V be a potentially semistable representation of $G_{\mathbf{Q}_p}$. Let

$$t_V(\mathbf{Q}_p) = \mathbf{D}_{\mathrm{st}}(V)/\mathrm{Fil}^0\mathbf{D}_{\mathrm{st}}(V)$$

denote the tangent space of V, and let $H^1_f(\mathbf{Q}_p, V)$ be the subgroup of $H^1(\mathbf{Q}_p, V)$ defined by (3.2). In [13], Bloch and Kato constructed a map

$$\exp_V \;:\; t_V(\mathbf{Q}_p) \to H^1_f(\mathbf{Q}_p, V) \tag{3.25}$$

using the fundamental exact sequence relating the rings of p-adic periods $\mathbf{B}_{\mathrm{cris}}$ and $\mathbf{B}_{\mathrm{dR}}$. It follows from Theorem 3.1 that $t_V(\mathbf{Q}_p)$ is canonically isomorphic to $t_{\mathbf{D}^\dagger_{\mathrm{rig}}(V)}(\mathbf{Q}_p)$ and that

$$H^1_f(\mathbf{Q}_p, \mathbf{D}^\dagger_{\mathrm{rig}}(V)) \simeq H^1_f(\mathbf{Q}_p, V)$$

(see [5], Proposition 1.4.2). Moreover, the diagram

$$\begin{array}{ccc} t_V(\mathbf{Q}_p) & \xrightarrow{\exp_V} & H^1_f(\mathbf{Q}_p,V) \\ \downarrow{=} & & \downarrow{=} \\ t_{\mathbf{D}^\dagger_{\mathrm{rig}}(V)}(\mathbf{Q}_p) & \xrightarrow{\exp_{\mathbf{D}^\dagger_{\mathrm{rig}}(V)}} & H^1_f(\mathbf{Q}_p,\mathbf{D}^\dagger_{\mathrm{rig}}(V)). \end{array}$$

commutes ([7], Section 2). Therefore our definition of the exponential map (3.24) agrees with the original one of Bloch and Kato in [13].

3.1.8 Cohomology of isoclinic modules

The results of this section are proved in [5] (see Proposition 1.5.9 and Section 1.5.10 of *op. cit.*). Let $\mathbf{D}$ be a semistable (φ,Γ)-module over $\mathscr{R}_E$ of rank d. Assume that $\mathcal{D}_{\mathrm{st}}(\mathbf{D})^{\varphi=1}=\mathcal{D}_{\mathrm{st}}(\mathbf{D})$ and that all Hodge–Tate weights of $\mathbf{D}$ are $\geqslant 0$. Since $N\varphi=p\varphi N$, this implies that $N=0$ on $\mathcal{D}_{\mathrm{st}}(\mathbf{D})$, and $\mathbf{D}$ is crystalline.

The canonical map $\mathbf{D}^\Gamma\to\mathscr{D}_{\mathrm{cris}}(\mathbf{D})$ is an isomorphism, and therefore

$$H^0(\mathbf{Q}_p,\mathbf{D})\simeq\mathscr{D}_{\mathrm{cris}}(\mathbf{D})=\mathbf{D}^\Gamma$$

is a E-vector space of dimension d. The Euler–Poincaré characteristic formula gives

$$\dim_E H^1(\mathbf{Q}_p,\mathbf{D})=d+\dim_E H^0(\mathbf{Q}_p,\mathbf{D})+\dim_E H^0(\mathbf{Q}_p,\mathbf{D}^*(\chi))=2d.$$

On the other hand, $\dim_E H^1_f(\mathbf{Q}_p,\mathbf{D})=d$ by (3.23). The group $H^1(\mathbf{Q}_p,\mathbf{D})$ has the following explicit description. The map

$$\begin{aligned} &i_{\mathbf{D}}\,:\,\mathscr{D}_{\mathrm{cris}}(\mathbf{D})\oplus\mathscr{D}_{\mathrm{cris}}(\mathbf{D})\to H^1(\mathbf{Q}_p,\mathbf{D}),\\ &i_{\mathbf{D}}(x,y)=\mathrm{cl}(-x,\,\log\chi(\gamma)\,y) \end{aligned}$$

is an isomorphism. (Remark that the sign -1 and $\log\chi(\gamma)$ are normalizing factors.) Let $i_{\mathbf{D},f}$ and $i_{\mathbf{D},c}$ denote the restrictions of $i_{\mathbf{D}}$ on the first and second summand, respectively. Then $\mathrm{Im}(i_{\mathbf{D},f})=H^1_f(\mathbf{Q}_p,\mathbf{D})$, and we set $H^1_c(\mathbf{Q}_p,\mathbf{D})=\mathrm{Im}(i_{\mathbf{D},c})$. Thus we have a canonical decomposition

$$H^1(\mathbf{Q}_p,\mathbf{D})\simeq H^1_f(\mathbf{Q}_p,\mathbf{D})\oplus H^1_c(\mathbf{Q}_p,\mathbf{D}). \tag{3.26}$$

Now consider the dual module $\mathbf{D}^*(\chi)$. It is crystalline, $\mathscr{D}_{\mathrm{cris}}(\mathbf{D}^*(\chi))^{\varphi=p^{-1}}=\mathscr{D}_{\mathrm{cris}}(\mathbf{D}^*(\chi))$, and the Hodge–Tate weights of $\mathbf{D}^*(\chi)$ are all $\leqslant 0$. Let

$$[\ ,\ ]_{\mathbf{D}}\,:\,\mathscr{D}_{\mathrm{cris}}(\mathbf{D}^*(\chi))\times\mathscr{D}_{\mathrm{cris}}(\mathbf{D})\to E$$

denote the canonical pairing. Define

$$i_{\mathbf{D}^*(\chi)} \,:\, \mathscr{D}_{\mathrm{cris}}(\mathbf{D}^*(\chi)) \oplus \mathscr{D}_{\mathrm{cris}}(\mathbf{D}^*(\chi)) \to H^1(\mathbf{Q}_p, \mathbf{D}^*(\chi))$$

by

$$i_{\mathbf{D}^*(\chi)}(\alpha, \beta) \cup i_{\mathbf{D}}(x, y) = [\beta, x]_{\mathbf{D}} - [\alpha, y]_{\mathbf{D}}.$$

As before, let $i_{\mathbf{D}^*(\chi),f}$ and $i_{\mathbf{D}^*(\chi),c}$ denote the restrictions of $i_{\mathbf{D}^*(\chi)}$ on the first and second summand respectively. From $H^1_f(\mathbf{Q}_p, \mathbf{D}^*(\chi)) = H^1_f(\mathbf{D})^{\perp}$ it follows that $\mathrm{Im}(i_{\mathbf{D}^*(\chi),f}) = H^1_f(\mathbf{Q}_p, \mathbf{D}^*(\chi))$, and we set $H^1_c(\mathbf{Q}_p, \mathbf{D}^*(\chi)) = \mathrm{Im}(i_{\mathbf{D}^*(\chi),c})$. Again we have a decomposition

$$H^1(\mathbf{Q}_p, \mathbf{D}^*(\chi)) \simeq H^1_f(\mathbf{Q}_p, \mathbf{D}^*(\chi)) \oplus H^1_c(\mathbf{Q}_p, \mathbf{D}^*(\chi)).$$

Now we will compute the Iwasawa cohomology of some isoclinic modules. Recall that the well-known computation of universal norms of $\mathbf{Q}_p(1)$ gives that

$$H^1_{\mathrm{Iw}}(\mathbf{Q}_p, \mathbf{Q}_p(1))_{\Gamma} \simeq p^{\mathbf{Z}} \tag{3.27}$$

under the Kummer map. The following proposition generalizes this result to (φ, Γ)-modules. It will be used in the proof of Proposition 3.13 below.

Proposition 3.8. *Let* $\mathbf{D}$ *be an isoclinic* (φ, Γ)*-module such that* $\mathscr{D}_{\mathrm{cris}}(\mathbf{D})^{\varphi=p^{-1}} = \mathscr{D}_{\mathrm{cris}}(\mathbf{D})$ *and* $\mathrm{Fil}^0 \mathscr{D}_{\mathrm{cris}}(\mathbf{D}) = 0$. *Then*

$$H^1_{\mathrm{Iw}}(\mathbf{D})_{\Gamma_0} = H^1_c(\mathbf{Q}_p, \mathbf{D}).$$

Proof. For any continuous character $\delta \,:\, \mathbf{Q}_p^* \to E^*$, let $\mathscr{R}_E(\delta)$ denote the (φ, Γ)-module of rank one $\mathscr{R}_E e_\delta$ such that $\varphi(e_\delta) = \delta(p) e_\delta$ and $\gamma(e_\delta) = \delta(\chi(\gamma)) e_\delta$. In [19], Proposition 3.1, Colmez proved that any (φ, Γ)-module of rank 1 is isomorphic to a unique module of the form $\mathscr{R}_E(\delta)$ for some δ.

Set $\mathbf{D}_m = \mathscr{R}_E(|x| x^m)$. Let e_m denote the canonical generator of $\mathbf{D}_m$. It is not difficult to see that an isoclinic (φ, Γ)-module $\mathbf{D}$ which satisfies the conditions of Proposition 3.8 is isomorphic to the direct sum $\bigoplus_{i=1}^{d} \mathbf{D}_{m_i}$ for some $m_i \geqslant 1$ ([5], Proposition 1.5.8). Therefore we can assume that $\mathbf{D} = \mathbf{D}_m$ for some $m \geqslant 1$. The E-vector space $H^1(\mathbf{D}_m)$ is generated by the cohomology classes $\mathrm{cl}(\alpha_m)$ and $\mathrm{cl}(\beta_m)$ defined by

$$\alpha_m = \partial^{m-1}\left(\frac{1}{X} + \frac{1}{2}, a\right) e_m, \quad (1-\varphi)\, a = (1 - \chi(\gamma)\gamma)\left(\frac{1}{X} + \frac{1}{2}\right),$$

$$\beta_m = \partial^{m-1}\left(b, \frac{1}{X}\right) e_m, \quad (1-\varphi)\left(\frac{1}{X}\right) = (1 - \chi(\gamma)\,\gamma)\, b,$$

where $\partial = (1+X)\,d/dX$ ([19], Sections 2.3–2.5). Moreover, $\mathrm{cl}(\alpha_m) \in H^1_f(\mathbf{D}_m)$ and $\mathrm{cl}(\alpha_m) \in H^1_c(\mathbf{D}_m)$ ([5], Theorem 1.5.7). For $m = 1$, the module $\mathbf{D}_1 = \mathscr{R}_E(\chi)$ is étale and corresponds to the p-adic representation $E(1)$. Using the formula

$$\frac{1}{X} = \sum_{i=0}^{p-1} (1+X)^i \varphi\left(\frac{1}{X}\right),$$

it can be checked directly that $\psi(1/X) = 1/X$. Thus $\psi\left(\frac{1}{X}e_1\right) = \frac{1}{X}e_1$, and we proved that $\frac{1}{X}e_1 \in \mathbf{D}_1^{\psi=1}$. Then easy induction (use the identity $\psi\partial = p\partial\psi$) shows that for each $m \geqslant 1$ one has $\partial^{m-1}\left(\frac{1}{X}\right)e_m \in \mathbf{D}_m^{\psi=1}$. Since the image of $\partial^{m-1}\left(\frac{1}{X}\right)e_m$ under the map $H^1_{\mathrm{Iw}}(\mathbf{D}_m) \to H^1_{\mathrm{Iw}}(\mathbf{D}_m)_{\Gamma_0} \subset H^1(\mathbf{D}_m)$ is β_m, this proves that $H^1_{\mathrm{Iw}}(\mathbf{D}_m)_{\Gamma_0} \subset H^1_c(\mathbf{D}_m)$. On the other hand, Theorem 3.7 iii) gives the Hochschild–Serre exact sequence

$$0 \to H^1_{\mathrm{Iw}}(\mathbf{D}_m)_{\Gamma_0} \to H^1(\mathbf{D}_m) \to H^2_{\mathrm{Iw}}(\mathbf{D}_m)^{\Gamma_0} \to 0.$$

By the same theorem, $H^2_{\mathrm{Iw}}(\mathbf{D}_m)$ is finite dimensional over E, and therefore

$$\dim_E H^2_{\mathrm{Iw}}(\mathbf{D}_m)^{\Gamma_0} = \dim_E H^2_{\mathrm{Iw}}(\mathbf{D}_m)_{\Gamma_0} = \dim_E H^2(\mathbf{D}_m) = 1.$$

Thus

$$\dim_E H^1_{\mathrm{Iw}}(\mathbf{D}_m)_{\Gamma_0} = \dim_E H^1(\mathbf{D}_m) - \dim_E H^2(\mathbf{D}_m) = 1,$$

and we proved that $\dim_E H^1_{\mathrm{Iw}}(\mathbf{D}_m)_{\Gamma_0} = \dim_E H^1_c(\mathbf{D}_m)$. This gives the proposition. □

Remark. It can be shown that the image of $p \in \mathbf{Q}_p^*$ under the Kummer map is $(1 - 1/p)\log(\chi)\mathrm{cl}(\beta_1)$ (see [4], Proposition 2.1.5).

3.2 The Main Conjecture

3.2.1 Regular submodules

In this section, we apply the theory of (φ, Γ)-modules to Iwasawa theory of p-adic representations. In particular, we propose a conjecture, which can be seen as a (weak) generalization of both Greenberg's [28] and Perrin-Riou's [51] main conjectures. Fix a prime number p and a finite set S of primes of $\mathbf{Q}$ containing p. For each number field F, we denote by $G_{F,S}$ the Galois group of the maximal algebraic extension of F unramified outside $S \cup \{\infty\}$. For any topological module M equipped with a continuous action of $G_{F,S}$, we write

$H^*_S(F,M)$ for the continuous cohomology of $G_{F,S}$ with coefficients in M. Let V be a p-adic representation of $G_{\mathbf{Q},S}$ with coefficients in a finite extension E of $\mathbf{Q}_p$. Recall that in Subsection 3.0.1 we defined the Bloch–Kato Selmer group $H^1_f(\mathbf{Q},V)$ and, for each $v\in S$, the subgroup $H^1_f(\mathbf{Q}_v,V)$ of $H^1(\mathbf{Q}_v,V)$.

The Poitou–Tate exact sequence gives the following exact sequence

$$0\to H^1_f(\mathbf{Q},V)\to H^1_S(\mathbf{Q},V)\to \bigoplus_{v\in S}\frac{H^1(\mathbf{Q}_v,V)}{H^1_f(\mathbf{Q}_v,V)}\to H^1_f(\mathbf{Q},V^*(1))^*\to$$
$$H^2_S(\mathbf{Q},V)\to \bigoplus_{v\in S}H^2(\mathbf{Q}_v,V)\to H^0_S(\mathbf{Q},V^*(1))^*\to 0 \quad (3.28)$$

(see [27], Proposition 2.2.1). Together with the well-known formula for the Euler characteristic, this implies that

$$\dim_E H^1_f(\mathbf{Q},V)-\dim_E H^1_f(\mathbf{Q},V^*(1))-\dim_E H^0_S(\mathbf{Q},V)+$$
$$+\dim_E H^0_S(\mathbf{Q},V^*(1))=\dim_E t_V(\mathbf{Q}_p)-d_+(V), \quad (3.29)$$

where $d_\pm(V)=\dim_E V^{c=\pm 1}$ and c denotes the complex conjugation.

In the rest of this section, we assume that V satisfies the following conditions:

C1) $H^0_S(\mathbf{Q},V)=H^0_S(\mathbf{Q},V^*(1))=0$.
C2) V is semistable at p.
C3) $\mathbf{D}_{\mathrm{cris}}(V)^{\varphi=1}=0$.
C4) V satisfies one of the following conditions
a) $H^1_f(\mathbf{Q},V^*(1))=0$.
b) V is a self dual representation, i.e. it is equipped with a non-degenerate bilinear form $V\times V\to E(1)$.

Note that the conditions **C1–3)** coincide with the conditions from Subsection 3.0.2. In support of these assumptions we notice that representations which we have in mind appear as p-adic realizations of pure motives over $\mathbf{Q}$. Let $X/\mathbf{Q}$ be a smooth proper variety having good reduction outside a finite set S of places containing p. Consider the motives $h^i(X)(m)$, where $0\leqslant i\leqslant 2\dim(X)$ and $m\in\mathbf{Z}$. Let $H^i_p(X)$ denote the p-adic étale cohomology of X. The p-adic realization of $h^i(X)(m)$ is $V=H^i_p(X)(m)$. The action of $\mathrm{Gal}(\overline{\mathbf{Q}}/\mathbf{Q})$ factors through $G_{\mathbf{Q},S}$. Moreover, the restriction of $H^i_p(X)$ on the decomposition group at p is semistable if X has semistable reduction at p (and potentially semistable in general) [21], [60]. The Poincaré duality and the hard Lefschetz theorem give a canonical isomorphism

$$H^i_p(X)^*\simeq H^i_p(X)(i), \quad (3.30)$$

and therefore

$$V^*(1) \simeq V(i+1-m).$$

The motive $h^i(X)(m)$ is pure of weight $w = i-2m$, and replacing, if necessary, V by $V^*(1)$ we can assume that $w \leqslant -1$. By the comparison isomorphism of the p-adic Hodge theory, [60] $\mathbf{D}_{\mathrm{st}}(H^i_p(X))$ is isomorphic to the log-crystalline cohomology $H^i_{\log-\mathrm{cris}}(X/\mathbf{Q}_p)$. The weight monodromy conjecture of Deligne–Jannsen [34] predicts that the absolute values of the eigenvalues of φ acting on $\mathbf{D}_{\mathrm{cris}}(H^i_p(X)) = \mathbf{D}_{\mathrm{st}}(H^i_p(X))^{N=0}$ are $\leqslant i/2$, and therefore $\mathbf{D}_{\mathrm{cris}}(V)^{\varphi=1}$ should be 0 if $w \leqslant -1$. If X has good reduction at p, we do not need the weight monodromy conjecture, and the nullity of $\mathbf{D}_{\mathrm{cris}}(V)^{\varphi=1}$ follows unconditionally from the result of Katz and Messing [37].

Assume that $w=-1$. Then i is odd, $m = \dfrac{i+1}{2}$ and the isomorphism (3.30) shows that V is self dual and therefore satisfies **C4b)**.

Assume that $w \leqslant -2$. One expects that

$$H^1_f(\mathbf{Q}, V^*(1)) = 0 \qquad \text{if } w \leqslant -2. \tag{3.31}$$

Namely, this follows from conjectural properties of the category $\mathcal{MM}$ of mixed motives over $\mathbf{Q}$. Consider the Yoneda groups

$$H^n(\mathbf{Q}, h^i(X)(m)) = \mathrm{Ext}^n_{\mathcal{MM}}(\mathbf{Q}(0), h^i(X)(m)), \quad n=0,1,$$

and denote by $H^1_f(\mathbf{Q}, h^i(X)(m))$ the subgroup of extensions having "good reduction". In [13], Bloch and Kato conjectured that the p-adic realization functor induces an isomorphism

$$R_p : H^1_f(\mathbf{Q}, h^i(X)(m)) \xrightarrow{\sim} H^1_f(\mathbf{Q}, H^i_p(X)(m)), \qquad \forall m \in \mathbf{Z} \tag{3.32}$$

(see also [34] and [24]). In particular, we should have an isomorphism

$$H^1_f(\mathbf{Q}, h^i(X)(i+1-m)) \xrightarrow{\sim} H^1_f(\mathbf{Q}, V^*(1)).$$

On the other hand, from the semisimplicity of the category of pure motives [35] it follows that $H^1(\mathbf{Q}, M) = 0$ for any pure motive M of weight $\geqslant 0$. In particular, $H^1(\mathbf{Q}, h^i(X)(i+1-m))$ should vanish if $-i-2+2m \geqslant 0$. Together with (3.32) this imples **C4b)**. To sum up, from the motivic point of view the conditions **C4a)** and **C4b)** correspond to the weight $\leqslant -2$ and the weight -1 cases respectively. We consider these two cases separately below.

The weight $\leqslant -2$ case. In this subsection, we assume that V satisfies the conditions **C1–3)** and **C4a)**. From **C3)** it follows that the exponential map $t_V(\mathbf{Q}_p) \to H^1_f(\mathbf{Q}_p, V)$ is an isomorphism, and we denote by $\log_V$ its inverse.

Compositing $\log_V$ with the localization map $H^1_f(\mathbf{Q}, V) \to H^1_f(\mathbf{Q}_p, V)$, we obtain a map

$$r_V : H^1_f(\mathbf{Q}, V) \to t_V(\mathbf{Q}_p).$$

If $V = H^i_p(X)(m)$ with $m \geqslant i/2 + 1$, this map is related to the syntomic regulator R_{syn} via the commutaive diagram

$$\begin{array}{ccc} H^1_f(\mathbf{Q}, h^i(X)(m)) & & \\ \downarrow R_p & \searrow R_{\mathrm{syn}} & \\ H^1_f(\mathbf{Q}, V) & \xrightarrow{r_V} & t_V(\mathbf{Q}_p) \end{array}$$

([47], see also [59]). Our next assumption reflects the hope that the syntomic regulator is an injective map.

C5a) The localization map

$$\mathrm{loc}_p : H^1_f(\mathbf{Q}, V) \to H^1_f(\mathbf{Q}_p, V)$$

is injective.

In support of this assumption consider the case where $V = H^i_p(X)(m)$ with $w = i - 2m \leqslant -2$. The weight monodromy conjecture of Deligne predicts that

$$H^0(\mathbf{Q}_v, H^i_p(X)(m)) = 0, \qquad \text{if } v \neq p \quad \text{and} \quad w = i - 2m \leqslant -2. \tag{3.33}$$

Since $\dim_E H^1_f(\mathbf{Q}_v, V) = \dim_E H^0(\mathbf{Q}_v, V)$, this implies that $H^1_f(\mathbf{Q}_v, V)$ should vanish for $v \neq p$. On the other hand, in [34], Jannsen conjectured that

$$H^2_S(\mathbf{Q}, H^i_p(X)(m)) = 0, \qquad \text{if } i + 1 < m \text{ or } i + 1 > 2m.$$

In particular, $H^2_S(\mathbf{Q}, V^*(1))$ should vanish. From the Poitou–Tate exact sequence we obtain an exact sequence

$$H^2_S(\mathbf{Q}, V^*(1)) \to H^1_f(\mathbf{Q}, V) \to \bigoplus_{v \in S} H^1_f(\mathbf{Q}_v, V).$$

Therefore the condition **C5a)** follows from Jannsen's conjecture together with the weight monodromy conjecture. Also note that if $w < -2$, then **C5a)** follows directly from (3.33) (see [34], Lemma 4 and Theorem 3).

We introduce the notion of regular submodule, which first appeared in Perrin-Riou's book [51] in the context of crystalline representations (see also [5]).

Definition (PERRIN-RIOU). Assume that V is a p-adic representation which satisfies the conditions **C1–3)**, **C4a)** and **C5a)**.

i) A (φ, N)-submodule D of $\mathbf{D}_{\mathrm{st}}(V)$ is regular if $D \cap \mathrm{Fil}^0\mathbf{D}_{\mathrm{st}}(V) = 0$ and the map

$$r_{V,D} : H^1_f(\mathbf{Q}, V) \to \mathbf{D}_{\mathrm{st}}(V)/(\mathrm{Fil}^0\mathbf{D}_{\mathrm{st}}(V) + D),$$

induced by r_V, is an isomorphism.

ii) A (φ, N)-submodule D of $\mathbf{D}_{\mathrm{st}}(V^*(1))$ is regular if

$$D + \mathrm{Fil}^0\mathbf{D}_{\mathrm{st}}(V^*(1)) = \mathbf{D}_{\mathrm{st}}(V^*(1))$$

and the map

$$D \cap \mathrm{Fil}^0\mathbf{D}_{\mathrm{st}}(V^*(1)) \to H^1_f(\mathbf{Q}, V)^*,$$

induced by the dual map $r_V^* : \mathrm{Fil}^0\mathbf{D}_{\mathrm{st}}(V^*(1)) \to H^1_f(V)^*$, is an isomorphism.

Remark. Assume that $H^1_f(\mathbf{Q}, V) = H^1_f(\mathbf{Q}, V^*(1)) = 0$. Then D is regular if the canonical projection $D \to t_V(\mathbf{Q}_p)$ is an isomorphism of vector spaces, and our definition agrees with the definition given in [5].

The weight -1 case. In this subsection we assume that V satisfies the conditions **C1–3)** and **C4b)**. Then formula (3.29) gives

$$\dim_E t_V(\mathbf{Q}_p) = d_+(V).$$

Definition (PERRIN-RIOU [51])**.** Assume that V is a p-adic representation which satisfies the conditions **C1–3)** and **C4b)**. A (φ, N)-submodule D of $\mathbf{D}_{\mathrm{st}}(V)$ is regular if the canonical map $D \to t_V(\mathbf{Q}_p)$ is an isomorphism.

In the both cases **C4a)** and **C4b)** from (3.29) it follows immediately that for a regular submodule D one has

$$\dim_E D = d_+(V). \tag{3.34}$$

Remark. Our definition of a regular module in the weight -1 case is slightly different from Perrin-Riou's definition. If we assume the non-degeneracy of p-adic height pairings $\langle\, ,\,\rangle_{V,D}$ (see Section 3.4), then our regular modules are regular also in the sense of [51], but the converse in not true.

3.2.2 Iwasawa cohomology

We keep previous notation and conventions. Let $\mathbf{Q}^{\mathrm{cyc}} = \underset{n\geqslant 1}{\cup} \mathbf{Q}(\mu_{p^n})$ and $\Gamma = \mathrm{Gal}(\mathbf{Q}^{\mathrm{cyc}}/\mathbf{Q})$. Then $\Gamma = \Delta \times \Gamma_0$ where $\Delta = \mathrm{Gal}(\mathbf{Q}(\mu_p)/\mathbf{Q})$ and $\Gamma_0 = \mathrm{Gal}(\mathbf{Q}^{\mathrm{cyc}}/\mathbf{Q}(\mu_p))$. Set $F_n = \mathbf{Q}(\mu_{p^{n+1}})^{\Delta}$ and $F_\infty = \underset{n\geqslant 1}{\cup} F_n$. The Galois group $\mathrm{Gal}(F_\infty/F)$ is canonically isomorphic to Γ_0. Let $\Lambda = \mathcal{O}_E[[\Gamma_0]]$ denote the Iwasawa algebra of Γ_0.

Let V be a p-adic representation of $G_{\mathbf{Q},S}$. Fix a $\mathcal{O}_E$-lattice T of V stable under the action of $G_{\mathbf{Q},S}$ and consider the Iwasawa cohomology groups $H^*_{\mathrm{Iw},S}(\mathbf{Q},T)$ and $H^*_{\mathrm{Iw}}(\mathbf{Q}_v,T)$ defined in Subsection 3.0.3. The main properties of these groups are summarized below (see [49]).

i) $H^i_{\mathrm{Iw},S}(\mathbf{Q},T)=0$ and $H^i_{\mathrm{Iw}}(\mathbf{Q}_v,T)=0$ if $i\neq 1,2$;

ii) If $v\neq p$, then $H^i_{\mathrm{Iw}}(\mathbf{Q}_v,T)$ are finitely generated Λ-torsion modules. In particular,
$$H^1_{\mathrm{Iw}}(\mathbf{Q}_v,T)\simeq H^1(\mathbf{Q}_v^{\mathrm{ur}}/\mathbf{Q}_v,(\Lambda\otimes_{\mathcal{O}_E}T^{I_v})^\iota). \tag{3.35}$$

iii) If $v=p$, then $H^2_{\mathrm{Iw}}(\mathbf{Q}_p,T)$ is a finitely generated Λ-torsion module. Moreover
$$\mathrm{rk}_\Lambda\left(H^1_{\mathrm{Iw}}(\mathbb{Q}_p,T)\right)=d,\qquad H^1_{\mathrm{Iw}}(\mathbf{Q}_p,T)_{\mathrm{tor}}\simeq H^0(\mathbf{Q}_p(\zeta_{p^\infty}),T).$$
Note that, by local duality, $H^2_{\mathrm{Iw}}(\mathbb{Q}_p,T)\simeq H^0(F_{\infty,p},V^*(1)/T^*(1))$, where $F_{\infty,p}$ is the completion of F_∞ at p.

The Λ-module structure of $H^i_{\mathrm{Iw},S}(\mathbf{Q},T)$ depends heavily on the following conjecture formulated by Greenberg [28].

Weak Leopoldt Conjecture. *Let V be a p-adic representation of $G_{\mathbf{Q},S}$ which is potentially semistable at p. Then*
$$H^2_S(F_\infty,V/T)=0.$$

We have the following result, proved in [51], Proposition 1.3.2.

Proposition 3.9. *Assume that the weak Leopoldt conjecture holds for V. Then $H^2_{\mathrm{Iw},S}(\mathbf{Q},T)$ is Λ-torsion, and*
$$\mathrm{rk}_\Lambda H^1_{\mathrm{Iw},S}(\mathbf{Q},T)=d_-(V).$$

Passing to projective limits in the Poitou–Tate exact sequence for V, we obtain an exact sequence
$$\begin{aligned}0\to H^2(F_\infty,V^*(1)/T^*(1))^\wedge &\to H^1_{\mathrm{Iw},S}(\mathbf{Q},T)\to\bigoplus_{v\in S}H^1_{\mathrm{Iw}}(\mathbf{Q}_v,T)\to\\ \to H^1_S(F_\infty,V^*(1)/T^*(1))^\wedge &\to H^2_{\mathrm{Iw},S}(\mathbf{Q},T)\to\bigoplus_{v\in S}H^2_{\mathrm{Iw}}(\mathbf{Q}_v,T)\to\\ &\to H^0_S(F_\infty,V^*(1)/T^*(1))^\wedge\to 0,\end{aligned}\tag{3.36}$$

where $(-)^\wedge$ denotes the Pontryagin dual. Therefore, if the weak Leopoldt conjecture holds for $V^*(1)$, we have an exact sequence

$$0 \to H^1_{\mathrm{Iw},S}(\mathbf{Q},T) \to \bigoplus_{v\in S} H^1_{\mathrm{Iw}}(\mathbf{Q}_v,T) \to H^1_S(F_\infty, V^*(1)/T^*(1))^\wedge \to$$
$$H^2_{\mathrm{Iw},S}(\mathbf{Q},T) \to \bigoplus_{v\in S} H^2_{\mathrm{Iw}}(\mathbf{Q}_v,T) \to H^0_S(F_\infty, V^*(1)/T^*(1))^\wedge \to 0.$$

3.2.3 The complexes $\mathbf{R}\Gamma_{\mathrm{Iw}}(V,D)$ and $\mathbf{R}\Gamma(V,D)$

In this section, we construct the Selmer complexes $\mathbf{R}\Gamma_{\mathrm{Iw}}(V,D)$ and $\mathbf{R}\Gamma(V,D)$, which play a central role in our approach to the Iwasawa theory. Nekovář's book [46] provides a detailed study of Selmer complexes associated to Greenberg's local conditions. For the purposes of this paper one should work with local conditions associated to general (φ,Γ)-submodules of $\mathbf{D}^\dagger_{\mathrm{rig}}(V)$. In this context, the general formalism of Selmer complexes was developed by Pottharst in [52], and we refer to *op. cit.* and [53] for further information and details.

Let V be a p-adic representation of the Galois group $G_{\mathbf{Q},S}$ with coefficients in E. We fix a $G_{\mathbf{Q},S}$-stable lattice T of V. Recall that we denote by $C^\bullet(G_{\mathbf{Q},S}, T\otimes\Lambda^\iota)$ and $C^\bullet(G_{\mathbf{Q}_v}, T\otimes\Lambda^\iota)$ the complexes of continuous cochaines of $G_{\mathbf{Q},S}$ and $G_{\mathbf{Q}_v}$ respectively with coefficients in $T\otimes\Lambda^\iota$ and by $\mathbf{R}\Gamma_{\mathrm{Iw},S}(\mathbf{Q},T)$ and $\mathbf{R}\Gamma_{\mathrm{Iw}}(\mathbf{Q}_v,T)$ these complexes viewed as objects of $D^{\mathrm{b}}(\Lambda)$. Recall that Shapiro's lemma gives canonical isomorphisms

$$\mathbf{R}^i\Gamma_{\mathrm{Iw},S}(\mathbf{Q},T) \simeq H^i_{\mathrm{Iw},S}(\mathbf{Q},T), \qquad \mathbf{R}^i\Gamma_{\mathrm{Iw}}(\mathbf{Q}_p,T) \simeq H^i_{\mathrm{Iw}}(\mathbf{Q}_p,T).$$

The derived version of the Poitou–Tate exact sequence for Iwasawa cohomology reads

$$\mathbf{R}\Gamma_{\mathrm{Iw},S}(\mathbf{Q},T) \to \bigoplus_{v\in S} \mathbf{R}\Gamma_{\mathrm{Iw}}(\mathbf{Q}_v,T) \to \mathbf{R}\Gamma(F_\infty, V^*(1)/T^*(1))^\wedge[-2]$$

(see [46], Proposition 8.4.22). Passing to cohomology in this exact triangle, we obtain (3.36).

Let $\overline{V} = V\otimes_{\mathbf{Q}_p}\mathscr{H}^\iota$. Then

$$\mathbf{R}\Gamma_S(\mathbf{Q},\overline{V}) \simeq \mathbf{R}\Gamma_{\mathrm{Iw},S}(\mathbf{Q},T)\otimes^{\mathbf{L}}_\Lambda \mathscr{H},$$
$$\mathbf{R}\Gamma(\mathbf{Q}_v,\overline{V}) \simeq \mathbf{R}\Gamma_{\mathrm{Iw}}(\mathbf{Q}_v,T)\otimes^{\mathbf{L}}_\Lambda \mathscr{H}.$$

Assume that V is potentially semistable. To each (φ,N)-submodule D of $\mathbf{D}_{\mathrm{st}}(V)$ we associate a complex $\mathbf{R}\Gamma_{\mathrm{Iw}}(V,D)$, which can be seen as a direct generalization of Selmer complexes determined by Greenberg's local conditions and studied in Chapters 7–8 of [46]. We define local conditions $U^{\bullet}_{\mathrm{Iw},v}(V,D)$ $(v\in S)$. For $v\neq p$ we set

$$U^{\bullet}_{\mathrm{Iw},v}(V,D)=\mathbf{R}\Gamma_{\mathrm{Iw},f}(\mathbf{Q}_v,T)\otimes_{\Lambda}\mathscr{H},$$

where

$$\mathbf{R}\Gamma_{\mathrm{Iw},f}(\mathbf{Q}_v,T)=\left[T^{I_v}\otimes\Lambda^{\iota}\xrightarrow{\mathrm{Fr}_v-1}T^{I_v}\otimes\Lambda^{\iota}\right].$$

Here I_v denotes the inertia group at v, Fr_v is the geometric Frobenius, and the first term is placed in degree 0. Let $H^i_{\mathrm{Iw},f}(\mathbf{Q}_v,T)$ denote the cohomology of this complex. From (3.35) it follows that $H^i_{\mathrm{Iw},f}(\mathbf{Q}_v,T)=0$ for $i\neq 1$, and $H^1_{\mathrm{Iw},f}(\mathbf{Q}_v,T)=H^1_{\mathrm{Iw}}(\mathbf{Q}_v,T)$ is Λ-torsion.

Now we define the local condition at p. Let $\mathbf{D}$ be the (φ,Γ)-submodule of $\mathbf{D}^{\dagger}_{\mathrm{rig}}(V)$ associated to D by Theorem 3.1. Set

$$U^{\bullet}_{\mathrm{Iw},p}(V,D)=\mathbf{R}\Gamma_{\mathrm{Iw}}(\mathbf{D})=\left[\mathbf{D}^{\Delta}\xrightarrow{\psi-1}\mathbf{D}^{\Delta}\right].$$

From Theorem 3.7 it follows that we have a canonical map $U^{\bullet}_{\mathrm{Iw},p}(V,D)\to\mathbf{R}\Gamma_{\mathrm{Iw}}(\mathbf{Q}_p,T)\otimes^{\mathbf{L}}_{\Lambda}\mathscr{H}$. This gives the following diagram in $\mathrm{D}^{\mathrm{b}}_{\mathrm{coad}}(\mathscr{H})$

$$\begin{array}{ccc}
\mathbf{R}\Gamma_{\mathrm{Iw},S}(\mathbf{Q},T)\otimes^{\mathbf{L}}_{\Lambda}\mathscr{H} & \longrightarrow & \bigoplus_{v\in S}\mathbf{R}\Gamma_{\mathrm{Iw}}(\mathbf{Q}_v,T)\otimes^{\mathbf{L}}_{\Lambda}\mathscr{H}\\
 & & \uparrow\\
 & & \bigoplus_{v\in S}U^{\bullet}_{\mathrm{Iw},v}(V,D),
\end{array}$$

and we denote by $\mathbf{R}\Gamma_{\mathrm{Iw}}(V,D)$ the associated Selmer complex

$$\mathbf{R}\Gamma_{\mathrm{Iw}}(V,D)=\mathrm{cone}\left(\left(\mathbf{R}\Gamma_{\mathrm{Iw},S}(\mathbf{Q},T)\otimes^{\mathbf{L}}_{\Lambda}\mathscr{H}\right)\bigoplus\bigoplus_{v\in S}U^{\bullet}_{\mathrm{Iw},v}(V,D)\to\right.$$
$$\left.\bigoplus_{v\in S}\mathbf{R}\Gamma_{\mathrm{Iw}}(\mathbf{Q}_v,T)\otimes^{\mathbf{L}}_{\Lambda}\mathscr{H}\right)[1]. \quad (3.37)$$

Now we define the complex $\mathbf{R}\Gamma(V,D)$. For $v\neq p$, let

$$\mathbf{R}\Gamma_f(\mathbf{Q}_v,V)=\left[V^{I_v}\xrightarrow{\mathrm{Fr}_v-1}V^{I_v}\right],$$

where the first term is placed in degree 0. It is clear that $\mathbf{R}^0\Gamma_f(\mathbf{Q}_v,V)=H^0(\mathbf{Q}_v,V)$ and $\mathbf{R}^1\Gamma_f(\mathbf{Q}_v,V)$ coincides with the group $H^1_f(\mathbf{Q}_v,V)$ defined by (3.2). Define

$$U_v^\bullet(V,D)=\begin{cases}\mathbf{R}\Gamma_f(\mathbf{Q}_v,V), & \text{if } v\neq p,\\ \mathbf{R}\Gamma(\mathbf{Q}_p,\mathbf{D}), & \text{if } v=p.\end{cases} \tag{3.38}$$

From Theorem 3.7 it follows that

$$U_v^\bullet(V,D)=U_{\mathrm{Iw},v}^\bullet(V,D)\otimes^{\mathbf{L}}_{\mathscr{H}}E, \qquad v\in S.$$

Consider the Selmer complex $\mathbf{R}\Gamma(V,D)$ associated to this data

$$\mathbf{R}\Gamma(V,D)=\mathrm{cone}\left(\mathbf{R}\Gamma_S(\mathbf{Q},T)\bigoplus\bigoplus_{v\in S}U_v^\bullet(V,D)\to\bigoplus_{v\in S}\mathbf{R}\Gamma(\mathbf{Q}_v,V)\right)[1]. \tag{3.39}$$

We denote by $H^*_{\mathrm{Iw}}(V,D)$ and $H^*(V,D)$ the cohomology of $\mathbf{R}\Gamma_{\mathrm{Iw}}(V,D)$ and $\mathbf{R}\Gamma(V,D)$ respectively. The main properties of our Selmer complexes are summarized below.

Proposition 3.10 (Pottharst). *i) The complex $\mathbf{R}\Gamma_{\mathrm{Iw}}(V,D)$ has cohomology concentrated in degrees $[1,3]$ consisting of coadmissible $\mathscr{H}$-modules.*

ii) One has

$$\mathrm{rk}_{\mathscr{H}}H^1_{\mathrm{Iw}}(V,D)=\mathrm{rk}_{\mathscr{H}}H^2_{\mathrm{Iw}}(V,D).$$

Moreover, $H^3_{\mathrm{Iw}}(V,D)=\left(T^*(1)^{H_{\mathbf{Q},S}}\right)^*\otimes_\Lambda\mathscr{H}$, *where* $H_{F,S}=\mathrm{Gal}(\mathbf{Q}^{(S)}/F_\infty)$.

iii) The complex $\mathbf{R}\Gamma(V,D)$ has cohomology concentrated in degrees $[0,3]$ consisting of finite dimensional E-vector spaces, and

$$\mathbf{R}\Gamma_{\mathrm{Iw}}(V,D)\otimes^{\mathbf{L}}_{\mathscr{H}}E\simeq\mathbf{R}\Gamma(V,D).$$

In particular, we have canonical exact sequences

$$0\to H^i_{\mathrm{Iw}}(V,D)_{\Gamma_0}\to H^i(V,D)\to H^{i+1}_{\mathrm{Iw}}(V,D)^{\Gamma_0}\to 0, \qquad i\in\mathbf{N}. \tag{3.40}$$

iv) There are canonical dualities

$$\mathrm{Hom}_E(\mathbf{R}\Gamma(V,D),E)\simeq\mathbf{R}\Gamma(V^*(1),D^\perp)[3],$$
$$\mathscr{D}\mathbf{R}\Gamma_{\mathrm{Iw}}(V,D)^\iota\simeq\mathbf{R}\Gamma_{\mathrm{Iw}}(V^*(1),D^\perp)[3].$$

In particular, we have canonical exact sequences

$$0\to\mathscr{D}^1H^{4-i}_{\mathrm{Iw}}(V,D)\to H^i_{\mathrm{Iw}}(V^*(1),D^\perp)^\iota\to\mathscr{D}^0H^{3-i}_{\mathrm{Iw}}(V,D)\to 0. \tag{3.41}$$

v) Assume that D satisfies the following conditions:

$$\mathrm{Fil}^0D=0 \quad \textit{and} \quad \mathrm{Fil}^0(\mathbf{D}_{\mathrm{st}}(V)/D)=\mathbf{D}_{\mathrm{st}}(V)/D;$$

$$(\mathbf{D}_{\mathrm{st}}(V)/D)^{\varphi=1,N=0}=0 \quad and \quad D/(ND+(p\varphi-1)D)=0.$$

Then $H^1(\mathbf{Q}_p,\mathbf{D})=H^1_f(\mathbf{Q}_p,V)$ *and* $H^1(V,D)=H^1_f(\mathbf{Q},V)$.

Proof. See [53], Theorem 4.1 and [54], Proposition 3.7. □

3.2.4 The Main Conjecture

In this subsection, we use the formalism of Selmer complexes to formulate a version of the Main Conjecture of Iwasawa theory. Let M be a pure motive over $\mathbf{Q}$ of weight $w \leqslant -1$. Since the category of pure motives is semisimple, we can assume that M is simple and $\neq \mathbf{Q}(m)$, $m\in\mathbf{Z}$. Let V denote the p-adic realization of M. Assume that V is semistable. One expects (see [51]) that for each regular (φ,N)-submodule D of $\mathbf{D}_{\mathrm{st}}(V)$ there exists a p-adic L-function of the form

$$L_p(M,D,s)=f_D(\chi(\gamma_0)^s-1),\qquad f_D\in\mathscr{H}$$

interpolating algebraic parts of special values of the complex L-function $L(M,s)$ (see Section 0.2). Assume that V satisfies the conditions **C1–4)** of Section 2.1. We propose the following conjecture.

Main Conjecture. *Let* $M/\mathbf{Q}$ *be a pure simple motive of weight* $\leqslant -1$. *Assume that* $M\neq \mathbf{Q}(m)$, $m\in\mathbf{Z}$. *Let* D *be a regular submodule of* $\mathbf{D}_{\mathrm{st}}(V)$. *Then*

i) $H^1_{\mathrm{Iw}}(V,D)=0$,
ii) $H^2_{\mathrm{Iw}}(V,D)$ *is* $\mathscr{H}$*-torsion and*

$$\mathrm{char}_{\mathscr{H}}(H^2_{\mathrm{Iw}}(V,D))=(f_D).$$

Remarks. 1) By Proposition 3.10, the nullity of $H^1_{\mathrm{Iw}}(V,D)$ implies that $H^2_{\mathrm{Iw}}(V,D)$ is $\mathscr{H}$-torsion.

2) The condition $M\neq\mathbf{Q}(m)$ implies that $(V^*(1))^{H_{\mathbf{Q},S}}=0$, and therefore $H^3_{\mathrm{Iw}}(V,D)$ should vanish by Proposition 3.10 ii). To sum up, we expect that under our assumptions the cohomology of $\mathbf{R}\Gamma_{\mathrm{Iw}}(V,D)$ is concentrated in degree 2.

3) Let $D^\perp$ denote the dual regular submodule. One can easily formulate the Main Conjecture for the dual pair $(M^*(1),D^\perp)$. Assume that $H^1_{\mathrm{Iw}}(V,D)=0$. From (3.41) it follows that

$$H^2_{\mathrm{Iw}}(V^*(1),D^\perp)^\iota\simeq H^2_{\mathrm{Iw}}(V,D).$$

This isomorphism reflects the conjectural functional equation relating $L_p(M^*(1),D^\perp,s)$ and $L_p(M,D,s)$.

4) Assume that V is ordinary at p, i.e. that the restriction of V on $G_{\mathbf{Q}_p}$ is equipped with an increasing exhaustive filtration F^iV such that for all $i \in \mathbf{Z}$ the inertia group I_p acts on $\mathrm{gr}_i(V) = F^iV/F^{i+1}V$ by χ^i. In [28], Greenberg works with the Selmer group defined as follows. For each place w of F_∞ including the unique place above p, fix a decomposition group H_w in $H_{F,S} = \mathrm{Gal}(\mathbf{Q}^{(S)}/F_\infty)$ and denote by I_w its inertia subgroup. Set $A = V/T$ and $F^1A = F^1V/F^1T$. Define

$$H^1_{\mathrm{Gr}}(F_{\infty,w}, A) = \begin{cases} H^1(H_w/I_w, A^{I_w}), & \text{if } w \neq p, \\ \ker\left(H^1(F_{\infty,p}, A) \to H^1(I_p, A/F^1A)\right) & \text{if } w = p. \end{cases} \tag{3.42}$$

The Greenberg Selmer group is defined as follows

$$S(F_\infty, V/T) = \ker\left(H^1_S(F_\infty, A) \to \bigoplus_{w \in S} \frac{H^1(F_{\infty,w}, A)}{H^1_{\mathrm{Gr}}(F_{\infty,w}, A)}\right).$$

It is well-known that each semistable representation is ordinary, and we set $D = \mathbf{D}_{\mathrm{st}}(F^1V)$. Then $\mathbf{R}\Gamma_{\mathrm{Iw}}(\mathbf{D}) = \mathbf{R}\Gamma_{\mathrm{Iw}}(\mathbf{Q}_p, F^1V) \otimes^{\mathbf{L}}_{\Lambda[1/p]} \mathscr{H}$, and directly from definition (3.37) it follows that the complex $\mathbf{R}\Gamma_{\mathrm{Iw}}(V, D)$ is isomorphic to $\widetilde{\mathbf{R}\Gamma}_{f,\mathrm{Iw}}(F_\infty/\mathbf{Q}, T) \otimes^{\mathbf{L}}_{\Lambda[1/p]} \mathscr{H}$, where $\widetilde{\mathbf{R}\Gamma}_{f,\mathrm{Iw}}(F_\infty/\mathbf{Q}, T)$ denotes Nekovář's Selmer complex associated to the local condition given by F^1V. In [46], Chapter 9, it is proved that under some technical conditions the characteristic ideal of the Pontryagin dual $S(F_\infty, V^*(1)/T^*(1))^\wedge$ of $S(F_\infty, V^*(1)/T^*(1))$ coincides with the characteristic ideal of the second cohomology group $\widetilde{H}^2_f(T)$ of $\widetilde{\mathbf{R}\Gamma}_{f,\mathrm{Iw}}(F_\infty/\mathbf{Q}, T)$. This allows us to compare our conjecture to the Main Conjecture from [28].

3.3 Local structure of p-adic representations

3.3.1 Filtration associated to a regular submodule

Let V be a semistable p-adic representation of $G_{\mathbf{Q}_p}$. In this subsection, we assume that V satisfies the following condition:

S) The endomorphism $\varphi : \mathbf{D}_{\mathrm{st}}(V) \to \mathbf{D}_{\mathrm{st}}(V)$ is semisimple at 1 and p^{-1}.

Note that if $X/\mathbf{Q}_p$ is a smooth proper variety having semistable reduction, then the semisimplicity of the Frobenius action on the log-crystalline cohomology $H^i_{\log-\mathrm{cris}}(X/\mathbf{Q}_p)$ is a well-known open question.

Let D be a (φ, N)-submodule of $\mathbf{D}_{\mathrm{st}}(V)$ such that $D \cap \mathrm{Fil}^0\mathbf{D}_{\mathrm{st}}(V) = \{0\}$. We associate to D an increasing filtration $(D_i)_{-2\leqslant i\leqslant 2}$ on $\mathbf{D}_{\mathrm{st}}(V)$ setting

$$D_i = \begin{cases} 0 & \text{if } i = -2, \\ (1 - p^{-1}\varphi^{-1})\, D + N(D^{\varphi=1}) & \text{if } i = -1, \\ D & \text{if } i = 0, \\ D + \mathbf{D}_{\mathrm{st}}(V)^{\varphi=1} \cap N^{-1}(D^{\varphi=p^{-1}}) & \text{if } i = 1, \\ \mathbf{D}_{\mathrm{st}}(V) & \text{if } i = 2. \end{cases}$$

It can be easily proved (see[5], Lemma 2.1.5) that $(D_i)_{i=-2}^2$ is the unique filtration on $\mathbf{D}_{\mathrm{st}}(V)$ such that

D1) $D_{-2} = 0$, $D_0 = D$ and $D_2 = \mathbf{D}_{\mathrm{st}}(V)$;
D2) $(\mathbf{D}_{\mathrm{st}}(V)/D_1)^{\varphi-1,N-0} = 0$ and $D_{-1} = (1 - p^{-1}\varphi^{-1})D_{-1} + N(D_{-1})$;
D3) $(D_0/D_{-1})^{\varphi=p^{-1}} = D_0/D_{-1}$ and $(D_1/D_0)^{\varphi=1} = D_1/D_0$.

In addition, for the dual regular submodule $D^{\perp}$, one has

$$D_i^{\perp} = \mathrm{Hom}_{\mathbf{Q}_p}(\mathbf{D}_{\mathrm{st}}(V)/D_{-i}, \mathbf{D}_{\mathrm{st}}(\mathbf{Q}_p(1))).$$

Let $(D_i)_{i=-2}^2$ be the filtration associated to D. By (3.14), it induces a filtration of $\mathbf{D}^{\dagger}_{\mathrm{rig}}(V)$, which we will denote by $(\mathbf{D}_i)_{i=2}^2$. Define

$$\mathbf{W} = \mathbf{D}_1/\mathbf{D}_{-1}, \quad \mathbf{W}_0 = \mathrm{gr}_0\mathbf{D}^{\dagger}_{\mathrm{rig}}(V), \quad \mathbf{W}_1 = \mathrm{gr}_0\mathbf{D}^{\dagger}_{\mathrm{rig}}(V).$$

Then we have an exact sequence

$$0 \to \mathbf{W}_0 \to \mathbf{W} \to \mathbf{W}_1 \to 0, \tag{3.43}$$

where, by **D1-3)**,

$$\mathcal{D}_{\mathrm{st}}(\mathbf{W}_0) = \mathcal{D}_{\mathrm{st}}(\mathbf{W}_0)^{\varphi=p^{-1}}, \quad \mathrm{Fil}^0\mathcal{D}_{\mathrm{st}}(\mathbf{W}_0) = 0, \tag{3.44}$$

$$\mathcal{D}_{\mathrm{st}}(\mathbf{W}_1) = \mathcal{D}_{\mathrm{st}}(\mathbf{W}_1)^{\varphi=1}. \tag{3.45}$$

Proposition 3.11. *Let D be a regular submodule of $\mathbf{D}_{\mathrm{st}}(V)$. Then*

i) There are canonical inclusions

$$H^1(\mathbf{D}_{-1}) \subset H^1_f(\mathbf{D}) \subset H^1(\mathbf{D}_1) \subset H^1(\mathbf{Q}_p, V);$$

ii) $H^1_f(\mathbf{D}_{-1}) = H^1(\mathbf{D}_{-1})$;
iii) The sequences

$$0 \to H^1(\mathbf{D}_{-1}) \to H^1(\mathbf{D}_1) \to H^1(\mathbf{W}) \to 0,$$
$$0 \to H^1(\mathbf{D}_{-1}) \to H^1_f(\mathbf{D}_1) \to H^1_f(\mathbf{W}) \to 0$$

are exact.

Proof. 1) By **D2)** one has

$$\mathrm{Hom}(D_{-1}, \mathbf{D}_{\mathrm{st}}(\mathbf{Q}_p(1)))^{\varphi=1,N=0} = 0,$$

and by the Poincaré duality

$$H^2(\mathbf{D}_{-1}) \simeq H^0(\mathbf{D}^*_{-1}(\chi))^* = 0. \tag{3.46}$$

Now (3.8) and (3.23) imply that $H^1_f(\mathbf{D}_{-1}) = H^1(\mathbf{D}_{-1})$, and ii) is proved.

2) The exact sequence

$$0 \to \mathbf{D}_1 \to \mathbf{D}^{\dagger}_{\mathrm{rig}}(V) \to \mathrm{gr}_2\mathbf{D}^{\dagger}_{\mathrm{rig}}(V) \to 0$$

gives an exact sequence

$$0 \to H^0\left(\mathrm{gr}_2\mathbf{D}^{\dagger}_{\mathrm{rig}}(V)\right) \to H^1(\mathbf{D}_1) \to H^1(\mathbf{Q}_p, V).$$

From **D2)** it follows that

$$H^0\left(\mathrm{gr}_2\mathbf{D}^{\dagger}_{\mathrm{rig}}(V)\right) = \mathrm{Fil}^0(\mathbf{D}_{\mathrm{st}}(V)/D_1)^{\varphi=1,N=0} = 0,$$

and the injectivity of $H^1(\mathbf{D}_1) \to H^1(\mathbf{Q}_p, V)$ is proved. Since $\mathrm{Fil}^0\mathscr{D}_{\mathrm{cris}}(\mathbf{D}) = 0$ and $\mathscr{D}_{\mathrm{cris}}(\mathbf{D})^{\varphi=1} = \mathbf{D}_{\mathrm{cris}}(V)^{\varphi=1} = 0$ by **C3)**, the exact sequence (3.22) induces a commutative diagram

$$\begin{array}{ccc} \mathscr{D}_{\mathrm{cris}}(\mathbf{D}) & \xrightarrow{\exp} & H^1_f(\mathbf{D}) \\ \downarrow & & \downarrow \\ t_V(\mathbf{Q}_p) & \xrightarrow{\exp} & H^1_f(\mathbf{Q}_p, V), \end{array}$$

where the exponential maps and the left vertical map are isomorphisms. Thus $H^1_f(\mathbf{D}) \to H^1_f(\mathbf{Q}_p, V)$ is an injection. The same argument, together with ii), proves that $H^1(\mathbf{D}_{-1}) \subset H^1_f(\mathbf{D})$. This implies i).

3) Since the sequence

$$0 \to H^1(\mathbf{D}_{-1}) \to H^1(\mathbf{D}_1) \to H^1(\mathbf{W}) \to H^2(\mathbf{D}_{-1})$$

is exact, iii) follows from ii) and (3.46). □

Assume now that the canonical projection $\mathcal{D}_{\mathrm{st}}(\mathbf{D}) \to t_V(\mathbf{Q}_p)$ is an isomorphism, i.e. that

$$\mathbf{D}_{\mathrm{st}}(V) = D \oplus \mathrm{Fil}^0\mathbf{D}_{\mathrm{st}}(V) \tag{3.47}$$

as E-vector spaces. In this case, the structure of $\mathbf{W}$ can be completely determined if we make the following additional assumption (see [29], [5]):

U) The (φ, Γ)-module $\mathbf{D}^{\dagger}_{\mathrm{rig}}(V)$ has no saturated *crystalline* subquotient U sitting in a non-split exact sequence of the form

$$0 \to \mathscr{R}_E(|x|x^k) \to U \to \mathscr{R}_E(x^m) \to 0. \tag{3.48}$$

Note that if $k \leqslant m$, then $H^1_f(\mathscr{R}_E(|x|x^{k-m})) = 0$, and there is no non-trivial crystalline extension (3.48). If $k > m$, it follows from (3.23) that $\dim_E H^1_f(\mathscr{R}_E(|x|x^{k-m})) = 1$, and therefore there exists a unique (up to isomorphism) crystalline (φ, Γ)-module U of the form (3.48).

Proposition 3.12. *Let D be a (φ, N)-submodule of $\mathbf{D}_{\mathrm{st}}(V)$ which satisfies* (3.47). *Assume that the condition* **U)** *holds. Then*

i) There exists a unique decomposition

$$\mathbf{W} \simeq \boldsymbol{A}_0 \oplus \boldsymbol{A}_1 \oplus \mathbf{M}, \tag{3.49}$$

where $\boldsymbol{A}_0$ and $\boldsymbol{A}_1$ are direct summands of $\mathbf{W}_0$ and $\mathbf{W}_1$ of ranks $\dim_E H^0(\mathbf{W}^*(\chi))$ and $\dim_E H^0(\mathbf{W})$ respectively. Moreover, $\mathbf{M}$ is inserted in an exact sequence

$$0 \to \mathbf{M}_0 \xrightarrow{f} \mathbf{M} \xrightarrow{g} \mathbf{M}_1 \to 0,$$

where $\mathbf{W}_0 \simeq \boldsymbol{A}_0 \oplus \mathbf{M}_0$, $\mathbf{W}_1 \simeq \boldsymbol{A}_1 \oplus \mathbf{M}_1$, and $\mathrm{rk}(\mathbf{M}_0) = \mathrm{rk}(\mathbf{M}_1)$.

ii) One has

$$\dim_E H^1(\mathbf{M}) = 2e, \quad \dim_E H^1_f(\mathbf{M}) = e, \quad \textit{where } e = \mathrm{rk}(\mathbf{M}_0) = \mathrm{rk}(\mathbf{M}_1).$$

iii) Consider the exact sequence

$$0 \to H^0(\mathbf{M}_1) \xrightarrow{\delta_0} H^1(\mathbf{M}_0) \xrightarrow{f_1} H^1(\mathbf{M}) \xrightarrow{g_1} H^1(\mathbf{M}_1) \xrightarrow{\delta_1} H^2(\mathbf{M}_0) \to 0.$$

Then $H^1(\mathbf{M}_0) \simeq \mathrm{Im}(\delta_0) \oplus H^1_f(\mathbf{M}_0)$, $\mathrm{Im}(f_1) = H^1_f(\mathbf{M})$, and $H^1(\mathbf{M}_1) \simeq \mathrm{Im}(g_1) \oplus H^1_f(\mathbf{M}_1)$.

Proof. See [5], Proposition 2.1.7 and Lemma 2.1.8. □

3.3.2 The weight $\leqslant -2$ case

We return to the study of the cohomology of p-adic representations. Let V be the p-adic realization of a pure simple motive $M \neq \mathbf{Q}(m)$. In this subsection, we assume that V satisfies the conditions **C1–4a)** of Subsection 3.2.1 and the conditions **S)** and **U)** of Subsection 3.3.1. Suppose D is a regular submodule of $\mathbf{D}_{\mathrm{st}}(V)$ and $\mathbf{D}$ is the associated (φ, Γ)-submodule of $\mathbf{D}^{\dagger}_{\mathrm{rig}}(V)$. In the remainder of this paper, we write $H^1(\mathbf{D})$ instead $H^1(\mathbf{Q}_p, \mathbf{D})$.

Define

$$H^1_{f,\{p\}}(\mathbf{Q}, V) = \ker \left(H^1_S(\mathbf{Q}, V) \to \bigoplus_{v \in S - \{p\}} \frac{H^1(\mathbf{Q}_v, V)}{H^1(\mathbf{Q}_v, V)} \right).$$

Then (3.28) gives an exact sequence

$$0 \to H^1_f(\mathbf{Q}, V) \to H^1_{f,\{p\}}(\mathbf{Q}, V) \to \frac{H^1(\mathbf{Q}_p, V)}{H^1_f(\mathbf{Q}_p, V)} \to 0. \tag{3.50}$$

Using the fact that D is regular, we obtain the following decomposition

$$H^1_f(\mathbf{Q}_p, V) = H^1_f(\mathbf{Q}, V) \oplus H^1_f(\mathbf{D}).$$

Therefore the restriction map induces an isomorphism

$$H^1_{f,\{p\}}(\mathbf{Q}, V) \simeq \frac{H^1(\mathbf{Q}_p, V)}{H^1_f(\mathbf{D})}.$$

Let $H^1_D(\mathbf{Q}, V)$ be the inverse image of $H^1(\mathbf{D}_1)/H^1_f(\mathbf{D})$ under this isomorphism. By Proposition 5 iii), we have an injection

$$\kappa_D : H^1_D(\mathbf{Q}, V) \hookrightarrow H^1(\mathbf{W}) \tag{3.51}$$

Since, by **D3)**, $\mathscr{D}_{\mathrm{cris}}(\mathbf{W}_0)^{\varphi = p^{-1}} = \mathscr{D}_{\mathrm{cris}}(\mathbf{W}_0)$ and $\mathrm{Fil}^0 \mathscr{D}_{\mathrm{cris}}(\mathbf{W}_0) = 0$, the results of Section 3.1.8 give a canonical decomposition

$$H^1(\mathbf{W}_0) = H^1_f(\mathbf{W}_0) \oplus H^1_c(\mathbf{W}_0). \tag{3.52}$$

Proposition 3.13. *Assume that the weak Leopoldt conjecture holds for $V^*(1)$ and that*

a) $H^0(\mathbf{W}) = 0$;
b) The map $H^1_c(\mathbf{W}_0) \to H^1(\mathbf{W})$ is injective;
c) One has

$$\mathrm{Im}(\kappa_D) \cap H^1_c(\mathbf{W}_0) = \{0\} \quad \text{in} \quad H^1(\mathbf{W}). \tag{3.53}$$

Then $H^1_{\mathrm{Iw}}(V, D) = 0$, and therefore $H^2_{\mathrm{Iw}}(V, D)$ is $\mathscr{H}$-torsion.

Proof. The reader can compare this proof to the proof of Theorem 5.1.3 of [7]. Recall that $\Gamma_0 = \mathrm{Gal}(\mathbf{Q}^{\mathrm{cyc}}/\mathbf{Q}(\mu_p))$, and that for each $\mathcal{H}$-module M we set $M_{\Gamma_0} = M \otimes_{\mathcal{H}} E$. We will use the following elementary lemma.

Lemma 3.14. *Let A and B be two submodules of a finitely-generated free $\mathcal{H}$-module M. Assume that the natural maps $A_{\Gamma_0} \to M_{\Gamma_0}$ and $B_{\Gamma_0} \to M_{\Gamma_0}$ are both injective. Then $A_{\Gamma_0} \cap B_{\Gamma_0} = \{0\}$ implies that $A \cap B = \{0\}$.*

Proof. This is Lemma 5.1.4.1 of [7]. □

We prove Proposition 3.13. Because the weak Leopoldt conjecture holds for $V^*(1)$, the group $H^1_{\mathrm{Iw},S}(\mathbf{Q},T)$ injects into $\oplus_{v\in S} H^1_{\mathrm{Iw}}(\mathbf{Q}_v,T)$. Since $H^1_{\mathrm{Iw},f}(\mathbf{Q}_v,T)=H^1_{\mathrm{Iw}}(\mathbf{Q}_v,T)$ for $v\neq p$, by the definition of the complex $\mathbf{R}\Gamma_{\mathrm{Iw}}(V,D)$ one has

$$H^1_{\mathrm{Iw}}(V,D)=\left(H^1_{\mathrm{Iw},S}(\mathbf{Q},T)\otimes_\Lambda \mathscr{H}\right)\cap H^1_{\mathrm{Iw}}(\mathbf{D}) \qquad \text{in} \quad H^1_{\mathrm{Iw}}(\mathbf{Q}_p,T)\otimes_\Lambda \mathscr{H}.$$

The Λ-torsion part of $H^1_{\mathrm{Iw},S}(\mathbf{Q},T)$ is isomorphic to $T^{H_{\mathbf{Q},S}}=0$, and therefore $A=H^1_{\mathrm{Iw},S}(\mathbf{Q},T)\otimes_\Lambda\mathscr{H}$ is a free $\mathscr{H}$-module. Moreover, $A_{\Gamma_0}\subset H^1_{f,\{p\}}(\mathbf{Q},V)$ Set $B=H^1_{\mathrm{Iw}}(\mathbf{D})/H^1_{\mathrm{Iw}}(\mathbf{D})_{\mathscr{H}-\mathrm{tor}}$. The $\mathscr{H}$-torsion part of $H^1_{\mathrm{Iw}}(\mathbf{D})$ is contained in $T^{H_{\mathbf{Q}_p}}\otimes_\Lambda\mathscr{H}$, and $\dim_E\left(V^{H_{\mathbf{Q}_p}}\right)_{\Gamma_0}=\dim_E\left(V^{H_{\mathbf{Q}_p}}\right)^{\Gamma_0}=0$, and therefore $B_{\Gamma_0}=H^1_{\mathrm{Iw}}(\mathbf{D})_{\Gamma_0}$.

Prove that $B_{\Gamma_0}=H^1_{\mathrm{Iw}}(\mathbf{D})_{\Gamma_0}$ injects into $H^1(\mathbf{D}_1)\subset H^1(\mathbf{Q}_p,V)$. The exact sequence

$$0\to\mathbf{D}\to\mathbf{D}_1\to\mathbf{W}_1\to 0$$

gives rise to an exact sequence $0\to H^0(\mathbf{W}_1)\to H^1(\mathbf{D})\to H^1(\mathbf{D}_1)$, and therefore it is sufficient to show that

$$H^0(\mathbf{W}_1)\cap H^1_{\mathrm{Iw}}(\mathbf{D})_{\Gamma_0}=0 \quad \text{in} \quad H^1(\mathbf{D}).$$

Set $Z=H^0(\mathbf{W}_1)\cap H^1_{\mathrm{Iw}}(\mathbf{D})_{\Gamma_0}$. Let $f\,:\,H^1(\mathbf{D})\to H^1(\mathbf{W}_0)$ denote the map induced by the natural projection $\mathbf{D}\to\mathbf{W}_0$. Since $H^0(\mathbf{W})=0$, the exact sequence $0\to\mathbf{W}_0\to\mathbf{W}\to\mathbf{W}_1\to 0$ induces an injection $H^0(\mathbf{W}_1)\hookrightarrow H^1(\mathbf{W}_0)$. Thus, $Z\cap\ker(f)=\{0\}$. On the other hand, by Proposition 3.8,

$$f(H^1_{\mathrm{Iw}}(\mathbf{D})_{\Gamma_0})\subset H^1_{\mathrm{Iw}}(\mathbf{W}_0)=H^1_c(\mathbf{W}_0).$$

From the exact sequence

$$0\to H^0(\mathbf{W}_1)\hookrightarrow H^1(\mathbf{W}_0)\to H^1(\mathbf{W})$$

and the injectivity of $H^1_c(\mathbf{W}_0)\to H^1(\mathbf{W})$ we obtain that

$$H^1_c(\mathbf{W}_0)\cap H^0(\mathbf{W}_1)=\{0\},$$

and therefore $f(Z)=0$. This proves that $Z=\{0\}$, and the map $H^1_{\mathrm{Iw}}(\mathbf{D})_{\Gamma_0}\to H^1(\mathbf{D}_1)$ is injective.

To complete the proof of Proposition it will be enough, by Lemma 3.14, to show that

$$H^1_{f,\{p\}}(\mathbf{Q},V)\cap H^1_{\mathrm{Iw}}(\mathbf{D})_{\Gamma_0}=0 \qquad \text{in} \quad H^1(\mathbf{Q}_p,V). \tag{3.54}$$

Since $H^1_{\mathrm{Iw}}(\mathbf{D})_{\Gamma_0} \subset H^1(\mathbf{D}_1)$ and the map κ_D is injective, (3.54) is equivalent to

$$H^1_D(\mathbf{Q}, V) \cap H^1_c(\mathbf{W}_0) = 0,$$

and the proposition is proved. □

We want to discuss how the above result is related to the phenomenon of trivial zeros of p-adic L-functions studied in [29], [5], [7]. We consider two cases.

A) Assume in addition that $\mathbf{D}_{\mathrm{st}}(V)^{\varphi=1} = 0$. Then $\mathbf{D}_1 = \mathbf{D}$ and $\mathbf{W} = \mathbf{W}_0$. By Proposition 3.11 iii), one has

$$H^1_D(\mathbf{Q}, V) \simeq H^1(\mathbf{D})/H^1_f(\mathbf{D}) \simeq H^1(\mathbf{W}_0)/H^1_f(\mathbf{W}_0).$$

Consider the commutative diagram

$$\begin{array}{ccc}
\mathscr{D}_{\mathrm{cris}}(\mathbf{W}_0) & \xrightarrow[\sim]{i_{D,f}} & H^1_f(\mathbf{W}_0) \\
\uparrow{\scriptstyle \rho_{D,f}} & & \uparrow{\scriptstyle p_{D,f}} \\
H^1_D(\mathbf{Q}, V) & \longrightarrow & H^1(\mathbf{W}_0) \\
\downarrow{\scriptstyle \rho_{D,c}} & & \downarrow{\scriptstyle p_{D,c}} \\
\mathscr{D}_{\mathrm{cris}}(\mathbf{W}_0) & \xrightarrow[\sim]{i_{D,c}} & H^1_c(\mathbf{W}_0),
\end{array}$$

where $p_{D,f}$ and $p_{D,c}$ are projections given by (3.52) and $\rho_{D,f}$ and $\rho_{D,c}$ are defined as the unique maps making this diagram commute.

Definition (see [7]). Let V be a p-adic representation which satisfies the conditions **C1–3)**, **C4a)**, **C5a)** from Subsection 3.2.1 and the conditions **S)** and **U)** from Subsection 3.3.1. Assume that $\mathbf{D}_{\mathrm{st}}(V)^{\varphi=1} = 0$. Let D be a regular submodule of $\mathbf{D}_{\mathrm{st}}(V)$. The determinant

$$\ell(V, D) = \det\left(\rho_{D,f} \circ \rho_{D,c}^{-1} \mid \mathscr{D}_{\mathrm{cris}}(\mathbf{W}_0)\right)$$

will be called the ℓ-invariant associated to V and D.

B) Assume that

$$H^1_f(\mathbf{Q}, V) = H^1_f(\mathbf{Q}, V^*(1)) = 0, \tag{3.55}$$

and that, in addition, V satisfies:

M) The condition **U)** of Section 3.3.1 holds and in the decomposition (3.49) $\mathbf{A}_0 = \mathbf{A}_1 = 0$.

Note that the typical example we have in mind is $W_f(k)$, where W_f is the p-adic representation associated to a split multiplicative newform f of weight $2k$ on $\Gamma_0(N)$ with $p \mid N$.

From Proposition 3.12 one has $H^1_f(\mathbf{M}_0) = H^1_f(\mathbf{M})$ and $H^1(\mathbf{M})/H^1_f(\mathbf{M}_0) \simeq H^1(\mathbf{M}_1)/H^1_f(\mathbf{M}_1)$. Thus, $H^1_D(\mathbf{Q}, V) \simeq H^1(\mathbf{M}_1)/H^1_f(\mathbf{M}_1)$ is a E-vector space of dimension $e = \mathrm{rank}(\mathbf{M}_0) = \mathrm{rank}(\mathbf{M}_1)$, and one has a commutative diagram

$$\begin{array}{ccccccccccc} 0 & \longrightarrow & H^0(\mathbf{M}_1) & \xrightarrow{\delta_0} & H^1(\mathbf{M}_0) & \xrightarrow{f_1} & H^1(\mathbf{M}) & \xrightarrow{g_1} & H^1(\mathbf{M}_1) & \xrightarrow{\delta_1} & H^2(\mathbf{M}_0) \longrightarrow 0, \\ & & & & & & \uparrow{\scriptstyle \kappa_D} & \nearrow{\scriptstyle \bar\kappa_D} & & & \\ & & & & & & H^1_D(\mathbf{Q},V) & & & & \end{array} \tag{3.56}$$

where the map $\bar{\kappa}_D$ is injective. As $H^0(\mathbf{M}) = H^2(\mathbf{M}) = 0$, the upper row is exact. This implies that $\mathrm{Im}(g_1)$ is a E-vector space of dimension e. Thus, $\mathrm{Im}(\bar{\kappa}_D) = \mathrm{Im}(g_1)$, and again one has a commutative diagram

$$\begin{array}{ccccc} & & \mathscr{D}_{\mathrm{cris}}(\mathbf{M}_1) & \xrightarrow[\sim]{i_{D,f}} & H^1_f(\mathbf{M}_1) \\ & & \uparrow{\scriptstyle \rho_{M,f}} & & \uparrow{\scriptstyle \rho_{D,f}} \\ H^1_D(\mathbf{Q},V) & \xrightarrow{\sim} & \mathrm{Im}(g_1) & \longrightarrow & H^1(\mathbf{M}_1) \\ & & \downarrow{\scriptstyle \rho_{M,c}} & & \downarrow{\scriptstyle \rho_{D,c}} \\ & & \mathscr{D}_{\mathrm{cris}}(\mathbf{M}_1) & \xrightarrow[\sim]{i_{D,c}} & H^1_c(\mathbf{M}_1). \end{array}$$

Definition (see[5]). Let V be a p-adic representation which satisfies the conditions **C1–3)** and (3.55). Let D be a regular submodule of $\mathbf{D}_{\mathrm{st}}(V)$. Assume that the condition **M)** holds. The determinant

$$\ell(V, D) = \det\left(\rho_{D,f} \circ \rho_{D,c}^{-1} \mid \mathscr{D}_{\mathrm{cris}}(\mathbf{M}_1)\right)$$

will be called the ℓ-invariant associated to V and D.

In particular, in this case the ℓ-invariant is local, i.e. depends only on the restriction of the representation V on the decomposition group at p.

Remark. In [5], the ℓ-invariant is defined in a slightly more general situation, where only $\mathbf{W}_0$ vanishes. We do not include it here to avoid additional technical complications in the formulation of our results.

The following conjecture was formulated in [5] and [7].

Conjecture (EXTRA ZERO CONJECTURE). *Let V be a p-adic representation which satisfies the conditions* **C1–4a)**. *Let D be a regular subspace of* $\mathbf{D}_{\mathrm{st}}(V)$, *and let* $e = \mathrm{rank}(\mathbf{W}_0)$. *Then, in both cases* **A)** *and* **B)**, *the p-adic L-function $L_p(V, D, s)$ has a zero of order e at $s = 0$ and*

$$\lim_{s\to 0} \frac{L_p(V, D, 0)}{s^e} = \ell(V, D)\,\mathcal{E}^+(V, D)\,\Omega_p(M, D)\frac{L(M, 0)}{\Omega_\infty(M)},$$

where $\mathcal{E}^+(V, D)$ is obtained from $\mathcal{E}(V, D)$ by excluding zero factors.

Recall (see Section 3.0.2) that $\Omega_p(M, D)$ denote the determinant of the regulator map $r_{V,D}$. Note that $\Omega_p(M, D) = 1$ if $H^1_f(\mathbf{Q}, V) = H^1_f(\mathbf{Q}, V^*(1)) = 0$. We refer to [5], [6] [7] for precise formulation of this conjecture and a survey of known cases. Note that the non-vanishing of $\ell(V, D)$ is a difficult open problem which is solved only for Dirichlet motives $\mathbf{Q}(\eta)$ [22] and for elliptic curves [1]. The following result shows that it is closely related to the expected vanishing of $H^1_{\mathrm{Iw}}(V, D)$.

Proposition 3.15. *Let V be a p-adic representation which satisfies the conditions* **C1-3)**. *Assume that the weak Leopoldt conjecture holds for $V^*(1)$. Then, in the both cases* **A)** *and* **B)** *above, the non-vanishing of $\ell(V, D)$ implies that $H^1_{\mathrm{Iw}}(V, D) = 0$, and therefore $H^2_{\mathrm{Iw}}(V, D)$ is $\mathscr{H}$-torsion.*

Proof. In the case **A)**, we have $\mathbf{W}_0 = \mathbf{W}$, and the statement follows directly from Proposition 3.13 and the definition of the ℓ-invariant. In the case **B)**, the diagram (3.56), together with Proposition 3.12, show that the injectivity of $H^1_c(\mathbf{M}_0) \to H^1(\mathbf{M})$ and the condition (3.53) are both equivalent to the following condition:

$$\mathrm{Im}(\delta_0) \cap H^1_c(\mathbf{M}_0) = \{0\}.$$

The statement follows now from [5], Proposition 2.2.4, where it is proved that $\ell(V, D)$ can be computed as the slope of the map $\delta_0 : H^0(\mathbf{M}_1) \to H^1(\mathbf{M})$ with respect to the decomposition $H^1(\mathbf{M}) \simeq H^1_f(\mathbf{M}) \oplus H^1_c(\mathbf{M})$. □

3.4 p-adic height pairings

3.4.1 Extended Selmer groups

Let V be a p-adic representation of $G_{\mathbf{Q},S}$ which satisfies the conditions **C1–3)** of Subsection 3.2.1. Fix a submodule D of $\mathbf{D}_{\mathrm{st}}(V)$ such that $\mathbf{D}_{\mathrm{st}}(V) = D \oplus$

$\mathrm{Fil}^0\mathbf{D}_{\mathrm{st}}(V)$. We will always assume that the condition $\mathbf{M}$ of Section 3.3.2 holds for D. Then, by Proposition 3.12, one has an exact sequence

$$0 \to \mathbf{M}_0 \to \mathbf{M} \to \mathbf{M}_1 \to 0,$$

where $\mathscr{D}_{\mathrm{cris}}(\mathbf{M}_0) = D/D_{-1}$, $\mathscr{D}_{\mathrm{cris}}(\mathbf{M}_1) = D_1/D$ and $\mathrm{rk}(\mathbf{M}_0) = \mathrm{rk}(\mathbf{M}_1)$.

Proposition 3.16. *There exists an exact sequence*

$$0 \to H^0(\mathbf{M}_1) \to H^1(V, D) \to H^1_f(\mathbf{Q}, V) \to 0, \tag{3.57}$$

where $H^1_f(\mathbf{Q}, V)$ is the Bloch–Kato Selmer group and

$$\dim_E H^0(\mathbf{M}_1) = \mathrm{rk}(\mathbf{M}_1).$$

Proof. By the definition of $\mathbf{R}\Gamma(V, D)$, the group $H^1(V, D)$ is the kernel of the map

$$H^1_S(\mathbf{Q}, V) \oplus \left(\bigoplus_{v \in S - \{p\}} H^1_f(\mathbf{Q}_v, V) \right) \oplus H^1(\mathbf{D}) \to \bigoplus_{v \in S} H^1(\mathbf{Q}_v, V).$$

The proposition follows directly from this description of $H^1(V, D)$ together with the following facts:

a) $H^0(\mathbf{M}_1) = \ker(H^1(\mathbf{D}) \to H^1(\mathbf{Q}_p, V))$;
b) $\mathrm{Im}(H^1(\mathbf{D}) \to H^1(\mathbf{Q}_p, V)) = H^1_f(\mathbf{Q}_p, V)$.

The proof of a) and b) can be extracted from the construction of the ℓ-invariant in [5], Section 2.2.1, but we recall the arguments for reader's convenience. Consider the exact sequence

$$0 \to \mathbf{D}_1 \to \mathbf{D}^{\dagger}_{\mathrm{rig}}(V) \to \mathrm{gr}_2\mathbf{D}^{\dagger}_{\mathrm{rig}}(V) \to 0.$$

In the proof of Proposition 3.11, we saw that $H^0\left(\mathrm{gr}_2\mathbf{D}^{\dagger}_{\mathrm{rig}}(V)\right) = 0$, and therefore the map $H^1(\mathbf{D}_1) \to H^1(\mathbf{Q}_p, V)$ is injective. Taking the long cohomology sequence associated to

$$0 \to \mathbf{D} \to \mathbf{D}_1 \to \mathbf{M}_1 \to 0,$$

we obtain

$$0 \to H^0(\mathbf{M}_1) \to H^1(\mathbf{D}) \to H^1(\mathbf{D}_1),$$

and a) is proved.

Using (3.22), it is not difficult to show that $H^1_f\left(\mathrm{gr}_2\mathbf{D}^{\dagger}_{\mathrm{rig}}(V)\right) = 0$, and therefore $H^1_f(\mathbf{D}_1) \simeq H^1_f(\mathbf{Q}_p, V)$. Consider the exact sequence

$$0 \to \mathbf{D}_{-1} \to \mathbf{D}_1 \to \mathbf{M} \to 0.$$

Since $H^0(\mathbf{M}) = 0$ and $H^1_f(\mathbf{D}_{-1}) = H^1(\mathbf{D}_{-1})$, by [5], Corollary 1.4.6, one has an exact sequence

$$0 \to H^1(\mathbf{D}_{-1}) \to H^1_f(\mathbf{D}_1) \to H^1_f(\mathbf{M}) \to 0.$$

On the other hand, from Proposition 3.12 it follows that

$$\mathrm{Im}(H^1(\mathbf{M}_0) \to H^1(\mathbf{M})) = H^1_f(\mathbf{M}).$$

Thus, $\mathrm{Im}(H^1(\mathbf{D}) \to H^1(\mathbf{D}_1)) = H^1_f(\mathbf{D}_1)$ and b) is proved. □

Definition. We will call $H^1(V, D)$ the extended Selmer group associated to D.

Remarks. 1) Extended Selmer groups associated to ordinary local conditions were studied in [45], [46].

2) If $\mathbf{M} = 0$, the group $H^1(V, D)$ coincides with the Bloch–Kato Selmer group. One expects that, if D is regular, the appearance of $H^0(\mathbf{M}_1)$ in the short exact sequence (3.57) reflects the presence of extra-zeros of the p-adic L-function $L_p(V, D, s)$. This question will be studied in [8].

3.4.2 p-adic height pairings

In [46], Nekovář found a new construction of the p-adic height pairing on extended Selmer groups defined by Greenberg's local conditions. In this section, we follow his approach *verbatim* working with local conditions defined by (φ, N)-submodules. Let V be a p-adic representation of $G_{\mathbf{Q},S}$ which satisfies the conditions **C1–3)** and the condition **C4b)** from Subsection 3.2.1. In particular, we assume that V is equipped with a bilinear form which induces an isomorphism $V \simeq V^*(1)$. We keep notation and conventions of Subsection 3.4.1. Set $A = \mathscr{H}/(X^2)$. Then $A = E[X]/(X^2)$, and one has an exact sequence

$$0 \to E \xrightarrow{X} A \to E \to 0. \qquad (3.58)$$

Set $V_A = V \otimes_E A^{\iota}$, and $D_A = D \otimes_E A^{\iota}$, and consider the complex $\mathbf{R}\Gamma(V_A, D_A)$. The sequence (3.58) induces a distinguished triangle

$$\mathbf{R}\Gamma(V, D) \to \mathbf{R}\Gamma(V_A, D_A) \to \mathbf{R}\Gamma(V, D),$$

which gives the coboundary map

$$H^1(V,D) \xrightarrow{\beta} H^2(V,D).$$

Let $\mathbf{D}^\perp \subset \mathbf{D}_{\mathrm{st}}(V^*(1))$ denote the dual submodule.

Definition. Let V be a p-adic representation which satisfies the conditions **C1–4b)** from Subsection 3.2.1, the condition **S)** from Subsection 3.4.1, and let D be a (φ,N)-submodule of $\mathbf{D}_{\mathrm{st}}(V)$ such that the condition **M)** holds for D. We define the p-adic height pairing associated to D as the bilinear map

$$\langle\,,\,\rangle_{V,D} \;:\; H^1(V,D)\times H^1(V^*(1),D^\perp)\to E \tag{3.59}$$

given by $\langle x,y\rangle_{V,D} = \beta(x)\cup y$, where $\cup\,:\,H^2(V,D)\times H^1(V^*(1),D^\perp)\to E$ denotes the duality defined in Proposition 3.10.

Remarks. 1) In the ordinary setting, it is possible to work over $\mathbf{Z}_p$, and Nekovář's descent machinery gives very general formulas of the Birch and Swinnerton-Dyer type. In the general setting, we are forced to work over $\mathbf{Q}_p$ if we want to use the theory of (φ,Γ)-modules.

2) The pairing (3.59) will be studied in detail in [8]. In particular, we compare our construction to the p-adic height pairing constructed by Nekovář in [45] and relate it to universal norms.

The following result can be seen as an analog of Proposition 3.15 in the weight -1 case.

Proposition 3.17. *Let V be a p-adic representation which satisfies the conditions* **C1–4b)** *from Subsection 3.2.1 and the condition* **S)** *from Subsection 3.3.1. Let D be a (φ,N)-submodule of $\mathbf{D}_{\mathrm{st}}(V)$ such that the condition* **M)** *holds for D. Assume that*

a) The weak Leopoldt conjecture holds for $V^*(1)$;
b) The pairing $\langle\,,\,\rangle_{V,D}$ is non-degenerate;
c) V does not contain subrepresentations of the form $\mathbf{Q}_p(m)$.

Then $H^1_{\mathrm{Iw}}(V,D)=0$, and therefore $H^2_{\mathrm{Iw}}(V,D)$ is $\mathscr{H}$-torsion.

Proof. If the p-adic height pairing is non-degenerate, the map β is injective. The diagram

$$\begin{array}{ccccccccc}
0 & \longrightarrow & \mathscr{H} & \xrightarrow{X} & \mathscr{H} & \longrightarrow & E & \longrightarrow & 0\\
 & & \downarrow & & \downarrow & & \downarrow & & \\
0 & \longrightarrow & E & \xrightarrow{X} & A & \longrightarrow & E & \longrightarrow & 0
\end{array}$$

gives rise to a commutative diagram with exact rows

$$\begin{array}{ccccc} 0 \longrightarrow H^1_{\mathrm{Iw}}(V,D)_{\Gamma_0} & \longrightarrow & H^1(V,D) & \longrightarrow & H^2_{\mathrm{Iw}}(V,D)^{\Gamma_0} \\ \downarrow & & \downarrow{=} & & \downarrow \\ H^1(V_A,D_A) & \longrightarrow & H^1(V,D) & \xrightarrow{\beta} & H^2(V,D). \end{array}$$

Thus $H^1_{\mathrm{Iw}}(V, D)_{\Gamma_0} = 0$. On the other hand, as the weak Leopoldt conjecture holds for $V^*(1)$ and $V^{H_{\mathbf{Q},S}} = 0$, the $\Lambda[1/p]$-module $H^1_{\mathrm{Iw},S}(\mathbf{Q}, V)$ is free and injects into $H^1_{\mathrm{Iw}}(\mathbf{Q}_p, V)$. Therefore, as in the proof of Proposition 3.13, one has

$$H^1_{\mathrm{Iw}}(V, D) = \left(H^1_{\mathrm{Iw},S}(\mathbf{Q}, T) \otimes_\Lambda \mathscr{H}\right) \cap H^1_{\mathrm{Iw}}(\mathbf{D}).$$

This implies that $H^1_{\mathrm{Iw}}(V, D)$ is $\mathscr{H}$-free, and the vanishing of $H^1_{\mathrm{Iw}}(V, D)_{\Gamma_0}$ gives $H^1_{\mathrm{Iw}}(V, D) = 0$. □

References

[1] K. Barré-Sirieix, G. Diaz, F. Gramain, G. Philibert, *Une preuve de la conjecture de Mahler-Manin,* Invent. Math. **124** (1996), 1–9.

[2] J. Bellaïche, G. Chenevier, *p-adic families of Galois representations and higher rank Selmer groups,* Astérisque **324** (2009), 314 pages.

[3] J. Bellaïche, *Ranks of Selmer groups in an analytic family,* Trans. Amer. Math. Soc. **364** (2012), 4735-4761.

[4] D. Benois, *On Iwasawa theory of p-adic representations,* Duke Math. J. **104** (2000), 211–267.

[5] D. Benois, *A generalization of Greenberg's $\mathscr{L}$-invariant,* Amer. J. Math. **133** (2011), 1573-1632.

[6] D. Benois, *Trivial zeros of p-adic L-functions at near central points,* J. Inst. Math. Jussieu **13** (2014), 561–598.

[7] D. Benois, *On extra zeros of p-adic L-functions: the crystalline case,* In: Iwasawa theory 2012. State of the Art and Recent Advances (T. Bouganis and O. Venjakob, eds.), Contributions in Mathematical and Computational Sciences, vol. 7, Springer, 2014 (to appear).

[8] D. Benois, *p-adic height pairings and p-adic Hodge theory,* preprint (2014), available at arXiv.org/pdf/1412.7305v1.pdf.

[9] L. Berger, *Représentations p-adiques et équations différentielles,* Invent. Math. **148** (2002), 219–284.

[10] L. Berger, *Bloch and Kato's exponential map: three explicit formulas,* Doc. Math., Extra Volume: Kazuya Kato's Fiftieth Birthday (2003), pp. 99–129.

[11] L. Berger, *Equations différentielles p-adiques et (φ, N)-modules filtrés,* In: Représentations p-adiques de groupes p-adiques I (L. Berger, P. Colmez, Ch. Breuil eds.), Astérisque **319** (2008), 13–38.

[12] L. Berger, P. Colmez, *Familles de représentations de de Rham et monodromie p-adique,* In: Représentations p-adiques de groupes p-adiques I (L. Berger, P. Colmez, Ch. Breuil eds.), Astérisque **319** (2008), 303–337.

[13] S. Bloch, K. Kato, *L-functions and Tamagawa numbers of motives,* In: The Grothendieck Festschrift, vol. 1 (P. Cartier, L. Illusie, N. M. Katz, G. Laumon, Yu. Manin, K. Ribet, eds.), Progress in Math. vol. 86, Birkhäuser Boston, Boston, MA, 1990, pp. 333–400.

[14] F. Cherbonnier, P. Colmez, *Représentations p-adiques surconvergentes,* Invent. Math., **133** (1998), 581–611.

[15] F. Cherbonnier, P. Colmez, *Théorie d'Iwasawa des représentations p-adiques d'un corps local,* J. Amer. Math. Soc., **12** (1999), 241–268.

[16] J. Coates, B. Perrin-Riou, *On p-adic L-functions attached to motives over* $\mathbb{Q}$, In: Algebraic Number Theory – in honor of K. Iwasawa (J. Coates, R. Greenberg, B. Mazur and I. Satake, eds.), Adv. Studies in Pure Math., **17** (1989), pp. 23–54.

[17] P. Colmez, *Fonctions L p-adiques,* In: Seminaire Bourbaki. Vol. 1998/99, Astérisque, **266** (2000), 21–58.

[18] P. Colmez, *Les conjectures de monodromie p-adiques,* In: Seminaire Bourbaki. Vol 2001/2002, Asterisque **290** (2003), 53–101.

[19] P. Colmez, *Représentations triangulines de dimension* 2, In: Représentations p-adiques de groupes p-adiques I (L. Berger, P. Colmez, Ch. Breuil eds.), Astérisque **319** (2008), 213-258.

[20] M. Flach, *A generalization of the Cassels–Tate pairing,* J. Reine Angew. Math. **412** (1990), 113–127.

[21] G. Faltings, *Crystalline cohomology and p-adic Galois representations,* In: Algebraic Analysis, Geometry and Number Theory (Baltimore, MD, 1988), John Hopkins University Press, Baltimore MD, 1989, pp. 25–80.

[22] B. Ferrero, R. Greenberg, *On the behavior of p-adic L-functions at s = 0,* Invent. Math. **50** (1978/79), 91–102.

[23] J.-M. Fontaine, *Représentations p-adiques des corps locaux,* In: The Grothendieck Festschrift, vol. 2 (P. Cartier, L. Illusie, N. M. Katz, G. Laumon, Yu. Manin, K. Ribet, eds.), Progress in Math. vol. 87, Birkhäuser Boston, Boston, MA, 1990, pp. 249–309.

[24] J.-M. Fontaine, *Valeurs spéciales de fonctions L des motifs,* In: Séminaire Bourbaki, Vol. 1991/1992, exposé 751, Astérisque, **206** (1992), 205–249

[25] J.-M. Fontaine, *Le corps des périodes p-adiques,* In: Périodes p-adiques (J.-M. Fontaine, ed.), Astérisque, **223** (1994), 59-102.

[26] J.-M. Fontaine, *Représentations p-adiques semi-stables,* In: Périodes p-adiques (J.-M. Fontaine, ed.), Astérisque, **223** (1994), 59-102.

[27] J.-M. Fontaine, B. Perrin-Riou, *Autour des conjectures de Bloch et Kato; cohomologie galoisienne et valeurs de fonctions L,* In: Motives (U. Jannsen, S. Kleiman, J.-P. Serre, eds.), Proc. Symp. in Pure Math., **55**, (1994), 599–706.

[28] R. Greenberg, *Iwasawa theory for p-adic representations,* In: Algebraic Number Theory – in honor of K. Iwasawa (J. Coates, R. Greenberg, B. Mazur and I. Satake, eds.), Adv. Studies in Pure Math., **17** (1989), 97–137.

[29] R. Greenberg, *Trivial zeros of p-adic L-functions,* In: p-Adic Monodromy and the Birch and Swinnerton-Dyer Conjecture (B. Mazur and G. Stevens, eds.), Contemp. Math. **165**, (1994), 149–174.

[30] R. Greenberg, *Iwasawa Theory and p-adic Deformations of Motives,* In: Motives (U. Jannsen, S. Kleiman, J.-P. Serre, eds.), Proc. of Symp. in Pure Math., **55**, (1994), 193–223.

[31] E. Hellmann, *Families of trianguline representations and nite slope spaces,* preprint available on http://arXiv:1202.4408v1.

[32] L. Herr, *Sur la cohomologie galoisienne des corps p-adiques,* Bull. Soc. Math. France **126** (1998), 563–600.

[33] L. Herr, *Une approche nouvelle de la dualité locale de Tate,* Math. Annalen **320** (2001), 307–337.

[34] U. Jannsen, *On the ℓ-adic cohomology of varieties over number fields and its Galois cohomology,* In: Galois Groups over $\mathbb{Q}$ (Y. Ihara, K. Ribet and J.-P. Serre, eds.) , Math. Sciences Research Inst. Publ., vol. 16, Springer, 1989, pp. 315–360.

[35] U. Jannsen, *Mixed motives and algebraic K-theory,* Lecture Notes in Math. vol. 1400, Springer, Berlin, 1990.

[36] K. Kato, p-adic Hodge theory and values of zeta-functions of modular forms, In: Cohomologies p-adiques et applications arithmétiques (III) (P. Berthelot, J.-M. Fontaine, L. Illusie, K. Kato, M. Rapoport, eds.), Astérisque **295** (2004), pp. 117-290.

[37] N. M. Katz, W. Messing, *Some consequences of the Riemann hypothesis for varieties over finite fields,* Invent. Math. **23** (1974), 73–77.

[38] K. Kedlaya, *A p-adic monodromy theorem,* Ann. of Math. **160** (2004), 94–184.

[39] K. Kedlaya, R. Liu, *On families of* (φ, Γ)*-modules,* Algebra and Number Theory, **4** (2010), 943967.

[40] K. Kedlaya, J. Pottharst, L. Xiao, *Cohomology of arithmetic families of* (φ, Γ)*-modules,* Preprint available on http://arxiv.org/abs/1203.5718.

[41] M. Lazard, *Les zéros des fonctions analytiques d'une variable sur un corps valué complet,* Publ. Math. IHES **14** (1962), 47–75.

[42] R. Liu, *Cohomology and Duality for* (φ, Γ)*-modules over the Robba ring,* Int. Math. Research Notices (2007), no. 3, 32 pages.

[43] Yu. Manin, *Periods of cusp forms and p-adic Hecke series,* Math USSR Sbornik **92** (1973), 371–393.

[44] B. Mazur, J.Tate, J. Teitelbaum, *On p-adic analogues of the conjectures of Birch and Swinnerton-Dyer,* Invent. Math. **84** (1986), 1–48.

[45] J. Nekovář, *On p-adic height pairing,* In: Séminaire de Théorie des Nombres, Paris, 1990/91 (S. David, ed.), Progress in Mathematics, vol. 108, Birkhäuser Boston, 1993, pp. 127202.

[46] J. Nekovář, *Selmer complexes,* Astérisque, **310** (2006), 559 pages.

[47] J. Nekovář, W. Nizioł, *Syntomic cohomology and p-adic regulators for varieties over p-adic fields,* Preprint available on http://arxiv.org/abs/1309.7620.

[48] T. Ochiai, *Control theorem for Bloch–Kato's Selmer groups of p-adic representations,* J. Number Theory **82** (1990), 69–90.

[49] B. Perrin-Riou, *Théorie d'Iwasawa et hauteurs p-adiques,* Invent. Math. **109** (1992), 137–185.

[50] B. Perrin-Riou, *Théorie d'Iwasawa des représentations p-adiques sur un corps local,* Invent. Math. **115** (1994), 81–149.

[51] B. Perrin-Riou, *Fonctions L p-adiques des représentations p-adiques,* Astérisque **229** (1995), 198 pages.

[52] J. Pottharst, *Analytic families of finite slope Selmer groups,* Algebra and Number Theory **7** (2013), 1571-1611.

[53] J. Pottharst, *Cyclotomic Iwasawa theory of motives,* Preprint available on http://math.bu.edu/people/potthars/.

[54] J. Pottharst, *Analytic families of finite slope Selmer groups,* Preprint available on math.bu.edu/people/potthars/writings/.

[55] J.-P. Serre, *Facteurs locaux des fonctions zêta des variétés algébriques (définitions et conjectures),* Séminaire Delange-Pisot-Poitou. Théorie des nombres, **11** no. 2 (1969-1970), Exp. No. 19, 15 pages.

[56] P. Schneider, *p-adic height pairings. I,* Invent. Math. **69** (1982), 401409.

[57] P. Schneider, J. Teitelbaum, *Algebras of p-adic distributions and ad- missible representations,* Invent. Math. **153** (2003), 145–196.

[58] C. Skinner, E. Urban, *The Iwasawa Main Conjectures for* GL_2, Invent. Math. **195** (2014), 1–277.

[59] G. Tamme, *Karoubi's relative Chern character, the rigid syntomic regulator, and the Bloch–Kato exponential map,* Preprint available on http://arxiv.org/abs/1111.4109.

[60] T. Tsuji, *p-adic etale cohomology and crystalline cohomology in the semistable reduction case,* Invent. Math. **137** (1999), 233–411.

[61] M. Vishik, *Non-archemedian measures connected with Dirichlet series,* Math. USSR Sbornik **28** (1976), 216–228.

[62] J.-P. Wintenberger, *Le corps des normes de certaines extensions infinies de corps locaux; applications,* Ann. Sc. ENS **16** (1983), 59–89.

4

A survey of applications of the circle method to rational points

T.D. Browning

School of Mathematics, University of Bristol, Bristol, BS8 1TW, UK

E-mail address: t.d.browning@bristol.ac.uk

Introduction

Given a number field k and a projective algebraic variety X defined over k, the question of whether X contains a k-rational point is both very natural and very difficult. In the event that the set $X(k)$ of k-rational points is not empty, one can also ask how the points of $X(k)$ are distributed. Are they dense in X under the Zariski topology? Are they dense in the set

$$X(\mathbf{A}_k) = \prod_{v \in \Omega_k} X(k_v)$$

of adèlic points under the product topology? Can one count the points in $X(k)$ of bounded height? In favourable circumstances, the Hardy–Littlewood circle method can systematically provide answers to all of these questions.

The focus of this survey will be upon the first and most basic of these questions: *when is* $X(k) \neq \emptyset$ *for a given variety* X *defined over* k*?* In considering what the circle method has to say about this we shall usually restrict our attention to varieties X defined over k that are geometrically integral, non-singular and projective. Recall that such a family is said to satisfy the Hasse principle if any variety in the family has a k-point as soon as it has a point in every completion of k. Sometimes we will drop the assumption on non-singularity, saying that the smooth Hasse principle holds for a family of such varieties when the Hasse principle holds for the smooth locus of X.

Date: April 2, 2014.
2010 *Mathematics Subject Classification.* 11P05 (11D72, 14G05).

Arithmetic and Geometry, ed. Luis Dieulefait *et al.* Published by Cambridge University Press.

Quadrics are among the first examples of families satisfying the Hasse principle. Beyond this simple setting, it is quite rare to find varieties which satisfy the Hasse principle. In dimension 2 we have the following counter-example.

Example 4.1. A generalised Châtelet surface over k is a proper smooth model of an affine surface

$$Y^2 - aZ^2 = f(T),$$

where $a \in k^* \setminus k^{*2}$ and $f \in k[T]$ is a polynomial. An infinite family of counter-examples to the Hasse principle was discovered by Colliot-Thélène, Coray and Sansuc [21, Prop. C]. This corresponds to taking $k = \mathbb{Q}$, $a = -1$ and $f(T) = (-T^2 + c)(T^2 - c + 1)$, where c is a positive integer congruent to 3 modulo 4. Letting $X_c/\mathbb{Q}$ denote this Châtelet surface, it follows from this work that $X_c(\mathbb{R}) \neq \emptyset$ and $X_c(\mathbb{Q}_p) \neq \emptyset$ for all primes p, but $X_c(\mathbb{Q}) = \emptyset$.

A systematic and unified explanation of various explicit failures of the Hasse principle was provided by Manin in his 1970 ICM address [51] in terms of the Brauer–Grothendieck group $\mathrm{Br}(X) = H^2_{\text{ét}}(X, \mathbb{G}_m)$, which can be associated to any geometrically integral, projective variety X. We shall discuss the Brauer–Manin obstruction to the Hasse principle in Section 4.3.1. The counter-examples $X_c/\mathbb{Q}$ were shown to conform to this explanation in [21]. It was later shown by Colliot-Thélène, Sansuc and Swinnerton-Dyer [24, 25] that this obstruction actually explains all counter-examples to the Hasse principle among generalised Châtelet surfaces, with $\deg f(T) \leqslant 4$, defined over any number field k. In particular, it follows from their work that such surfaces satisfy the Hasse principle when f is irreducible over k, which is the generic situation. In Section 4.3.3 we discuss a partial extension of this result when $k = \mathbb{Q}$.

The topic of rational points on varieties is *vast* and we shall only touch upon those aspects which have involved the circle method intrinsically (and even then we will not be able to give many details). This survey will focus on the following subjects:

§1 The circle method and complete intersections
§2 The circle method and diagonalisation
§3 The circle method and descent

Acknowledgements. The author is grateful to Jean-Louis Colliot-Thélène, Boris Moroz, Sean Prendiville, Damaris Schindler, Alexei Skorobogatov and

Efthymios Sofos for comments on an earlier draft of this paper. While working on this paper the author was supported by ERC grant 306457.

4.1 The circle method and complete intersections

4.1.1 The circle method over $\mathbb{Q}$

Let us begin with some generalities about the circle method over $\mathbb{Q}$. Suppose that we are given a system of polynomials $\mathbf{f} = (f_1, \dots, f_s)$, with $f_i \in \mathbb{Z}[x_1, \dots, x_n]$ for $1 \leqslant i \leqslant s$. Let $\mathfrak{B} \subset \mathbb{R}^n$ be a fixed box, with side lengths parallel to the coordinate axes, and write $B\mathfrak{B}$ for the dilation of this region by a parameter $B \geqslant 1$. Define the counting function

$$N(B) = \#\{\mathbf{x} \in \mathbb{Z}^n \cap B\mathfrak{B} : f_i(\mathbf{x}) = 0 \text{ for } 1 \leqslant i \leqslant s\},$$

for $B \geqslant 1$. When $f_1, \dots, f_s$ are homogeneous and cut out a non-singular complete intersection $X \subset \mathbb{P}^{n-1}$, with $\dim(X) \geqslant 3$, this amounts to counting integral points of bounded height on the universal torsor over X. (Note that the affine cone over X in $\mathbb{A}^n \setminus \{\mathbf{0}\}$ is the unique universal torsor over X up to isomorphism.)

In order to show that the variety $f_1 = \cdots = f_s = 0$ has an integral point it suffices to show $N(B) > 0$ if B is taken to be sufficiently large and $\mathfrak{B}$ is chosen suitably. Let $\mathrm{e}(z) = e^{2\pi i z}$ for any $z \in \mathbb{R}$. By orthogonality one has

$$N(B) = \int_{(0,1]^s} S(\boldsymbol{\alpha})\mathrm{d}\boldsymbol{\alpha}, \tag{4.1}$$

where

$$S(\boldsymbol{\alpha}) = \sum_{\mathbf{x} \in \mathbb{Z}^n \cap B\mathfrak{B}} \mathrm{e}\left(\alpha_1 f_1(\mathbf{x}) + \cdots + \alpha_s f_s(\mathbf{x})\right).$$

The classical circle method is concerned with an asymptotic analysis of this integral, as $B \to \infty$, based on a suitable dissection of the range of integration.

Suppose that $\deg f_i = d_i$ for $1 \leqslant i \leqslant s$. Let Q satisfy $1 \leqslant Q \leqslant B^{\max_i d_i}$. For given integers a_i, q such that $1 \leqslant a_i \leqslant q \leqslant Q$, we define

$$\mathfrak{M}(q, \mathbf{a}; Q) = \left\{\boldsymbol{\alpha} \in (0,1]^s : \left|\alpha_i - \tfrac{a_i}{q}\right| \leqslant QB^{-d_i} \text{ for } 1 \leqslant i \leqslant s\right\}.$$

The set of major arcs of level Q is the union

$$\mathfrak{M}(Q) = \bigcup_{q \leqslant Q} \bigcup_{\substack{1 \leqslant a_i \leqslant q \\ \gcd(\mathbf{a}, q) = 1}} \mathfrak{M}(q, \mathbf{a}; Q).$$

The general philosophy behind the circle method is that the contribution from integrating over the major arcs should produce a main term in our asymptotic formula for $N(B)$. Generally, this main term takes the shape

$$B^{n-\sum_{1\leqslant i\leqslant s} d_i} \cdot \prod_{v\in\Omega_{\mathbb{Q}}} \sigma_v, \tag{4.2}$$

where σ_v are local densities for the variety $f_1 = \cdots = f_s = 0$. More precisely,

$$\sigma_\infty = \int_{\mathbb{R}^s} \int_{\mathfrak{B}} \mathrm{e}(z_1 f_1(\mathbf{x}) + \cdots + z_s f_s(\mathbf{x}))\, \mathbf{dx}\, \mathbf{dz}$$

is the real density and

$$\sigma_p = \lim_{t\to\infty} p^{-t(n-s)} \# \left\{ \mathbf{x} \in (\mathbb{Z}/p^t\mathbb{Z})^n : f_i(\mathbf{x}) \equiv 0 \,(\mathrm{mod}\, p^t) \text{ for } 1 \leqslant i \leqslant s \right\}$$

is the p-adic density, for each prime p.

On the other hand, the minor arcs, which are defined to be

$$\mathfrak{m}(Q) = (0, 1]^s \setminus \mathfrak{M}(Q),$$

should produce a smaller overall contribution, with

$$\int_{\mathfrak{m}(Q)} S(\boldsymbol{\alpha})\mathrm{d}\boldsymbol{\alpha} = o(B^{n-\sum_{1\leqslant i\leqslant s} d_i}).$$

We refer the reader to Davenport [27] or Vaughan [71] for a detailed account of the circle method. The fundamental tool here is Weyl's inequality and its generalisations. One such generalisation, which was used to spectacular effect by Heath-Brown [40] to analyse cubic forms in 14 variables, is the method of van der Corput differencing.

Suppose for the moment that $s = 1$ and let $f \in \mathbb{Z}[x_1, \ldots, x_n]$ be a polynomial of degree d. Let $H \in \mathbb{Z} \cap [1, B]$. It will be convenient to write, temporarily,

$$w(\mathbf{x}) = \begin{cases} \mathrm{e}(\alpha f(\mathbf{x})), & \text{if } \mathbf{x} \in B\mathfrak{B}, \\ 0, & \text{otherwise.} \end{cases}$$

Let $\mathscr{H} = \mathbb{Z}^n \cap (0, H]^n$. Then the kernel of the van der Corput method is the observation that

$$\#\mathscr{H} S(\alpha) = \sum_{\mathbf{h}\in\mathscr{H}} \sum_{\mathbf{x}\in\mathbb{Z}^n} w(\mathbf{x}+\mathbf{h}) = \sum_{\mathbf{x}\in\mathbb{Z}^n} \sum_{\mathbf{h}\in\mathscr{H}} w(\mathbf{x}+\mathbf{h}).$$

An application of Cauchy's inequality yields

$$\begin{aligned} H^{2n}|S(\alpha)|^2 &\ll B^n \sum_{\mathbf{x}\in\mathbb{Z}^n} \Big| \sum_{\mathbf{h}\in\mathscr{H}} w(\mathbf{x}+\mathbf{h})\Big|^2 \\ &= B^n \sum_{\mathbf{h}_1\in\mathscr{H}} \sum_{\mathbf{h}_2\in\mathscr{H}} \sum_{\mathbf{x}\in\mathbb{Z}^n} w(\mathbf{x}+\mathbf{h}_1)\overline{w(\mathbf{x}+\mathbf{h}_2)} \\ &= B^n \sum_{\mathbf{h}_1\in\mathscr{H}} \sum_{\mathbf{h}_2\in\mathscr{H}} \sum_{\mathbf{y}\in\mathbb{Z}^n} w(\mathbf{y}+\mathbf{h}_1-\mathbf{h}_2)\overline{w(\mathbf{y})} \\ &= B^n \sum_{\substack{\mathbf{h}\in\mathbb{Z}^n\\ |\mathbf{h}|\leqslant H}} N(\mathbf{h}) \sum_{\mathbf{y}\in\mathbb{Z}^n} w(\mathbf{y}+\mathbf{h})\overline{w(\mathbf{y})}, \end{aligned}$$

where $N(\mathbf{h}) = \#\{\mathbf{h}_1, \mathbf{h}_2 \in \mathscr{H} : \mathbf{h} = \mathbf{h}_1 - \mathbf{h}_2\} \ll H^n$. We therefore conclude that

$$|S(\alpha)|^2 \ll H^{-n}B^n \sum_{\mathbf{h}} |T_{\mathbf{h}}(\alpha)| \ll \frac{B^{2n}}{H^n} + \frac{B^n}{H^n} \sum_{0<|\mathbf{h}|\leqslant H} |T_{\mathbf{h}}(\alpha)|, \tag{4.3}$$

where

$$T_{\mathbf{h}}(\alpha) = \sum_{\substack{\mathbf{x}\in\mathbb{Z}^n\cap B\mathfrak{B}\\ \mathbf{x}+\mathbf{h}\in B\mathfrak{B}}} e\big(\alpha(f(\mathbf{x}+\mathbf{h}) - f(\mathbf{x}))\big).$$

The significance of this is that, for each non-zero $\mathbf{h} \in \mathbb{Z}^n$, the exponential sum $T_{\mathbf{h}}(\alpha)$ is an exponential sum involving the polynomial $f(\mathbf{x}+\mathbf{h}) - f(\mathbf{x})$ of degree $d-1$. Taking $H = B$ recovers the first step in the proof of Weyl's inequality

$$|S(\alpha)|^2 \ll B^n + \sum_{0<|\mathbf{h}|\leqslant B} |T_{\mathbf{h}}(\alpha)|. \tag{4.4}$$

The presence of a free parameter H in (4.3) brings a lot of extra flexibility into the analysis, compared with (4.4). This process can be repeated, using (4.3) or (4.4) at each step, until one arrives at exponential sums of smaller degree that can be estimated efficiently using other methods.

Suppose that $X \subset \mathbb{P}^{n-1}$ is a complete intersection cut out by s hypersurfaces of degrees $d_1, \dots, d_s$, such that X is geometrically integral, non-singular and projective. Assuming that one is fortunate enough to show that the major and minor arc contributions behave as above, with $\prod_v \sigma_v$ absolutely convergent, one can conclude that the Hasse principle holds for X. Indeed, if $X(\mathbb{R}) \neq \emptyset$, one takes $\mathfrak{B}$ to be a small box centred on a (necessarily non-singular) real point. This is enough to ensure that $\sigma_\infty > 0$. Likewise, one can prove that $\prod_p \sigma_p > 0$ if $X(\mathbb{Q}_p) \neq \emptyset$ for every p. Hence $N(B) > 0$ if B is sufficiently large, which implies that $X(\mathbb{Q}) \neq \emptyset$, as desired.

As $\boldsymbol{\alpha}$ runs through elements of $(0, 1]^s$ one might expect the numbers

$$e(\alpha_1 f_1(\mathbf{x}) + \cdots + \alpha_s f_s(\mathbf{x}))$$

to be randomly scattered around the unit circle as we vary over $\mathbf{x} \in \mathbb{Z}^n$. This leads via the central limit theorem to the expectation that $S(\boldsymbol{\alpha})$ should usually be of order roughly $B^{n/2}$. On the other hand, as in (4.2), one heuristically expects the main term in any asymptotic formula to be of order $B^{n-\sum_{1\leqslant i\leqslant s} d_i}$. Thus we can only hope to succeed with the circle method when

$$n > 2 \sum_{1\leqslant i\leqslant s} d_i. \tag{4.5}$$

We call this the square root barrier.

Example 4.2. In most applications we remain very far from the square root barrier. One notable exception is given by considering varieties defined by norm forms. Let $k/\mathbb{Q}$ be a number field of degree $d \geqslant 2$ and fix a basis $\{\omega_1, \dots, \omega_d\}$ for k as a $\mathbb{Q}$-vector space. We denote the norm form by

$$\mathbf{N}_{k/\mathbb{Q}}(x_1, \dots, x_d) = N_{k/\mathbb{Q}}(x_1\omega_1 + \cdots + x_d\omega_d),$$

where $N_{k/\mathbb{Q}}$ denotes the field norm. This is a homogeneous polynomial of degree d defined over $\mathbb{Q}$. Let $X \subset \mathbb{P}^{2d}$ denote the hypersurface

$$cx_0^d + \mathbf{N}_{k_1/\mathbb{Q}}(x_1, \dots, x_d) + \mathbf{N}_{k_2/\mathbb{Q}}(x_{d+1}, \dots, x_{2d}) = 0,$$

for any $c \in \mathbb{Q}^*$. This variety is geometrically integral and projective, but it is singular. Birch, Davenport and Lewis [5] established the smooth Hasse principle for X. Note that this is at the square-root barrier, since the number of variables is $2d + 1$ and the degree is d.

The Birch–Davenport–Lewis result has been generalised to arbitrary number fields independently by Schindler and Skorobogatov [62] and by Swarbrick Jones [68]. Their work covers the hypersurfaces

$$cx_0^d + \mathbf{N}_{K_1/k}(x_1, \dots, x_d) + \mathbf{N}_{K_2/k}(x_{d+1}, \dots, x_{2d}) = 0, \tag{4.6}$$

where K_1, K_2 are degree d extensions of any number field k and $c \in k^*$.

4.1.2 The circle method over k

We proceed to discuss one aspect of the circle method over a general number field k, in the context of hypersurfaces (i.e. $s = 1$). When $k = \mathbb{Q}$ a key innovation, due to Kloosterman [47], involves decomposing $(0, 1]$ in (4.1) using a Farey dissection to keep track of the precise end points of the intervals.

This allows one to introduce non-trivial averaging over the numerators of the approximating fractions a/q, an approach that is usually called the "Kloosterman refinement". This method is not immediately available to us when passing to general k, since aside from work of Cassels, Ledermann and Mahler [17] about the imaginary quadratic fields $\mathbb{Q}(i)$ and $\mathbb{Q}(\sqrt{-3})$, no generalisation is known of the Farey dissection to the number field analogue of (0, 1].

An alternative approach is provided via the smooth δ-function technology of Duke, Friedlander and Iwaniec [32], as revisited by Heath-Brown [39, Thm. 1] and recently extended to general number fields by Browning and Vishe [14]. Suppose $\mathfrak{o}$ is the ring of integers of our number field k, which we assume to have degree d. The ideal norm will be designated $\mathrm{N}\mathfrak{a} = \#\mathfrak{o}/\mathfrak{a}$ for any integral ideal $\mathfrak{a} \subseteq \mathfrak{o}$. In line with our description of the Hardy–Littlewood circle method over $\mathbb{Q}$, we would like to use Fourier analysis to find an alternative description of the indicator function $\delta_k : \mathfrak{o} \to \{0, 1\}$, given by

$$\delta_k(\alpha) = \begin{cases} 1, & \text{if } \alpha = 0, \\ 0, & \text{otherwise.} \end{cases}$$

The following result is a special case of [14, Thm. 1.2].

Theorem 4.3. *Let $Q \geqslant 1$ and let $\alpha \in \mathfrak{o}$. Then there exists a positive constant c_Q and an infinitely differentiable function $h(x, y) : (0, \infty) \times \mathbb{R} \to \mathbb{R}$ such that*

$$\delta_k(\alpha) = \frac{c_Q}{Q^{2d}} \sum_{(0) \neq \mathfrak{b} \subseteq \mathfrak{o}} \sideset{}{^*}\sum_{\sigma \pmod{\mathfrak{b}}} \sigma(\alpha) h\left(\frac{\mathrm{N}\mathfrak{b}}{Q^d}, \frac{N_{k/\mathbb{Q}}(\alpha)}{Q^{2d}}\right),$$

where the notation $\sum^*_{\sigma \pmod{\mathfrak{b}}}$ means that the sum is taken over primitive additive characters modulo $\mathfrak{b}$. The constant c_Q satisfies

$$c_Q = 1 + O_N(Q^{-N}),$$

for any $N > 0$. Furthermore, we have $h(x, y) \ll x^{-1}$ for all y and $h(x, y) \neq 0$ only if $x \leqslant \max\{1, 2|y|\}$.

In [14, Thm 1.1] this result is used to establish that $X(k) \neq \emptyset$, for any non-singular cubic hypersurface $X \subset \mathbb{P}^{n-1}$ defined over k, with $n \geqslant 10$.

Question 4.4. Can one apply Theorem 4.3 to generalise Heath-Brown's work on quadratic forms [39] to arbitrary number fields k?

4.1.3 Complete intersections

Throughout this section let $X_{d,s} \subset \mathbb{P}^{n-1}$ be a complete intersection over a number field k cut out by s hypersurfaces $F_i = 0$ of degree d, with forms $F_i \in k[x_1, \ldots, x_n]$. We will always assume that our complete intersection is geometrically integral and non-degenerate, but it need not be non-singular. When $X_{d,s}$ is non-singular it has been conjectured by Colliot-Thélène (see [58, App. A] and [8, App.]) that the Hasse principle holds when $n + 1 \geqslant ds$ (i.e. $X_{d,s}$ is Fano) and $n - 1 - s \geqslant 3$ (i.e. $X_{d,s}$ has dimension at least 3). We will summarise conditions on d, s and the structure of $X_{d,s}$ under which the smooth Hasse principle has been established for $X_{d,s}$ using the circle method.

The local problem

With the Hasse principle in place for $X = X_{d,s}$ one is naturally led to seek further conditions under which $X(k_v) \neq \emptyset$ for every completion k_v of k. One might expect this purely local problem to be easier, but apart from a few exceptions, the bounds required on n needed to ensure local solubility are rather poor. Let us discuss this briefly in the case $s = 1$ of hypersurfaces. When $d = 2$ we have $X(k_v) \neq \emptyset$ for every non-archimedean place $v \in \Omega_k$, provided only that $n \geqslant 5$. When $d = 3$ work of Lewis [49] (and Dem'yanov [29] when the characteristic of the residue field is distinct from 3) secures the same conclusion provided that $n \geqslant 10$.

For every non-archimedean completion k_v of k and for every $d \geqslant 2$ it is fairly easy to construct examples of hypersurfaces of degree d in d^2 variables which do not have k_v-points. This shows that the above results for $d \in \{2, 3\}$ are best possible. The situation for $d \geqslant 4$ is much less satisfactory. Specialising to hypersurfaces X defined over $\mathbb{Q}$, we know that there is a number v_d such that $X(\mathbb{Q}_p) \neq \emptyset$ for every prime p, as soon as $n > v_d$. Brauer [7] achieved the first result in this direction using an elementary argument based on multiple nested inductions. The resulting value of v_d was too large to write down, but the underlying ideas have since been revisited and improved upon by Wooley [74], with the outcome that we may take $v_d \leq d^{2^d}$ in general. An account of recent developments concerning the local solubility problem for forms in many variables can be found in Heath-Brown's 2010 ICM address [41].

The global problem

We begin by summarising two conditions that any application of the circle method naturally imposes on the complete intersection $X_{d,s}$. The first involves the h-invariant $h(F)$ of a form $F \in k[x_1, \ldots, x_n]$. This is defined to be the least positive integer h such that F can be written identically as

$$A_1B_1 + \cdots + A_hB_h,$$

for forms $A_i, B_i \in k[x_1, \ldots, x_n]$ of positive degree. We define the h-invariant $h(X_{d,s})$ of $X_{d,s}$ to be $\min\{h(F)\}$, with the minimum taken over all forms F of the rational pencil $c_1F_1 + \cdots + c_sF_s$, with $[c_1, \ldots, c_s] \in \mathbb{P}^{s-1}(k)$. A rich seam of results is founded on $h(X_{d,s})$ being "sufficiently large", in terms of d, n and s. The most important contribution adopting this point of view is due to Schmidt [63].

A second point of view requires the singular locus of $X_{d,s}$ to be "sufficiently small". Define the Birch singular locus for $X_{d,s}$ to be the projective variety

$$\{[\mathbf{x}] \in \mathbb{P}^{n-1} : \operatorname{rank}(J(\mathbf{x})) < s\},$$

where $J(\mathbf{x})$ is the Jacobian matrix of size $s \times n$ formed from the gradient vectors $\nabla F_1(\mathbf{x}), \ldots, \nabla F_s(\mathbf{x})$. We define $\Delta \in \{-1, 0 \ldots, n-1\}$ to be the dimension of the Birch singular locus, with $\Delta = -1$ if and only if the variety is empty. There are several places in the literature (see Peyre [55, Cor. 5.4.9], for example) where it has been erroneously assumed that $\Delta = -1$ when $X_{d,s}$ is non-singular. However, this is not true and in general one only has

$$\Delta \leqslant s - 1 + \dim \operatorname{sing}(X_{d,s}),$$

where $\operatorname{sing}(X_{d,s})$ is the singular locus of $X_{d,s}$. This can be seen by intersecting the Birch singular locus with $X_{d,s}$, before applying the projective dimension theorem. When $X_{d,s}$ is non-singular it is possible to improve this inequality slightly. Thus it follows from work of Browning and Heath-Brown [10, Lemma 3.1] that

$$\Delta \leqslant s - 2$$

when $X_{d,s}$ is non-singular. In particular, when $s = 1$ this shows that $\Delta = -1$ when $X_{d,1}$ is non-singular. It is easy to show, moreover, that $\Delta = s - 2$ for any complete intersection cut out by diagonal hypersurfaces.

At this point we should observe that independent works of Dietmann [31] and Schindler [61] allow one to replace Δ by an alternative invariant, which we denote temporarily by Δ'. One can show in complete generality that $\Delta' \leq \Delta$, but that Δ' can be strictly less than Δ in appropriate cases. We shall work with Δ in this survey, however.

We are now ready to recall the *gold standard*, by which all subsequent results have been measured. Let us put

$$\nu_{d,s} = \Delta + 2 + s(s+1)(d-1)2^{d-1}.$$

The following result is due to Birch [4] when $k = \mathbb{Q}$ and to Skinner [65] for general number fields. (See Aleksandrov and Moroz [1] for more on the role of

Table 4.1 *Improvements (via the circle method) to the Birch bound over $\mathbb{Q}$*

(d, s)	$\nu_{d,s}$	$n \geqslant ?$	Conditions?	Who?
(2, 1)	5	3	X non-singular	Heath-Brown [39]
(3, 1)	$\Delta + 18$	16	—	Davenport [26]
		14	—	Heath-Brown [40]
		13	X splits off a form	Browning [8]
		11	X splits off two forms	Xue and Dai [75]
		10	X non-singular	Heath-Brown [38]
		9	X non-singular	Hooley [44]
		9	X contains isolated double points	Hooley [45]
(4, 1)	49	41	X non-singular	Browning and Heath-Brown [9]
		40	X non-singular	Hanselmann [36]
(5, 1)	129	111	X non singular	Browning and Prendiville [13]
(2, 2)	$\Delta + 14$	9	X contains 2 conjugate double points over $\mathbb{Q}(i)$	Browning and Munshi [12]
	14	11	X non-singular	Munshi [54]

Δ in this result.) It is heavily influenced by the pioneering work of Davenport [26] on cubic forms.

Theorem 4.5. *The smooth Hasse principle holds for $X_{d,s}$ if $n \geqslant \nu_{d,s}$.*

We say that a form $F \in k[x_1, \dots, x_n]$ splits off a form if F can be written identically as $A_1 + A_2$, for forms $A_1 \in k[x_1, \dots, x_m]$ and $A_2 \in k[x_{m+1}, \dots, x_n]$ of positive degree, for some $m \in \{1, \dots, n\}$. In Table 4.1 we summarise the improvements that have been made to Theorem 4.5 in the special case $k = \mathbb{Q}$, using the circle method.

Remark. Some of the results in Table 4.1 pertain to non-singular complete intersections $X_{d,s} \subset \mathbb{P}^{n-1}$ with $n \geqslant n_0$, for appropriate n_0 depending on d and s. It is easy to extend these results to establish the smooth Hasse principle for complete intersections in $n \geqslant n_0+1+\sigma$ variables, where σ is the dimension of the singular locus. To do so one argues by induction on $\sigma \geqslant -1$, the case $\sigma = -1$ being represented in the table. If $\sigma \geqslant 0$ one intersects the variety with a generic $\mathbb{Q}$-rational linear space of dimension $n - 2 - \sigma$. Bertini's theorem ensures that the resulting variety is a non-singular complete intersection of codimension s and degree d in $\mathbb{P}^{n-2-\sigma}$, from which the claim easily follows.

The analogue of Table 4.1 for complete intersections over arbitrary number fields k is much smaller, with Theorem 4.5 still providing the best bounds in most cases. One notable exception concerns cubic hypersurfaces $X \subset \mathbb{P}^{n-1}$ defined over k. In this setting a folklore conjecture predicts that $X(k) \neq \emptyset$ as soon as $n \geqslant 10$. When no constraints are placed on the singular locus of X, the work of Pleasants [57] shows that $n \geqslant 16$ variables are needed to achieve this conclusion. When X is non-singular, building on Heath-Brown's [38] breakthrough result when $k = \mathbb{Q}$, Skinner [64] showed that $n \geqslant 13$ variables suffice. The loss of 3 variables is entirely due to the lack of a suitable Kloosterman methodology, a situation that is remedied by Theorem 4.3. Using this, Browning and Vishe [14] have established the conjecture for all non-singular cubic hypersurfaces over any number field. Over $\mathbb{Q}$, as indicated in Table 4.1, Heath-Brown has shown that $n \geqslant 14$ suffices when no restrictions are placed on X, which suggests the following.

Question 4.6. Can Heath-Brown's [40] work on cubic forms in 14 variables be generalised to arbitrary number fields?

So far we have only talked about complete intersections over number fields k which can be handled using the circle method. For certain cases discussed in Table 4.1 significant improvements are available through other channels. Using descent theory, for example, Colliot-Thélène and Salberger [19] have shown that the Hasse principle holds for any cubic hypersurface over k in $\mathbb{P}^{n-1}$ containing a set of three conjugate singular points, provided only that $n \geqslant 4$. As suggested by Example 4.2, the circle method can also be brought to bear on cubics of this type, but the number of variables required is greater. The case $(d, s) = (2, 2)$ of pairs of quadrics is a further important class of varieties that can be handled via descent methods. Assuming that $X \subset \mathbb{P}^{n-1}$ is the common zero locus of two quadratic forms $q_1, q_2 \in k[x_1, \dots, x_n]$, which we assume to be geometrically integral and not a cone, the work of Colliot-Thélène, Sansuc and Swinnerton-Dyer [24] demonstrates that the smooth Hasse principle holds if $n \geqslant 9$. This can be reduced to $n \geqslant 8$ when X is non-singular (see Heath-Brown [42]) or $n \geqslant 5$ when X contains a pair of conjugate singular points and does not belong to a certain explicit class of varieties for which the Hasse principle is known to fail (see [24, Thm. A]).

4.1.4 Recent developments

The technology that lies behind Theorem 4.5 is remarkably robust. In this section we summarise some of the most interesting ways in which it has recently been refined and extended.

Cubics

It turns out that when the singular locus is not too large, cubic hypersurfaces are most efficiently handled by conjoining Theorem 4.3 with Poisson summation, rather than using the division into major and minor arcs from Section 4.1.1 (and the squaring and differencing approach based on (4.3) and (4.4)). This leads to estimates for complete exponential sums over finite fields, which can be estimated very efficiently using Deligne's [28] resolution of the Riemann Hypothesis for varieties over finite fields (however, the methods become less effective as the dimension of the singular locus grows). Deligne's bounds handle sums to prime, or square-free, moduli, but sums to prime power moduli remain a considerable problem. Circumventing these difficulties, Hooley [46] has recently established the Hasse principle for non-singular cubic hypersurfaces over $\mathbb{Q}$ in only 8 variables, provided that one assumes Hypothesis HW. This amounts to some technical (as yet unproved) assumptions about the analytic properties of certain associated Hasse–Weil L-functions.

Improvements for every degree

In Table 4.1 a saving of 9 variables is recorded for non-singular quartic hypersurfaces. Given a quartic form $f \in \mathbb{Z}[x_1, \dots, x_n]$, the idea behind the proof in [9] is to first carry out van der Corput differencing, as in (4.3) (the saving of an extra variable comes through an *averaged* version of this in [36]). The exponential sum $T_{\mathbf{h}}(\alpha)$ now involves the cubic polynomial $f(\mathbf{x}+\mathbf{h}) - f(\mathbf{x})$, which has cubic part equal to $\mathbf{h}.\nabla f(\mathbf{x})$. The idea is then to estimate these exponential sums directly, using Poisson summation, which appears to be the only other weapon we have in our arsenal.

Let $f \in \mathbb{Z}[x_1, \dots, x_n]$ be a non-singular form of degree $d \geqslant 5$. It is natural to ask whether a similar method can be adapted to improve Theorem 4.5 in the setting of hypersurfaces, which we recall establishes the Hasse principle for $n \geqslant \nu_{d,1} = 1+(d-1)2^d$. This is answered affirmatively in forthcoming work of Browning and Prendiville [13], where it is shown that one can take

$$n \geqslant \left(d - \frac{\sqrt{d}}{2}\right) 2^d.$$

The proof of this result involves throwing every available tool at the problem, depending on the Diophantine approximation properties of the number α in the exponential sum $S(\alpha)$. For generic α it is most efficient to apply (4.3) $d-3$ times, producing a family of cubic exponential sums with associated parameters $H_1, \dots, H_{d-3} \in \mathbb{Z} \cap [1, B]$, which one estimates using Poisson summation as in [9]. For other α it is more advantageous to apply Weyl differencing (4.4) multiple times, and for others still, one is better off applying a mixture of the approaches. It turns out that the method is most efficient when d is small. When $d = 5$, as in Table 4.1, one is able to save 18 variables over Birch's theorem.

Forms of differing degree

Theorem 4.5 only applies to complete intersections cut out by hypersurfaces of the same degree. It is not entirely clear how this method can be adapted to handle a system of forms of differing degree, since the process of Weyl differencing (4.4) involved in the proof eradicates the presence of the lower degree forms. Schmidt encounters the same problem in the work [63] cited above. A more efficient treatment of this issue is a key ingredient in work of Browning, Dietmann and Heath-Brown [15], which shows how the circle method can be used to analyse non-singular varieties cut out by a cubic and quadric hypersurface in $n \geqslant 29$ variables. In fact p-adic solubility is guaranteed for n in this range for every p and so one gets $\mathbb{Q}$-points provided only that there are $\mathbb{R}$-points.

Browning and Heath-Brown [10] have looked at generalising Birch's result to complete intersections cut out by hypersurfaces of arbitrary degree. The final result is too complicated to state here, but it has a number of particularly succinct corollaries. First, for a non-singular complete intersection $X \subset \mathbb{P}^{n-1}$ cut out by a hypersurface of degree D and a hypersurface of degree E, where $D > E \geq 2$, the Hasse principle holds for X whenever

$$n > (2 + E)(D - 1)2^{D-1} + E2^{E-1}.$$

In the case of one quadratic and one cubic we find that $n \geq 37$ suffices, which is superseded by the result of Browning, Dietmann and Heath-Brown [15, Thm. 1.3].

Second, suppose that $V \subseteq \mathbb{P}^m$ is a geometrically integral, non-singular and projective variety over $\mathbb{Q}$. Then V satisfies the Hasse principle provided only that

$$\dim(V) \geqslant (\deg(V) - 1)2^{\deg(V)}.$$

Note that when V is a hypersurface this result reduces to Theorem 4.5 over $\mathbb{Q}$. It is worth emphasising here that our hypotheses make no reference to the shape of the defining equations for V. In particular, V is not required to be a complete intersection!

Complete intersections in biprojective space

Birch's theorem is concerned with complete intersections embedded in projective space $\mathbb{P}^{n-1}$. As formalised by Peyre [56], one expects that the methods can be generalised to handle varieties which are embedded inside more general toric varieties. Schindler [60] has carried out this plan for complete intersections in $\mathbb{P}^{m-1} \times \mathbb{P}^{n-1}$. Whereas Birch works with forms of total degree d and applies (4.4) $d-1$ times to obtain linear exponential sums, Schindler used the structure of bihomogeneous forms $F_i(\mathbf{x};\mathbf{y})$ of bidegree (d_1, d_2) to obtain results in which fewer variables are needed than would arise through the most naive adaptation of Birch's work. Indeed, Schindler profits by observing that it suffices to apply the Weyl differencing process $d_1 - 1$ times for the $\mathbf{x}$-variables and $d_2 - 1$ times for the $\mathbf{y}$-variables, leading to a total of only $d_1 + d_2 - 2$ differencing steps.

Question 4.7. Can one generalise Birch's theorem to complete intersections embedded in weighted projective space $\mathbb{P}(a_1, \ldots, a_n)$, where $a_1, \ldots, a_n$ are pairwise coprime positive integers?

Linear spaces on complete intersections

So far we have been concerned with the Hasse principle for complete intersections $X \subset \mathbb{P}^{n-1}$ over number fields k. The Hasse principle is concerned with the existence of k-points on X, or equivalently, the k-rational linear spaces of dimension 0 contained in X. It is natural to ask whether a local–global principle is valid for k-rational linear spaces of arbitrary given dimension $m \geqslant 1$. Brandes [6] has shown how Birch's result can be adapted to handle this kind of question when n is sufficiently large.

Complete intersections over $\mathbb{F}_q[t]$

One can also ask whether the circle method carries over to other global fields. Suppose that $X_{d,s} \subset \mathbb{P}^{n-1}$ is defined over the function field $K = \mathbb{F}_q[t]$. It follows from the Lang–Tsen theorem that $X_{d,s}(K) \neq \emptyset$ whenever $n > sd^2$. Thus the interest here lies with the quantitative distribution of K-points on X. Lee [48] has shown that the proof of Theorem 4.5 can be extended to address this question in the function field setting.

4.2 The circle method and diagonalisation

In this survey we have concentrated on varieties cut out by general polynomials. When the underlying equations have extra additive structure (such as being diagonal) one can usually obtain much sharper results. In this section we content ourselves with summarising what is known for diagonal cubic hypersurfaces $X \subset \mathbb{P}^{n-1}$ defined over a number field k by

$$c_1 x_1^3 + \cdots + c_n x_n^3 = 0, \tag{4.7}$$

for $c_1, \ldots, c_n \in k^*$. We would like conditions on n and the coefficients under which the Hasse principle holds for X. When $k = \mathbb{Q}$ this problem was solved for $n \geqslant 7$, for any non-zero coefficients, by Baker [2] (using the version of the circle method developed by Vaughan [70]). Using completely different methods, when $n = 4$ and k is general (but doesn't contain the primitive cube roots of unity), it follows from work of Swinnerton-Dyer [69] that the Hasse principle holds, subject to (very mild) conditions on the coefficients and the (rather strong) assumption that the Tate–Shafarevich group of every relevant elliptic curve is finite. Using the fibration method, moreover, Colliot-Thélène, Kanevsky and Sansuc [23, Prop. 7] have shown that in order to deduce that the Hasse principle holds for (4.7) for $k = \mathbb{Q}$ and any $n \geqslant 5$, it suffices to show that the Brauer–Manin obstruction is the only obstruction to the Hasse principle for $k = \mathbb{Q}$ and $n = 4$.

As pioneered by Birch [3], it turns out that for any general system of polynomial equations over a number field k one can carry out a diagonalisation process to reduce its analysis to that of an appropriate system of diagonal equations. This idea was refined substantially by Wooley [73], in such a way as to incorporate ideas of Lewis and Schulze-Pillot [50], enabling him to avoid the previous inductive approach. Given positive integers d, s, let $\mu_{d,s}(k)$ denote the least integer (if any such integer exists) with the property that whenever $n > \mu_{d,s}(k)$, and $f_1, \ldots, f_s \in k[x_1, \ldots, x_n]$ are forms of degree d, then the variety $f_1 = \cdots = f_s = 0$ has a k-point. If no such integer exists, we define $\mu_{d,s}(k) = \infty$. Then it follows from [73, Thms. 1 and 2] that

$$\mu_{3,s}(\mathbb{Q}) < (90s)^8 \log^5(27s) \quad \text{and} \quad \mu_{5,s}(\mathbb{Q}) = o(e^{s^{6}}).$$

The bounds for $\mu_{d,s}(\mathbb{Q})$ for odd $d > 5$ are *much* weaker.

As we saw in Section 4.1.3, a great deal of work has been invested in understanding $\mu(k) = \mu_{3,1}(k)$ in the case $(d, s) = (3, 1)$ of cubic hypersurfaces. The best bounds that we have are $\mu(k) < 16$ for any number field k (Pleasants [57]) and $\mu(\mathbb{Q}) < 14$ (Heath-Brown [40]).

Question 4.8. Can one (conditionally) reduce the bound for $\mu(k)$ by combining an efficient diagonalisation argument with the work of Swinnerton-Dyer [69]?

4.3 The circle method and descent

4.3.1 The Brauer–Manin obstruction

We now recall the Brauer–Manin obstruction to the Hasse principle, as introduced by Manin in 1970 using class field theory. Let X be a variety over a number field k that is geometrically integral, non-singular and projective. Let $\mathrm{Br}(X) = H^2_{\text{ét}}(X, \mathbb{G}_m)$ be the associated Brauer group. For any place $v \in \Omega_k$ let

$$\mathrm{inv}_v : \mathrm{Br}(k_v) \to \mathbb{Q}/\mathbb{Z}$$

be the invariant map from local class field theory. For any field $K \supseteq k$, each $\mathscr{A} \in \mathrm{Br}(X)$ gives rise to an evaluation map $\mathrm{ev}_{\mathscr{A}} : X(K) \to \mathrm{Br}(K)$. Global class field theory (the Hasse reciprocity law) gives the exact sequence

$$0 \longrightarrow \mathrm{Br}(k) \longrightarrow \bigoplus_v \mathrm{Br}(k_v) \xrightarrow{\sum_v \mathrm{inv}_v} \mathbb{Q}/\mathbb{Z} \longrightarrow 0. \tag{4.8}$$

We then have the commutative diagram

$$\begin{array}{ccccc} X(k) & \longrightarrow & X(\mathbf{A}_k) & & \\ \downarrow{\scriptstyle \mathrm{ev}_{\mathscr{A}}} & & \downarrow{\scriptstyle \mathrm{ev}_{\mathscr{A}}} & & \\ \mathrm{Br}(k) & \longrightarrow & \bigoplus_v \mathrm{Br}(k_v) & \longrightarrow & \mathbb{Q}/\mathbb{Z} \end{array}$$

where $X(\mathbf{A}_k) = \prod_v X(k_v)$. Let $\Theta_{\mathscr{A}} : X(\mathbf{A}_k) \to \mathbb{Q}/\mathbb{Z}$ denote the composed map. Then it follows that the image of $X(k)$ in $X(\mathbf{A}_k)$ under the diagonal embedding is contained in $\ker \Theta_{\mathscr{A}}$ for all $\mathscr{A} \in \mathrm{Br}(X)$. We call

$$X(\mathbf{A}_k)^{\mathrm{Br}} = \bigcap_{\mathscr{A} \in \mathrm{Br}(X)} \ker \Theta_{\mathscr{A}}$$

the Brauer set. This gives an obstruction to the Hasse principle, since if $X(\mathbf{A}_k)^{\mathrm{Br}}$ is empty we cannot expect X to have k-rational points. We say that the Brauer–Manin obstruction to the Hasse principle is the only one if $X(\mathbf{A}_k)^{\mathrm{Br}} \neq \emptyset$ implies that $X(k) \neq \emptyset$.

In seeking to determine whether there is a Brauer–Manin obstruction it is enough to consider elements $\mathscr{A}$ belonging to the quotient $\mathrm{Br}(X)/\mathrm{Br}(k)$. Indeed it follows from the exact sequence (4.8) that elements of $\mathrm{Br}(k)$ are orthogonal to any adelic point under the pairing $X(\mathbf{A}_k) \times \mathrm{Br}(X) \to \mathbb{Q}/\mathbb{Z}$.

Example 4.9. Taking up the example of generalised Châtelet surfaces X from Example 4.1, let $p(T)$ be any monic irreducible factor of $f(T)$. If a is a square in $k_p = k[T]/(p(T))$ then X is k-rational and it follows that $\mathrm{Br}(X)/\mathrm{Br}(k) = 0$. Let n denote the number of monic irreducible factors $p(T)$ of $f(T)$ such that $a \notin k_p^2$. Then $\mathrm{Br}(X)/\mathrm{Br}(k) \cong (\mathbb{Z}/2\mathbb{Z})^{n-1}$ if either all such $p(T)$ have even degree and $\deg f(T)$ is even or else if $\deg f(T)$ is odd. Alternatively, if $\deg f(T)$ is even, but some $p(T)$ with $a \notin k_p^2$ has odd degree, then $\mathrm{Br}(X)/\mathrm{Br}(k) \cong (\mathbb{Z}/2\mathbb{Z})^{n-2}$.

In particular, $\mathrm{inv}_v\,(\mathrm{ev}_{\mathscr{A}}(M_v))$ takes the values 0 or 1/2 in $\mathbb{Q}/\mathbb{Z}$, for any $(M_v) \in X(\mathbf{A}_k)$. An obstruction to the Hasse principle arises if and only if $X(\mathbf{A}_k)^{\mathrm{Br}} = \emptyset$, that is to say, if and only if for any $(M_v) \in X(\mathbf{A}_k)$ there exists $\mathscr{A} \in \mathrm{Br}(X)$ such that

$$\sum_v \mathrm{inv}_v\,(\mathrm{ev}_{\mathscr{A}}(M_v)) = \frac{1}{2}. \tag{4.9}$$

In order to obtain counter-examples to the Hasse principle, it will be necessary for $\mathrm{inv}_v\,(\mathrm{ev}_{\mathscr{A}}(M_v))$ to be constant for every valuation v. Indeed, if there is a valuation v for which $\mathrm{inv}_v\,(\mathrm{ev}_{\mathscr{A}}(M_v))$ takes both values 0 and 1/2, as a function of M_v, then (4.9) is clearly impossible. This can be used to explain the failure of the Hasse principle for the surfaces $X_c/\mathbb{Q}$ considered in Example 4.1, where $\mathrm{Br}(X_c)/\mathrm{Br}(\mathbb{Q}) \cong \mathbb{Z}/2\mathbb{Z}$ and a generator is given by the quaternion algebra $(-1, T^2 - c + 1)$ (see [21, §5] for details).

The main conjecture in this field is the following one due to Colliot-Thélène (see [18]).

Conjecture 4.10. *The Brauer–Manin obstruction is the only obstruction to the Hasse principle for any geometrically integral, non-singular and projective variety which is geometrically rationally connected.*

The class of geometrically rationally connected varieties covers all Fano varieties and all generalised Châtelet surfaces. It would appear that the circle method has not yet been made to say anything significant about varieties which are not geometrically rationally connected.

4.3.2 Descent varieties

A powerful and very general descent machinery has been developed by Colliot-Thélène and Sansuc [20]. In favourable circumstances, this allows one to reduce the task of establishing Conjecture 4.10 to proving the Hasse principle for some associated descent varieties. We refer the reader to Skorobogatov [67] for a general account of the theory. A fruitful development in the last decade has been the use of the circle method to analyse these descent varieties. It is this aspect that we proceed to discuss here.

One particularly rich seam of varieties arises through the Diophantine equations

$$P(t) = \mathbf{N}_{K/k}(x_1, \ldots, x_n), \tag{4.10}$$

defined over a number field k, where $P(t) \in k[t]$ is a polynomial and $\mathbf{N}_{K/k}$ is a norm form associated to an arbitrary degree n extension K/k (as in Example 4.2). Let X be a smooth and projective model of the affine variety in $\mathbb{A}^{n+1}$ that (4.10) defines. Then X is geometrically rational and so is covered by Conjecture 4.10. Note that when K/k is quadratic this is an example of the generalised Châtelet surface that we met in Example 4.1. In particular we have already seen that X need not satisfy the Hasse principle in general.

It turns out that understanding the Hasse principle for X becomes harder as the number of distinct roots of $P(t)$ grows larger. When $P(t)$ has at most one distinct root (so that either $P(t)$ is constant or else $P(t) = ct^d$ for some $c \in k^\times$ and $d \in \mathbb{Z}_{>0}$), the affine variety (4.10) is a principal homogeneous space for an algebraic k-torus, and the work of Sansuc [59] and Voskresenskiĭ [72] establishes Conjecture 4.10 for X. When $P(t)$ has precisely two distinct roots, both of which are defined over k, it can be written

$$P(t) = ct^{d_1}(t-1)^{d_2}, \tag{4.11}$$

after a possible change of variables. In this case there are obvious rational points on the affine variety (4.10), coming from the roots of $P(t)$. However, these will be singular as soon as $\min\{d_1, d_2\} > 1$. When $P(t)$ is given by (4.11), it turns out that the descent varieties involve equations of the shape (4.6), which can be analysed using the circle method. This programme was first carried out by Heath-Brown and Skorobogatov [43] when $k = \mathbb{Q}$ and $\gcd(n, d_1, d_2) = 1$, but both of these conditions have subsequently been removed (see [22, 62, 68]). For other work on the arithmetic of (4.10), together with a survey of the literature, we refer the reader to Derenthal, Smeets and Wei [30] and the references therein.

We proceed to record some descent varieties associated to two particular families of varieties. Example 4.11 has recently been solved using additive combinatorics (see Section 4.3.3). Example 4.12, on the other hand, is still wide open.

Example 4.11. Working over $\mathbb{Q}$, suppose that P takes the form

$$P(t) = c\prod_{i=1}^{r}(t-e_i)^{m_i},$$

for $c \in \mathbb{Q}^*$, $m_1, \ldots, m_r \in \mathbb{Z}_{>0}$ and pairwise distinct $e_1, \ldots, e_r \in \mathbb{Q}$. This generalises the example (4.11) that we met above. Let π be the projection of the affine variety given by (4.10) to $\mathbb{A}^1$, and let $U_0 \subset \mathbb{A}^1$ be the open subset on which $\prod_{i=1}^r (t-e_i) \neq 0$. Let $U = \pi^{-1}(U_0)$ and let $T = R^1_{K/\mathbb{Q}}$ be the torus given by the affine equation $\mathbf{N}_{K/\mathbb{Q}}(\mathbf{x}) = 1$. It is possible to construct a partial compactification Y of U and show that vertical torsors $\mathscr{T} \to Y$ exist (see Colliot-Thélène, Harari and Skorobogatov [22] and Schindler and Skorobogatov [62, §2]). These are torsors $\mathscr{T} \to Y$ whose type is the injective map of $\mathrm{Gal}(\overline{\mathbb{Q}}/\mathbb{Q})$-modules $\widehat{T}^r \to \mathrm{Pic}(\overline{Y})$. It follows from [62, Lemma 2.2] that the restriction $\mathscr{T}_U$ of $\mathscr{T}$ to $U \subset Y$ is $E \times V$, where E is a principal homogeneous space for T and $V \subset \mathbb{A}^{nr+1}$ is defined by

$$t - e_i = \lambda_i \mathbf{N}_{K/\mathbb{Q}}(\mathbf{x}_i) \neq 0, \quad (1 \leqslant i \leqslant r),$$

for $\lambda_1, \dots, \lambda_r \in \mathbb{Q}^*$. This relies crucially on the local description of torsors laid down by Colliot-Thélène and Sansuc [20, Thm. 2.3.1]. Finally, it follows from [62, Thm. 2.1] that in order to establish Conjecture 4.10 for a smooth and projective model of (4.10), it suffices to show that V satisfies the Hasse principle. But V is isomorphic to the variety cut out by the system of equations

$$e_1 - e_i = \lambda_i \mathbf{N}_{K/\mathbb{Q}}(\mathbf{x}_i) - \lambda_1 \mathbf{N}_{K/\mathbb{Q}}(\mathbf{x}_1), \quad (2 \leqslant i \leqslant r).$$

By an obvious change of variables it suffices to establish the Hasse principle for the variety in $\mathbb{A}^{nr+r}$ defined by the system of equations

$$0 \neq (e_1 - e_i)\mathbf{N}_{K/\mathbb{Q}}(\mathbf{y}) = \lambda_i \mathbf{N}_{K/\mathbb{Q}}(\mathbf{x}_i) - \lambda_1 \mathbf{N}_{K/\mathbb{Q}}(\mathbf{x}_1), \quad (2 \leqslant i \leqslant r).$$

But this variety is isomorphic to the non-singular variety $W \subset \mathbb{A}^{n(r+1)+2}$ given by

$$\begin{aligned} v &= \mathbf{N}_{K/\mathbb{Q}}(\mathbf{y}) \neq 0, \\ u - e_i v &= \lambda_i \mathbf{N}_{K/\mathbb{Q}}(\mathbf{x}_i) \neq 0, \quad (1 \leqslant i \leqslant r). \end{aligned} \tag{4.12}$$

We shall return to this example in Section 4.3.3.

Example 4.12. Suppose that $X \subset \mathbb{P}^3$ is a non-singular diagonal cubic surface defined over k, with equation

$$a_1 x_1^3 + \cdots + a_4 x_4^3 = 0,$$

for $a_1, \dots, a_4 \in k^*$. When $k = \mathbb{Q}$, Colliot-Thélène, Kanevsky and Sansuc [23] have undertaken an extensive investigation of the Brauer–Manin obstruction for X. Assuming without loss of generality that $a_1, \dots, a_4 \in \mathbb{Z}$ are cube-free, it follows from this work that the Brauer–Manin obstruction to the Hasse principle is empty if there is a prime p which divides precisely one of the coefficients.

It is still an open problem to establish Conjecture 4.10 unconditionally for diagonal cubic surfaces X. One approach would involve establishing the Hasse principle for some associated intermediate torsors (see [23, Prop. 10]). These possess a Zariski open subset which is $\mathbb{Q}$-isomorphic to an intersection of two cubics in $\mathbb{A}^9$ involving norm forms (see [23, §10] for details). Over $\overline{\mathbb{Q}}$ these take the shape

$$ax_1x_2x_3 + by_1y_2y_3 + cz_1z_2z_3 = 0$$
$$dx_1y_1z_1 + ex_2y_2z_2 + fx_3y_3z_3 = 0,$$

for non-zero $a, \ldots, f \in \overline{\mathbb{Q}}$. Alas this system is beyond the square-root barrier (4.5) and so it seems unlikely that the circle method could be used to say anything useful in its present form.

Let $\mathscr{C}$ denote the family of all systems of Diophantine equations over the ring of integers $\mathfrak{o}_k$ of a number field k, for which the circle method has proved the Hasse principle for the system. Let $\mathscr{T}$ denote the family of varieties which are either affine space or a principal homogeneous space for a torus.

Question 4.13. What is the most general class of geometrically integral, non-singular and projective varieties, for which universal torsors exist and are k-birational to a product of varieties from $\mathscr{C}$ and $\mathscr{T}$?

In Example 4.11, according to [22] and [62], any smooth and projective model of (4.10) with $P(t) = ct^{d_1}(t-1)^{d_2}$ belongs to the class of varieties considered in Question 4.13. Indeed, the universal torsor in this case is k-birational to a product of V and a principal homogeneous space for the torus $\mathbf{N}_{K/k}(\mathbf{x}) = 1$.

4.3.3 The nilpotent circle method

Let $h : \mathbb{Z} \to \mathbb{R}$ be an arithmetic function and let $L_1, \ldots, L_r \in \mathbb{Z}[u, v]$ be a system of binary linear forms, no two of which are proportional. A powerful new development concerns the ability (at least in principle) to produce asymptotic formulae for sums of the shape

$$\sum_{(u,v)\in\mathbb{Z}^2\cap B\mathfrak{B}} h(L_1(u, v)) \ldots h(L_r(u, v)), \tag{4.13}$$

as $B \to \infty$, for any bounded convex region $\mathfrak{B} \subset \mathbb{R}^2$. The underlying technology comes from additive combinatorics and is due to Green and Tao [33] and Green, Tao and Ziegler [35], where it is used to tackle this problem when $h = \Lambda$ is the von Mangoldt function. The term nilpotent circle method originates from [34] and refers to the methods and concepts that were developed in the course of this work.

So far, the analysis of (4.13) has been carried out for a fairly restrictive family of arithmetic functions h. Matthiesen [52, 53] has done so when $h(N)$ is the function which counts the number of (inequivalent) representations of $N \in \mathbb{Z}$ by a given binary quadratic form. It was noticed by Browning, Matthiesen and Skorobogatov [16] that, once unravelled in terms of counting solutions to a system of underlying Diophantine equations, this is exactly what is needed to establish the Hasse principle for the variety W in (4.12), in the special case $n = 2$ of quadratic extensions $K/\mathbb{Q}$. This therefore yields a proof of Conjecture 4.10 over $\mathbb{Q}$, for the generalised Châtelet surfaces considered in Examples 4.1 and 4.9, provided that the polynomial $f(T)$ decomposes completely as a product of linear polynomials with coefficients in $\mathbb{Q}$. The significance of this is that there was previously no *unconditional* proof of Conjecture 4.10 for generalised Châtelet surfaces in which $f(T)$ has $r > 4$ distinct roots (the case $r \leqslant 4$ is covered by [24, 25]).

Very recently, Browning and Matthiesen [11] have used the nilpotent circle method to generalise this considerably. Here the Hasse principle is established for the variety W in (4.12), without any restriction on the degree or type of the number field K (note that when $K/\mathbb{Q}$ is cyclic, Harpaz, Skorobogatov and Wittenberg [37] have shown how to arrive at the same conclusion by combining the fibration method with the work of Green and Tao [33] and Green, Tao and Ziegler [35] on the generalised Hardy–Littlewood primes conjecture). A consequence of this is that Conjecture 4.10 holds for the varieties X considered in Example 4.11 for *any* $r \geqslant 2$ (generalising the case $r = 2$ handled in [43]). Using descent, Skorobogatov [66] has shown how the main result in [11] can be used to study the Diophantine equation (4.10) when the norm form on the right hand side is replaced by a product of norm forms.

Eliminating u, v from (4.12) the variety W is seen to be $\mathbb{Q}$-birational to the system of equations

$$\mathbf{N}_{K/\mathbb{Q}}(\mathbf{x}_i) + b_i\mathbf{N}_{K/\mathbb{Q}}(\mathbf{y}) = c_i\mathbf{N}_{K/\mathbb{Q}}(\mathbf{z}), \quad (1 \le i \le r-1),$$

for suitable $b_i, c_i \in \mathbb{Q}$. This can be viewed as a variety in $\mathbb{P}^{(r+1)n-1}$ cut out by $r-1$ hypersurfaces of degree n. The main result in [11] shows that the smooth Hasse principle holds for this variety. By contrast, the inequality (4.5) becomes $(r+1)n > 2n(r-1)$, which is true if and only if $3 > r$. The significance of [11], therefore, is that it goes *beyond* the square-root barrier.

References

[1] A.G. Aleksandrov and B.Z. Moroz. Complete intersections in relation to a paper of B. J. Birch. *Bull. London Math. Soc.* **34** (2002), 149–154.

[2] R.C. Baker, Diagonal cubic equations, II. *Acta Arith.* **53** (1989) 217–250.

[3] B.J. Birch, Homogeneous forms of odd degree in a large number of variables. *Mathematika* **4** (1957), 102–105.

[4] B.J. Birch, Forms in many variables. *Proc. Roy. Soc. Ser. A* **265** (1961), 245–263.

[5] B.J. Birch, H. Davenport and D.J. Lewis, The addition of norm forms. *Mathematika* **9** (1962), 75–82.

[6] J. Brandes, Forms representing forms and linear spaces on hypersurfaces. *Proc. London Math. Soc.* **108** (2014), 809–835.

[7] R. Brauer, A note on systems of homogeneous algebraic equations. *Bull. Amer. Math. Soc.* **51** (1945), 749–755.

[8] T.D. Browning, Rational points on cubic hypersurfaces that split off a form. *Compositio Math.* **146** (2010), 853–885.

[9] T.D. Browning and D.R. Heath-Brown, Rational points on quartic hypersurfaces. *J. reine angew. Math.* **629** (2009), 37–88.

[10] T.D. Browning and D.R. Heath-Brown, Forms in many variables and differing degrees. *J. Eur. Math. Soc.*, to appear.

[11] T.D. Browning and L. Matthiesen, Norm forms for arbitrary number fields as products of linear polynomials. *Submitted*, 2013. (`arXiv:1307.7641`)

[12] T.D. Browning and R. Munshi, Rational points on singular intersections of quadrics. *Compositio Math.* **149** (2013), 1457–1494.

[13] T.D. Browning and S. Prendiville, Improvements in Birch's theorem on forms in many variables. *J. reine angew. Math.*, to appear.

[14] T.D. Browning and P. Vishe, Cubic hypersurfaces and a version of the circle method for number fields. *Duke Math. J.* **183** (2014), 1825–1883.

[15] T.D. Browning, R. Dietmann, D.R. Heath-Brown, Rational points on intersections of cubic and quadric hypersurfaces. *J. Inst. Math. Jussieu*, to appear.

[16] T.D. Browning, L. Matthiesen and A.N. Skorobogatov, Rational points on pencils of conics and quadrics with many degenerate fibres. *Annals of Math.* **180** (2014), 381–402.

[17] J.W.S. Cassels, W. Ledermann and K. Mahler, Farey section in $k(i)$ and $k(\varrho)$. *Philos. Trans. Roy. Soc. London. Ser. A.* **243** (1951), 585–626.

[18] J.-L. Colliot-Thélène. Points rationnels sur les fibrations. *Higher dimensional varieties and rational points (Budapest, 2001)*, 171–221, Springer-Verlag, 2003.

[19] J.-L. Colliot-Thélène and P. Salberger, Arithmetic on some singular cubic hypersurfaces. *Proc. London Math. Soc.* **58** (1989), 519–549.

[20] J.-L. Colliot-Thélène and J.-J. Sansuc, La descente sur les variétés rationnelles, II. *Duke Math. J.* **54** (1987), 375–492.

[21] J.-L. Colliot-Thélène, D. Coray and J.-J. Sansuc, Descente et principe de Hasse pour certaines variétés rationnelles. *J. reine angew. Math.* **320** (1980), 150–191.

[22] J.-L. Colliot-Thélène, D. Harari and A.N. Skorobogatov, Valeurs d'un polynôme à une variable représentés par une norme. *Number theory and algebraic geometry*, 69–89, London Math. Soc. Lecture Note Ser. **303** Cambridge University Press, 2003.

[23] J.-L. Colliot-Thélène, D. Kanevsky and J.-J. Sansuc, Arithmétique des surfaces cubiques diagonales. *Diophantine approximation and transcendence theory*, 1–108, Springer Lecture Notes Math. **1290** Springer-Verlag, 1987.

[24] J.-L. Colliot-Thélène, J.-J. Sansuc and P. Swinnerton-Dyer, Intersections of two quadrics and Châtelet surfaces, I. *J. reine angew. Math.* **373** (1987), 37–107.

[25] J.-L. Colliot-Thélène, J.-J. Sansuc and P. Swinnerton-Dyer, Intersections of two quadrics and Châtelet surfaces, II. *J. reine angew. Math.* **374** (1987), 72–168.

[26] H. Davenport, Cubic forms in 16 variables. *Proc. Roy. Soc. A* **272** (1963), 285–303.

[27] H. Davenport, *Analytic methods for Diophantine equations and Diophantine inequalities*. 2nd ed., edited by T.D. Browning, Cambridge University Press, 2005.

[28] P. Deligne, La conjecture de Weil, I. *Inst. Hautes Études Sci. Publ. Math.* **43** (1974), 273–307.

[29] V.B. Dem'yanov, On cubic forms in discretely normed fields. (Russian) *Doklady Akad. Nauk SSSR* **74** (1950), 888–891.

[30] U. Derenthal, A. Smeets and D. Wei, Universal torsors and values of quadratic polynomials represented by norms. *Math. Annalen*, to appear.

[31] R. Dietmann, Weyl's inequality and systems of forms, *Quart. J. Math.* **66** (2015), 97–110.

[32] W. Duke, J.B. Friedlander and H. Iwaniec, Bounds for automorphic L-functions. *Invent. Math.* **112** (1993), 1–8.

[33] B. Green and T. Tao, Linear equations in primes. *Annals of Math.* **171** (2010), 1753–1850.

[34] B. Green and T. Tao, The Möbius function is strongly orthogonal to nilsequences. *Annals of Math.* **175** (2012), 541–566.

[35] B. Green, T. Tao and T. Ziegler, An inverse theorem for the Gowers $U_{s+1}[N]$-norm. *Annals of Math.* **176** (2012), 1231–1372.

[36] M. Hanselmann, Rational points on quartic hypersurfaces. *Ph.D. thesis*, Ludwig Maximilians Universität Munchen, 2012.

[37] J. Harpaz, A.N. Skorobogatov and O. Wittenberg, The Hardy–Littlewood conjecture and rational points. *Compositio Math.* **150** (2014), 2095–2111.

[38] D.R. Heath-Brown, Cubic forms in ten variables. *Proc. London Math. Soc.* **47** (1983), 225–257.

[39] D.R. Heath-Brown, A new form of the circle method, and its application to quadratic forms. *J. reine angew. Math.* **481** (1996), 149–206.

[40] D.R. Heath-Brown, Cubic forms in 14 variables. *Invent. Math.* **170** (2007), 199–230.

[41] D.R. Heath-Brown, Artin's conjecture on zeros of p-adic forms. *Proceedings of the International Congress of Mathematicians. Vol. II*, 249–257, Hindustan Book Agency, New Delhi, 2010.

[42] D.R. Heath-Brown, Zeros of pairs of quadratic forms. *Submitted*, 2013. (`arXiv:1304.3894`)

[43] D.R. Heath-Brown and A.N. Skorobogatov, Rational solutions of certain equations involving norms. *Acta Math.* **189** (2002), 161–177.

[44] C. Hooley, On nonary cubic forms. *J. reine angew. Math.* **386** (1988), 32–98.

[45] C. Hooley, On nonary cubic forms: IV. *J. reine angew. Math.* **680** (2013), 23–39.

[46] C. Hooley, On octonary cubic forms. *Proc. London Math. Soc.* **109** (2014), 241–281.

[47] H.D. Kloosterman, On the representation of numbers in the form $ax^2 + by^2 + cz^2 + dt^2$. *Acta. Math.* **49** (1926), 407–464.

[48] S.-L. A. Lee, Birch's theorem in function fields. *Submitted*, 2012. (arXiv:1109.4953)

[49] D.J. Lewis, Cubic homogeneous polynomials over $\mathfrak{p}$-adic fields. *Annals of Math.* **56** (1952), 473–478.

[50] D.J. Lewis and R. Schulze-Pillot, Linear spaces on the intersection of cubic hypersurfaces. *Monatsh. Math.* **97** (1984), 277–285.

[51] Y.I. Manin, Le groupe de Brauer–Grothendieck en géométrie diophantienne. *Actes du Congrès International des Mathématiciens (Nice, 1970)*. Tome 1, 401–411, Gauthier-Villars, Paris, 1971.

[52] L. Matthiesen, Linear correlations amongst numbers represented by positive definite binary quadratic forms. *Acta Arith.* **154** (2012), 235–306.

[53] L. Matthiesen, Correlations of representation functions of binary quadratic forms. *Acta Arith.* **158** (2013), 245–252.

[54] R. Munshi, Pairs of quadrics in 11 variables. *Compositio Math.*, to appear.

[55] E. Peyre, Hauteurs et mesures de Tamagawa sur les variétiés de Fano. *Duke Math. J.* **79** (1995), 101–218.

[56] E. Peyre, Torseurs universels et méthode du cercle. *Rational points on algebraic varieties*, 221–274, Progr. Math. **199**, Birhäuser, 2001.

[57] P.A.B. Pleasants, Cubic polynomials over algebraic number fields. *J. Number Theory* **7** (1975), 310–344.

[58] B. Poonen and J.F. Voloch, Random Diophantine equations. *Arithmetic of higher-dimensional algebraic varieties (Palo Alto, CA, 2002)*, 175–184, Progr. Math. **226**, Birkhäuser, 2004.

[59] J.-J. Sansuc, Groupe de Brauer et arithmétique des groupes algébriques linéaires sur un corps de nombres. *J. reine angew. Math.* **327** (1981), 12–80.

[60] D. Schindler, Bihomogeneous forms in many variables. *J. Théorie Nombres Bordeaux* **26** (2014), 483–506.

[61] D. Schindler, A variant of Weyl's inequality for systems of forms and applications. To be submitted, arXiv:1403.7156.

[62] D. Schindler and A.N. Skorobogatov, Norms as products of linear polynomials. *J. London Math. Soc.* **89** (2014), 559–580.

[63] W. Schmidt, The density of integer points on homogeneous varieties. *Acta Math.* **154** (1985), 243–296.

[64] C.M. Skinner, Rational points on nonsingular cubic hypersurfaces. *Duke Math. J.* **75** (1994), 409–466.

[65] C.M. Skinner, Forms over number fields and weak approximation. *Compositio Math.* **106** (1997), no. 1, 11–29.

[66] A.N. Skorobogatov, Descent on toric fibrations. This volume, pp. 422–435.

[67] A.N. Skorobogatov, *Torsors and rational points*. Cambridge University Press, 2001.

[68] M. Swarbrick Jones, A note on a theorem of Heath-Brown and Skorobogatov. *Quart. J. Math.* **64** (2013), 1239–1251.

[69] P. Swinnerton-Dyer, The solubility of diagonal cubic surfaces. *Annales Sci. École Normale Sup.* **34** (2001), 891–912.

[70] R.C. Vaughan, On Waring's problem for cubes. *J. reine angew. Math.* **365** (1986) 122–170.

[71] R.C. Vaughan, *The Hardy-Littlewood method.* 2nd ed., Cambridge Tracts in Mathematics **125**, Cambridge University Press, 1997.

[72] V.E. Voskresenskiĭ, *Algebraic groups and their birational invariants.* Translations of Math. Monographs **179**, American Math. Soc., 1998.

[73] T.D. Wooley, Forms in many variables. *Analytic number theory (Kyoto, 1996)*, 361–376, London Math. Soc. Lecture Note Ser. **247**, Cambridge University Press, 1997.

[74] T.D. Wooley, On the local solubility of Diophantine systems. *Compositio Math.* **111** (1998), 149–165.

[75] B. Xue and H. Dai, Rational points on cubic hypersurfaces that split off two forms. *Bull. London Math. Soc.* **46** (2014), 169–184.

5

Arithmetic differential equations of Painlevé VI type

Alexandru Buium[1] *and Yuri I. Manin*[2]
[1]University of New Mexico, Albuquerque
[2]Max–Planck–Institut für Mathematik, Bonn, Germany

ABSTRACT. Using the description of Painlevé VI family of differential equations in terms of a universal elliptic curve, going back to R. Fuchs (cf. [Ma96]), we translate it into the realm of Arithmetic Differential Equations (cf. [Bu05]), where the role of derivative "in the p-adic direction" is played by a version of Fermat quotient.

Introduction and brief summary

This article is dedicated to the study of differential equations of the Painlevé VI type "with p-adic time", in which the role of time derivative is played by a version of p-Fermat quotient. Richness of p-adic differential geometry was already demonstrated in many papers of the first-named author: see in particular the monograph [Bu05].

Applicability of this technique to the Painlevé equations is ensured by the combination of two approaches: R. Fuchs's treatment of the classical case modernized in [Ma96], and constructions of p-adic differential characters in [Bu95].

In sec. 5.1 below we present a short introduction to the p-adic differential geometry. In sec. 5.2 and 5.3 we introduce several versions of p-adic PVI. Sec. 5.4 is dedicated to the problem of transposing into p-adic domain of main features of Hamiltonian formalism. We have not found a definitive answer to this problem, and Painlevé equations serve here as a very stimulating testing ground. Finally, sec. 5.5 treats another problem, suggested by comparison of several versions of p-adic PVI, but presenting an independent interest.

Arithmetic and Geometry, ed. Luis Dieulefait *et al.* Published by Cambridge University Press.

Acknowledgement. This article was conceived during the Spring 2013 Trimester program "Arithmetic and Geometry" of the Hausdorff Institute for Mathematics (HIM), Bonn. The authors are grateful to HIM for a stimulating atmosphere and working conditions.

5.1 Arithmetic differential equations: background

We start with a brief summary of relevant material from [Bu05], Chapters 2, 3. Fix a prime p; in our applications it will be assumed that $p \geq 5$.

5.1.1 p-derivations

Recall that, in the conventional commutative algebra, given a ring A and an A-module N, *a derivation of A with values in N* is any map $\partial : A \to N$ such that the map $A \to A \times N : a \mapsto (a, \partial a)$ is a ring homomorphism, where $A \times N$ is endowed with the structure of commutative ring with componentwise addition, and multiplication $(a, m) \cdot (b, n) := (ab, an + bm)$. Notice that $\{0\} \times N$ is the ideal with square zero in $A \times N$.

Similarly, in arithmetic geometry *a p-derivation of A with values in an A-algebra B*, $f : A \to B$, is a map $\delta_p : A \to B$ such that the map $A \to B \times B : a \mapsto (f(a), \delta_p(a))$ is a ring homomorphism $A \to W_2(B)$ where $W_2(B)$ is the ring of p-typical Witt vectors of length 2. Again, if $pB = \{0\}$, Witt vectors of the form $(0, b)$ form the ideal of square zero.

Explicitly, this means that $\delta_p(1) = 0$, and

$$\delta_p(x + y) = \delta_p(x) + \delta_p(y) + C_p(x, y), \tag{5.1}$$

$$\delta_p(xy) = f(x)^p \cdot \delta_p(y) + f(y)^p \cdot \delta_p(x) + p \cdot \delta_p(x) \cdot \delta_p(y), \tag{5.2}$$

where

$$C_p(X, Y) := \frac{X^p + Y^p - (X + Y)^p}{p} \in \mathbf{Z}[X, Y]. \tag{5.3}$$

In particular, this implies that for any p-derivation $\delta_p : A \to B$ the respective map $\phi_p : A \to B$ defined by $\phi_p(a) := f(a)^p + p\delta_p(a)$ is a ring homomorphism satisfying $\phi_p(x) \equiv f(x)^p \bmod p$, that is "a lift of the Frobenius map applied to f".

Conversely, having such a lift of Frobenius, we can uniquely reconstruct the respective derivation δ_p *under the condition that B has no p-torsion.*

We will often work with p-derivations $A \to A$ with respect to the identity map $A \to A$ and keep p fixed; then it might be kept off the notation. Such a pair (A, δ) is called a δ-ring. Morphisms of δ-rings are algebra morphisms compatible with their p-derivations.

Moreover, our rings (and more generally, schemes) will be R-algebras where $R = W(k)$ (ring of infinite p-typical Witt vectors) is the completion of the maximal unramified extension of $\mathbf{Z}_p$, with residue field $k :=$ an algebraic closure of $\mathbf{Z}/p\mathbf{Z}$. By $\phi : R \to R$ we denote the automorphism acting as Frobenius $x \mapsto x^p$ on k, and by δ the respective p-derivation: $\delta(x) = (\phi(x) - x^p)/p$. The R-algebra structure on a δ-ring is always assumed to be compatible with this p-derivation.

5.1.2 Prolongation sequences and p-jet spaces

In the classical situation invoked in 1.1, there exists an universal derivation

$$d : A \to \Omega^1(A) \tag{5.4}$$

with values in the A-module of differentials.

For p-derivations, (5.4) might be replaced by the following construction (however, see the Remark 5.1 below).

Let A be an R-algebra. A *prolongation sequence* for A consists of a family of p-adically complete R-algebras $A^i, i \geq 0$, where $A^0 = A\hat{}$ is the p-adic completion of A, and of maps $\varphi_i, \delta_i : A^i \to A^{i+1}$ satisfying the following conditions:

a) *φ_i are ring homomorphisms, each δ_i is a p-derivation with respect to φ_i, compatible with δ on R.*
b) *$\delta_i \circ \varphi_{i-1} = \varphi_i \circ \delta_{i-1}$ for all $i \geq 1$.*

Prolongation sequences form a category with evident morphisms, ring homomorphisms $f_i : A^i \to B^i$ commuting with φ_i and δ_i, and in its subcategory with fixed A^0 *there exists an initial element, defined up to unique isomorphism* (cf. [Bu05], Chapter 3). It can be called the universal prolongation sequence.

In the geometric language, if $X = \operatorname{Spec} A$, the formal spectrum of the i-th ring A^i in the universal prolongation sequence is denoted $J^i(X)$ and called *the i-th p-jet space of X*. Conversely, $A^i = \mathcal{O}(J^i(X))$, the ring of global functions.

The geometric morphisms (of formal schemes over $\mathbf{Z}$) corresponding to ϕ_i are denoted $\phi^i : J^i(X) \to J^0(X) =: X\hat{}$ (formal p-adic completion of X).

This construction is compatible with localisation so that it can be applied to the non-necessarily affine schemes: cf. [Bu05], Chapter 3.

Remark 5.1. Classically, (5.4) extends to the universal map of A to the differential graded algebra $\Omega^*(A)$, and there is a superficial similarity of this map with the one, say of A to the inductive limit of its universal prolongation sequence in the p-adic arithmetics context.

However, the classical differential acts in $\mathbf{Z}_2$-graded supercommutative algebras and is an *odd* operator with square zero, whereas δ_p are even.

The differential geometry of smooth schemes *in characteristic* $p > 0$ suggests a perspective worth exploring. Namely, the sheaf of differential forms on such a scheme is endowed with the so-called *Cartier operator* C, which is dual to the Frobenius operator $F : \partial \mapsto \partial^p$ acting upon vector fields. This operator C is F^{-1}-linear. One could consider studying p-adic lifts of the Cartier operator from the closed fibre of the relevant scheme to its p-adic completion, following the lead of [Bu05]. For a recent survey of F^{-1}-linear maps cf. [BlSch12] and references therein.

5.1.3 Flows

Let now X be a smooth affine scheme over $R = W(k)$. Each element of $\mathcal{O}(J^r(X))$ induces a function $f : X(R) \mapsto R$. Such functions are called δ-functions of order r on X, and we may and will identify them with respective elements of $\mathcal{O}(J^r(X))$. For $r = 0$, $\mathcal{O}(J^0(X)) = \mathcal{O}(X)\widehat{}$, the p-adic completion of $\mathcal{O}(X)$.

Definition 5.2. a) A system of arithmetic differential equations of order r on X is a subset $\mathcal{E}$ of $\mathcal{O}(J^r(X))$.

b) A solution of $\mathcal{E}$ is an R-point $P \in X(R)$ such that $f(P) = 0$ for all $f \in \mathcal{E}$. The set of solutions of $\mathcal{E}$ is denoted $Sol(\mathcal{E}) \subset X(R)$.

c) A prime integral of $\mathcal{E}$ is a function $\mathcal{H} \in \mathcal{O}(X)\widehat{}$ such that $\delta(\mathcal{H}(P)) = 0$ for all $P \in Sol(\mathcal{E})$.

We will also denote by $Z^r(\mathcal{E})$ the closed formal subscheme of $J^r(X)$ generated by $\mathcal{E}$.

Now, let δ_X be a p-derivation of $\mathcal{O}(X)\widehat{}$. From the universality of the jet sequence explained in 1.2, it follows that such derivations are in a bijection with the sections of the canonical morphism $J^1(X) \to J^0(X)$.

Definition 5.3. The δ-flow associated to δ_X is the system of arithmetic differential equations of order 1 which is the ideal in $\mathcal{O}(J^1(X))$ generated by elements of the form $\delta f_i - \delta_X f_i$ where $f_i \in \mathcal{O}(X)$ generate $\mathcal{O}(X)$ as R-algebra.

We use the word "flow" in this context in order to suggest that in our main applications we consider the p-adic axis as an arithmetic version of the time axis.

The derivation δ_X is completely determined by its δ-flow. If δ_X corresponds to the section $s : J^0(X) \to J^1(X)$ of $J^1(X) \to J^0(X)$ then $Z(\mathcal{E}(\delta_X)) \subset J^1(X)$ coincides with the image of ths section. One easily checks that if $\mathcal{H} \in \mathcal{O}(X)\widehat{}$ is such that $\delta_X \mathcal{H} = 0$ then $\mathcal{H}$ is a prime integral for $\mathcal{E}(\delta_X)$. All of the above can be transposed to the case when X a p-formal scheme, locally a p-adic completion of a smooth scheme over R.

In what follows we choose a smooth affine scheme Y and apply the constructions discussed above to $X := J^1(Y)$. In this case one can define a special class of δ-flows on $J^1(Y)$ which will be called *canonical δ-flows.*

Definition 5.4. A canonical δ-flow is a δ-flow $\mathcal{E}(\delta_{J^1(Y)})$ with the property that the composition of $\delta_{J^1(Y)} : \mathcal{O}(J^1(Y)) \to \mathcal{O}(J^1(Y))$ with the pull back map $\mathcal{O}(Y) \to \mathcal{O}(J^1(Y))$ equals the universal p-derivation $\delta : \mathcal{O}(Y) \to \mathcal{O}(J^1(Y))$.

Notice that in view of the universality property of p-jet spaces, one gets a natural closed embedding $\iota : J^2(Y) \to J^1(J^1(Y))$. This induces an injective map (which we view as an identification) from the set of sections of $J^2(Y) \to J^1(Y)$ to the set of sections of $J^1(J^1(Y)) \to J^1(Y)$. The sections of $J^2(Y) \to J^1(Y)$ are in a natural bijection with canonical δ-flows on $J^1(Y)$, whereas the sections of $J^1(J^1(Y)) \to J^1(Y)$ are in a bijection with (not necessarily canonical) δ-flows on $J^1(Y)$.

Finally, consider a system of arithmetic differential equations of order 2, $\mathcal{F} \subset \mathcal{O}(J^2(Y))$.

Definition 5.5. $\mathcal{F}$ defines a δ-flow on $J^1(Y)$ if the map $Z^2(\mathcal{F}) \to J^1(Y)$ is an isomorphism.

In this case then $Z^2(\mathcal{F}) \to J^1(Y)$ defines a section of $J^2(Y) \to J^1(Y)$ and hence a canonical δ-flow $\mathcal{E}(\delta_{J^1(Y)})$ on $J^1(Y)$ such that $\iota(Z^2(\mathcal{F})) = Z^1(\mathcal{E}(\delta_{J^1(Y)}))$.

The differential algebra counterpart of the above definition yields the natural concept of flow on the (co)tangent space defining a second-order differential equation.

5.2 Painlevé VI and differential characters of elliptic curves

5.2.1 Classical case

The family of sixth Painlevé equations depends on four arbitrary constants $(\alpha, \beta, \gamma, \delta)$ and is classically written as

$$\frac{d^2X}{dt^2} = \frac{1}{2}\left(\frac{1}{X} + \frac{1}{X-1} + \frac{1}{X-t}\right)\left(\frac{dX}{dt}\right)^2 - \left(\frac{1}{t} + \frac{1}{t-1} + \frac{1}{X-t}\right)\frac{dX}{dt} + \\ + \frac{X(X-1)(X-t)}{t^2(t-1)^2}\left[\alpha + \beta\frac{t}{X^2} + \gamma\frac{t-1}{(X-1)^2} + \delta\frac{t(t-1)}{(X-t)^2}\right]. \tag{5.5}$$

As R. Fuchs remarked in 1907, (5.5) can be rewritten as the differential equation for a (local) section $P := (X(t), Y(t))$ of the generic elliptic curve $E = E(t) : Y^2 = X(X-1)(X-t)$:

$$t(1-t)\left[t(1-t)\frac{d^2}{dt^2} + (1-2t)\frac{d}{dt} - \frac{1}{4}\right]\int_{\infty}^{(X,Y)} \frac{dx}{\sqrt{x(x-1)(x-t)}} = \\ = \alpha Y + \beta\frac{tY}{X^2} + \gamma\frac{(t-1)Y}{(X-1)^2} + (\delta - \frac{1}{2})\frac{t(t-1)Y}{(X-t)^2} \tag{5.6}$$

The l.h.s. of (5.6) can be called *the additive differential character* μ *of order two* of E: it is a non-linear differential expression in coordinates of P such that $\mu(P+Q) = \mu(P) + \mu(Q)$ where $P+Q$ means addition of points of the generic elliptic curve E, with infinity as zero. In particular, $\mu(Q) = 0$ for points of finite order. The point is that the integral in the l.h.s. of (5.6) already has such additivity property, but it is defined only modulo periods, and the latter are annihilated by the Gauss differential operator.

Thus $\mu(P)$ is defined up to multiplication by an invertible function of t. If we choose a differential of the first kind ω on the generic curve and the symbol of the Picard–Fuchs operator of the second-order annihilating periods of ω, the character will be defined uniquely. In particular, if we pass to the analytic picture replacing the algebraic family of curves $E(t)$ by the analytic one $E_\tau := \mathbf{C}/(\mathbf{Z} + \mathbf{Z}\tau) \mapsto \tau \in H$, and denote by z a fixed coordinate on $\mathbf{C}$, then (5.5) and (5.6) can be equivalently written in the form

$$\frac{d^2z}{d\tau^2} = \frac{1}{(2\pi i)^2}\sum_{j=0}^{3} \alpha_j \wp_z(z + \frac{T_j}{2}, \tau) \tag{5.7}$$

where $(\alpha_0, \ldots, \alpha_3) := (\alpha, -\beta, \gamma, \frac{1}{2} - \delta)$ and

$$\wp(z,\tau) := \frac{1}{z^2} + \sum_{(m,n)\neq(0.0)}\left(\frac{1}{(z+m\tau+n)^2} - \frac{1}{(m\tau+n)^2}\right). \tag{5.8}$$

Moreover, we have

$$\wp_z(z,\tau)^2 = 4(\wp(z,\tau) - e_1(\tau))(\wp(z,\tau) - e_2(\tau))(\wp(z,\tau) - e_3(\tau)) \tag{5.9}$$

where

$$e_i(\tau) = \wp(\frac{T_i}{2}, \tau), \ (T_0, \dots, T_3) = (0, 1, \tau, 1+\tau) \tag{5.10}$$

so that $e_1 + e_2 + e_3 = 0$.

For a more geometric description, cf. sec. 5.4.3 below.

5.2.2 p-adic differential additive characters

The form (5.7) is more suggestive than (5.5) for devising a p-adic version of PVI. In purely algebraic terms, z can be described as the logarithm of the formal group law, and the r.h.s. of (5.7) is simply a linear combination of shifts of $\wp_z$- (or Y-)coordinate by sections of the second order.

More precisely, using basic conventions of [Bu05], let $p \geq 5$ be a prime, k an algebraic closure of $\mathbf{Z}/p\mathbf{Z}$, $R = W(k)$ the ring of p-typical Witt vectors., as in 1.1.

Consider a smooth projective curve of genus one E over R, with four marked and numbered R-sections P_i, $i = 0, \dots, 3$, such that all divisors $2(P_i - P_j)$ are principal ones. Choosing, say, P_0 as zero section, we may and will identify E with its Jacobian and represent E as the closure of the affine curve $y^2 = 4x^3 + ax + b = 4(x - e_1)(x - e_2)(x - e_3)$, with $e_i \in R$, corresponding to $P_i - P_0$. Assume that E is *not a canonical lift of its (good) reduction*, that is, does not admit a lift of the Frobenius morphism of $E \otimes k$.

Choose the differential $\omega := dx/y$. Whenever we work with several elliptic curves simultaneously, we will write $x_{E,\omega}$, $y_{E,\omega}$ in place of former x, y etc.

The curve E has a canonical p-differential character $\psi_{E,\omega}$ of order 2 ([Bu05], pp. 201 and 197), which corresponds to $(2\pi i)^2 d^2 z/d\tau^2$ in (5.7).

5.2.3 Painlevé p-adic equation and the problem of constants

Now we can directly write a p-adic version of (5.7) as

$$\psi_{E,\omega}(Q) = \sum_{j=0}^{3} \alpha_j s_j^*(y(Q)) \tag{5.11}$$

where Q is a variable section of E/R, and $s_j : E_j \to E_j$ is the shift by P_j.

At this point we have to mention two problems.

(A) In (5.7), α_j must be absolute constants rather than, say, functions of t. Directly imitating this condition, we have to postulate that in (5.11), α_j must be roots of unity or zero, i.e. δ_p-constants. It is desirable to find a justification of such a requirement (or a version of it) in a more extended p-adic theory of

Painlevé VI, e.g. tracing its source to the arithmetic analog of isomonodromy deformations.

(B) The relevance and non-triviality of the problem of "constants" in the p-adic differential equations context is also implicit in our exclusion of those E that are canonical lifts of their reductions.

A formal reason for this exclusion was the fact that for such E the basic differential character ψ has order 1 rather than 2, thus being outside the framework of Painlevé VI. But the analogy with functional case suggests that canonical lifts should be morally considered as analogs of families with constant j-invariant in the functional case.

This agrees also with J. Borger's philosophy that Frobenius lift(s) should be treated as descent data to an "*algebraic geometry below* Spec $\mathbf{Z}$" (cf. [Bo09]). In our particular case the latter might be called "geometry over p-typical field with one element" (Borger's suggestion in e-mail to Yu. M. of April 22, 2013).

Indeed, canonical lifts X in such a geometry are endowed with an isomorphism $X^{\phi} \to X^{(p)}$ that can be seen as a categorification of the identity $c^{\phi} = c^p$ defining roots of unity.

In the p-adic case, however, if we decide to declare j-invariants of canonical lifts "constants" in some sense, this will require a revision of the latter notion. These invariants generally are not roots of unity: cf. Finotti's papers in http://www.math.utk.edu/~finotti/.

5.3 Symmetries and variants

Lemma 5.6. *(i) (Landin's transform). In the notations of 2.3, denote for each $i = 1, 2, 3$ by $\pi_i : E \to E_i := E/\langle P_i \rangle$ the respective isogeny. Let ω_i be the 1-form on E_i such that $\pi^*(\omega_i) = \omega$. Then*

$$\pi_i^*(y_{E_i,\omega_i}) = y_{E,\omega} + s_i^*(y_{E,\omega}). \tag{5.12}$$

(ii) We have

$$\pi_i^*(\psi_{E_i,\omega_i}) = \psi_{E,\omega}. \tag{5.13}$$

Proof. (i) If we choose an embedding of R in $\mathbf{C}$ identifying $E(\mathbf{C})$ with $\mathbf{C}/(\mathbf{Z}+\mathbf{Z}\tau)$ in such a way that P_i becomes the point $\tau/2$ (modulo periods), then (5.12) turns into the classical Landin identity (cf. [Ma96], sec. 1.6):

$$\wp(z, \tau/2) = \wp(z, \tau) + \wp(z + \tau/2, \tau)$$

(ii) The identity (5.13) can be stated and proved in wider generality. Namely, let E, E' be two smooth elliptic curves over R, not admitting lifts of

Frobenius. Let $\pi : E \to E'$ be an isogeny of degree prime to p and and ω, ω' such bases of 1-forms on E, E' that $\pi^*(\omega') = \omega$. Then

$$\pi^*(\psi_{E',\omega'}) = \psi_{E,\omega}. \tag{5.14}$$

In fact from [Bu05], Theorem 7.34, it follows that $\pi^*(\psi_{E',\omega'}) = c\psi_{E,\omega}$ for some $c \in R$.

Now consider the second jet space $J^2(E)$. Then $\psi = \psi_{E,\omega}$ is an element of $\mathcal{O}(J^2(E))$ (cf. [Bu05], p. 201). Denote by $\phi^i : J^i(E) \to J^0(E)$ the map defined at the end of sec. 5.1.2. Put

$$\omega^{(i)} := \frac{(\phi^i)^*(\omega)}{p^i}, \ i = 0, 1, 2. \tag{5.15}$$

Then in view of [Bu05], p. 203 (where our $\omega^{(i)}$ were denoted ω_i), we have

$$d\psi = p\omega^{(2)} + \lambda_1\omega^{(1)} + \lambda_0\omega^{(0)} \tag{5.16}$$

with $\lambda_1 \in R$, $\lambda_1 \in R^\times$ depending on (E, ω). Notice that here d means the usual differential taken in the "geometric", or "vertical" direction, that is $dc = 0$ for any $c \in R$. Moreover, in [Bu97] it was proved that if E is defined over $\mathbf{Z}_p \subset R$, then $\lambda_0 = 1, \lambda_1 = -a_p$ where a_p is the trace of Frobenius on the reduction E mod p.

Returning to the proof of (5.14), we can now compare the $\omega^{(2)}$-contributions to $d\psi_{E',\omega'}$ in two different ways. First, we have

$$\pi^*(d\psi_{E',\omega'}) = \pi^*(p(\omega')^{(2)} + \dots) = p\,(\pi^*(\omega')^{(2)}) + \dots$$

Second,

$$\pi^*(d\psi_{E',\omega'}) = d\pi^*(\psi_{E',\omega'}) = d(c\,\psi_{E,\omega}) = c\,(p\omega^{(2)} + \dots).$$

This shows that $c = 1$.

5.3.1 Two versions of PVI

Put $Y := E \setminus \cup_{i=0}^3 P_i$. Denote by $r := \sum_{j=0}^3 \alpha_j s_j^*(y) \in \mathcal{O}(Y)$ (cf. (5.11)).

Below we will denote by ρ *either* r, *or* $\phi(r) \in \mathcal{O}(J^1(Y))$. Although the introduction of the version with $\phi(r)$ is not motivated at this point, we will see in sec. 5.4 below that exactly this version admits a "p-adic Hamiltonian" description.

The character ψ induces the map of sets, that we will also denote $\psi : E(R) \to R$; similarly, ρ induces the map of sets $\rho : Y(R) \to R$.

Proposition 5.7. *Denote by the symbol* $PVI(E,\omega,P_0,P_1,P_2,P_3,\alpha_0,\alpha_1,\alpha_2,\alpha_3)$ *any of the two equations*

$$\psi(P)=\rho(P). \tag{5.17}$$

Let $\pi:=\pi_2: E\to E':=E/\langle P_2\rangle$ *as in Lemma 5.6. Put* $P_0'=\pi(P_0)$ *(zero point),* $P_1'=\pi(P_1)$*, and choose remaining two points of order two* P_2',P_3'*. Assume that* $\pi^*(\omega')=\omega$*.*

In this case for any solution Q *to (5.17), the point* $Q':=\pi(Q)$ *will be a solution to*

$$PVI'=PVI(E',\omega',P_0',P_1',P_2',P_3',\alpha_0,\alpha_1,0,0), \tag{5.18}$$

and conversely, if Q' *is a solution to* PVI'*, then* Q *is a solution to* PVI*.*

Proof. Let $\rho=\phi^j(r)$, $j=0$ or 1. Our statement results from the following calculation, using (5.12), (5.13):

$$\begin{gathered}\psi_{E',\omega'}(\pi(Q))=\psi_{E,\omega}(Q)=\\ \phi^j(\alpha_0(y_{E,\omega}(Q)+y_{E,\omega}(Q+P_2))+\alpha_1(y_{E,\omega}(Q+P_1)+\\ y_{E,\omega}(Q+P_1+P_2)))=\\ \phi^j(\alpha_0 y_{E',\omega'}(\pi(Q))+\alpha_1 y_{E',\omega'}(\pi(Q)+\pi(P_1))).\end{gathered}$$

Of course, similar statement will hold if π_2 is replaced by π_1 or π_3.

5.3.2 Two more versions of PVI

The character ψ is of course not algebraic, but the equation (5.16) shows that it has an algebraic vertical differential. The same is obviously true for the r.h.s. of (5.11): one easily sees that if E is given in the Weierstrass form $y^2=f(x)$, then

$$dr=-\frac{1}{2}\sum_{j=0}^{3}\alpha_j s_j^*(f'(x)))\,\omega. \tag{5.19}$$

Hence we may consider two more versions of the arithmetic PVI: the condition of vanishing of one of the following 1-forms on $J^2(E)$:

$$d\psi-dr=p\omega^{(2)}+\lambda_1\omega^{(1)}+\left(\lambda_0-\frac{dr}{\omega}\right)\omega^{(0)}, \tag{5.20a}$$

$$d\psi-d\phi(r)=p\omega^{(2)}+\left(\lambda_1-p\phi\left(\frac{dr}{\omega}\right)\right)\omega^{(0)}+\lambda_0\omega^{(0)}. \tag{5.20b}$$

But in the p-adic situation solutions to (5.17) and (5.20a,b) respectively are not related to each other in the way we would expect by analogy with usual calculus.

Indeed, let us consider a smooth function $f(x, y, y', \dots, y^{(r)})$ in $r+2$ variables defined on $\mathbf{R}^{r+2}$, where the latter is viewed as the r-th jet space of the first projection $\mathbf{R}^2 \to \mathbf{R}$, $(x, y) \mapsto x$. Let $u = u(x)$ be an unknown smooth function $u : \mathbf{R} \to \mathbf{R}$. Then the equation

$$f(x, u(x), u'(x), \dots, u^{(r)}(x)) = 0 \tag{5.21}$$

is related to the 1-form df on $\mathbf{R}^{r+2}$ as follows. If u solves (5.21), then taking the derivative of (5.21) with respect to x we get

$$\nabla^r(u)^*(df) = 0 \tag{5.22}$$

where $\nabla^r(u) : \mathbf{R} \to \mathbf{R}^{r+2}$ is given by $\nabla^r(u)(x) = (x, u(x), \dots, u^{(r)}(x))$.

However in the case of arithmetic jet spaces the situation is different.

Indeed, let $f \in \mathcal{O}(J^r(\mathbf{A}^1)) = R[y, y', \dots, y^{(r)}]\hat{}$ where $J^r(\mathbf{A}^1)$ is the arithmetic jet space of the affine line over R. Furthermore, let $u \in R$ be a solution of

$$f(u, \delta u, \dots, \delta^r u) = 0 \tag{5.23}$$

where $\delta = \delta_p : R \to R$ is the standard p-derivation on R. Applying δ to (5.23) we get:

$$\begin{aligned}&\frac{1}{p}(f^{(\phi)}(u^p + p\delta u, (\delta u)^p + p\delta^2 u, \dots, (\delta^r u)^p + p\delta^{r+1} u) \\ &\quad - f(u, \delta u, \dots, \delta^r u)^p) = 0\end{aligned} \tag{5.24}$$

where $f^{(\phi)}$ is f with coefficients twisted by the Frobenius ϕ. To compare this latter equation with (5.22), one can apply the Taylor formula and get

$$\frac{f^{(\phi)}(\nabla^r(u)^p) - f(\nabla^r(u))^p}{p} + \sum_{i=0}^{r} \frac{\partial f^{(\phi)}}{\partial y^{(i)}}(\nabla^r(u)^p)(\delta^{i+1} u) + M = 0, \tag{5.25}$$

where $\nabla^r(u) = (u, \delta u, \dots, \delta^r u)$.

The first term here is an arithmetic version of the pullback of the term $\frac{\partial f}{\partial x} dx$ in df. The second term clearly involves df. However, M involves higher partial derivatives of f; it is highly non-linear, it is divisible by p, and "the more non-linear its terms are the more they are divisible by p". In some sense, since p is small, one can view (5.25) as a non-linear deformation of (5.22).

We will continue discussion of this formalism in sec. 5.5 below. The reader may wish to skip the next section, or to postpone reading it.

5.4 Hamiltonian formalism

5.4.1 Classical case

The classical PVI equation written in the form (5.7) can be represented as a Hamiltonian flow on the variable two-dimensional phase space (twisted cotangent spaces to a versal family of elliptic curves, cf. [Ma96] and the end of this section), with time-dependent Hamiltonian:

$$\frac{dz}{d\tau} = \frac{\partial \mathcal{H}}{\partial y}, \ \frac{dy}{d\tau} = -\frac{\partial \mathcal{H}}{\partial z}, \tag{5.26}$$

where

$$\mathcal{H} := \mathcal{H}(\alpha_0, \dots, \alpha_3) := \frac{y^2}{2} - \frac{1}{(2\pi i)^2} \sum_{j=0}^{3} \alpha_j \wp(z + \frac{T_j}{2}, \tau). \tag{5.27}$$

In more geometric terms, this means that solutions to the PVI become leaves of the null-foliation of the following closed two-form:

$$\omega = \omega(\alpha_0, \dots, \alpha_3) := 2\pi i (dy \wedge dz - d\mathcal{H} \wedge d\tau) =$$

$$= 2\pi i (dy \wedge dz - y dy \wedge d\tau) + \frac{1}{2\pi i} \sum_{j=0}^{3} \alpha_j \wp_z(z + \frac{T_j}{2}, \tau) dz \wedge d\tau. \tag{5.28}$$

The extra factor $2\pi i$ makes ω defined over $\mathbf{Q}[\alpha_i]$ on a natural algebraic model of (twisted) relative cotangent bundle to the respective versal family of elliptic curves.

In the expression (5.28), the summand $2\pi i\, dy \wedge dz$ is the canonical fibrewise symplectic form on the relative cotangent bundle. The terms involving $d\tau$ uniquely determine the differential of the (time-dependent) Hamiltonian.

Moreover, ω is not just closed, but is a global differential: $\omega = d\nu$ where the form

$$\nu = \nu(\alpha_0, \dots, \alpha_3) := 2\pi i\, (y dz - \frac{1}{2} y^2 d\tau) + d \log \theta(z, \tau) + 2\pi i\, G_2(\tau) d\tau +$$

$$+ \frac{1}{2\pi i} \sum_{j=0}^{3} \alpha_j \wp(z + \frac{T_j}{2}, \tau) d\tau \tag{5.29}$$

also descends to an appropriate algebraic model, and the Hamiltonian $\mathcal{H}$ is again encoded in the $d\tau$-part of ν. Here is a convenient way to represent this encoding:

$$\mathcal{H}(\alpha_0, \dots \alpha_3) = i_{\partial_\tau} \left(\frac{y^2}{2} d\tau + \frac{1}{2\pi i} (\nu(0, \dots, 0) - \nu(\alpha_0, \dots, \alpha_3)) \right), \tag{5.30}$$

where $\partial_\tau := \frac{\partial}{\partial \tau}$.

Finally, the last summand in (5.28) is simply the additive differential character $2\pi i \frac{d^2z}{d\tau^2}$ that is generally denoted ψ in the p-adic case.

In the following subsections, we try to imitate this description of Hamiltonian structure for p-adic PVI equations. The reader should be aware that our treatment is somewhat *ad hoc*, and must be considered as a tentative step towards a more coherent vision of Hamiltonian flows with p-adic time. In fact, we do not yet have an appropriate version of dp replacing $d\tau$ and generally do not know what are differential forms involving "differentials in the arithmetical direction".

5.4.2 Arithmetical case: preparation

Let Y be a formal affine scheme over $R = W(k)$. Modules of vertical differential forms on Y are defined as

$$\Omega_Y = \lim \operatorname{inv} \Omega_{Y_n/R_n}$$

where $R_n = R/p^{n+1}R$, $Y_n = Y \otimes_R R_n$.

Let now $Z \subset J^n(Y)$ be a closed formal subscheme defined by the ideal $I_Z \subset \mathcal{O}(J^n(Y))$. Put

$$\Omega'_Z := \frac{\Omega_{J^n(Y)}}{\langle I_Z \Omega_{J^n(Y)}, dI_Z \rangle} \tag{5.31}$$

Given a system of arithmetic differential equations $\mathcal{F} \subset \mathcal{O}(J^r(Y))$, denote by $Z^r := Z^r(\mathcal{F})$ the ideal generated by $\mathcal{F}$. For each $s \leq r$, there is a natural map $\pi_{r,s} : Z^r \to J^s(Y)$.

Generally, the natural maps ϕ^* respect degrees of differential forms, one can define natural maps $\phi^*/p^i : \Omega^i_{J^{r-1}(Y)} \to \Omega^i_{J^r(Y)}$ and, for $f \in \mathcal{O}(J^2(Y))$, they induce maps which we will denote

$$\frac{\phi^*_Z}{p^i} : \Omega^i_{J^1(Y)} \to \Omega'^i_{Z^2(f)} \tag{5.32}$$

Definition 5.8. We say that $\mathcal{F} \subset \mathcal{O}(J^r(Y))$ defines a generalized canonical δ-flow on $J^s(Y)$, if the induced map

$$\pi^*_{r,s}\Omega_{J^s(Y)} \to \Omega'_{Z^r}$$

is injective, and its cokernel is annihilated by a power of p.

The cokernel here intuitively measures "how singular" $\mathcal{F}$ is on the closed fibre of Y.

Definition 5.9. a) Let X be a smooth surface over R (or the p-adic completion of such a surface).
A symplectic form on X is an invertible 2-form on X.
A contact form on X is an 1-form on X such that $d\nu$ is symplectic.
b) Let Y be a smooth curve over R. An 1-form on $X := J^1(Y)$ is called canonical, if $\nu = f\beta$, where $f \in \mathcal{O}(X)$ and β is an 1-form lifted from Y.

Notice that any closed canonical 1-form on $X = J^1(Y)$ is lifted from Y.

We now come to the main definition.

Definition 5.10. Let Y be a smooth affine curve over R and let $f \in \mathcal{O}(J^2(Y))$ be a function defining a generalized canonical δ-flow on $J^1(Y)$.

a) The respective generalized δ-flow is called Hamiltonian with respect to the symplectic form η on $J^1(Y)$, if $\phi_Z^*\eta = \mu \cdot \eta$ in $\Omega'^2_{Z^2(f)}$ for some $\mu \in pR$ called the eigenvalue.
b) Assume that moreover $\eta = d\nu$ for some canonical 1-form ν on $J^1(Y)$. Then we call

$$\epsilon := \frac{\phi_Z \nu - \mu\nu}{p} \in \Omega'_{Z^2(f)} \tag{5.33}$$

an Euler–Lagrange form.

We consider (5.33) as an (admittedly, half-baked) arithmetical analog of the expression $i_{\partial_\tau}(ydz - \mathcal{H}d\tau)$ (cf. (5.30)) in the same sense as the p-derivation

$$\delta_p(x) = \frac{\phi(x) - x^p}{p}$$

is an analog of ∂_τ.

Now we pass to the arithmetical PVI. Let again E be an elliptic curve over R that does not admit a lift of Frobenius and let $\psi \in \mathcal{O}(J^2(E))$ be the canonical δ-character of order 2 attached to an invertible 1-form ω on E. Consider the symplectic form $\eta = \omega^{(0)} \wedge \omega^{(1)}$ on $J^1(E)$: cf. (5.15). Let $Y \subset E$ be an affine open set and let $r \in \mathcal{O}(Y)$. Assume in addition that Y has an étale coordinate. (The basic example is E with sections of the second order deleted.)

Denoting such an étale coordinate by T, put $\mathcal{A}_2 = K[[T, T']]$, $\mathcal{A}_3 = K[[T, T', T'']]$, where K is the quotient ring of R.

Proposition 5.11. *The following assertions hold:*

1) The function $f = \psi - \phi(r)$ defines a generalized canonical δ-flow on $J^1(Y)$ which is Hamiltonian with respect to η.

2) *There exists a canonical* 1*-form* ν *on* X *such that* $d\nu = \eta$*; in particular the symplectic form* η *is exact and if* $\epsilon := \frac{1}{p}(\phi_Z^*\nu - \mu\nu)$ *is the Euler–Lagrange form then* $p\epsilon$ *is closed.*
3) *Let* ϵ *be Euler–Lagrange form and* $f = \psi - \phi(r)$*. Then we have the following equality in* $\Omega_{\mathcal{A}_3}$*:*

$$\epsilon = f\omega^{(1)} - \frac{1}{p}(\phi^* - \mu)\nu.$$

4) *Let* $r_1, r_2 \in \mathcal{O}(Y)$ *be such that* $r_2 - r_1 = \partial s$*, for some* $s \in \mathcal{O}(Y)$*, where* ∂ *is the derivation on* E *dual to* ω*. (This holds for two right-hand sides of any two PVI equations.) Consider the equations* $\psi - \phi(r_1)$ *and* $\psi - \phi(r_2)$ *respectively. Then there exists a canonical* 1*-form* ν *on* $J^1(Y)$ *such that* $d\nu = \eta$ *and such that, if* ϵ_1, ϵ_2 *are the corresponding Euler–Lagrange forms, then:*

$$\epsilon_1 - \epsilon_2 = \frac{1}{p}d\phi(s) \in \Omega_{\mathcal{A}_2}.$$

Remark. We will deduce from this Proposition below (cf. Corollary 5.16) that in fact p-adic PVI in the form (5.17) defines a generalized canonical δ-flow. However, it does not define a δ-flow in the sense of Definition 5.5. This motivated our Definition 5.8.

Proof. From (5.15), we get the following equality in $\Omega^2_{J^2(Y)}$:

$$\frac{\phi^*\eta}{p^2} = \frac{\phi^*\omega^{(0)}}{p} \wedge \frac{\phi^*\omega^{(1)}}{p} = \omega^{(1)} \wedge \omega^{(2)}.$$

Recall the formula (5.20b):

$$df = p\omega^{(2)} + (\lambda_1 - p\phi(\partial r))\omega^{(1)} + \lambda_0\omega^{(0)}$$

in $\Omega_{J^2(E)}$ where $\lambda_1 \in R$, $\lambda_0 \in R^\times$. Hence, if we keep notation $\omega^{(0)}, \omega^{(1)}$ also for the images of the respective forms in $\Omega'_{Z^2(f)}$, in view of $\omega^{(1)} \wedge df = 0$ in $\Omega'_{Z^2(f)}$, we have the following equality in $\Omega'^2_{Z^2(f)}$:

$$\phi_Z^*\eta = -p \cdot \omega^{(1)} \wedge ((\lambda_1 - p\phi(dr))\omega^{(1)} + \lambda_0\omega^{(0)}) = p\lambda_0\eta,$$

This completes the proof of 1).

Write now $\omega = dL = \frac{dL}{dT}dT$ where $L = L(T) \in TK[[T]]$. (For instance, we can take L to be the formal logarithm of E.) Then

$$\mathcal{A}_3 = K[[T, \phi(T), \phi^2(T)]] = K[[L, \phi(L), \phi^2(L)]].$$

So the image of ψ in $\mathcal{A}_3$ is

$$\psi = \frac{1}{p}(\phi^2(L) + \lambda_1\phi(L) + p\lambda_0 L) + \lambda_{-1}$$

for some $\lambda_{-1} \in R$. Therefore the maps

$$\Omega_{\mathcal{A}_2} \to \frac{\Omega_{\mathcal{A}_3}}{\langle f\Omega_{\mathcal{A}_3}, df\rangle}, \quad \Omega^2_{\mathcal{A}_2} \to \frac{\Omega^2_{\mathcal{A}_3}}{\langle f\Omega^2_{\mathcal{A}_3}, df \wedge \Omega_{\mathcal{A}_3}\rangle}$$

are isomorphisms and so we have induced Frobenii maps $\phi^*_f : \Omega_{\mathcal{A}_2} \to \Omega_{\mathcal{A}_2}$ and $\phi^*_f : \Omega^2_{\mathcal{A}_2} \to \Omega^2_{\mathcal{A}_2}$. Since T is étale the lift of Frobenius $T \mapsto T^p$ on $\mathbf{A}^1$ extends to a lift of Frobenius $\phi_0 : \widehat{Y} \to \widehat{Y}$ of the p-adic completion of Y. Also the derivation $\frac{d}{dT}$ on $R[T]$ extends to a derivation still denoted by $\frac{d}{dT}$ on $\mathcal{O}(\widehat{Y})$. We claim that

$$\frac{\phi(L) - \phi_0(L)}{p},$$

which a priori is an element of $\mathcal{A}_2$, actually belongs to $\mathcal{O}(J^1(Y))$. Indeed we have the following expansion in $\mathcal{A}_2$:

$$\begin{aligned}
\frac{\phi(L) - \phi_0(L)}{p} &= \frac{L^{(\phi)}(T^p + pT') - L^{(\phi)}(T^p)}{p} \\
&= \sum_{i=1}^{\infty} \frac{p^{i-1}}{i!}\frac{d^i L^{(\phi)}}{dT^i}(T^p)(T')^i = \sum_{i=1}^{\infty} \frac{p^{i-1}}{i!}\phi_0\left(\frac{d^i L}{dT^i}\right)(T')^i \\
&= \sum_{i=1}^{\infty} \frac{p^{i-1}}{i!}\phi_0\left(\left(\frac{d}{dT}\right)^{i-1}\left(\frac{\omega}{dT}\right)\right)(T')^i \in \mathcal{O}(J^1(Y)),
\end{aligned}$$

where the superscript (ϕ) means twisting coefficients by ϕ. The latter inclusion follows because $T' = \delta T \in \mathcal{O}(J^1(Y))$, $\frac{\omega}{dT} \in \mathcal{O}(Y) \subset \mathcal{O}(\widehat{Y})$, and the latter is stable under $\frac{d}{dT}$ and ϕ_0. Now set

$$\nu := -\frac{\phi(L) - \phi_0(L)}{p}\omega \in \mathcal{O}(J^1(Y))\,\omega \in \Omega_{J^1(Y)}.$$

Then

$$\begin{aligned}
d\nu &= -d\left(\frac{\phi(L) - \phi_0(L)}{p}\right) \wedge \omega \\
&= -d\left(\frac{\phi(L)}{p}\right) \wedge \omega + d\left(\frac{\phi_0(L)}{p}\right) \wedge \omega \\
&= -\omega^{(1)} \wedge \omega^{(0)} = \eta.
\end{aligned}$$

This completes the proof of 2).

Next for $f = \psi - \phi(r)$ we have the following computation in $\Omega_{\mathcal{A}_3}$:

$$\begin{aligned} p\epsilon &= \phi_f^* \nu - \mu\nu \\ &= -\phi_f^* \left(\frac{\phi(L) - \phi_0(L)}{p} \omega \right) + \mu \frac{\phi(L) - \phi_0(L)}{p} \omega \\ &= -\phi_f \phi(L) \omega^{(1)} + \phi\phi_0(L)\omega^{(1)} + \mu \frac{\phi(L) - \phi_0(L)}{p} \omega \\ &= (\lambda_1 \phi(L) + p\lambda_0 L + \phi\phi_0(L) - p\phi(r) + p\lambda_{-1})\omega^{(1)} + \mu \frac{\phi(L) - \phi_0(L)}{p} \omega^{(0)} \\ &= (\lambda_1 \phi(L) + p\lambda_0 L + \phi\phi_0(L) - p\phi(r) + p\lambda_{-1})\omega^{(1)} \\ &\quad + \phi^* \left(\frac{\phi(L) - \phi_0(L)}{p} \omega^{(0)} \right) - (\phi^* - \mu)\nu \\ &= \left(\phi^2(L) + \lambda_1 \phi(L) + p\lambda_0 L - p\phi(r) + p\lambda_{-1} \right) \omega^{(1)} - (\phi^* - \mu)\nu \\ &= pf\omega^{(1)} - (\phi^* - \mu)\nu. \end{aligned}$$

This ends the proof of assertion 3). Assertion 4) follows from the fact that

$$\epsilon_1 - \epsilon_2 = \phi(\partial s)\omega^{(1)} = \frac{1}{p} d(\phi(s)).$$

Remarks. a) Some of our arguments above break down if $\psi - \phi(r)$ is replaced by $\psi - r$. This is our main motivation for studying $\psi - \phi(r)$.

b) Assertion 3) implies that if $[\epsilon]_2, [f\omega^{(1)}]_2 \in \frac{\Omega_{\mathcal{A}_3}}{(\phi^* - \mu)\Omega_{\mathcal{A}_2}}$ are the images of $\epsilon, f\omega^{(1)}$ then

$$[\epsilon]_2 = [f\omega^{(1)}]_2.$$

This justifies our suggestion that f is the "Euler–Lagrange equation attached to our Hamiltonian data". Assertion 4) says that the Euler–Lagrange forms of various PVI equations differ by exact forms.

c) Finally, we could treat in this way also the multicomponent version of PVI and the degeneration PV as they are described in [Ta01].

5.4.3 Geometry of PVI in various categories

Below we essentially reproduce from [Ma96] a geometric description of the natural habitats of various forms of PVI including its arithmetic version.

Our series of constructions starts with a non–constant family ("pencil") of elliptic curves over one-dimensional base. in one of several natural categories: schemes over a field of characteristic zero or a ring of algebraic integers, or a p-adic completen of the latter, analytic spaces etc.

a. Let $(\pi : E \to B; D_0, \ldots, D_3)$ be a pencil of compact smooth curves of genus one, with variable absolute invariant, endowed with four labelled sections D_i such that if any one of them is taken as zero, the others will be of order two.

We will call E *a configuration space* of PVI (common for all values of parameters.) Solutions to all equations will be represented by some multisections of π.

b. Let $\mathcal{F}$ be the subsheaf of the sheaf of vertical 1-forms $\Omega^1_{E/B}(D_3)$ on E with pole at D_3 and residue 1 at this pole. It is an affine twisted version of $\Omega^1_{E/B}$ which is the sheaf of sections of the relative cotangent bundle $T^*_{E/B}$. Similarly, $\mathcal{F}$ itself "is" the sheaf of sections of an affine line bundle $F = F_{E/B}$ on E. More precisely, we can construct a bundle $\lambda : F \to E$ and a form $\nu_F \in \Gamma(F, \Omega^1_{F/B}(\lambda^{-1}(D_3))$ such that the map

$$\{\text{local section } s \text{ of } F\} \mapsto s^*(\nu_F)$$

identifies the sheaf of sections of F/E with $\mathcal{F}$.

We will call F *a phase space* for PVI (again, common for all parameter values). It is this space that carries a canonical symplectic form (relative over the base) rather than the usual cotangent bundle.

c. E carries a distinguished family of algebraic/arithmetic curves transversal to the fibers of E: considered as multisections of E/B they are of finite order (if any of D_i is chosen as zero). It is important that each curve of this family has a canonical lifting to F (for its description, see [Ma96], especially (2.12) and (2.29)).

d. F carries a closed two-form ω which can be characterised by the following two properties:

i). The vertical part of ω, i.e. its restriction to the fibers of $\pi \circ \lambda : F \to B$, coincides with $d_{F/B}(\nu_F)$.

ii). Any canonical lift to F of a connected multisection of finite order of $E \to B$, referred to above, is a leaf of the null-foliation of ω.

e. E also carries four distinguished closed two-forms $\omega_0, \ldots, \omega_3$. They are determined, up to multiplication by a constant, by the following properties.

i). The divisor of ω_i is $\dfrac{D_j D_k D_l}{D_i^3}$ where $\{i, j, k, l\} = \{0, 1, 2, 3\}$.

ii). Identify the sheaves Ω^2_E and $\pi^*(\Omega^1_{E/B})^{\otimes 3}$ on E using the Kodaira–Spencer isomorphism $\pi^*(\Omega^1_B) \cong (\Omega^1_{E/B})^{\otimes 2}$ and the exact sequence $0 \to \pi^*(\Omega^1_B) \to \Omega^1_E \to \Omega^1_{E/B} \to 0$. Then the image of ω_i in $\pi^*(\Omega^1_{E/B})^{\otimes 3}$

considered in the formal neighborhood of D_i is the cube of a vertical 1-form with a constant residue along D_i.

Notice that up to introduction of ω, all constructions were valid both in geometric and arithmetic cases. It is precisely the absence of the differential in arithmetic direction was the object of our concerns in 4.2–4.4 above.

The affine space $P_0 := \omega + \sum_{i=0}^{3} \mathbf{C}\lambda^*(\omega_i)$ of closed two-forms on F is our version of the *moduli space of the PVI equations* replacing the classical $(\alpha, \beta, \gamma, \delta)$-space.

We can now summarize our geometric definition of PVI equations and their solutions.

Definition 5.12. a) A Painlevé two-form on F is a point $\Omega \in P_0$.
b) The Painlevé foliation corresponding to Ω is the null-foliation of Ω.
c) The solutions to the respective Painlevé equation are the leaves of this foliation (in the Hamiltonian description).

The form ω corresponds to $(\alpha_0, \alpha_1, \alpha_2, \alpha_3) = (0, 0, 0, 0)$.

5.5 Pfaffian congruences

5.5.1 Notation

We will continue here the discussion started in 3.3.

For $u \in R$, define the following infinite vectors in $R \times R \times \dots$:

$$\nabla(u) := (u, \delta u, \delta^2 u, \dots),$$
$$\delta\nabla(u) := (\delta u, \delta^2 u, \delta^3 u, \dots),$$
$$\nabla(u)^p := (u^p, (\delta u)^p, (\delta^2 u)^p, \dots).$$

For $g \in R[y, y', \dots, y^{(r)}]\widehat{\ }$, denote by $g^{(\phi)}$ be the series obtained from g by applying ϕ to the coefficients of g. Moreover, put

$$dg := \sum_{i=0}^{r} \frac{\partial g}{\partial y^{(i)}} dy^{(i)},$$

$$\frac{\partial g}{\partial p} := \frac{1}{p}(g^{(\phi)}(y^p, \dots (y^{(r)})^p) - g(y, \dots, y^{(r)})^p),$$

$$\frac{\partial}{\partial \nabla(y)} := \left(\frac{\partial}{\partial y}, \frac{\partial}{\partial y'}, \frac{\partial}{\partial y''} \dots\right),$$

Denoting by $\langle , \rangle$ the basic pairing between 1-forms and vector fields, and writing $\left(\frac{\partial}{\partial \nabla(y)}\right)^t$ for the column transpose of $\frac{\partial}{\partial \nabla(y)}$, we get for any $f \in R[y, y', \dots, y^{(r)}]\widehat{\ }$ and $i \geq 0$:

$$\delta^{i+1} f(u) = \delta\delta^i f(u)$$
$$= \frac{1}{p}((\delta^i f)^{(\phi)}(u^p + p\delta u, (\delta u)^p + p\delta^2 u, \ldots) - (\delta^i f)(u, \delta u, \ldots)^p)$$
$$\equiv \frac{\partial \delta^i f}{\partial p}(\nabla(u)^p) + \sum_{j=0}^{r} \frac{\partial(\delta^i f)^{(\phi)}}{\partial y^{(j)}}(\nabla(u)^p)\delta^{j+1} u \bmod p$$
$$\equiv \frac{\partial \delta^i f}{\partial p}(\nabla(u)^p) + \langle d[(\delta^i f)^{(\phi)}](\nabla(u)^p), \delta\nabla(u) \cdot \left(\frac{\partial}{\partial \nabla(y)}\right)^t \rangle \bmod p.$$

It is known ([Bu05], Lemma 3.20), that if $b \in R$ then $b = 0$ if and only if

$$\delta^i b \equiv 0 \bmod p \quad \text{for all } i \geq 0.$$

Combining the above facts we get:

Proposition 5.13. *An element $u \in R$ is a solution to $f(u, \delta u, \ldots, \delta^r u) = 0$ if and only if the following hold:*

a) $f(\nabla(u)) \equiv 0 \bmod p$,

b) $\frac{\partial(\delta^i f)}{\partial p}(\nabla(u)^p) \equiv -\langle d[(\delta^i f)^{(\phi)}](\nabla(u)^p), \delta\nabla(u) \cdot \left(\frac{\partial}{\partial \nabla(y)}\right)^t \rangle \bmod p,$ $i \geq 0$.

Moreover, u is a solution to the equation $\delta f(u, \delta u, \ldots, \delta^r u) = 0$ if and only if b) above holds.

The above discussion can be obviously generalized to the case when y is a tuple of variables and f is a tuple of equations. It shows that the equation $f(u, \delta u, \ldots, \delta^r u) = 0$ is controlled by a system of "Pfaffian" congruences involving the 1-forms

$$d((\delta^i f)^{(\phi)}) = d(\delta^i(f^{(\phi)})), \quad i \geq 0.$$

If f has $\mathbf{Z}_p$-coefficients then the above forms are, of course,

$$d(\delta^i f), \quad i \geq 0.$$

In what follows we analyse such forms relevant for PVI equations.

5.5.2 The forms $\delta^i(\psi - \rho)$

We start by defining inductively certain universal δ-polynomials.

Consider a family of commuting free variables $z = (z_0, z_1, \ldots, z_r)$ and let w be another variable. As in the jet theory in [Bu05], denote by $z', z'', \ldots$, resp. $w', w'', \ldots$ new (families of) independent variables indexed additionally by the formal order of derivative. Consider the ring $\mathbf{Z}[z, z', z'', \dots, w, w', w'', \dots]$, equipped with the tautological p-derivation $\delta z_i^{(k)} = z_i^{(k+1)}$ etc., and hence with the lift of Frobenius $\phi(F) = F^p + p\delta F$. Define the elements $A_{m,i} = A_{m,i}(z, z', z'', \dots, w, w', w' \dots)$ of this ring ($m \geq 0$, $i = 0, \dots, m+r$) by induction:

$$A_{0,i} := z^{(i)}, \quad i = 0, \dots, r$$
$$A_{m+1,0} := -(w^{(m)})^{p-1} A_{m,0}, \quad m \geq 0 \tag{5.34}$$
$$A_{m+1,i} := \phi(A_{m,i-1}) - (w^{(m)})^{p-1} A_{m,i}, \; i = 1, \dots, m+r, \; m \geq 0$$
$$A_{m+1,m+r+1} := \phi(A_{m,m+r}), \; m \geq 0.$$

It is easy then to check that

$$A_{m,0} = (-1)^{m-1}(ww' \dots w^{(m-1)})^{p-1} z^{(0)}, \quad m \geq 1,$$
$$A_{m,m+r} = \phi^m(z^{(r)}), \quad m \geq 0,$$
$$A_{m,m+r-1} \equiv \phi^m(z^{(r-1)}) \bmod (z^{(r)}, \phi(z^{(r)}), \dots, \phi^{m-1}(z^{(r)})), \quad m \geq 0 \tag{5.35}$$
$$A_{m,m+i} \equiv \phi^m(z^{(i)}) \bmod (w, w', \dots, w^{(m-1)}), \quad m \geq 1, \quad i = 0, \dots, r,$$
$$A_{m,i} \equiv 0 \bmod (w, w', \dots, w^{(m-1)}), \quad i = 0, \dots, m-1.$$

In the following statement Y is an affine smooth curve over R and $f \in \mathcal{O}(J^r(Y))$. Then by [Bu05], there exist $a_0, \dots, a_r \in \mathcal{O}(J^r(Y))$ such that

$$df = \sum_{i=0}^{r} a_i \omega^{(i)}.$$

Put also $a = (a_1, \dots, a_r)$. Then we have:

Proposition 5.14. *Put $a_{m,i} = A_{m,i}(a, \delta a, \delta^2 a, \dots, f, \delta f, \delta^2 f, \dots)$. Then*

$$d(\delta^m f) = \sum_{i=0}^{m+r} a_{m,i} \omega^{(i)}.$$

Proof. We proceed by induction on $m \geq 0$. The case $m = 0$ is trivial. The passage from m to $m+1$ runs as follows:

$$d(\delta^{m+1} f) = d(\delta\delta^m f) = d\left(\frac{\phi(\delta^m f) - (\delta^m f)^p}{p}\right)$$
$$= \frac{\phi^*}{p}(d(\delta^m f)) - (\delta^m f)^{p-1} d(\delta^m f)$$

$$= \frac{\phi^*}{p}(\sum_{i=0}^{m+r} a_{m,i}\omega^{(i)}) - (\delta^m f)^{p-1}(\sum_{i=0}^{m+r} a_{m,i}\omega^{(i)})$$

$$= (\sum_{i=0}^{m+r} \phi(a_{m,i})\omega^{(i+1)}) - (\delta^m f)^{p-1}(\sum_{i=0}^{m+r} a_{m,i}\omega^{(i)}) = \sum_{i=0}^{m+1+r} a_{m+1,i}\omega^{(i)}.$$

This ends the proof.

Now we apply this to the case when Y is an affine open subset of the elliptic curve E over $\mathbf{Z}_p$ without lift of Frobenius. Denote by $a_p \in \mathbf{Z}$ the trace of Frobenius on the reduction E mod p.

Corollary 5.15. *Let $f = \psi - \rho$, $\rho = \phi(r)$, where $r \in \mathcal{O}(Y)$, ψ the canonical δ-character of order 2 attached to ω. Let $d(\delta^m f) = \sum_{i=0}^{m+2} a_{m,i}\omega^{(i)}$, $m \geq 0$, $a_{m,i} \in \mathcal{O}(J^{m+2}(Y))$. So*

$$a_{0,2} = p,\ a_{0,1} = -(a_p + p\phi(\frac{dr}{\omega})),\ a_{0,0} = 1.$$

Then, for $m \geq 1$:

1) $a_{m,m+2} = p$;
2) $a_{m,m+1} \equiv -a_p \bmod p$;
3) $a_{m,m+1} \equiv -(a_p + p\phi^{m+1}(\frac{dr}{\omega})) \bmod (f, \delta f, \ldots, \delta^{m-1} f)$;
4) $a_{m,m} \equiv 1 \bmod (f, \delta f, \ldots, \delta^{m-1} f)$;
5) $a_{m,i} \equiv 0 \bmod (f, \delta f, \ldots, \delta^{m-1} f)$ for $i = 0, \ldots, m-1$;
6) $a_{m,0} = (-1)^{m-1}(f \cdot \delta f \cdots \delta^{m-1} f)^{p-1}$, for $m \geq 1$.
In particular if E has ordinary reduction then $a_{m,m+1}$ is invertible in $\mathcal{O}(J^{m+2}(Y))$.

One can prove a similar statement for $\psi - r$ in place of $\psi - \phi(r)$.

Corollary 5.16. *In the same situation, let $f = \psi - \rho$, where $\rho = r$ or $\rho = \phi(r)$. Let ψ be the canonical δ-character of order 2 attached to ω.*

Denote by $Z^{m+2} = Z^{m+2}(f, \delta f, \ldots, \delta^m f)$ the closed formal subscheme of $J^{m+2}(Y)$ defined by the ideal generated by $f, \delta f, \ldots, \delta^m f$. Let $\pi_m : Z^{m+2} \to J^1(Y)$ be the canonical projection.

Then the map

$$\pi_m^* \Omega_{J^1(Y)} \to \Omega'_{Z^{m+2}}$$

is injective with cokernel annihilated by p^{m+1}. In particular, for any $m \geq 0$, the system $\{f, \delta f, \ldots, \delta^m f\} \subset \mathcal{O}(J^{m+2}(Y))$ defines a generalised δ-flow on

$J^1(Y)$. If moreover E has ordinary reduction then the above cokernel is a cyclic $\mathcal{O}(Z^{m+2})$-module generated by the class of $\omega^{(m+2)}$.

Proof. Consider the case $\rho = \phi(r)$; a similar argument holds for $\rho = r$. Recall that $\Omega_{J^n(Y)}$ is a free $\mathcal{O}(J^n(Y))$-module generated by $\omega^{(0)}, \dots, \omega^{(n)}$. Also, by definition,

$$\mathcal{O}(Z^{m+2}) = \frac{\mathcal{O}(J^{m+2}(Y))}{(f, \delta f, \dots, \delta^m f)}$$

and

$$\Omega'_{Z^{m+2}} = \frac{\mathcal{O}(J^{m+2}(Y))\omega^{(0)} \oplus \dots \oplus \mathcal{O}(J^{m+2}(Y))\omega^{(m+2)}}{\langle(\delta^i f)\omega^{(j)}, d(\delta^i f)\rangle},$$

where $\langle\ \rangle$ means $\mathcal{O}(J^{m+2}(Y))$-linear span and $j = 0, \dots, m+2, i = 0, \dots, m$. By Corollary 5.15 we have:

$$\Omega'_{Z^{m+2}} = \frac{\mathcal{O}(Z^{m+2})\omega^{(0)} \oplus \dots \oplus \mathcal{O}(Z^{m+2})\omega^{(m+2)}}{\langle p\omega^{(2+i)} - (a_p + p\phi^{i+1}(\frac{dr}{\omega}))\omega^{(1+i)} + \omega^{(i)}\rangle}$$

where $i = 0, \dots, m$ and $\langle\ \ \rangle$ means here $\mathcal{O}(Z^{m+2})$-linear span. Note that $\pi_m^*\Omega_{J^1(Y)}$ is a free $\mathcal{O}(J^{m+2}(Y))$-module with basis $\omega^{(0)}, \omega^{(1)}$. So in order to prove that the map

$$\pi_m^*\Omega_{J^1(Y)} \to \Omega'_{Z^{m+2}} \tag{5.36}$$

is injective we need to check that no $\mathcal{O}(Z^{m+2})$-linear combination of $\omega^{(0)}$, $\omega^{(1)}$ can be a $\mathcal{O}(Z^{m+2})$-linear combination of elements $p\omega^{(2+i)} - (a_p + p\phi^{i+1}(\frac{dr}{\omega}))\omega^{(1+i)} + \omega^{(i)}$ which is clear. For $j = 2, \dots, m+2$ let

$$\overline{\omega^{(j)}} \in \frac{\Omega'_{Z^{m+2}}}{\langle\omega^{(0)}, \omega^{(1)}\rangle}$$

be the image of $\omega^{(j)}$. Clearly $p\overline{\omega^{(2)}} = 0$ hence $p^2\overline{\omega^{(3)}} = 0$, etc. We conclude that p^{m+1} annihilates the cokernel of (5.36).

Finally assume E has ordinary reduction. We want to show that the above cokernel is generated by $\overline{\omega^{(m+2)}}$. It is enough to show that for all $j = 1, \dots, m+2$ we have

$$\overline{\omega^{(j-1)}} \in p\,\mathcal{O}(Z^{m+2})\,\overline{\omega^{(j)}}.$$

We proceed by induction on j. For $j = 1$ this is clear. Now assume the above is true for some $1 \le j < m+2$. We have

$$p\overline{\omega^{(j+1)}} - (a_p + p\phi^j(\frac{dr}{\omega}))\overline{\omega^{(j)}} + \overline{\omega^{(j-1)}} = 0.$$

By induction $\overline{\omega^{(j-1)}} = p\,c\,\overline{\omega^{(j)}}$ for some $c \in \mathcal{O}(Z^{m+2})$. Since a_p is invertible in R it follows that $a_p - pc(1 + \phi^{j-1}(dr/\omega))$ is invertible hence

$$\overline{\omega^{(j)}} = p(a_p + p\phi^j(\frac{dr}{\omega}) - pc)^{-1}\overline{\omega^{(j+1)}},$$

which ends the proof.

Remark 5.17. Let Y be a smooth affine scheme over R and let $\mathcal{F} \subset \mathcal{O}(J^r(Y))$ be a system of arithmetic differential equations. Consider the set

$$\{\mathcal{F}, \delta\mathcal{F}, \ldots, \mathcal{F}\} \subset \mathcal{O}(J^{r+m}(Y))$$

which can be referred to as the m-th prolongation of $\mathcal{F}$. Let

$$Z^{r+m} := Z^{r+m}(\mathcal{F}, \delta\mathcal{F}, \ldots, \delta^m\mathcal{F}) \subset J^{r+m}(Y)$$

be the closed subscheme defined by this prolongation and let

$$Z^{\infty} = Z^{\infty}(\delta^{\infty}\mathcal{F})$$

be defined by

$$Z^{\infty} = \lim_{\leftarrow} Z^{r+m},$$

the projective limit (defined as the Spf of the p-adic completion of the inductive limit of $\mathcal{O}(Z^{r+m})$ as m varies; this is a generally non-Noetherian formal scheme). Then Z^{∞} is a closed horizontal formal subscheme of

$$J^{\infty}(Y) := \lim_{\leftarrow} J^n(Y);$$

by horizontal we mean here that its ideal is sent into itself by δ. So there is an induced p-derivation $\delta_{Z^{\infty}}$ on $\mathcal{O}(Z^{\infty})$.

If Z^{∞} happens to be a smooth formal scheme then $\delta_{Z^{\infty}}$ defines a (genuine) δ-flow $\mathcal{E}(\delta_{Z^{\infty}}) \subset \mathcal{O}(J^1(Z^{\infty}))$ on Z^{∞} (in the sense of our previous definition). In case Z^{∞} is not necessarily a smooth formal scheme one can still define the affine formal scheme $J^1(Z^{\infty})$ and the δ-flow $\mathcal{E}(\delta_{Z^{\infty}})$ on Z^{∞} by copying the definitions from the smooth case.

The following problem needs to be investigated. Assume that $\mathcal{F} \subset \mathcal{O}(J^2(Y))$ defines a δ-flow $\mathcal{E}(\delta_{J^1(Y)}) \subset \mathcal{O}(J^1(J^1(Y)))$ on $J^1(Y)$. Consider the formal scheme $Z^{\infty} = Z^{\infty}(\delta^{\infty}\mathcal{F})$ as above. Prove that Z^{∞} is naturally isomorphic to $J^1(Y)$ and $\mathcal{E}(\delta_{J^1(Y)})$ is naturally identified with $\mathcal{E}(\delta_{Z^{\infty}})$. This would show the naturality of the above definitions.

References

[BlSch12] M. Blickle, K. Schwede. *p^{-1}-linear maps in algebra and geometry.* arXiv:1205.4577

[Bo09] J. Borger. *Lambda-rings and the field with one element.* arXiv:0906.3146

[Bu95] A. Buium. *Differential characters of abelian varieties over p-adic fields.* Inv. Math., vol. 122 (1995), 309–340.

[Bu97] A. Buium. *Differential characters and characteristic polynomial of Frobenius.* J. reine u. angew. Math. 485 (1997), 209–219.

[Bu05] A. Buium. *Arithmetic Differential Equations.* Math. Surveys and Monographs, 118, MS, Providence RI, 2005. xxxii+310 pp.

[Ma96] Yu. Manin. *Sixth Painlevé equation, universal elliptic curve, and mirror of* $\mathbf{P}^2$. In: Geometry of Differential Equations, ed. by A. Khovanskii, A. Varchenko, V. Vassiliev. Amer. Math. Soc. Transl. (2), vol. 186 (1998), 131–151. Preprint alg–geom/9605010.

[Ma08] Yu. Manin. *Cyclotomy and analytic geometry over F_1.* In: Quanta of Maths. Conference in honour of Alain Connes. Clay Math. Proceedings, vol. 11 (2010), 385–408. Preprint math.AG/0809.2716.

[Ta01] K. Takasaki. *Painlevé–Calogero correspondence revisited.* Journ. Math. Phys., vol. 42, Nr 3 (2001), 1443–1473.

6

Differential calculus with integers

Alexandru Buium
Department of Mathematics and Statistics,
University of New Mexico,
Albuquerque, NM 87131, USA
E-mail address: buium@math.unm.edu

ABSTRACT. Ordinary differential equations have an arithmetic analogue in which functions are replaced by numbers and the derivation operator is replaced by a Fermat quotient operator. In this survey we explain the main motivations, constructions, results, applications, and open problems of the theory.

The main purpose of these notes is to show how one can develop an arithmetic analogue of differential calculus in which differentiable functions $x(t)$ are replaced by integer numbers n and the derivation operator $x \mapsto \frac{dx}{dt}$ is replaced by the Fermat quotient operator $n \mapsto \frac{n-n^p}{p}$, where p is a prime integer. The Lie–Cartan geometric theory of differential equations (in which solutions are smooth maps) is then replaced by a theory of "arithmetic differential equations" (in which solutions are integral points of algebraic varieties). In particular the differential invariants of groups in the Lie–Cartan theory are replaced by "arithmetic differential invariants" of correspondences between algebraic varieties. A number of applications to diophantine geometry over number fields and to classical modular forms will be explained.

This program was initiated in [11] and pursued, in particular, in [12]–[35]. For an exposition of some of these ideas we refer to the monograph [16]; cf. also the survey paper [58]. We shall restrict ourselves here to the *ordinary differential* case. For the *partial differential* case we refer to [20, 21, 22, 7]. Throughout these notes we assume familiarity with the basic concepts of algebraic geometry and differential geometry; some of the standard material is being reviewed, however, for the sake of introducing notation, and "setting the stage". The notes are organized as follows. The first section presents some

Arithmetic and Geometry, ed. Luis Dieulefait *et al.* Published by Cambridge University Press.

classical background, the main concepts of the theory, a discussion of the main motivations, and a comparison with other theories. The second section presents a sample of the main results. The third section presents a list of open problems.

Acknowledgements. The author is indebted to HIM for support during part of the semester on Algebra and Geometry in Spring 2013. These notes are partially based on lectures given at the IHES in Fall 2011 and MPI in Summer 2012 when the author was partially supported by IHES and MPI, respectively. Partial support was also received from the NSF through grant DMS 0852591.

6.1 Main concepts

6.1.1 Classical analogies

The analogies between functions and numbers have played a key role in the development of modern number theory. Here are some classical analogies. All facts in this subsection are well known and entirely classical; we review them only in order to introduce notation and put things in perspective.

6.1.1.1 Polynomial functions

The ring $\mathbb{C}[t]$ of polynomial functions with complex coefficients is analogous to the ring $\mathbb{Z}$ of integers. The field of rational functions $\mathbb{C}(t)$ is then analogous to the field of rational numbers $\mathbb{Q}$. In $\mathbb{C}[t]$ any non-constant polynomial is a product of linear factors. In $\mathbb{Z}$ any integer different from $0, \pm 1$ is up to a sign a product of prime numbers. To summarize

$$\mathbb{C} \subset \mathbb{C}[t] \subset \mathbb{C}(t)$$

are analogous to

$$\{0, \pm 1\} \subset \mathbb{Z} \subset \mathbb{Q}$$

6.1.1.2 Regular functions

More generally rings $\mathcal{O}(T)$ of regular functions on complex algebraic affine non-singular curves T are analogous to rings of integers $\mathcal{O}_F$ in number fields F. Hence curves T themselves are analogous to schemes $Spec\ \mathcal{O}_F$. Compactifications

$$T \subset \overline{T} = T \cup \{\infty_1, \ldots, \infty_n\} \simeq \text{(compact Riemann surface of genus } g)$$

are analogous to "compactifications"

$$Spec\ \mathcal{O}_F \subset \overline{Spec\ \mathcal{O}_F} = (Spec\ \mathcal{O}_F) \cup \frac{Hom(F, \mathbb{C})}{\text{conjugation}}$$

6.1.1.3 Formal functions

The inclusions

$$\mathbb{C} \subset \mathbb{C}[[t]] \subset \mathbb{C}((t))$$

(where $\mathbb{C}[[t]]$ is the ring of power series and $\mathbb{C}((t))$ is the ring of Laurent series) are analogous to the inclusion

$$\{0\} \cup \mu_{p-1} = \{c \in \mathbb{Z}_p;\ c^p = c\} \subset \mathbb{Z}_p \subset \mathbb{Q}_p$$

(where $\mathbb{Z}_p$ is ring of p-adic integers and $\mathbb{Q}_p = \mathbb{Z}_p[1/p]$). Recall that

$$\mathbb{Z}_p = \lim_{\leftarrow} \mathbb{Z}/p^n\mathbb{Z} = \{\sum_{n=0}^{\infty} c_i p^i;\ c_i \in \{0\} \cup \mu_{p-1}\}$$

So $\{0\} \cup \mu_{p-1}$ plays the role of "constants" in $\mathbb{Z}_p$. Sometimes we need more "constants" and we are led to consider, instead, the inclusions:

$$\{0\} \cup \bigcup_{\nu} \mu_{p^\nu - 1} \subset \widehat{\mathbb{Z}_p^{ur}} \subset \widehat{\mathbb{Z}_p^{ur}}[1/p]$$

where

$$\widehat{\mathbb{Z}_p^{ur}} = \mathbb{Z}_p[\zeta;\ \zeta^{p^\nu - 1} = 1, \nu \geq 1]\hat{} = \{\sum_{i=0}^{\infty} c_i p^i;\ c_i \in \{0\} \cup \bigcup_{\nu} \mu_{p^\nu - 1}\}.$$

Here the upper hat on a ring A means its p-adic completion:

$$\widehat{A} := \lim_{\leftarrow} A/p^n A.$$

So in the latter case the monoid $\{0\} \cup \bigcup_\nu \mu_{p^\nu - 1}$ should be viewed as the set of "constants" of $\widehat{\mathbb{Z}_p^{ur}}$; this is consistent with the "philosophy of the field with one element" to which we are going to allude later. Let us say that a ring is a *local p-ring* if it is a discrete valuation ring with maximal ideal generated by a prime $p \in \mathbb{Z}$. Then $\mathbb{Z}_p$ and $\widehat{\mathbb{Z}_p^{ur}}$ are local p-rings. Also for any local p-ring R we denote by $k = R/pR$ the residue field and by $K = R[1/p]$ the fraction field of R. Sometimes we will view local p-rings as analogues of rings $\mathbb{C}\{x\}$ of germs of analytic functions on Riemann surfaces and even as analogues of rings of global analytic (respectively C^∞ functions) on a Riemann surface T (respectively on a 1-dimensional real manifold T, i.e. on a circle S^1 or $\mathbb{R}$).

6.1.1.4 Topology

Fundamental groups of complex curves (more precisely Deck transformation groups of normal covers $T' \to T$ of Riemann surfaces) have, as analogues, Galois groups $G(F'/F)$ of normal extensions number fields $F \subset F'$. The genus of a Riemann surface has an analogue for number fields defined in terms of ramification. All of this is very classical. There are other, less classical, topological analogies like the one between primes in $\mathbb{Z}$ and nodes in 3-dimensional real manifolds [61].

6.1.1.5 Divisors

The group of divisors

$$Div(\overline{T}) = \{\sum_{P \in \overline{T}} n_P P; n_P \in \mathbb{Z}\}$$

on a non-singular complex algebraic curve $\overline{T}$ is analogous to the group of divisors

$$Div(\overline{Spec\ \mathcal{O}_F}) = \{\sum \nu_P P;\ \nu_P \in \mathbb{Z} \text{ if } P \text{ is finite, } \nu_P \in \mathbb{R} \text{ if } P \text{ is infinite}\}$$

One can attach divisors to rational functions f on $\overline{T}$ ($Div(f)$ is the sum of poles minus the sum of zeroes); similarly one can attach divisors to elements $f \in F$. In both cases one is lead to a "control" of the spaces of fs that have a "controlled" divisor (the Riemann–Roch theorem). One also defines in both settings divisor class groups. In the geometric setting the divisor class group of $\overline{T}$ is an extension of $\mathbb{Z}$ by the Jacobian

$$Jac(\overline{T}) = \mathbb{C}^g/(\text{period lattice of } \overline{T})$$

where g is the genus of $\overline{T}$. In the number theoretic setting divisor class groups can be interpreted as "Arakelov class groups"; one recaptures, in particular, the usual class groups $Cl(F)$. Exploring usual class groups "in the limit", when one adjoins roots of unity, leads to Iwasawa theory. We will encounter Jacobians later in relation, for instance, to the Manin–Mumford conjecture. This conjecture (proved by Raynaud) says that if one views $\overline{T}$ as embedded into $Jac(\overline{T})$ (via the "Abel–Jacobi map") the the intersection of $\overline{T}$ with the torsion group of $Jac(\overline{T})$ is a finite set. This particular conjecture does not seem to have an analogue for numbers.

6.1.1.6 Families

Maps

$$M \to T$$

of complex algebraic varieties, complex analytic, or real smooth manifolds, where $\dim T = 1$, are analogous to arithmetic schemes, i.e. schemes of finite type

$$X \to Spec\ R$$

where R is either the ring of integers $\mathcal{O}_F$ in a number field F or a complete local p-ring respectively. Note however that in this analogy one "goes arithmetic only half way": indeed for $X \to Spec\ R$ the basis is arithmetic yet the fibers are still geometric. One can attempt to "go arithmetic all the way" and find an analogue of $M \to T$ for which both the base and the fiber are "arithmetic"; in particular one would like to have an analogue of $T \times T$ which is "arithmetic" in two directions. This is one of the main motivations in the search for $\mathbb{F}_1$, the "field with one element".

6.1.1.7 Sections

The set of sections

$$\Gamma(M/T) = \{s : T \to M; \pi \circ \sigma = 1\}$$

of a map $\pi : M \to T$ is analogous to the set

$$X(R) = \{s : Spec\ R \to X; \pi \circ s = 1\}$$

of R-points of X where $\pi : X \to Spec\ R$ is the structure morphism. This analogy suggests that finiteness conjectures for sets of the form $X(R)$, which one makes in Diophantine geometry, should have as analogues finiteness conjectures for sets of sections $\Gamma(M/T)$. A typical example of this phenomenon is the Mordell conjecture (Faltings' theorem) saying that if X is an algebraic curve of "genus" ≥ 2, defined by polynomials with coefficients in a number field F, then the set $X(F)$ is finite. Before the proof of this conjecture Manin [55] proved a parallel finiteness result for $\Gamma(M/T)$ where $M \to T$ is a "non-isotrivial" morphism from an algebraic surface to a curve, whose fibers have genus ≥ 2. Manin's proof involved the consideration of differential equations with respect to vector fields on T. Faltings' proof went along completely different lines. This raised the question whether one can develop a theory of differential equations in which one can differentiate numbers.

All these examples of analogies are classical; cf. work of Dedekind, Hilbert, Hensel, Artin, Weil, Lang, Tate, Iwasawa, Grothendieck, and many others.

6.1.2 Analogies proposed in [11]–[35]

One thing that seems to be missing from the classical picture is a counterpart, in number theory, of the differential calculus (in particular of differential equations) for functions. Morally the question is whether one can meaningfully consider (and successfully use) "arithmetic differential equations" satisfied by numbers. In our research on the subject [11]–[35] we proposed such a theory based on the following sequence of analogies:

6.1.2.1 Derivatives

The derivative operator $\delta_t = \frac{d}{dt} : \mathbb{C}[t] \to \mathbb{C}[t]$ is analogous to the *Fermat quotient operator* $\delta = \delta_p : \mathbb{Z} \to \mathbb{Z}$, $\delta_p a = \frac{a - a^p}{p}$. More generally the derivative operator $\delta_t = \frac{d}{dt} : C^\infty(\mathbb{R}) \to C^\infty(\mathbb{R})$ (with t the coordinate on $\mathbb{R}$) is analogous to the operator $\delta = \delta_p : R \to R$ on a complete local p-ring R, $\delta_p \alpha = \frac{\phi(\alpha) - \alpha^p}{p}$ (where $\phi : R \to R$ is a fixed homomorphism lifting the p-power Frobenius map on R/pR). The map $\delta = \delta_p$ above is, of course, not a derivation but, rather, it satisfies the following conditions:

$$\begin{array}{lcl} \delta(1) & = & 0 \\ \delta(a+b) & = & \delta(a) + \delta(b) + C_p(a,b) \\ \delta(ab) & = & a^p \delta(b) + b^p \delta(a) + p\delta(a)\delta(b), \end{array}$$

where $C_p(x, y) \in \mathbb{Z}[x, y]$ is the polynomial $C_p(x, y) = p^{-1}(x^p + y^p - (x + y)^p)$. Any set theoretic map $\delta : A \to A$ from a ring A to itself satisfying the above axioms will be referred to a *p-derivation*; such operators were introduced independently in [49, 11] and they implicitly arise in the theory of Witt rings. For any such δ, the map $\phi : A \to A$, $\phi(a) = a^p + p\delta a$ is a ring homomorphism lifting the p-power Frobenius on A/pA; and vice versa, given a p-torsion free ring A and a ring homomorphism $\phi : A \to A$ lifting Frobenius the map $\delta : A \to A$, $\delta a = \frac{\phi(a) - a^p}{p}$ is a p-derivation.

6.1.2.2 Differential equations

Differential equations $F(t, x, \delta_t x, \dots, \delta_t^n x) = 0$ (with F smooth) satisfied by smooth functions $x \in C^\infty(\mathbb{R})$ are replaced by *arithmetic differential equations* $F(\alpha, \delta_p \alpha, \dots, \delta_p^n \alpha) = 0$ with $F \in R[x_0, x_1, \dots, x_n]\hat{}$ satisfied by numbers $\alpha \in R$. As we shall see it is crucial to allow F to be in the p-adic completion of the polynomial ring rather than in the polynomial ring itself; indeed if one restricts to polynomial F's the main interesting examples of the theory are left out.

6.1.2.3 Jet spaces

More generally, the Lie–Cartan geometric theory of differential equations has an arithmetic analogue which we explain now. Let $M \to T$ be a submersion of smooth manifolds with $\dim T = 1$, and let

$$J^n(M/T) = \{J_t^n(s); s \in \Gamma(M/T), t \in T\}$$

be the space of n-jets $J_t^n(s)$ of smooth sections s of $M \to T$ at points $t \in T$. Cf. [1, 48, 62] for references to differential geometry. If $M = \mathbb{R}^d \times \mathbb{R}$, $T = \mathbb{R}$, $M \to T$ is the second projection, and $x = (x_1, \dots, x_d)$, t are global coordinates on $\mathbb{R}^d$ and $\mathbb{R}$ respectively then $J^n(M/T) = \mathbb{R}^{(n+1)d} \times \mathbb{R}$ with *jet coordinates* $x, x', \dots, x^{(n)}, t$. So for general $M \to T$ the map $J^n(M/T) \to M$ is a fiber bundle with fiber $\mathbb{R}^{nd}$ where $d + 1 = \dim(M)$. One has the total derivative operator

$$\delta_t : C^\infty(J^n(M/T)) \to C^\infty(J^{n+1}(M/T))$$

which in coordinates is given by

$$\delta_t = \frac{\partial}{\partial t} + \sum_{j=0}^{n} \sum_{i=1}^{d} x_i^{(j+1)} \frac{\partial}{\partial x_i^{(j)}}.$$

In the arithmetic theory the analogues of the manifolds $J^n(M/T)$ are certain formal schemes (called *arithmetic jet spaces* or *p-jet spaces*) $J^n(X) = J^n(X/R)$ defined as follows. Assume X is affine, $X = Spec \frac{R[x]}{(f)}$ with x, f tuples; the construction that follows is easily globalized to the non-affine case. Let $x', \dots, x^{(n)}, \dots$ be new tuples of variables, consider the polynomial ring $R\{x\} := R[x, x', x'', \dots]$, let $\phi : R\{x\} \to R\{x\}$ be the unique ring homomorphism extending ϕ on R and sending $\phi(x) = x^p + px'$, $\phi(x') = (x')^p + px''$,..., and let

$$\delta = \delta_p : R\{x\} \to R\{x\}$$

be the p-derivation

$$\delta F = \frac{\phi(F) - F^p}{p}.$$

Then one defines

$$J^n(X) = Spf \frac{R[x, x', \dots, x^{(n)}]\hat{}}{(f, \delta f, \dots, \delta^n f)}.$$

6.1.2.4 Differential equations on manifolds

Usual *differential equations* are defined geometrically as elements of the ring $C^\infty(J^n(M/T))$; alternatively such elements are referred to as (time dependent) *Lagrangians* on M. Their analogue in the arithmetic theory, which we call *arithmetic differential equations* [16], are the elements of the ring $\mathcal{O}^n(X) := \mathcal{O}(J^n(X))$. For group schemes the following concept [11] plays an important role: if G is a group scheme of finite type over R then we may consider the R-module $\mathcal{X}^n(G) = Hom_{gr}(J^n(G), \widehat{\mathbb{G}}_a) \subset \mathcal{O}^n(G)$.

Going back to arbitrary schemes of finite type X/R note that $\delta_p : R\{x\} \to R\{x\}$ induces maps $\delta = \delta_p : \mathcal{O}^n(X) \to \mathcal{O}^{n+1}(X)$ which can be viewed as arithmetic analogues of the total derivative operator $\delta_t : C^\infty(J^n(M/T)) \to C^\infty(J^{n+1}(M/T))$. The latter is a "generator" of the Cartan distribution defined by $dx_i - x_i' dt$, $dx_i' - x_i'' dt$, etc. Note however that the forms in differential geometry defining the Cartan distribution do not have a direct arithmetic analogue; for one thing there is no form "dp" analogous to dt. On the other hand in the arithmetic case we have induced ring homomorphisms $\phi : \mathcal{O}^n(X) \to \mathcal{O}^{n+1}(X)$ which have no analogue in differential geometry.

6.1.2.5 Differential functions

Any differential equation $F \in C^\infty(J^n(M/T))$ defines a natural *differential function* $F_* : \Gamma(M/T) \to C^\infty(T)$; in coordinates sections $s \in \Gamma(M/T)$ correspond to functions $x = x(t)$ and then F_* sends $x(t)$ into $F_*(x(t)) = F(x(t), \delta_t x(t), \dots, \delta_t^n x(t))$. Analogously any arithmetic differential equation $f \in \mathcal{O}(J^n(X))$ defines a map of sets $f_* : X(R) \to R$, referred to as a δ-*function*, which in affine coordinates sends $\alpha \in X(R) \subset R^N$ into $f_*(\alpha) := F(\alpha, \delta_p \alpha, \dots, \delta_p^n \alpha) \in R$ if $F \in R[x, x', \dots, x^{(n)}]\hat{}$ represents f. If $X = G$ is in addition a group scheme and $\psi \in \mathcal{X}^n(G)$ then $\psi_* : G(R) \to \mathbb{G}_a(R) = R$ is a group homomorphism called a δ-*character* of G. For X/R smooth and R/pR algebraically closed f is uniquely determined by f_* so one can identify f with f_* and ψ with ψ_*.

6.1.2.6 Prolongations of vector fields

For any vertical vector field

$$\xi := \sum_{i=1}^{d} a_i(t, x) \frac{\partial}{\partial x_i}$$

on M/T one can consider the canonical prolongations

$$\xi^{(n)} := \sum_{j=0}^{n} \sum_{i=1}^{d} (\delta_t^j a_i(t, x)) \frac{\partial}{\partial x_i^{(j)}}$$

on $J^n(M/T)$. The map

$$\xi^{(n)} : C^\infty(J^n(M/T)) \to C^\infty(J^n(M/T))$$

is the unique $\mathbb{R}$-derivation whose restriction to $C^\infty(M)$ is ξ and which commutes with the total derivative operator δ_t. The above construction has an arithmetic analogue that plays a key technical role in the development of the theory. Indeed for any affine smooth X/R and any R-derivation $\xi : \mathcal{O}(X) \to \mathcal{O}(X)$ the canonical prolongation

$$\xi^{(n)} : \mathcal{O}^n(X)[1/p] \to \mathcal{O}^n(X)[1/p]$$

is defined as the unique $K = R[1/p]$-derivation whose restriction to $\mathcal{O}(X)$ is ξ and which commutes with ϕ. This construction then obviously globalizes.

6.1.2.7 Infinitesimal symmetries of differential equations

Some Galois theoretic concepts based on prolongations of vector fields have arithmetic analogues. Indeed recall that if $\mathcal{L} \subset C^\infty(J^n(M/T))$ is a linear subspace of differential equations then a vertical vector field ξ on M/T is called an *infinitesimal symmetry* of $\mathcal{L}$ if $\xi^{(n)}\mathcal{L} \subset \mathcal{L}$; it is called a *variational infinitesimal symmetry* if $\xi^{(n)}\mathcal{L} = 0$. Similarly given an R-submodule $\mathcal{L} \subset \mathcal{O}^n(X)$ an *infinitesimal symmetry* of $\mathcal{L}$ is an R-derivation $\xi : \mathcal{O}(X) \to \mathcal{O}(X)$ such that $\xi^{(n)}\mathcal{L} \subset \mathcal{L}[1/p]$. One says ξ is a *variational infinitesimal symmetry* of $\mathcal{L}$ if $\xi^{(n)}\mathcal{L} = 0$.

6.1.2.8 Total differential forms on manifolds

Recall that a *total differential form* ([62], p. 351) on M/T is an expression that in coordinates looks like a sum of expressions

$$F(t, x, x', \dots, x^{(n)})dx_{j_1} \wedge \dots \wedge dx_{j_i},$$

It is important to introduce an arithmetic analogue of this which we now explain. We consider the case of top forms ($i = d$) which leads to what we will call *δ-line bundles*. Denote by $\mathcal{O}^n$ the sheaf $U \mapsto \mathcal{O}^n(U)$ on X for the Zariski topology. Define a *δ-line bundle* of order n on X to be a locally free $\mathcal{O}^n$-module of rank 1. Integral powers of bundles need to be generalized as follows. Set $W = \mathbb{Z}[\phi]$ (ring of polynomials with $\mathbb{Z}$-coefficients in the symbol ϕ). For $w = \sum a_s\phi^s$ write $deg(w) = \sum a_s$. Also let W_+ be the set of all $w = \sum a_s\phi^s \in W$ with $a_s \geq 0$ for all s. If L is a line bundle on X given by a cocycle (g_{ij}) and $w = \sum_{s=0}^n a_s\phi^s \in W$, $w \neq 0$, $a_n \neq 0$, then define a δ-line bundle L^w of order n by the cocycle (g_{ij}^w), $g_{ij}^w = \prod_s \phi^s(g_{ij})^{a_s}$. With all these definitions in place we may define the following rings which, by the way, are the main objects of the theory:

$$R_\delta(X, L) = \bigoplus_{0 \neq w \in W_+} H^0(X, L^w).$$

Note that the above is a graded ring without unity. The homogeneous elements of $R_\delta(X, L)$ can be viewed as arithmetic analogues of Lagrangian densities and can also be referred to as *arithmetic differential equations* [16].

6.1.2.9 Differential forms on jet spaces and calculus of variations

The spaces

$$\Omega^i_\uparrow(J^n(M/T))$$

of vertical smooth i-forms on $J^n(M/T)$ (generated by i-wedge products of forms $dx, dx', \dots, dx^{(n)}$) play an important role in the calculus of variations [62]. These spaces fit into a deRham complex where the differential is the vertical exterior differential

$$d_\uparrow : \Omega^i_\uparrow(J^n(M/T)) \to \Omega^{i+1}_\uparrow(J^n(M/T))$$

with respect to the variables $x, x', \dots, x^{(n)}$. On the other hand we have unique operators

$$\delta_t : \Omega^i_\uparrow(J^n(M/T)) \to \Omega^i_\uparrow(J^{n+1}(M/T))$$

that commute with $d_\uparrow$, induce a derivation on the exterior algebra, and for $i = 0$ coincide with the total derivative operators. Then one can define spaces of *functional forms* ([62], p. 357)

$$\Omega^i_*(J^n(M/T)) = \frac{\Omega^i_\uparrow(J^n(M/T))}{Im(\delta_t)}$$

and the (vertical part of the) *variational complex* ([62], p. 361) with differentials

$$d : \Omega^i_*(J^n(M/T)) \to \Omega^{i+1}_*(J^n(M/T)).$$

The class of $\omega \in \Omega^i_\uparrow(J^n(M/T))$ in $\Omega^i_*(J^n(M/T))$ is denoted by $\int \omega dt$. For $i = 0$, $\omega = F$, have the formula $d(\int F dt) = \int EL(F)dt$ in $\Omega^i_*(J^{2n}(M/T))$ where $EL(F) = \sum_{i=1}^d F_i dx_i$ is the *Euler–Lagrange* total differential form,

$$F_i = \sum_{j=0}^n (-1)^j \delta_t^j \left(\frac{\partial F}{\partial x_i^{(j)}} \right).$$

Noether's theorem then says that for any vertical vector field ξ on M/T we have the formula

$$\langle \xi, EL(F) \rangle - \xi^{(n)}(F) = \delta_t G$$

for some $G \in C^\infty(J^{2n-1}(M/T))$; G is unique up to a constant. If ξ is a variational infinitesimal symmetry of F then G is referred to as the *conservation law* attached to this symmetry; in this case if $x(t)$ is a solution of all $F_i = 0$ then G evaluated at $x(t)$ will be a constant.

Analogously, for X/R smooth and affine, one can consider the modules $\Omega^i_{\mathcal{O}^n(X)/R}$ defined as the exterior powers of the inverse limit of the Käher differentials

$$\Omega_{\mathcal{O}^n(X)\otimes(R/p^nR)/(R/p^nR)},$$

and the exterior differential

$$d : \Omega^i_{\mathcal{O}^n(X)/R} \to \Omega^{i+1}_{\mathcal{O}^n(X)/R}.$$

Also one may consider the operators

$$\phi^* : \Omega^i_{\mathcal{O}^n(X)/R} \to \Omega^i_{\mathcal{O}^{n+1}(X)/R};$$

again ϕ^* and d commute. Also note that any element in the image of ϕ^* is uniquely divisible by p^i; for any $\omega \in \Omega^i_{\mathcal{O}^n(X)/R}$ and $r \geq 0$ we then set $\omega_r = p^{-ir}\phi^{*r}\omega$. The operation $p^{-ir}\phi^{*r}$ is a characteristic zero version of the inverse Cartier operator. For any element $\mu \in R$ (which we refer to as *eigenvalue*) one can define groups

$$\Omega^i_*(J^n(X)) = \frac{\Omega^i_{\mathcal{O}^n(X)/R}}{Im(\phi^* - \mu)}$$

that fit into a *variational complex* with differentials

$$d : \Omega^i_*(J^n(X)) \to \Omega^{i+1}_*(J^n(X)).$$

The class of $\omega \in \Omega^i_{\mathcal{O}^n(X)/R}$ in $\Omega^i_*(J^n(X))$ is denoted by $\int \omega dp$. For $i = 0$, $\mu = 1$, $\omega = F \in \mathcal{O}^n(X)$ we have $d(\int F dp) = \int \{\sum_{i=1}^d F_i \omega^i_n\} dp$ in $\Omega^1_*(J^{2n}(X))$ where $F_i \in \mathcal{O}^{2n}(X)$, ω^i is a basis of $\Omega_{\mathcal{O}(X)/R}$, and $\omega^i_n = p^{-n}\phi^{*n}\omega^i$. Also an analogue of the Noether theorem holds in this context with $\epsilon(F) := \sum_{i=1}^d F_i \omega^i_n$ playing the role of the Euler–Lagrange form; indeed for any vector field ξ on X there exists $G \in \mathcal{O}^{2n-1}(X)$ such that

$$\langle \xi^{(n)}, \epsilon(F) \rangle - \xi^{(n)}(F) = G^\phi - G.$$

If ξ is a variational infinitesimal symmetry of F then G can be referred to as a *conservation law*; in this case if $P \in X(R)$ is a point which is a solution to all $F_{i*}(P) = 0$ then $G_*(P)^\phi = G_*(P)$.

6.1.2.10 Flows and Hamiltonian formalism [31]

We place ourselves in either the smooth or the complex analytic setting. Assume one is given a section $\sigma : M \to J^1(M/T)$ of the projection $J^1(M/T) \to M$. To give such a σ is equivalent to giving a vector field δ_M on M lifting δ_t; such a vector field can be referred to as a δ_t-flow on M/T. In coordinates, if σ is defined by $(t, x) \mapsto (t, x, s(t, x))$, then

$$\delta_M = \sum s_i(t, x) \frac{\partial}{\partial x_i}.$$

The composition

$$L_{\delta_M} : \Omega^i_\uparrow(M/T) \xrightarrow{\delta_t} \Omega^i_\uparrow(J^1(M/T)) \xrightarrow{\sigma^*} \Omega^i_\uparrow(M/T)$$

induces a derivation on the exterior algebra of vertical forms on M/T, commutes with exterior vertical differentiation $d_\uparrow$, and coincides with δ_M for $i = 0$; one can refer to L_{δ_M} as the "Lie derivative" attached to the vector field δ_M (it is a relative version of the usual Lie derivative).

We consider now the "time-dependent" Hamiltonian formalism. The classical treatment of this formalism (e.g. [56]) usually involves the differential of time, dt. Since "dp" does not exist in the arithmetic setting what we will do will be to first find a presentation of the classical time dependent Hamiltonian formalism without reference to dt and then we shall transpose the latter to the arithmetic case. Assume for simplicity that M has dimension 3 so the fibers of $M \to T$ are 2-dimensional. A vertical symplectic form on M is a nowhere vanishing vertical 2-form on M. A vertical contact form on M is a vertical 1-form ν on M such that $d\nu$ is symplectic. We shall drop the word "vertical" in what follows. Recall that a δ_t-flow on M is a vector field δ_M on M that lifts the vector field δ_t on T; and recall that to give a δ-flow on M is the same as to give a section of the projection $J^1(M/T) \to M$. Given a δ-flow δ_M as above we may consider the relative Lie derivative L_{δ_M} on the vertical forms $\Omega^i(M/T)$. A δ_t-flow δ_M will be called *Hamiltonian* with respect to a symplectic form η on M if $L_{\delta_M}\eta = \mu \cdot \eta$ for some function μ on T. Note that if $\delta_t\gamma = -\mu \cdot \gamma$ for some function γ on T and if $\eta_1 = \gamma\eta$ then $L_{\delta_M}\eta_1 = 0$. If ν is a contact form, $\eta = d_\uparrow\nu$, and δ_M is a δ_t-flow which is Hamiltonian with respect to η (hence $L_{\delta_M}\eta = \mu\eta$) then $\epsilon := L_{\delta_M}\nu - \mu\nu$ is closed; the latter can be referred to as an Euler–Lagrange form attached to the symplectic form η and the δ_t-flow δ_M. If ϵ happens to be exact, i.e. $L_{\delta_M}\nu - \mu\nu = d\mathfrak{L}$ for some function $\mathfrak{L}$ on M, then $\mathfrak{L}$ can be referred to as a Lagrangian attached to η and δ_M. (This Euler–Lagrange formalism is related to, but does not coincide with, the Euler–Lagrange formalism in 6.1.2.9.) In a series of interesting applications (especially in the complex analytic setting of the above, e.g. to Painlevé equations [56]) M is taken to be $J^1(N/T)$ where $N \to T$ is a family of curves; in this case a 1-form on M is called canonical if it is a function on M times the pull-back of a 1-form on N. Also one is interested, in this situation, in δ_t-flows defined by sections of $J^1(M/T) \to M = J^1(N/T)$ that arise from sections of $J^2(N/T) \to J^1(N/T)$ (where $J^2(N/T)$ is naturally embedded into $J^1(M/T)$); such δ_t-flows are referred to as *canonical* and are defined by order 2 differential equations (functions on $J^2(N)$).

A justification for the above terminology comes, for instance, from the following local algebraic considerations. Let F be a field equipped with a derivation $\delta : F \to F$ (e.g. $F = \mathbb{C}((t))$, $\delta = \delta_t$) and let A be a ring isomorphic to a power series ring in 2 variables with coefficients in F. Choose formal coordinates $x, y \in A$ (i.e. $A = F[[x, y]]$), and set $\Omega_A = A \cdot dx + A \cdot dy$,

$\Omega^2_A = A \cdot dx \wedge dy$. A form $\eta = u dx \wedge dy \in \Omega^2_A$ is called invertible if $u(0,0) \neq 0$. (These definitions are independent of the choice of the formal coordinates x, y.) The following is a translation into the language of differential algebra of the classical Hamiltonian/Lagrangian formalism in 2 analytic coordinates. Assume one is given a derivation $\delta : A \to A$ such that $\delta F \subset F$ and sending the maximal ideal of A into itself. Then the following are easy to check:

1) Assume we are given an invertible $\eta \in \Omega^2_A$. Then there exist formal coordinates $x, y \in A$ such that $\eta = dy \wedge dx$. Furthermore if $\delta\eta = 0$ then and there exists an element $\mathcal{H} \in A$ such that
$$\delta y = -\frac{\partial \mathcal{H}}{\partial x}, \quad \delta x = \frac{\partial \mathcal{H}}{\partial y}.$$
2) Assume there exist formal coordinates $x, y \in A$, and a series $\mathcal{H} \in A$ such that $\delta y = -\frac{\partial \mathcal{H}}{\partial x}$ and $\delta x = \frac{\partial \mathcal{H}}{\partial y}$. Let $\eta = dy \wedge dx$, $\nu = y dx$, $\epsilon = \delta\nu$, and $\mathcal{L} = y\frac{\partial \mathcal{H}}{\partial y} - \mathcal{H} = y\delta x - \mathcal{H}$. Then
$$d\nu = \eta, \quad \delta\eta = 0, \quad d\mathcal{L} = \epsilon.$$
3) Assume the notation in 2) and assume the series $\frac{\partial^2 \mathcal{H}}{\partial y^2}$ is invertible in $F[[x, y]]$. Then there exists a unique F-derivation denoted by $\frac{\partial}{\partial \delta x}$ on $F[[x, y]]$ sending $x \mapsto 0$ and $\delta x \mapsto 1$. Moreover we have
$$\frac{\partial \mathcal{L}}{\partial \delta x} = y, \quad \frac{\partial \mathcal{L}}{\partial x} = -\frac{\partial \mathcal{H}}{\partial x}, \quad \delta\left(\frac{\partial \mathcal{L}}{\partial \delta x}\right) = \frac{\partial \mathcal{L}}{\partial x}.$$

We want to introduce an arithmetic analogue of the above formalism; cf. [31]. To simplify our discussion let X be a smooth surface over R (or the p-adic completion of such a surface). Recall that by an i-form on X we understand a global section of Ω^i_X (projective limit of usual Kähler differentials of X mod p^n). A symplectic form on X is an invertible 2-form on X. A contact form on X is a 1-form on X such that $d\nu$ is symplectic. By a δ-flow on X we will understand a section $X \to J^1(X)$ of the projection $J^1(X) \to X$. Let us further specialize the situation to the case when $X = J^1(Y)$ where Y is a curve (affine, to simplify). A 1-form ν on $X = J^1(Y)$ is called canonical if $\nu = g\beta$ with $g \in \mathcal{O}(X)$ and β a 1-form on Y. If ν is a canonical 1-form on $X = J^1(Y)$ and is closed, i.e. $d\nu = 0$, then ν is (the pull-back of) a 1-form on Y. Note that $J^2(Y)$ naturally embeds into $J^1(J^1(Y))$. A δ-flow on $X = J^1(Y)$ will be called *canonical* if the section $J^1(Y) = X \to J^1(X) = J^1(J^1(Y))$ factors through a section $J^1(Y) \to J^2(Y)$. This concept is, however, too restrictive for applications. We will need to relax the concept of canonical δ-flow as follows. To simplify the discussion we restrict to the case of order 2 differential equations. Let $f \in \mathcal{O}(J^2(Y))$ and let $Z = Z^2(f) \subset J^2(Y)$ be the closed formal subscheme defined by f. For $i \geq 1$ we define

$$\Omega'^i_Z = \frac{\Omega^i_{J^n(Y)}}{\langle f\Omega^i_{J^n(Y)}, df \wedge \Omega^{i-1}_{J^n(Y)}\rangle}.$$

So $\Omega'^1_Z = \Omega'_Z$. Also we have natural wedge products

$$\wedge : \Omega'^i_Z \times \Omega'^j_Z \to \Omega'^{i+j}_Z,$$

and exterior differentiation maps

$$d : \Omega'^i_Z \to \Omega'^{i+1}_Z.$$

We say that f defines a generalized canonical δ-flow on $J^1(Y)$ if the natural map

$$\pi^*\Omega_{J^1(Y)} \to \Omega'_Z$$

is injective and has a cokernel annihilated by a power of p. Note now that the natural maps $\phi^* : \Omega^i_{J^1(Y)} \to \Omega^i_{J^2(Y)}$ induce natural maps denoted by

$$\phi^*_Z : \Omega^i_{J^1(Y)} \to \Omega'^i_Z.$$

We say that the generalized canonical δ-flow is *Hamiltonian* with respect to a sympletic form η on $J^1(Y)$ if $\phi^*_Z\eta = \mu \cdot \eta$ in $\Omega'^2_{Z^2(f)}$ for some $\mu \in pR$. If in addition $\eta = d\nu$ for some canonical 1-form ν on $J^1(Y)$ we call $\epsilon := \frac{1}{p}(\phi^*_Z\nu - \mu\nu) \in \Omega'_{Z^2(f)}$ an Euler–Lagrange form. (This arithmetic Euler–Lagrange formalism is related to, but does not coincide with, the arithmetic Euler–Lagrange formalism in 6.1.2.9.) In the notation above $p\epsilon$ is closed in the sense that $d(p\epsilon) = 0$ in $\Omega'^2_{Z^2(f)}$. The above is the framework for the main results in [31] about the arithmetic analogue of Painlevé VI.

6.1.2.11 "Category of differential equations"

A categorical framework for differential equations on manifolds can be introduced (cf. [1], for instance). In one variant of this the objects are locally isomorphic to projective systems of submanifolds of $J^n(M/T)$ compatible with the total derivative operator and morphisms are smooth maps between these, again compatible with the total derivative operator. This categorical framework has an arithmetic analogue as follows. Note first that ϕ acts naturally on $R_\delta(X, L)$ but δ does not. Nevertheless for any homogeneous $s \in R_\delta(X, L)$ of degree v the ring $R_\delta(X, L)_{(s)}$ of all fractions f/s^w with f homogeneous of degree wv has a natural p-derivation δ on it. This inductive system $X_\delta(L) = (R_\delta(X, L)_{(s)}, \delta)_s$ of rings equipped with p-derivations δ is an object of a natural category underlying a geometry more general than algebraic geometry which we refer to as *δ-geometry* [16]. This geometry is an arithmetic analogue of the categorical setting in [1] and also an arithmetic analogue of the Ritt–Kolchin δ-algebraic geometry [52, 9]. If one restricts to étale maps of smooth schemes we have a functor $X \mapsto X_\delta = X_\delta(K^\nu)$ from "algebraic geometry" to our "δ-geometry" by taking $L = K^\nu$ to be a fixed power of the canonical bundle; the natural choice later in the theory turns out to be the anti-canonical bundle $L = K^{-1}$. In δ-geometry X_δ should be viewed as an infinite dimensional object.

6.1.2.12 Differential invariants

Recall the concept of *differential invariant* which plays a key role in the "Galois theoretic" work of Lie and Cartan. Assume that a group G acts on M and T such that $M \to T$ is G-equivariant. (In this discussion assume $d := \dim T$ is arbitrary.) Then G acts on $J^n(M/T)$ and the ring $C^\infty(J^n(M/T))^G$ of G-invariant elements in the ring $C^\infty(J^n(M/T))$ is called the ring of *differential invariants*. There are two extreme cases of special interest: the case when G acts trivially on T and the case when G acts transitively on T. A special case of the first extreme case mentioned above is that in which $M = F \times T$, with G a Lie group acting trivially on T and transitively on F; this leads to the context of Cartan's moving frame method and of Cartan connections. For $\dim T = 1$, T is the "time" manifold, F is the "physical space", and sections in $\Gamma(M/T)$ correspond to particle trajectories. A special case of the second extreme case mentioned above is the situation encountered in the study of "geometric structures" and in the formulation of field theories, in which G is the group Diff(T) of diffeomorphisms of T, T is viewed as the "physical space" or "physical space-time", and M is a *natural bundle* over T, i.e. a quotient $M = \Gamma\backslash\mathrm{Rep}_n(T)$ of the bundle $\mathrm{Rep}_n(T) \to T$ of n-jets of frames $(\mathbb{R}^d, 0) \to T$ by a Lie subgroup Γ of the group $\mathrm{Aut}_n(\mathbb{R}^d, 0)$ of n-jets of diffeomorphisms of $(\mathbb{R}^d, 0)$; cf. [1], pp. 150-153 and 183. (E.g. the Riemannian metrics on T identify with the sections in $\Gamma(M_{1,O(d)}/T)$ where $O(d) < GL(d) = \mathrm{Aut}_1(\mathbb{R}^d, 0)$ is the orthogonal group.) In the case when the action of G on T is trivial the ring of differential invariants above turns out to have an interesting arithmetic analogue, namely the ring of *δ-invariants of a correspondence*; cf. the discussion below. The special case of Cartan connections has also an arithmetic analogue: the *arithmetic logarithmic derivative* attached to a *δ-flow*. The latter has a flavor different from that of δ-invariants of correspondences and will be discussed later. There are also interesting candidates for arithmetic analogues of the situation when $G = \mathrm{Diff}(T)$; cf. our discussion of Lie groupoids below.

We next explain the ring of δ-invariants of a correspondence. Pursuing an analogy with usual geometric invariant theory assume we are given (X, L), a *correspondence* on X (i.e. a morphism $\sigma = (\sigma_1, \sigma_2) : \tilde{X} \to X \times X$), and a *linearization* of L (i.e. an isomorphism $\beta : \sigma_1^* L \simeq \sigma_2^* L$). Then we may define the ring of *δ-invariants of σ* by

$$R_\delta(X, L)^\sigma = \{f \in R_\delta(X, L);\ \beta\sigma_1^* f = \sigma_2^* f\}.$$

(This ring, again, has no unity!) The homogeneous elements s of $R_\delta(X, L)^\sigma$ can be viewed as arithmetic analogues of total differential forms invariant under appropriate symmetries. The inductive system of rings

$$(R_\delta(X, L)^\sigma_{\langle s \rangle}, \delta)$$

equipped with p-derivations δ can be viewed as an incarnation of the quotient space "X_δ/σ_δ" in δ-geometry (and, under quite general hypotheses, is indeed the categorical quotient of X_δ by σ_δ); note that, in most interesting examples (like the ones in Theorem 6.6 below), the quotient X/σ does not exist in usual algebraic geometry (or rather the categorical quotient in algebraic geometry reduces to a point).

It is worth revisiting the sequence of ideas around differential invariants. Classical Galois theory deals with algebraic equations (satisfied by numbers). Lie and Cartan extended Galois' ideas, especially through the concept of differential invariant, to the study of differential equations (satisfied by functions); roughly speaking they replaced numbers by functions. In the theory presented here functions are replaced back by numbers. But we did *not* come back to where things started because we have added new structure, the operator $\delta = \delta_p$. In particular the δ-invariants mentioned above, although arithmetic in nature, and although attached to algebraic equations (defining X, $\tilde{X}$, σ), are nevertheless *not* "Galois theoretic" in any classical sense.

6.1.2.13 Lie groupoids

Let us go back to the (interrelated) problems of finding arithmetic analogues of $T \times T$ and of Diff(T). The arithmetic analogue of $S := T \times T$ is usually referred to as the hypothetical tensor product "$\mathbb{Z} \otimes_{\mathbb{F}_1} \mathbb{Z}$" over the "field with one element"; cf. [57] for some history of this. The arithmetic analogue of Diff(T) could be thought of as the Galois group "Gal$(\mathbb{Q}/\mathbb{F}_1)$". According to a suggestion of Borger [6] one should take "$\mathbb{Z} \otimes_{\mathbb{F}_1} \mathbb{Z}$" to be, by definition, the big Witt ring of $\mathbb{Z}$. Since we are here in a local arithmetic situation it is convenient, for our purposes, to take, as an arithmetic analogue of $S = T \times T$, the schemes $\Sigma_m = Spec\ W_m(R)$, where W_m is the p-typical functor of Witt vectors of length $m + 1$ (we use Borger's indexing) and R is a complete local p-ring. On the other hand an infinitesimal analogue of Diff(T) is the Lie groupoid of T defined as the projective system of groupoids $\mathcal{G}_n(T) := J^n(S/T)^*$ (where the upper $*$ means taking "invertible" elements and $S = T \times T \to T$ is the second projection.) So an arithmetic analogue of the system $J^n(S/T)^*$ (which at the same time would be an infinitesimal analogue of "Gal($\mathbb{Q}/\mathbb{F}_1$)") would be the system $J^n(\Sigma_m)$ equipped with the natural maps induced by the comonad map. This system is a rather non-trivial object [29]. It is worth noting that a good arithmetic analogue of $J^n(S/T)^*$ could also be the "usual" jet spaces (in the sense of [8]) of $W_m(R)/R$ (which in this case can be constructed directly from the module of Käher differentials $\Omega_{W_m(R)/R}$). By the way $\Omega_{W_m(R)/R}$ is also the starting point for the construction of the deRham–Witt complex [44]. However Ω involves usual derivations (rather than Fermat quotients) so taking Ω as a path to an arithmetic analogue of $J^n(S/T)^*$ seems, again, like "going arithmetic halfway". On the contrary taking the system $J^n(\Sigma_m)$ as the analogue of $J^n(S/T)^*$ seems to achieve, in some sense, the task of "going arithmetic all the way".

We end by remarking that if we denote π_1 and π_2 the source and target projections from $\mathcal{G}_n(T)$ into T then there are natural "actions" ρ of the groupoids $\mathcal{G}_n(T)$ on all natural bundles $M_{n,\Gamma} := \Gamma\backslash \text{Rep}_n(T) \to T$ fitting into diagrams:

$$\begin{array}{ccc} M_{n,\Gamma} \times_{T,\pi_1} \mathcal{G}_n(T) & \stackrel{\rho}{\longrightarrow} & M_{n,\Gamma} \\ p_2 \downarrow & & \downarrow \\ \mathcal{G}_n(T) & \stackrel{\pi_2}{\longrightarrow} & T \end{array}$$

where p_2 is the second projection. The above induce "actions"

$$\begin{array}{ccc} J^n(M_{n,\Gamma}/T) \times_{T,\pi_1} \mathcal{G}_n(T) & \stackrel{\rho}{\longrightarrow} & J^n(M_{n,\Gamma}/T) \\ p_2 \downarrow & & \downarrow \\ \mathcal{G}_n(T) & \stackrel{\pi_2}{\longrightarrow} & T \end{array}$$

where p_2 is again the second projection. One can consider rings of differential invariants

$$C^\infty(J^n(M_{n,\Gamma}/T))^\rho := \{F \in C^\infty(J^n(M_{n,\Gamma}/T));\ F \circ \rho = F \circ p_1\}$$

where $p_1 : J^n(M_{n,\Gamma}/T) \times_{T,\pi_1} \mathcal{G}_n(T) \to J^n(M_{n,\Gamma}/T)$ is the fist projection. There should be arithmetic analogues of the above "actions" and rings of differential invariants. One can argue that the analogue of $\text{Rep}_n(T)$ is, again, the system $J^n(\Sigma_m)$. Then for $\Gamma = 1$ the analogue of $J^n(M_{n,\Gamma}/T)$ might be $J^n(J^n(\Sigma_m))$; the analogue, for $\Gamma = 1$, of the "action" ρ above could then be the map $J^n(J^n(\Sigma_{m',m''})) \to J^n(J^n(\Sigma_{m'+m''}))$ where $\Sigma_{m',m''} = Spec\ W_{m'}(W_{m''}(R))$. A challenge would then be to find arithmetic analogues of non-trivial Γs.

6.1.2.14 Differential Galois theory of linear equations

This subject is best explained in a complex (rather than real) situation. Classically (following Picard-Vessiot and Kolchin [51]) one starts with a differential field $\mathcal{F}$ of meromorphic functions on a domain D in the complex plane $\mathbb{C}$ and one fixes an $n \times n$ matrix $A \in \mathfrak{gl}_n(\mathcal{F})$ in the Lie algebra of the general linear group $GL_n(\mathcal{F})$. The problem then is to develop a "differential Galois theory" for equations of the form

$$\delta_z U = A \cdot U$$

where $U \in GL_n(\mathcal{G})$, $\mathcal{G}$ a field of meromorphic functions on a subdomain of D, and z a coordinate on D. The start of the theory is as follows. One fixes a solution U and introduces the differential Galois group $G_{U/\mathcal{F}}$ of $U/\mathcal{F}$ as the group of all $\mathcal{F}$−automorphisms of the field $\mathcal{F}(U)$ that commute with d/dz. One can ask for an arithmetic analogue of this theory. There is a well developed difference algebra analogue of linear differential equations [65]; but the arithmetic differential theory is still in its infancy [33, 34, 35]. What is being proposed in loc. cit. in the arithmetic theory is to fix a matrix $\alpha \in \mathfrak{gl}_n(R)$ and define δ-linear equations as equations of the form

$$\delta u = \alpha \cdot u^{(p)}$$

where $u = (u_{ij}) \in GL_n(R)$, $\delta u := (\delta u_{ij})$, and $u^{(p)} := (u_{ij}^p)$. Note that the above equation is equivalent to $\phi(u) = \epsilon \cdot u^{(p)}$ where $\epsilon = 1 + p\alpha$ which is *not* a difference equation for ϕ in the sense of [65]; indeed difference equations for ϕ have the form $\phi(u) = \epsilon \cdot u$. To define the δ-Galois group of such an equation start with a δ-subring $\mathcal{O} \subset R$ and let $u \in GL_n(R)$ be a solution of our equation. Let $\mathcal{O}[u] \subset R$ the ring generated by the entries of u; clearly $\mathcal{O}[u]$ is a δ-subring of R. We define the *δ-Galois group* $G_{u/\mathcal{O}}$ of $u/\mathcal{O}$ as the subgroup of all $c \in GL_n(\mathcal{O})$ for which there exists an $\mathcal{O}$-algebra automorphism σ of $\mathcal{O}[u]$ such that $\sigma \circ \delta = \delta \circ \sigma$ on $\mathcal{O}[u]$ and such that $\sigma(u) = uc$. The theory starts from here.

6.1.2.15 Maurer–Cartan connections

The Maurer–Cartan connection attached to a Lie group G is a canonical map $T(G) \to L(G)$ from the tangent bundle $T(G)$ to the Lie algebra $L(G)$; for $G = GL_n$ it is given by the form $dg \cdot g^{-1}$ and its algebraic incarnation is Kolchin's logarithmic derivative map [51] $GL_n(\mathcal{G}) \to \mathfrak{gl}_n(\mathcal{G})$, $u \mapsto \delta_z u \cdot u^{-1}$. In our discussion of linear equations above the arithmetic analogue of the Kolchin logarithmic derivative map is the map $GL_n(R) \to \mathfrak{gl}_n(R)$,

$$u \mapsto \delta u \cdot (u^{(p)})^{-1}.$$

This map is naturally attached to the lift of Frobenius $\phi_{GL_n,0} : \widehat{GL_n} \to \widehat{GL_n}$ whose effect on the ring $\mathcal{O}(\widehat{GL_n}) = R[x, \det(x)^{-1}]\hat{}$ is $\phi_{GL_n,0}(x) = x^{(p)}$. The latter lift of Frobenius behaves well (in a precise sense to be explained later) with respect to the maximal torus $T \subset GL_n$ of diagonal matrices and with respect to the Weyl group $W \subset GL_n$ of permutation matrices but it behaves "badly" with respect to other subgroups like the classical groups SL_n, SO_n, Sp_n. (This phenomenon does not occur in the geometry of Lie groups where the Maurer–Cartan form behaves well with respect to *any* Lie subgroup of GL_n, in particular with respect to the classical groups.) In order to remedy the situation one is lead to generalize the above constructions by replacing $\phi_{GL_n,0}$ with other lifts of Frobenius $\phi_{GL_n} : \widehat{GL_n} \to \widehat{GL_n}$ that are adapted to each of these classical groups. Here is a description of the resulting framework.

First we define an arithmetic analogue of the Lie algebra $\mathfrak{gl}_n$ of GL_n as the set $\mathfrak{gl}_n$ of $n \times n$ matrices over R equipped with the non-commutative group law $+_\delta : \mathfrak{gl}_n \times \mathfrak{gl}_n \to \mathfrak{gl}_n$ given by

$$a +_\delta b := a + b + pab,$$

where the addition and multiplication in the right hand side are those of $\mathfrak{gl}_n$, viewed as an associative algebra. There is a natural "δ-adjoint" action $\star_\delta$ of GL_n on $\mathfrak{gl}_n$ given by

$$a \star_\delta b := \phi(a) \cdot b \cdot \phi(a)^{-1}.$$

Assume now one is given a ring endomorphism ϕ_{GL_n} of $\mathcal{O}(\widehat{GL_n})$ lifting Frobenius, i.e. a ring endomorphism whose reduction mod p is the p-power Frobenius on $\mathcal{O}(GL_n)\hat{}/(p) = k[x, \det(x)^{-1}]$; we still denote by $\phi_{GL_n} : \widehat{GL_n} \to \widehat{GL_n}$ the induced morphism of p-formal schemes and we refer to it as a lift of Frobenius on $\widehat{GL_n}$ or simply as a δ-*flow* on $\widehat{GL_n}$. Consider the matrices $\Phi(x) = (\phi_{GL_n}(x_{ij}))$ and $x^{(p)} = (x_{ij}^p)$ with entries in $\mathcal{O}(GL_n)\hat{}$; then $\Phi(x) = x^{(p)} + p\Delta(x)$ where $\Delta(x)$ is some matrix with entries in $\mathcal{O}(GL_n)\hat{}$. Furthermore given a lift of Frobenius ϕ_{GL_n} as above one defines, as usual, a p-derivation δ_{GL_n} on $\mathcal{O}(GL_n)\hat{}$ by setting $\delta_{GL_n}(f) = \frac{\phi_{GL_n}(f) - f^p}{p}$.

Assume furthermore that we are given a smooth closed subgroup scheme $G \subset GL_n$. We say that G is ϕ_{GL_n}-horizontal if ϕ_{GL_n} sends the ideal of G into itself; in this case we have a lift of Frobenius endomorphism ϕ_G on $\widehat{G}$, equivalently on $\mathcal{O}(G)\hat{}$.

Assume the ideal of G in $\mathcal{O}(GL_n)$ is generated by polynomials $f_i(x)$. Then recall that the Lie algebra $L(G)$ of G identifies, as an additive group, to the subgroup of the usual additive group $(\mathfrak{gl}_n, +)$ consisting of all matrices a satisfying

$$\text{“}\epsilon^{-1}\text{”} f_i(1 + \epsilon a) = 0,$$

where $\epsilon^2 = 0$. Let $f_i^{(\phi)}$ be the polynomials obtained from f_i by applying ϕ to the coefficients. Then we define the δ-Lie algebra $L_\delta(G)$ as the subgroup of $(\mathfrak{gl}_n, +_\delta)$ consisting of all the matrices $a \in \mathfrak{gl}_n$ satisfying

$$p^{-1} f_i^{(\phi)}(1 + pa) = 0.$$

The analogue of Kolchin's logarithmic derivative (or of the Maurer–Cartan connection) will then be the map, referred to as the *arithmetic logarithmic derivative*, $l\delta : GL_n \to \mathfrak{gl}_n$, defined by

$$l\delta a := \frac{1}{p}\left(\phi(a)\Phi(a)^{-1} - 1\right) = (\delta a - \Delta(a))(a^{(p)} + p\Delta(a))^{-1}.$$

For G a ϕ_{GL_n}-horizontal subgroup $l\delta$ above induces a map $l\delta : G \to L_\delta(G)$. Now any $\alpha \in L_\delta(G)$ defines an equation of the form $l\delta u = \alpha$, with unknown $u \in G$; such an equation will be referred to as a δ-linear equation (with respect to our δ-flow). Later in the paper we will explain our results about existence of δ-flows on GL_n compatible with the classical groups. These δ-flows will produce, as explained above, corresponding δ-linear equations.

6.1.2.16 Symmetric spaces

In the classical Cartan theory of symmetric spaces one starts with a (real) Lie group G equipped with a Lie group automorphism $x \mapsto x^\tau$ of order 1 or 2 (which we refer to as an involution). One defines the fixed group of the involution

$$G^+ := \{a \in G; a^\tau = a\} \tag{6.1}$$

and homogeneous spaces G/S where $(G^+)^\circ \subset S \subset G^+$ are Lie groups intermediate between the identity component $(G^+)^\circ$ and G^+. Then τ acts on the Lie algebra $L(G)$ and one denotes by $L(G)^+$ and $L(G)^-$ the $+$ and $-$ eigenspaces of this action. We have a decomposition $L(G) = L(G)^+ \oplus L(G)^-$ which can be referred to as a Cartan decomposition corresponding to τ. One can also consider the closed subset of G defined by

$$G^- := \{a \in G; a^\tau = a^{-1}\}. \tag{6.2}$$

This is not a subgroup of G. Set $a^{-\tau} = (a^\tau)^{-1}$. The map $\mathcal{H} : G \to G^-$, $\mathcal{H}(a) := a^{-\tau}a$ identifies G/G^+ with a subset of G^- and there is an obvious compatibility between the tangent map to $\mathcal{H}$ and the Cartan decomposition.

All of the above has an arithmetic analogue as follows. We start with a linear smooth group scheme G over R and an involution τ on G. We define G^+ and G^- as in 6.1 and 6.2 respectively. One also has an action of τ on the δ-Lie algebra $L_\delta(G)$. One considers the subgroup $L_\delta(G)^+ \subset L_\delta(G)$ of all $b \in L_\delta(G)$ such that $b^\tau = b$. One also considers the closed subscheme $L_\delta(G)^- \subset L_\delta(G)$ whose points b satisfy $b^\tau +_\delta b = 0$; $L_\delta(G)^-$ is not a subgroup. For $G = GL_n$ one easily proves that the map $+_\delta : L_\delta(G)^+ \times L_\delta(G)^- \to L_\delta(G)$ is a bijection on points; this can be viewed as an analogue of the Cartan decomposition. Later we will state results having the above as background.

The classical theory of symmetric spaces also considers bilinear (positive definite) forms B on $L(G)$ such that G^+ acts on $L(G)$ via the adjoint action by orthogonal transformations with respect to B; it would be interesting to find an arithmetic analogue of this condition.

6.1.3 Main task of the theory

At this point we may formulate the main technical tasks of the theory. Let, from now on, $R = \widehat{\mathbb{Z}_p^{ur}}$. First given a specific scheme X (or group scheme G) the task is to compute the rings $\mathcal{O}^n(X)$ (respectively the modules $\mathcal{X}^n(G)$). More generally given a specific pair (X, L) we want to compute the ring $R_\delta(X, L)$. Finally given (X, L), a correspondence σ on X, and a linearization of L, we want to compute the ring $R_\delta(X, L)^\sigma$. *The main applications of the theory (cf. the subsection below on motivations) arise as a result of the presence of interesting/unexpected elements and relations in the rings $\mathcal{O}^n(X)$, $R_\delta(X, L)$, $R_\delta(X, L)^\sigma$.*

6.1.4 Motivations of the theory

There are a number of motivations for developing such a theory.

6.1.4.1 Diophantine geometry

Usual differential equations satisfied by functions can be used to prove diophantine results over function fields (e.g. the function field analogues of the Mordell conjecture [55] and of the Lang conjecture [8]). In the same vein one can hope to use "arithmetic differential equations" satisfied by numbers to prove diophantine results over number fields. This idea actually works in certain situations, as we will explain below. Cf. [12, 19]. The general strategy is as follows. Assume one wants to prove that a set S of points on an algebraic variety is finite. What one tries to do is find a system of arithmetic differential equations $F_i(\alpha, \delta_p\alpha, \ldots, \delta_p^n\alpha) = 0$ satisfied by all $\alpha \in S$; then one tries, using algebraic operations and the "differentiation" δ_p, to eliminate $\delta_p\alpha, \ldots, \delta_p^n\alpha$ from this system to obtain another system $G_j(\alpha) = 0$ satisfied by all $\alpha \in S$, where G_j do not involve the "derivatives" anymore. Finally one proves, using usual algebraic geometry that the latter system has only finitely many solutions.

6.1.4.2 "Impossible" quotient spaces

If X is an algebraic variety and $\sigma : \tilde{X} \to X \times X$ is a correspondence on X then the categorical quotient X/σ usually reduces to a point in the category of algebraic varieties. In some sense this is an illustration of the limitations of classical algebraic geometry and suggests the challenge of creating more general geometries in which X/σ becomes interesting. (The non-commutative geometry of A. Connes [38] serves in particular this purpose.) As explained above we proposed, in our work, to replace the algebraic equations of usual algebraic geometry by "arithmetic differential equations"; the resulting new geometry is called δ_p-geometry (or simply δ-geometry). Then it turns out that important class of quotients X/σ that reduce to a point in usual algebraic geometry become interesting in δ-geometry (due to the existence of interesting δ-invariants). A general principle seems to emerge according to which this class coincides with the class of "analytically uniformizable" correspondences. Cf. [16].

6.1.4.3 "Impossible" liftings to characteristic zero

A series of phenomena belonging to algebraic geometry in characteristic p, which do not lift to characteristic 0 in algebraic geometry, can be lifted nevertheless to characteristic 0 in δ-geometry. This seems to be a quite general principle with various incarnations throughout the theory (cf. [15, 16, 25]) and illustrates, again, how a limitation of classical algebraic geometry can be overcome by passing to δ-geometry.

6.1.5 Comparison with other theories

It is worth noting that the paradigm of our work is quite different from the following other paradigms:

1) Dwork's theory of p-adic differential equations [41] (which is a theory about δ_t acting on functions in $\mathbb{Q}_p[[t]]$ and not about δ_p acting on numbers; also Dwork's theory is a theory of linear differential equations whereas the theory here is about analogues of non-linear differential equations),

2) Vojta's jet spaces [66] (which, again, although designed for arithmetic purposes, are nevertheless constructed from Hasse–Schmidt derivations "$\frac{1}{n!}\delta_t^n$" acting on functions and not from operators acting on numbers),
3) Ihara's differentiation of integers [47] (which, although based on Fermat quotients, goes from characteristic zero to characteristic p and hence, unlike our δ_p, cannot be iterated),
4) the point of view of Kurokawa et al. [53] (which uses an operator on numbers very different from δ_p namely $\frac{\partial \alpha}{\partial p} := np^{n-1}\beta$ for $\alpha = p^n\beta \in \mathbb{Z}$, $p \not| \beta$),
5) the theory of Drinfeld modules [39](which is entirely in characteristic p),
6) the difference geometry in the work of Cohn, Hrushovski-Chatzidakis [37], and others (in which the jet spaces are n-fold products of the original varieties as opposed to the jet spaces in our work which are, as we shall see, bundles over the original varieties with fibers affine spaces),
7) Raynaud's deformation to Witt vectors $W_2(k)$ over a field k of characteristic p [64] (which again leads to operators from characteristic zero to characteristic p which cannot be iterated; loosely speaking $W_2(k)$ in Raynaud's approach is replaced in our theory by $W_2(W(k))$).
8) the work of Soulé, Deitmar, Connes, Berkovich, and many others on the "geometry over the field $\mathbb{F}_1$ with one element". In their work passing from the geometry over $\mathbb{Z}$ to the geometry over $\mathbb{F}_1$ amounts to *removing* part of the structure defining commutative rings, e.g. removing addition and hence considering multiplicative monoids instead of rings. On the contrary our theory can be seen as a tentative approach to $\mathbb{F}_1$ (cf. the Introduction to [16]) that passes from $\mathbb{Z}$ to $\mathbb{F}_1$ by *adding* structure to the commutative rings, specifically adding the operator(s) δ_p. This point of view was independently proposed (in a much more systematic form) by Borger [6]. Borger's philosophy is global in the sense that it involves all the primes (instead of just one prime as in our work) and it also proposes to see "positivity" as the corresponding story at the "infinite" prime; making our theory fit into Borger's larger picture is an intriguing challenge.
9) the work of Joyal [49] and Borger [4, 5] on the Witt functor; the Witt functor is a right adjoint to the forgetful functor from "δ-rings" to rings as opposed to the arithmetic jet functor which is a left adjoint to the same forgetful functor. As it is usually the case the left and right stories turn out to be rather different.
10) the theory of the Greenberg transform, cf. Lang's thesis and [42] (which attaches to a scheme X/R varieties $G^n(X)$ over k; one can show [12] that $G^n(X) \simeq J^n(X) \otimes_R k$ so the arithmetic jet spaces are certain remarkable liftings to characteristic zero of the Greenberg transforms. The operators

δ on $\mathcal{O}^n(X)$ do not survive after reduction mod p as operators on the Greenberg transforms.)

11) the work on the deRham–Witt complex, cf., e.g. [44] (which has as its starting point the study of Kähler differential of Witt vectors; on the contrary, what our research suggests, cf. [29], is to push arithmetization one step further by analyzing instead the *arithmetic* jet spaces of Witt vectors.)
12) the theory ϕ-modules (which is a theory about linear equations as opposed to the theory here which is non-linear).

6.2 Main results

We present in what follows a sample of our results. We always set $\overline{A} = A \otimes_{\mathbb{Z}} \mathbb{F}_p = A/pA, \overline{X} = X \otimes_{\mathbb{Z}} \mathbb{F}_p$ for any ring A and any scheme X respectively. Recall that we denote by $\widehat{A}$ the p-adic completion of A; for X Noetherian we denote by $\widehat{X}$ the p-adic completion of X. Also, in what follows, $R := \widehat{\mathbb{Z}_p^{ur}}$, $k := R/pR$. We begin with completely general facts:

6.2.1 Affine fibration structure of p-jet spaces

Theorem 6.1. [11]

1) If X/R is a smooth scheme of relative dimension d then X has an affine covering X_i such that $J^n(X_i) \simeq \widehat{X_i} \times \widehat{\mathbb{A}^{nd}}$ in the category of p-adic formal schemes.

2) If G/R is a smooth group scheme of relative dimension d, with formal group law $\mathcal{F} \in R[[T_1, T_2]]^d$ (T_1, T_2 d-tuples), then the kernel of $J^n(G) \to \widehat{G}$ is isomorphic to $\widehat{\mathbb{A}^{nd}}$ with composition law obtained from the formal series

$$\delta\mathcal{F}, \dots, \delta^n\mathcal{F} \in R[[T_1, T_2, \dots, T_1^{(n)}, T_2^{(n)}]]^d$$

by setting $T_1 = T_2 = 0$.

Note that after setting $T_1 = T_2 = 0$ the series $\delta^n\mathcal{F}$ become restricted, i.e. elements of $(R[T_1', T_2', \dots, T_1^{(n)}, T_2^{(n)}]\hat{\ })^d$ so they define morphisms in the category of p-adic formal schemes. Assertion 1) in the theorem makes p-jet spaces resemble the usual jet spaces of the Lie–Cartan theory. Note however that, even if G is commutative, the kernel of $J^n(G) \to \widehat{G}$ is not, in general the additive group raised to some power. Here is the idea of the proof of 1) for $n = 1$. We may assume $X = Spec\ A$, $A = R[x]/(f)$, and there is an étale map $R[T] \subset A$ with T a d-tuple of indeterminates. Consider the unique ring homomorphism $R[T] \to W_1(A[T'])$ sending $T \mapsto (T, T')$ where W_1 is the functor of Witt vectors of length 2 and T' is a d-tuple of indeterminates. Using the fact that $R[T] \subset A$ is formally étale and the fact that the first projection $W_1(A[T']/(p^n)) \to A[T']/(p^n)$ has a nilpotent kernel, one constructs a compatible sequence of homomorphisms $A \to W_1(A[T']/(p^n))$, $T \mapsto (T, T')$. Hence one gets a homomorphism $A \to W_1(A[T']\hat{})$, $a \mapsto (a, \delta a)$. Then one defines a homomorphism $R[x, x']\hat{}/(f, \delta f) \to A[T']\hat{}$ by sending $x \mapsto a := \text{class}(x) \in A$, $x' \mapsto \delta a$. Conversely one defines a homomorphism $A[T']\hat{} \to R[x, x']\hat{}/(f, \delta f)$ by sending $T' \mapsto \delta T$. The two homomorphisms are inverse to each other which ends the proof of 1) for $n = 1$.

6.2.2 δ-functions and δ-characters on curves

The behavior of the rings $\mathcal{O}^n(X)$ for smooth projective curves X depends on the genus of X as follows:

Theorem 6.2. [11, 12, 32] *Let X be a smooth projective curve over R of genus g.*

1) *If $g = 0$ then $\mathcal{O}^n(X) = R$ for all $n \geq 0$.*
2) *Let $g = 1$. If X is not a canonical lift then $\mathcal{O}^1(X) = R$ (hence $\mathcal{X}^1(X) = 0$) and $\mathcal{X}^2(X)$ is a free module of rank 1; in particular $\mathcal{O}^2(X) \neq R$. If, on the other hand, X is a canonical lift then $\mathcal{O}^1(X) = \mathcal{O}(\widehat{\mathbb{A}^1})$ and $\mathcal{X}^1(X)$ is free of rank one.*
3) *If $g \geq 2$ then $J^n(X)$ is affine for $n \geq 1$; in particular $\mathcal{O}^1(X)$ separates the points of $X(R)$.*

The proof of 1) is a direct computation. The idea of proof of the statements about $\mathcal{X}^n$ in 2) is as follows. Let $N^n = Ker(J^n(X) \to \widehat{X})$. Then one first proves (using Theorem 6.1) that $Hom(N^n, \widehat{\mathbb{G}}_a)$ has rank at least n over R and one computes the ranks of $\mathcal{X}^n(X)$ by looking at the exact squence

$$Hom(J^2(X), \widehat{\mathbb{G}}_a) \to Hom(N^2, \widehat{\mathbb{G}}_a) \to H^1(X, \mathcal{O}).$$

Here $Hom = Hom_{gr}$. The proof of 3) is based on representing $\overline{J^1(X)}$ as $\mathbb{P}(\mathcal{E}) \backslash D$ where $\mathcal{E}$ is a rank 2 vector bundle on $\overline{X}$, and D is an ample divisor.

If $g = 1$ and X is not a canonical lift then a basis ψ for $\mathcal{X}^2(X)$ can be viewed as an analogue of the "Manin map" of an elliptic fibration [55]. Also note that assertion 3) in Theorem 6.2 implies the effective version of the Manin–Mumford conjecture [12]. Indeed Manin and Mumford conjectured that if X

is a complex curve of genus ≥ 2 embedded into its Jacobian A then $X \cap A_{tors}$ is a finite set. This was proved by Raynaud [64]. Mazur later asked [59] if $\sharp(X \cap A_{tors}) \leq C(g)$ where $C(g)$ is a constant that depends only on the genus g of X. Using 3) in Theorem 6.2 one can prove:

Theorem 6.3. [12] *For a smooth projective complex curve X in its Jacobian A we have $\sharp(X \cap A_{tors}) \leq C(g, p)$ where $C(g, p)$ is a constant that depends only on the genus g and (in case X is defined over $\overline{\mathbb{Q}}$) on the smallest prime p of good reduction of X.*

Roughly speaking the idea of proof is as follows. First one can replace the complex numbers by R and A_{tors} by its prime to p torsion subgroup $\Gamma < A(R)$. Then one embeds $X(R) \cap \Gamma$ (via the "jet map") into the the set of k-points of $\overline{J^1(X)} \cap p\overline{J^1(A)}$. But the latter is a finite set because $\overline{J^1(X)}$ is affine, $p\overline{J^1(A)}$ is projective, and both are closed in $\overline{J^1(A)}$. Moreover the cardinality of this finite set can be bounded using Bézout's theorem.

One can ask for global vector fields on a smooth projective curve X/R that are infinitesimal symmetries for given δ-functions on X. The only non-trivial care is that of genus 1 (elliptic curves); indeed for genus 0 there are no non-constant δ-functions (cf. Theorem 6.2) while for genus ≥ 2 there are no non-zero vector fields. Here is the result:

Theorem 6.4. [11, 16] *Let X be an elliptic curve over R with ordinary reduction.*

1) *If X has Serre–Tate parameter $q(X) \not\equiv 1 \mod p^2$ then there exists a non-zero global vector field on X which is a variational infinitesimal symmetry for all the modules $\mathfrak{X}^n(X)$, $n \geq 1$.*
2) *If X is a canonical lift (equivalently has Serre–Tate parameter $q(X) = 1$) then there is no non-zero global vector field on X which is a variational infinitesimal symmetry of $\mathfrak{X}^1(X)$.*

Finally here is a computation of differentials of δ-characters.

Theorem 6.5. [13, 16] *Assume X is an elliptic curve over R which is not a canonical lift and comes from an elliptic curve $X_{\mathbb{Z}_p}$ over $\mathbb{Z}_p$. Let $a_p \in \mathbb{Z}$ be the trace of Frobenius of the reduction mod p of $X_{\mathbb{Z}_p}$ and let ω be a basis for the global 1-forms on $X_{\mathbb{Z}_p}$. Then there exists an R-basis ψ of $\mathfrak{X}^2(X)$ such that*

$$p \cdot d\psi = (\phi^{*2} - a_p\phi^* + p)\omega.$$

In particular for the eigenvalue $\mu = 1$ we have

$$p \cdot d\left(\int \psi dp\right) = (1 - a_p + p) \cdot \int \omega dp.$$

A similar result holds in case X is a canonical lift.

6.2.3 δ-invariants of correspondences

Here is now a (rather roughly stated) result about δ-invariants of correspondences on curves; for precise statements we refer to [16].

Theorem 6.6. [16] *The ring $R_\delta(X, K^{-1})^\sigma$ is "δ-birationally equivalent" to the ring $R_\delta(\mathbb{P}^1, \mathcal{O}(1))$ if the correspondence σ on X "comes from" one of the following cases:*

1) *(spherical case) The standard action of $SL_2(\mathbb{Z}_p)$ on $\mathbb{P}^1$.*
2) *(flat case) A dynamical system $\mathbb{P}^1 \to \mathbb{P}^1$ which is post-critically finite with (orbifold) Euler characteristic zero.*
3) *(hyperbolic case) The action of a Hecke correspondence on a modular (or Shimura) curve.*

Here by saying that σ "comes from" a group action on X (respectively from an endomorphism of X) we mean that ("up to some specific finite subschemes" for which we refer to [16]) $\tilde{X}$ is the disjoint union of the graphs of finitely many automorphisms generating the action (respectively $\tilde{X}$ is the graph of the endomorphism). Also δ-birational equivalence means isomorphism (compatible with the actions of δ) between the p-adic completions of the rings of homogeneous fractions of degree zero with denominators not divisible by p. The proofs behind the spherical case involve direct computations. The proofs behind the flat case use the arithmetic Manin map, i.e. the space $\mathcal{X}^2(X)$ in Theorem 6.2, plus an induction in which canonical prolongations of vector fields play a crucial role. The proofs behind the hyperbolic case of Theorem 6.6 are based on a theory of *δ-modular forms* which we quickly survey next.

6.2.4 δ-modular forms

Cf. [15, 2, 16, 26, 17, 25]. Let $X_1(N)$ be the complete modular curve over R of level $N > 4$ and let $L_1(N) \to X_1(N)$ be the line bundle with the property that the sections of its various powers are the classical modular forms on $\Gamma_1(N)$ of various weights. Let X be $X_1(N)$ minus the cusps and the supersingular locus (zero locus of the Eisenstein form E_{p-1} of weight $p-1$). Let $L \to X$ be the restriction of the above line bundle and let V be L with the zero section removed. So $L^2 \simeq K$. The elements of $M^n = \mathcal{O}^n(V)$ are called δ-modular functions. Set $M^\infty = \cup M^n$. The elements of $\mathcal{O}^n(X)$ are called δ-modular forms of weight 0. For $w \in W$, $w \neq 0$, the elements of $H^0(X, L^w) \subset M^\infty$ are called δ-modular forms of weight w. We let $\sigma = (\sigma_1, \sigma_2) : \tilde{X} \to X \times X$ be the union of all the (prime to p) Hecke correspondences. Any δ-modular function $f \in M^n$ has a "δ-Fourier expansion" in $R((q))[q', \ldots, q^{(n)}]\hat{}$; setting $q' = q'' = \ldots = 0$ in this δ-Fourier expansion one gets a series in $R((q))\hat{}$ called the Fourier expansion of f. Finally let a_4 and a_6 be variables; then consideration of the elliptic curve $y^2 = x^3 + a_4 x + a_6$ yields an R-algebra map

$$R[a_4, a_6, \Delta^{-1}] \to M^0$$

where $\Delta = \Delta(a_4, a_6)$ is the discriminant polynomial. By universality we have induced homomorphisms

$$R[a_4, a_6, \ldots, a_4^{(n)}, a_6^{(n)}, \Delta^{-1}]\hat{} \to M^n$$

that are compatible with δ.

Example 6.7. [17, 19] For $f = \sum a_n q^n$ a classical newform over $\mathbb{Z}$ of weight 2, we get a δ-modular form of weight 0 and order 2

$$f^\sharp : J^2(X) \subset J^2(X_1(N)) \stackrel{J^2(\Phi)}{\to} J^2(E_f) \stackrel{\psi}{\to} \widehat{\mathbb{G}_a}$$

where $\Phi : X_1(N) \to E_f$ is the Eichler–Shimura map to the corresponding elliptic curve E_f (assumed for simplicity to be non-CM), and ψ the "unique" δ-character of order 2. Then the δ-Fourier expansion of $f^\sharp$ is congruent mod p to

$$\sum_{(n,p)=1} \frac{a_n}{n} q^n - a_p \left(\sum a_m q^{mp}\right) \frac{q'}{q^p} + \left(\sum a_m q^{mp^2}\right) \left(\frac{q'}{q^p}\right)^p ;$$

hence the Fourier expansion of $f^\sharp$ is congruent mod p to $\sum_{(n,p)=1} \frac{a_n}{n} q^n$.

Example 6.8. [15] There exists a unique δ-modular form $f^1 \in H^0(X, L^{-\phi-1})$ with δ-Fourier expansion

$$\sum_{n \geq 1} (-1)^{n-1} \frac{p^{n-1}}{n} \left(\frac{q'}{q^p}\right)^n = \frac{q'}{q^p} + p(\ldots).$$

Hence the Fourier expansion of f^1 is 0. By the way the above δ-Fourier expansion has the same shape as a certain function playing a role in "explicit" local class field theory. Here is the rough idea for the construction of f^1. One considers the universal elliptic curve $E = \bigcup U_i \to Spec\ M^{\infty}$, one considers sections $s_i : \widehat{U}_i \to J^1(U_i)$ of the projection, one considers the differences

$$s_i - s_j : \widehat{U}_i \cap \widehat{U}_j \to N^1 \simeq \widehat{\mathbb{G}}_a,$$

and one considers the cohomology class $[s_i - s_j] \in H^1(\widehat{E}, \mathcal{O}) = H^1(E, \mathcal{O})$. Then $f^1 \in M^{\infty}$ is defined as the cup product of this class with the canonical generator of the 1-forms; in fact f^1 is the image of some element

$$f^1(a_4, a_6, a_4', a_6') \in R[a_4, a_6, a_4', a_6', \Delta^{-1}]\hat{}.$$

By the way it is a result of Hurlburt [45] that

$$f^1(a_4, a_6, a_4', a_6') \equiv E_{p-1}\frac{2a_4^p a_6' - 3a_6^p a_4'}{\Delta^p} + f_0(a_4, a_6) \mod p$$

where $f_0 \in R[a_4, a_6, \Delta^{-1}]$ and we recall that E_{p-1} is the Eisenstein form of weight $p-1$. (The polynomial f_0 is related to the Kronecker modular polynomial mod p^2.) This plus the δ-Fourier expansion of f^1 should be viewed as an arithmetic analogue of the fact (due to Ramanujan) that

$$\frac{2a_4 da_6 - 3a_6 da_4}{\Delta} \mapsto \frac{dq}{q}$$

via the map between the Kähler differentials induced by the Fourier expansion map $\mathbb{C}[a_4, a_6, \Delta^{-1}] \to \mathbb{C}((q))$.

Example 6.9. [2] There exists unique δ-modular forms $f^{\partial} \in H^0(X, L^{\phi-1})$ and $f_{\partial} \in H^0(X, L^{1-\phi})$ with δ-Fourier expansions 1 (and hence Fourier expansions 1). They satisfy $f^{\partial} \cdot f_{\partial} = 1$. The form f^{∂} is constructed by applying the "top part" of the canonical prolongation of the "Serre operator" to the form f^1. More precisely f^{∂} is a constant times the image of

$$(72\phi(a_6)\frac{\partial}{\partial a_4'} - 16\phi(a_4)^2\frac{\partial}{\partial a_6'} - p\cdot\phi(P))(f^1(a_4, a_6, a_4', a_6'))$$

where $P \in R[a_4, a_6, \Delta^{-1}, E_{p-1}^{-1}]\hat{}$ is the Ramanujan form.

Theorem 6.10. [2, 15, 16] *The ring $R_\delta(X, K^{-1})^{\sigma}$ is "δ-generated" by f^1 and f^{∂}.*

Note that f^1 and f^{∂} do not actually belong to $R_\delta(X, K^{-1})$ so the above theorem needs some further explanation which we skip here; essentially, what happens is that f^1 and f^{∂} belong to a ring slightly bigger than $R_\delta(X, K^{-1})$ and they "δ-generate" that ring. We also note the following structure theorem for the kernel and image of the δ-Fourier expansion map, in which the forms f^1 and f^∂ play a key role:

Theorem 6.11. [25]

1) *The kernel of the Fourier expansion map $M^\infty \to R((q))\hat{}$ is the p-adic closure of the smallest δ-stable ideal containing f^1 and $f^\partial - 1$.*
2) *The p-adic closure of the image of the Fourier expansion map $M^\infty \to R((q))\hat{}$ equals Katz' ring $\mathbb{W}$ of generalized p-adic modular forms.*

The proof of the Theorem above is rather indirect and heavily Galois-theoretic. Statement 2) in Theorem 6.11 says that all Katz' divided congruences between classical modular forms can be obtained by taking combinations of "higher p-derivatives" of classical modular forms. Statement 1) above is a lift to characteristic zero of the Serre and Swinnerton–Dyer theorem about the kernel of the classical Fourier expansion map for classical modular forms mod p. Theorem 6.10 can also be viewed as a lift to characteristic zero of results of Ihara [46] about the Hasse invariant in characteristic p. These mod p results do not lift to characteristic zero in usual algebraic geometry but do lift, as we see, to characteristic zero in δ-geometry.

We mention the following remarkable infinitesimal symmetry property; recall the classical Serre operator $\partial : \mathcal{O}(V) \to \mathcal{O}(V)$. Also consider the Euler derivation operator $\mathcal{D} : \mathcal{O}(V) \to \mathcal{O}(V)$ given by multiplication by the degree on each graded component of $\mathcal{O}(V)$. Finally let P be the Ramanujan form (in the degree 2 component of $\mathcal{O}(V)$) and let $\theta : \mathcal{O}(V) \to \mathcal{O}(V)$ be the derivation $\theta = \partial + P\mathcal{D}$.

Theorem 6.12. [16] *The operator θ is an infinitesimal symmetry of the R-module generated by f^1, f^∂, and f_∂ in $M^1 = \mathcal{O}^1(V)$. Also θ is a variational infinitesimal symmetry of the R-module generated by f^∂ and f_∂ in $M^1 = \mathcal{O}^1(V)$.*

Here is a calculation of differentials of the forms $f^1, f^\partial, f_\partial$.

Theorem 6.13. [16] *Let ω, α be the basis of $\Omega_{\mathcal{O}(V)/R}$ dual to $\theta, \mathcal{D}$. Then*

$$\begin{aligned} d(f^\partial) &= f^\partial \cdot (\phi^*\alpha - \alpha) \\ d(f_\partial) &= -f_\partial \cdot (\phi^*\alpha - \alpha) \\ d(f^1) &= -f^1 \cdot (\phi^*\alpha + \alpha) - f_\partial \cdot \omega + f^\partial \cdot p^{-1}\phi^*\omega. \end{aligned}$$

In particular, for the eigenvalue $\mu = 1$ we have

$$\int \{\frac{d(f^\partial)}{f^\partial}\}dp = \int \{\frac{d(f_\partial)}{f_\partial}\}dp = 0.$$

By the way, the forms $f^\sharp$ and f^1 introduced above can be used to prove some interesting purely diophantine results. For instance we have the following:

Theorem 6.14. [19] *Assume that $\Phi : X = X_1(N) \to A$ is a modular parametrization of an elliptic curve. Let p be a sufficiently large "good" prime and let $Q \in X(R)$ be an ordinary point. Let S be the set of all rational primes that are inert in the imaginary quadratic field attached to Q. Let C be the S-isogeny class of Q in $X(R)$ (consisting of points corresponding to isogenies of degrees only divisible by primes in S). Then there exists a constant c such that for any subgroup $\Gamma \leq A(R)$ with $r := rank(\Gamma) < \infty$ the set $\Phi(C) \cap \Gamma$ is finite of cardinality at most cp^r.*

Other results of the same type (e.g. involving Heegner points) were proved in [19]. In particular an analogue of the above Theorem is true with C replaced by the locus CL of canonical lifts. To have a rough idea about the arguments involved assume we want to prove that $\Phi(CL) \cap \Gamma$ is finite (and to bound the cardinality of this set) in case Γ is the torsion group of $A(R)$. One considers the order 2 δ-modular form $f^\sharp : X_1(N)(R) \to R$ and one constructs, using f^1, a δ-function of order 1, $f^\flat : X(R) \to R$, on an open set $X \subset X_1(N)$ which vanishes exactly on $CL \cap X(R)$. Then any point P in the intersection $X(R) \cap \Phi(CL) \cap \Gamma$ satisfies the system of "differential equations of order ≤ 2 in 1 unknown"

$$\begin{cases} f^\sharp(P) = 0 \\ f^\flat(P) = 0 \end{cases}$$

One can show that this system is "sufficiently non-degenerate" to allow the elimination of the "derivatives" of the unknown; one is left with a differential equation $f^0(P) = 0$ "of order 0" which has, then, only finitely many solutions (by Krasner's theorem). By that theorem one can also bound the number of solutions.

We end the discussion here by noting that the main players in the theory above enjoy a certain remarkable property which we call δ-overconvergence. Morally this is a an overconvergence property (in the classical sense of Dwork, Monsky, Washnitzer) "in the direction of the variables $x', x'', \ldots, x^{(n)}$" (but not necessarily in the direction of x). We prove:

Theorem 6.15. [27] *The δ-functions $f^\sharp$, f^1, f^∂ are δ-overconvergent.*

6.2.5 δ-Hecke operators

Next we discuss the Hecke action on δ-modular forms. For $(n, p) = 1$ the Hecke operators $T_m(n)$ act naturally on δ-series (i.e. series in $R((q))[q', \dots, q^{(r)}]\hat{}$) by the usual formula that inserts roots of unity of order prime to p which are all in R. However no naive definition of $T_m(p)$ seems to work. Instead we consider the situation mod p and make the following definition. Let x be a p-tuple $x_1, \dots, x_p$ of indeterminates and let s be the p-tuple $s_1, \dots, s_p$ of fundamental symmetric polynomials in x. An element $f \in k[[q]][q', \dots, q^{(r)}]$ is called δ-p-symmetric mod p if the sum

$$f(x_1, \dots, x_1^{(r)}) + \dots + f(x_p, \dots, x_p^{(r)}) \in k[[x]][x', \dots, x^{(r)}]$$

is the image of an element

$$f_{(p)} = f_{(p)}(s_1, \dots, s_p, \dots, s_1^{(r)}, \dots, s_p^{(r)}) \in k[[s]][s', \dots, s^{(r)}].$$

For f that is δ-p-symmetric mod p define

$$\begin{aligned} \text{“}pT_m(p)\text{''} f = {} & f_{(p)}(0, \dots, 0, q, \dots, 0, \dots, 0, q^{(r)}) \\ & + p^m f(q^p, \dots, \delta^r(q^p)) \in k[[q]][q', \dots, q^{(r)}]. \end{aligned}$$

Eigenvectors of “$pT_m(p)$'' will be automatically understood to be δ-p-symmetric. Also let us say that a series in $k[[q]][q', \dots, q^{(r)}]$ is primitive if the series in $k[[q]]$ obtained by setting $q' = \dots = q^{(r)} = 0$ is killed by the classical U-operator. Then one can give a complete description (in terms of classical Hecke eigenforms mod p) of δ-“eigenseries” mod p of order 1 which are δ-Fourier expansions of δ-modular forms of arbitrary order and weight:

Theorem 6.16. [26] *There is a* $1-1$ *correspondence between:*

1) *Series in* $k[[q]]$ *which are eigenvectors of all* $T_{m+2}(n), T_{m+2}(p), (n, p) = 1$*, and which are Fourier expansions of classical modular forms over* k *of weight* $\equiv m+2 \bmod p-1$.
2) *Primitive series in* $k[[q]][q']$ *which are eigenvectors of all* $nT_m(n)$*, “*$pT_m(p)$*”,* $(n, p) = 1$*, and which are* δ*-Fourier expansions of* δ*-modular forms of some order* ≥ 0 *with weight* w*,* $deg(w) = m$.

Note that the δ-Fourier expansion of the form $f^\sharp$ discussed in Example 6.7 is an example of series in 2) of Theorem 6.16. (Note that $f^\sharp$ has order 2 although its δ-Fourier expansion reduced mod p has order 1!) More generally the series in 1) and 2) of Theorem 6.16 are related in an explicit way, similar to the way f and $f^\sharp$ of Example 6.7 are related. The proof of Theorem 6.16 involves a careful study of the action of δ-Hecke operators on δ-series plus the use of the canonical prolongations of the Serre operator acting on δ-modular forms.

6.2.6 δ-functions on finite flat schemes

The p-jet spaces of finite flat schemes over R seem to play a key role in many aspects of the theory. These p-jet spaces are neither finite nor flat in general and overall they seem quite pathological. There are two remarkable classes of examples, however, where some order seems to be restored in the limit; these classes are finite flat p-group schemes that fit into p-divisible groups and finite length p-typical Witt rings. Recall that for any ring A we write $\overline{A} = A/pA$. Then for connected p-divisible groups we have:

Theorem 6.17. [28] *Let $\mathcal{F}$ be a formal group law over R in one variable x, assume $\mathcal{F}$ has finite height, and let $\mathcal{F}[p^\nu]$ be the kernel of the multiplication by p^ν viewed as a finite flat group scheme over R. Then*

$$\lim_{\vec{n}} \overline{\mathcal{O}^n(\mathcal{F}[p^\nu])} \simeq \frac{k[x, x', x'', \dots]}{(x^{p^\nu}, (x')^{p^\nu}, (x'')^{p^\nu}, \dots)}$$

sending $x, \delta x, \delta^2 x, \dots$ into the classes of $x, x', x'', \dots$.

A similar result is obtained in [28] for the divisible group $E[p^\nu]$ of an ordinary elliptic curve; some of the components of $J^n(E[p^\nu])$ will be empty and exactly which ones are so is dictated by the valuation of $q - 1$ where q is the Serre–Tate parameter. The components that are non-empty (in particular the identity component) behave in the same way as the formal groups examined in Theorem 6.17 above.

In the same spirit one can compute p-jet spaces of Witt rings. Let us consider the ring $W_m(R)$ of p-typical Witt vectors of length $m+1$, $m \geq 1$, and denote by $\Sigma_m = Spec\ W_m(R)$ its spectrum. Set $v_i = (0, \dots, 0, 1, 0, \dots, 0) \in W_m(R)$, (1 preceded by i zeroes, $i = 1, \dots, m$), set $\pi = 1 - \delta v_1 \in \mathcal{O}^1(\Sigma_m)$, and let $\Omega_m = \{1, \dots, m\}$. The following is a description of the identity component of the limit of p-jet spaces mod p:

Theorem 6.18. [29] *For $n \geq 2$ the image of π^p in $\overline{\mathcal{O}^n(\Sigma_m)}$ is idempotent and we have an isomorphism*

$$\lim_{\vec{n}} \overline{\mathcal{O}^n(\Sigma_m)_\pi} \simeq \frac{k[x_i^{(r)}; i \in \Omega_m; r \geq 0]}{(x_i x_j, (x_i^{(r)})^p; i, j \in \Omega_m, r \geq 1)}$$

sending each $\overline{\delta^r v_i}$ into the class of the variable $x_i^{(r)}$.

A similar description is obtained in [29] for the p-jet maps induced by the Witt comonad maps. We recall that the data consisting of $\mathcal{O}^n(\Sigma_m)$ and the maps induced by the comonad maps should be viewed as an arithmetic analogue of the Lie groupoid of the line.

6.2.7 δ-Galois groups of δ-linear equations

Recall that for any solution $u \in GL_n(R)$ of a δ-linear equation

$$\delta u = \alpha \cdot u^{(p)}$$

(where $\alpha \in \mathfrak{gl}_n(R)$) and for any δ-subring $\mathcal{O} \subset R$ we defined the δ-Galois group $G_{u/\mathcal{O}} \subset GL_n(\mathcal{O})$. We want to explain a result proved in [34]. Consider the maximal torus $T \subset GL_n(R)$ of diagonal matrices, the Weyl group $W \subset GL_n(R)$ of permutation matrices, the normalizer $N = WT = TW$ of T in $GL_n(R)$, and the subgroup N^δ of N consisting of all elements of N whose entries are in the monoid of constants R^δ. We also use below the notation K^a for the algebraic closure of the fraction field K of R; the Zariski closed sets Z of $GL_n(K^a)$ are then viewed as varieties over K^a. The next result illustrates some "generic" features of our δ-Galois groups; assertion 1) of the next theorem shows that the δ-Galois group is generically "not too large". Assertions 2) and 3) show that the δ-Galois group are generically "as large as possible". As we shall see presently, the meaning of the word *generic* is different in each of the 3 situations: in situation 1) *generic* means *outside a Zariski closed set*; in situation 2) *generic* means *outside a thin set* (in the sense of diophantine geometry); in situation 3) *generic* means *outside a set of the first category* (in the sense of Baire category).

Theorem 6.19. *1) There exists a Zariski closed subset $\Omega \subset GL_n(K^a)$ not containing 1 such that for any $u \in GL_n(R)\backslash\Omega$ the following holds. Let $\alpha = \delta u \cdot (u^{(p)})^{-1}$ and let $\mathcal{O}$ be a δ-subring of R containing α. Then $G_{u/\mathcal{O}}$ contains a normal subgroup of finite index which is diagonalizable over K^a.*

2) Let $\mathcal{O} = \mathbb{Z}_{(p)}$. There exists a thin set $\Omega \subset \mathbb{Q}^{n^2}$ such that for any $\alpha \in \mathbb{Z}^{n^2}\backslash\Omega$ there exists a solution u of the equation $\delta u = \alpha u^{(p)}$ with the property that $G_{u/\mathcal{O}}$ is a finite group containing the Weyl group W.

3) There exists a subset Ω of the first category in the metric space

$$X = \{u \in GL_n(R);\, u \equiv 1 \ mod \ p\}$$

such that for any $u \in X\backslash\Omega$ the following holds. Let $\alpha = \delta u \cdot (u^{(p)})^{-1}$. Then there exists a δ-subring $\mathcal{O}$ of R containing R^δ such that $\alpha \in \mathfrak{gl}_n(\mathcal{O})$ and such that $G_{u/\mathcal{O}} = N^\delta$.

The groups W and N^{δ} should be morally viewed as "incarnations" of the groups "$GL_n(\mathbb{F}_1)$" and "$GL_n(\mathbb{F}_1^a)$" where "$\mathbb{F}_1$" and "$\mathbb{F}_1^a$" are the "field with element" and "its algebraic closure" respectively [6]. This suggests that the δ-Galois theory we are proposing here should be viewed as a Galois theory over "$\mathbb{F}_1$".

6.2.8 δ-flows for the classical groups

The main results in [34] concern the existence of certain δ-flows on $\widehat{GL_n}$ that are adapted, in a certain precise sense, to the various classical subgroups GL_n, SL_n, SO_n, Sp_n.

Let $G = GL_n$ and let H be a smooth closed subgroup scheme of G. We say that a δ-flow ϕ_G is left (respectively right) compatible with H if H is ϕ_G-horizontal and the left (respectively right) diagram below is commutative:

$$\begin{array}{ccc} \widehat{H}\times\widehat{G} & \to & \widehat{G} \\ \phi_H\times\phi_G\downarrow & & \downarrow\phi_G \\ \widehat{H}\times\widehat{G} & \to & \widehat{G} \end{array} \qquad \begin{array}{ccc} \widehat{G}\times\widehat{H} & \to & \widehat{G} \\ \phi_G\times\phi_H\downarrow & & \downarrow\phi_G \\ \widehat{G}\times\widehat{H} & \to & \widehat{G} \end{array}$$

where $\phi_H : \widehat{H} \to \widehat{H}$ is induced by ϕ_G and the horizontal maps are given by multiplication.

Next recall that by an involution on G we understand an automorphism $\tau : G \to G$ over R, $x \mapsto x^{\tau}$, of order 1 or 2. For each such τ one may consider the subgroup $G^+ = \{a \in G; a^{\tau} = a\}$. The identity component $(G^+)^{\circ}$ is referred to as the subgroup defined by τ. Also set $\mathcal{H} : G \to G$, $\mathcal{H}(x) = x^{-\tau}x$ and $\mathcal{B} : G \times G \to G$, $\mathcal{B}(x, y) = x^{-\tau}y$.

Let us fix now a lift of Frobenius $\phi_{G,0}$ on $\widehat{G}$. A lift of Frobenius ϕ_G on $\widehat{G}$ is said to be $\mathcal{H}$-horizontal (respectively $\mathcal{B}$-symmetric) with respect to $\phi_{G,0}$ if the left (respectively right) diagram below is commutative:

$$\begin{array}{ccc} \widehat{G} & \stackrel{\phi_G}{\longrightarrow} & \widehat{G} \\ \mathcal{H}\downarrow & & \downarrow\mathcal{H} \\ \widehat{G} & \stackrel{\phi_{G,0}}{\longrightarrow} & \widehat{G} \end{array} \qquad \begin{array}{ccc} \widehat{G} & \stackrel{\phi_{G,0}\times\phi_G}{\longrightarrow} & \widehat{G}\times\widehat{G} \\ \phi_G\times\phi_{G,0}\downarrow & & \downarrow\mathcal{B} \\ \widehat{G}\times\widehat{G} & \stackrel{\mathcal{B}}{\longrightarrow} & \widehat{G} \end{array}$$

If this is the case and $\phi_{G,0}(1) = 1$ then the group $S := (G^+)^{\circ}$ defined by τ is ϕ_G-horizontal; in particular there is an induced lift of Frobenius ϕ_S on $\widehat{S}$. Also note that if we set $\phi_{G,0}(x) = x^{(p)}$, viewing $\mathcal{H}$ as a matrix $\mathcal{H}(x)$ with entries in $R[x, \det(x)^{-1}]\hat{}$, we have that $\delta_G\mathcal{H} = 0$, which can be interpreted as saying that $\mathcal{H}$ is a *prime integral* for our δ-flow ϕ_G.

The basic split classical groups GL_n, SO_n, Sp_n are defined by involutions on $G = GL_n$ as follows. We start with GL_n itself which is defined by $x^{\tau} = x$.

We call τ the canonical involution defining GL_n. We also recall that if $T \subset G$ is the maximal torus of diagonal matrices, N is its normalizer in G, and $W \subset G$ is the group of permutation matrices then $N = TW = WT$ and W is isomorphic to the Weyl group N/T. Throughout our discussion we let $\phi_{G,0}(x)$ be the lift of Frobenius on $\widehat{GL_n}$ defined by $\phi_{G,0}(x) := x^{(p)}$; one can prove that this $\phi_{G,0}(x)$ is the unique lift of Frobenius on $\widehat{G}$ that is left and right compatible with N and extends to a lift of Frobenius on $\widehat{\mathfrak{gl}_n}$ (where we view $\widehat{GL_n}$ as an open set of $\widehat{\mathfrak{gl}_n}$). On the other hand the groups $Sp_{2r}, SO_{2r}, SO_{2r+1}$ are defined by the involutions on $G = GL_n$ given by $x^\tau = q^{-1}(x^t)^{-1}q$ where q is equal to

$$\begin{pmatrix} 0 & 1_r \\ -1_r & 0 \end{pmatrix}, \begin{pmatrix} 0 & 1_r \\ 1_r & 0 \end{pmatrix}, \begin{pmatrix} 1 & 0 & 0 \\ 0 & 0 & 1_r \\ 0 & 1_r & 0 \end{pmatrix},$$

$n = 2r, 2r, 2r+1$ respectively, x^t is the transpose, and 1_r is the $r \times r$ identity matrix. We call this τ the canonical involution defining $Sp_{2r}, SO_{2r}, SO_{2r+1}$, respectively. All these groups are smooth over R. If S is any of these groups and we set $T_S = T \cap S$ and $N_S = N \cap S$ then these groups are smooth, T_S is a maximal torus in S and N_S is the normalizer of T_S in S. Call a root of one of these groups *abnormal* if it is a shortest root of a group SO_n with n odd.

Theorem 6.20. *Let S be any of the groups GL_n, SO_n, Sp_n and let τ be the canonical involution on $G = GL_n$ defining S. Then the following hold.*

1) *(Compatibility with involutions) There exists a unique lift of Frobenius ϕ_G on $\widehat{G}$ that is $\mathcal{H}$-horizontal and $\mathcal{B}$-symmetric with respect to $\phi_{G,0}$. In particular if $l\delta, l\delta_0 : GL_n \to \mathfrak{gl}_n$ are the arithmetic logarithmic derivative associated to $\phi_G, \phi_{G,0}$ then for all $a \in S$ we have that the Cartan decomposition*

$$l\delta_0(a) = (l\delta_0(a))^+ +_\delta (l\delta_0(a))^-$$

of $l\delta_0(a)$ satisfies $(l\delta_0(a))^+ = l\delta(a)$.

2) *(Compatibility with the normalizer of maximal torus.) ϕ_G is right compatible with N and left and right compatible with N_S. In particular for all $a \in N_S$ and $b \in GL_n$ (alternatively for all $a \in GL_n$ and $b \in N$) we have*

$$l\delta(ab) = a \star_\delta l\delta(b) +_\delta l\delta(a).$$

3) *(Compatibility with root groups.) If χ is a root of S which is not abnormal then the corresponding root group $U_\chi \simeq \mathbb{G}_a$ is ϕ_{GL_n}-horizontal.*

Note that a similar result can be proved for SL_n; the involution τ lives, in this case, on a cover of GL_n rather than on GL_n itself. Note also that the exception in assertion 3 of the Theorem (occurring in case χ is a shortest root of SO_n with n odd) is a curious phenomenon which deserves further understanding.

6.2.9 Arithmetic Painlevé VI

The classical Painlevé VI equation has an elegant geometric description due to Manin [56]. In [31] an arithmetic analogue of this equation has been introduced and studied. We present below the main result in [31]. Recall our framework in 6.1.2.10 above.

Theorem 6.21. [31] *Let E be an elliptic curve over R which does not have a lift of Frobenius and let $\psi \in \mathcal{O}(J^2(E))$ be δ-character of order 2 attached to an invertible 1-form ω on E. Consider the symplectic form $\eta = \omega \wedge \frac{\phi^*\omega}{p}$ on $J^1(E)$. Let $Y = E \backslash E[2]$ and let $r \in \mathcal{O}(Y)$. Then the following hold:*

1) $f = \psi - \phi(r)$ defines a generalized canonical δ-flow on $J^1(Y)$ which is Hamiltonian with respect to η.
2) There exists a canonical 1-form ν on X such that $d\nu = \eta$; in particular η is exact and if ϵ is the associated Euler–Lagrange form then $p\epsilon$ is closed.

Note that f in the Theorem does not define a δ-flow on $J^1(E)$ (equivalently, $J^1(E)$ has no lift of Frobenius, equivalently the projection $J^1(J^1(E)) \to J^1(E)$ does not have a section):

Theorem 6.22. [32]. *Let E be an elliptic curve over R which does not have a lift of Frobenius. Then $J^1(E)$ does not have a lift of Frobenius.*

6.3 Problems

Problem 6.1. *Study the arithmetic jet spaces $J^n(X)$ of curves X (and more general varieties) with bad reduction.*

This could be applied, in particular, to tackle Mazur's question [59] about bounding the torsion points on curves unifromly in terms of the genus; in other words replacing $C(g, p)$ by $C(g)$ in Theorem 6.3. Our proof in [12] is based on the study of the arithmetic jet space of $J^1(X)$ at a prime p of good reduction. A study of the arithmetic jet space of curves at primes of bad reduction might lead to dropping the dependence of $C(g, p)$ on p. Evidence that arithmetic jet spaces can be handled in the case of bad reduction comes in particular from the recent paper [27].

Problem 6.2. *Study the δ-modular forms that vanish on arithmetically interesting Zariski dense subsets of Shimura varieties (such as CM loci or individual non-CM isogeny classes). Compute δ-invariants of higher dimensional correspondences.*

This could be applied to extend results in [19], e.g. Theorem 6.14 above. A deeper study of differential modular forms may allow one, for instance, to replace S-isogeny class with the full isogeny class. The arguments might then be extended to higher dimensional contexts and to the global field rather than the local field situation. That such a deeper study is possible is shown by papers like [25], for instance. For the higher dimensional case the theory in [3] might have to be further developed to match the one dimensional theory in [15, 16]. In a related direction one might attempt to use the methods in [19] to tackle Pink's conjectures in [63]. In [19] it was shown that behind finiteness theorems in diophantine geometry one can have reciprocity maps that are somehow inherited from δ-geometry (and that provide effective bounds); a similar picture might hold for (cases of) Pink's conjecture. The first case to look at for such reciprocity maps would be in the case of the intersection between a multisection X of an abelian (or semiabelian) scheme $G \to S$ over a curve S with the set of torsion points lying in special (CM or otherwise) fibers; more general situations, in which torsion points are replaced by division points of a group generated by finitely many sections, can be considered. Results of André, Ribet, and Bertrand are pertinent to this question.

Problem 6.3. *Compare the δ-geometric approach to quotient spaces with the approach via non-commutative geometry.*

The quotients X/σ for the correspondences appearing in Theorem 6.6 do not exist, of course, in usual algebraic geometry. As Theorem 6.6 shows these quotients exist, however, and are interesting in δ-geometry. Remarkably such quotients also exist and are interesting in non-commutative geometry [60]. More precisely the 3 cases (spherical, flat, hyperbolic) of Theorem 6.6 are closely related to the following 3 classes of examples studied in non-commutative geometry:

1) (spherical) $\frac{\mathbb{P}^1(\mathbb{R})}{SL_2(\mathbb{Z})}$, non-commutative boundary of the classical modular curve;

2) (flat) $\frac{S^1}{\langle e^{2\pi i\tau}\rangle}$ ($\theta \in \mathbb{R}\backslash\mathbb{Q}$): non-commutative elliptic curves;
3) (hyperbolic) Non-commutative space Sh^{nc} containing the classical Shimura variety Sh (2-dimensional analogue of Bost–Connes systems).

It would be interesting to understand why these 3 classes appear in both contexts (δ-geometry and non-commutative geometry); also one would like to see whether there is a connection, in the case of these 3 classes, between the 2 contexts.

Note that non-commutative geometry can also tackle the dynamics of rational functions that are not necessarily post-critically finite of Euler characteristic zero. It is conceivable that some post-critically finite polynomials of non-zero Euler characteristic possess δ-invariants for some particular primes (with respect to the anticanonical bundle or other bundles). A good start would be to investigate the δ-invariants of $\sigma(x) = x^2 - 1$. Another good start would be to investigate δ-invariants of post-critically finite polynomials with Euler characteristic zero that are congruent modulo special primes to post-critically finite polynomials with Euler characteristic zero.

Problem 6.4. *Study the de Rham cohomology of arithmetic jet spaces. Find arithmetic analogues of Kähler differentials* Ω *and* $\mathcal{D}$*-modules. Find an object that is to* $\mathcal{D}$ *what* $\mathcal{O}^1$ *is to* $Sym(\Omega)$.

The study of de Rham cohomology of arithmetic jet spaces was started in [7] where it is shown that the de Rham cohomology of $J^n(X)$ carries information about the arithmetic of X. The de Rham computations in [7] are probably shadows of more general phenomena which deserve being understood. Also the de Rham setting could be replaced by an overconvergent one; overconvergence is known to give an improved picture of the de Rham story and, on the other hand, as already mentioned, it was proved in [27] that most of the remarkable δ-functions occurring in the theory possess a remarkable overconvergence property in the "δ-variables" called *δ-overconvergence*. Finally one is tempted to try to relate the de Rham cohomology of arithmetic jet spaces $J^n(X)$ to the de Rham–Witt complex of X in characteristic p and in mixed characteristic. Note further that since the arithmetic jet space $J^1(X)$ is an analogue of the (physical) tangent bundle $T(X)$ of X it follows that the sheaf $\mathcal{O}^1$ is an arithmetic analogue of the sheaf $Sym(\Omega_{X/R})$, symmetric algebra on the Kähler differentials. But there is no obvious arithmetic analogue of the sheaf $\Omega_{X/R}$ itself. Also there is no obvious arithmetic analogue of the sheaf $\mathcal{D}_X$ of differential operators on X and of $\mathcal{D}_X$-modules. The absence of immediate analogues of Ω and $\mathcal{D}$ is of course related to the intrinsic non-linearity of p-derivations. It would be interesting to search for such analogues. It is on the other hand conceivable that there is a sheaf in the arithmetic theory that is to $\mathcal{D}$ what $\mathcal{O}^1$ is to $Sym(\Omega)$. Recall that the associated graded algebra of $\mathcal{D}$ is canonically isomorphic to the algebra of functions on the (physical) cotangent bundle $\mathcal{O}(T^*(X))$ (and not on the tangent bundle); this looks like a discrepancy but actually the arithmetic jet space $J^1(X)$ has a sort of intrinsic self-duality (cf. [16]) that is missing in the classical algebro-geometric case where the tangent bundle $T(X)$ and the cotangent bundle $T^*(X)$ and not naturally dual (unless, say, a symplectic structure is given).

The δ-overconvergence property mentioned in Problem 6.4 may hold the key to:

Problem 6.5. *Define and study the/a maximal space of δ-modular forms on which Atkin's U operator can be defined.*

Indeed the Hecke operators $T(n)$ with $p \not| n$ are defined on δ-modular forms and have a rich theory in this context [15]. In contrast to this $T(p)$ and hence U are still mysterious in the theory of δ-modular forms. A step in understanding U was taken in [26] where the theory mod p for series of order 1 was given a rather definitive treatment. However the theory in characteristic zero seems elusive at this point. There are two paths towards such a theory so far. One path is via *δ-symmetry* [24, 26]; this is a characteristic 0 analogue of the concept of δ-p-symmetry mod p discussed above. Another path is via δ-overconvergence [27]. The two paths seem to lead into different directions and this discrepancy needs to be better understood. Assuming that a good theory of U is achieved, this might lead to a Hida-like theory of families of differential modular forms, including Galois representations attached to such forms. It is conceivable that families in this context are not power series but Witt vectors. Part of the quest for a U theory of δ-modular forms is to seek a δ-analogue of Eisentein series. It is conceivable that the rings $\mathcal{O}^n(X_1(N))$ contain functions that do not vanish at the cusps and are eigenvectors of the Hecke operators; such functions could be viewed as "δ-Eisenstein" forms of weight zero.

Problem 6.6. *Interpret information contained in the arithmetic jet spaces $J^n(X)$ as an arithmetic Kodaira-Spencer "class" of X.*

Indeed some of these arithmetic Kodaira–Spencer classes (e.g in the case of elliptic curves or, more generally, abelian schemes) were studied in [15, 16] and lead to interesting δ-modular forms. For general schemes (e.g for curves of higher genus) these classes were explored in Dupuy's thesis [40]. They are non-abelian cohomology classes with values in the sheaf of automorphisms of p-adic affine spaces $\widehat{\mathbb{A}}^d$ (in the case of curves $d = 1$). These classes arise from comparing the local trivializations of arithmetic jet spaces. In this more general case these classes may hold the key to a "deformation theory over the field with one element". On the other hand Dupuy proved in [40] that if X is a smooth projective curve of genus ≥ 2 over R then $J^1(X)$ is a torsor for some line bundle over X; this is rather surprising in view of the high non-linearity of the theory. One should say that the line bundle in question is still mysterious and deserves further investigation.

Problem 6.7. *Further develop the partial differential theory in* [20, 21, 22].

Indeed in spite of the extensive work done in [20, 21, 22] the arithmetic *partial* differential theory is still in its infancy. The elliptic case of that theory [20] (which, we recall, involves operators δ_{p_1} and δ_{p_2} corresponding to two primes p_1 and p_2) is directly related to the study of the de Rham cohomology of arithmetic jet spaces [7]; indeed one of the main results in [7] shows that the arithmetic Laplacians in [20] are formal primitives (both p_1-adically and p_2-adically) of global 1-forms on the arithmetic jet spaces (these forms being not exact, although formally exact, and hence closed). By the way analogues of these results in [7] probably exist in the case of modular curves; in the one prime case a beginning of such a study was undertaken in [16], where some of the main δ-modular forms of the theory were shown to satisfy some remarkable systems of Pfaff equations. The hyperbolic/parabolic case of the theory [21, 22] (which, we recall, involves a p-derivation δ_p with respect to a prime p and a usual derivation operator δ_q) could be further developed as follows. One could start by "specializing" the variable q in δ_q to elements π in *arbitrarily ramified* extensions of $\mathbb{Z}_p$. This might push the theory in the "arbitrarily ramified direction" which would be extremely desirable for arithmetic–geometric applications. Indeed our ordinary arithmetic differential theory is, at present, a non-ramified (or at most "boundedly ramified") theory. A further idea along these lines would be to use the solutions of the arithmetic partial differential equations in [21, 22] to let points "flow" on varieties defined over number fields. Some of the solutions in [21, 22] have interesting arithmetic features (some look like hybrids between quantum exponentials and Artin–Hasse exponentials, for instance) so the "flows" defined by them might have arithmetic consequences. The challenge is to find (if at all possible) "special values" of these solutions that are algebraic. One should also mention that the arithmetic hyperbolic and parabolic equations in [21, 22] have, in special cases, well defined "indices" that seem to carry arithmetic information; the challenge would be to make the index machinery work in general situations and to study the variation of indices in families.

Problem 6.8. *Find an arithmetic analogue of Sato hyperfunction solutions of both "ordinary" and "partial" arithmetic differential equations.*

Indeed Sato's hyperfunctions, in their simplest incarnation, are pairs $(f(x), g(x))$ of functions on the unit disk (corresponding to the distribution $f(x) - g(x^{-1})$) modulo (c, c), c a constant. The derivative of a pair is then $(\frac{df}{dx}(x), -x^2\frac{dg}{dx}(x))$.) One could then try to consider, in the arithmetic case, pairs (P, Q) of points of algebraic groups with values in δ-rings modulo an appropriate equivalence relation and with an appropriate analogue of differentiation with respect to p; this framework could be the correct one for "non-analytic" solutions of the equations in [20, 21, 22].

Problem 6.9. *Construct p-adic measures from δ-modular forms.*

Indeed one of the main ideas in Katz's approach to p-adic interpolation [50] was to lift some of the remarkable $\mathbb{Z}_p$-valued (p-adic) measures of the theory to $\mathbb{W}$-valued measures where $\mathbb{W}$ is the ring of (generalized) p-adic modular forms. One can hope that some of these $\mathbb{W}$-valued measures of Katz can be further lifted to measures with values in the p-adic completion of the ring of δ-modular functions M^∞. Indeed recall from Theorem 6.11 that there is a canonical homomorphism $M^\infty \to \mathbb{W}$ whose image is p-adically dense, hence the "lifting" problem makes sense. These lifted $\widehat{M^\infty}$-valued measures could then be evaluated at various elliptic curves defined over δ-rings to obtain new $\mathbb{Z}_p$-valued measures (and hence new p-adic interpolation results) in the same way in which Katz evaluated his measures at special elliptic curves. Another related idea would be to interpret the solutions in $\mathbb{Z}_p[[q]]$ of the arithmetic partial differential equations in Problem 6.7 above as measures (via Iwasawa's representation of measures as power series). There is a discrepancy in this approach in that the derivation of interest in Iwasawa's theory is $(1+q)\frac{d}{dq}$ whereas the derivation of interest in Problem 6.7 is $q\frac{d}{dq}$; nevertheless one should pursue this idea and understand the discrepancy.

Problem 6.10. *Find arithmetic analogues of classical theorems in the theory of differential algebraic groups and further develop the δ-Galois theory in* [35]. *Also further develop the arithmetic analogue of the theory of symmetric spaces in* [34], *especially on the Riemannian side.*

Indeed the theory of groups defined by (usual) differential equations ("differential algebraic groups") is by now a classical subject: it goes back to Lie and Cartan and underwent a new development, from a rather new angle, through the work of Cassidy and Kolchin [36, 52]. It is tempting to seek an arithmetic analogue of this theory: one would like to understand, for instance, the structure of all subgroups of $GL_n(R)$ that are defined by arithmetic differential equations. The paper [14] proves an arithmetic analogue of Cassidy's theorem about Zariski dense subgroups of simple algebraic groups over differential fields; in [14] the case of Zariski dense mod p groups is considered. But Zariski dense groups such as $GL_n(\mathbb{Z}_p)$ are definitely extremely interesting (and lead to interesting Galois theoretic results such as in [16], Chapter 5). So a generalization of [14] to the case of Zariski dense (rather than Zariski dense mod p) groups, together with a generalization of the Galois theoretic results in [16], would be very desirable. For instance it would be interesting to classify all δ-subgroups of the multiplicative group $\mathbb{G}_m(R) = R^\times$ (or more generally of $GL_2(R)$) and find the invariants of such groups acting on $\widehat{R\{x\}_{(p)}}$ (where $R\{x\} = R[x, x', x'', \ldots])$. Also remark that, as in the case of usual derivations, there are interesting ("unexpected") homomorphisms in δ-geometry between $GL_1 = \mathbb{G}_m$ and GL_2, for instance

$$\mathbb{G}_m(R) \to GL_2(R), \quad a \mapsto \begin{pmatrix} a & a\psi_*(a) \\ 0 & a \end{pmatrix}$$

where ψ_* is a δ-character, i.e. $\psi \in \mathcal{X}^n(\mathbb{G}_m)$. Cf. [43] for interesting developments into this subject. The main open problem in the δ-Galois theory of GL_n [35] seems at this point to decide if the δ-Galois groups always contain a subgroup of the diagonal matrices as a subgroup of finite index. Other problems are: to establish a Galois correspondence; to understand the relation (already hinted at in [35]) between δ-Galois groups and Galois problems arising from the dynamics on $\mathbb{P}^n$; to generalize the theory by replacing GL_n with an arbitrary reductive group. With regards to the analogue in [34] of the theory of symmetric spaces the main open problems are to find analogues of the Riemannian metrics and curvature. An analogue of Lie brackets can, by the way, be developed; it would be interesting to use it to develop an "arithmetic Riemannian theory".

Problem 6.11. *Compose some of the basic δ-functions of the theory in* [16] *(e.g. the δ-modular forms on Shimura curves) with p-adic uniformization maps (e.g. with Drinfeld's uniformization map of Shimura curves).*

Indeed this might shed a new light (coming from the analytic world) on δ-geometry. These composed maps would belong to a "δ-rigid geometry" which has yet to be developed in case these maps are interesting enough to require it. By the way it is not at all clear that the functor that attaches to a formal p-adic scheme its arithmetic jet space can be prolonged to a functor in the rigid category. The problem (which seems to boil down to some quite non-trivial combinatorially flavored calculation) is to show that the arithmetic jet space functor sends blow-ups with centers in the closed fiber into (some version of) blow-ups.

In the spirit of the last comments on Problem 6.11 one can ask:

Problem 6.12. *Study the morphisms $J^n(X) \to J^n(Y)$ induced by non-étale finite flat covers of smooth schemes $X \to Y$.*

Indeed note that if $X \to Y$ is finite and étale then it is well known that $J^n(X) \to J^n(Y)$ are finite and étale; indeed $J^n(X) \simeq J^n(Y) \times_Y X$. But note that if $X \to Y$ is only finite and flat then $J^n(X) \to J^n(Y)$ is neither finite nor flat in general. One of the simplest examples which need to be investigated (some partial results are available [28], cf. Theorem 6.17) is that of the covers $[p^\nu] : G \to G$ of smooth group schemes (or formal groups) given by multiplication by p^ν. The geometry of the induced endomorphisms of the arithmetic jet spaces is highly complex and mysterious. Understanding it might be, in particular, another path towards introducing/understanding the Atkin operator U on δ-modular forms referred to in Problem 6.5. An obviously closely related problem is to understand the p-jet spaces of p-divisible groups; cf. Theorem 6.17. Yet another example of interest is the study of $J^n(X) \to J^n(X/\Gamma)$ for Γ a finite group acting on a smooth X; even the case $X = Y^n$ with Y a curve and $\Sigma = S_n$ the symmetric group acting naturally is still completely mysterious. This latter case is related to the concept of δ-symmetry mentioned in Problem 6.5 and appeared in an essential manner in the paper [24]: the failure of invariants to commute with formation of jet spaces (in the ramified case) is directly responsible for the existence (and indeed abundance) of δ-functions on smooth projective curves which do not arise from δ-characters of the Jacobian.

Problem 6.13. *Understand which analytic functions $X(\mathbb{Z}_p) \to \mathbb{Z}_p$ for smooth schemes $X/\mathbb{Z}_p$ are induced by δ-functions.*

Indeed it was proved in [23] that a function $f : \mathbb{Z}_p \to \mathbb{Z}_p$ is analytic if, and only if, there exists m such that f can be represented as $f(x) = F(x, \delta x, \ldots, \delta^m x)$, where $F \in \mathbb{Z}_p[x_0, x_1, \ldots, x_m]\hat{}$ is a restricted power series with $\mathbb{Z}_p$-coefficients in $m+1$ variables. This can be viewed as a "differential interpolation result": indeed $f(x)$ is given by a finite family of (unrelated) power series $F_i(x)$ convergent on disjoint balls B_i that cover $\mathbb{Z}_p$ and the result says that one can find a single power series $F(x, \delta x, \ldots, \delta^m x)$ that equals $F_i(x)$ on B_i for each i. One can ask for a generalization of this by asking which analytic functions $f : X(\mathbb{Z}_p) \to \mathbb{Z}_p$ defined on the $\mathbb{Z}_p$-points of a smooth scheme $X/\mathbb{Z}_p$ come from a δ-function $\tilde{f} \in \mathcal{O}^n(X)$ (i.e. $f = \tilde{f}_*$); of course such a $\tilde{f}$ cannot be, in general, unique. If $X = \mathbb{A}^1$ is the affine line then the result in [23] says that any f comes from some $\tilde{f}$. This is probably still the case if X is any affine smooth scheme. On the other hand this fails if $X = \mathbb{P}^1$ is the projective line simply because there are no non-constant δ-functions in $\mathcal{O}^n(\mathbb{P}^1)$ [12] but, of course, there are plenty of non-constant analytic functions $\mathbb{P}^1(\mathbb{Z}_p) \to \mathbb{Z}_p$. There should be a collection of cohomological obstructions to lifting f to some $\tilde{f}$ that should reflect the global geometry of X. This seems to us a rather fundamental question in understanding the relation between p-adic analytic geometry and δ-geometry.

In the light of Theorem 6.17 one can ask:

Problem 6.14. *Let $\alpha_0, \ldots, \alpha_\nu$ be elements in the algebraic closure of the fraction field of R which are integral over R. Let $X_i = Spec\ R[\alpha_i]$ be viewed as a closed subscheme of the line $\mathbb{A}^1$ over R. Compute/understand the arithmetic jet spaces $J^n(\cup_{i=0}^{\nu} X_i)$.*

This is a problem "about the interaction" of algebraic numbers. Indeed the case $\nu = 0$ is clear; for instance if $R[\alpha]$ is totally ramified over R (and $\neq R$) then the arithmetic jet spaces are empty: $J^n(Spec\ R[\alpha]) = \emptyset$ for $n \geq 1$. However, for $\nu \geq 1$, an interesting new phenomenon occurs. Indeed if $\alpha_i = \zeta_{p^i}$ (primitive p^i-th root of unity) then although $J^n(Spec\ R[\zeta_{p^i}]) = \emptyset$ for $n \geq 1$ and $i \geq 1$ we have that $J^n(\cup_{i=0}^{\nu} Spec\ R[\zeta_{p^i}]) = J^n(\mu_{p^\nu})$ is non-empty and indeed extremely interesting; cf. Theorem 6.17.

Problem 6.15. *Find arithmetic analogues of Hamiltonian systems and of algebraically completely integrable systems. Find arithmetic analogues of the formal pseudo-differential calculus.*

This problem is motivated by the link (due to Fuchs and Manin [56]) between the Painlevé VI equation (which has a Hamiltonian structure) and the Manin map of an elliptic fibration. Painlevé VI possesses an arithmetic analogue whose study was begun in [31]; this study is in its infancy and deserves further attention. More generally there is an intriguing possibility that other physically relevant differential equations (especially arising in Hamiltonian contexts, especially in the completely integrable situation, both finite and infinite dimensional) have arithmetic analogues carrying arithmetic significance. Finally, it is conceivable that a meaningful arithmetic analogue of the formal pseudo-differential calculus in one variable [62]. p. 318 can be developed; in other words one should be able to bring into the picture negative powers of the p-derivation $\delta : \mathcal{O}^n(X) \to \mathcal{O}^{n+1}(X)$ in the same way in which formal pseudo-differential calculus brings into the picture the negative powers of the total derivative operator acting on functions on jet spaces.

Problem 6.16. *Find an arithmetic analogue of the differential groupoids/Lie pseudogroups in the work of Lie, Cartan, Malgrange* [39].

Indeed recall that one can view the p-jet spaces $J^n(\Sigma_m)$ of $\Sigma_m = Spec\ W_m(R)$ and the jet maps induced by the comonad maps as an arithmetic analogue of the Lie groupoid $J^n(\mathbb{R} \times \mathbb{R}/\mathbb{R})^*$ of the line; cf. [29] and Theorem 6.18 above for results on this. It would be interesting to investigate subobjects of the system $J^n(\Sigma_m)$ that play the role of analogues of differential sub-groupoids and to find arithmetic analogues of the differential invariants of diffeomorphism groups acting on natural bundles arising from frame bundles. A theory of $J^n(\Sigma_m)$, suitably extended to other rings and to arithmetic analogues of poly-vector fields, could also be interpreted as an arithmetic analogue of the theory of the deRham–Witt complex [44] because it would replace the usual Kähler differentials Ω by constructions involving the operators δ_p.

Problem 6.17. *Compute $\mathcal{O}^n(\mathbb{T}(N, \kappa))$ where $\mathbb{T}(N, \kappa)$ is the Hecke $\mathbb{Z}$-algebra attached to cusp forms on $\Gamma_1(N)$ of level κ.*

This could lead to a way of associating differential modular forms $f^\sharp$ to classical newforms f of weight $\neq 2$. (Cf. Theorem 6.16.) This might also lead to a link between δ-geometry and Galois representations. The problem may be related to the study of double coset sets of GL_2 (or more generally GL_n) and a link of this to the study in [33, 34, 35] is plausible.

We end by stating two of the most puzzling concrete open problems of the theory.

Problem 6.18. *Compute $\mathcal{O}^n(X)$ for an elliptic curve X.*

Note that $\mathcal{O}^1(X)$ and $\mathcal{X}^n(X)$ have been computed (Theorem 6.2) and one expects $\mathcal{X}^n(X)$ to "generate" $\mathcal{O}^n(X)$. In particular assume X is not a canonical lift and ψ is a basis of $\mathcal{X}^2(X)$; is it true that $\mathcal{O}^n(X) = R[\psi, \delta\psi, \dots, \delta^{n-1}\psi]\hat{}$? The analogue of this in differential algebra is true; cf. [9].

Problem 6.19. *Compute $\mathcal{X}^n(G)$ for G an extension of an elliptic curve by $\mathbb{G}_m$.*

One expects that this module depends in an arithmetically interesting way on the class of the extension. The problem is directly related to that of understanding the cohomology of arithmetic jet spaces.

References

[1] D. V. Alexeevski, V. V. Lychagin, A. M. Vinogradov, *Basic ideas and concepts of differential geometry*, in: Geometry I, R. V. Gamkrelidze, Ed., Encyclopedia of Mathematical Sciences, Volume 28, Springer 1991.

[2] M.Barcau, *Isogeny covariant differential modular forms and the space of elliptic curves up to isogeny*, Compositio Math., 137 (2003), 237-273.

[3] M. Barcau, A. Buium, *Siegel differential modular forms*, International Math. Res. Notices, 28 (2002), 1459-1503.

[4] J. Borger, *The basic geometry of Witt vectors, I*: the affine case, Algebra and Number Theory, 5 (2011), No. 2, 231-285.

[5] Borger, J., *The basic geometry of Witt vectors, II: Spaces*, Mathematische Annalen, 351 (2011), No. 4, 877-933.

[6] J. Borger, *Λ-rings and the field with one element*, arXiv:0906.3146v1

[7] J. Borger, A. Buium, *Differential forms on arithmetic jet spaces*, Selecta Math., 17 (2011), no. 2, 301-335.

[8] A. Buium, *Intersections in jet spaces and a conjecture of S.Lang*, Annals of Math. 136 (1992), 557-567.

[9] A. Buium, *Differential algebra and diophantine geometry*, Hermann, 1994.

[10] A. Buium, *On a question of B.Mazur*, Duke Math. J., 75 (1994), no. 3, 639-644.

[11] A. Buium, *Differential characters of Abelian varieties over p-adic fields*, Invent. Math., 122 (1995), 309-340.

[12] A. Buium, *Geometry of p-jets*, Duke J. Math., 82, (1996), no. 2, 349-367.

[13] A. Buium, *Differential characters and characteristic polynomial of Frobenius*, Crelle J. 485 (1997), 209-219.

[14] A. Buium, *Differential subgroups of simple algebraic groups over p-adic fields*, Amer. J. Math. 120 (1998), 1277-1287.

[15] A. Buium, *Differential modular forms*, J. reine angew. Math., 520 (2000), 95-167.

[16] A. Buium, *Arithmetic differential equations*, Math. Surveys and Monographs, 118, American Mathematical Society, Providence, RI, 2005. xxxii+310 pp.

[17] A. Buium, *Differential eigenforms*, J. Number Theory, 128 (2008), 979-1010.

[18] A. Buium, B. Poonen, *Independence of points on elliptic curves arising from special points on modular and Shimura curves, I: global results*, Duke Math. J., 147 (2009), No. 1, 181-191.

[19] A. Buium, B. Poonen, *Independence of points on elliptic curves arising from special points on modular and Shimura curves, II: local results*, Compositio Math., 145 (2009), 566-602.

[20] A. Buium, S.R. Simanca, *Arithmetic Laplacians*, Adv. Math., 220 (2009), 246-277.

[21] A. Buium, S.R. Simanca, *Arithmetic partial differential equations, I*, Advances in Math. 225 (2010), 689-793.

[22] A. Buium, S.R. Simanca, *Arithmetic partial differential equations II*, Advances in Math., 225 (2010), 1308-1340.

[23] A. Buium, C. Ralph, S.R. Simanca, *Arithmetic differential operators on $\mathbb{Z}_p$*, J. Number Theory, 131 (2011), 96-105.

[24] A. Buium, *Differential characters on curves*, in: Number Theory, Analysis and Geometry: In Memory of Serge Lang, D. Goldfeld et al. Editors, Springer, 2011, pp. 111-123.

[25] A. Buium, A. Saha, *The ring of differential Fourier expansions*, J. Number Theory, 132 (2012), 896-937.

[26] A. Buium, A. Saha, *Hecke operators on differential modular forms mod p*, J. Number Theory, 132 (2012), 966-997

[27] A. Buium, A. Saha, *Differential overconvergence*, in: Algebraic methods in dynamical systems; volume dedicated to Michael Singer's 60th birthday, Banach Center Publications, Vol 94 (2011), 99-129.

[28] A. Buium, *p-jets of finite algebras, I: p-divisible groups*, Documenta Math., 18 (2013), 943-969.

[29] A. Buium, *p-jets of finite algebras, II: p-typical Witt rings*, Documenta Math., 18 (2013), 971-996.

[30] A. Buium, *Galois groups arising from arithmetic differential equations*, to appear in: Proceedings of a Conference in Luminy, Seminaires et Congrès, Soc. Math. France.

[31] A. Buium, Yu. I. Manin, *Arithmetic differential equations of Painlevé VI type*, arXiv:1307.3841

[32] A. Buium, A. Saha, *The first p-jet space of an elliptic curve: global functions and lifts of Frobenius*, arXiv:1308.0578, to appear in Math. Res. Letters.

[33] A. Buium, T. Dupuy, *Arithmetic differential equations on GL_n, I: differential cocycles*, arXiv:1308.0748v1.

[34] A. Buium, T. Dupuy, *Arithmetic differential equations on GL_n, II: arithmetic Lie theory*, arXiv:1308.0744.

[35] A. Buium, T. Dupuy, *Arithmetic differential equations on GL_n, III: Galois groups*, arXiv:1308.0747

[36] P. Cassidy, *Differential algebraic groups*, Amer. J. Math, 94 (1972), 891-954.

[37] Z. Chatzidakis, E. Hrushovski, *Model theory of difference fields*, Trans. AMS, 351 (1999), 2997-3071.

[38] A. Connes, *Non-commutative Geometry*, Academic Press, 1994.

[39] V. G. Drinfeld, *Elliptic modules*, Math. Sbornik 94 (1974), 594-627.

[40] T. Dupuy, Arithmetic deformation theory of algebraic curves, PhD Thesis (UNM).

[41] B. Dwork, G. Gerotto, F. J. Sullivan, *An introduction to G-functions*. Annals of Mathematics Studies, 133. Princeton University Press, Princeton, NJ, 1994. xxii+323 pp.

[42] M.Greenberg, *Schemata over local rings*, Annals of Math., 73 (1961), 624-648.

[43] A. Herras-Llanos, PhD thesis (UNM), in preparation.

[44] L. Hesselholt, The big deRham–Witt complex, preprint, 2011.

[45] C. Hurlburt, *Isogeny covariant differential modular forms modulo p*, Compositio Math., 128, (2001), No. 1, 17-34.

[46] Y. Ihara, *An invariant multiple differential attached to the field of elliptic modular functions of characteristic p*, Amer. J. Math., XCIII (1971), No. 1, 139-147.

[47] Y. Ihara, *On Fermat quotient and differentiation of numbers*, RIMS Kokyuroku 810 (1992), 324-341, (In Japanese). English translation by S. Hahn, Univ. of Georgia preprint.

[48] T. A. Ivey, J. M. Landsberg, *Cartan for beginners: differential geometry via moving frames and exterior differential systems*, GTM 61, AMS 2003.

[49] A. Joyal, *δ–anneaux et vecteurs de Witt*, C.R. Acad. Sci. Canada, VII, (1985), No. 3, 177-182.

[50] N. Katz, *p-adic interpolation of real Eisenstein series*, Ann. of Math. 104 (1976), 459-571.

[51] E.R. Kolchin, *Differential algebra and algebraic groups*. Pure and Applied Mathematics, Vol. 54. Academic Press, New York-London, 1973. xviii+446 pp.

[52] E.Kolchin, *Differential Algebraic Groups*, Academic Press, 1985.

[53] N. Kurokawa, H. Ochiai, M. Wakayama, *Absolute derivations and zeta functions*, Documenta Math.,Extra Volume Kato (2003), 565-584.

[54] B. Malgrange, Le groupoïde de Galois d'un feuilletage, Monographie 28 Vol 2, L'enseignement Mathématique (2001).

[55] Yu. I. Manin, *Rational points on algebraic curves over function fields*, Izv. Acad. Nauk USSR, 27 (1963), 1395-1440.

[56] Yu. I. Manin, *Sixth Painlevé equation, universal elliptic curve, and mirror of* $\mathbb{P}^2$, arXiv:alg-geom/9605010.

[57] Yu. I. Manin, *Cyclotomy and analytic geometry over* $\mathbb{F}_1$, arXiv:0809.1564.

[58] Yu. I. Manin, *Numbers as functions*, *p*-Adic Numbers, Ultrametric Analysis, and Applications, 5 (2013), No. 4, 313-325.

[59] B.Mazur, *Arithmetic on curves*, Bull. Amer. Math. Soc., 14, 2 (1986), 207-259.

[60] M.Marcolli, *Lectures on Arithmetic Non-commutative Geometry*, Univ. Lecture Series 36, AMS, 2005.

[61] M. Morishita, *Knots and primes*, Springer, 2012.

[62] P. J. Olver, *Appications of Lie Groups to Differential Equations*, GTM 107, Springer, 2000.

[63] R. Pink, *A combination of the conjectures of Mordell-Lang and André-Oort*, in: Progress in Math 235, Birkhauser 2005, pp. 251-282.

[64] M.Raynaud, *Courbes sur une variété abélienne et points de torsion*, Invent. Math. 71 (1983), 207-235.

[65] M. Singer, M. van der Put, *Galois theory of difference equaltions*, LNM, Springer 1997.

[66] P. Vojta, Jets via Hasse-Schmidt derivations, in: Diophantine geometry, pp. 335-361, CRM Series, 4, U. Zannier Ed., Sc. Norm., Pisa, 2007.

7

Un calcul de groupe de Brauer et une application arithmétique

Jean-Louis Colliot-Thélène
C.N.R.S., Université Paris Sud, Mathématiques, Bâtiment 425,
91405 Orsay Cedex, France
E-mail address : jlct@math.u-psud.fr

RÉSUMÉ. Browning et Matthiesen ont récemment démontré le principe de Hasse et l'approximation faible pour une certaine classe de variétés algébriques. Skorobogatov a utilisé ce résultat pour montrer que sur une deuxième classe, plus large, de variétés, les points rationnels sont denses dans l'ensemble de Brauer–Manin. On considère ici une troisième classe de variétés, intermédiaire entre la première et la seconde, pour laquelle on établit, de façon purement algébrique, que le groupe de Brauer non ramifié est trivial. Le résultat de Skorobogatov donne alors le principe de Hasse et l'approximation faible pour ces variétés.

7.1 Un calcul algébrique

Soient k un corps, $K = \overline{k}$ une clôture séparable de k et $G = \mathrm{Gal}(K/k)$ le groupe de Galois absolu. Pour toute k-variété X, on note $X_K = X \times_k K$.

On a la proposition bien connue (cf. [**CTS**, (1.5.0)]) :

***Proposition* 7.1** *Pour toute k-variété X, on a une suite exacte naturelle*

$$0 \to H^1(G, K[X]^\times) \to \mathrm{Pic}\, X \to (\mathrm{Pic}\, X_K)^G \to H^2(G, K[X]^\times) \to$$
$$\mathrm{Ker}[\mathrm{Br}\, X \to \mathrm{Br}\, X_K] \to H^1(G, \mathrm{Pic}\, X_K)$$

Arithmetic and Geometry, ed. Luis Dieulefait *et al.* Published by Cambridge University Press.

C'est la suite exacte des termes de bas degré de la suite spectrale de Hochschild-Serre pour le faisceau $\mathbb{G}_m$ pour la topologie étale, et le revêtement $X_K \to X$.

Soient $L_i(x_1, \ldots, x_n)$, $i = 1, \ldots, m$ des formes linéaires à coefficients dans k, *linéairement indépendantes deux à deux*, et soient K_i/k, $i = 1, \ldots, m$, des k-algèbres finies étales.

Soit Y la k-variété affine définie par le système d'équations affines

$$L_i(x_1, \ldots, x_n) = \mathrm{Norm}_{K_i/k}(\Xi_i),\ i = 1, \ldots, m.$$

Ici Ξ_i varie dans la k-variété $R_{K_i/k}\mathbf{A}^1$ définie par la restriction à la Weil de K_i à k de la droite affine $\mathbf{A}^1_{K_i}$.

Soit $p : Y \to \mathbf{A}^n_k$ la projection sur l'espace affine de dimension n définie par les coordonnées $(x_1, \ldots, x_n)$. Soit $F \subset \mathbf{A}^n$ le complémentaire de la réunion des fermés de codimension 2 définis par $L_i = L_j = 0,\ i < j$. Soit $U \subset \mathbf{A}^n_k$ l'ouvert complémentaire de F et soit $X = Y_U \subset Y$ l'ouvert image inverse de U dans Y.

Proposition 7.2 *(i) La k-variété X est lisse et géométriquement intègre.*

(ii) Toute fonction inversible sur X est constante : $k^\times = k[X]^\times$.

(iii) Le groupe de Picard Pic X *est nul.*

(iv) Pour toute compactification lisse Z de X, le groupe de Picard Pic Z_K *est un G-réseau de permutation.*

(v) Supposons k de caractéristique zéro. Pour toute k-variété Z projective, lisse, géométriquement intègre k-birationnelle à Y, l'appplication naturelle de groupes de Brauer $\mathrm{Br}\, k \to \mathrm{Br}\, Z$ *est un isomorphisme.*

Démonstration Pour les fonctions inversibles, on a $k[X]^\times = (K[X]^\times)^G$. Le théorème 90 de Hilbert donne $H^1(G, K^\times) = 0$. En utilisant la proposition 7.1, on voit que pour établir les énoncés (i), (ii) et (iii) on peut se limiter au cas $k = K$ séparablement clos, ce que nous supposons jusqu'à nouvel ordre.

Chaque norme $\mathrm{Norm}_{K_i/k}(\Xi_i)$ est alors un produit de variables indépendantes :

$$\mathrm{Norm}_{K_i/k}(\Xi_i) = \prod_{r \in I_i} \Xi_{i,r}.$$

Au-dessus de l'ouvert $V \subset U \subset \mathbf{A}^n_k$ défini par $\prod_i L_i \neq 0$, la fibration $p : X_V \to V$ est clairement lisse, les fibres sont des tores. Plus précisément, pour $i = 1, \ldots, m$, on a

$$\Xi_{i,1} = L_i(x_1, \ldots, x_n). \prod_{r \in I_i,\ r > 1} \Xi_{i,r}^{-1}$$

si bien que X_V est le produit de $V \subset \mathbf{A}_k^n$ et d'un k-tore de groupe des caractères ayant pour base les $\Xi_{i,r}$ pour $r > 1$. Il est alors clair que l'on a $\operatorname{Pic} X_V = 0$. Le groupe des fonctions inversibles sur X_V est engendré par les L_i et les $\Xi_{i,r}$, où, pour i donné, r varie dans I_i. Il est donc engendré par les $\Xi_{i,r}$. Ces éléments sont indépendants multiplicativement.

Comme L_1 n'est pas identiquement nul, la k-variété définie par

$$L_1(x_1, \ldots, x_n) = \operatorname{Norm}_{K_1/k}(\Xi_1)$$

est isomorphe à un espace affine, donc intègre et lisse sur k. La projection de la variété définie par

$$0 \neq L_i(x_1, \ldots, x_n) = \operatorname{Norm}_{K_i/k}(\Xi_i), \quad i = 2, \ldots, m,$$

sur $\mathbf{A}_k^n$ est un morphisme lisse. La k-variété $X_1 \subset X$ définie par les équations

$$L_1(x_1, \ldots, x_n) = \operatorname{Norm}_{K_1/k}(\Xi_1)$$

$$0 \neq L_i(x_1, \ldots, x_n) = \operatorname{Norm}_{K_i/k}(\Xi_i), \quad i = 2, \ldots, m,$$

On voit aisément qu'elle est intègre. Définissant X_i, pour chaque $i = 2, \ldots, m$ de façon analogue à X_1, on voit que la k-variété X, qui est la réunion des ouverts X_i, est une k-variété lisse (cf. [**Sk1**, Lemma 4.4.5]). Notons $\Delta_{i,r}$ le diviseur de X défini par $\Xi_{i,r} = 0$, qui est inclus dans $L_i = 0$. Un élément $f \in k[X]^\times$ produit d'éléments $\Xi_{i,r}^{n_{i,r}}$ avec l'un des $n_{i,r} \in \mathbb{Z}$ non nul n'est pas inversible. Ceci établit $k^\times = k[X]^\times$.

Le complémentaire de X_V dans X est la réunion des diviseurs $\Delta_{i,r}$, et chacun d'eux est principal sur X. Comme on a $\operatorname{Pic} X_V = 0$, ceci implique $\operatorname{Pic} X = 0$.

Sur tout corps k, désormais quelconque et de clôture séparable K, on a donc établi les points (i) à (iii), et la suite exacte de la proposition 7.1 donne de plus un isomorphisme :

$$\operatorname{Br} k \xrightarrow{\simeq} \operatorname{Ker}[\operatorname{Br} X \to \operatorname{Br} X_K].$$

Pour Z comme en (iv), le groupe $\operatorname{Div}_\infty Z_K$ des diviseurs à support dans le complémentaire de X_K dans Z_K est un G-réseau de permutation (possédant une base globalement respectée par l'action de G), et ce groupe est G-isomorphe à $\operatorname{Pic} Z_K$, comme le montre la suite exacte

$$K[Z]^\times \to K[X]^\times \to \operatorname{Div}_\infty Z_K \to \operatorname{Pic} Z_K \to \operatorname{Pic} X_K.$$

Ceci établit (iv).

On a vu plus haut que l'ouvert $X_{V,K}$ de X_K est le produit d'un ouvert de $\mathbf{A}_K^n$ et d'un K-tore. La K-variété X_K est donc K-birationnelle à un espace projectif.

Supposons désormais k de caractéristique zéro. Soit Z une k-variété propre, lisse, géométriquement intègre k-birationnelle à X. Il existe donc un ouvert $W \subset X$ contenant tous les points de codimension 1 de X et un k-morphisme birationnel $f : W \to Z$, d'où l'on déduit par les propriétés standards du groupe de Brauer des injections $\mathrm{Br}\, Z \hookrightarrow \mathrm{Br}\, W$ (lissité des k-variétés), puis $\mathrm{Br}\, X \xrightarrow{\simeq} \mathrm{Br}\, W$ (pureté pour le groupe de Brauer en caractéristique zéro, [**Gr**, §6]), donc $\mathrm{Br}\, Z \hookrightarrow \mathrm{Br}\, X$. L'invariance birationnelle du groupe de Brauer en caractéristique zéro [**Gr**, §7], sa nullité sur l'espace affine $\mathbf{A}^n_K$ en caractéristique zéro et la K-rationalité de la K-variété propre et lisse Z_K donnent $\mathrm{Br}\, Z_K = 0$. On a donc

$$\mathrm{Br}\, Z = \mathrm{Ker}[\mathrm{Br}\, Z \to \mathrm{Br}\, Z_K] \hookrightarrow \mathrm{Ker}[\mathrm{Br}\, X \to \mathrm{Br}\, X_K],$$

et comme la flèche $\mathrm{Br}\, k \to \mathrm{Ker}[\mathrm{Br}\, X \to \mathrm{Br}\, X_K]$ est un isomorphisme, il en résulte que la flèche $\mathrm{Br}\, k \to \mathrm{Br}\, Z$ est un isomorphisme, ce qui établit (v). □

***Remarque* 7.3** Le calcul ci-dessus est une généralisation du calcul fait au Lemme 2.6.1 de [**CTS**].

7.2 Application arithmétique

Le calcul du paragraphe précédent a été motivé par le théorème 1.3 de l'article [**BM**] de T. Browning et L. Matthiesen. Pour toute $\mathbb{Q}$-variété définie par un système

$$0 \neq L_i(x_1, \dots, x_n) = \mathrm{Norm}_{K_i/\mathbb{Q}}(\Xi_i),\ i = 1, \dots, m$$

comme à la proposition 7.2, les algèbres K_i *étant de plus supposées être des corps*, les auteurs montrent que le principe de Hasse et l'approximation faible valent pour les points rationnels. Il ne saurait donc y avoir d'obstruction de Brauer–Manin associée au groupe de Brauer d'une compactification lisse.

On eût pu difficilement imaginer un tel théorème si le groupe de Brauer d'une telle compactification n'avait pas été réduit au groupe de Brauer du corps de base, ce qu'établit la proposition 7.2.

A. Skorobogatov [**Sk2**] a récemment combiné la théorie de la descente et le théorème de Browning et Matthiesen. Il a ainsi établi que l'obstruction de Brauer–Manin est la seule obstruction au principe de Hasse pour certaines variétés plus générales que celles considérées par Browning et Matthiesen. La conjonction de [**Sk2**, Prop. 3.5] et de la proposition 7.2 ci-dessus donne l'énoncé suivant, qui étend directement celui de Browning et Matthiesen.

***Théorème* 7.4** *Soient* $L_i(x_1, \ldots, x_n)$, $i = 1, \ldots, m$ *des formes linéaires à coefficients dans* $\mathbb{Q}$, *linéairement indépendantes deux à deux, et soient* $K_i/\mathbb{Q}$, $i = 1, \ldots, m$, *des* $\mathbb{Q}$*-algèbres finies étales, c'est-à-dire des produits d'extensions finies de* $\mathbb{Q}$.

Soit Y *la* $\mathbb{Q}$*-variété affine définie par le système d'équations affines*

$$0 \neq L_i(x_1, \ldots, x_n) = \mathrm{Norm}_{K_i/k}(\Xi_i), \quad i = 1, \ldots, m.$$

Ici Ξ_i *varie dans la* $\mathbb{Q}$*-variété* $R_{K_i/\mathbb{Q}}\mathbf{A}^1$ *définie par la restriction à la Weil de* K_i *à* $\mathbb{Q}$ *de la droite affine* $\mathbf{A}^1_{K_i}$.

Le principe de Hasse et l'approximation faible valent pour les points rationnels de Y. □

Ce travail, stimulé par un séjour à l'Institut Hausdorff à Bonn en avril 2013, a bénéficié d'une aide de l'Agence Nationale de la Recherche portant la référence ANR-12-BL01-0005.

Références

[BM] T. Browning et L. Matthiesen, Norm forms for arbitrary number fields as products of linear polynomials, `http://arxiv.org/abs/1307.7641`

[CTS] J.-L. Colliot-Thélène et J.-J. Sansuc, La descente sur les variétés rationnelles, II, Duke Math. J. **54** (1987) 375–492.

[Gr] A. Grothendieck, Le groupe de Brauer III, in *Dix exposés sur la cohomologie des schémas*, Masson & North-Holland, 1968.

[Sk1] A. N. Skorobogatov, *Torsors and rational points*, Cambridge tracts in mathematics **144**, Cambridge University Press, 2001.

[Sk2] A. N. Skorobogatov, Descent on toric fibrations. This volume, pp. 422–435.

8

Connectedness of Hecke algebras and the Rayuela conjecture: a path to functoriality and modularity

Luis Dieulefait[1] *and Ariel Pacetti*[2]

[1]Departament d'Àlgebra i Geometria, Facultat de Matemàtiques, Universitat de Barcelona, Gran Via de les Corts Catalanes, 585. 08007 Barcelona
E-mail address: ldieulefait@ub.edu

[2]Departamento de Matemática, Facultad de Ciencias Exactas y Naturales, Universidad de Buenos Aires and IMAS, CONICET, Argentina
E-mail address: apacetti@dm.uba.ar

A la memoria de Julio Cortazar, autor de "Rayuela", al cumplirse hoy cien años de su nacimiento.

ABSTRACT. Let ρ_1 and ρ_2 be a pair of residual, odd, absolutely irreducible two-dimensional Galois representations of a totally real number field F. In this article we propose a conjecture asserting existence of "safe" chains of compatible systems of Galois representations linking ρ_1 to ρ_2. Such conjecture implies the generalized Serre's conjecture and is equivalent to Serre's conjecture under a modular version of it. We prove a weak version of the modular variant using the connectedness of certain Hecke algebras, and we comment on possible applications of these results to establish some cases of Langlands functoriality.

8.1 Introduction

Let F be a totally real number field. In [KK03] Khare and Kiming conjectured that given ρ_1 and ρ_2 two odd, absolutely irreducible, two-dimensional residual representations of the absolute Galois group of F with values on finite fields of

Date: August 26, 2014
2010 *Mathematics Subject Classification.* 11F33
Keywords and phrases. Base Change, Rayuela Conjecture
L.D. partially supported by MICINN grants MTM2012-33830 and by an ICREA Academia Research Prize.
AP was partially supported by CONICET PIP 2010-2012 GI and FonCyT BID-PICT 2010-0681.

Arithmetic and Geometry, ed. Luis Dieulefait *et al.* Published by Cambridge University Press.

different prime characteristics ℓ and ℓ', satisfying certain local compatibilities (for all primes q different from ℓ and ℓ', there exists a Weil-Deligne parameter and a choice of integral model such that their reductions modulo ℓ and ℓ' are isomorphic to the restrictions of ρ_1 and ρ_2 to a decomposition group at q), there exists a modular form which residually coincides with ρ_1 at ℓ and ρ_2 at ℓ'. This is a very strong statement and no evidence for its truth is known. In particular it implies that given two Hilbert modular forms f and g over F, and two prime numbers ℓ and ℓ', if we assume suitable local compatibilities, then there exists a Hilbert modular form h which is congruent to f modulo ℓ and to g modulo ℓ'.

The purpose of this article is to state a weaker conjecture called **The Rayuela Conjecture** involving "chains of congruent systems" connecting two given Galois representations. The idea is not to link any given pair of residual Galois representations via a single compatible system of Galois representations but via a chain of such systems, meaning that each of them has to be congruent to the next one in the chain, modulo a suitable prime, and it will also be required that some reasonable properties are being satisfied at each of these congruences (this will be done in Section 8.2). Such conjecture has important consequences, like Serre's conjectures for totally real number fields or base change for totally real number fields (the second one follows from a weaker version that we will explain later). The conjecture for the field of rational numbers follows from the results proved by Dieulefait in [Die12a] (the reader can easily check that the method used to prove base change in loc. cit. implies, in fact is based on, the fact that this conjecture is true over $\mathbb{Q}$ at least for modular Galois representations, and this combined with the truth of Serre's conjecture over $\mathbb{Q}$ implies the conjecture).

Although we are not able to prove the conjecture itself, in Sections 8.4 and 8.5 we prove that some (weak) variant of it holds in the modular world. This is a generalization of a Theorem of Mazur, a result that he proved for the case of weight 2 modular forms of prime level N.

In this article, for general weights and levels, and for any totally real number field, we extend Mazur's result, and the control we get in the level and weights of the modular forms involved might be used to prove the modular version of the conjecture (although there are still some ingredients missing to get the full statement).

Also we will show that our Conjecture is equivalent to the generalized Serre's conjecture proposed in [BDJ10], at least if some strengthening of the available Modularity Lifting Theorems is assumed. What we show is how one can manipulate a pair of modular Galois representations to end up in a controlled situation where both representations have the same Hodge-Tate weights and ramification in a small controlled set of primes (and then this is combined with Mazur's result).

Combining the ideas of the present article with some nowadays standard tricks (like the use of Micro Good Dihedral primes), one can prove for some small real quadratic fields a Base Change theorem, all that is left is a finite computational check (to ensure that some Hecke Algebra of known level and weight is connected in a good way). This is part of a work in progress of the authors but more will be said at the end of the present article.

For most of the chains constructed in this paper, the construction carries on for the case of abstract Galois representations (the reader should keep in mind that the possibility of constructing congruences between abstract Galois representations where some local information changes and the existence of compatible systems containing most geometric p-adic Galois representations are the two main technical ingredients in the proofs of Serre's conjecture over $\mathbb{Q}$), and then again the use of Micro Good Dihedral primes combined with the results in this paper is enough to reduce the proof of Serre's conjecture over a given small real quadratic field to some special cases of Galois representations with small invariants (Serre's weights and conductor). Thus it is perfectly conceivable that one can give a complete proof of Serre's conjecture over a small real quadratic field F, by just completing this process with a few extra steps designed to remove the Micro Good Dihedral prime from the level (eventually relying on results of Skinner and Wiles [SW99] if the residually reducible case has to be considered) and end up in some "base case for modularity" over F, such as the modularity/non-existence of semistable abelian varieties of conductor 3 over $\mathbb{Q}(\sqrt{5})$ proved by Schoof ([Sch12]). We plan to check over which real quadratic fields, such proof of Serre's conjecture can be completed in a future work.

Section 8.3 contains the Modularity Lifting Theorems (that will be denoted MLT) used and needed in the present article. We have included a mixed case that has only been proved in weight 2 situations, yet we assume its truth in more general situations. We believe that such a result can be deduced from the techniques in [BLGGT], but we haven't formally checked this.

Conventions and notations: If F is a number field, we denote by $\mathcal{O}_F$ its ring of integers. By G_F we denote the Galois group $\mathrm{Gal}(\bar{F}/F)$. All Galois representations (residual or ℓ-adic) are assumed to be continuous. Let $\mathfrak{p}$ be a prime ideal in $\mathcal{O}_F$. We denote by $\mathrm{Frob}_{\mathfrak{p}}$ a Frobenius element over $\mathfrak{p}$.

To ease notation, instead of specifying a prime π in the field of coefficients of a Galois representation and reducing mod π, sometimes we will simply say that we reduce "mod p", where p is the rational prime below π. We expect that this will be no cause of confusion.

Concerning local types of strictly compatible systems of Galois representations, we will use the words "Steinberg", "principal series" and "supercuspidal", to denote the local representations they correspond to under local Langlands.

Acknowledgment: we want to thank Fred Diamond for useful comments and remarks and for the proof of Lemma 8.32.

8.2 The Rayuela Conjecture

There are nowadays many MLT that can be applied to propagate modularity through a congruence between two ℓ-adic Galois representations. For us, the main interest in congruences between Galois representations is in the case where a MLT holds in both directions (i.e. modularity of either of the two representations implies modularity of the other one), so we will call a congruence an MLT congruence if the hypothesis of some MLT theorem are fulfilled in both directions. Since MLT hypothesis are becoming less restrictive with time, some of the proofs we give make use of tricks that are required to reduce to situations in which some of the available MLT applies, but some of these tricks are very likely to become obsolete in the near future (when new MLT are proved). There are also steps in the chains that we are going to build in our attempt to connect two given modular or abstract Galois representations that involve congruences that are not known to be MLT. This is the unique reason why we are not able to prove any strong result in this paper, such as relative base change (see the discussion at the end of the article).

Let us now state our conjecture for abstract Galois representations.

Conjecture 8.1. The Rayuela Conjecture*: Let F be a totally real number field, ℓ_0, ℓ_∞ prime numbers and*

$$\rho_i : \mathrm{Gal}(\bar{F}/F) \to \mathrm{GL}_2(\overline{\mathbb{F}}_{\ell_i}), \qquad i = 0, \infty,$$

two absolutely irreducible odd Galois representations. Then there exists a family of odd, absolutely irreducible, two-dimensional strictly compatible systems of Galois representations $\{\rho_{i,\lambda}\}_{i=1}^{n}$ such that:

- $\rho_0 \equiv \rho_{1,\lambda_0} \pmod{\lambda_0}$*, with* $\lambda_0 \mid \ell_0$.
- $\rho_\infty \equiv \rho_{n,\lambda_\infty} \pmod{\lambda_\infty}$*, with* $\lambda_\infty \mid \ell_\infty$.

- *for* $i = 1, \dots, n-1$ *there exist* λ_i *prime such that* $\rho_{i,\lambda_i} \equiv \rho_{i+1,\lambda_i} \pmod{\lambda_i}$,
- *all the congruences involved are MLT.*

Remark 8.2. The primes involved always are taken to be in the field of coefficient of the corresponding system of representations (for the first and second condition) or in the compositum of the two relevant such fields (for the third conditions). From now on, this remark applies to all congruence between compatible systems appearing in this paper.

The last hypothesis of the conjecture implies that modularity of one of the given representations can be propagated through the chain of congruences produced by the conjecture allowing to prove modularity of the other one. It is thus clear that Conjecture 8.1 implies Serre's generalized conjecture over F by just taking ρ_∞ to be an irreducible residual representation attached to any cuspidal Hilbert modular form over F.

At this time we think Conjecture 8.1 is out of reach, but since it does not involve directly modular forms, it might lead to a different attack to Serre's conjecture. Note that in the hypothesis of the Conjecture, one can put ρ_0 inside a strictly compatible system of Galois representations as done by [Die04] for representations over $\mathbb{Q}$ and by [Sno09] for totally real fields (using Theorem 7.6.1 and Corollary 1.1.2). The main difficulty for abstract representations is to connect them. During this work we will show how to manipulate abstract representations such that if we start with any pair of them we can connect both, through suitable chains of MLT congruences, to representations having common values for their Serre's level and weights, but we cannot go any further so far.

For different purposes, such as applications to Langlands functoriality, it is interesting to study the previous conjecture in the modular setting. In this case we are able to prove part of it, namely, we are able to build the chain of congruences but we can not ensure that congruences are MLT at all steps. Still, we find this interesting because advances in MLT theorems may eventually lead to a proof of this modular variant of the conjecture, and this will be evidence for the truth of the Rayuela conjecture. In fact it will give a proof that this conjecture is equivalent to the generalized Serre's conjecture over F (the other implication being trivial as already remarked). The modular variant is the following:

Conjecture 8.3. *Let* F *be a totally real number field,* ℓ_0, ℓ_∞ *prime numbers and*

$$\rho_i : \mathrm{Gal}(\bar{F}/F) \to \mathrm{GL}_2(\overline{\mathbb{F}}_{\ell_i}), \qquad i = 0, \infty,$$

two absolutely irreducible odd Galois representations attached to Hilbert modular newforms f_0 and f_∞, respectively. Then there exists a family of odd, absolutely irreducible, two-dimensional strictly compatible systems of Galois representations $\{\rho_{i,\lambda}\}_{i=1}^n$ such that:

- $\rho_0 \equiv \rho_{1,\lambda_0} \pmod{\lambda_0}$, *with* $\lambda_0 \mid \ell_0$.
- $\rho_\infty \equiv \rho_{n,\lambda_\infty} \pmod{\lambda_\infty}$, *with* $\lambda_\infty \mid \ell_\infty$.
- *for* $i = 1, \dots, n-1$ *there exist* λ_i *prime such that* $\rho_{i,\lambda_i} \equiv \rho_{i+1,\lambda_i} \pmod{\lambda_i}$,
- *all the congruences involved are MLT.*

The aim of this article is to prove part of Conjecture 8.3 and show some implications of it related to modularity and functoriality. It is clear that all the representations appearing in Conjecture 8.3 are modular. Clearly we have:

Theorem 8.4. *Serre's generalized conjecture + Conjecture 8.3 imply Conjecture 8.1.*

Proof. This is clear, since if ρ_0 and ρ_∞ are any two Galois representation and if Serre's conjecture holds, they are modular, and then they are connected in the right way by Conjecture 8.3. □

Conjecture 8.3 can be proved using standard arguments mainly due to Mazur, if we remove the condition that the congruences are MLT.

Theorem (Mazur). *Let F be a totally real number field and f_0, f_∞ two Hilbert modular forms, whose weights are congruent modulo 2. Then there exists Hilbert modular forms $\{h_i\}_{i=1}^n$ such that $h_1 = f_0$, $h_n = f_\infty$ and for $i = 1, \dots, n-1$ there exist λ_i prime such that $\rho_{h_i,\lambda_i} \equiv \rho_{h_{i+1},\lambda_i} \pmod{\lambda_i}$.*

Remark 8.5. Although connectedness of the Hecke algebra was proved by Mazur for classical modular forms of weight 2 and prime level N using the curve $X_0(N)$, the proof we will give is strongly based in his argument.

Remark 8.6. If modularity would propagate via any congruence, the previous Theorem would be equivalent to Conjecture 8.3. Unfortunately this is not clear even for classical modular forms.

The main result of the present article is the following:

Theorem 8.7. *Let F be a totally real number field, ℓ_f, ℓ_g prime numbers and*

$$\rho_f : \mathrm{Gal}(\bar{F}/F) \to \mathrm{GL}_2(\overline{\mathbb{F}}_{\ell_f}), \qquad \rho_g : \mathrm{Gal}(\bar{F}/F) \to \mathrm{GL}_2(\overline{\mathbb{F}}_{\ell_g}),$$

two absolutely irreducible odd Galois representations attached to Hilbert modular newforms f and g, respectively. Then there exists prime ideals $\mathfrak{q} \subset \mathcal{O}_F$ and $p \in \mathbb{Q}$ which splits completely in F, Hilbert modular forms f_∞ and g_∞ of parallel weight 2 and level $\Gamma_0(p\mathfrak{q}^2)$ and two families of odd, absolutely irreducible, 2-dimensional strictly compatible systems of Galois representations $\{\rho^f_{i,\lambda}\}_{i=1}^{n_f}$, $\{\rho^g_{i,\lambda}\}_{i=1}^{n_g}$, such that:

- $\rho_f \equiv \rho^f_{1,\lambda_f} \pmod{\lambda_f}$*, with* $\lambda_f \mid \ell_f$.
- $\rho_{f_\infty,\lambda_\infty} \equiv \rho^f_{n_f,\lambda_\infty} \pmod{\lambda_\infty}$*, for some prime ideal* λ_∞.
- *for* $i = 1, \ldots, n-1$ *there exist* λ_i *prime such that* $\rho^f_{i,\lambda_i} \equiv \rho^f_{i+1,\lambda_i} \pmod{\lambda_i}$,
- $\rho_g \equiv \rho^g_{1,\lambda_g} \pmod{\lambda_g}$*, with* $\lambda_g \mid \ell_g$.
- $\rho_{g_\infty,\lambda_\infty} \equiv \rho^g_{n_g,\lambda_\infty} \pmod{\lambda_\infty}$*, for some prime ideal* λ_∞.
- *for* $i = 1, \ldots, n-1$ *there exist* λ_i *prime such that* $\rho^g_{i,\lambda_i} \equiv \rho^g_{i+1,\lambda_i} \pmod{\lambda_i}$,
- *all the congruences involved are MLT.*

Remark 8.8. The importance of our main result is that we can translate Conjecture 8.3 to a situation where the level and the weight are known. Some of the primes involved are not explicit, since they come from some application of Tchebotarev density theorem. See Section 8.6.

8.3 MLT theorems used

In this section we just enumerate the MLT theorems that will be used during this work. By E we denote a finite extension of $\mathbb{Q}_\ell$.

Theorem 8.9 (MLT1). *Let F be a totally real number field, and $\ell \geq 5$ a prime number which splits completely in F. Let $\rho : G_F \to \mathrm{GL}_2(E)$ be a continuous irreducible representation such that:*

- ρ *ramifies only at finitely many primes,*
- $\overline{\rho}$ *is odd,*
- $\rho|_{G_{F_v}}$ *is potentially semi-stable for any* $v \mid \ell$ *with different Hodge-Tate weights.*
- *The restriction* $\overline{\rho}|_{G_{F(\xi_\ell)}}$ *is absolutely irreducible.*
- $\bar{\rho} \sim \bar{\rho}_f$ *for a Hilbert modular form* f.

Then ρ is automorphic.

Proof. This is Theorem 6.4 of [HT13]. □

Theorem 8.10 (MLT2 - ordinary case)**.** *Let F be a totally real field, ℓ and odd prime and $\rho : G_F \to \mathrm{GL}_2(E)$ be a continuous irreducible representation such that:*

- *ρ is unramified at all but finitely many primes.*
- *ρ is de Rham at all primes above ℓ.*
- *The reduction $\bar{\rho}$ is irreducible and $\bar{\rho}(G_{F(\xi_\ell)}) \subset \mathrm{GL}_2(\overline{\mathbb{F}_\ell})$ is adequate.*
- *ρ is ordinary at all primes above ℓ.*
- *$\bar{\rho}$ is ordinarily automorphic.*

Then ρ is ordinarily automorphic. If ρ is also crystalline (resp. potentially crystalline), then ρ is ordinarily automorphic of level prime to ℓ (resp. potentially level prime to ℓ).

Proof. This is just Theorem 2.4.1 of [BLGGT]. □

Theorem 8.11 (MLT3 - pot. diagonalizable case)**.** *Let F be a totally real field, $\ell \geq 7$ be and odd prime and $\rho : G_F \to \mathrm{GL}_2(E)$ be a continuous irreducible representation such that:*

- *ρ is unramified at all but finitely many primes.*
- *ρ is de Rham at all primes above ℓ, with different Hodge-Tate numbers.*
- *$\rho|_{G_{F_\lambda}}$ is potentially diagonalizable for all $\lambda \mid \ell$.*
- *The restriction $\bar{\rho}|_{G_{F(\xi_\ell)}}$ is irreducible.*
- *$\bar{\rho}$ is either ordinarily automorphic or potentially diagonalizable automorphic.*

Then ρ is potentially diagonalizable automorphic (of level potentially prime to ℓ).

Proof. This is Theorem 4.2.1 of [BLGGT]. Note that although it is stated only for CM fields, one can chose a suitable CM extension and get the same result for totally real fields using solvable base change. □

Remark 8.12. The hardest condition to check in the previous Theorem is that of ρ being potentially diagonalizable, but this is satisfied if, in particular, ℓ is unramified in F, ρ is crystalline at all primes above ℓ and the Hodge-Tate weights are in the Fontaine-Laffaille interval.

Remark 8.13. The last result is stated for $\ell \geq 7$, BUT if ℓ is odd and $\overline{\rho}|G_{F(\xi_\ell)}$ is adequate, the same holds (see [DG12]).

We also need a mixed variant. Recall the following property.

Lemma 8.14. *Let F be a totally real field, and $\{\rho_\lambda\}$ be a strictly compatible system of continuous, odd, irreducible, parallel weight 2 representations, then ρ_λ is potentially Barsotti-Tate or ordinary at λ.*

Proof. If the restriction to λ is not potentially crystalline, then it is potentially semistable non-crystalline, in which case ordinariness is known. In fact, using potential modularity and the semistability at λ, the system is known to correspond to an abelian variety with potentially semistable reduction at λ, and potentially (over a suitable extension) to a parallel weight 2 Hilbert modular form which is Steinberg at λ. □

The following variant of the current MLT is not yet known:

Assumption 8.15 (MLT4 - mixed case)**.** *Let F be a totally real field, ℓ be and odd prime and $\rho : G_F \to \mathrm{GL}_2(E)$ be a continuous irreducible representation such that:*

- *ρ is unramified at all but finitely many primes.*
- *ρ is de Rham at all primes above ℓ, with different Hodge-Tate numbers.*
- *$\rho|G_{F_\lambda}$ is potentially diagonalizable for some $\lambda \mid \ell$ and is ordinary at the others.*
- *The restriction $\bar{\rho}|_{G_{F(\xi_\ell)}}$ is irreducible and adequate.*
- *$\bar{\rho}$ is either ordinarily automorphic or potentially diagonalizably automorphic at the same places as $\bar{\rho}$.*

Then ρ is potentially automorphic (of level potentially prime to ℓ).

Remark 8.16. We believe that using the tools developed in [BLGGT] this result is accessible. Not only we consider the above result accessible, but also if we restrict to the case where both Galois representations involved are of parallel weight 2 (thus, because of the previous Lemma, locally potentially Barsotti-Tate or ordinary at all primes above p), it is a Theorem as proved in [BD], Theorem 3.2.2.

Remark 8.17. We will need also need Assumption 8.15 to hold for $p = 2$ and parallel weight 2.

Remark 8.18. To apply the previous MLTs we need $\overline{\rho}(G_{F(\xi_\ell)})$ to be irreducible and to be adequate. When $\ell \geq 7$, adequacy is equivalent to irreducibility, so this imposes no extra hypothesis. Nevertheless, when $\ell = 3$ and $\ell = 5$ this is not the case. Theorem 1.5 of [Gur12] implies that if $\ell \geq 3$ and

$SL_2(\mathbb{F}_{p^r}) \subset \text{Im}(\overline{\rho})$ for $r > 1$ then the residual image of the representation (and the same for the image of its restriction to any abelian extension) is adequate as well. In what follows, we will only apply any of the previous MLTs in characteristics 3 or 5 AFTER having added a good dihedral prime (which is added modulo a prime greater than 5) and from this it follows that this result of Guralnick can be applied ensuring adequacy (see [Die12a] for more details).

8.4 Abstract representations

As mentioned in the introduction, most of the level/weight manipulations that we are going to perform, work not only for modular representation but for abstract ones as well. In this section we will work in the greatest generality possible, and in the next section we will restrict to modular representations putting emphasis on the results that are nowadays only known for modular representations. The results in this section are enough to prove Mazur connectedness Theorem. We begin by recalling the following well-known definition (see [Ser98] for example). If λ_i is a prime in a number field K, we denote by ℓ_i the rational prime below λ_i and by L_{ℓ_i} the set of primes in K dividing ℓ_i (which clearly contains λ_i).

Definition 8.19. A *compatible system* of Galois representations over F is a family of continuous Galois representations

$$\rho_\lambda : \text{Gal}(\overline{F}/F) \to GL_2(K_\lambda),$$

where K is a finite extension of $\mathbb{Q}$ and λ runs through the prime ideals of $\mathcal{O}_K$, which satisfy:

(1) There exists a finite set of primes S (independent of λ) such that ρ_λ is unramified outside $S \cup L_\ell$.
(2) For each pair of prime ideals (λ_1, λ_2) in $\mathcal{O}_K$ and for each prime ideal $\mathfrak{p} \notin S \cup L_{\ell_1} \cup L_{\ell_2}$, the characteristic polynomials $Q_\mathfrak{p}(x)$ of $\rho_{\lambda_i}(\text{Frob}_\mathfrak{p})$ lie in $K[x]$ and are equal.

The most important examples of such families are the ones arising from the étale cohomology of a variety defined over F. In this case we also have some control on the roots of the characteristic polynomials, and some control on the λ-adic representation at primes in L. This motivate the following definition (see [BLGGT], Section 5).

Definition 8.20. A rank 2 *strictly compatible system* of Galois representations $\mathcal{R}$ of G_F defined over K is a 5-tuple

$$\mathcal{R} = (K, S, \{Q_{\mathfrak{p}}(x)\}, \{\rho_\lambda\}, \{H_\tau\}),$$

where

(1) K is a number field.
(2) S is a finite set of primes of F.
(3) for each prime $\mathfrak{p} \notin S$, $Q_{\mathfrak{p}}(x)$ is a degree 2 polynomial in $K[x]$.
(4) For each prime λ of K, the representation

$$\rho_\lambda : G_F \to GL_2(K_\lambda),$$

is a continuous semi-simple representation such that:
- If $\mathfrak{p} \notin S$ and $\mathfrak{p} \nmid \ell$, then ρ_λ is unramified at $\mathfrak{p}$ and $\rho_\lambda(\mathrm{Frob}_{\mathfrak{p}})$ has characteristic polynomial $Q_{\mathfrak{p}}(x)$.
- If $\mathfrak{p} \mid \ell$, then $\rho|_{G_{F_{\mathfrak{p}}}}$ is de Rham and in the case $\mathfrak{p} \notin S$, crystalline.

(5) for $\tau : F \hookrightarrow \overline{K}$, H_τ contains 2 different integers such that for any $\overline{K} \hookrightarrow \overline{K}_\lambda$ over K, we have that $\mathrm{HT}_\tau(\rho_\lambda) = H_\tau$.
(6) For each finite place $\mathfrak{p}$ of F there exists a Weil-Deligne representation $\mathrm{WD}_{\mathfrak{p}}(\mathcal{R})$ of $W_{F_{\mathfrak{p}}}$ over $\overline{K}$ such that for each place λ of K not dividing the residue characteristic of $\mathfrak{p}$ and every K-linear embedding $\iota : \overline{K} \hookrightarrow \overline{K}_\lambda$, the push forward $\iota\,\mathrm{WD}_{\mathfrak{p}}(\mathcal{R}) \simeq \mathrm{WD}(\rho_\lambda|_{G_{F_{\mathfrak{p}}}})^{K\text{-ss}}$.

Remark 8.21. If one starts with a two-dimensional continuous, odd, Galois representations over a totally real number field F, under some minor hypothesis (which are exactly the hypothesis for an MLT theorem to hold), one can prove that such representations is potentially modular. In particular, this implies that the representations is part of a strictly compatible system. Since all the congruences we will work with are where an MLT theorem works in both directions, without loss of generality, we will assume that all the representations come in strictly compatible systems.

Definition 8.22. Let $\{\rho_\lambda\}$ be a strictly compatible system of Galois representations. We say that the system is *dihedral* if the images are compatible dihedral groups, i.e. if there exists a quadratic extension L/F (independent of λ) such that ρ_λ is induced from a λ-adic character of L.

Lemma 8.23. *The family $\{\rho_\lambda\}$ is dihedral if and only if one representation is dihedral.*

Proof. It is clear that if the whole family is dihedral, in particular any of them is dihedral. For the converse, let λ be a prime ideal of $\mathcal{O}_F$, and suppose that ρ_{λ_0} is dihedral. Then there exists a quadratic extension L/F and a λ-adic character $\chi_{\lambda_0} : \mathrm{Gal}(\overline{L}/L) \to K_{\lambda_0}$ such that $\rho_{\lambda_0} = \mathrm{Ind}_{G_L}^{G_F} \chi_{\lambda_0}$. We know that χ_{λ_0} is part of a strictly compatible system of one-dimensional representations $\{\chi_\lambda\}$, so we are led to prove that $\rho_\lambda \simeq \mathrm{Ind}_{G_L}^{G_F} \chi_\lambda$ for all primes $\lambda \subset \mathcal{O}_F$. This comes from a straightforward computation, since the values of the traces (an even the whole characteristic polynomial) of $\rho_{\lambda_0}(\mathrm{Frob}_\mathfrak{p})$ are given in terms of the values of χ_{λ_0}. For split primes in the extension L/F, the trace of $\rho_{\lambda_0}(\mathrm{Frob}_\mathfrak{p})$ equals $-\chi_{\lambda_0}(\mathfrak{p}_1) - \chi_{\lambda_0}(\mathfrak{p}_2)$, where $\mathfrak{p}\mathcal{O}_L = \mathfrak{p}_1\mathfrak{p}_2$ and for inert primes, the trace is zero. In particular, the same happens to $\rho_\lambda(\mathrm{Frob}_\mathfrak{p})$, for all λ so ρ_λ and $\mathrm{Ind}_{G_L}^{G_F} \chi_\lambda$ have the same trace and are thus isomorphic. □

To apply most MLTs theorems we will need to have some control of the image of our residual Galois representations. In particular we will need the image of its restriction to a cyclotomic extension to be adequate. To avoid checking this particular condition at each step of our chain of congruences, we will move to families with "big image", in the sense that for all but finitely many primes p (and in most steps, for all primes of bounded size), the residual representation has an image containing $\mathrm{SL}_2(\mathbb{F}_p)$.

Proposition 8.24. *Let ρ_λ be a strictly compatible system of odd dihedral representations. Then there exists a strictly compatible system of odd non-dihedral representations $\{\rho_{2,\lambda}\}$ and a prime $\mathfrak{p}$ such that $\rho_\mathfrak{p} \equiv \rho_{2,\mathfrak{p}} \pmod{p}$ and MLT holds in both directions.*

Proof. It is well known that dihedral compatible families do not have Steinberg primes in the level, so we just need to add a Steinberg prime to our representation by some raising the level argument. Let λ be a prime over a prime $p > 5$ and such that:

- p splits completely in F.
- $\lambda \notin S$, i.e. $\rho_\mathfrak{q}$ is unramified at λ if $\mathfrak{q} \neq \lambda$.
- L and $F(\xi_p)$ are disjoint, i.e. $F(\xi_p) \cap L = F$.

We want to add a Steinberg prime $\mathfrak{q}$ modulo λ, and by the previous choice, we are in the hypothesis of Theorem 7.2.1 of [Sno09] which says that it is enough to raise the level locally. The local problem is standard, and can be achieve by Tchebotarev's Theorem as follows: take $\mathfrak{q}$ inert in the extension L/F, so that $a_q = 0$. Then the local raising the level condition becomes $\mathcal{N}\mathfrak{q} \equiv -1 \pmod{p}$, so we chose any such prime and get a global representation with the desired properties. The existence of a strictly compatible system attached to such global representation follows from Taylor's potential modularity result (see [BLGGT]) plus the argument from [Die04], as generalized in [BLGGT]. That MLT holds in both directions comes from Theorem 8.9. □

Proposition 8.25. *Let $\{\rho_\lambda\}$ be a strictly compatible system of odd, non-dihedral representations. Then there exists an integer B, such that if $\mathcal{N}(\lambda) > B$ then* $\mathrm{SL}_2(\mathbb{F}_p) \subset \mathrm{Im}(\overline{\rho}_\lambda)$.

Proof. According to Dickson's classifications of subgroups of $\mathrm{PGL}_2(\mathbb{F}_\lambda)$, when we consider the residual representations, they might:

(1) contain $\mathrm{PSL}_2(\mathbb{F}_\lambda)$,
(2) be reducible,
(3) be dihedral,
(4) be isomorphic to A_4, S_4 or A_5.

We want to prove under the running hypothesis, there are only finitely many primes where the first case does not hold. But the second case is exactly Lemma 5.4 of [CG11], the third case Corollary 5.2 of [CG11] (here we use the assumption that the system is not dihedral), and the last case is Lemma 5.3 of [CG11]. □

Remark 8.26. Another way to prove the last Proposition (following the classical approach of Ribet) it to first apply some potential automorphy result (like in [BLGGT]) to deduce that the system is potentially modular. Then its restriction is isomorphic to that of a Hilbert modular form. Furthermore, since our abstract representations are not dihedral, the respective Hilbert modular form has no CM. But for Hilbert modular forms, such result is proven in [Dim05] Proposition 0.1.

By the above considerations, from now on we will only consider non-dihedral families, so we will skip writing this hypothesis in the next results.

Proposition 8.27. *Let $\{\rho_{1,\lambda}\}$ be a strictly compatible system of odd irreducible Galois representations. Then there exists a compatible system $\{\rho_{2,\lambda}\}$ of parallel weight 2 representations and a prime $\mathfrak{p}$ such that $\rho_{1,p} \equiv \rho_{2,p} \pmod{p}$ and MLT holds in both directions.*

Proof. Let $p > 5$ be a prime which splits completely in F, does not divide the level of ρ, is larger than all weights of the system and such that the image of $\overline{\rho}_{1,\mathfrak{p}}$ is large. Let $\det(\overline{\rho}) = \overline{\psi}\overline{\chi_p}$, where $\overline{\chi_p}$ is the reduction of the cyclotomic character, and let ψ be any lift of $\overline{\psi}$. Then Theorem 7.6.1 of [Sno09] implies that $\overline{\rho}$ admits a weight two lift to $\overline{\mathbb{Q}}_p$ which ramifies at the same primes as ρ and ψ, with determinant $\psi\chi_p$. MLT hold in both directions by the same proof as the previous Proposition 8.24. □

8.4.1 Adding a good dihedral prime

As already mentioned, while working with MLT one needs to ensure that residual images are big enough to be in the hypothesis of such a theorem. Sometimes, one needs the restriction of the residual representation to the cyclotomic extension of p-th roots of unity to have adequate image, but for most MLT requiring irreducibility of this restriction is enough (and for $p > 5$ it is known that both properties are equivalent). A way to get this property guaranteed at most steps is by introducing to the level an extra prime which forces the image modulo all primes up to a certain bound to be "non-exceptional", i.e. it is irreducible, and its projectivization is not dihedral, nor the exceptional groups A_4, S_4, A_5. A way to get this, is by adding a "good dihedral prime" (with respect to the given bound) as was introduced by Khare and Wintenberger in their work on Serre's conjecture.

The difference with the classical setting is that since we work with two strictly compatible systems at the same time, we need to add the same good dihedral prime to both of them.

Proposition 8.28. *Let $\{\rho_{1,\lambda}\}$ and $\{\rho_{2,\lambda}\}$ be two strictly compatible systems of continuous, odd, irreducible representations of parallel weight 2. Fix B a positive integer, larger than 5 and than all primes in the conductors of both systems. Let $p \equiv 1 \pmod 4$ be a rational prime such that:*

- *p is bigger than B.*
- *p splits completely in the compositum of the coefficient fields of ρ_1 and ρ_2.*
- *p is relatively prime to the conductors of both systems.*
- *$\mathrm{Im}(\overline{\rho}_{i,\mathfrak{p}}) = \mathrm{GL}_2(\mathbb{F}_p)$, $i = 1, 2$, for some prime $\mathfrak{p}$ over p.*

Then there exists a prime q not dividing the conductor of the systems such that:

- *$q \equiv -1 \pmod p$.*
- *q splits completely in the extension F' given by the compositum of all quadratic extensions of F ramified only at primes above rational primes $\ell < B$.*

- $q \equiv 1 \pmod 8$.
- *There exists a prime ideal* $\mathfrak{q}$ *in* F *over* q *such that the image of* $\overline{\rho}_{1,\mathfrak{p}}(\mathrm{Frob}_{\mathfrak{q}})$ *and* $\overline{\rho}_{2,\mathfrak{p}}(\mathrm{Frob}_{\mathfrak{q}})$ *both have eigenvalues* 1 *and* -1.

With this choice of primes, there exists two strictly compatible systems of continuous representations $\{\varrho_{1,\lambda}\}$ *and* $\{\varrho_{2,\lambda}\}$ *of parallel weight* 2 *such that:*

(1) $\overline{\rho}_{1,\mathfrak{p}} \simeq \overline{\varrho}_{1,\mathfrak{p}}$.
(2) $\overline{\rho}_{2,\mathfrak{p}} \simeq \overline{\varrho}_{2,\mathfrak{p}}$.
(3) $\{\varrho_{i,\lambda}\}$, *for* $i = 1, 2$, *is locally good dihedral at* $\mathfrak{q}$ *(w.r.t. the bound* B*).*
(4) $\varrho_{i,\mathfrak{p}}$, *is Barsotti-Tate at all primes dividing* p *and has the same type as* $\rho_{i,\mathfrak{p}}$ *locally at any prime other that* $\mathfrak{q}$ *for* $i = 1, 2$.
(5) *The congruences are MLT.*

Remark 8.29. We will not reproduce here the precise definition of good-dihedral prime (it can be found in [KW09] and [Die12b]) or what it means for a compatible system to be locally good dihedral at a prime w.r.t. certain bound B. Let us just recall that a good-dihedral prime is a supercuspidal prime in a compatible system such that the ramification at this prime forces the system to have residual images containing $SL_2(\mathbb{F}_p)$ every time the residual characteristic p is bounded by B (and a bit more for $p = 2$ or 3: also in these characteristics the image is forced to be non-solvable).

Remark 8.30. We do not want to make precise what condition (4) of the last statement means (the local type), since it will not be important for our purposes, but what we prove is the following: for abstract representations, we will use Theorem 7.2.1 of [Sno09], where a "type" is understood as a Weil type (no information on the monodromy), while for modular representations (actually residually modular ones), we will use Theorem 3.2.2 of [BD], which uses the complete notion of type.

Proof. The existence of the primes $\mathfrak{p}$ and $\mathfrak{q}$ follows with almost the same arguments as [Die12b] (Lemma 3.3). The only difference is that we need to consider the compositum of $\mathbb{Q}(i)$, F and the coefficient field of ρ_1 and ρ_2. Take p big enough (for the images to be large) and split in such extension. Then q is chosen using Tchebotarev density Theorem, with the condition that it hits complex conjugation in the same suitable field.

We are lead to prove the existence of a lift of $\bar{\rho}_{1,\mathfrak{p}}$ with the desired properties. By Theorem 7.2.1 of [Sno09], or by Theorem 3.2.2 of [BD] (see the last remark), we know that a global representation with the desired properties exists if and only if locally the corresponding lifts do exist, so we only need to show which are the local deformation conditions:

- At the primes $\mathfrak{l} \neq \mathfrak{q}$, that of $\rho_i|_{G_{\mathfrak{l}}}$.
- $\varrho_i|_{D_{\mathfrak{q}}} = \mathrm{Ind}_{\mathbb{Q}_q}^{\mathbb{Q}_q[\sqrt{\epsilon}]}(\chi)$, where $\mathbb{Q}_q[\sqrt{\epsilon}]$ is the unique quadratic unramified extension of $\mathbb{Q}_q$, and χ is a character with order p (this is called type C in [Sno09]).

This proves the existence of $\varrho_{i,\mathfrak{p}}$. The congruence is MLT in both directions because of Theorem 8.9 (in this case both forms are of parallel Hodge-Tate weight 2 which is smaller than the prime p). Since we are in the hypothesis of an MLT theorem, we can put such representation into a strictly compatible system and get the result. □

Remark 8.31. Although we stated the result for representations of parallel weight 2, in general we can first move to parallel weight 2 (using Proposition 8.27) and then add the good dihedral prime.

With this result, we can prove Mazur's Theorem.

Theorem (Mazur). *Let F be a totally real number field, f_0 and f_∞ two Hilbert modular forms, whose weights are congruent modulo 2, then there exists Hilbert modular forms $\{h_i\}_{i=1}^n$ such that $h_1 = f_0$, $h_n = f_\infty$ and such that for $i = 1, \ldots, n-1$ there exist λ_i prime such that $\rho_{h_i,\lambda_i} \equiv \rho_{h_{i+1},\lambda_i}$ (mod λ_i).*

Proof. By the previous results, we can assume that both forms are of parallel weight 2, and do not have complex multiplication (or we move to such a situation using Proposition 8.24 and Proposition 8.27), for some common level $\Gamma_1(\mathfrak{n})$ (of course they might not be new at the same level). The idea now is to find both modular forms in the cohomology of a Shimura curve, and for this purpose we need to add an auxiliary prime to the level where both forms are not principal series (in case $[F : \mathbb{Q}]$ is odd). So what we do is to raise the level of both forms at an auxiliary prime $\mathfrak{p}$ as in Proposition 8.28 (we could also add a Steinberg prime). Now we consider the Shimura curve $X^{\mathfrak{p}}(\mathfrak{n})$ ramified at all infinite places of F but one, and at the auxiliary prime $\mathfrak{p}$ if needed (depending whether $[F : \mathbb{Q}]$ is odd or even) and with level $\mathfrak{n}$.

We apply Mazur's argument just as in [Maz77] (proposition 10.6, page 98) with some minor adjustments. The main idea of his proof is the following: if the Jacobian J is a product of two abelian varieties $A \times B$, since J decomposes (up to isogeny) as a product of simple factors with multiplicity one, there are no nontrivial homomorphisms from A to $\hat{B}$ nor from B to $\hat{A}$. Then the principal polarization of J induces principal polarizations in A and B, but a

Jacobian cannot decompose as a nontrivial direct product of principally polarized abelian varieties (which follows from the irreducibility of its θ-divisor, see [ACGH85]). From this it follows that Spec $\mathbb{T}$ is connected, where $\mathbb{T}$ denotes the Hecke algebra acting on J.

In his original article Mazur was dealing with the curve $X_0(N)$, with N a prime number, so there are no old forms appearing in $J_0(N)$. To use the same argument in our context, we have to deal with old forms as well, and the problem is that the abelian varieties A_f corresponding to old forms do not appear with multiplicity one in the decomposition (up to isogeny) of the Jacobian of a modular or a Shimura curve. But this is not a problem if we observe that for what we want it is not necessary to prove the connectedness of Spec $\mathbb{T}$, it is enough to show that the anemic Hecke algebra $\mathbb{T}_0$ generated only by the Hecke operators with index prime to the level is connected. Therefore, what we need is to discard the cases where the Jacobian of $X^{\mathfrak{p}}(\mathfrak{n})$ decomposes as a product of abelian varieties $A \times B$ with every simple factor in A and every simple factor in B being orthogonal (recall that now these simple factors need not appear with multiplicity one). In such case, the same proof as in Mazur's article applies, and gives the connectedness we are looking for. □

8.5 Killing the level

Now that we have a good dihedral prime, which controls the image of the residual representations in the families of the two representations for small primes, we want to connect them at a chosen level. From know on, we will only consider modular representations, pointing out in each case the needed result for abstract ones. To lower the level we will need the following result, whose proof is due to Fred Diamond.

Lemma 8.32. *Let* $\rho_f : \mathrm{Gal}(\overline{F}/F) \to \mathrm{GL}_2(K_{\mathfrak{p}})$ *be a modular representation of level dividing* $\mathfrak{n}p^r$ *(where* $\mathfrak{p} \mid p$ *and* $\mathfrak{p} \nmid \mathfrak{n}$*) which satisfies the following hypothesis:*

- *The form* f *has a supercuspidal prime* $\mathfrak{q} \nmid p$.
- *The residual image* $\overline{\rho_f}$ *is absolutely irreducible and not bad dihedral.*
- *The determinant* $\det(\overline{\rho_f}|_{I_v}) = \chi_p^n$ *for all* $v|p$ *and some* n *independent of* v.

Then $\overline{\rho}$ *has a modular lift of level dividing* $\mathfrak{n}$*. Moreover we can assume the weight of the form is parallel and the character has order prime to* p.

Proof. We do not assume that p is unramified in F because the results we need from [BDJ10] can be generalized to the ramified case in a straightforward way. We will follow the notation of the aforementioned article.

By Proposition 2.10 and Corollary 2.11 of [BDJ10], $\overline{\rho}$ is modular of some weight $\sigma = \otimes_{v|p}\sigma_v$ and central character $\otimes_{v|p}(\mathcal{N}_{k_v/F_p})^{n-1}$. The results of [Roz12] show that σ is a Jordan-Holder constituent of the reduction of $\otimes_\tau \mathrm{Symm}^{k-1}\mathcal{O}_E^2$ for some sufficiently large k, where the tensor product is over embeddings $\tau : F \to E$ for a sufficiently large number field E, viewed as contained both in $\mathbb{C}$ and in $\overline{\mathbb{Q}_p}$. Another application of Proposition 2.10 of [BDJ10] shows that $\overline{\rho}$ is modular of parallel weight k and level prime to p. Moreover the presence of a supercuspidal prime $\mathfrak{q}$ allows us to use an indefinite quaternion algebra of discriminant dividing $\mathfrak{q}$ and hence assume the open compact subgroup has level dividing $\mathfrak{n}$ in the first application of Proposition 2.10 of [BDJ10]. Moreover $\overline{\rho}$ is not badly dihedral in the sense of Lemma 4.11 of [BDJ10], so the arguments there ensure that we can assume the divisibility of $\mathfrak{n}$ by the level is preserved in the second application of Proposition 2.10 as well. We can similarly ensure the conclusion on the central character. □

For further details and more general results along these lines, see forthcoming work of Diamond and Reduzzi.

8.5.1 Modifying the non-Steinberg primes

From now on we work under the assumption that MLFMT (Assertion 8.15) is true. We call an abstract strictly compatible system of continuous, odd, irreducible Galois representations *modular* if there exists a Hilbert modular form, whose attached Galois representations matches the abstract one.

Theorem 8.33. *Let $\{\rho_{1,\lambda}\}$ be a strictly compatible system of continuous, odd, irreducible, Galois representations attached to a Hilbert newform f over a totally real number field F of parallel weight 2, containing in its conductor a locally good dihedral prime $\mathfrak{q}$ (w.r.t. some sufficiently large bound B). Then:*

- *There exists a strictly compatible system of continuous, odd, irreducible, parallel weight 2 representations $\{\rho_{2,\lambda}\}$, which is semistable at all primes except the same good dihedral prime $\mathfrak{q}$ and such that the Steinberg ramified primes are bounded in norm by B.*
- *There exists a chain of congruences of compatible systems linking $\{\rho_{1,\lambda}\}$ and $\{\rho_{2,\lambda}\}$ such that all congruences involved occur in residual characteristics bounded by B and are MLT.*

In particular the system $\{\rho_{2,\lambda}\}$ is also modular.

Proof. Let λ be a prime which is supercuspidal or principal series. Let p be a prime number dividing the order of the character corresponding to ramification at λ. We consider two different cases: if p is relative prime to λ, we call it the *tamely ramified case*, while the other case we call it the *wildly ramified case*.

The tamely ramified case. Let $\mathfrak{p}$ be a prime ideal in K (the coefficient field) dividing p, and consider the residual mod $\mathfrak{p}$ representation. Then the p-part of the ramification is lost, so we take a minimal lift with the same parallel weight 2 (it exists by Theorem 3.2.2 in [BD]). Observe that at this step we are not only modifying the ramification type at λ, but also at any other prime supercuspidal or principal series in the prime-to-p part of the conductor with ramification given by a character of order divisible by p. Then by Lemma 8.14, we might be in a mixed situation of potentially Barsotti-Tate and ordinary representations. Using Theorem 3.2.2 of [BD] (recall that this is a special case of our Assumption 8.15), we get rid of this p-part of the inertia with an MLT congruence. Iterating this process, we end up killing all tamely ramified ramification given by characters, i.e. the prime-to-p part of the ramification at primes dividing p is killed. So we are reduced to the case where all primes in the conductor are either Steinberg or with ramification given by a prime order character whose order is divisible by the ramified prime.

The wildly ramified case. In this case, we will move the wildly ramified primes (up to twist) to tamely ramified primes, so the previous argument ends our proof. For a prime $\mathfrak{t}$ in the conductor of the wildly ramified case, let us call t the rational prime below $\mathfrak{t}$ and consider a mod t congruence, up to twist by some finite order character ψ, with the Galois representation corresponding to a Hilbert newform H of parallel weight 2 with at most $\mathfrak{t}$ to the first power in the level (i.e. Γ_1 at $\mathfrak{t}$), and the same for the other primes dividing t. The existence of such a form comes from Lemma 8.32.

By level-lowering, we can assume that the only extra primes in the level of H are those primes that have been introduced to the residual conductor while twisting by a character ψ. It is easy to see that such character can be chosen such that at primes other than those dividing t it has square-free conductor. To this congruence, Theorem 3.2.2 of [BD] applies (the conditions for this theorem are preserved by twisting, and modularity too) so we are reduced to a case where we have a system that is either Steinberg or tamely ramified principal series at primes dividing t, and tamely ramified principal series at all extra primes introduced by ψ. If we iterate this process at all wildly ramified primes, we end up with a system with no wildly ramified primes. We repeat the previous case procedure of killing all ramification given by tamely ramified characters, but now in the absence of wild ramification we finish with a compatible system such that all its ramified primes other than the good-dihedral prime are Steinberg. It is not hard to see that in all this process, for a suitably chosen bound B, all auxiliary primes can be taken to be smaller than B. □

Remark 8.34. Except for the application of Corollary 2.12 of [BDJ10] at a key point, and a better control of the local types (which seems reasonable for abstract representations), all congruences in the above proof are known to exist for abstract Galois representation, so an analogue of the above result for abstract compatible systems can be proved assuming that this result from [BDJ10] generalizes to the abstract setting (thus relating the modularity of any geometric compatible system to that of a system with only Steinberg primes).

8.5.2 Killing the Steinberg primes

This part of the process is a little more delicate, and the MLT theorems are more restrictive, so what we do first is move the Steinberg primes to primes which split completely in F.

Theorem 8.35. *Suppose that Assumption 8.15 is true. Let $\{\rho_{1,\lambda}\}$ be a system of modular continuous, odd, irreducible, parallel weight 2 representations with a big locally good dihedral prime $\mathfrak{q}$. Let $\mathcal{L} \mid \ell$ ($\ell \in \mathbb{Z}$) be a Steinberg prime of the system which does not split completely in F. We also assume that the system is either unramified or Steinberg at all primes dividing ℓ. Then:*

- *there exists a strictly compatible system $\{\rho_{2,\lambda}\}$ of continuous, odd, irreducible, parallel weight 2 representations, which has the same ramification behavior at all primes except those dividing ℓ, where it is unramified, and has at most a set of extra Steinberg primes, all of them dividing the same rational prime which splits completely in F.*
- *there exists a chain of MLT congruences linking the two systems.*

In particular, the system $\{\rho_{2,\lambda}\}$ is also modular.

Proof. We look at $\rho_{1,\lambda'}$ for a prime λ' dividing ℓ and we reduce it modulo ℓ. We can construct then a modular lift which is unramified at $\mathcal{L}$ and at all primes dividing ℓ, and with weights among those predicted by the Serre's weights of the residual representation. This follows from iterated applications of Lemma 8.32 since the determinant locally at primes above ℓ is a fixed power of the cyclotomic character, and the form (without ℓ in the level) has parallel weight as well. Note that this congruence is MLT, at least under Assumption 8.15, which applies because at Steinberg primes, weight 2 representations are ordinary, and the crystalline lift of higher weight can also be taken to be ordinary (observe that to deduce modularity of the weight 2 family from the other one we can apply Theorem 3.2.2 in [BD]).

Let p be a big prime (bigger than the weights of this second family) which splits completely in F, then by [Sno09] we can construct a third family with parallel weight 2 by looking modulo p, and this congruence is MLT because of the results in [BLGGT] (we are comparing a Fontaine-Laffaille with a potentially Barsotti-Tate representation, so they are both potentially diagonalizable). We apply the previous section method to this new family to end up with a representation which is at most Steinberg at all primes dividing p as desired. Observe that in this last step no extra ramified primes are introduced because we do not have wild ramification at p (the parallel weight 2 lift implies tamely ramified principal series at all primes above p). □

Now that we have only totally split Steinberg primes in our family, we can get rid of the Steinberg primes.

Theorem 8.36. *Let $\{\rho_{1,\lambda}\}$ be a system of modular, irreducible, parallel weight 2 representations, whose ramification consists of a big locally good dihedral prime $\mathfrak{q}$ and Steinberg primes which split completely in F. Then:*

- *There exists a strictly compatible system $\{\rho_{2,\lambda}\}$ of continuous, odd, irreducible representations which are only ramified at the locally good dihedral prime $\mathfrak{q}$.*
- *There exists a chain of MLT congruences linking the two systems.*

In particular, the system $\{\rho_{2,\lambda}\}$ is also modular.

Proof. The procedure is quite similar to the one in the previous Theorem, but now we can use Theorem 8.9. For each Steinberg prime of residual characteristic ℓ we look at $\overline{\rho}_{1,\lambda}$ with $\lambda \mid \ell$. There exists a minimal lift which is crystalline at all primes of residual characteristic ℓ by Lemma 8.32. Now, since the prime is split, the congruence is MLT by the cited Theorem. The only delicate point here is that to apply Theorem 8.9 we need the residual characteristic to be different from 2 and 3, so if we have such a small Steinberg prime, we first apply Theorem 8.35 to transfer the ramification to Steinberg ramification at some larger split prime. □

Remark: In the previous two Theorems, modularity of the given system was only used to ensure existence of the lift corresponding to a system with no ℓ in its conductor, and this was deduced from results of [BDJ10]. In the case of abstract Galois representations, after computing the Serre's weights of the residual representation one should be able to propose locally at all primes above ℓ a crystalline lift of the residual local representation, but then the problem is that in order to apply results such as those in [BLGGT] that guarantee existence of a global lift with such local conditions, one should be in a potentially diagonalizable case. In particular, under the conjecture that all potentially crystalline representations are potentially diagonalizable, the previous two theorems shall generalize to the abstract setting (of course, the conclusion that the last system is modular shall also be removed).

Proof of Theorem 8.7. With the machinery developed in the previous chapters, starting with two modular representations, we can take each of them to representations which are only ramified at the good dihedral prime $\mathfrak{q}$. The only problem is that while killing the Steinberg primes, we lost control over the weights, so to take the families to parallel weight 2, we chose an auxiliary prime p which splits completely in F and consider a parallel weight 2 lift of each family modulo p, with the same local type at $\mathfrak{q}$. Note that we cannot assure that the forms we get at the end of the process will be newforms for $\Gamma_0(p\mathfrak{q}^2)$, because at some of the prime ideals of F dividing p our representations could be unramified. □

8.6 Further developments

As mentioned before, although we have some good control on the level of the forms we started with, the primes in the level are not explicit. One can go further and change the primes in the level for smaller and concrete ones, so as to check whether the connectedness of the Hecke algebras in Mazur's Theorem corresponds to a chain where all congruences are MLT or not. For this purpose, one can use the notion of *micro good dihedral primes*. Adding a micro good dihedral prime (which is chosen asking some splitting behavior in the base field) one can get rid of the good dihedral prime, and also bound the Steinberg primes in the level. These ideas, although standard (see for example [Die12b]) are more delicate, and involve some technicalities that we prefer to avoid in the present article. With this control, one can give an algorithm that given a totally real number field, checks whether our approach implies Base Change over that field via a finite computation involving Hilbert modular forms. See [DP] for more details.

Concerning Serre's conjecture over a specified small real quadratic field, one should carefully check that the chain of congruences constructed carries on to the case of abstract representations, and then once having reduced the problem to the case of representations of concrete small invariants (those where the above process concludes, after the introduction of the micro good dihedral prime) one should connect such representations to some "base case" where modularity or residual reducibility is known (applying for example the result of Schoof recalled in the introduction), checking on the way that all congruences are MLT (with the advantage that over quadratic fields there are also MLT of Skinner and Wiles that deal with the residually reducible case, under suitable assumptions). We plan to check in a future work if this strategy succeeds in giving a proof of Serre's conjecture over some small real quadratic field.

References

[ACGH85] E. Arbarello, M. Cornalba, P. A. Griffiths, and J. Harris. *Geometry of algebraic curves. Vol. I*, volume 267 of *Grundlehren der Mathematischen Wissenschaften [Fundamental Principles of Mathematical Sciences]*. Springer-Verlag, New York, 1985.

[BD] Christophe Breuil and Fred Diamond. Formes modulaires de Hilbert modulo p et valeurs d'extensions entre caractères galoisiens. *Ann. Scient. de l'E.N.S., to appear.*

[BDJ10] Kevin Buzzard, Fred Diamond, and Frazer Jarvis. On Serre's conjecture for mod ℓ Galois representations over totally real fields. *Duke Math. J.*, 155(1):105–161, 2010.

[BLGGT] Thomas Barnet-Lamb, Toby Gee, David Geraghty, and Richard Taylor. Potential automorphy and change of weight. *Ann. of Math., to appear.*

[CG11] Frank Calegari and Toby Gee. Irreducibility of automorphic galois representations of $\mathrm{GL}(n)$, n at most 5. *arXiv:1104.4827 [math.NT]*, 2011.

[DG12] Luis Dieulefait and Toby Gee. Automorphy lifting for small ℓ - appendix b to "Automorphy of $Symm^5(GL(2))$ and base change". *Journal de Math. Pures et Appl.*, to appear

[Die04] Luis V. Dieulefait. Existence of families of Galois representations and new cases of the Fontaine-Mazur conjecture. *J. Reine Angew. Math.*, 577:147–151, 2004.

[Die12a] Luis Dieulefait. Automorphy of $Symm^5(\mathrm{GL}(2))$ and base change. *Journal de Math. Pures et Appl.*, to appear.

[Die12b] Luis Dieulefait. Langlands base change for GL(2). *Ann. of Math. (2)*, 176(2):1015–1038, 2012.

[Dim05] Mladen Dimitrov. Galois representations modulo p and cohomology of Hilbert modular varieties. *Ann. Sci. École Norm. Sup. (4)*, 38(4):505–551, 2005.

[DP] Luis Dieulefait and Ariel Pacetti. Examples of base change for real quadratic fields. *In preparation.*

[Gur12] Robert Guralnick. Adequacy of representations of finite groups of lie type – appendix a to "automorphy of $symm^5(gl(2))$ and base change". *Journal de Math. Pures et Appl.*, to appear

[HT13] Yongquan Hu and Fucheng Tan. The Breuil-Mezard conjecture for non-scalar split residual representations. *Ann. Sci. ENS*, to appear.

[KK03] Chandrashekhar Khare and Ian Kiming. Mod pq Galois representations and Serre's conjecture. *J. Number Theory*, 98(2):329–347, 2003.

[KW09] Chandrashekhar Khare and Jean-Pierre Wintenberger. Serre's modularity conjecture. I. *Invent. Math.*, 178(3):485–504, 2009.

[Maz77] B. Mazur. Modular curves and the Eisenstein ideal. *Inst. Hautes Études Sci. Publ. Math.*, (47):33–186 (1978), 1977.

[Roz12] Sandra Rozensztajn. Asymptotic values of modular multiplicities for GL_2. *arXiv:1209.5666, J. Théorie des Nombres Bordeaux*, 2012, *to appear*.

[Sch12] René Schoof. Semistable abelian varieties with good reduction outside 15. *Manuscripta Math.*, 139(1-2):49–70, 2012.

[Ser98] Jean-Pierre Serre. *Abelian l-adic representations and elliptic curves*, volume 7 of *Research Notes in Mathematics*. A K Peters Ltd., Wellesley, MA, 1998. With the collaboration of Willem Kuyk and John Labute, Revised reprint of the 1968 original.

[Sno09] Andrew Snowden. On two dimensional weight two odd representations of totally real fields. *arXiv:0905.4266v1 [math.NT]*, 2009.

[SW99] C. M. Skinner and A. J. Wiles. Residually reducible representations and modular forms. *Inst. Hautes Études Sci. Publ. Math.*, (89):5–126 (2000), 1999.

9

Big image of Galois representations and congruence ideals

Haruzo Hida and Jacques Tilouine

Department of Mathematics, UCLA, Los Angeles, CA 90095-1555, U.S.A., Université Paris 13, Sorbonne Paris-Cité, LAGA, CNRS (UMR 7539), 99 av. J.-B. Clément, F-93430, Villetaneuse

E-mail address: hida@math.ucla.edu, tilouine@math.univ-paris13.fr

Contents

Date: August 16, 2014
The first author is partially supported by the NSF grant: DMS 0753991. The second author is partially supported by the ANR grant: ArShiFo ANR-10-BLAN-0114

Arithmetic and Geometry, ed. Luis Dieulefait *et al.* Published by Cambridge University Press.

9.1 Introduction

Let $n \geq 1$; we consider the $n \times n$ antidiagonal unit matrix $s = (\delta_{i,n+1-j})_{1\leq i,j\leq n}$ and the $2n \times 2n$ antisymmetric matrix $J = \begin{pmatrix} 0 & s \\ -s & 0 \end{pmatrix}$; we denote by $G = \mathrm{GSp}_{2n}$ the Chevalley group of symplectic similitude matrices for J and by $B = TN$ its standard Borel consisting of upper triangular matrices in G; let ρ be the half-sum of positive roots for (G, B, T) and $\lambda \in X^*(T)$ a dominant weight. Let π be a cuspidal automorphic representation on $G(\mathbf{A}_{\mathbb{Q}})$ of level $M \geq 1$, whose infinity type is in the discrete series, with infinitesimal character $\lambda + \rho$. It occurs in the cohomology of the genus n Siegel variety of level M, with coefficients in the local system of highest weight λ. It implies that there exists a number field K_0 whose ring of integers $\mathcal{O}_0$ contains all Hecke eigenvalues of π (for prime to M Hecke operators). Let $G_{\mathbb{Q}} = \mathrm{Gal}(\overline{\mathbb{Q}}/\mathbb{Q})$ and $\widehat{G}$ be the dual Chevalley group for (G, B, T); one has $\widehat{G} = \mathrm{GSpin}_{2n+1}$ (see below); we assume that the compatible system of Galois representations $\rho_{\pi,\iota_\ell} \colon G_{\mathbb{Q}} \to \widehat{G}(\overline{\mathbb{Q}}_\ell)$ associated to π and to embeddings $\iota_\ell \colon \overline{\mathbb{Q}} \hookrightarrow \overline{\mathbb{Q}}_\ell$ is constructed (for the moment, it is known for $n = 1, 2$). Let $p \nmid M$ be a rational prime. Assume that π is ordinary for ι_p; there exists an n-variable Hida family passing by π; more precisely, let Λ_n be the n-variable Iwasawa algebra and $\mathbf{T}^M$ be the ordinary prime-to-M Hida-Hecke algebra [MT02], [H02], [Pi12]; it is finite torsion-free over Λ_n. A Hida family is a Λ_n-algebra homomorphism $\theta \colon \mathbf{T}^M \to \mathbb{I}$ where $\mathbb{I}$ is an integrally closed finite torsion-free extension of Λ_n.

By interpolating the pseudo-representations associated to the arithmetic specializations of θ (see Sect. 4.2), one constructs a continuous homomorphism $\rho_\theta \colon G_{\mathbb{Q}} \to \widehat{G}(\mathbb{I})$ associated to θ (see Lemma 9.12 in the text). We assume throughout that the residual representation is irreducible. Following the ideas of [H13b], we prove first, under some assumptions (mostly $\mathbb{I} = \Lambda_n$ and $\mathbb{Z}_p$-regularity), that if the Hida family θ is "generic", the image of ρ_θ is "Λ_n-full" (see 9.17); more precisely, there is a non-zero ideal $\mathfrak{l}$ of Λ_n such that the image of Galois contains the principal congruence subgroup $\Gamma_{\widehat{G}'}(\mathfrak{l})$ in $\widehat{G}'(\Lambda_n)$, where $\widehat{G}'$ denotes the derived group of $\widehat{G}$. An important point of the proof is to create a Λ_n-structure on the submodule of a Lie algebra associated to $\mathrm{Im}\,\rho_\theta$; in the case $n = 1$, Hida [H13b] used conjugation by certain elements of the inertia; a suitable generalization using repeated use of Poisson brackets provides the desired structure of Λ_n-module. Moreover, we ask the relation between the maximal such ideal $\mathfrak{l}_\theta$, called the Galois level of θ, and the schematic intersection of the irreducible component Spec $(\mathbb{I})$ with other components, more precisely, at least if $\mathbb{I} = \Lambda_n$, we ask whether the set of primes containing $\mathfrak{l}_\theta$ coincides with the union of the zero loci in Λ_n of the greatest common divisor of the congruence ideal of θ and those of other non-generic families congruent to θ. The equality of these has been established for $n = 1$ by Hida [H13b]. We prove this equality for $n = 2$ in the case of twisted Yoshida lift congruence. In a subsequent paper, we plan to formulate more precisely this relation (including exponents of height one prime ideals) and generalize it to more general groups. The question of replacing Λ_n-fullness by $\mathbb{I}$-fullness, or rather $\mathbb{I}'$-fullness for a suitable subring of $\mathbb{I}$ is currently investigated, by a student of one of the authors[1].

9.2 Galois representations associated to Siegel modular forms

Let $G = \mathrm{GSp}_{2n}$ be the Chevalley group of symplectic similitude matrices for J. Let $\nu : G \to \mathbb{G}_m$ be the similitude factor character. Its kernel Sp_{2n} coincides with the derived group G' of G. We denote by $B = TU$ its standard Borel (consisting of upper triangular matrices in G).

Let $S_k^{(n)}(M)$ be the space of genus n holomorphic Siegel cusp forms of weight $k = (k_1, \dots, k_n)$ with $k_1 \geq k_2 \geq \dots \geq k_n \geq n + 1$, and level M (say, for the principal congruence subgroup of level M). Let $\mathbf{T}_k^M$ be the M-spherical $\mathbb{Z}[\frac{1}{M}]$-algebra generated by the Hecke operators $T_{\ell,i}$ for all rational primes ℓ prime to M and all $i = 1, \dots, n$, acting on $S_k^{(n)}(M)$. Recall that $T_{\ell,i}$ denotes the action of the double class $\Gamma(M) \cdot d_{\ell,i} \cdot \Gamma(M)$ where $d_{\ell,1} = \mathrm{diag}(1_n, \ell \cdot 1_n)$ and $d_{\ell,i} = \mathrm{diag}(1_{n+1-i}, \ell \cdot 1_{2i}, \ell^2 \cdot 1_{n+1-i})$ for $i = 2, \dots n$. Let $f \in S_k^{(n)}(M)$ be an eigenform for $\mathbf{T}_k^M$; let K_0 be a number field containing its eigenvalues; let $\mathcal{O}_0$ be its ring of integers. It gives rise to a character $\theta_f : \mathbf{T}_k^M \to \mathcal{O}_0$. For any prime ℓ not dividing M, we denote by $P_{f,\ell}(X)$ the Hecke polynomial (see [FC90] Chapt.VII, Sect. 1); it is monic of degree 2^n.

Fix an odd prime p relatively prime to M, an embedding $\iota_p : \overline{\mathbb{Q}} \hookrightarrow \overline{\mathbb{Q}}_p$ and assume that f is ordinary for ι_p. It means that the multiset of p-adic valuations of the (images by ι_p of the) roots of the Hecke polynomial $P_{f,p}(X)$ coincides with the multiset $\{\sum_{i \in I}(k_i - i); I \subset [1, n]\}$ of Hodge weights.

To the data (G, B, T), one can associate a dual reductive group over $\mathbb{Z}[\frac{1}{2}]$ with a standard Borel and a standard torus $(\widehat{G}, \widehat{B}, \widehat{T})$, with an identification $X^*(T) = X_*(\widehat{T})$. By comparing the root data (with épinglage, see [MT02] Sect. 3.2.2), one has a canonical identification $\widehat{G} = \mathrm{GSpin}_{2n+1}$ over $\mathbb{Z}[\frac{1}{2}]$. Therefore $\widehat{G}$ is endowed with a linear representation $\mathrm{spin} : \widehat{G} \to \mathrm{GL}_{2^n}$ and an orthogonal representation $\pi : \widehat{G} \to \mathrm{SO}_{2n+1}$ whose kernel is the center $\mathbb{G}_m$ of $\widehat{G}$ (the central inclusion $\mathbb{G}_m \to \widehat{G}$ is dual to the similitude factor $G \to \mathbb{G}_m$). It also carries a similitude factor $\widehat{G} \to \mathbb{G}_m$, dual to the central inclusion $\mathbb{G}_m \to G$; its kernel coincides with the derived group $\widehat{G}' = \mathrm{Spin}_{2n+1}$.

[1] H. Hida

For any number field F, we put $G_F = \mathrm{Gal}(\overline{F}/F)$ and for any finite place v of F, we denote by D_v a decomposition subgroup at v in G_F; let $\epsilon\colon G_F \to \mathbb{Z}_p^\times$ be the p-adic cyclotomic character. One still denotes by ϵ its restriction to D_v. We assume that there exists a Galois representation $\rho_f\colon G_{\mathbb{Q}} \to \widehat{G}(\overline{\mathbb{Q}}_p)$ unramified outside Mp and such that for any rational prime ℓ not dividing Mp, the characteristic polynomial $\mathrm{Char}(\mathrm{spin} \circ \rho_f(\mathrm{Frob}_\ell))$ coincides with $\iota_p(P_{f,\ell}(X))$. It is called the Galois representation associated to f. The integer $w = k_1 + \cdots + k_n - \frac{n(n+1)}{2}$ is called the motivic weight of ρ_f.

The existence of ρ_f is known for $n = 1$ (Deligne) and for $n = 2$ (due to R. Taylor, Laumon and Weissauer, in [Ta93], [We05] and [Lau05]). By compactness, there exists a p-adic field $K \subset \overline{\mathbb{Q}}_p$ such that ρ_f is defined over K. We may assume that $\iota_p(K_0) \subset K$, hence if $\mathcal{O}$ denotes its valuation ring, we have $\iota_p(\mathcal{O}_0) \subset \mathcal{O}$. One can choose an $\mathcal{O}$-lattice stable in the orthogonal representation $\pi \circ \rho_f$. We assume it is unimodular. It implies that ρ_f takes values in $\widehat{G}(\mathcal{O})$. Let ϖ be a uniformizing parameter of $\mathcal{O}$ and $\mathbb{F} = \mathcal{O}/\varpi\mathcal{O}$ its residue field. Let $\overline{\rho}_f\colon G_{\mathbb{Q}} \to \widehat{G}(\mathbb{F})$ be the residual representation.

Under the assumption of "automorphic ordinarity", it is conjectured that the following "Galois ordinarity" holds. Let $\widetilde{\rho} = \sum_{i=1}^n \varpi_i$ be the character of T sum of all the fundamental weights of G; it is given by ρ on $T' = T \cap G'$ and by $z \mapsto z^{\frac{n(n+1)}{2}}$ on the center. Let us view $\lambda + \widetilde{\rho}$ as a cocharacter of $\widehat{T}$ via $X^*(T) = X_*(\widehat{T})$. For $\widehat{g}, \widehat{h} \in \widehat{G}$, let us put $\widehat{h}^{\widehat{g}} = \widehat{g}^{-1}\widehat{h}\widehat{g}$. Write D_p for a (chosen) decomposition group at p inside $\mathrm{Gal}(\overline{\mathbb{Q}}/\mathbb{Q})$ with the inertia subgroup I_p.

(GO) There exists $\widehat{g} \in \widehat{G}(\mathcal{O})$ such that $\rho_f(D_p) \subset \widehat{g} \cdot \widehat{B}(\mathcal{O}) \cdot \widehat{g}^{-1}$ and for any $\sigma \in I_p$, $\rho_f^{\widehat{g}}(\sigma) \pmod{\widehat{N}(\mathcal{O})}$ is given by $(\lambda + \widetilde{\rho}) \circ \epsilon(\sigma)$.

"Automorphic ordinarity" implies "Galois ordinarity" is known for $n = 1$ (due to Wiles) and for $n = 2$ (due to Urban [Ur05]). We assume in the sequel it holds for any n. It would follow from Katz-Messing theorem if one knew that $\mathrm{spin} \circ \rho_f$ is motivic.

In order to motivate the assumptions we shall make in genus n, let us prove some results in genus 2, where the results have less conditions.

9.3 Fullness of the image for Galois representations in GSp(4)

9.3.1 Irreducibility and open image

Let $n = 2$ and f be a genus 2 cusp eigenform of weight $k = (k_1, k_2)$, $k_1 \geq k_2 \geq 3$, level M, as above. Fix an odd prime p prime to M at which f is ordinary; let $P_{f,p}(X) = (X - \alpha)(X - \beta)(X - \gamma)(X - \delta)$ be the Hecke polynomial of f at p, with roots ordered with increasing p-adic valuation (so that $\mathrm{ord}_p(\alpha) = 0$, $\mathrm{ord}_p(\beta) = k_2 - 2$, $\mathrm{ord}_p(\gamma) = k_1 - 1$ and

$\mathrm{ord}_p(\alpha) = k_1 + k_2 - 3$); we assume that all the Hecke eigenvalues, together with α, β, γ and δ, are contained in $\mathcal{O}$ and that the Galois representation associated to f ([Ta93], [We05] and [Lau05]) is defined over $\mathcal{O}$: $\rho_f : G_{\mathbb{Q}} \to \mathrm{GSp}_4(\mathcal{O})$ where the group of symplectic similitudes is relative to the matrix $J = \begin{pmatrix} 0 & 0 & 0 & 1 \\ 0 & 0 & 1 & 0 \\ 0 & -1 & 0 & 0 \\ -1 & 0 & 0 & 0 \end{pmatrix}$. Recall that the spin representation provides an isomorphism $\mathrm{GSpin}_5 \cong \mathrm{GSp}_4 \subset \mathrm{GL}_4$. Let (L, ψ) be the $\mathcal{O}$-module of the representation endowed with a unimodular symplectic pairing ψ; we fix a symplectic $\mathcal{O}$-basis (e_1, e_2, e_3, e_4) of L with $\psi(e_1, e_4) = \psi(e_2, e_3) = 1$ so that we have an identification $\mathrm{GSp}(L, \psi) = \mathrm{GSp}_4(\mathcal{O})$; let $V = L[\frac{1}{p}]$ be the K-vector space spanned by L. We have

$$\rho_f|_{D_p} \sim \begin{pmatrix} \epsilon^{k_1+k_2-3} ur(\frac{\delta}{p^{k_1+k_2-3}}) & * & * & * \\ 0 & \epsilon^{k_1-1} ur(\frac{\gamma}{p^{k_1-1}}) & * & * \\ 0 & 0 & \epsilon^{k_2-2} ur(\frac{\beta}{p^{k_2-2}}) & * \\ 0 & 0 & 0 & ur(\alpha) \end{pmatrix}.$$

Here $ur(x) : D_p \to \mathcal{O}^\times$ is the unramified character sending Frob_p to $x \in \mathcal{O}^\times$. We shall assume in the sequel that the conjugation needed for this description is the identity. We also put $\overline{\rho}\colon G_{\mathbb{Q}} \to \mathrm{GSp}_4(\mathbb{F})$ its reduction modulo ϖ.

We first show

Proposition 9.1. *If f is neither CAP nor endoscopic, and if either*

(i) Langlands transfer holds from GSp(4) *to* GL(4) *(no need of the ordinarity assumption then),*

(ii) $p > 7$, the semisimplification $(\overline{\rho}|_{G_{\mathbb{Q}(\zeta_p)}})^{ss}$ contains a copy of $\mathrm{SL}_2(\mathbb{F}_p)$, and the p-adic units $\frac{\delta}{p^{k_1+k_2-3}}$, $\frac{\gamma}{p^{k_1-1}}$, $\frac{\beta}{p^{k_2-2}}$ and α are distinct modulo ϖ, then ρ_f is absolutely irreducible.

In the first case, one can use [CaGe13]. The transfer has been established by Arthur for Sp(4), but not yet for GSp(4), unless f admits a generic form with the same eigenvalues.

In the second case, we proceed by case inspection. If there is a stable line $D = \langle e'_2 \rangle$ in L, let $P_1 = \langle e'_2, e'_3 \rangle$ be a hyperbolic plane of L containing D and $P_2 = \langle e'_1, e'_4 \rangle$ its orthogonal, so that $P_1 \perp P_2 = L$ be a decomposition of L into two orthogonal hyperbolic planes (it is possible by unimodularity of the symplectic pairing on L). By taking (e'_1, e'_2, e'_3, e'_4) as symplectic basis of L adapted to this decomposition, we see in this basis that

$$\rho_f = \begin{pmatrix} * & 0 & \bullet & * \\ \bullet & \chi & \bullet & \bullet \\ 0 & 0 & \chi' & 0 \\ * & 0 & \bullet & * \end{pmatrix}.$$

hence,

- ρ_f admits a degree 1 subrepresentation χ which is locally algebraic at p and finitely ramified, hence it is given by an (A_0)-type Hecke character of $\mathbb{Q}$, which cannot be of finite order because $\chi(p)$ cannot be of archimedean absolute value one by Deligne purity for α, β, γ and δ, so that χ cannot have Hodge-Tate weight 0: it must be $k_2 - 2$ or $k_1 - 1$
- ρ_f has a degree 2 subquotient $\sigma = \left(\begin{smallmatrix} * & * \\ * & * \end{smallmatrix}\right)$, which is odd because its determinant is the similitude factor of ρ, which is odd.

Moreover, σ is p-ordinary with two distinct Hodge-Tate weights (0 and $k_1 + k_2 - 3$) and its residual representation is irreducible over $G_{\mathbb{Q}(\zeta_p)}$ by assumption. Therefore, the representation σ is modular: $\sigma = \sigma_g$ for an ordinary form g of weight $\ell \geq 2$ and level prime to p by Emerton's theorem on Fontaine-Mazur Conjecture ([Em14, Cor.1.2.2]); this implies by the classification of Vogan-Zuckerman (see [Ta93] Section 1 or [Ti09] Section 6) that, $k_1 = k_2 = k$, $\ell = 2k - 2$, $\chi = \epsilon^{k-1} \cdot (finite)$, and that the L-function of f is of the form $L(\chi, s)L(g, s)L(\chi, s - 1)$, hence f is CAP of Siegel type (i.e. is a Saito-Kurokawa lift), contrary to the assumption.

Similarly, if there is a stable isotropic plane, ρ_f contains a two-dimensional subrepresentation; it is p-ordinary with distinct Hodge-Tate weights, $p > 7$, and the image of its reduction modulo ϖ restricted to $G_{\mathbb{Q}(\zeta_p)}$ contains $\mathrm{SL}_2(\mathbb{F})$, it is odd by Cor.1.3 of Calegari [Ca13]; so again by [Em14, Cor.1.2.2], Fontaine-Mazur Conjecture holds and shows that f is CAP of Klingen type, contradiction. Finally, if there is a stable hyperbolic plane, we see by a similar argument (but without using Calegari's argument, hence without assuming $p > 7$ and the $\mathrm{SL}_2(\mathbb{F}_p)$ condition, because the oddness is obvious in this case), that f is a Yoshida lift, so it is endoscopic. QED.

Proposition 9.2. *Assume that the adjoint representation of ρ_f on the Lie algebra $\mathfrak{sp}_4$ is irreducible. Then (without assuming ordinarity), the image of Galois is full, i.e. contains a conjugate in $G(\overline{\mathbb{Q}}_p)$ of a congruence subgroup of $G(\mathbb{Z}_p)$.*

We first show that the image of Galois is open in its Zariski closure. Say that $\mathrm{Im}\,\rho_f \subset \mathrm{GSp}_4(E)$ for a finite extension $E/\mathbb{Q}_p$, and let $G = \mathrm{Res}_{E/\mathbb{Q}_p}\mathrm{GSp}_4$. Let $\mathfrak{g}$, resp. $\mathfrak{G}$, be the Lie algebra over $\mathbb{Q}_p$ of image of Galois (with its natural structure of p-adic Lie group), resp. of its $\mathbb{Q}_p$-Zariski closure in G. For any embedding $\sigma : E \to \overline{\mathbb{Q}}_p$, let $\mathfrak{G}_\sigma$ resp. $\mathfrak{g}_\sigma$ be the σ-projection of $\mathfrak{G} \otimes \overline{\mathbb{Q}}_p$ resp. $\mathfrak{g} \otimes \overline{\mathbb{Q}}_p$; By a theorem of Chevalley [Che51, p.177, Th.15], we see that, for any σ, the derived Lie algebras $\mathfrak{g}'_\sigma$ and $\mathfrak{G}'_\sigma$ coincide. By absolute irreducibility of ρ_f, the center of $\mathfrak{g}_\sigma$ has rank at most one. Moreover, since the similitude factor

$\nu \circ \rho$ is a non-trivial power of the p-adic cyclotomic character, the rank of the center is exactly one for any σ, so that $\mathfrak{g} \otimes \overline{\mathbb{Q}}_p = \mathfrak{G} \otimes \overline{\mathbb{Q}}_p$.

The possibilities for $\mathfrak{G}' \otimes \overline{\mathbb{Q}}_p$ are listed below

(F) $\mathfrak{G}' \otimes \overline{\mathbb{Q}}_p \sim \mathfrak{sp}_4$.
(Y) $\mathfrak{G}' \otimes \overline{\mathbb{Q}}_p \sim \mathfrak{sl}(2) \times \mathfrak{sl}(2)$
(K) $\mathfrak{G}' \otimes \overline{\mathbb{Q}}_p \sim \mathfrak{sl}(2)$ with $\mathfrak{sl}(2)$ embedded into a Klingen parabolic subalgebra.
(C) $\mathfrak{G}' \otimes \overline{\mathbb{Q}}_p \sim \mathfrak{sl}(2)$ via the symmetric cube representation of SL(2),
(S) $\mathfrak{G}' \otimes \overline{\mathbb{Q}}_p \sim \mathfrak{sl}(2)$ with $\mathfrak{sl}(2)$ embedded into a Siegel parabolic subalgebra.
(T) $\mathfrak{G}' \otimes \overline{\mathbb{Q}}_p \sim \{1\}$

Let us show that in all cases except (F), the adjoint representation of ρ_f on $\mathfrak{gsp}_4$ (denoted by adding the superscript "ad") would be reducible. Indeed, in all cases except (F), we see that $\mathfrak{G}^{ad}$ is reducible (note that $\left(\mathrm{Symm}^3\, \mathfrak{sl}(2)\right)^{ad}$ is also reducible). This shows that the $\mathbb{Q}_p$-Lie algebra $\mathfrak{g}$ of $\mathrm{Im}\,\rho_f$ is a $\mathbb{Q}_p$-form of $\mathfrak{gsp}_4(\mathbb{Q}_p)$ contained in $\mathfrak{gsp}_4(\overline{\mathbb{Q}}_p)$. Since $\mathrm{Aut}(\mathfrak{gsp}_4) = \mathrm{Int}(\mathfrak{gsp}_4)$, it implies that there exists $\alpha \in G(\overline{\mathbb{Q}}_p)$ such that

$$\mathfrak{g} = \alpha \cdot \mathfrak{gsp}_4(\mathbb{Q}_p) \cdot \alpha^{-1}.$$

In order to get a more complete result without assuming adjoint irreducibility, one needs to analyze the connected components of the Galois image.

9.3.2 Connected components of the image of Galois

Let $\rho = \{\rho_l\}_l$ be an n-dimensional compatible system of l-adic representations of $\mathrm{Gal}(\overline{\mathbb{Q}}/k)$ satisfying the assertions of [P98a, Theorems 3.2–3]. Write $V_l \cong \mathbb{Q}_l^n$ for the space of ρ_l. Replacing ρ_l by its semi-simplification, we may assume that ρ_l is semi-simple for all l. Define an algebraic group $G_{p/\mathbb{Q}_p}$ over $\mathbb{Q}_p$ by the $\mathbb{Q}_p$-Zariski closure in $\mathrm{GL}(V_{p/\mathbb{Q}_p})$ of $\mathrm{Im}(\rho_p)$ for a prime p, and let G_p° be its connected component. Since ρ_p is semi-simple, G_p is reductive. Then by a result of Serre, G_p/G_p° and $\mathrm{rank}\, G_p^\circ$ are independent of p (see [P98a, Theorem 3.6]). Let k^{conn} be the fixed field of $\rho_p^{-1}(G_p^\circ(\mathbb{Q}_p))$. We fix a prime p and write $\mathfrak{l}$ for a prime ideal of k^{conn} prime to p. Let $\rho_p(\mathrm{Frob}_\mathfrak{l})^{ss}$ be the semi-simplification of $\rho_p(\mathrm{Frob}_\mathfrak{l})$ inside $G_p(\mathbb{Q}_p)$. Write $T_\mathfrak{l}$ for the smallest algebraic subgroup over $\mathbb{Q}_p$ containing $\rho(\mathrm{Frob}_\mathfrak{l})^{ss}$ inside $G_p^\circ(\mathbb{Q}_p)$. Then the connected component $T_\mathfrak{l}^\circ$ is a torus, and it is isomorphic to a base-change to $\mathbb{Q}_p$ of a $\mathbb{Q}$-torus in $\mathrm{GL}(n)$. If $T_\mathfrak{l}$ is a maximal torus of G_p°, then $\rho_p(\mathrm{Frob}_\mathfrak{l})$ is already semi-simple.

Lemma 9.3 (Serre). *Let p be a prime. The algebraic group G_p° has a dense Zariski open subset U stable under conjugation such that T_l is connected and maximal if $\rho_p(\mathrm{Frob}_l)^{ss}$ is in $U(\mathbb{Q}_p)$. In particular, for any prime $\mathfrak{l}$ in a density one subset of primes of k^{conn}, $T_\mathfrak{l}$ is maximal.*

This follows from the argument in [Ch92, Theorem 3.7]. Though Chi assumes that ρ comes from an abelian variety, his argument works in general.

9.3.3 Connected semi-simple subgroups of GSp(4)

Let $\rho = \{\rho_\mathfrak{l}\}_\mathfrak{l}$ be a strict compatible system of symplectic semi-simple representations of $\mathrm{Gal}(\overline{k}/k)$ into $\mathrm{GSp}_4(T_\mathfrak{l})$ for a number field T (associated to a motive). Assume that T is the smallest coefficient field (i.e. it is generated by trace of Frobenii). Here $\mathfrak{l}$ runs over primes of T. Let l be the residual characteristic of $\mathfrak{l}$. Let $\mathrm{Res}\,\rho = \{\rho_l\}_l$ for $\rho_l = \prod_{\mathfrak{l}|l} \rho_\mathfrak{l}$ regarded to have values in $\mathrm{GSp}_4(T \otimes_\mathbb{Q} \mathbb{Q}_l)$. Then $\mathrm{Res}\,\rho$ is a compatible system with coefficients in $\mathbb{Q}$ acting on the $\mathbb{Q}_l$-vector spaces $V_l = (T \otimes_\mathbb{Q} \mathbb{Q}_l)^4$. Consider the $\mathbb{Q}_l$-Zariski closure G_l of $\mathrm{Im}(\rho_l)$ in $\mathrm{GL}(V_l)$.

As mentioned earlier, G_l is a reductive subgroup of $\mathrm{GL}(V_l)$ defined over $\mathbb{Q}_l$, $\mathrm{rank}\, G_l^\circ$ is independent of l and G_l/G_l° is independent of l and is isomorphic to $\mathrm{Gal}(k^{conn}/k)$ for the fixed field k^{conn} of $\rho_l^{-1}(G_l^\circ)$

Since G_l commutes with the action of T on V_l, for each prime factor $\mathfrak{l}$, we have its projection $G_\mathfrak{l}$ in $\mathrm{GSp}_4(T_\mathfrak{l})$. Let G_l° be the connected component of G_l (and $G_\mathfrak{l}^\circ$ for the $\mathfrak{l}$-component of G_l°; so, $G_\mathfrak{l}^\circ$ is the connected component of $G_\mathfrak{l}$).

Let $G'_\mathfrak{l}$ be the derived group of $G_\mathfrak{l}^\circ$. We may regard $G'_\mathfrak{l} \subset \mathrm{Res}_{T_\mathfrak{l}/\mathbb{Q}_l}\mathrm{Sp}(4)$. Thus the projections of $G'_\mathfrak{l}(\overline{\mathbb{Q}}_l)$ to each simple component $\mathrm{Sp}_4(\overline{\mathbb{Q}}_l)$ of $(\mathrm{Res}_{T_\mathfrak{l}/\mathbb{Q}_l}\mathrm{Sp}(4))(\overline{\mathbb{Q}}_l)$ are mutually isomorphic. We write $G'_\mathfrak{l} \sim G'$ for a semi-simple connected subgroup G' of $\mathrm{Sp}(4)_{/\overline{\mathbb{Q}}_l}$ if its projections to any simple factor $\mathrm{Sp}_4(\overline{\mathbb{Q}}_l)$ of $\mathrm{Res}_{T_\mathfrak{l}/\mathbb{Q}_l}\mathrm{Sp}(4)_{/\overline{\mathbb{Q}}_l}$ are isomorphic to G' over $\overline{\mathbb{Q}}_l$. Assume that ρ has weight w, i.e. $\det\rho$ is, up to finite character, given by $\{\epsilon_l^w\}_\mathfrak{l}$ for the l-adic cyclotomic character ϵ_l.

Let $\mathcal{G}$ be a connected semi-simple subgroup of $\mathrm{Sp}(4)_{/k}$ for a field k, and write $\mathfrak{g}$ for its Lie algebra. Write V_n for the symplectic space of dimension $2n$ on which $\mathrm{Sp}(2n)$ acts. Then by [LS98, Theorem 1], the largest proper semi-simple connected subgroup $\mathcal{G}$ of $\mathrm{Sp}(4)$ is isomorphic to $\mathrm{SL}(2) \times \mathrm{SL}(2)$ over an algebraic closure $\overline{k}$. Making an identification $V_2 \cong V_1 \oplus V_1$ as symplectic spaces, this group $\mathcal{G}$ acts diagonally on $V_1 \oplus V_1$. As for smaller semi-simple connected subgroups, we have two possibilities $\mathcal{G} = 1$ and $\mathcal{G} \cong \mathrm{SL}(2)$. The second possibility includes the case where the isomorphism $\mathrm{SL}(2) \cong \mathcal{G}$ is given by the symmetric cube representation of $\mathrm{SL}(2)$. Thus irreducibility of

$\mathcal{G}$-module $\mathfrak{g}$ (under the adjoint action) is satisfied if the derived group $\mathcal{G}$ is either SL(2) (the image of symmetric cube) or Sp(4).

For a prime $\mathfrak{P}$ over p of k, we suppose to have

$$\rho|_{I_{\mathfrak{P}}} \sim \begin{pmatrix} \epsilon^{w_1} & * & * & * \\ 0 & \epsilon^{w_2} & * & * \\ 0 & 0 & \epsilon^{w_3} & * \\ 0 & 0 & 0 & \epsilon^{w_4} \end{pmatrix} \quad (w_1 > w_2 > w_3 > w_4) \qquad \text{(Reg)}$$

up to finite error for the p-adic cyclotomic character $\epsilon = \epsilon_p$ of the inertia group $I_{\mathfrak{P}}$ of $\mathfrak{P}$. Thus we have six possibilities of the derived group $G'_{\mathfrak{p}}$ (the group $G^\circ_{\mathfrak{p}}$ is split over $T_{\mathfrak{p}}$ by (Reg)):

(F) $G'_{\mathfrak{p}} \sim \mathrm{Sp}(4)$.
(Y) $G'_{\mathfrak{p}} \sim \mathrm{SL}(2) \times \mathrm{SL}(2)$ with $\rho^{conn}_{\mathfrak{p}} = \rho_{1,\mathfrak{p}} \oplus \rho_{2,\mathfrak{p}}$, where $\rho_{1,\mathfrak{p}}$ is not twist equivalent to $\rho_{2,\mathfrak{p}}$.
(K) $G'_{\mathfrak{p}} \sim \mathrm{SL}(2)$ with $\rho^{conn}_{\mathfrak{p}} = \chi \oplus \chi' \oplus \rho_{1,\mathfrak{p}}$ with irreducible two-dimensional $\rho_{1,\mathfrak{p}}$.
(C) $G'_{\mathfrak{p}} \sim \mathrm{SL}(2)$ via the symmetric cube representation of SL(2).
(S) $G'_{\mathfrak{p}} \sim \mathrm{SL}(2)$ with $\rho^{conn}_{\mathfrak{p}} = \rho_{1,\mathfrak{p}} \oplus (\rho_{1,\mathfrak{p}} \otimes \chi)$ with absolutely irreducible two-dimensional $\rho_{1,\mathfrak{p}}$.
(T) $G'_{\mathfrak{p}} \sim \{1\}$ with $\rho_{\mathfrak{p}} = \mathrm{Ind}^{\mathbb{Q}}_F \chi$ for a degree 4 CM field $F/\mathbb{Q}$.

Here we write $\rho^{conn}_{\mathfrak{p}}$ for the restriction of $\rho_{\mathfrak{p}}$ to $\mathrm{Gal}(\overline{\mathbb{Q}}/k^{conn})$.

9.3.4 $\mathbb{Z}_p$-Fullness for GSp(4) and its subgroups

Note that $\mathrm{Sp}_4 \cong \mathrm{Spin}_5$ over $\mathbb{Z}_p$ (for $p > 2$). Suppose that ρ is the (semi-simple) system associated to a Siegel cusp form (so $k = \mathbb{Q}$) with regular weight and that $\rho_{\mathfrak{p}}$ (for $\mathfrak{p}|p$) is ordinary as in Proposition 9.1. Recall that $G_{\mathfrak{p}}$ is the Zariski closure of $\mathrm{Im}(\rho_{\mathfrak{p}})$, that $G^\circ_{\mathfrak{p}}$ is the identity connected component of $G_{\mathfrak{p}}$ and that $G'_{\mathfrak{p}}$ is the derived group of $G^\circ_{\mathfrak{p}}$ for a prime $\mathfrak{p}|p$ of T. For a semi-simple $\mathbb{Q}$-split subgroup G of Sp(4) and a semi-simple $T_{\mathfrak{p}}$ subgroup H, we write $H \sim G$ if $H \cong G \times_{\mathbb{Q}} T_{\mathfrak{p}}$. Then as seen in Sect. 9.3.3, we have the following six possibilities of $G'_{\mathfrak{p}}$:

(F) $G'_{\mathfrak{p}} \sim G$ for $G = \mathrm{Sp}(4)$.
(Y) $G'_{\mathfrak{p}} \sim G$ for $G = \mathrm{SL}(2) \times \mathrm{SL}(2)$ with $\rho^{conn}_{\mathfrak{p}} = \rho_{1,\mathfrak{p}} \oplus \rho_{2,\mathfrak{p}}$, where $\rho_{1,\mathfrak{p}}$ is not twist equivalent to $\rho_{2,\mathfrak{p}}$.
(K) $G'_{\mathfrak{p}} \sim G$ for $G = \mathrm{SL}(2)$ with $\rho^{conn}_{\mathfrak{p}} = \chi \oplus \chi' \oplus \rho_{1,\mathfrak{p}}$ with irreducible two-dimensional $\rho_{1,\mathfrak{p}}$.
(C) $G'_{\mathfrak{p}} \sim G$ for $G = \mathrm{SL}(2)$ via the symmetric cube representation of SL(2).

(S) $G'_{\mathfrak{p}} \sim G$ for $G = \mathrm{SL}(2)$ with $\rho_{\mathfrak{p}}^{conn} = \rho_{1,\mathfrak{p}} \oplus (\rho_{1,\mathfrak{p}} \otimes \chi)$ with absolutely irreducible two-dimensional $\rho_{1,\mathfrak{p}}$.
(T) $G'_{\mathfrak{p}} \sim \{1\}$ with $\rho_{\mathfrak{p}} = \mathrm{Ind}_F^{\mathbb{Q}} \chi$ for a degree 4 CM field $F/\mathbb{Q}$.

Here $\mathrm{Im}(\rho_{j,\mathfrak{p}})$ $(j = 1, 2)$ contains an open subgroup of $\mathrm{SL}_2(\mathbb{Z}_p)$ by the definition of $G'_{\mathfrak{p}}$, and χ, χ' are characters. In the automorphic side, Case (Y) corresponds to the Yoshida lift from an automorphic form $\mathrm{GL}(2) \times \mathrm{GL}(2)_{/\mathbb{Q}}$ or $\mathrm{Res}_{F/\mathbb{Q}}\,\mathrm{GL}(2)$ for a real quadratic field F, under the hypothesis that the starting automorphic form is not endo-scopic (i.e. non-CM). Cases (K) and (S) are either CAP or Eisenstein associated to the Siegel parabolic (S) and Klingen parabolic (K), assuming the corresponding GL(2) automorphic form is not endoscopic. Case (C) is the symmetric cube lift from $\mathrm{GL}(2)_{/\mathbb{Q}}$.

If G is absolutely simple, let $\mathfrak{g}$ be the Lie algebra of G and put $\mathfrak{g}_{\mathfrak{p}} = \mathfrak{g} \otimes_{\mathbb{Q}} T_{\mathfrak{p}}$ and regard it as the Lie algebra of $G'_{\mathfrak{p}}$. Since $\mathrm{Im}(\rho_{\mathfrak{p}})$ is Zariski dense in $G'_{\mathfrak{p}}$, the adjoint action of $\rho_{\mathfrak{p}}$ is absolutely irreducible. Write $Ad(\rho_{\mathfrak{p}})$ for the adjoint representation acting on $\mathfrak{g}_{\mathfrak{p}}$, and let $E_{\mathfrak{p}}$ be the subfield of $T_{\mathfrak{p}}$ generated over $\mathbb{Q}_p$ by $\mathrm{Tr}(Ad(\rho_{\mathfrak{p}}))$ on the Galois group. Let W be the $\mathfrak{p}$-adic integer ring of $E_{\mathfrak{p}}$. If $G = \mathrm{SL}(2) \times \mathrm{SL}(2)$ (i.e. we are in Case (Y)), we consider $Ad(\rho_{j,\mathfrak{p}})$ acting on $\mathfrak{sl}_2(T_{\mathfrak{p}})$. We then define E_j by the subfield of $T_{\mathfrak{p}}$ generated over $\mathbb{Q}_p$ by the values of $\mathrm{Tr}(Ad(\rho_{j,\mathfrak{p}}))$ over $\mathrm{Gal}(\overline{\mathbb{Q}}/k^{conn})$. Then we write W_j for the $\mathfrak{p}$-adic integer ring of E_j.

Proposition 9.4. *Let the notation be as above. Suppose* $p > 3$. *If* G *is absolutely simple,* $\mathrm{Im}(\rho_{\mathfrak{p}})$ *contains an open subgroup of* $G(W)$. *If* $G = \mathrm{SL}(2) \times \mathrm{SL}(2)$ *(i.e. in Case* (Y)*), then* $\mathrm{Im}(\rho_{\mathfrak{p}})$ *contains an open subgroup of* $\mathrm{SL}_2(W_1) \times \mathrm{SL}_2(W_2)$.

Proof. Suppose first that G is absolutely simple (i.e. we are not in Case (Y)); so, G is either SL(2) or Sp(4). Thus the adjoint representation of G on its Lie algebra $\mathfrak{g}$ is absolutely irreducible. We regard $Ad(\rho_{\mathfrak{p}})$ restricted to $\mathrm{Gal}(\overline{\mathbb{Q}}/k^{conn})$ as having values in G^{ad} (the adjoint group of G). Since $\mathrm{Im}(\rho_{\mathfrak{p}}^{conn})$ is Zariski dense in G, $Ad(\rho_{\mathfrak{p}}^{conn})$ is absolutely irreducible. Then by a theorem of Weisfeiler (see [P98b, Theorem 0.7]), there exists a linear algebraic group H defined over $E_{\mathfrak{p}}$ in $T_{\mathfrak{p}}$ such that $H \times_{E_{\mathfrak{p}}} T_{\mathfrak{p}} = G^{ad} \times_{\mathbb{Q}} T_{\mathfrak{p}}$ with $\mathrm{Im}(Ad(\rho_{\mathfrak{p}}^{conn})$ being an open subgroup of $H(W)$. Since the Siegel cusp form has regular weight, $\mathrm{Im}(\rho_{\mathfrak{p}}^{conn})$ contains p-inertia regular element g such that any of its power g^n $(0 < n \in \mathbb{Z})$ has a split torus of maximal rank in G as its centralizer. Thus H is split, and the result follows as $G \to G^{ad}$ is a central isogeny.

Now we assume that we are in Case (Y). Applying the above argument to $\rho_{j,\mathfrak{p}}$, $\mathrm{Im}(\rho_{j,\mathfrak{p}})$ contains an open subgroup of $\mathrm{SL}_2(W_j)$. Thus we need to show that $\mathrm{Ad}(\rho_1) \times \mathrm{Ad}(\rho_2)$ contains an open subgroup of $\mathrm{PGL}_2(W_1) \times \mathrm{PGL}_2(W_2)$. Let PG be PGL(2) defined over F; so, $PG(F) = \mathrm{PGL}_2(T_{\mathfrak{p}}) \times \mathrm{PGL}_2(T_{\mathfrak{p}})$. By [P98b, Main Theorem 0.2 (a)], there exist a semi-simple $\mathbb{Q}_p$-subalgebra E of $F := T_{\mathfrak{p}} \oplus T_{\mathfrak{p}}$ and a connected adjoint group $H_{/E}$ and an isogeny $\varphi : H \times_E F \to PG$ such that $\mathrm{Im}(\rho_{\mathfrak{p}}^{conn}))$ is an open subgroup of $\varphi(H(E))$. If E is a field, we have $E = E_1 = E_2$, and $\rho_{1,\mathfrak{P}} \cong \rho_{2,\mathfrak{p}}$, and hence we are in Case (S), a contradiction. Thus $E = E_1 \oplus E_2$, and $H(E) = \mathrm{PGL}_2(E_1) \times \mathrm{PGL}_2(E_2)$ as desired. □

9.3.5 Pink's Lie algebra theory

Let G be a split (smooth) connected reductive $\mathbb{Z}_p$-group with maximal split torus T. Let G' be the derived group of G, and put $\Gamma_A^G(\mathfrak{c})$ to be the kernel of the reduction map $G'(A) \twoheadrightarrow G'(A/\mathfrak{c})$ for a non-zero ideal $\mathfrak{c}$ of A for a $\mathbb{Z}_p$-algebra A. For a $\mathbb{Z}_p$-algebra $B \subset A$, a subgroup H of $G(A)$ is B-full if $H \supset \Gamma_B^G(\mathfrak{c})$ for a non-zero B-ideal $\mathfrak{c}$. Let $\rho : \mathrm{Gal}(\overline{\mathbb{Q}}/K) \to G(A)$ be a continuous representation for a finite extension $K/\mathbb{Q}$. Suppose A is an integral domain with quotient field $Q(A)$. We call ρ *B-full with respect to G* if after replacing ρ by a conjugate of ρ under an element of $G(Q(A))$, $\mathrm{Im}(\rho)$ is B-full. Suppose that B is a local ring with maximal ideal $\mathfrak{m}_B$. We call ρ *B-regular* if $\mathrm{Im}(\rho)$ contains a regular element b in the torus $T(B)$ such that $\alpha(b) \not\equiv \beta(b) \bmod \mathfrak{m}_B$ for all distinct roots α and β.

Via Pink's theory of tight correspondence between p-profinite Lie subalgebras of $\mathfrak{sl}(2)$ and p-profinite subgroups, we want to show that if a p-profinite subgroup of Spin_{2n+1} contains sufficiently large unipotent subgroups, it is full. Let us recall Pink's theory briefly. To study a general p-profinite subgroup $\mathcal{G}$ of $\mathrm{SL}_2(A)$ for a general p-profinite ring A, we want to have an explicit relation between p-profinite subgroups $\mathcal{G}$ of the form $\mathrm{SL}_2(A) \cap (1 + X)$ for a Lie $\mathbb{Z}_p$-subalgebra $X \subset \mathfrak{gl}_2(A)$. Assuming $p > 2$, Pink [P93] found a functorial explicit relation between closed subgroups in $\mathrm{SL}_2(A)$ and Lie subalgebras of $\mathfrak{gl}_2(A)$ (valid even for A of characteristic p). We call subgroups of the form $\mathrm{SL}_2(A) \cap (1 + X)$ (for a p-profinite Lie $\mathbb{Z}_p$-subalgebra X of $\mathfrak{gl}_2(A)$) *basic subgroups* following Pink.

We prepare some notation to quote here the results in [P93]. Let A be a semi-local p-profinite ring (not necessarily of characteristic p). Since Pink's result allows semi-local p-profinite algebra, we do not assume A to be local in the exposition of his result. We assume $p > 2$. Define maps $\Theta : \mathrm{SL}_2(A) \to \mathfrak{sl}_2(A)$ and $\zeta : \mathrm{SL}_2(A) \to Z(A)$ for the center $Z(A)$ of the algebra $M_2(A)$ by

$$\Theta(x) = x - \frac{1}{2}\mathrm{Tr}(x)\begin{pmatrix}1 & 0\\0 & 1\end{pmatrix} \text{ and } \zeta(x) = \frac{1}{2}(\mathrm{Tr}(x) - 2)\begin{pmatrix}1 & 0\\0 & 1\end{pmatrix}.$$

For each p-profinite subgroup $\mathcal{G}$ of $\mathrm{SL}_2(A)$, define L by the closed additive subgroup of $\mathfrak{sl}_2(A)$ topologically generated by $\Theta(x)$ for all $x \in \mathcal{G}$. Then we put $C = \mathrm{Tr}(L \cdot L)$. Here $L \cdot L$ is the closed additive subgroup of $M_2(A)$ generated by $\{xy|x, y \in L\}$ for the matrix product xy, similarly L^j is the closed additive subgroup generated by iterated products (j times) of elements in L. We then define $L_1 = L$ and inductively $L_{j+1} = [L, L_j]$; so, $L_2 = [L, L]$, where $[L, L_j]$ is the closed additive subgroup generated by Lie bracket $[x, y] = xy - yx$ for $x \in L$ and $y \in L_n$. Then by [P93, Proposition 3.1], we have

$$[L, L] \subset L,\ C \cdot L \subset L,\ L = L_1 \supset \cdots \supset L_j \supset L_{j+1} \supset \cdots \text{ and } \bigcap_{j \geq 1} L_n = \bigcap_{j \geq 1} L^j = 0. \quad (9.1)$$

In particular, L is a Lie $\mathbb{Z}_p$-subalgebra of $\mathfrak{sl}_2(A)$. Put $\mathcal{M}_j(\mathcal{G}) = C\left(\begin{smallmatrix}1&0\\0&1\end{smallmatrix}\right) \oplus L_j \subset M_2(A) = \mathfrak{gl}_2(A)$, which is a closed Lie $\mathbb{Z}_p$-subalgebra by (9.1). Define

$$\mathcal{H}_j = \{x \in \mathrm{SL}_2(A) | \Theta(x) \in L_j,\ \mathrm{Tr}(x) - 2 \in C\} \text{ for } j \geq 1.$$

If $x \in \mathcal{H}_n$, then $x = \Theta(x) + \zeta(x) + \left(\begin{smallmatrix}1&0\\0&1\end{smallmatrix}\right)$; thus, $\mathcal{H}_j \subset \mathrm{SL}_2(A) \cap (1 + \mathcal{M}_j(\mathcal{G}))$. If we pick $x \in \mathrm{SL}_2(A) \cap (1 + \mathcal{M}_j(\mathcal{G}))$, then $x = 1 + c \cdot 1 + y$ with $y \in L_n$ and $c \in C$. Thus $\mathrm{Tr}(x) - 2 = 2c \in C$ and $\Theta(x) = \left(\begin{smallmatrix}1&0\\0&1\end{smallmatrix}\right) + c \cdot \left(\begin{smallmatrix}1&0\\0&1\end{smallmatrix}\right) + y - \frac{1}{2}(2 + 2c) \cdot \left(\begin{smallmatrix}1&0\\0&1\end{smallmatrix}\right) = y$. This shows

$$\mathcal{H}_j = \mathrm{SL}_2(A) \cap (1 + \mathcal{M}_j(\mathcal{G})).$$

Here is a result of Pink (Theorem 3.3 combined with Theorem 2.7 both in [P93]):

Theorem 9.5 (Pink)**.** *Let the notation be as above. Suppose $p > 2$, and let A be a semi-local p-profinite commutative ring with identity. Take a p-profinite subgroup $\mathcal{G} \subset \mathrm{SL}_2(A)$. Then we have*

(1) *$\mathcal{G}$ is a normal closed subgroup of $\mathcal{H}_1$ (defined as above for $\mathcal{G}$),*
(2) *$\mathcal{H}_n$ is a p-profinite subgroup of $\mathrm{SL}_2(A)$ inductively given by $\mathcal{H}_{j+1} = (\mathcal{H}_1, \mathcal{H}_j)$ which is the closed subgroup topologically generated by commutators (x, y) with $x \in \mathcal{H}_1$ and $y \in \mathcal{H}_j$,*
(3) *$\{\mathcal{H}_j\}_{j \geq 2}$ coincides with the descending central series of $\{\mathcal{G}_j\}_{j \geq 2}$ of $\mathcal{G}$, where $\mathcal{G}_{j+1} = (\mathcal{G}, \mathcal{G}_j)$ starting with $\mathcal{G}_1 = \mathcal{G}$.*

In particular, we have

(P) *The topological commutator subgroup $\mathcal{G}'$ of $\mathcal{G}$ is the subgroup given by $\mathrm{SL}_2(A) \cap (1 + \mathcal{M}_2(\mathcal{G}))$ for the closed Lie subalgebra $\mathcal{M}_2(\mathcal{G}) \subset M_2(A)$ as above.*

Recall that G is a split (smooth) connected reductive $\mathbb{Z}_p$-group with maximal split torus T. Let G' be the derived group of G. Now let $\mathcal{G}$ be a p-profinite subgroup of $G'(A)$. Then on $\mathcal{G}_1 = \mathcal{G} \cap \Gamma_A^G(p)$ the logarithm $\sum_{m=1}^{\infty}(-1)^{m+1}\frac{(x-1)^m}{m}$ converges to an element $\mathrm{Log}(x) \in Lie(G)(A)$. Note that in general $\mathrm{Log}(\mathcal{G}) = \{\mathrm{Log}(x)|x \in \mathcal{G}\}$ is neither an abelian group nor a Lie algebra. Fix a Borel subgroup $B_{/\mathbb{Z}_p}$ of G containing T with $B = TU^+$ for the unipotent radical U^+. Write U^- for the unipotent subgroup opposite to U^+; so, TU^- is the opposite Borel subgroup of B.

Lemma 9.6. *Suppose that A is a p-profinite complete local integral domain over $\mathbb{Z}_p$ for $p > 2$. If $\mathcal{G}_2$ contains $\Gamma_A^{U^+}(\mathfrak{a})$ and $\Gamma_A^{U^-}(\mathfrak{b})$ for non-zero ideals $\mathfrak{a}$ and $\mathfrak{b}$ of A, then $\mathcal{G}_2$ (and hence $\mathcal{G}$) contains $\Gamma_A^G(\mathfrak{a}^2\mathfrak{b}^2)$.*

Proof. We write $\mathfrak{u}^+$ (resp. $\mathfrak{u}^-$, $\mathfrak{t}$) for the Lie subalgebra of $Lie(G)$ corresponding to U^+ (resp. U^-, T). We start with the case where $G = \mathrm{SL}(2)$. Note that $\mathrm{GSpin}_3 \cong \mathrm{SL}(2)$.

In our setting, we assume that A is local. Then by the definition as above, under the notation in Theorem 9.5, we have $\mathrm{Log}(\mathcal{G}_1 \cap U^+(A)) = \mathcal{M}_1(\mathcal{G}_1) \cap U^+(A)$ and $\mathrm{Log}(\mathcal{G}_1 \cap U^-(A)) = \mathcal{M}_1(\mathcal{G}_1) \cap U^-(A)$. Thus $\mathcal{M}_1(\mathcal{G}_1)$ contains $\mathfrak{u}^+(\mathfrak{a}) = \left\{\left(\begin{smallmatrix} 0 & a \\ 0 & 0 \end{smallmatrix}\right) \middle| a \in \mathfrak{a}\right\}$ and $\mathfrak{u}^-(\mathfrak{a}') = \left\{\left(\begin{smallmatrix} 0 & 0 \\ a & 0 \end{smallmatrix}\right) \middle| a \in \mathfrak{a}'\right\}$. Thus $[\mathfrak{u}^+(\mathfrak{a}), \mathfrak{u}^-(\mathfrak{a}')] \supset \mathfrak{t}(\mathfrak{a}\mathfrak{a}') := \left\{\left(\begin{smallmatrix} a & 0 \\ 0 & -a \end{smallmatrix}\right) \middle| a \in \mathfrak{a}\mathfrak{a}'\right\}$ is contained in $\mathcal{M}_1(\mathcal{G}')$. In other words, $\mathcal{M}_1(\mathcal{G}_1) \supset \mathcal{M}_1(\Gamma_A^G(\mathfrak{a}\mathfrak{a}'))$. This implies $\mathcal{G}'_1$ contains the derived group of $\Gamma_A^G(\mathfrak{a}\mathfrak{a}')$ which contains $\Gamma_A^G(\mathfrak{a}^2\mathfrak{a}'^2)$. This finishes the proof for $n = 1$.

For general G, we pick a positive root α of T with root subgroup $U_\alpha \subset U^+$. Then we have an embedding $i_\alpha : \mathrm{SL}(2) \hookrightarrow G$ sending the upper (resp. lower) triangular unipotent subgroup of $\mathrm{SL}(2)$ to U_α (resp. $U_{-\alpha} \subset U^-$). Applying the argument in the case of $\mathrm{SL}(2)$ to $\mathcal{G}_\alpha := i_\alpha^{-1}(\mathcal{G})$, we have $\mathcal{G}_\alpha \supset \Gamma_A^{\mathrm{SL}(2)}(\mathfrak{a}^2\mathfrak{b}^2)$. It is easy to see that $\{i_\alpha(\Gamma_A^{\mathrm{SL}(2)}(\mathfrak{a}^2\mathfrak{b}^2))\}_\alpha$ generates $\Gamma_A^G(\mathfrak{a}^2\mathfrak{b}^2)$, and hence the desired assertion holds. □

9.3.6 Λ-fullness for the endoscopic cases

We return to the setting of Sect. 9.3.4. We now study Λ-fullness in the one variable case of $\Lambda = \Lambda_1 = \mathbb{Z}_p[[T]]$ in the endoscopic cases (Y), (K), (C) and (S) below. In each endoscopic case, we have a factor $\mathrm{GL}(2)$ of G. Even if the component of the Hecke algebra has two variables, the Galois image of the projected factor falls in $\mathrm{GL}_2(\mathbb{I})$ for $\mathbb{I}$ finite over Λ (for a suitable quotient Λ of Λ_2). Thus we do not lose any generality assuming that $\mathbb{I}$ is finite over one variable Λ. To describe the quotient map $\Lambda_2 \to \Lambda$ explicitly, we normalize first $\Lambda_2 = \mathbb{Z}_p[[T_1, T_2]]$ so that the Galois image to contain

$$\mathcal{T} = \{\mathrm{diag}[(t_1t_2)^s, t_1^s, t_2^s, 1] \in T(\Lambda_2) | s \in \mathbb{Z}_p, t_j = 1 + T_j\}$$

as the wild p-inertia image. Then we normalize the Galois representation attached an $\mathbb{I}$-adic ordinary Hecke eigenform form as follows depending on cases for $\rho = \rho_{\mathbb{I}}$.

- In Case (Y), we write $\rho \cong \phi \times \varphi$ with $\phi(\sigma) = \begin{pmatrix} a(\sigma) & b(\sigma) \\ c(\sigma) & d(\sigma) \end{pmatrix}$, $\varphi(\sigma) = \begin{pmatrix} \alpha(\sigma) & \beta(\sigma) \\ \gamma(\sigma) & \delta(\sigma) \end{pmatrix}$ and

$$\rho(\sigma) = \begin{pmatrix} a(\sigma) & 0 & 0 & b(\sigma) \\ 0 & \alpha(\sigma) & \beta(\sigma) & 0 \\ 0 & \gamma(\sigma) & \delta(\sigma) & 0 \\ c(\sigma) & 0 & 0 & d(\sigma) \end{pmatrix} \quad \begin{array}{l} \text{and } \Lambda = \Lambda_2/(t_1t_2 - t_1) \\ \text{with } t_1 \mapsto t = 1 + T. \end{array}$$

- In Case (K), $\rho \cong \chi \oplus \chi' \oplus \phi$ with $\phi(\sigma) = \begin{pmatrix} a(\sigma) & b(\sigma) \\ c(\sigma) & d(\sigma) \end{pmatrix}$ and

$$\rho(\sigma) = \begin{pmatrix} a(\sigma) & 0 & 0 & b(\sigma) \\ 0 & \chi(\sigma) & 0 & 0 \\ 0 & 0 & \chi'(\sigma) & 0 \\ c(\sigma) & 0 & 0 & d(\sigma) \end{pmatrix} \quad \begin{array}{l} \text{and } \Lambda = \Lambda_2/(t_1 - t_2) \\ \text{with } t_1t_2 \mapsto t = 1 + T. \end{array}$$

- In Case (S), $\rho \cong (\phi^\iota \otimes \chi) \oplus \phi$; i.e.,

$$\rho(\sigma) = \begin{pmatrix} \chi(\sigma)s^t\phi(\sigma)^\iota s^{-1} & 0 \\ 0 & \phi(\sigma) \end{pmatrix} \text{ and } \Lambda = \Lambda_2/(t_1 - 1) \text{ with } t_2 \mapsto t = 1 + T.$$

- In Case (C), and $\Lambda = \Lambda_2/(t_1 - t_2^2)$ with $t_2 \mapsto t = 1 + T$.

Suppose

(y) In Case (Y), $n = 2$, $\det \varphi = \det \rho = \nu\rho$ and for $\sigma \in I_p$,

$$\phi(\sigma) = \begin{pmatrix} (t_1t_2)^{\log_p(\epsilon(\sigma)/\log_p(1+p)} & * \\ 0 & 1 \end{pmatrix},$$

$$\varphi(\sigma) = \begin{pmatrix} t_1^{\log_p(\epsilon(\sigma)/\log_p(1+p)} & * \\ 0 & t_2^{\log_p(\epsilon(\sigma)/\log_p(1+p)} \end{pmatrix},$$

and ϕ has values in a finite extension $\mathbb{I}_\phi \subset \mathbb{I}$ of $\Lambda_\phi = \mathbb{Z}_p[[T_\phi]]$ for $T_\phi = t_1t_2 - 1$ and $\varphi \otimes \kappa^{-1}$ has values in a finite extension $\mathbb{I}_\varphi \subset \mathbb{I}$ of $\Lambda_\varphi = \mathbb{Z}_p[[T_\varphi]]$ for $T_\varphi = t_1t_2^{-1} - 1$, where $\kappa : G_\mathbb{Q} \to \Lambda_\varphi^\times$ is a character such that $\kappa(\sigma) = t_2^{\log_p(\epsilon(\sigma)/\log_p(1+p)}$.

(k) In Case (K), $n = 2$ and for $\sigma \in I_p$, $\chi(\sigma) = t_1^{\log_p(\epsilon(\sigma)/\log_p(1+p)}$, $\chi'(\sigma) = t_2^{\log_p(\epsilon(\sigma)/\log_p(1+p)}$ and

$$\phi(\sigma) = \begin{pmatrix} (t_1 t_2)^{\log_p(\epsilon(\sigma)/\log_p(1+p)} & * \\ 0 & 1 \end{pmatrix},$$

and ϕ has values in a finite extension $\mathbb{I}_\phi \subset \mathbb{I}$ of $\Lambda_\phi = \mathbb{Z}_p[[T_\phi]]$ for $T_\phi = t_1 t_2 - 1$.

(s) In Case (S), $n = 2$ and $\chi(\sigma) = t_1^{\log_p(\epsilon(\sigma)/\log_p(1+p)}$ and $\phi(\sigma) = \begin{pmatrix} t_2^{\log_p(\epsilon(\sigma)/\log_p(1+p)} & * \\ 0 & 1 \end{pmatrix}$ for $\sigma \in I_p$, and ϕ has values in a finite extension $\mathbb{I}_\phi \subset \mathbb{I}$ of $\Lambda_\phi = \mathbb{Z}_p[[T_\phi]]$ for $T_\phi = t_2 - 1$.

(c) In Case (C), $n = 1$ and for $\sigma \in I_p$,

$$\rho(\sigma) = \begin{pmatrix} t_1{}^{3\log_p(\epsilon(\sigma)/\log_p(1+p)} & * & * & * \\ 0 & t_1{}^{2\log_p(\epsilon(\sigma)/\log_p(1+p)} & * & * \\ 0 & 0 & t_1{}^{\log_p(\epsilon(\sigma)/\log_p(1+p)} & * \\ 0 & 0 & 0 & 1 \end{pmatrix}.$$

Note here $\mathrm{GSpin}_3 \cong \mathrm{SL}(2)$ and $\mathrm{GSpin}_5 = \mathrm{GSp}(4)$ over $\mathbb{Z}_p$ (for p odd). Let $\mathbb{I}$ be a finite flat normal extension of Λ as before, and $\rho_{\mathbb{I}} : \mathrm{Gal}(\overline{\mathbb{Q}}/\mathbb{Q}) \to \mathrm{GSpin}_{2n+1}(\mathbb{I})$ be a Galois representation. Suppose

(G0) $G \supset T$; so, $\mathrm{rank}\, G = \mathrm{rank}\, \mathrm{GSpin}_{2n+1}$,

(G1) there exists a point $P \in \mathrm{Spec}\,(\Lambda)(\overline{\mathbb{Q}}_p)$ such that $\rho_P = (\rho_{\mathbb{I}} \mod P)$ is $\mathbb{Z}_p$-full with respect to G, regarding $\mathbb{I}/P\mathbb{I}$ as a $\mathbb{Z}_p$-algebra.

(G2) the semi-simplification of $\rho_{\mathbb{I}}$ restricted to the wild p-inertia group I_p^w is isomorphic to the diagonal representation $\mathrm{diag}[\kappa^{2n-1}, \kappa^{2n-2}, \cdots, 1]$ for the character $\kappa : I_p^w \to \Lambda^\times$ such that $\kappa([1+p, \mathbb{Q}_p]) = t = 1 + T$.

(G3) $(\nu \circ \rho_{\mathbb{I}})/\kappa^{n(2n-1)}$ has finite order.

(G4) We have $\mathrm{diag}[\kappa^{2n-1}, \kappa^{2n-2}, \cdots, 1]$ in $\rho_{\mathbb{I}}(I_p^w)$.

Conjecture 9.7. *Assume $p > 2n - 1 > 2$ in addition to* (G0–4). *Then $\rho_{\mathbb{I}}$ is Λ-full with respect to G.*

Theorem 9.8. *Suppose that $n = 1, 2$ and that we are not in Case* (F) *nor in Case* (T) *(so we are in the endoscopic cases). Let $\rho_{\mathbb{I}}$ be associated to an $\mathbb{I}$-adic ordinary Hecke eigen cusp form. Assume $\mathbb{Z}_p$-regularity if $\mathbb{I} \neq \Lambda$ or in Case* (Y). *Then the above conjecture holds.*

We prove below the Λ-fullness including $p = 3$.

Proof. When $n = 1$, this follows from [H13b, Theorem I]. If $\mathbb{I} = \Lambda$, by [Z14], we have a non-trivial upper unipotent element in $\rho_{\mathbb{I}}(I_p)$ and also a lower one; so, (G4) is actually not necessary if $\mathbb{I} = \Lambda$. If $\mathbb{I} \neq \Lambda$, $\mathbb{Z}_p$-regularity is used in [H13b] to show non-trivial unipotents can be found in $\mathrm{SL}_2(\Lambda)$ (not just in $\mathrm{SL}_2(\mathbb{I})$ shown in [Z14]). Since we are in the cases different from (F) and (T), G' is either split SL(2) or SL(2) $\times$ SL(2). This implies that, by GL(2)-theory, we can find an arithmetic prime $P \in \mathrm{Spec}\ (\Lambda)$ (sufficiently regular) such that $\rho := \rho_{\mathfrak{P}}$ for every prime $\mathfrak{P} \in \mathrm{Spec}\ (\mathbb{I})$ above P belongs to the same case except for Case (F). Then by Proposition 9.4 applied to ρ, $\rho_{j,\mathfrak{p}}$ is $\mathbb{Z}_p$-full, and hence the argument in [H13b] works.

Since we have one simple component SL(2) except for Case (Y), if we are not in Cases (Y) and (F), there is no difference in the argument proving the theorem from the one given in [H13b]. Thus we give more details in Case (Y). In this case, $\rho_{\mathbb{I}}$ has values in $\mathrm{GL}_2(\mathbb{I}) \times \mathrm{GL}_2(\mathbb{I})$. We write $\rho_L, \rho_R : \mathrm{Gal}(\overline{\mathbb{Q}}/\mathbb{Q}) \to \mathrm{GL}_2(\mathbb{I})$ for the projection of $\rho_{\mathbb{I}}$ to the left and right factor $\mathrm{GL}_2(\mathbb{I})$. Each projection $\rho_{\mathbb{I},j}$ $(j = 1, 2)$ to each simple component GL(2) is Λ-full by the argument in [H13b]. Pink's theory can be also applied to semi-local p-profinite rings, like $A = \mathbb{I} \oplus \mathbb{I}$. Put $\mathcal{G} = \mathrm{Im}(\rho_{\mathbb{I}}) \cap \Gamma^{\mathrm{SL}(2)}_{\mathbb{I}\oplus\mathbb{I}}(\mathfrak{m}_{\mathbb{I}})$ for the maximal ideal $\mathfrak{m}_{\mathbb{I}}$. and $\mathcal{G}_P = \mathrm{Im}(\rho_P) \cap \Gamma^{\mathrm{SL}(2)}_{\mathbb{I}/P\oplus\mathbb{I}/P}(\mathfrak{m}_{\mathbb{I}/P})$. Since the reduction map modulo P is onto for Pink's Lie algebra, we have a surjective morphism $\mathcal{M}_j(\mathcal{G}) \twoheadrightarrow \mathcal{M}_j(\mathcal{G}_P)$. Adding the superscript "+" (resp. "−", "0") to Pink's Lie algebra, we indicates the upper nilpotent subalgebra (resp. the lower nilpotent subalgebra, the diagonal subalgebra). Then by $\mathbb{Z}_p$-regularity, $\mathcal{M}_j(\mathcal{G}) = \mathcal{M}_j^+(\mathcal{G}) \oplus \mathcal{M}_j^0(\mathcal{G}) \oplus \mathcal{M}_j^-(\mathcal{G})$ and the same for $\mathcal{M}_j(\mathcal{G}_P)$. Thus the reduction map modulo P induces a sujection $\mathcal{M}_j^*(\mathcal{G}) \to \mathcal{M}_j^*(\mathcal{G}_P)$ for $* = \pm, 0$.

Let 1_L (resp. 1_R) be the idempotent of the left (resp. the right) factor $\mathbb{I}$ of A. Since $\mathcal{G}_P$ is $\mathbb{Z}_p$-full, for $\mathcal{M}_j(\mathcal{G}_P)$, $1_L\mathcal{M}_j^*(\mathcal{G}_P) \neq 0$ and $1_R\mathcal{M}_j^*(\mathcal{G}_P) \neq 0$ for $* = \pm, 0$. Take $g \in \mathrm{Im}(\rho_{\mathbb{I}})$ giving the $\mathbb{Z}_p$-regularity, $Ad(g)$ acts on $1_L\mathcal{M}_j^\pm(\mathcal{G}_P) \cap \mathcal{M}_j^\pm(\mathcal{G}_P)$ and $1_R\mathcal{M}_j^\pm(\mathcal{G}_P) \cap \mathcal{M}_j^\pm(\mathcal{G}_P)$ by the different set of eigenvalues. Thus $1_?\mathcal{M}_j^\pm(\mathcal{G}_P) \cap \mathcal{M}_j^\pm(\mathcal{G}_P) \neq 0$ for $? = L, R$. We have eigenspace decompositions under the action of $Ad(g)$: $\mathcal{M}_j^\pm(\mathcal{G}_P) = (1_L\mathcal{M}_j^\pm(\mathcal{G}_P) \cap \mathcal{M}_j^\pm(\mathcal{G}_P)) \oplus (1_L\mathcal{M}_j^\pm(\mathcal{G}_P) \cap \mathcal{M}_j^\pm(\mathcal{G}_P))$ and $\mathcal{M}_j^\pm(\mathcal{G}) = (1_L\mathcal{M}_j^\pm(\mathcal{G}) \cap \mathcal{M}_j^\pm(\mathcal{G})) \oplus (1_L\mathcal{M}_j^\pm(\mathcal{G}) \cap \mathcal{M}_j^\pm(\mathcal{G}))$. This implies $1_L\mathcal{M}_j^\pm(\mathcal{G}) \cap \mathcal{M}_j^\pm(\mathcal{G}) \neq 0$ and $1_R\mathcal{M}_j^\pm(\mathcal{G}) \cap \mathcal{M}_j^\pm(\mathcal{G}) \neq 0$ as the modulo P reduction map of $1_?\mathcal{M}_j^\pm(\mathcal{G}) \cap \mathcal{M}_j^\pm(\mathcal{G})$ to $1_?\mathcal{M}_j^\pm(\mathcal{G}_P) \cap \mathcal{M}_j^\pm(\mathcal{G}_P)$ is surjective for $? = L, R$. This implies that $(1_?\mathcal{M}_j^\pm(\mathcal{G}) \cap \mathcal{M}_j^\pm(\mathcal{G})) = 1_?\mathcal{M}_j^\pm(\mathcal{G})$ for $? = L, R$ and for all $j > 0$. After replacing $\rho_{\mathbb{I}}$ by a conjugate under $G(Q(\mathbb{I}))$, by [H13b, Lemma 2.9], that $\mathrm{Im}(\rho_L) \supset \Gamma^{\mathrm{SL}(2)}_{\Lambda}(\mathfrak{c}_L)$ and $\mathrm{Im}(\rho_R) \supset \Gamma^{\mathrm{SL}(2)}_{\Lambda}(\mathfrak{c}_R)$ for non-zero

Λ-ideals $\mathfrak{c}_L$ and $\mathfrak{c}_R$. In particular, by Lemma 9.6, we get $\mathrm{Im}(\rho_{\mathbb{I}}) \supset \Gamma_A^{\mathrm{SL}(2)}(\mathfrak{c})$ for $\mathfrak{c} = \mathfrak{c}_L^2 \cap \mathfrak{c}_R^2$ as desired. □

9.3.7 $\mathbb{Z}_p[[S_1]] \oplus \mathbb{Z}_p[[S_2]]$-fullness in Case (Y)

We can think of fullness property in a two-variable setting. Case (F) will be treated in the following section in the general setting of GSpin representations as $\mathrm{GSpin}_5 \cong \mathrm{GSp}(4)$. In the cases other than (F) and (Y), the abelian part consumes one variable out of the two, we do not have a properly two-variable Galois representation $\rho_{\mathbb{I}}$. Thus we assume that there exists a finite extension $k/\mathbb{Q}$ such that $\rho := \rho_{\mathbb{I}}|_{\mathrm{Gal}(\overline{\mathbb{Q}}/k)}$ has values in

$$G_2 = \{(x, y) \in \mathrm{GL}(2) \times \mathrm{GL}(2) | \det(x) = \det(y)\}.$$

We suppose that G_2 is embedded into GSp(4) in the following way:

$$G_2 \ni \left(\begin{pmatrix} a & b \\ c & d \end{pmatrix}, \begin{pmatrix} \alpha & \beta \\ \gamma & \delta \end{pmatrix}\right) \mapsto \begin{pmatrix} a & 0 & 0 & b \\ 0 & \alpha & \beta & 0 \\ 0 & \gamma & \delta & 0 \\ c & 0 & 0 & d \end{pmatrix} \in \mathrm{GSp}(4).$$

As before, we suppose that $\rho(I_p^w)$ for the wild p-inertia group I_p^w contains the following split torus

$$\mathcal{T} = \left\{\mathrm{diag}[t_1^s t_2^s, t_1^s, t_2^s, 1] \in T | s \in \mathbb{Z}_p\right\}$$

for $t_j = 1 + T_j$ and $\Lambda_2 = \mathbb{Z}_p[[T_1, T_s]]$. We make the following variable change $s_1 = t_1 t_2$ and $s_2 = t_1/t_2$; so, $\mathbb{Z}_p[[S_1, S_2]] = \Lambda_2$ for $S_j = s_j - 1$. We put $A_j = \mathbb{Z}_p[[S_j]]$; so, $\Lambda_2 = \mathbb{Z}_p[[S_1, S_2]] = A_1 \widehat{\otimes}_{\mathbb{Z}_p} A_2$.

Proposition 9.9. *Let $\rho_{\mathbb{I}}$ is associated to $\mathbb{I}$-adic ordinary Hecke eigenform on* GSp(4) *for an algebra $\mathbb{I}$ finite torsion-free over Λ_2. Suppose $\mathbb{Z}_p$-regularity for $\rho_{\mathbb{I}}$ and that for an arithmetic point $P \in \mathrm{Spec}\,(\Lambda_2)(\overline{\mathbb{Q}}_p)$ and for all primes $\mathfrak{P} \in \mathrm{Spec}\,(\mathbb{I})(\overline{\mathbb{Q}}_p)$ above P, $\rho_{\mathfrak{P}} = \rho \bmod \mathfrak{P}$ belongs to Case* (Y) *in Section 9.3.4. Then, if $p \geq 3$, replacing ρ by a conjugate of ρ by an element in $\mathrm{GL}_2(Q(A_1)) \times \mathrm{GL}_2(Q(A_2))$ if necessary, there exist non-zero A_j-ideals $\mathfrak{c}_j$ such that* $\mathrm{Im}(\rho)$ *contains*

$$\Gamma_{A_1}^{\mathrm{SL}(2)}(\mathfrak{c}_1) \times \Gamma_{A_2}^{\mathrm{SL}(2)}(\mathfrak{c}_2) \subset \mathrm{SL}_2(A_1) \times \mathrm{SL}_2(A_2) \subset G_2(\Lambda_2).$$

Proof. Let $\{i, j\} = \{1, 2\}$. We can write $P = P_i \otimes P_j$ for $P_? \in \mathrm{Spec}\,(A_?)(\overline{\mathbb{Q}})$ for $? = i, j$. Consider the projection $P_i \times \mathrm{id} : \Lambda_2 := A_i \widehat{\otimes}_{\mathbb{Z}_p} A_j \to W_i \otimes_{\mathbb{Z}_p} A_j = W_i[[S_j]]$ for $W_i = A_i/P_i$. Take a prime $\widetilde{\mathfrak{P}}_i \in \mathrm{Spec}\,(\mathbb{I})$ above

$\mathrm{Ker}(P_i \times \mathrm{id})$. Let $\rho_j = \rho \mod \widetilde{\mathfrak{P}}_i$. Then by Theorem 9.8 (or more precisely, by the same argument proving the theorem), replacing ρ_j by its conjugate under an element of $\mathrm{GL}_2(Q(A_j))$, $\mathrm{Im}(\rho_j)$ contains $\Gamma_{A_j}^{\mathrm{SL}(2)}(\mathfrak{a}_j)$ for a non-zero A_j-ideal $\mathfrak{a}_j$.

Put $\mathcal{G} = \mathrm{Im}(\rho) \cap \Gamma_{\Lambda_2}^{\mathrm{Sp}(4)}(\Lambda_2)$, and consider Pink's Lie algebra $\mathcal{M}_2(\mathcal{G})$ with respect to $A = A_1 \oplus A_2$ regarding $\mathrm{SL}_2(A_1) \times \mathrm{SL}_2(A_2) \subset G_2(\Lambda_2)$ as $\mathrm{SL}_2(A_1 \oplus A_2)$. Write $\mathfrak{u}_j^+$ for the upper nilpotent Lie algebra of the j-th component of $\mathrm{SL}(2) \times \mathrm{SL}(2)$, and let $\mathfrak{u}_j^-$ be the opposite algebra of $\mathfrak{u}_j^+$. By $\mathbb{Z}_p$-regularity of ρ, we have $\sigma \in \mathrm{Gal}(\overline{\mathbb{Q}}/K)$ normalizing $\mathcal{G}$ whose adjoint action $Ad(\rho(\sigma)$ has distinct $\mathbb{Z}_p$-eigenvalues on $\mathfrak{u}_j^{\pm}$ for $j = 1, 2$.

Since $\mathrm{Im}(\rho_j)$ contains $\Gamma_{A_j}^{\mathrm{SL}(2)}(\mathfrak{a}_j)$, $\mathfrak{n}_j^{\pm} := \mathcal{M}_2(\mathcal{G}) \cap \mathfrak{u}_j^{\pm}$ is non-trivial. By the adjoint action of $\mathcal{T}$, $\mathfrak{n}_j^{\pm}$ is a A_j-module. Then by [H13b, Lemma 2.9] applied to each projection of ρ to GL(2), we conclude a suitable conjugate of ρ has image containing the product of congruence subgroups as in the proposition. □

9.4 Fullness of the Galois image in general Spin groups over Iwasawa algebras

9.4.1 Spin groups

Let us first recall the definition and the basic properties of the $\mathbb{Z}[\frac{1}{2}]$-group scheme GSpin_{2n+1} of spinorial similitudes. Consider the quadratic form $Q(\underline{x}) = 2x_1x_{2n+1} + 2x_2x_{2n} + \ldots + 2x_{n-1}x_{n+1} + x_n^2$ on $L = \bigoplus_{i=1}^{2n+1} \mathbb{Z}[\frac{1}{2}] \cdot e_i$. Let $C = C(L, Q)$ be the Clifford algebra associated to (L, Q) and $C = C^+ \oplus C^-$ be its decomposition as $\mathbb{Z}/2\mathbb{Z}$-graded algebra into even and odd Clifford elements; C^+ is a subalgebra. The main (anti-)involution $*$ sends a pure element $v_1 \cdot \ldots \cdot v_r$ to $v_r \cdot \ldots \cdot v_1$. It leaves C^+ stable; the module C, resp. C^+, is locally free of rank 2^{2n+1}, resp. 2^{2n}; L embeds in C^- by $v \mapsto v$. Then $\mathrm{GSpin}_{2n+1} = \{x \in (C^+)^\times; x \cdot L \cdot x^* = L\}$ and $\mathrm{Spin}_{2n+1} = \{x \in (C^+)^\times; x \cdot L \cdot x^* = L, x^*x = 1\}$. We have two exact sequences of group schemes

$$1 \to \mathrm{Spin}_{2n+1} \to \mathrm{GSpin}_{2n+1} \xrightarrow{\mu} \mathbb{G}_m \to 1$$

by $\mu \colon x \mapsto x^*x$, and

$$1 \to \mathbb{G}_m \to \mathrm{GSpin}_{2n+1} \to \mathrm{SO}(L, Q) \to 1$$

by $x \mapsto (v \mapsto xvx^{-1})$. Since 2 is invertible, we can write $L = W \oplus W^\vee \oplus U$ where W, resp. $W^\vee$, is the standard lagrangian generated by $e_1, \ldots, e_n$, resp. its dual, identified to the direct factor generated by

$e_{n+2}, \ldots, e_{2n+1}$, and $U = \mathbb{Z}e_{n+1}$. There is a natural isomorphism $C^+ \cong \mathrm{End}(\bigwedge^\bullet W)$ (Fulton-Harris, Lemma 20.16, the proof works with $\mathbb{C}$ replaced by $\mathbb{Z}[\frac{1}{2}]$); by restriction to GSpin, it induces the spin representation $\mathrm{spin}\colon \mathrm{GSpin} \to \mathrm{GL}(\bigwedge^\bullet W) = \mathrm{GL}_{2^n}$. Let T_{spin} be the diagonal maximal torus of GSpin_{2n+1}, and put $T'_{\mathrm{spin}} = T_{\mathrm{spin}} \cap \mathrm{Spin}_{2n+1}$. We denote by $(t_1, \ldots, t_n; \mu) \mapsto [t_1, \ldots, t_n; \mu]$ the standard isomorphism $\mathbb{G}_m^n \times \mathbb{G}_m \cong T_{\mathrm{spin}}$, with $\mu([t_1, \ldots, t_n; \mu]) = \mu$. In particular $(t_1, \ldots, t_n) \mapsto [t_1, \ldots, t_n; 1]$ is an isomorphism $\mathbb{G}_m^n \cong T'_{\mathrm{spin}}$. We define the upper, resp. lower, triangular Borel subgroup $B_{\mathrm{spin}} = T'_{\mathrm{spin}} \cdot U_{\mathrm{spin}}$ resp. ${}^tB_{\mathrm{spin}} = T'_{\mathrm{spin}} \cdot {}^tU_{\mathrm{spin}}$, of Spin_{2n+1} as $\mathrm{spin}^{-1}(B_{2^n})$,resp. $\mathrm{spin}^{-1}({}^tB_{2^n})$, where $B_{2^n} = T_{2^n} \cdot U_{2^n}$ denotes the upper triangular Borel subgroup of GL_{2^n}.

The triple $(\mathrm{GSpin}_{2n+1}, B_{\mathrm{spin}}, T_{\mathrm{spin}})$ identifies to the dual $(\widehat{G}, \widehat{B}, \widehat{T})$ of the triple (G, B, T) (see [MT02, Sect. 3.2]). In particular, the parametrization of (the semi-simple part of) the standard torus $T'_{\mathrm{spin}} = \widehat{T}$ of $\widehat{G} = \mathrm{GSpin}_{2n+1}$ by $[t_1, \ldots, t_n; 1]$ can be written as $\prod_i [e_i - \frac{1}{2}f](t_i)$ via $\widehat{T} = \mathbb{G}_m \otimes X^*(T)$, where $e_i\colon (x_1, \ldots, x_n, \nu x_n^{-1}, \ldots, \nu x_1^{-1}) \mapsto x_i$ and $f\colon (x_1, \ldots, x_n, \nu x_n^{-1}, \ldots, \nu x_1^{-1}) \mapsto \nu$.

9.4.2 Genus *n* Hida families

Let $n \geq 1$ and $d = \frac{n(n+1)}{2}$. Let $\mathfrak{H}$ be the genus n Siegel half space; it is a d-dimensional hermitian symmetric domain. Let $G = \mathrm{GSp}_{2n}$ with its standard Borel subgroup B and its diagonal torus T as before. For any torsion-free compact open subgroup $K \subset G(\mathbb{A}^\infty)$, we form the d-dimensional Shimura Siegel variety $X_K = G(\mathbb{Q})\backslash(\mathfrak{H} \times G(\mathbb{A}^\infty)/K)$. These are non-connected d-dimensional complex varieties; the inclusions $K' \subset K$ induce a projective system of finite coverings $\pi_{K',K}\colon X_{K'} \to X_K$. These varieties and morphisms are algebraic and admit quasi-projective canonical models over $\mathbb{Q}$ which we still denote by X_K and $\pi_{K',K}$. Let $\lambda \in X^*(T)$ be a dominant weight for (G, B, T) and V_λ the irreducible representation of G over $\mathbb{Q}$ associated to λ. It defines a projective system of étale local systems $V_{\lambda,K}(\overline{\mathbb{Q}}_p)$ over the X_K's: $\pi^*_{K',K} V_{\lambda,K}(\overline{\mathbb{Q}}_p) = V_{\lambda,K'}(\overline{\mathbb{Q}}_p)$. Let $M \geq 3$ and $K(M)$ be the level M principal congruence subgroup of $G(\widehat{\mathbb{Z}})$. Let us fix an odd prime p not dividing M and consider the subgroups $I_m^+ \subset I_m$ of $K_p = G(\mathbb{Z}_p)$ where $I_m = \{g \in K_p; g \pmod{p^m} \in B(\mathbb{Z}/p^m\mathbb{Z})\}$ and $I_m^+ = \{g \in K_p; g \pmod{p^m} \in N(\mathbb{Z}/p^m\mathbb{Z})\}$ denotes its (normal) pro-p-Sylow. Let $K_m = K(M)^p \times I_m^+$. Consider the middle degree cohomology

$$\mathbb{H} = \lim_m H^d(X_{K_m} \times \overline{\mathbb{Q}}, V_{\lambda,K_m}(\overline{\mathbb{Q}}_p))$$

Note that any $\mathbb{Z}_p$-model V_λ of the representation of G defines a compatible system of integral étale sheaves $V_{\lambda,K_m}(\overline{\mathbb{Z}}_p))$, hence an integral structure $\mathbb{H}_{\mathbb{Z}_p}$ of $\mathbb{H}$: $\mathbb{H} = \mathbb{H}_{\mathbb{Z}_p} \otimes \mathbb{Q}_p$; note that $\mathbb{H}_{\mathbb{Z}_p}$ may have torsion. Let $\epsilon = (1, \ldots, 1; 0) \in X_*(T)$; we put $w = \langle \lambda + \rho, \epsilon \rangle$; it is the motivic weight of these cohomology groups.

For any prime ℓ not dividing M, let $d_{\ell,1} = \mathrm{diag}(1_n, \ell \cdot 1_n)$ and for any $i = 2 \ldots, n$, let $d_{\ell,i} = \mathrm{diag}(1_{n+1-i}, \ell \cdot 1_{2n-i}, \ell^2 \cdot 1_{n+1-i})$. Consider the groups $K'_m = d_{\ell,i} K_m d_{\ell,i}^{-1}$ and $K''_m = d_{\ell,i}^{-1} K_m d_{\ell,i}$, we have an isomorphism $\pi(d_{\ell,i})\colon X_{K'_m} \to X_{K''_m}$. We define an action of the double class $K_m d_{\ell,i} K_m$ on the pair $(X_{K_m}, V_{\lambda,K_m}(\overline{\mathbb{Q}}_p))$ by using the two finite morphisms $\pi_{K'_m,K_m}$, $\pi_{K''_m,K_m}$ and the morphism of sheaves

$$(\pi(d_{\ell,i}), d^*_{\ell,i})\colon V_{\lambda,K''_m}(\overline{\mathbb{Q}}_p) \to V_{\lambda,K'_m}(\overline{\mathbb{Q}}_p)$$

where for $t \in T(\mathbb{Q}_p)$, we write $t^* = \nu(t) \cdot t^{-1} \in T(\mathbb{Q}_p)$; and we consider its action on $V_\lambda(\mathbb{Q}_p)$. Note that for any $\ell \neq p$, $d^*_{\ell,i} = d^*_{\ell,i} = 1$ on $V_\lambda(\mathbb{Q}_p)$, in particular, it preserves any integral structure of V_λ. For $\ell = p$, we see that $d^*_{\ell,i}$ defines an isomorphism of $V_\lambda(\mathbb{Q}_p)$, that it preserves $V_\lambda^{\max}(\mathbb{Z}_p) = \mathrm{Ind}_{B^-}^G \lambda$, but has non-trivial cokernel; we won't make use of this integral structure in the sequel. We define for any prime ℓ not dividing M, and any $i = 1, \ldots, n$:

$$[K_m d_{\ell,i} K_m] = \pi_{K'_m,K_m,*} \circ (\pi d_{\ell,i}, d^*_{\ell,i}))^* \circ \pi^*_{K''_m,K_m}.$$

These actions are compatible when m varies and define $\mathbb{Q}_p$-linear endomorphisms on $\mathbb{H}$, denoted by $T_{\ell,i}$ for ℓ prime to Mp, resp. by $U_{p,i}$ for $\ell = p$. These operators are the Hecke correspondences acting on $\mathbb{H}$. The space $\mathbb{H}$ carries a Galois action which commutes to the $T_{\ell,i}$s and $U_{p,i}$s. These actions preserve the integral structures.

It carries also a normal action of the torus $T(\mathbb{Z}_p)$ viewed as the quotient $B(\mathbb{Z}_p)/N(\mathbb{Z}_p) = \varprojlim I_m/I_m^+$. This action is continuous. One can decompose $T(\mathbb{Z}_p)$ as $T_f \times T_1$ where T_f is the torsion subgroup and T_1 is the pro-p-Sylow, kernel of the reduction modulo p. We fix a topological basis $(u_1, \ldots, u_n)$ of T_1 by fixing $u = 1 + p$ as topological basis of $1 + p\mathbb{Z}_p$ and defining u_i as the image of u by the cocharacter $t \mapsto \mathrm{diag}(E_i(t), E_{n+1-i}(t^{-1}))$ where $E_i(t) = \mathrm{diag}(1, \ldots, t, \ldots, 1)$ (t is the i-th component). By this choice, we identify the completed group algebra $\mathbb{Z}_p[[T_1]]$ to the n-variable Iwasawa algebra $\Lambda_n = \mathbb{Z}_p[[X_1, \ldots, X_n]]$ via $u_i \mapsto X_i + 1$. $\mathbb{H}$ is therefore a $\Lambda_n[T_f]$-module. This action commutes to the $T_{\ell,i}$ and the $U_{p,i}$s; moreover it preserves any integral structure given by a $\mathbb{Z}_p$-model V_λ of the representation of G. We consider the largest subspace $H_0^d(X_{K_m} \times \overline{\mathbb{Q}}, V_{\lambda,K_m}(\overline{\mathbb{Q}}_p))$, resp. $\mathbb{H}_0$, of $H^d(X_{K_m} \times \overline{\mathbb{Q}}, V_{\lambda,K_m}(\overline{\mathbb{Q}}_p))$, resp. of $\mathbb{H}$, on which the eigenvalues of the operators $U_{p,i}$ are p-adic units; it is the sum of all the generalized eigenspaces for such eigenvalues. We denote by $\mathbf{T}^M_\lambda(K_m)$ the $\mathbb{Z}_p$-algebra generated by the $T_{\ell,i}$s, the $U_{p,i}$s and the group action of $T(\mathbb{Z}/p^m\mathbb{Z})$ on $H^d_0(X_{K_m} \times \overline{\mathbb{Q}}, V_{\lambda,K_m}(\overline{\mathbb{Q}}_p))$; similarly, let $\mathbf{T}^M$ be the Λ_n-subalgebra of $\mathrm{End}_{\Lambda_n}\mathbb{H}_0$ generated by the $T_{\ell,i}$s, the $U_{p,i}$s and the action of T_f. Recall ([M04] Prop. 6.4.1) that $\mathbb{H}_0$ is a finite Λ_n-module. Therefore, $\mathbf{T}^M$ is a finite Λ_n-algebra. Note however that it is not cuspidal, in general (this would mean that all specializations at cohomological weights would be cuspidal). It is conjectured that cuspidality holds for the localization of $\mathbf{T}^M$ at a "weakly non-Eisenstein" maximal ideal (assuming the existence of the $\widehat{G}$-valued p-adic Galois representation and its ordinarity at p, "weak non-Eisenstein-ness" means that the corresponding residual Galois representation is absolutely irreducible).

However, another obstacle in this definition (due to the presence of torsion in the cohomology groups) is that the exact control theorem has not been established for all cohomological weights, unless one makes some strong extra assumptions. This seems to prevent to show in full generality that $\mathbf{T}^M$ is torsion-free as a Λ_n-algebra. Let $\lambda' \in X^*(T)$ be a dominant weight such that $\lambda'|_{T_f} = \lambda|_{T_f}$, its restriction to T_1 defines by linearity and continuity an algebra homomorphism $\lambda' \colon \Lambda_n \to \mathbb{Z}_p$; we denote by $P_{\lambda'}$ the prime ideal of Λ_n kernel of λ'.

Let $C > 0$ be a real number. The set of C-regular weights is the set of λ''s such that $\langle \lambda', \alpha \rangle > C$ for all simple roots α of G.

Definition 9.10. We say that control theorem, resp. $\mathfrak{m}$-control theorem, holds for the weight λ' (congruent to λ mod. $p-1$), if the surjective morphism $\mathbf{T}^M \otimes_{\Lambda_n,\lambda'} \mathbb{Z}_p \to \mathbf{T}^M_{\lambda'}(K_1)$, resp. $\mathbf{T}^M_{\mathfrak{m}} \otimes_{\Lambda_n,\lambda'} \mathbb{Z}_p \to \mathbf{T}^M_{\lambda'}(K_1)_{\mathfrak{m}}$, has finite kernel.

If one assumes that $p-1 > w$ (for the motivic weight w) and if the maximal ideal $\mathfrak{m}$ is "strongly non-Eisenstein" (in the sense of [MT02] Sect. 9: it means that the image of the residual representation contains the normalizer of the $\mathbb{F}_p$-points of the standard maximal torus $\widehat{T}$) then cuspidality and Λ_n-torsion-freeness hold for the $\mathfrak{m}$-localized Hecke algebra and $\mathfrak{m}$-control theorem do hold for all cohomological weights λ' congruent to λ mod. $p-1$ (see Th.10 of [MT02]), otherwise, without assuming the existence of the Galois representations and without localization, there exists a constant C depending on the data (λ, p) such that control theorem holds for all C-regular weights (but it does not imply the torsion-freeness of the Hecke algebra, except for Siegel-Hilbert varieties of genus 2, see [TU99], or for unitary groups in three variables, see [M04]).

In order to obtain a control theorem for all weights without localization, one needs to follow another approach via coherent cohomology, started by Hida [H02]. We identify the character group $X^*(T)$ to the index two sublattice of $\mathbb{Z}^{n+1}$ by sending the character

$$\lambda\colon \mathrm{diag}(t_1,\ldots,t_n,\nu t_n^{-1},\ldots,\nu t_1^{-1}) \mapsto t_1^{k_1}\ldots t_n^{k_n}\nu^{(c-k_1-\ldots-k_n)/2}$$

to $(k_1,\ldots,k_n;c)$. Let $t=(1,\ldots,1;1)$; we fix the cohomological weight $k=\lambda+(n+1)t=(k_1,\ldots,k_n;k_1+\cdots+k_n)$ and the sheaf ω^k_{cusp} over X_{K_1} (see [H02, Sect. 2.3] and [Pi12, Sect. 4]). These objects are now viewed on the rigid analytic site over $\mathbb{Q}_p$. One starts from a holomorphic cusp form $f\in H^0(X_{K_1},\omega^k_{\mathrm{cusp}})$; instead of the whole tower of finite coverings $X_{K_m}\to X_{K_1}$, one forms the Igusa tower $T_m=\mathrm{Isom}_{X^0_{K_1}}(\mu^n_{p^m},A[p^m]^\circ)$ over the ordinary locus $X^0_{K_1}$ of X_{K_1} (here, $A[p^m]^\circ$ denotes the connected component of the p^m-torsion of the universal abelian variety over $X^0_{K_1}$); T_m can be viewed as the rigid tube of the "multiplicative type" connected component of the ordinary locus in the reduction modulo p of X_{K_m}. Consider the $\mathbb{Q}_p$-vector space $\mathbb{HC}=\lim_m H^0(T_m,\omega^k_{\mathrm{cusp}})$; one defines similarly action of the Hecke operators $T_{\ell,i}$s and $U_{p,i}$s $(i=1,\ldots,n)$; these are compatible with the transition morphisms of the Igusa tower; the Galois group of the Igusa tower is $T(\mathbb{Z}_p)$ which therefore acts on $H^0(T_m,\omega^k_{\mathrm{cusp}})$ and $\mathbb{HC}$. One defines the $\mathbb{Z}_p$-Hecke algebras $\mathbf{T}^M(T_m)$, resp. $\mathbf{T}^M(T_\infty)$ as generated by $T_{\ell,i}$s, $U_{p,i}$s and the image of $T(\mathbb{Z}_p)$ acting on $H^0(T_m,\omega^k)$, resp. $\mathbb{HC}$. Similarly, one could define $\mathbf{T}^{M,\mathrm{holcusp}}$ as the $\mathbb{Z}_p$-algebra acting on the ordinary part $\mathbb{H}_0^{\mathrm{holcusp}}$ of $\mathbb{H}^{\mathrm{holcusp}}=\lim_m H^0(X_{K_m},\omega^k_{\mathrm{cusp}})$; by density of classical Siegel modular cusp forms in p-adic modular cusp forms, the natural morphism $\mathbf{T}^M(T_\infty)\to\mathbf{T}^{M,\mathrm{holcusp}}$ is an isomorphism of Λ_n-algebras (note though that one has only surjections $\mathbf{T}^M(T_m)\to\mathbf{T}^{M,\mathrm{holcusp}}(K_m)$). By Hodge–Tate decomposition (see Faltings–Chai Chapter VI, using BGG theory), $\mathbf{T}^{M,\mathrm{holcusp}}$ is a quotient of $\mathbf{T}^M$ (it cannot be injective since the left-hand side contains non-cuspidal families, but after localization at any weakly non-Eisenstein maximal ideal should be an isomorphism).

By [Pi12, Th.6.7 and Th.1.2], and [PiSt, Th.6.3.1], $\mathbf{T}^{M,\mathrm{holcusp}}$ is finite torsion-free over Λ_n and (cuspidal) Control Theorem holds without localization (and without assuming the existence of the Galois representations); in other words, for any dominant weight λ' such that $\lambda'|_{T_f}=\lambda|_{T_f}$, the surjective morphism $\mathbf{T}^{M,\mathrm{holcusp}}\otimes_{\Lambda_n,\lambda'}\mathbb{Z}_p\to\mathbf{T}^{M,\mathrm{holcusp}}_{\lambda'}(K_1)$ has finite kernel.

Let $\mathbb{I}$ be the normal closure of an irreducible component of Spec $(\mathbf{T}^{M,\mathrm{holcusp}})$. Let $\theta\colon\mathbf{T}^{M,\mathrm{holcusp}}\to\mathbb{I}$ be the corresponding Hida family. Assuming the existence of the Galois representations $\rho_f\colon G_{\mathbb{Q}}\to\widehat{G}(\overline{\mathbb{Z}}_p)$ associated to the holomorphic cusp forms of level K_1 occurring in the family, let us construct the Galois representation $G_{\mathbb{Q}} \to \widehat{G}(\mathbb{I})$ associated to θ. We say that a cusp eigenform f' of level K_1, weight $k' = \lambda' + (n+1)t$ and eigensystem $\theta_{f'}$ occurs in the family if there exists a prime P' of $\mathbb{I}$ above $P_{\lambda'}$ such that, via the control theorem, one has $\theta_{f'} = \theta \pmod{P'}$.

Definition 9.11. We say that the irreducible component $\mathbb{I}$ is weakly non-Eisenstein if for some (hence all) cusp eigenform f' occurring in the family, the residual representation composed with spin is absolutely irreducible.

Lemma 9.12. *For any weakly non-Eisenstein irreducible component $\mathbb{I}$, there exists a continuous homomorphism $\rho_\theta \colon G_{\mathbb{Q}} \to \widehat{G}(\mathbb{I})$ such that for any dominant weight λ' such that $\lambda'|_{T_f} = \lambda|_{T_f}$, and for any cusp eigenform f' of weight $k' = \lambda' + (n+1)t$ in the family, the representation $\rho_{f'}$ is conjugate in $\widehat{G}(\overline{\mathbb{Q}}_p)$ to $\rho_\theta \pmod{P'}$.*

Proof. We give the proof only for $\mathbb{I} = \Lambda_n$. Let us consider the $\mathbb{Z}_p$-morphism $\mathrm{Ad} \colon \widehat{G} \to \mathrm{SO}_{2n+1}$; note that Ker(Ad) is the center $\widehat{Z} \cong \mathbb{G}_m$ of $\widehat{G}$. We first construct a representation $R \colon G_{\mathbb{Q}} \to \mathrm{SO}_{2n+1}(\mathbb{I})$ and then lift it to $\widehat{G}(\mathbb{I})$. For any arithmetic prime P' of $\mathbb{I}$ above a dominant weight λ' such that $\lambda'|_{T_f} = \lambda|_{T_f}$, let $f_{P'}$ be a cusp eigenform occurring in the family with this weight; by residual irreducibility of spin $\circ \rho_{f_{P'}}$ and Carayol's theorem [Car94], we can assume that this representation is defined over $\mathbb{I}/P'$; by injectivity of spin, it implies that $\rho_{P'} = \rho_{f_{P'}}$ takes values in $\widehat{G}(\mathbb{I}/P')$: $\rho_{P'} \colon G_{\mathbb{Q}} \to \widehat{G}(\mathbb{I}/P')$.

On the other hand, by residual irreducibility, it follows from Carayol's theorem [Car94] that $\mathrm{Ad} \circ \overline{\rho}$ lifts to a continuous representation $R \colon G_{\mathbb{Q}} \to \mathrm{GL}_{2n+1}(\mathbb{I})$. Since $\mathrm{Ad} \circ \overline{\rho}$ is orthogonal, and each specialization modulo an arithmetic prime is orthogonal, we see that R is orthogonal. Since $\mathbb{I}$ is profinite and by formal smoothness of the group $\widehat{G}$, there exists a continuous lift $\widetilde{R} \colon G_{\mathbb{Q}} \to \widehat{G}(\mathbb{I})$, which gives rise to a 2-cocyle $c_{\sigma,\tau} \in Z^2(G_{\mathbb{Q}}, \mathbb{I}^\times)$.

For any prime P', the existence of the representation $\rho_{P'}$ implies that the image of $c_{\sigma,\tau}$ is a coboundary: $c_{\sigma,\tau} = b_{\sigma\tau} b_\sigma^{-1} b_\tau^{-1}$over $(\mathbb{I}/P')^\times$. Take $\mathbb{I} = \Lambda_n$ and consider the Iwasawa ideal $\Omega_{n,\underline{k}}$ generated by the $(1+T_i)^{p^n} - u^{k_i p^n}$, $= 1, \ldots, n$. The quotient ring $\Lambda_{n,\underline{k}} = \Lambda/\Omega_{n,\underline{k}}$ embeds in the product of cyclotomic rings $\Lambda(\zeta_1, \ldots, \zeta_n) = \Lambda_n / I(\zeta_1, \ldots, \zeta_n)$ where

$$I(\zeta_1, \ldots, \zeta_n) = (1 + T_i - u^{k_i}\zeta_i, \quad i = 1 \ldots, n)$$

where ζ_i is a p^n-th root of unity. The condition for a family $(b(\zeta_1, \ldots, \zeta_n))_{\zeta_1,\ldots,\zeta_n}$ to be in the image of the embedding is that for any pair

$(\zeta_1,\dots,\zeta_n)$ and $(\zeta'_1,\dots,\zeta'_n)$ the components $b(\zeta_1,\dots,\zeta_n)$ and $b(\zeta'_1,\dots,\zeta'_n)$ are congruent modulo $I(\zeta_1,\dots,\zeta_n)+I(\zeta'_1,\dots,\zeta'_n)$. Let us check it is the case for the family of 1-cochains $(b(\zeta_1,\dots,\zeta_n)_\sigma)$ associated to the specialization $c^{n,\underline{k}}_{\sigma,\tau}$ of $c_{\sigma,\tau}$ over $(\Lambda_{n,\underline{k}})^\times$. Indeed, let us compare the reductions modulo $I(\zeta_1,\dots,\zeta_n)+I(\zeta'_1,\dots,\zeta'_n)$ of the representations $\mathrm{spin}\circ\widetilde{R}(\sigma)b(\zeta_1,\dots,\zeta_n)_\sigma$ and $\mathrm{spin}\circ\widetilde{R}(\sigma)b(\zeta'_1,\dots,\zeta'_n)_\sigma$; they have the same characteristic polynomial and they both lift the irreducible representation $\mathrm{spin}\circ\overline{\rho}$. By Carayol's theorem [Car94], they are conjugate, so that $b(\zeta_1,\dots,\zeta_n)_\sigma=b(\zeta'_1,\dots,\zeta'_n)_\sigma$ in $\Lambda_n/I(\zeta_1,\dots,\zeta_n)+I(\zeta'_1,\dots,\zeta'_n)$.

From this, we conclude that there exists a 1-cochain $b^{n,\underline{k}}_\sigma$ with values in $(\Lambda_{n,\underline{k}})^\times$ such that $c^{n,\underline{k}}_{\sigma,\tau}=b^{n,\underline{k}}_{\sigma\tau}b^{n,\underline{k}-1}_\sigma b^{n,\underline{k}-1}_\tau$. This construction is compatible when n grows and gives rise to a 1-cochain b_σ with values in $\Lambda_n^\times$ as desired. The resulting map $\sigma\mapsto\widetilde{R}(\sigma)b_\sigma$ is the desired representation $G_{\mathbb{Q}}\to\widehat{G}(\Lambda_n)$. □

Moreover ρ_θ is ordinary at p. More precisely, recall that ϵ denotes the p-adic cyclotomic character; we define for any $\sigma\in\mathrm{Gal}(\overline{\mathbb{Q}}/\mathbb{Q})$ the cyclotomic logarithm $\ell(\sigma)=\frac{\log_p(\epsilon(\sigma))}{\log_p(1+p)}$ (for the p-adic cyclotomic character ϵ). Let $t_i=1+X_i$ $(i=1,\dots,n)$, then

Lemma 9.13. *Assume $p-1>w$. The representation ρ_θ is ordinary: there exists $\widehat{g}\in\widehat{G}(\mathbb{I})$ such that $\rho_\theta(D_p)\subset\widehat{g}\widehat{B}(\mathbb{I})\widehat{g}^{-1}$ and for any $\sigma\in I_p$,*

$$\widehat{g}^{-1}\rho_\theta(\sigma)\widehat{g}\equiv[\epsilon(\sigma)^{-1}t_1^{\ell(\sigma)},\epsilon(\sigma)^{-2}t_2^{\ell(\sigma)},\dots,\epsilon(\sigma)^{-n}t_n^{\ell(\sigma)};\epsilon(\sigma)^{-d}]\pmod{U(\mathbb{I})}$$

where $d=\frac{n(n+1)}{2}$.

Proof. Since all Borel subgroups are conjugated in $\widehat{G}$, it is enough to apply the spin representation and to find a D_p-stable flag in $\mathbb{I}^{2^n}$ with action of I_p on the graded pieces given by the mutually distinct characters $\prod_{i\in I}\epsilon(\sigma)^{-i}t_i^{\ell(\sigma)}$ for all subset Is of $\{1,\dots,n\}$. Over $\mathbb{J}=\prod_{\mathfrak{P}'}\mathcal{O}_{\mathfrak{P}'}$ as in the previous proof, we have such a flag (in $\mathbb{J}^{2^n}$). If we assume $p-1>w$ so that all the characters above take values in $\mathbb{I}$ and are mutually distinct modulo p, we see that the intersection of this flag with $\mathbb{I}^{2^n}$ induces a flag as desired. □

9.4.3 Λ_n-fullness

In the sequel, we consider an irreducible component $\mathbb{I}$ of $\mathbf{T}^M$ such that $\mathbb{I}=\Lambda_n=\mathbb{Z}_p[[X_1,\dots X_n]]$ (with $p>2$). Let $u=1+p$. Let $\mathbb{Q}_p^{\mathrm{cyc}}$ be the p-power-cyclotomic extension of $\mathbb{Q}_p$. For any weight $\underline{k}=(k_1,\dots,k_n)$

dominant for (G, B, T), and cohomological, we denote by $P_{\underline{k}}$ the point of Spec Λ_n given by $P_{\underline{k}} = (1 + X_1 - u^{k_1}, \ldots, 1 + X_n - u^{k_n})$. the point

We define for any $\sigma \in \mathrm{Gal}(\overline{\mathbb{Q}}/\mathbb{Q})$ the cyclotomic logarithm $\ell(\sigma) = \frac{\log_p(\epsilon(\sigma))}{\log_p(1+p)}$ (for the p-adic cyclotomic character ϵ). We put, as above, $t_i = 1+X_i$ $(i = 1, \ldots, n)$, and we assume that the semi-simplification of ρ restricted to the p-inertia group I_p is

$$[\epsilon(\sigma)^{-1} t_1^{\ell(\sigma)}, \epsilon(\sigma)^{-2} t_2^{\ell(\sigma)} \ldots, \epsilon(\sigma)^{-n} t_n^{\ell(\sigma)}; \epsilon(\sigma)^{-d} (t_1 \cdot \ldots \cdot t_n)^{\ell(\sigma)}],$$

where $d = \frac{n(n+1)}{2}$. This is conjecturally the case if ρ is associated to an n-variable p-adic Hida family of Siegel forms of genus n, passing by a form f of level prime to p and weight $\underline{k}^0 = (k_1^0, \ldots, k_n^0)$, with $k_1^0 \geq \ldots \geq k_n^0 \geq n+1$. It is actually proven if $n = 2$.

We consider the Galois representation $\rho = \rho_\theta : \mathrm{Gal}(\overline{\mathbb{Q}}/\mathbb{Q}) \to \mathrm{GSpin}_{2n+1}(\Lambda_n)$ associated to the family $\theta\colon \mathbf{T}^{M,\mathrm{holcusp}} \to \mathbb{I} = \Lambda_n$; the representation is ρ ordinary at p; After conjugation, we can and will assume in the sequel that $\rho(D_p) \subset \widehat{B}(\Lambda_n)$. For any root α of $(\widehat{G}, \widehat{T})$, we denote by U_α the one-dimensional Lie subgroup of Spin_{2n+1}, by $u_\alpha\colon \mathbf{G}_a \to U_\alpha$ the corresponding one-parameter subgroup and by $\mathfrak{u}_\alpha = \mathrm{Lie}(U_\alpha) \subset \mathfrak{spin}_{2n+1}$ the corresponding Lie subalgebra (say over $\mathbb{Z}_p$ or any base change thereof). Because of these notations, it will be convenient to write U instead of $\widehat{N}$ for the unipotent radical of the standard Borel subgroup $\widehat{B} = \widehat{T}\widehat{N}$. We have the following key lemma

Lemma 9.14. *Let $\rho : \mathrm{Gal}(\overline{\mathbb{Q}}/\mathbb{Q}) \to \mathrm{GSpin}_{2n+1}(\Lambda_n)$ be the Galois representation associated to a family $\theta\colon \mathbf{T}^{M,holcusp} \to \mathbb{I} = \Lambda_n$. Suppose*

(1) *there is a cohomological weight k^0 for which the specialization $\rho_{P_{k^0}} = \rho_f$ has $\mathbb{Z}_p$-full image; in other words,* $\mathrm{Im}(\rho_{P_{k^0}})$ *contains an open subgroup of* $\mathrm{Spin}_{2n+1}(\mathbb{Z}_p)$,
(2) strong regularity, (compare to 9.3.5) $\mathrm{Im}(\rho)$ *contains an element $\delta \in \widehat{T}(\Lambda_n)$ such that for any two roots $\alpha \neq \alpha'$ of* SO_{2n+1}, $\alpha(\mathrm{Ad}(\delta)) \not\equiv \alpha'(\mathrm{Ad}(\delta)) \pmod{\mathfrak{m}_{\Lambda_n}}$,

Then for any root α of $(\widehat{G}, \widehat{T})$, $\mathrm{Im}\,\rho \cap U_\alpha(\Lambda_n) \neq 0$

Remarks. 1) Note that the roots of $(\widehat{G} = \mathrm{GSpin}_{2n+1}, \widehat{T})$ are the same as those of $(\mathrm{SO}_{2n+1}, T_{\mathrm{SO}_{2n+1}})$.

2) If instead of Assumption 2 we only assume that there exists an element $\delta \in \widehat{B}(\Lambda_n)$ such that its projection $\delta_1 \in \widehat{T}(\Lambda_n)$ is strongly regular mod. $\mathfrak{m}_{\Lambda_n}$, by which we mean that for any two roots $\alpha \neq \alpha'$ of SO_{2n+1},

$\alpha(\mathrm{Ad}(\delta_1)) \not\equiv \alpha'(\mathrm{Ad}(\delta_1)) \pmod{\mathfrak{m}_{\Lambda_n}}$, then we can modify it by conjugations by elements $u_\alpha(x)$ for positive roots α roots and suitable elements x so that it belongs to $\widehat{T}(\Lambda_n)$ and it satisfies the exact statement of Assumption 2.

3) If we assume furthermore that $p-1 > 2w$ and $\underline{k}^0 = (k_1^0, \dots, k_n^0)$ is such that for any $i \neq j$, $k_i^0 - i \not\equiv \pm(k_j^0 - j) \pmod{p-1}$, we see by Lemma 4.4 above that if we choose $\sigma \in I_p$ such that $\kappa(\sigma) = \omega(\sigma)$ is of exact order $p-1$, we can take $\delta = \rho(\sigma)$.

Proof. Since the argument is the same for positive and negative roots, we restrict ourselves to the "upper triangular" unipotent radical U (corresponding to positive roots α) of $\mathrm{Spin}(2n+1)$. Let $K = \mathrm{Im}\,\rho \subset \widehat{G}(\Lambda_n)$. We write $P = P_{\underline{k}^0}$ for the arithmetic point such that the specialization ρ_P of the representation ρ is $\mathbb{Z}_p$-full. Fix a positive root α and define

$$\begin{aligned}\Gamma_{U_\alpha}(P^j) &= \{x \in \mathrm{Spin}_{2n+1}(\Lambda_n) | x \mod P^j \in U_\alpha(\Lambda_n/P^j)\}\\ \Gamma_{B_\alpha}(P^j) &= \{x \in \mathrm{Spin}_{2n+1}(\Lambda_n) | x \mod P^j \in \widehat{T}(\Lambda_n/P^j)U_\alpha(\Lambda_n/P^j)\}\end{aligned} \tag{9.2}$$

and

$$\begin{aligned}K_{U_\alpha}(P^j) &= K \cap \Gamma_{U_\alpha}(P^j)\\ K_{B_\alpha}(P^j) &= K \cap \Gamma_{B_\alpha}(P^j)\end{aligned} \tag{9.3}$$

Since $U(\Lambda_n)$ and $\Gamma_{\Lambda_n}(P)$ are p-profinite, the groups $\Gamma_{U_\alpha}(P^j)$ and $K_{U_\alpha}(P^j)$ for all $j \geq 1$ are also p-profinite. Recall

$$\Gamma_{\Lambda_n}(P^i) = \{x \in \mathrm{Spin}_{2n+1}(\Lambda_n) | (x \mod P^i) = 1\}$$

is the kernel of the reduction morphism $\pi_j\colon \mathrm{Spin}_{2n+1}(\Lambda_n) \to \mathrm{Spin}_{2n+1}(\Lambda_n/P^j)$. Recall the embedding $\iota_\alpha : \mathrm{SL}(2) \to \mathrm{Spin}_{2n+1}$ associated to the root α (i.e. the upper (resp. lower) unipotent subgroup $U^+_{\mathrm{SL}(2)}$ (resp. $U^-_{\mathrm{SL}(2)}$) of $\mathrm{SL}(2)$ is sent to U_α (resp. $U_{-\alpha}$) by ι_α). Thus, we see

$$\Gamma_{U_\alpha}(P^j) = \Gamma_{\Lambda_n}(P^j)\iota_\alpha(U_{\mathrm{SL}(2)}(\Lambda_n)).$$

Note that

$$\left[\begin{pmatrix} a & b\\ c & -a\end{pmatrix}, \begin{pmatrix} e & f\\ g & -e\end{pmatrix}\right] = \begin{pmatrix} bg-cf & 2(af-be)\\ 2(ce-ag) & cf-bg\end{pmatrix}.$$

From this, we have

Lemma 9.15. *If $X, Y \in \mathfrak{sl}_2(\Lambda_n) \cap \begin{pmatrix} P^j & P^k\\ P^i & P^j\end{pmatrix}$ with $i \geq j \geq k$, $[X,Y] \in \begin{pmatrix} P^{i+k} & P^{j+k}\\ P^{i+j} & P^{i+k}\end{pmatrix}$.*

This tells us, for the topological commutator subgroup $D\Gamma_{U_\alpha}(P^j) := (\Gamma_{U_\alpha}(P^j), \Gamma_{U_\alpha}(P^j))$ that

$$D\Gamma_{U_\alpha}(P^j) \subset \Gamma_{B_\alpha}(P^{2j}) \cap \Gamma_{U_\alpha}(P^j). \tag{9.4}$$

Using now the $\mathbb{Z}_p$-regularity assumption (in Sect. 9.3.5), we consider an element $\delta \in \widehat{T}(\Lambda_n) \in K$, with distinct root values modulo $\mathfrak{m}_{\Lambda_n}$. Replacing δ by $\lim_{m\to\infty} \delta^{p^m}$, we may assume that $\delta \in \widehat{T}(\mathbb{Z}_p)$ has finite order a. The order a is prime to p (indeed it is a factor of $p-1$). Note that δ normalizes $K_{U_\alpha}(P^j)$ and $\Gamma_{B_\alpha}(P^j)$. Since H is p-profinite, $x \in H$ has unique a-th root inside H. We define

$$\Delta_\alpha(x) = \sqrt[a]{x \cdot (\delta x \delta^{-1})^{\alpha(\delta)^{-1}} \cdot (\delta^2 x \delta^{-2})^{\alpha(\delta)^{-2}} \cdots (\delta^{a-1} x \delta^{1-a})^{\alpha(\delta)^{1-a}}} \in H.$$

Lemma 9.16. *If $u \in \Gamma_{U_\alpha}(P^j)$ ($j \geq 1$), then $\Delta_\alpha^2(u) \in \Gamma_{U_\alpha}(P^{2j})$ and $\pi_j(\Delta_\alpha(u)) = \pi_j(u)$.*

Proof. If $u \in \Gamma_{U_\alpha}(P^j)$, we have $\pi_j(\Delta_\alpha(u)) = \pi_j(u)$ as Δ_α is the identity map on $U_\alpha(\Lambda_n/P^j)$. Let $D\Gamma_{U_\alpha}(P^j)$ be the topological commutator subgroup of $\Gamma_{U_\alpha}(P^j)$. Since Δ_α induces the projection of the $\mathbb{Z}_p$-module $\Gamma_{U_\alpha}(P^j)/D\Gamma_{U_\alpha}(P^j)$ onto its $\alpha(\delta)$-eigenspace for $\mathrm{Ad}(\delta)$, it is a projection onto $U_\alpha(\Lambda_n)D\Gamma_{U_\alpha}(P^j)/D\Gamma_{U_\alpha}(P^j)$. The fact that this is exactly the $\alpha(\delta)$-eigenspace comes from the Iwahori decomposition of $\Gamma_{U_\alpha}(P^j)$ which shows it is generated by the $U_\beta(P^j\Lambda_n)$ for all roots $\beta \neq \alpha$ and by $U_\alpha(\Lambda_n)$, hence a similar direct sum decomposition holds in the abelianization $\Gamma_{U_\alpha}(P^j)/D\Gamma_{U_\alpha}(P^j)$.

By (9.4) $D\Gamma_{U_\alpha}(P^j) \subset \Gamma_{B_\alpha}(P^{2j}) \cap \Gamma_{U_\alpha}(P^j)$. Since the $\alpha(\delta)$-eigenspace of $\Gamma_{U_\alpha}(P^j)/D\Gamma_{U_\alpha}(P^j)$ is inside $\Gamma_{B_\alpha}(P^{2j})$, Δ_α projects $u\Gamma_{U_\alpha}(P^j)$ to

$$\overline{\Delta}_\alpha(u) \in (\Gamma_{B_\alpha}(P^{2j}) \cap \Gamma_{U_\alpha}(P^j))/D\Gamma_{U_\alpha}(P^j).$$

In particular, $\Delta_\alpha(u) \in \Gamma_{B_\alpha}(P^{2j}) \cap \Gamma_{U_\alpha}(P^j)$. Again apply Δ_α. Since $\Gamma_{B_\alpha}(P^{2j})/\Gamma_{\Lambda_n}(P^{2j})$ is sent to to $\Gamma_{U_\alpha}(P^{2j})/\Gamma_{\Lambda_n}(P^{2j})$ by Δ_α, we get $\Delta_\alpha^2(u) \in \Gamma_{U_\alpha}(P^{2j})$ as desired. □

We can now prove the key Lemma 9.14. Pick $0 \neq \overline{u}_\alpha \in U_\alpha(A/P) \cap \mathrm{Im}(\rho_P)$. By $\mathbb{Z}_p$-fullness of ρ_P, $\overline{u}_\alpha$ exists. Since the reduction map $\mathrm{Im}(\rho) \to \mathrm{Im}(\rho_1)$ is surjective, we have $u_\alpha \in \mathrm{Im}(\rho)$ such that $u_\alpha \mod P = \overline{u}_\alpha$. Take $v_1 \in U_\alpha(\Lambda_n)$ such that $v_1 \mod P = \overline{u}_\alpha$. Such a v_1 exists as the reduction map $U_\alpha(\Lambda_n) \to U_\alpha(\Lambda_n/P)$ is onto. Then $u_\alpha v_1^{-1} \in \Gamma_{\Lambda_n}(P)$, we find that $u_\alpha \in K_{U_\alpha}(P)$.

By the compactness of $K_{U_\alpha}(P)$, by Lemma 9.16, starting with u_α as above, $\lim_{m\to\infty} \Delta_\alpha^m(u_\alpha)$ converges P-adically to $\Delta_\alpha^\infty(u_\alpha) \in U_\alpha(\Lambda_n) \cap \mathrm{Im}(\rho)$ with

$\Delta_\alpha^\infty(u_\alpha)\Gamma_{\Lambda_n}^{\mathrm{Spin}(2n+1)}(P) = u_\alpha\Gamma_{\Lambda_n}^{\mathrm{Spin}(2n+1)}(P)$ (i.e., $\overline{u}_\alpha = (\Delta_\alpha^\infty(u_\alpha) \mod P)$). Thus $\Delta_\alpha^\infty(u_\alpha)$ is a non-trivial unipotent element in $U_\alpha(\Lambda_n) \cap \mathrm{Im}(\rho)$. □

From this lemma we can deduce the following main theorem (we still assume that $\rho(D_p) \subset \widehat{B}(\Lambda_n)$).

Theorem 9.17. *Let* $\rho : \mathrm{Gal}(\overline{\mathbb{Q}}/\mathbb{Q}) \to \mathrm{GSpin}_{2n+1}(\Lambda_n)$ *be a continuous representation. Suppose*

(1) there is a cohomological weight k^0 *for which the specialization* $\rho_{P_{k^0}} = \rho_f$ *has* $\mathbb{Z}_p$*-full image; in other words,* $\mathrm{Im}(\rho_{P_{k^0}})$ *contains an open subgroup of* $\mathrm{Spin}_{2n+1}(\mathbb{Z}_p)$,

(2) $\mathrm{Im}(\rho)$ *contains an element* $\delta \in \widehat{B}(\Lambda_n)$*, whose projection* $\delta_1 \in \widehat{T}(\Lambda_n)$ *is* $\mathfrak{spin}_{2n+1}$*-regular mod.* $\mathfrak{m}_{\Lambda_n}$*, that means that for any two roots* $\alpha \neq \alpha'$ *of* SO_{2n+1}, $\alpha(\mathrm{Ad}(\delta_1)) \not\equiv \alpha'(\mathrm{Ad}(\delta_1)) \pmod{\mathfrak{m}_{\Lambda_n}}$,

Then $\mathrm{Im}\ \rho$ *contains a principal congruence subgroup of* $\mathrm{Spin}_{2n+1}(\Lambda_n)$

Definition 9.18. The largest ideal of Λ_n such that $\Gamma_{\Lambda_n}(\mathfrak{l}_\theta) \subset \mathrm{Im}\ \rho_\theta$ is called the Galois level of θ, and is denoted by $\mathfrak{l}_\theta$.

There are two proofs. It can either be deduced from Pink's theory using Lemma 9.6, as the only missing point was the non-triviality of $U_\alpha(\Lambda_n)\cap\mathrm{Im}(\rho)$ for all αs (which follows from Lemma 9.14), or it can be proven using Lie algebras as follows.

Let $H = \mathrm{Im}\ \rho\cap\Gamma_{\Lambda_n}(p)$. We have seen that for any root α, $\mathrm{Im}\ \rho\cap U_\alpha(\Lambda_n) \neq \{1\}$; therefore, by taking possibly a p-th power of the non-trivial element in this intersection, we can assume that for any root α, $H \cap U_\alpha(\Lambda_n) \neq \{1\}$. The series defining Log still provides a bijection between H and the set $\mathrm{Log}\ H$ which is contained in $p \cdot \mathfrak{spin}_{2n+1}(\Lambda_n)$, but may not be stable by addition; however, the Log map induces group isomorphisms from $H \cap U_\alpha(\Lambda_n)$ to the (possibly infinite dimensional) $\mathbb{Z}_p$-Lie algebras $\mathrm{Log}(H) \cap \mathfrak{u}_\alpha(\Lambda)$, which are therefore non-trivial $\mathbb{Z}_p$-modules for all roots α. Recall that the $\mathbb{Q}_p \cdot \mathrm{Log}(H)$ is a $\mathbb{Q}_p$-Lie algebra, denoted $\mathrm{Log}_{\mathbb{Q}_p} H$ below.

Using the adjoint action of δ, we have a decomposition into root spaces

$$(*) \quad \mathrm{Log}_{\mathbb{Q}_p} H = (\mathrm{Log}_{\mathbb{Q}_p} H)_0 \oplus \bigoplus_{\alpha\in\Phi} (\mathrm{Log}_{\mathbb{Q}_p} H)_\alpha$$

compatible with the root decomposition

$$\mathfrak{spin}_{2n+1} = \mathrm{Lie}(\widehat{T}) \oplus \bigoplus_{\alpha} \mathfrak{u}_\alpha$$

which is valid over Λ_n, hence also over $\Lambda_n[\frac{1}{p}]$

Let $[t_1, \dots, t_n] = \prod_i [e_i - \frac{1}{2} f](t_i)$ denote the parametrization of (the semi-simple part of) the standard torus $\mathcal{T}' = \widehat{T}$ of $\widehat{G} = \mathrm{GSpin}_{2n+1}$ via $\widehat{T} = \mathbb{G}_m \otimes X^*(T)$, where $e_i \colon (x_1, \dots, x_n, \nu x_n^{-1}, \dots, \nu x_1^{-1}) \mapsto x_i$ and $f \colon (x_1, \dots, x_n, \nu x_n^{-1}, \dots, \nu x_1^{-1}) \mapsto \nu$. For any $\sigma \in I_p$, we put

$$t(\sigma) = [(1+T_1)^{\ell(\sigma)}, \dots, (1+T_n)^{\ell(\sigma)}]$$

We know that for any $\sigma \in I_p$,

$$\rho'(\sigma) \equiv t(\sigma) \pmod{U(\Lambda_n)}$$

The simple roots of $\widehat{G}$ (for the standard "upper triangular" Borel) are $\alpha_i \colon [t_1, \dots, t_n] \mapsto t_i/t_{i+1}$ $(i = 1, \dots, n-1)$ and $\alpha_n \colon [t_1, \dots, t_n] \mapsto t_n$. The positive roots are then t_i/t_j for $1 \le i < j \le n$, t_i, $i = 1, \dots, n$, and , $t_i t_j$ for $1 \le i < j \le n$; they can be written, respectively, as $\alpha_i + \ldots + \alpha_{j-1}$, $\alpha_i + \ldots + \alpha_n$, and $\alpha_i + \ldots + \alpha_{j-1} + 2(\alpha_j + \ldots + \alpha_n)$. Conjugation by $\rho(\sigma)$ acts on $\mathrm{Log}_{\mathbb{Q}_p} H_\alpha = \mathrm{Log}_{\mathbb{Q}_p} H \cap \mathfrak{u}_\alpha(\Lambda_n)$ by $\alpha(t(\sigma))$; one can choose $\sigma \in I_p$ such that $\ell(\sigma) = 1$. We find that we can let $\frac{1+X_i}{1+X_j}$, resp. $1+X_i$, resp. $(1+X_i)(1+X_j)$ act on $(\mathrm{Log}_{\mathbb{Q}_p} H)_{t_i/t_j}$ resp. $\mathrm{Log}_{\mathbb{Q}_p} H_{t_i}$, resp. $(\mathrm{Log}_{\mathbb{Q}_p} H)_{t_i t_j}$. We have an action of $1 + X_n$ on the non-zero subgroup $(\mathrm{Log}_{\mathbb{Q}_p} H)_{\alpha_n}$ and of $\frac{1+X_{n-1}}{1+X_n}$ on the non-zero subgroup $(\mathrm{Log}_{\mathbb{Q}_p} H)_{\alpha_{n-1}}$; by forming the Poisson bracket of these two submodules, we find a non-zero modules of $(\mathrm{Log}_{\mathbb{Q}_p} H)_{\alpha_{n-1}+\alpha_n}$ on which both $(1+X_n)$ and $\frac{1+X_{n-1}}{1+X_n}$ act; therefore $(1+X_n)$ and $1+X_{n-1}$ both act on this non-zero subgroup of $(\mathrm{Log}_{\mathbb{Q}_p} H)_{\alpha_{n-1}+\alpha_n}$. By repeating this procedure, we find, for $\alpha = \alpha_1 + \ldots + \alpha_n$, a non-zero submodule of $(\mathrm{Log}_{\mathbb{Q}_p} H)_\alpha$ on which all $1+X_i$s act. This is a non-zero Λ_n-submodule of $U_\alpha(\Lambda_n)$, contained in $(\mathrm{Log}\, H)_\alpha$. We then use successive Poisson brackets with $(\mathrm{Log}_{\mathbb{Q}_p} H)_{-\alpha_1}, \dots, (\mathrm{Log}_{\mathbb{Q}_p} H)_{-\alpha_j}$ to construct inside all $(\mathrm{Log}_{\mathbb{Q}_p} H)_{\alpha_i}$'s a non-trivial Λ_n-submodule of $\mathfrak{u}_{\alpha_i}(\Lambda_n)$. Once we have non-zero Λ_n-submodules in $(\mathrm{Log}_{\mathbb{Q}_p} H)_{\alpha_i}$ for all the simple roots, one can construct non-trivial Λ_n-submodules in $(\mathrm{Log}_{\mathbb{Q}_p} H)_\alpha$ for all the roots α. After multiplying by a power of p, we obtain non-trivial Λ_n-submodules in $(\mathrm{Log}_{\mathbb{Q}_p} H)_\alpha$ for all roots. Taking the exponential, we see that there exists a non-zero ideal $\mathfrak{l}'$ of Λ_n such that $U_\alpha(\mathfrak{l}') \subset \mathrm{Im}\,\rho$. Since these subgroups generate a group containing a principal congruence subgroup of $\mathrm{Spin}_{2n+1}(\Lambda_n)$, we see that there exists a non-zero ideal $\mathfrak{l}$ of Λ_n such that $\Gamma_{\Lambda_n}(\mathfrak{l}) \subset \mathrm{Im}\,\rho$ as desired. □

We give a lemma which follows from the proof above. For any simple root α_i of $\mathfrak{spin}_{2n+1}$, let $\overline{\mathfrak{u}}_{\alpha_i} = \mathfrak{u}_{\alpha_i}(\Lambda_n[\frac{1}{p}]) \cap \mathrm{Lie}_{\mathbb{Q}_p} \mathrm{Im}\,\rho_\theta$. For any characteristic zero quotient A of Λ_n, let ρ_A be the specialization of ρ_θ to A and let $\overline{\mathfrak{u}}^A_{\alpha_i}$ be the image of $\overline{\mathfrak{u}}_{\alpha_i}$ in $\mathfrak{spin}_{2n+1}(A[\frac{1}{p}])$. Let $\overline{\mathfrak{u}}^{A,j}_{\alpha_\bullet}$ the sequence defined by $\overline{\mathfrak{u}}^{A,1}_{\alpha_\bullet} = \overline{\mathfrak{u}}^{A}_{\alpha_1}$, $\overline{\mathfrak{u}}^{A,2}_{\alpha_\bullet} = [\overline{\mathfrak{u}}^{A}_{\alpha_1}, \overline{\mathfrak{u}}^{A}_{\alpha_2}]$, and for $j \leq n-1$:

$$\overline{\mathfrak{u}}^{A,j+1}_{\alpha_\bullet} = [\overline{\mathfrak{u}}^{A,j}_{\alpha_\bullet}, \overline{\mathfrak{u}}^{A}_{\alpha_{j+1}}]$$

Note that by calculations in the proof of Theorem 9.17, $\overline{\mathfrak{u}}^{A,n}_{\alpha_\bullet}$ is an $A[\frac{1}{p}]$-module.

Lemma 9.19. *For any $n \geq 1$, assume ρ_θ is full over Λ_n. For any characteristic zero quotient A of Λ_n, there exists a unique Lie subalgebra Lie_A of the $\mathbb{Q}_p$-Lie algebra $Lie_{\mathbb{Q}_p}$ Im ρ_A which is an $A \otimes \mathbb{Q}_p$-Lie algebra, such that $Lie_A \cap \overline{\mathfrak{u}}^{A}_{\sum_i \alpha_i} = \overline{\mathfrak{u}}^{A,n}_{\alpha_\bullet}$ and which is stable by the adjoint action of* Im ρ_A. *If $\Lambda_n \to A \to B$ are two characteristic zero quotients of Λ_n, then $Lie_A \to Lie_B$ is surjective.*

Note however that for some quotients A, this algebra may be zero because $\overline{\mathfrak{u}}^{A,n}_{\alpha_\bullet}$ may be zero, although it is not the case over Λ_n. This is why we'll need a modified construction later (see Lemma 9.24).

Proof. To simplify notations, we prove it for $\mathfrak{sp}_4$. Let $\alpha_1 = t_1 t_2^{-1}$ and $\alpha_2 = t_2^2$ be the two simple roots. We form $\overline{\mathfrak{u}}^{A,2}_{\alpha_\bullet} = [\overline{\mathfrak{u}}^{A}_{\alpha_1}, \overline{\mathfrak{u}}^{A}_{\alpha_2}]$ which is an A-module by calculations at the end of the proof of Theorem 9.17. If it is non-zero, by Poisson bracket of $\overline{\mathfrak{u}}^{A,2}_{\alpha_\bullet}$ with all the $\overline{\mathfrak{u}}^{A}_{\gamma}$s for all the root (positive or negative) of $\mathfrak{sp}_4$, we construct $A[\frac{1}{p}]$-submodules $\overline{\mathfrak{u}}^{A,\prime}_{\alpha_1+\alpha_2+\gamma}$ of $\overline{\mathfrak{u}}^{A}_{\alpha_1+\alpha_2+\gamma}$. Note that in particular, $\overline{\mathfrak{u}}^{A,\prime}_{\alpha_1+\alpha_2} = \overline{\mathfrak{u}}^{A}_{\alpha_1+\alpha_2}$. We also define a Cartan subalgebra $\overline{\mathfrak{t}}^{A,\prime}$ by $[\overline{\mathfrak{u}}^{A,\prime}_{\alpha_1}, \overline{\mathfrak{u}}^{A,\prime}_{-\alpha_1}] \oplus [\overline{\mathfrak{u}}^{A,\prime}_{\alpha_2}, \overline{\mathfrak{u}}^{A,\prime}_{-\alpha_2}]$, and we then define an $A \otimes \mathbb{Q}_p$-module by

$$\mathrm{Lie}_A = \overline{\mathfrak{t}}^{A,\prime} \oplus \bigoplus_{\gamma} \overline{\mathfrak{u}}^{A,\prime}_{\gamma}$$

Using the Jacobi identity for the Poisson bracket, one sees easily that it is an $A \otimes \mathbb{Q}_p$-Lie algebra which has all the desired properties. □

It follows from the construction above that if ρ_θ is full over Λ_n; then for all roots γ, $\overline{\mathfrak{u}}'_\gamma$ which is the $\gamma(\delta)$-eigenspace for the action of δ as in Theorem 9.17) on Lie_{Λ_n}, is non-zero. More precisely,

Corollary 9.20. *If ρ_θ is full over Λ_n, there exists a non-zero largest $\Lambda_n[\frac{1}{p}]$-ideal $\mathfrak{l}'_\theta$ such that*

$$\mathfrak{l}'_\theta \cdot \mathfrak{sp}_4(\Lambda_n \otimes \mathbb{Q}_p) \subset Lie_{\Lambda_n}$$

We have $\mathfrak{l}'_\theta \subset \mathfrak{l}_\theta \otimes \mathbb{Q}_p$ and these two ideals have the same radical: $\sqrt{\mathfrak{l}'_\theta} = \sqrt{\mathfrak{l}_\theta \otimes \mathbb{Q}_p}$ in $\Lambda_n[\frac{1}{p}]$.

9.5 Galois level and congruence ideals

9.5.1 Congruence ideals

Let $\theta\colon \mathbf{T}^{M,\mathrm{holcusp}} \to \mathbb{I}_0$ be an irreducible component of Spec $\mathbf{T}^{M,\mathrm{holcusp}}$; we denote by $\mathbb{I}$ the normalization of $\mathbb{I}_0$ and by $\mathcal{K}$ its field of fractions; let $\mathbf{T}_{\mathbb{I}} = \mathbf{T}^{M,\mathrm{holcusp}} \otimes_{\Lambda_n} \mathbb{I}$ and $\theta_{\mathbb{I}} = m_{\mathbb{I}} \circ \theta \otimes \mathrm{Id}_{\mathbb{I}}$ the composition of the extension of scalars with the multiplication on $\mathbb{I}$; by the diagonalisability of $\mathbf{T}^{M,\mathrm{holcusp}}$ over $\mathcal{K}$, the character $\theta_{\mathbb{I}}\colon \mathbf{T}_{\mathbb{I}} \to \mathbb{I}$ splits over $\mathcal{K}$, and there is an isomorphism of $\mathcal{K}$-algebras:

$$\mathbf{T}_{\mathbb{I}} \otimes_{\mathbb{I}} \mathcal{K} \cong \mathcal{K} \times \mathbf{T}'_{\mathcal{K}}$$

with the first projection pr_1 given by $\theta_{\mathcal{K}} = \theta_{\mathbb{I}} \otimes \mathcal{K}$.

(Mult 1) We assume that $\theta_{\mathbb{I}}$ does not factor through pr_2. In other words, the family $\theta_{\mathbb{I}}$ occurs with multiplicity 1 in $\mathbf{T}_{\mathbb{I}}$.

If $n = 1$, this assumption amounts to saying that the family $\theta_{\mathbb{I}}$ is M-new (i.e. all the forms occurring in the family are M-new); if $n = 2$, using the Roberts-Schmidt theory of newforms, and assuming a folklore conjecture which says that automorphic forms on GSp_4 which are neither CAP nor endoscopic satisfy multiplicity one, one can prove that it holds for M-new families of squarefree level M. For $n \geq 3$, we simply assume (Mult 1).

Let $\mathbf{T}'$ be the image of $\mathbf{T}_{\mathbb{I}}$ by the second projection pr_2. We define then $\mathfrak{c}_1 = \mathbf{T}_{\mathbb{I}} \cap (\mathbb{I} \times \{0\})$, which is an ideal of $\mathbf{T}_{\mathbb{I}}$ and of $\mathbb{I} \times \{0\}$, and $\mathfrak{c}_2 = \mathbf{T}_{\mathbb{I}} \cap (\{0\} \times \mathbf{T}')$ which is an ideal of $\mathbf{T}_{\mathbb{I}}$ and of $\times \mathbf{T}' \times \{0\}$. Note that $\mathfrak{c}_1 = \mathrm{Ker}\,\mathrm{pr}_2$ and $\mathfrak{c}_2 = \mathrm{Ker}\,\mathrm{pr}_1$, and that for $i = 1, 2$, one can identify $\mathfrak{c}_i$ with its image $\mathrm{pr}_i(\mathfrak{c}_i)$. We also define $\mathfrak{c} = \mathfrak{c}_1 + \mathfrak{c}_2$. The projections pr_i induce isomorphisms of $\mathbb{I}$-modules

$$\overline{\mathrm{pr}}_1\colon \mathbf{T}_{\mathbb{I}}/\mathfrak{c} \to \mathbb{I}/\mathfrak{c}_1 \text{ and } \quad \overline{\mathrm{pr}}_2\colon \mathbf{T}_{\mathbb{I}}/\mathfrak{c} \to \mathbf{T}'/\mathfrak{c}_2$$

We thus obtain the congruence isomorphism $\overline{\mathrm{pr}}_2 \circ \overline{\mathrm{pr}}_1^{-1}\colon \mathbb{I}/\mathfrak{c}_1 \cong \mathbf{T}'/\mathfrak{c}_2$.

Definition 9.21. The ideal $\mathfrak{c}_\theta = \mathfrak{c}_1$ is called the congruence ideal associated to the family θ.

For any prime P of $\mathbb{I}$ containing $\mathfrak{c}_\theta$, there exists another family $\theta'\colon \mathbf{T}_{\mathbb{I}} \to \mathbb{J}$ (for some finite normal extension $\mathbb{J}$ of $\mathbb{I}$) and a prime Q of $\mathbb{J}$ above P such that $\theta_{\mathbb{I}} \cong \theta' \pmod{Q}$.

Let $\theta_2\colon \mathbf{T}^{M,\mathrm{holcusp}} \to \mathbb{J}$ be a second family, different from θ, hence it factors through pr_2. Let $\mathfrak{c}_{\theta_2} \subset \mathbb{J}$ be its congruence ideal; if Q is a prime of $\mathbb{J}$ containing both the ideals $\mathfrak{c}_\theta$ and $\mathfrak{c}_{\theta_2}$, and if $P = \mathbb{I} \cap Q$, the congruence isomorphism induces a congruence injection $\mathbb{I}/P \cong \mathbb{J}/Q$ which expresses a congruence between the two specific families θ and θ_2 modulo Q. Therefore the intersection locus $V(\mathfrak{c}_\theta + \mathfrak{c}_{\theta_2})$ in Spec $\mathbf{T}^{M,\mathrm{holcusp}}$ of the (normalizations of the) two irreducible components Spec $\mathbb{I}$ and Spec $\mathbb{J}$ is also the congruence locus of the two families θ and θ_2. Note that this locus is empty unless both families factor through the same local factor of the semilocal algebra $\mathbf{T}^{M,\mathrm{holcusp}}$; in other words a necessary condition for this locus to be non empty is that the residual representations $\overline{\rho}_\theta$ and $\overline{\rho}_{\theta_2}$ coincide.

9.5.2 Comparison between Galois level and congruence ideals

In this section we assume again that the family $\theta\colon \mathbf{T}^{M,\mathrm{holcusp}} \to \mathbb{I}$ is Λ_n-valued, i.e. $\mathbb{I} = \Lambda_n$. We also assume that its residual Galois representation $\overline{\rho}_\theta$ is irreducible and that $\rho = \rho_\theta$ satisfies the assumptions of Theorem 9.17: there is an arithmetic specialization for which Im $\rho_{P_{k0}}$ is full and that there is a $\mathbb{Z}_p$-regular element δ in Im ρ. By Theorem 9.17, the representation ρ_θ is full; we denote by $\mathfrak{l}_\theta$ the Galois level of θ, that is, the largest ideal of Λ_n such that $\Gamma_{\Lambda_n}(\mathfrak{l}_\theta) \subset \mathrm{Im}\ \rho_\theta$.

Let us assume that there exists another family θ_2 in the same local component of $\mathbf{T}^{M,\mathrm{holcusp}}$ as θ (so that its residual representation coincides with that of θ, in particular it is irreducible) but such that ρ_{θ_2} is not full. Let $\mathfrak{c}(\theta,\theta_2) = \mathfrak{c}_\theta \cdot \mathbb{J} + \mathfrak{c}_{\theta_2}$ be the ideal of $\mathbb{J}$ defining the congruence locus between θ and θ_2 and let $c(\theta) = \bigcap_{\theta_2} \mathfrak{c}(\theta,\theta_2) \cap \Lambda_n$, where the intersection of the $\mathfrak{c}(\theta,\theta_2)$s is taken over all the families θ_2 congruent to θ but not full. Let Q be a prime ideal of $\mathbb{J}$ containing $\mathfrak{c}(\theta,\theta_2)$, and let $P = Q \cap \Lambda_n$. Then, we have a congruence $\rho_\theta \equiv \rho_{\theta_2} \pmod{Q}$; then $\rho_\theta \pmod{P}$ is not full; hence we must have $\mathfrak{l}_\theta \subset Q$. Since for any prime P of Λ_n containing $c(\theta)$ there exists a prime Q above P in a finite extension which contains $\bigcap_{\theta_2} \mathfrak{c}(\theta,\theta_2)$, we see that any prime ideal P containing $c(\theta)$ contains $\mathfrak{l}_\theta$. For any ideal $\mathfrak{a}$ of Λ_n, let $V(\mathfrak{a})$ be the set of prime ideals containing $\mathfrak{a}$. Recall that in the case $n = 1$, Hida has proved in [H13b] that if the family θ is not CM but there is another family θ_2 which is CM for a given imaginary quadratic field, then the sets $V(\mathfrak{l}_\theta)$ and $V(c(\theta))$ coincide; actually under slightly stronger assumptions (see [H13b, Th.7.2]), he could use Pink's Lie algebra theory for SL_2 to compare the exponents of the height one primes occurring in (the reflexive envelopes of) the two ideals $\mathfrak{l}_\theta$ and $c(\theta)$, see [H13b, Th.8.5.].

For general n, a natural question is to ask whether the converse is true, namely that if a prime P divides $\mathfrak{l}_\theta$, there exists a non full family θ_2 congruent to θ at a prime Q above P. Assume again $n = 2$, recall that given a real quadratic field F of discriminant D, and a Hilbert modular cusp form for F of level $\mathfrak{n}$ and weight (m_1, m_σ) with $m_1 = m_\sigma + 2r$ with $r \geq 1$ and

$m_\sigma \geq 2$, eigen for the Hecke operators and p-ordinary, one can define its theta lift to GSp_4 (see [Y79] and [R01, Th.8.6]), which we call its twisted Yoshida lift. It gives rise to an eigensystem of $\mathbf{T}^{M,\mathrm{holcusp}}$ with $M = N(\mathfrak{n})D^2$ of weight (k_1, k_2) with $k_1 = m_\sigma + r$ and $k_2 = r + 2$, (so, $k_1 \geq k_2 \geq 3$), see for instance [MT02, Sect. 7.3]. Assuming moreover that $\mathfrak{n}$ is prime to p and that g is p-ordinary, there exists a unique two-variable p-adic family of p-ordinary Hilbert modular forms $\mathbf{g}$ of auxiliary level $\mathfrak{n}$ passing through g (after p-stabilization). It defines a homomorphism of Λ_2-algebras $\theta_{\mathbf{g}} : \mathbf{T}_F^{\mathfrak{n}} \to \mathbb{J}$ of the ordinary Hida Hecke algebra $\mathbf{T}_F^{\mathfrak{n}}$ for Hilbert cusp forms over F, of auxiliary level $\mathfrak{n}$. For any place w of F prime to $\mathfrak{n}p$, this homomorphism sends the Hecke operator T_w to the eigenvalue $\theta_{\mathbf{g}}(T_w)$ on $\mathbf{g}$. The map $(m_1, m_\sigma) \mapsto (k_1, k_2)$ defines an automorphism of Λ_2, denoted by τ.

Lemma 9.22. *Given $(F, \mathbf{g})$ of a real quadratic field and a two-variable family of p-ordinary cusp forms $\theta_{\mathbf{g}} : \mathbf{T}_F^{\mathfrak{n}} \to \mathbb{J}$, then, for $M = N(\mathfrak{n})D^2$, there exists a two-variable p-adic family $\theta_2 \colon \mathbf{T}^{M,holcusp} \to \mathbb{J}^\tau$, which is a homomorphism of Λ_2-algebras via τ: $\theta_2(\alpha \cdot T) = \tau(\alpha) \cdot \theta_2(T)$, such that for every prime q not dividing Mp,*

$$\theta_2(T_{q,1}) = \lambda_q(\mathbf{g}) \text{ and } \quad \theta_2(T_{q,2}) = \mu_q(\mathbf{g})$$

where the Hecke eigenvalues $\lambda_q(\mathbf{g})$ and $\mu_q(\mathbf{g})$ are defined as follows

- *If q splits in F, let w_1 and w_2 the places above q, then $\lambda_q(\mathbf{g}) = \theta_{\mathbf{g}}(T_{w_1}) + \theta_{\mathbf{g}}(T_{w_2})$ and $q \cdot \mu_q(\mathbf{g}) = q^2 + q\theta_{\mathbf{g}}(T_{w_1})\theta_{\mathbf{g}}(T_{w_2}) - 1$*
- *If q is inert, $\lambda_q(\mathbf{g}) = 0$ and $q \cdot \mu_q(\mathbf{g}) = -(q^2 + q\theta_{\mathbf{g}}(T_w) + 1)$*

We call $\theta_2 = \theta_2(F, \mathbf{g})$ the twisted Yoshida lift of $(F, \mathbf{g})$

The proof is by comparing for each prime q not dividing Mp the characteristic polynomial of $\mathrm{Ind}_{G_F}^{G_{\mathbb{Q}}} \rho_h$ for all classical Hilbert forms h in the family $\mathbf{g}$ and the GSp_4-Hecke polynomial at q of its Yoshida lift.

From now on, we fix a real quadratic field F of discriminant D, an auxiliary level $\mathfrak{n}$ in F prime to p and we pose $M = N(\mathfrak{n})D^2$; we give ourselves a Λ_2-valued family $\theta \colon \mathbf{T}^{M,\mathrm{holcusp}} \to \Lambda_2$ which satisfies the assumptions of Sect. 9.4.3, so that its Galois representation $\rho_\theta \colon G_{\mathbb{Q}} \to \mathrm{GSp}_4(\Lambda_2)$ is full.

In the following theorem, we assume there exists a Hilbert modular form g of weight $m_1 = m_\sigma + 2r$ with $r \geq 1$ (hence the form g does not descend to $\mathbb{Q}$) and $m_\sigma \geq 2$, such that

- $\overline{\rho}_g$ is "full": that is, $\mathrm{SL}_2(\mathbb{F}_p) \subset \mathrm{Im}\, \overline{\rho}_g$,
- there exists no character $\xi \colon G_{\mathbb{Q}} \to \overline{\mathbb{F}}_p^\times$ such that $\overline{\rho}_{g^\sigma} \cong \overline{\rho}_g \otimes \xi$.

We finally assume that $\overline{\rho}_\theta = \mathrm{Ind}_{G_F}^{G_\mathbb{Q}} \overline{\rho}_g$. By the previous lemma, there exists a "twisted Yoshida lift" family $\theta_2 = \theta_2(F, \mathbf{g})$ in the same local component of $\mathbf{T}^{M,\mathrm{holcusp}}$ as θ. Note that if there is another quadratic field F' and a Hilbert form g' on F' such that $\overline{\rho}_\theta = \mathrm{Ind}_{G_{F'}}^{G_\mathbb{Q}} \overline{\rho}_{g'}$ then $F' = F$. Indeed, if this equality holds and $F' \neq F$, we have an isomorphism of $G_\mathbb{Q}$-modules $\overline{\rho}_\theta \otimes \xi \cong \overline{\rho}_\theta$ where ξ is the (non-trivial) quadratic character associated to F'; hence, by irreducibility of the G_F-modules $\overline{\rho}_g$ and $\overline{\rho}_{g^\sigma}$, we have an isomorphism of G_F-modules $\overline{\rho}_g \otimes \xi \cong \overline{\rho}_g$ (which implies that $\overline{\rho}_g \cong \mathrm{Ind}_{FF'}^{F} \varphi$ for a character φ and is impossible by fullness of $\overline{\rho}_g$) or $\overline{\rho}_g \otimes \xi \cong \overline{\rho}_{g^\sigma}$ which is impossible by assumption.

Theorem 9.23. *For $n = 2$, assume that ρ_θ satisfies the assumptions of* Theorem 9.17, *hence is full, and that there is a Hilbert cusp form g for F as above such that $\overline{\rho}_\theta = \mathrm{Ind}_{G_F}^{G_\mathbb{Q}} \overline{\rho}_g$, then we have $V(c(\theta)[\frac{1}{p}]) = V(\mathfrak{l}_\theta[\frac{1}{p}])$.*

Note that we are not yet able to study the prime (p); more precisely, we can't prove that if $(p) \in V(\mathfrak{l}_\theta)$, then $(p) \in V(c(\theta))$. We need first a variant of 9.19 and 9.20, because in our case 9.19 applied to the quotient A of interest for us yields a trivial $A \otimes \mathbb{Q}_p$-Lie algebra. Since ρ_θ is congruent to $\mathrm{Ind}_{G_F}^{G_\mathbb{Q}} \rho_g$ and since its image contains the diagonal element δ with distinct eigenvalues, we see that the diedral group generated by δ and by the element w of the Weyl group which exchanges t_1 and t_2 in $\mathrm{diag}(t_1, t_2, t_2^{-1}, t_1^{-1})$ lifts to characteristic zero as a subgroup of Im ρ_θ (by a theorem of P. Hall, as the kernel of Im $\rho_\theta \to$ Im $\mathrm{Ind}_{G_F}^{G_\mathbb{Q}} \rho_g$ is pro-p); hence Im ρ_θ contains the element w of the Weyl group which exchanges t_1 and t_2. Let $\alpha_1 = t_1 t_2^{-1}$ and $\alpha_2 = t_2^2$ be the two simple roots. We recall the notations $\overline{\mathfrak{u}}_\gamma^A = \mathfrak{u}_\gamma(A) \cap \mathrm{Lie}_{\mathbb{Q}_p}$ Im ρ_A for all roots γs.

Lemma 9.24. *For any characteristic zero quotient A of Λ_2, there exists a Lie subalgebra Lie_A of the $\mathbb{Q}_p$-Lie algebra $Lie_{\mathbb{Q}_p}$* Im ρ_A *which is an $A \otimes \mathbb{Q}_p$-Lie algebra, which contains $\overline{\mathfrak{u}}_{\alpha_2}^A$ and which is stable by the adjoint action of* Im ρ_A. *If $\Lambda_2 \to A \to B$ are two characteristic zero quotients of Λ_2, then $Lie_A \to Lie_B$ is surjective.*

Proof. By action of the inertia as in the proof of Theorem 9.17, we see that $\overline{\mathfrak{u}}_{\alpha_2}$, resp. $\overline{\mathfrak{u}}_{2\alpha_1+\alpha_2}$, carries an action of $(1+T_1)(1+T_2)^{-1}$, resp. of $(1+T_1)(1+T_2)$; moreover $\overline{\mathfrak{u}}_{\alpha_2} = w(\overline{\mathfrak{u}}_{2\alpha_1+\alpha_2})$ hence $\overline{\mathfrak{u}}_{\alpha_2}$ carries a structure of A-module by transport of structure. We construct Lie_A as the Lie algebra generated by $\overline{\mathfrak{u}}_{\alpha_2}^A$, $\overline{\mathfrak{u}}_{2\alpha_1+\alpha_2}$ and their Poisson brackets by the $\overline{\mathfrak{u}}_{\alpha_2}^A$s for all the root (positive or negative) of $\mathfrak{sp}_4$. This defines the desired $A \otimes \mathbb{Q}_p$-Lie algebra. □

Again, it follows from the construction above that if ρ_θ is full over Λ_2; then for all roots γ, the $\gamma(\delta)$-eigenspace for the action of δ as in Theorem 9.17) on Lie_{Λ_2}, is non-zero. More precisely,

Corollary 9.25. *If ρ_θ is full over Λ_2, there exists a non-zero largest $\Lambda_n[\frac{1}{p}]$-ideal $\mathfrak{l}'_\theta$ such that*

$$\mathfrak{l}'_\theta \cdot \mathfrak{sp}_4(\Lambda_n[\frac{1}{p}]) \subset Lie_{\Lambda_2}$$

We have $\mathfrak{l}'_\theta \subset \mathfrak{l}_\theta \otimes \mathbb{Q}_p$ and these two ideals have the same radical: $\sqrt{\mathfrak{l}'_\theta} = \sqrt{\mathfrak{l}_\theta \otimes \mathbb{Q}_p}$.

Proof. Simply review the construction of Lie_{Λ_2} starting from $\mathfrak{u}_\gamma(\mathfrak{l}_\theta) \subset \overline{\mathfrak{u}}_\gamma$ for all γs, and take Poisson brackets. □

Proof of the theorem. Let us prove that if P is a characteristic zero prime containing $\mathfrak{l}_\theta$, then it contains $c(\theta)$. We can assume that P is an isolated component in the primary decomposition of the ideal $\mathfrak{l}'_\theta$ defined in the previous corollary. Let $\mathfrak{q}$ be the primary component of $\mathfrak{l}'_\theta$ whose radical is P. Let Λ_P be the localization of Λ_2 at P, $\kappa_P = \mathrm{Frac}(\Lambda_2/P)$ its residue field. Let $\widetilde{\Lambda} = \Lambda_2/\mathfrak{q}$ and $\widetilde{\Lambda}_P$ its localization at P; it is an artinian Λ_P-algebra with residue field κ_P.

Let $\widetilde{\rho}$ be the reduction of ρ_θ modulo $\mathfrak{q}$; it takes values in $G(\widetilde{\Lambda})$. We consider an ideal $\mathfrak{a}$ of Λ_2 containing $\mathfrak{q}$ and $\widetilde{\mathfrak{a}} = \mathfrak{a}/\mathfrak{q}$, such that $\widetilde{\mathfrak{a}}_P$ is one-dimensional over κ_P. Consider the $\widetilde{\Lambda} \otimes \mathbb{Q}_p$-Lie algebra $\mathrm{Lie}_{\widetilde{\Lambda}}$ of Lemma 9.24 in $\mathfrak{gsp}_4(\widetilde{\Lambda})$ and the Lie subalgebra $\mathfrak{s} = \widetilde{\mathfrak{a}} \cdot \mathfrak{sp}_4(\widetilde{\Lambda}) \cap \mathrm{Lie}_{\widetilde{\Lambda}}$ inside the Lie algebra $\widetilde{\mathfrak{a}} \cdot \mathfrak{sp}_4(\widetilde{\Lambda})$, after localization at P, the latter is isomorphic to $\mathfrak{sp}_4(\kappa_P)$. So we can identify $\mathfrak{s}_P$ to a κ_P-Lie subalgebra of $\mathfrak{sp}_4(\kappa_P)$. Let us prove that $\mathfrak{s}_P \neq \mathfrak{sp}_4(\kappa_P)$. Indeed if equality holds, we have $\widetilde{\mathfrak{a}} \cdot \mathfrak{sp}_4(\widetilde{\Lambda}) \subset \mathrm{Lie}_{\widetilde{\Lambda}}$. hence $\mathfrak{a} \cdot \mathfrak{sp}_4(\Lambda_2) \subset \mathrm{Lie}_{\Lambda_2} + \mathfrak{q} \cdot \mathfrak{sp}_4(\Lambda_2)$ so after localizing at P, noting that $\mathfrak{l}'_{\theta,P} = \mathfrak{q}_P$ and that $\mathfrak{l}'_\theta \cdot \mathfrak{sp}_4(\Lambda_2) \subset \mathrm{Lie}_{\Lambda_2}$, we have $\mathfrak{a}_P \cdot \mathfrak{sp}_4(\Lambda_P) \subset \mathrm{Lie}_{\Lambda_P}$. Consider the ideal $\mathfrak{b} = \mathfrak{l}'_\theta + \mathfrak{a}$, it is strictly larger than $\mathfrak{l}'_\theta$ and we have $\mathfrak{b} \cdot \mathfrak{sp}_4(\Lambda_2) \subset \mathrm{Lie}_{\Lambda_2}$; this contradicts the definition of $\mathfrak{l}'_\theta$.

In conclusion, $\mathfrak{s}_P$ is a strict κ_P-Lie subalgebra of $\mathfrak{sp}_4(\kappa_P)$. It is semi-simple because $\widetilde{\rho}$ is irreducible. Hence it is either 0, or an $\mathfrak{sl}_2$ in the Levi of a maximal parabolic, or $\mathrm{Sym}^3\,\mathfrak{sl}_2$, or the endoscopic $\mathfrak{sl}_2 \times \mathfrak{sl}_2$. Since it is normalized by Im ρ_P (recall ρ_P is the representation over Λ_2/P), Im ρ_P is contained in the corresponding normalizer. However the second and third cases are impossible. Indeed, for any root γ occurring in $\mathfrak{sl}_2 \times \mathfrak{sl}_2$, one can show using the projection Δ_γ as Lemma 9.14 and Lemma 9.16, that $U_\gamma(\widetilde{\Lambda}) \cap \mathrm{Im}\,\widetilde{\rho}$ maps surjectively to $U_\gamma(\mathbb{F}_p) \cap \mathrm{Im}\,\overline{\rho}$. Hence by the assumption of fullness of $\overline{\rho}_g$ and

$\overline{\rho}_{g^\sigma}$, these groups are non-trivial. For instance for $\gamma = \alpha_2$, by taking a power of p and taking the Log of a non-trivial element, we find a non-trivial element of $\overline{\mathfrak{u}}_{\alpha_2}^{\widetilde{\Lambda}} \subset \mathrm{Lie}_{\widetilde{\Lambda}}$, since both sides are $\widetilde{\Lambda}$-modules, one can assume that this non-zero element is in $\overline{\mathfrak{u}}_{\alpha_2}^{\widetilde{\Lambda}}(\widetilde{\mathfrak{a}}) \subset \mathrm{Lie}_{\widetilde{\Lambda}}$, hence $\mathfrak{s}_P \neq 0$. The only remaining case is the endoscopic $\mathfrak{sl}_2 \times \mathfrak{sl}_2$ and $\mathrm{Im}\,\rho_P \subset N(\mathrm{GL}_2 \times \mathrm{GL}_2)^\circ)$. Comparing with the reduction modulo the maximal ideal, we find a two-dimensional Galois representation of G_F over Λ_2/P congruent to $\overline{g}$ modulo the maximal ideal of Λ_2/P.

From this, one can deduce, using an $R_F = \mathbf{T}_F$ of Skinner-Wiles [SW01], that there exists a Hilbert family $\mathbf{g}$ such that the family $\theta_2 \colon \mathbf{T}^{M,\mathrm{holcusp}} \to \mathbb{J}$ associated to $(F, \mathbf{g})$ which is congruent to θ modulo a prime Q of $\mathbb{J}$ above P. □

Questions: 1) By this theorem, we can ask more precisely about the primary ideals (for minimal primes) occurring in $c(\theta)[\frac{1}{p}]$ and $\mathfrak{l}_\theta[\frac{1}{p}]$. Are they equal?

2) Note that for a Hilbert form g on a real quadratic field F, the L function of $\mathrm{Ad}\,\mathrm{Ind}_{G_F}^{G_\mathbb{Q}} \rho_g$ can be decomposed as

$$L(\mathrm{Ad}\,\mathrm{Ind}_{G_F}^{G_\mathbb{Q}} \rho_g, s) = L(\mathrm{Ind}_{G_F}^{G_\mathbb{Q}} \mathrm{Ad}\rho_g, s) L(R_g, s)$$

where R_g is the extension to $G_\mathbb{Q}$ of the representation $\rho_g \otimes \rho_{g^\sigma}^\vee$ of G_F appearing as a factor of $\mathrm{Ad}\,\mathrm{Ind}_{G_F}^{G_\mathbb{Q}} \rho_g$.

By a theorem of [Di05], the normalized special value at 1 of the first factor control congruences between the Hilbert form g and other Hilbert forms, hence a similar normalized special value at 1 of the second factor should control the congruences between the Yoshida lift of g and other Siegel forms which are not theta lifts of Hilbert forms. The p-adic L function interpolating the values of the first factor for g varying in a Hida family $\mathbf{g}$ has been defined by Hida. It might be possible to define the p-adic L function interpolating the second factor for g varying in a Hida family. Then one could ask how it is related to the determination of $\mathfrak{l}_\theta$ for a family θ congruent to $\Theta(\mathbf{g})$.

3) More generally, we hope to treat other cases of congruences between a "full" family and other families of automorphic forms with "small but irreducible Galois image", for GSp(4) and other groups.

References

[Ca13] F. Calegari, Even Galois representations and the Fontaine-Mazur conjecture, Invent math. (DOI 10.1007/s00222-010-0297-0)

[CaGe13] F. Calegari, T. Gee, Irreducibility of automorphic Galois representations of $\mathrm{GL}(n)$, n at most 5, to appear in Ann. Inst. Fourier

[Car94] H. Carayol, Formes modulaires et représentations galoisiennes à valeurs dans un anneau local complet, in "*p-adic monodromy and the Birch and Swinnerton-Dyer conjecture*", Contemp. Math. vol.165, AMS 1994, pp. 213-237

[Che51] C. Chevalley, Théorie des groupes de LIE, II et III, Publ. Inst. Math. Univ. Nancago, Paris, Hermann 1951 et 1955.

[Ch92] W.-C. Chi, l-adic and λ-adic representations associated to abelian varieties defined over number fields. Amer. J. Math. **114** (1992), 315–353.

[Di05] M. Dimitrov, On Ihara's Lemma for Hilbert modular varieties, Compositio Mathematica 145, Issue 05 (2009), 1114-1146.

[Em14] M. Emerton, Local-global compatibility in the p-adic Langlands programme for $\mathrm{GL}_{2/\mathbb{Q}}$, preprint

[FC90] G. Faltings, C.-L. Chai, *Degeneration of Abelian Varieties*, Erg. Math. Series 3-22, Springer Verlag 1990

[H02] H. Hida, Control theorems of coherent sheaves on Shimura varieties of PEL type, J. Inst. Math. Jussieu, **1**, 2002, pp.1-76

[H13a] H. Hida, Image of Λ-adic Galois representations modulo p, Invent. Math. **194** (2013), 1–40.

[H13b] H. Hida, Big Galois representations and p-adic L-functions, preprint, 2012, 51 pages, to appear in Compositio Math.

[Lau05] G. Laumon, Fonctions zêta des variétés de Siegel de dimension trois,*in* Formes Automorphes (II), le cas du groupe GSp(4), Astérisque **302**, SMF, 2005

[LS98] Martin W. Liebeck and Gary M. Seitz, On the subgroup structure of classical groups, Invent. math. **134** (1998), 427–453

[M04] D. Mauger, Algèbres de Hecke quasi-ordinaires universelles, Ann. Sci. École Norm. Sup. (4) **37** (2004), 171–222.

[MT02] A. Mokrane, J. Tilouine, Cohomology of Siegel varieties with p-adic integral coefficients and applications, *in* Asterisque 280, 2002

[Pi12] V. Pilloni, Sur la théorie de Hida pour le groupe GSp_{2g}, Bulletin de la SMF **140** (2012), 335-400

[PiSt] V. Pilloni, B. Stroh, Surconvergence et classicité: le cas déployé, submitted

[P93] R. Pink, Classification of pro-p subgroups of SL_2 over a p-adic ring, where p is an odd prime, Compositio Math. **88** (1993), 251–264

[P98a] R. Pink, l-adic algebraic monodromy groups, cocharacters, and the Mumford-Tate conjecture, J. reine angew. Math. **495** (1998), 187–237

[P98b] R. Pink, Compact subgroups of linear algebraic groups. J. Algebra **206** (1998), 438–504.

[R01] B. Roberts, Global L-packets for GSp_2 and Theta lifts, Doc. Math. **6** (2001), pp.247-314

[SW01] C. Skinner, A. Wiles, Nearly ordinary deformations of irreducible residual representations, Ann. Fac. Sci. Toulouse Math. (6), **10** (1):185-215, 2001

[Ta93] R. Taylor, On the cohomology of Siegel threefolds, Invent. Math. **114** (1993), pp.289-310

[TU99] J. Tilouine, E. Urban, Several variable p-adic families of Siegel-Hilbert cusp eigensystems and their Galois representations, Ann. Sci. E.N.S., 4 série, t. **32**, p. 499-574, 1999

[Ti09] J. Tilouine, Cohomologie des variétés de Siegel et représentations galoisiennes associées aux représentations cuspidales cohomologiques de GSp(4), Publ. Math. Besançon, Algèbre et Théorie des Nombres, 2009

[Ur05] E. Urban, Sur les représentations p-adiques associées aux représentations cuspidales de $GSp_4(\mathbb{Q})$, *in* Formes Automorphes (II) , le cas du groupe GSp(4), pp. 151-176, Astérisque **302**, SMF, 2005

[Ur11] E. Urban, Eigenvarieties for Reductive Groups, Annals of Math. **174** (2011), 1685-1784

[We05] R. Weissauer, Four-dimensional Galois representations, *in* Formes Automorphes (II) , le cas du groupe GSp(4), pp., Astérisque **302**, SMF, 2005

[Y79] H. Yoshida, Weil's representations and Siegel's modular forms. Lectures on harmonic analysis on Lie groups and related topics (Strasbourg, 1979), pp. 319-341, Lectures in Math., **14**, Kinokuniya Book Store, Tokyo, 1982.

[Z14] B. Zhao, Local indecomposability of Hilbert modular Galois representations, to appear in Ann. Inst. Fourier (Grenoble) (posted in web: arXiv:1204.4007v1 [math.NT])

10

The skew-symmetric pairing on the Lubin–Tate formal module

M. A. Ivanov[1] *and S. V. Vostokov*[2]

[1]Saint Petersburg state university, 198504, Universitetsky prospekt, 28, Peterhof, St. Petersburg, Russia
E-mail address: micliva@gmail.com

[2]Saint Petersburg state university, 198504, Universitetsky prospekt, 28, Peterhof, St. Petersburg, Russia
E-mail address: sergei.vostokov@gmail.com

§10.1 Introduction

The classical Hilbert symbol determines a skew-symmetric pairing on the multiplicative group of a local field. Explicit formulae for the Hilbert symbol were obtained independently by Brückner ([3]) and Vostokov ([2]) in 1978. Subsequently, these formulae were generalized to multi-dimensional local fields (see [7]). In this note, we shall generalize these results to the formal modules.

Let F be a commutative formal group over the ring of integers $\mathfrak{o}_k$ of a local field k; let $k_0 \subset k$ be the subfield such that $\mathfrak{o}_{k_0}$ is the endomorphism ring of F. Let $k'|k$ be a finite extension and let $\mathfrak{M} := \pi\,\mathfrak{o}_{k'}$, where π is a prime element of k'. Our goal is to construct a skew-symmetric bilinear pairing $(\alpha, \beta) : F(\mathfrak{M}) \times F(\mathfrak{M}) \to \ker[\pi^n]_F$.

Due to the short exact sequence

$$0 \to F_E(\mathfrak{M}) \to E_0(k) \to E_{ns}(\overline{k}) \to 0,$$

where E is an elliptic curve, E_0 is the set of points with non-singular reduction, E_{ns} is the set of non-singular points, and F_E is a formal group, associated to E (cf. [9, Ch. 7]), our pairing defines a pairing on the group $E_0(k)$ of the elliptic curve E with non-singular reduction. In the multiplicative case this

The authors are grateful to the Hausdorff Research Institute for Mathematics and Max Planck Institute for Mathematics in Bonn (Germany) for the excellent working conditions. This work has been partially supported by the RFFI (grant 11-01-00588-a)

Arithmetic and Geometry, ed. Luis Dieulefait *et al.* Published by Cambridge University Press.

construction determines the wild Hilbert symbol (a pairing on $E_{ns}(\overline{k})$ gives the tame symbol).

We shall begin with the simplest case of formal groups, that is with the Lubin–Tate formal group, corresponding to an elliptic curve with complex multiplication. The endomorphism ring of such a group coincides with $\mathfrak{o}_k$ and the kernel of the isogeny is generated by the unique element denoted by ξ.

§10.2 The construction

We introduce the basic notation of this article:

- k_0 is a local field (a finite extension of $\mathbb{Q}_p$), $p > 2$,
- $\mathfrak{o}_0$ is the ring of integers of the field k_0,
- π_0 is a prime element of $\mathfrak{o}_0$,
- $F(X, Y) \in \mathfrak{o}_0[[X, Y]]$ is the formal series, defining a Lubin–Tate formal group over $\mathfrak{o}_0$,
- $\lambda(X)$ is the logarithm of F (cf. [8, Ch. 8]),
- ξ is a fixed primitive root of the isogeny $[\pi_0^n](X)$,
- $k|k_0$ is a finite extension with $\xi \in k$,
- π is a prime element of k,
- e is the ramification index of the extension $k|k_0$,
- the Lubin–Tate formal group F defines a formal module $F(\mathfrak{M})$ on the maximal ideal $\mathfrak{M} = (\pi)$ of k,
- T is the inertia subfield in k/k_0,
- $\mathfrak{o}_T$ is the ring of integers of T,
- Δ is the Frobenius endomorphism of $T|k_0$,
- $\mathfrak{R}$ is the Teichmüller system in T,
- for $\alpha \in \mathfrak{M}$, let $\underline{\alpha}(X) \in X\mathfrak{o}_T[[X]]$ be such that $\underline{\alpha}(\pi) = \alpha$.

Our goal is to define a bilinear skew-symmetric pairing

$$(\cdot, \cdot) : F(\mathfrak{M}) \times F(\mathfrak{M}) \to \kappa_n := \ker[\pi_0^n]_F. \tag{10.1}$$

At first, we shall introduce a pairing on the rings of formal series

$$\langle \cdot, \cdot \rangle : X\mathfrak{o}_T[[X]] \times X\mathfrak{o}_T[[X]] \to \mathfrak{o}_T/\pi_0^n\mathfrak{o}_T. \tag{10.2}$$

The pairing (2) allows us to define a pairing on the formal module

$$(\alpha, \beta) = [\langle \underline{\alpha}(X), \underline{\beta}(X) \rangle](\xi), \tag{10.3}$$

As usual, we define the action of the Frobenius automorphism Δ on the series $\phi = \sum d_i X^i$ (lying either in $\mathfrak{o}_T((X))$ or in $\mathfrak{o}_T\{\{X\}\}$) by

$$\Delta\phi = \phi^\Delta = \sum d_i^\Delta X^{qi}. \tag{10.4}$$

Let us also define the Artin–Hasse logarithm of the Lubin–Tate formal group:

$$\ell(\phi) = \left(1 - \tfrac{\Delta}{\pi_0}\right)\lambda(\phi), \qquad \text{for } \phi \in X\mathfrak{o}_T[[X]],$$

and the Artin–Hasse exponent (inverse to $\ell(\phi)$):

$$E(\psi) = \lambda^{-1}(1 + \tfrac{\Delta}{\pi_0} + \tfrac{\Delta^2}{\pi_0^2} + \ldots)(\psi).$$

For $\langle\cdot,\cdot\rangle$ we take the pairing constructed similarly to the multiplicative case:

$$\langle \underline{\alpha}(X), \underline{\beta}(X)\rangle = \operatorname{Tr}\operatorname{res}\left[\Phi(\underline{\alpha}(X), \underline{\beta}(X))\frac{1}{s(X)}\right] \tag{10.5}$$

with $s(X) = [\pi_0^n](\underline{\xi}(X))$, $\operatorname{Tr} = \operatorname{Tr}_{T/k_0}$, and

$$\begin{aligned}&\Phi(\underline{\alpha}(X), \underline{\beta}(X))\\ &\quad = \ell(\underline{\alpha}(X))\,\mathrm{d}\ell(\underline{\beta}(X)) - \ell(\underline{\alpha}(X))\,\mathrm{d}\lambda(\underline{\beta}(X)) + \ell(\underline{\beta}(X))\,\mathrm{d}\lambda(\underline{\alpha}(X)).\end{aligned}$$

Simple properties

The pairing (5) is *well defined* because $\Phi(\alpha, \beta) \in \mathfrak{o}_T[[X]]$ ($\ell(X) \in \mathfrak{o}_T[[X]]$ and $\mathrm{d}\lambda(X) \in \mathfrak{o}_T[[X]]$ (see [4, §2.3])).

The *bilinearity* of the pairing (5) (and of the pairing (3) as consequence of it) follows from the properties of $\lambda(X)$ and $\ell(X)$ (see [4, §0.3, §2.3]):

$$\begin{aligned}\langle \alpha_1 +_F \alpha_2,\, \beta\rangle &= \langle \alpha_1,\, \beta\rangle + \langle \alpha_2,\, \beta\rangle,\\ \langle \alpha,\, \beta_1 +_F \beta_2\rangle &= \langle \alpha,\, \beta_1\rangle + \langle \alpha,\, \beta_2\rangle\\ \langle [a]_F\alpha,\, \beta\rangle &= a\langle \alpha,\, \beta\rangle,\\ \langle \alpha,\, [c]_F\beta\rangle &= c\langle \alpha,\, \beta\rangle,\end{aligned} \tag{10.6}$$

$$\begin{aligned}(\alpha_1 +_F \alpha_2,\, \beta) &= (\alpha_1,\, \beta) +_F (\alpha_2,\, \beta),\\ (\alpha,\, \beta_1 +_F \beta_2) &= (\alpha,\, \beta_1) +_F (\alpha,\, \beta_2)\\ ([a]_F\alpha,\, \beta) &= [a]_F(\alpha,\, \beta),\\ (\alpha,\, [c]_F\beta) &= [c]_F(\alpha,\, \beta)\end{aligned} \tag{10.7}$$

with $a, c \in \mathfrak{o}_0$.

By Lemma 10.2 below,

$$\operatorname{res} d(\ell(\alpha)\,\ell(\beta))\frac{1}{s} \equiv 0 \mod \pi_0^n;$$

on the other hand,

$$\langle \underline{\alpha}, \underline{\beta} \rangle + \langle \underline{\beta}, \underline{\alpha} \rangle = \operatorname{res}[\Phi(\alpha, \beta) + \Phi(\beta, \alpha)]\frac{1}{s} = \operatorname{res} d(\ell(\alpha)\,\ell(\beta))\frac{1}{s},$$

therefore the pairing $\langle \cdot, \cdot \rangle$ is *skew-symmetry*.

The Shafarevich basis

A suitable basis of the formal module $F(\mathfrak{M})$ can viewed as a generalization the Shafarevich canonical basic of the group of principal units of a local field (see [1, 2]). It can be described as follows. Let $\alpha \in F(\mathfrak{M})$, then

$$\alpha = \sum_i {}_{(F)} E_F(a_i X^i)|_{X=\pi} +_F \omega(a), \qquad a_i, a \in \mathfrak{o}_T,$$

where $1 \leqslant i < qe/(q-1)$, $q \nmid i$, and $\omega(a) = E_F(as(X))|_{X=\pi} \in \mathfrak{o}_T$ is a π_0^n-primary element (see [4]). Besides the $\mathfrak{o}_{k_0}$-module Ω of the π_0^n-primary elements is generated by $\omega(a_0)$, so that $a_0 \in \mathfrak{o}_T$ and $\operatorname{Tr} a_0 \notin \pi\,\mathfrak{o}_{k_0}$.

An element α belongs to $[\pi_0^n]_F F(\mathfrak{M})$ if and only if $\operatorname{Tr} a \in \pi^n \mathfrak{o}_k$ and $a_i \in \pi^n \mathfrak{o}_T$. There are also other generalizations of the Shafarevich basis [4, 6].

To compare pairing (3) with the known pairings on the multiplicative group [2] and on the product of the multiplicative group and the formal Lubin–Tate module [4], let us remark that in those two cases the proof of the explicit formulae depends on the study of the pairs (π, β), where π is the uniformizing element and β is a principal unit. Such a method does not work for pairing (3). Loosely speaking, we define a pairing on the product of two groups of "principal units".

§10.3 Invariance of the pairing

Let us prove that our pairing (5) is invariant under a change of variable and therefore the pairing (3) does not depend on the choice of the prime element π. We refer to [2, §2.4] for analogous considerations in the multiplicative case.

It suffices to establish invariance of $\langle E(aX^u), E(bX^v)\rangle$ with $a, b \in \mathfrak{R}$ because any element in $\mathfrak{o}_T$ is a linear combination of some elements in $\mathfrak{R}$, the pairing (5) is bilinear, and $\varepsilon = E(\ell(\varepsilon))$ for $\varepsilon \in X\mathfrak{o}_T[\![X]\!]$, where E is an additive bilinear function. Let

$$X = g(Y) = c\psi(Y) \text{ with } c \in \mathfrak{R} \text{ and } \psi(0) = 1.$$

Since $\triangle$ and E depend on the choice of X, let us write $\triangle_X$ and $\triangle_Y$ instead of $\triangle$ when appropriate, and similarly for E and res.

In these notations,

$$E_X(aX^u) = E_Y\big((1 - \tfrac{\triangle_Y}{\pi_0})S\big), \tag{10.8}$$

where

$$a \in \mathfrak{R}, \text{ and } S = \sum_{r=0}^{\infty} \frac{(ag^u)^{q^r}}{\pi_0^r},$$

for

$$E_Y\big((1 - \tfrac{\triangle_Y}{\pi_0})S\big) = \lambda^{-1}(1 + \tfrac{\triangle_Y}{\pi_0} + \ldots)(1 - \tfrac{\triangle_Y}{\pi_0})S = \lambda^{-1}S =$$
$$= \lambda^{-1}\Big(ag^m + \frac{(ag^m)^q}{\pi_0} + \ldots\Big) = \lambda^{-1}\Big(aX^m + \frac{(aX^m)^q}{\pi_0} + \ldots\Big) = E_X(aX^m).$$

Let us note that the series $(1 - \frac{\triangle_Y}{\pi_0})S$ has integer coefficients.

Lemma 10.1. *We have*

$$\frac{h^{q^k} - h^{q^{k-1}\triangle}}{\pi_0^k} \in \mathfrak{o}_T[[X]],$$

for $h \in \mathfrak{o}_T[[X]]$.

Proof. This Lemma can be easily proved by induction on k.

Lemma 10.2. *We have*

$$\operatorname{res} dU \frac{1}{s} \equiv 0 \mod \pi_0^n,$$

for $U \in \mathfrak{o}_T[[X]]$.

Proof. See [4, (18)].

Proposition 10.3. *For $p \neq 2$ the following congruence holds true:*

$$\langle E_X(aX^u), E_X(bX^v)\rangle_X \equiv \langle E_Y\big((1 - \tfrac{\triangle_Y}{\pi_0})S_1\big), E_Y\big((1 - \tfrac{\triangle_Y}{\pi_0})S_2\big)\rangle_Y \mod \pi_0^n.$$

Proof. It follows from (5) that

$$\langle E_X(aX^u), E_X(bX^v)\rangle_X = \operatorname{Tr}\operatorname{res}_X \Phi(X)\frac{1}{s(X)},$$

where

$$\Phi(X) = aX^u\, \mathrm{d}(bX^v) - aX^u\, \mathrm{d}\Big(bX^v + \frac{(bX^v)^q}{\pi_0} + \ldots\Big)$$
$$+ bX^v\, \mathrm{d}\Big(aX^u + \frac{(aX^u)^q}{\pi_0} + \ldots\Big).$$

Moreover,

$$\langle E_Y\big((1-\tfrac{\Delta_Y}{\pi_0})S_1\big), E_Y\big((1-\tfrac{\Delta_Y}{\pi_0})S_2\big)\rangle_Y = \operatorname{Tr}\operatorname{res}_Y \Psi(Y)\frac{1}{s(g(Y))},$$

where

$$\begin{aligned}\Psi(Y) &= (1-\tfrac{\Delta_Y}{\pi_0})S_1\,\mathrm{d}(1-\tfrac{\Delta_Y}{\pi_0})S_2 - (1-\tfrac{\Delta_Y}{\pi_0})S_1\,\mathrm{d}S_2 + (1-\tfrac{\Delta_Y}{\pi_0})S_2\,\mathrm{d}S_1 = \\ &= ag^u\,\mathrm{d}(bg^v) - ag^u\,\mathrm{d}\Big(bg^v + \frac{(bg^v)^q}{\pi_0} + \dots\Big) + bg^v\,\mathrm{d}\Big(ag^u + \frac{(ag^u)^q}{\pi_0} + \dots\Big) + \\ &\quad + \sum_{r\geqslant 1}\sigma_r + \sum_{k,s\geqslant 1}\tau_{r,s} = \sum_{r\geqslant 1}\sigma_r + \sum_{k,s\geqslant 1}\tau_{r,s} + \Phi(g(Y))\,\mathrm{d}g(Y)\end{aligned}$$

and

$$\begin{aligned}\sigma_k &= ag^u\,\mathrm{d}\Big(\frac{(bg^v)^{q^k}}{\pi_0^k} - \frac{(bg^v)^{q^{k-1}\Delta_Y}}{\pi_0^k}\Big) + \\ &+ \mathrm{d}(ag^u)\cdot\Big(\frac{(bg^v)^{q^k}}{\pi_0^k} - \frac{(bg^v)^{q^{k-1}\Delta_y}}{\pi_0^k}\Big) = \mathrm{d}\Big(ag^u\cdot\frac{(bg^v)^{q^k} - (bg^v)^{q^{k-1}\Delta_Y}}{\pi_0^k}\Big).\end{aligned} \tag{10.9}$$

$$\begin{aligned}\tau_{k,s} &= \mathrm{d}\Big(\frac{(ag^u)^{q^s}}{\pi_0^s}\Big)\cdot\frac{(bg^v)^{q^k} - (bg^v)^{q^{k-1}\Delta_Y}}{\pi_0^k} - \\ &- \frac{(ag^u)^{q^s} - (ag^u)^{q^{s-1}\Delta_Y}}{\pi_0^s}\cdot\mathrm{d}\Big(\frac{(bg^v)^{q^{k-1}\Delta_Y}}{\pi_0^k}\Big).\end{aligned}$$

The series

$$ag^u\cdot\frac{(bg^v)^{q^k} - (bg^v)^{q^{k-1}\Delta_Y}}{\pi_0^k}$$

has integer coefficients (see Lemma 10.1), and therefore

$$\sum_{r=1}^{\infty}\sigma_r = \mathrm{d}\varphi_1(Y), \quad \varphi_1(Y)\in\mathfrak{o}_T[[X]].$$

It follows also that

$$\begin{aligned}\tau_{k,s} &= \mathrm{d}\left(\frac{uq^s}{uq^s+vq^k}\frac{1}{\pi_0^{s+k}}\Big((ag^u)^{q^s}(bq^v)^{q^k} - (ag^u)^{q^{s-1}\Delta_y}(bg^v)^{q^{r-1}\Delta_Y}\Big)\right) - \\ &\quad - \mathrm{d}\left(\frac{(bq^v)^{q^{k-1}\Delta_Y}}{\pi_0^{s+k}}\Big((ag^u)^{q^s} - (ag^u)^{q^{s-1}\Delta_y}\Big)\right)\end{aligned}$$

Let us verify that the power series under the differential have integer coefficients. In view of

$$g^{uq^s - uq^{s-1}\Delta_Y} = \psi^{uq^s - uq^{s-1}\Delta_Y} = \exp\log(\psi^{uq^s(1-\frac{\Delta_Y}{q})}) = \exp(uq^s\,\ell_m(\psi)),$$

one obtains

$$\frac{uq^s}{uq^s+vq^k}\frac{1}{\pi_0^{s+k}}(\exp((uq^s+vq^k)\,\ell_m\,\psi)-1)-\frac{1}{\pi_0^{s+k}}(\exp(uq^s\,\ell_m\,\psi)-1\,)=$$
$$=\sum_{n\geqslant 2}(\ell_m\,\psi)^n\frac{uq^s(uq^s+vq^k)^{n-1}-(uq^s)^n}{\pi_0^{s+k}n!}.$$

It is not hard to see that for $q \neq 2$ all the coefficients are integers, hence

$$\sum_{k,s}^{\infty}\tau_{k,s} = \mathrm{d}\varphi_2(Y), \quad \varphi_2(Y) \in \mathfrak{o}_T[[X]].$$

The above calculations, when combined with Lemma 10.2, show that

$$\operatorname{res}_Y \Psi(Y)\frac{1}{s(g(Y))} \equiv \operatorname{res}_Y\left[\left(\Phi(g(Y))\,\mathrm{d}g(Y)+\mathrm{d}\varphi_1+\mathrm{d}\varphi_2\right)\frac{1}{s(g(Y))}\right] \equiv$$
$$\equiv \operatorname{res}_X\left[\Phi(X)\frac{1}{s(X)}\right] \mod \pi_0^n.$$

The proposition is proved.

From now on, let us fix the prime π and define the pairing $(\cdot,\cdot)$ by (3). We have just proved that the pairing $(\cdot,\cdot)$ is bilinear, skew-symmetric, and invariant under the choice of the prime element π.

Moreover, it can be checked in a standard way (see, for instance, [5]) that the pairing $(\cdot,\cdot)$ does not depend on the expansions of $\underline{\alpha}$ and $\underline{\beta}$ of α and β.

§10.4 The kernel of the pairing

Let us restrict ourselves, for simplicity, to the totally ramified extensions.

Proposition 10.4. 1) *The module of the π_0^n-primary elements is contained in the kernel of the pairing.*

2) *The pairing*

$$F(\mathfrak{M})/\Omega \times F(\mathfrak{M})/\Omega \to \ker[\pi_0^n] \tag{10.10}$$

is nondegenerate modulo $[\pi_0^n]$*, that is, if* $(\alpha, \beta) = 0$ *for all* $\beta \in F(\mathfrak{M})/\Omega$*, then* $\alpha \in [\pi_0^n](F(\mathfrak{M}))$.

Proof. To prove 1), let us calculate the pairing $\langle \omega(a), \varepsilon \rangle$. We have

$$\langle \omega(a), \varepsilon \rangle = \operatorname{Tr}\operatorname{res}\left[-as(X)\,\mathrm{d}(\tfrac{\Delta}{\pi_0}\lambda(s(X)))\frac{1}{s(X)} \right] + \\ + \operatorname{Tr}\operatorname{res}\left[\ell(\varepsilon(X))\,\mathrm{d}(1 + \tfrac{\Delta}{\pi_0} + \tfrac{\Delta^2}{\pi_0^2} + \ldots)(as(X))\frac{1}{s(X)} \right]$$

Obviously, the first term is equal to zero; as for the second term, it suffices to note that

$$\mathrm{d}\frac{s^{\Delta^r}}{\pi_0^r} = X^{q^r-1}\frac{q^r}{\pi_0^r}(\mathrm{d}s(X))^{\Delta^r} \equiv 0 \mod \pi_0^n,$$

see, for example, [4, §1.4]. This proves 1).

For $\alpha \in F(M)/\Omega$, let us constuct an element β such that $\langle \alpha, \beta \rangle \neq 0$. We have:

$$\alpha = \underline{\alpha}(X)|_{X=\pi} = E(w(X))|_{X=\pi},$$

with $w(X) = w_1 X^{m_1} + \ldots + w_t X^{m_t} \in \mathfrak{o}_T[[X]]$, where $p \nmid m_i$ and $m_i < \frac{pe}{p-1}$ for $1 \leqslant i \leqslant t$.

Let us suppose first that α is not divisible by $[\pi_0]$, that is, there exists j such that $a_j \in \mathfrak{o}_T^*$. Let $j_0 = j_0(\alpha)$ be the minimal such index. Then

$$\Phi(\underline{\alpha}(X), E(X^{\frac{pe}{p-1}-j_0})) \equiv X^{\frac{pe}{p-1}-j_0}\,\mathrm{d}w(X) \equiv \\ \equiv w_{j_0} j_0 X^{\frac{pe}{p-1}-1} \mod (\pi_0, X^{\frac{pe}{p-1}}),$$

and one can prove [6, §10.25] that

$$\frac{1}{s(X)} \equiv \frac{c_0}{X^{\frac{pe}{p-1}}} + \frac{c_1}{X^{\frac{pe}{p-1}-p}} + \ldots \mod \pi_0, \quad c_0 \in \mathfrak{o}_T^*,$$

It follows from those two congruences that

$$\langle \alpha, E(X^{\frac{pe}{p-1}-j_0})|_{\pi_0} \rangle \equiv \operatorname{Tr} c_0 w_{j_0} j_0 \not\equiv 0 \mod \pi_0.$$

In the general case, if

$$\alpha \in [\pi_0^k]F(\mathfrak{M}) \text{ and } \alpha \notin [\pi_0^{k+1}]F(\mathfrak{M}),$$

then there is α' such that $\alpha = [\pi_0^k]\alpha'$. Let $\beta = E(X^{\frac{pe}{p-1}-j_0(\alpha')})|_{\pi_0}$, then it follows from our calculation that

$$\langle \alpha, \beta \rangle = [\pi_0^k]\langle \alpha', \beta \rangle \not\equiv 0 \mod \pi_0^k.$$

This proves 2).

References

[1] I. R. Shafarevich, *A general reciprocity law*, Mat. Sbornik N.S. **26(68)** (1950), 113–146 (Russian). MR0031944 (11,230f)

[2] S. V. Vostokov, *An explicit form of the reciprocity law*, Izv. Akad. Nauk SSSR Ser. Mat. **42** (1978), no. 6, 1288–1321, 1439 (Russian); English transl., Math. USSR-Izv. **13** (1979), no. 3, 557–588. MR522940 (80f:12014)

[3] H. Brückner, *Explizites Reziprozitätsgesetz und Anwendungen*, Vorlesungen aus dem Fachbereich Mathematik der Universität Essen [Lecture Notes in Mathematics at the University of Essen], vol. 2, Universität Essen Fachbereich Mathematik, Essen, 1979 (German). MR533354 (80m:12015)

[4] S. V. Vostokov, *A norm pairing in formal modules*, Izv. Akad. Nauk SSSR Ser. Mat. **43** (1979), no. 4, 765–794, 966 (Russian); English transl., Math. USSR-Izv. **15** (1980), no. 1, 25–51. MR548504 (81b:12017)

[5] ______, *The Hilbert symbol in a discretely valued field*, Zap. Nauchn. Sem. Leningrad. Otdel. Mat. Inst. Steklov. (LOMI) **94** (1979), 50–69, 150 (Russian); English transl., Journal of Soviet Mathematics **19** (1982), no. 1, 1006–1019. Rings and modules, 2. MR571515 (81g:12019)

[6] ______, *Symbols on formal groups*, Izv. Akad. Nauk SSSR Ser. Mat. **45** (1981), no. 5, 985–1014, 1198 (Russian); English transl., Math. USSR-Izv. **19** (1982), no. 2, 261–284. MR637613 (83c:12016)

[7] ______, *Explicit construction of the class field theory for a multidimensional local field*, Izv. Akad. Nauk SSSR Ser. Mat. **49** (1985), no. 2, 283–308, 461 (Russian); English transl., Math. USSR-Izv. **22** (1986), no. 2, 263–288. MR791304 (86m:11096)

[8] I. B. Fesenko and S. V. Vostokov, *Local fields and their extensions*, 2nd ed., Translations of Mathematical Monographs, vol. 121, American Mathematical Society, Providence, RI, 2002. With a foreword by I. R. Shafarevich. MR1915966 (2003c:11150)

[9] Joseph H. Silverman, *The arithmetic of elliptic curves*, 2nd ed., Graduate Texts in Mathematics, vol. 106, Springer, Dordrecht, 2009. MR2514094 (2010i:11005)

11

Equations in matrix groups and algebras over number fields and rings: prolegomena to a lowbrow noncommutative Diophantine geometry

Boris Kunyavskiĭ

Department of Mathematics, Bar-Ilan University, 5290002 Ramat Gan, ISRAEL
E-mail address: kunyav@macs.biu.ac.il

In memory of Valentin Evgenyevich Voskresenskiĭ

ABSTRACT. We give a brief survey of some problems related to the solvability of word equations in matrix groups and polynomial equations in matrix algebras, focusing on the cases where the matrix entries belong to a global field or its ring of integers.

> *What is needed is a corpus of explicit concrete cases and a middlebrow arithmetic theory which would provide both a practicable means to obtain them and a framework to understand any unexpected regularities.*
>
> J. W. S. Cassels, E. V. Flynn, *Prolegomena to a Middlebrow Arithmetic of Curves of Genus 2*, Cambridge, 1996

11.1 Introduction

In the present survey, by a noncommutative Diophantine problem we mean the question on the existence of a solution to an equation of the form

$$F(A_1, \dots, A_m, X_1, \dots, X_d) = 0, \tag{11.1}$$

where the A_i are fixed elements of an associative or Lie algebra $\mathcal{A}$, or of a group G, F is an associative or Lie polynomial P, or a group word w, and

Arithmetic and Geometry, ed. Luis Dieulefait *et al.* Published by Cambridge University Press.

solutions are d-uples $(X_1, \ldots, X_d)$ sought in $\mathcal{A}^d$ or G^d, respectively (in the group case we change 0 to 1 in the right-hand side of (11.1)). We restrict our attention to a particular class of equations:

$$F(X_1, \ldots, X_d) = A, \tag{11.2}$$

where A is fixed and only *scalar* constants are permitted in the left-hand side of (11.2).

If the algebra $\mathcal{A}$ or the group G admits a faithful matrix representation over a field k or a ring R, then we can regard the matrix entries as *variables* and thus reduce our *noncommutative* Diophantine problem to a *commutative* one (in the group case, this reduction may require either cancelling denominators arising from negative powers of X_i's in group words, or using representations in SL rather than in GL, which can easily be achieved by embedding GL_n into SL_{n+1}).

This "abelianization" has several drawbacks. First, we replace a *single* (though noncommutative) polynomial equation with a *system* of polynomial equations which may be really huge (though commutative), require computer investigation, and go far beyond software limitations. Second, this abelianization process is somehow forgetful: essential symmetries built in the original algebra or group are hidden in the arising polynomial system, and it is not an easy task to reveal them.

However, sometimes one should be ready to pay this price, having in mind a potential reward: full access to a mill driven by arithmetic-geometric power, allowing one to grind hard grain. Such a reward is particularly achievable in the *finite group* case, when one can use representations over *finite fields* and apply to the resulting problem of the existence of rational points on algebraic varieties over finite fields the whole strength of high-tech homological algebra developed over the twentieth century (Weil–Grothendieck–Deligne). Perhaps the first, quite impressive instance of such tour de force is due to Bombieri [Bom1] (see also a follow-up in [Bom2]), who ingeniously applied elimination techniques to polynomial systems over finite fields which arise in classification of finite simple groups of Ree type (note that computer calculations played a critically important role). Over the past decade, some new applications of arithmetic-geometric methods appeared, which resulted in solving long-standing problems in the theory of finite [BGGKPP1]–[BGGKPP2] and infinite [BS1]–[BS2] groups; see surveys [Sh1]–[Sh3], [GKP], [BGaK] for more details (including the discussion of the famous Ore problem [Mall], where the impressive final stop was recently put in [LOST]).

In the present paper, our focus is, however, somewhat different: we are going to discuss the situation when matrix representations described above are defined over some *infinite* field (or ring) of arithmetic nature. Our primary interest is addressed to the field $\mathbb{Q}$ of rational numbers and the ring $\mathbb{Z}$ of rational integers. Here, comparing to the finite field case, achievements are much more modest, to put it mildly. We will give a brief survey of some of those, trying to put the relevant problems in a general context of local–global considerations, traditional for classical "commutative" Diophantine geometry.

Note that perhaps the only existing general result, Borel's theorem [Bor], guaranteeing, in the case where G is a connected semisimple linear algebraic group over an algebraically closed field and w is any non-identity group word, the solvability of equation (11.2) with a "general" (in the sense of the Zariski topology) right-hand side, gives very little for smaller fields: A. Thom showed that even over reals, if G is compact, the solutions of the relevant equation may, for an appropriately chosen w, be included in an arbitrarily small (in the sense of real topology) neighbourhood of the identity [Th]. Once again, it is worth quoting the book of Cassels and Flynn [CF], the title of which inspired (and was shamelessly plagiarised by) the author of this survey: 'Modern geometers take account of the ground field, but regard a field extension as a cheap manoeuvre. For practical computations it is desperately expensive: the theory we seek must avoid them at almost all costs.'

The author has to apologise that the corpus of presented explicit examples, as well as the theory brought for explaining some regularities, seem, at the current stage, too poor. What is badly needed is some replacement of the missing algebraic chain in Manin's triad appearing in the title of [Man]: by now, we are not aware of any reasonable analogues of Galois-cohomological machinery and descent methods, which proved their power in clarifying ties between geometric and arithmetic properties of varieties considered in [Man] and [CF] (and in many other sources). This explains why the term "middlebrow" has been downgraded in the title of this paper.

11.2 Local obstructions

First, it should be emphasised once again that in the present text we restrict our attention to considering word equations in *simple algebraic groups*, as well as to polynomial equations in *simple matrix algebras* (and close to such). Apart from the reason mentioned in the introduction (that we want to use algebraic–geometric machinery), there are other arguments in favour of such limitations: roughly, in a "random" group a "random" word equation has little chance to be solvable; even if we drop one of the assumptions on the group (either simplicity, or algebraicity), even the simplest word equations, such as the commutator equation $[x, y] = g$, may turn out to be nonsolvable; see [KBKP] for relevant discussions and counter-examples.

So throughout below all word (resp. polynomial) equations will be considered in $\mathcal{G}(k)$, where $\mathcal{G}$ is a connected simple linear algebraic k-group (resp. in the matrix algebra $\mathrm{M}_n(k)$).

As our focus is on arithmetic properties of matrix equations, k will typically stand for a number field (usually $k = \mathbb{Q}$) or its ring of integers.

We are going to discuss the traditional local–global approach to solving matrix equations over a number field or its ring of integers, so the first question is the existence of *local obstructions*. Here we notice the first unusual phenomenon: even an innocent looking matrix equation may have no solutions over $\mathbb{C}$. Below are the simplest instances.

11.2.1 Power maps

Let $w\colon G \to G$ be defined by $w(x) = x^m$ where $m \geq 2$ is an integer. Looking at the equation

$$x^2 = \begin{pmatrix} -1 & 1 \\ 0 & -1 \end{pmatrix} \tag{11.3}$$

in $G = \mathrm{SL}_2(\mathbb{C})$, we immediately notice that it has no solutions (apply Jordan's theorem). This obstruction can easily be removed by factoring out the centre: indeed, one can show that in $\mathrm{PSL}_2(\mathbb{C})$ any power equation $x^m = a$ is solvable. However, this observation is somewhat misleading because for simple algebraic groups of types other than **A** the obstruction over $\mathbb{C}$ does exist:

Theorem 11.1 ([Ch1], [Ch2], [Ste]). *The map $x \mapsto x^m$ is surjective on $\mathcal{G}(k)$ (k is an algebraically closed field of characteristic exponent p, $\mathcal{G}$ is a connected semisimple algebraic k-group) if and only if m is prime to prz, where z is the order of the centre of $\mathcal{G}$ and r is the product of "bad" primes.*

So even in characteristic zero (when $p = 1$), even for groups of adjoint type (when $z = 1$), the power map $x \mapsto x^m$ is not surjective whenever m is divisible by a bad prime r, which can be 2,3 or 5, depending on the type of the Dynkin diagram of $\mathcal{G}$.

Other local obstructions (p-adic and real) to the surjectivity of power maps are also completely described, see [Ch3], [Ch4].

Let us now go over to more complicated maps.

11.2.2 Group case

As above, here $\mathcal{G}$ stands for a connected semisimple algebraic k-group, $G = \mathcal{G}(k)$, w is a word in d letters. We denote by the same letter the map $w\colon G^d \to G$, $(g_1, \dots, g_d) \mapsto w(g_1, \dots, g_d)$.

Assume that $\mathcal{G}$ is of *adjoint* type, so G has no centre. Here is a brief account of what is known by now.

(i) For $k = \mathbb{C}$ it is a challenging open question whether or not there exist a nonpower word w (i.e. w cannot be represented as a proper power of another word v) and a group $\mathcal{G}$ such that the map w is not surjective; see [KBKP] for a survey of known surjectivity results.
(ii) For $k = \mathbb{R}$, A. Thom [Th] constructed words w in two variables such that the image of the corresponding word map on the unitary group lies within a neighbourhood of the identity matrix, and this neighbourhood can be made as small as we wish by an appropriate choice of w.
(iii) This construction can easily be extended to an arbitrary compact simple real group and also (using a theorem of Bruhat–Tits–Rousseau, see [Pr] for a short proof) to an arbitrary anisotropic p-adic simple group (which is necessarily of type **A**).

If examples as in (i) above do exist, one can ask whether the surjectivity can be tested on real points:

Question 11.2. Do there exist a word w and a simple algebraic $\mathbb{R}$-group $\mathcal{G}$ such that the map $w_{\mathbb{R}}\colon (\mathcal{G}(\mathbb{R}))^d \to \mathcal{G}(\mathbb{R})$ is surjective but the map $w_{\mathbb{C}}\colon (\mathcal{G}(\mathbb{C}))^d \to \mathcal{G}(\mathbb{C})$ is not?

Similar questions can be asked for justification of "global-to-local" surjectivity principles:

Question 11.3. Do there exist a word w and a simple algebraic $\mathbb{Q}$-group $\mathcal{G}$ such that the map $w_{\mathbb{Q}}\colon (\mathcal{G}(\mathbb{Q}))^d \to \mathcal{G}(\mathbb{Q})$ is surjective but the map $w_{\mathbb{R}}\colon (\mathcal{G}(\mathbb{R}))^d \to \mathcal{G}(\mathbb{R})$ is not?

Question 11.4. Do there exist a word w, a simple algebraic $\mathbb{Q}$-group $\mathcal{G}$ and a prime number p such that the map $w_{\mathbb{Q}}\colon (\mathcal{G}(\mathbb{Q}))^d \to \mathcal{G}(\mathbb{Q})$ is surjective but the map $w_p\colon (\mathcal{G}(\mathbb{Q}_p))^d \to \mathcal{G}(\mathbb{Q}_p)$ is not?

The author is not aware of any results in this direction.

11.2.3 Algebra case

For the solvability of polynomial equations (11.2) in matrix algebras there are several obstructions: obvious, such as the case where F is an identical relation of the algebra under consideration, or the case where F is a Lie polynomial and hence its image consists of trace zero matrices; less obvious but known for decades, such as the case where all values of F lie in the centre of the algebra; more recent, described in [KBMR1]–[KBMR3], [Sp] (for matrices over an algebraically closed or, more generally, a quadratically closed field) and [Male] (for matrices over a real closed field); see these papers, as well as the survey [KBKP] for more details and some open problems.

The case of Lie polynomials on simple Chevalley Lie algebras was treated in some detail in [BGKP], where among other things, one can find counter-examples to the surjectivity as well. However, the cases of real and p-adic algebras remain almost unexplored, for both associative and Lie algebras. In particular, the author does not know anything regarding analogues of Questions 11.2–11.4.

The next questions refer to an eventual analogue of Thom's construction described in 12.4(ii):

Questions 11.5. Do there exist a simple real Lie algebra $\mathfrak{g}$ and a Lie polynomial P such that the image of the map $P\colon \mathfrak{g}^d \to \mathfrak{g}$ lies within an open (in the real topology) neighbourhood of zero U (different from $\mathfrak{g}$)? Can one choose U as small as we wish by an appropriate choice of P? Can one arrange this for $\mathfrak{g} = \mathfrak{su}(n)$, the algebra of skew-hermitian trace zero matrices? Can one do the same for some simple p-adic Lie algebra?

11.3 Local–global principles

In light of Questions 11.2–11.4, it seems reasonable to formulate local–global principles for the surjectivity of word (or polynomial) maps as follows. Let $\mathcal{G}$ denote a simple linear algebraic $\mathbb{Q}$-group. For $p \leq \infty$ we denote by $\mathbb{Q}_p$ the completion of $\mathbb{Q}$ at p where $\mathbb{Q}_\infty := \mathbb{R}$.

Question 11.6. Let $w(x_1, \ldots, x_d) = g$ (resp. $P(X_1, \ldots, X_d) = a$) be a word (resp. polynomial) equation in the group $\mathcal{G}(\mathbb{Q})$ (resp. in the associative algebra $\mathrm{M}_n(\mathbb{Q})$). Suppose that for every $g \in \mathcal{G}(\mathbb{Q})$ (resp. for every $a \in \mathrm{M}_n(\mathbb{Q})$) the equation is solvable in $\mathcal{G}(\mathbb{Q}_p)$ (resp. in $\mathrm{M}_n(\mathbb{Q}_p)$) for all $p \leq \infty$. Is it true that for every $g \in \mathcal{G}(\mathbb{Q})$ (resp. for every $a \in \mathrm{M}_n(\mathbb{Q})$) it is solvable in $\mathcal{G}(\mathbb{Q})$ (resp. in $\mathrm{M}_n(\mathbb{Q})$)?

If the answer to this question is "yes", we say that the pair (w, G) (resp. (P, M_n)) satisfies *weak Grunwald–Wang principle* (see the next section for justification of this terminology).

One can also consider a variant of this principle, where the local requirements are weakened so that the global solvability is implied by the local solvability at *all but finitely many* p. In this case we say that the pair under consideration satisfies *strong Grunwald–Wang principle.*

Naturally, both versions can be discussed when $\mathbb{Q}$ is replaced with any number field (or, more generally, any global field).

As in the case of obstructions over $\mathbb{C}$, let us start with power maps.

11.3.1 Power maps

Variants of Question 11.6 can be posed for $\mathcal{G}$ other than simple linear algebraic groups, say, for commutative algebraic groups. In this context, the case of power words $w = x^m$ was thoroughly investigated. For the multiplicative group $\mathbb{G}_{\mathrm{m}}$, the answer follows from the Grunwald–Wang theorem [AT, Chapter X], [NSW, Chapter IX], [Co, Appendix A] (see [Roq] for historical details). Given a global field k and a natural number m, it says whether there is an element of k which is an mth power everywhere (or almost everywhere) locally but not globally. In particular, if m is divisible by 8, there are counter-examples to strong Grunwald–Wang principle for $(x^m, \mathbb{G}_{\mathrm{m},\mathbb{Q}})$. For $(x^m, \mathbb{G}_{\mathrm{m},\mathbb{Q}(\sqrt{7})})$ there are well-known counter-examples to weak Grunwald–Wang principle: 16 is not an 8th power in $\mathbb{Q}(\sqrt{7})$ though it is an 8th power everywhere locally.

For power maps on more general commutative algebraic groups, perhaps the first counter-example to weak Grunwald–Wang principle was constructed by Cassels and Flynn [CF, 6.9]. In their example, $m = 2$ and $\mathcal{G}$ is an abelian $\mathbb{Q}$-surface. Over the past decade, both versions of Grunwald–Wang principles (weak and strong) for power maps have been studied in detail, which led to almost conclusive results for algebraic tori [DZ1], [Il], elliptic curves [Wo], [DZ1]–[DZ3], [Pa1]–[Pa3], [PRV], [Cr2], and higher-dimensional abelian varieties [Wo], [Cr1]. The interested reader is referred to [Cr1], [Cr2] for a detailed account of these results as well as for various ramifications of local–global principles for higher cohomology.

These considerations lead to counter-examples to weak Grunwald–Wang principle for power maps on *simple* linear algebraic groups defined over number fields such as $\mathbb{Q}(\sqrt{7})$: indeed, the diagonal 2×2 matrix with diagonal entries 16 and 1/16 cannot be an 8th power of a matrix from $\mathrm{SL}_2(\mathbb{Q})$ though it is an 8th power everywhere locally.

As to more general words, the author is not aware of any counter-examples to Grunwald–Wang principles.

11.3.2 Digression on relations with rationality and approximation problems

It is well known that counter-examples to Grunwald–Wang principles are closely related to counter-examples to E. Noether's problem on the rationality of the field of invariants $\mathbb{Q}(x_1, \dots, x_n)^G$ with respect to the permutation action of a finite group G on the x_i. Perhaps this relation was first noticed by Saltman [Sal1], [Sal2] (see also [Sw] for a survey of these results), who provided an argument based on the existence (or, more precisely, non-existence) of generic G-extensions and described relations with lifting and approximation problems. There is another approach, where a bridge between the Grunwald–Wang and Noether problems goes through birational invariants of certain algebraic tori.

Namely, according to Sansuc [San, §§1–2], the cohomological obstruction to the classical *strong* Grunwald–Wang principle is of purely *algebraic* nature: the deviation from this principle is measured by the finite abelian group

$$Ш^1_\omega(k, \mu_m) = \ker\left[H^1(\Gamma, \mu_m) \to \bigoplus_{\langle\gamma\rangle\subset\Gamma} H^1(\langle\gamma\rangle, \mu_m)\right], \qquad (11.4)$$

where μ_m is the group of mth roots of unity, $\Gamma = \mathrm{Gal}(K/k)$ is the Galois group of the splitting field K of μ_m, and the sum in the right-hand side is taken over all cyclic subgroups $\langle\gamma\rangle$ of Γ. Note that the deviation from the *weak* Grunwald–Wang principle is measured by the subgroup $Ш^1(k, \mu_m)$ of $Ш^1_\omega(k, \mu_m)$, which is of *arithmetic* nature, as well as their quotient, measuring the gap between the weak and strong principles (see [San, Lemme 1.4 and Sec. 2a]).

On the other hand, one can show (see [CTS]) that the same finite abelian group $Ш^1_\omega(k, \mu_m)$ provides a Galois-cohomological obstruction to the rationality of the field of invariants in Noether's problem, and give another interpretation of this obstruction. Namely, following Voskresenskiĭ (see [Vo1], [Vo2, Chapter VIII]), consider, for any finite abelian group C (say, for $C = \mathbb{Z}/m$), the k-torus $Q_C = S/C$ where S is a quasitrivial torus (say, for $C = \mathbb{Z}/m$ we have $S = R_{K/k}\mathbb{G}_{\mathrm{m}}$ where K is the splitting field of the character group $\hat{C} = \mu_m$, i.e. the fixed field with respect to the action of $\mathrm{Gal}(k)$ on $\hat{C}$). Then the field of invariants in Noether's problem for C is none other than the field of rational functions $k(Q_C)$. From the exact sequence of Γ-modules

$$0 \to \hat{Q}_C \to \hat{S} \to \hat{C} \to 0,$$

using the fact that $\hat{S}$ is a permutation Γ-module and hence the group

$$\text{Ш}^2_\omega(k, \hat{S}) := \ker \left[H^2(\Gamma, \hat{S}) \to \bigoplus_{\langle\gamma\rangle \subset \Gamma} H^2(\langle\gamma\rangle, \hat{S}) \right]$$

is trivial [San, Lemme 1.9], we obtain that $\text{Ш}^1_\omega(k, \hat{C}) \cong \text{Ш}^2_\omega(k, \hat{Q}_C)$. This latter group is a birational invariant of the k-torus Q_C. It is isomorphic to $\mathrm{Br}(X)/\mathrm{Br}(k)$, where X is a smooth projective k-variety containing Q_C as an open subset [CTS, Proposition 9.5(ii)] (the unramified Brauer group of $k(Q_C)/k$, using an alternative terminology), so whenever this invariant does not vanish, one can conclude that the field in Noether's problem for C is not rational. Thus any counter-example to strong Grunwald–Wang principle yields a counter-example to Noether's problem.

It is worth noting that the abelian group $\text{Ш}^1_\omega(k, \mu_m)/\text{Ш}^1(k, \mu_m)$, mentioned above (or, more precisely, its Pontryagin dual), carries somewhat different arithmetical information, which is in a sense complementary to weak Grunwald–Wang principle. Namely, whenever this group vanishes, the classical version of the Grunwald–Wang theorem for the field k and the cyclic group $C = \mathbb{Z}/m$ holds: for any finite set S of places of k and any collection of local field extensions $K^{(v)}/k_v$ with Galois group Γ_v ($v \in S$) embeddable into C, there exists an extension K/k with Galois group C such that for every $w|v$ one has $K_w \cong K^{(v)}$ (see [NSW, IX.2], [Co], [Ne] for other versions). This property of the triple (k, C, S) is sometimes called *weak approximation* with respect to S. The pair (k, C) is said to satisfy weak (resp. very weak) approximation if the triple (k, C, S) satisfies weak approximation for every finite S (resp. for every finite S disjoint from some finite set S_0 depending from k and C). The classical Grunwald–Wang theory establishes very weak approximation for all abelian groups C and all global fields k and describes all cases when weak approximation holds. These properties also make sense for nonabelian finite groups C and are equivalent to weak (resp. very weak) approximation for quotients SL_n/C (see, e.g., [Ha2]). According to a well-known observation due to Colliot-Thélène, very weak approximation for (k, C) implies that the inverse Galois problem for C over k is solvable; moreover, it is solvable under a weaker assumption of so-called hyper-weak approximation [Ha2]; a recent paper [DG] describes other interrelations between Grunwald–Wang and inverse Galois problems.

For nonabelian groups C, conjectural relations of weak approximation properties to the unramified Brauer group are much more tricky comparing with the abelian case. We leave them aside referring the interested reader to [Ha2] and more recent papers [De], [BDH], [LA], [Ne]. In the latter paper there has been defined another cohomological invariant, detecting the failure of weak approximation for some semi-direct products, and there has been introduced a finer subdivision of Grunwald problems into tame (where all primes dividing the order of C are outside S) and wild ones. It was shown in [Ne] that for every C there are wild problems with negative solution. For tame problems no such examples are known.

To finish with this digression, let us mention a recent generalization of the Grunwald–Wang theorem, viewed as a statement on characters of $\mathbb{G}_{\mathrm{m}}$, to automorphic representations of semisimple groups [So, §3].

11.3.3 Back to general words

It is not clear to what extent power words can serve as a prototypical toy-model for investigating local–global properties in the general case. It is a common understanding that, say, commutator words are in a certain sense opposite to powers (cf., e.g., [HSVZ, Section 6.5]). Moreover, looking at equidistribution properties of fibres of word maps on simple algebraic groups over finite fields, one can observe that the behaviour of all nonpower words is in striking contrast with the case of power words [BK1]. However, the author believes that the lack of experimental evidence does not prevent from raising various questions of local–global nature, such as Question 11.6, for nonpower words as well. Note that collecting experimental data is not an easy task: it is not obvious how to produce a nonpower word w such that for some simple algebraic $\mathbb{Q}$-group $\mathcal{G}$ the map $w\colon \mathcal{G}(\mathbb{Q})^d \to \mathcal{G}(\mathbb{Q})$ is not surjective. Here is an example of such a word.

Proposition 11.7. *For* $w(x, y) = x^2[x^{-2}, y^{-1}]^2$, *the map* $w\colon \mathrm{SL}_2(\mathbb{Q})^2 \to \mathrm{SL}_2(\mathbb{Q})$ *is not surjective.*

The assertion is an immediate consequence of results obtained by Jambor, Liebeck and O'Brien [JLO]. We revisit their proof, adjusting it to our setting, to use this opportunity for demonstrating the power of the *trace method*.

This method dates back to classical works of Vogt, Fricke and Klein. The needed results are quoted below from [Ho] (see [BGrK], [BGG], [BG], [BGaK] for more details and applications).

Let $\mathcal{F}_d$ denote the free group on d generators. For $G = \mathrm{SL}_2(k)$ (k is any commutative ring with 1) and for any $u \in \mathcal{F}_d$ denote by $\mathrm{tr}(u)\colon G^d \to G$ the trace character, $(g_1, \ldots, g_d) \mapsto \mathrm{tr}(u(g_1, \ldots, g_d))$.

Theorem 11.8. [Ho] *If w is an arbitrary element of $\mathcal{F}_d$, then the character of w can be expressed as a polynomial*

$$\mathrm{tr}(w) = P_w(t_1, \dots, t_d, t_{12}, \dots, t_{12\dots d})$$

with integer coefficients in the $2^d - 1$ characters $t_{i_1 i_2 \dots i_\nu} = \mathrm{tr}(x_{i_1} x_{i_2} \dots x_{i_\nu})$, $1 \le \nu \le d$, $1 \le i_1 < i_2 < \dots < i_\nu \le d$.

Let $G = \mathrm{SL}_2(k)$, and let $\pi : G^d \to \mathbb{A}^{2^d-1}$ be defined by

$$\pi(g_1, \dots, g_d) = (t_1, \dots, t_d, t_{12}, \dots, t_{12\dots d})$$

in the notation of Theorem 11.8.

Let $Z_d := \pi(G^d) \subset \mathbb{A}^{2^d-1}$. Let $w \colon G^d \to G$ be a word map. It follows from Theorem 11.8 that for every d there exists a polynomial map $\psi : \mathbb{A}^{2^d-1} \to \mathbb{A}^1$ such that the following diagram commutes:

$$\begin{array}{ccc} G^d & \xrightarrow{w} & G \\ {\scriptstyle\pi}\downarrow & & \downarrow{\scriptstyle\mathrm{tr}} \\ Z_d(k) & \xrightarrow{\psi} & \mathbb{A}^1. \end{array} \tag{11.5}$$

Moreover, for small d we have a more precise information: one can take $Z_2 = \mathbb{A}^3$ and $Z_3 \subset \mathbb{A}^7$ an explicitly given hypersurface. This diagram allows one to reduce the study of the image and fibres of w to the corresponding problems for ψ, which may be much simpler. To ease the notation, we denote the coordinates of $Z_2 = \mathbb{A}^3$ by (s, t, u). For any prime p we denote by $P_{w,p}(s, t, u)$ the reduction of the integer polynomial $P_w(s, t, u)$ modulo p.

Proof of the proposition. We will use diagram (11.5) for $k = \mathbb{Q}$ and $k = \mathbb{F}_p$. Corollary 3 of [JLO] implies that for infinitely many primes p the word map $w \colon \mathrm{SL}_2(\mathbb{F}_p)^2 \to \mathrm{SL}_2(\mathbb{F}_p)$ induced by the word w from the statement of the proposition is not surjective. More precisely, looking at the proof of this corollary, one can see that for these primes the image of ψ does not contain 0. In other words, for any such p all values of the polynomial $P_{w,p}$ are nonzero. Let $\mathcal{P}$ denote the set of all these primes.

Let us show that the image of the map $w \colon \mathrm{SL}_2(\mathbb{Q})^2 \to \mathrm{SL}_2(\mathbb{Q})$ does not contain matrices with trace zero. Assume to the contrary that there exist $A, B \in \mathrm{SL}_2(\mathbb{Q})$ such that the trace of $w(A, B)$ equals zero. Let $s = \mathrm{tr}(A)$, $t = \mathrm{tr}(B)$, $u = \mathrm{tr}(AB)$. From diagram (11.5) we have $P_w(s, t, u) = 0$. Since the set of prime divisors of denominators of rational numbers s, t, u is finite and $\mathcal{P}$ is infinite, there exists $p \in \mathcal{P}$ such that the reductions $\bar{s}, \bar{t}, \bar{u} \in \mathbb{F}_p$ are well-defined and $P_{w,p}(\bar{s}, \bar{t}, \bar{u}) = 0$, contradiction. □

11.3.4 Approximation problems

By analogy with the commutative case, along with local–global problems such as Question 11.6, one can consider various approximation problems. We restrict ourselves to stating them in the group case.

For a word $w = w(x_1, \dots, x_d)$ and $g \in \mathcal{G}(\mathbb{Q})$, where $\mathcal{G}$, as above, stands for a simple linear algebraic $\mathbb{Q}$-group embedded into an affine $\mathbb{Q}$-space $\mathbb{A}^N$, denote by $X_{w,g}$ the affine algebraic variety defined over $\mathbb{Q}$ by the system of polynomial equations in $\mathbb{A}^N$ arising from the matrix equation $w(x_1, \dots, x_d) = g$. Assume that for every $g \in \mathcal{G}(\mathbb{Q})$ the variety $X_{w,g}$ has a $\mathbb{Q}$-point.

Question 11.9. (i) Is it true that for any finite set S of primes $p \leq \infty$ and for every $g \in \mathbb{G}(\mathbb{Q})$ the set $X_{w,g}(\mathbb{Q})$ is dense in $\prod_{p \in S} X_{w,g}(\mathbb{Q}_p)$?

(ii) Does there exist a finite set S_0 such that for any finite set S of primes $p \leq \infty$ disjoint from S_0 and for every $g \in \mathbb{G}(\mathbb{Q})$ the set $X_{w,g}(\mathbb{Q})$ is dense in $\prod_{p \in S} X_{w,g}(\mathbb{Q}_p)$?

If Question 11.9(i) (resp. (ii)) is answered in the affirmative, we say that the word w satisfies weak (resp. very weak) approximation in $\mathcal{G}$.

Although Question 11.9 does not seem to be easy for an arbitrary $\mathcal{G}$, there is a hope that the trace method may be helpful in the case $\mathcal{G} = \mathrm{SL}_2$. Say, for $d = 2$ diagram (11.5) then induces a fibration $\pi : X_{w,g} \to Y_{w,g}$ where the surface $Y_{w,g} \subset \mathbb{A}^3_{\mathbb{Q}}$ is given by the equation $\psi(s, u, t) = \mathrm{tr}(g)$. For the commutator word $w = [x, y]$ this approach will hopefully lead to (at least very) weak approximation in SL_2; using [Hal], one can show that in this case for $X_{w,g}$ the Brauer–Manin obstruction to weak approximation is the only obstruction. Details will follow in a forthcoming paper [BK2].

11.3.5 Strong approximation and local–global principles over integers

Diophantine questions similar to those discussed above can be posed in the case where $\mathcal{G}$ is a group scheme over $\mathbb{Z}$ (or, more generally, over the ring of integers O_K of a global field K) such that its generic fibre is a simple linear algebraic group over $\mathbb{Q}$ (or K).

Little is known in such a set-up. Even local obstructions are not completely understood though some recent results give certain hope for conclusive answers, at least for $\mathcal{G} = \mathrm{SL}_n$. Say, for the commutator word $w = [x, y]$ the situation is roughly as follows: obstructions for representing an element of $\mathbb{Z}_p$ as a commutator vanish (after excluding necessary congruence obstructions) as soon as p is sufficiently large with respect to the Lie rank of $\mathcal{G}$ (see [GS, Theorem 5], [AGKS, Theorem 3.5]). Here is an explicit instance of such a phenomenon [AGKS, Theorem 3.8]: if n is a proper divisor of $p-1$, then every element of $\mathrm{PSL}_n(\mathbb{Z}_p)$ is a commutator. The reader is referred to [AGKS] for more results in this direction, some of those answer the questions raised in [Sh2].

As to the global case, practically nothing is known, even for simplest words and groups. Here is a tempting question raised by Shalev [Sh2], [Sh3]: is every element of $\mathrm{SL}_n(\mathbb{Z})$ $(n > 3)$ a commutator? For $n = 2$ the answer to a similar question is obviously negative. However, the author is not aware of any answer to the following question.

Question 11.10. Let $A \in \mathrm{SL}_2(\mathbb{Z})$ be representable as a commutator in all groups $\mathrm{SL}_2(\mathbb{Z}_p)$. Is A a commutator in $\mathrm{SL}_2(\mathbb{Z})$?

(Apparently, only $p = 2$ and $p = 3$ provide nontrivial local obstructions.)

Recent developments in understanding strong approximation for varieties admitting nice fibrations [CTX], [CTH] give some hope for successful application of the trace method, as in approximation problems mentioned above.

One can consider analogues of Question 11.10 for other arithmetically interesting cases, such as Bianchi groups $\mathrm{PSL}_2(O_d)$, where O_d is the ring of integers of the imaginary quadratic field $\mathbb{Q}(\sqrt{-d})$. Higher-dimensional analogues, such as Picard modular groups $\mathrm{SU}(n, 1; O_d)$, look even more challenging.

11.3.6 Case of simple algebras

In the case where $P(X_1, \dots, X_d)$ is an associative (or Lie) polynomial, the author is not aware of any example of failure of analogues of Grunwald–Wang principles for the map of simple associative (or Lie) algebras $\mathcal{A}^d \to \mathcal{A}$ induced by P (apart from the power maps discussed above), neither over global fields, nor over their rings of integers. Some natural cases to be considered are, say, the following ones:

(i) $P = X_1^2 + \cdots + X_d^2$;
(ii) additive commutator $P = [X_1, X_2] = X_1X_2 - X_2X_1$.

Some interesting results for case (i) were obtained in [Va1], [KG], [Pu], however, local–global questions as above seem to remain open. For the additive commutator in matrix algebras, the induced map $\mathcal{A}^2 \to \mathcal{A}_0$, where $\mathcal{A}_0$ denotes the collection of all trace zero elements of $\mathcal{A}$, is surjective in each of the following cases (see [KBKP], [Sta] for more details):

(i) $\mathcal{A} = \mathrm{M}_n(D)$, where D is a central division algebra over a field k and $n \geq 2$ [AR];
(ii) for $n = 1$ this is only known in the case where k is a local field and the degree of D is prime [Ros];
(iii) $\mathcal{A} = \mathrm{M}_2(R)$, where R is a principal ideal domain [Li], [Va2], [RR];
(iv) $\mathcal{A} = \mathrm{M}_n(\mathbb{Z})$ [LR] or, more generally, $\mathcal{A} = \mathrm{M}_n(R)$, where R is a principal ideal ring [Sta].

Remark 11.11. Regarding case (iv), for $\mathcal{A} = \mathrm{M}_n(R)$ where $n \geq 3$ and R is an arbitrary Dedekind ring, the question on the surjectivity of the commutator map $\mathcal{A}^2 \to \mathcal{A}_0$ remains open, see a discussion in [Sta].

Remark 11.12. If the commutator map as above is regarded as a map of Lie algebras $\mathfrak{sl}(n, R) \times \mathfrak{sl}(n, R) \to \mathfrak{sl}(n, R)$, the situation becomes more subtle: it is not clear whether in the Laffey–Reams–Stasinski theorem on representing any trace zero matrix in the form $AB - BA$ one can choose A and B to be trace zero matrices. Say, this is not always possible in the case $n = 2$, $R = \mathbb{Z}$: if it were the case, it would also be possible in $\mathfrak{g} = \mathfrak{sl}(2, \mathbb{F}_2)$, but $[\mathfrak{g}, \mathfrak{g}] \neq \mathfrak{g}$. (I thank N. Gordeev for this remark.)

It is tempting to use the local–global approach to the surjectivity problem for this map as well. Naturally, similar problems make sense for any Lie polynomial $P(X_1, \dots, X_d)$, when one can ask about the surjectivity of the induced map of Lie rings $\mathfrak{g}(R)^d \to \mathfrak{g}(R)$, where $\mathfrak{g}$ is a Chevalley Lie algebra over $\mathbb{Z}$ and R is the ring of integers of a global field.

Remark 11.13. Perhaps, the first class of polynomials to which one can try to extend results known for additive commutator are additive Engel polynomials $E_n(X_1, X_2) = [[[X_1, X_2], \dots, X_2]]$. Nothing is known about the surjectivity of induced maps. Both the associative and Lie case are completely open, even over fields, even for the simplest polynomial $E_2(X_1, X_2) = [[[X_1, X_2], X_2]]$. Note, however, that for a somewhat similar polynomial, generalised commutator $P(X_1, X_2, X_3) = X_1X_2X_3 - X_3X_2X_1$, the surjectivity of the induced map $\mathcal{A}^3 \to \mathcal{A}$ is known for any division algebra $\mathcal{A}$ [GKKL] and for any matrix algebra $\mathcal{A} = \mathrm{M}_n(R)$ over a principal ideal domain R [KL].

Remark 11.14. To conclude, it is worth emphasizing the importance of pursuing parallels between matrix equations in groups and algebras, especially in the Lie algebra case that can be viewed as a bridge between groups and associative algebras. Certain considerations in [AGKS] and [BRL] give some hope for making this into a working tool.

Acknowledgements. The author was supported in part by the Israel Science Foundation, grant 1207/12, and by the Minerva Foundation through the Emmy Noether Research Institute for Mathematics; a part of this work was done when he participated in the trimester program “Arithmetic and Geometry” in the Hausdorff Research Institute for Mathematics (Bonn).

The author thanks T. Bandman, N. Gordeev, D. Neftin and A. Shalev for useful discussions and correspondence and T. Bauer for providing reference [Sta].

References

[AR] S. A. Amitsur, L. H. Rowen, *Elements of reduced trace 0*, Israel J. Math. **87** (1994), 161–179.

[AT] E. Artin, J. Tate, *Class Field Theory*, AMS Chelsea Publishing, Providence, RI, 2009.

[AGKS] N. Avni, T. Gelander, M. Kassabov, A. Shalev, *Word values in p-adic and adelic groups*, Bull. London Math. Soc. **45** (2013), 1323–1330.

[BG] T. Bandman, S. Garion, *Surjectivity and equidistribution of the word $x^a y^b$ on* $\mathrm{PSL}(2,q)$ *and* $\mathrm{SL}(2,q)$, Internat. J. Algebra Comput. **22** (2012), no. 2, 1250017, 33 pp.

[BGG] T. Bandman, S. Garion, F. Grunewald, *On the surjectivity of Engel words on* $\mathrm{PSL}(2,q)$, Groups Geom. Dyn. **6** (2012), 409–439.

[BGaK] T. Bandman, S. Garion, B. Kunyavskiĭ, *Equations in simple matrix groups: algebra, geometry, arithmetic, dynamics*, Central European J. Math. **12** (2014), 175–211.

[BGKP] T. Bandman, N. Gordeev, B. Kunyavskiĭ, E. Plotkin, *Equations in simple Lie algebras*, J. Algebra **355** (2012), 67–79.

[BGGKPP1] T. Bandman, G.-M. Greuel, F. Grunewald, B. Kunyavskiĭ, G. Pfister, E. Plotkin, *Two-variable identities for finite solvable groups*, C. R. Acad. Sci. Paris, Sér. A **337** (2003), 581–586.

[BGGKPP2] T. Bandman, G.-M. Greuel, F. Grunewald, B. Kunyavskiĭ, G. Pfister, E. Plotkin, *Identities for finite solvable groups and equations in finite simple groups*, Compos. Math. **142** (2006), 734–764.

[BGrK] T. Bandman, F. Grunewald, B. Kunyavskiĭ (with an appendix by N. Jones), *Geometry and arithmetic of verbal dynamical systems on simple groups*, Groups Geom. Dyn. **4** (2010), 607–655.

[BK1] T. Bandman, B. Kunyavskiĭ, *Criteria for equidistribution of solutions of word equations on* $\mathrm{SL}(2)$, J. Algebra **382** (2013), 282–302.

[BK2] T. Bandman, B. Kunyavskiĭ, *Local-global properties of word varieties*, in preparation.

[Bom1] E. Bombieri (with appendices by A. Odlyzko and D. Hunt), *Thompson's problem* ($\sigma^2 = 3$), Invent. Math. **58** (1980), 77–100.

[Bom2] E. Bombieri, *Sastry automorphisms*, J. Algebra **257** (2002), 222–243.

[Bor] A. Borel, *On free subgroups of semisimple groups*, Enseign. Math. **29** (1983) 151–164; reproduced in Œuvres - Collected Papers, vol. IV, Springer-Verlag, Berlin–Heidelberg, 2001, pp. 41–54.

[BS1] A. Borisov, M. Sapir, *Polynomial maps over finite fields and residual finiteness of mapping tori of group endomorphisms*, Invent. Math. **160** (2005), 341–356.

[BS2] A. Borisov, M. Sapir, *Polynomial maps over p-adics and redisual properties of mapping tori of group endomorphisms*, Intern. Math. Res. Notices **2009**, 3002–3015.

[BDH] M. Borovoi, C. Demarche, D. Harari, *Complexes de groupes de type multiplicatif et groupe de Brauer non ramifié des espaces homogènes*, Ann. Sci. Éc. Norm. Supér. **46** (2013), 651–692.

[BRL] K. Bou-Rabee, M. Larsen, *Linear groups with Borel's property*, J. Europ. Math. Soc., to appear.

[CF] J. W. S. Cassels, E. V. Flynn, *Prolegomena to a Middlebrow Arithmetic of Curves of Genus 2*, London Math. Soc. Lecture Note Ser., vol. 230, Cambridge Univ. Press, Cambridge, 1996.

[Ch1] P. Chatterjee, *On the surjectivity of the power maps of algebraic groups in characteristic zero*, Math. Res. Lett. **9** (2002), 741–756.

[Ch2] P. Chatterjee, *On the surjectivity of the power maps of semisimple algebraic groups*, Math. Res. Lett. **10** (2003), 625–633.

[Ch3] P. Chatterjee, *On the power maps, orders and exponentiality of p-adic algebraic groups*, J. reine angew. Math. **629** (2009), 201–220.

[Ch4] P. Chatterjee, *Surjectivity of power maps of real algebraic groups*, Adv. Math. **226** (2011), 4639–4666.

[CTH] J.-L. Colliot-Thélène, D. Harari, *Approximation forte en famille*, J. reine angew. Math., to appear, `doi:10.1515/crelle-2013-0092`.

[CTS] J.-L. Colliot-Thélène, J.-J. Sansuc, *Principal homogeneous spaces under flasque tori: applications*, J. Algebra **106** (1987), 148–205.

[CTX] J.-L. Colliot-Thélène, F. Xu, *Strong approximation for total space of certain quadric fibrations*, Acta Arith. **157** (2013), 169–199.

[Co] B. Conrad, *Lifting global representations with local properties*, preprint, 2011, available at `http://math.stanford.edu/~conrad/papers/locchar.pdf`.

[Cr1] B. Creutz, *A Grunwald–Wang type theorem for abelian varieties*, Acta Arith. **154** (2012), 353–370.

[Cr2] B. Creutz, *On the local–global principle for divisibility in the cohomology of elliptic curves*, `arXiv:1305.5881`.

[DG] P. Dèbes, N. Ghazi, *Galois covers and the Hilbert–Grunwald property*, Ann. Inst. Fourier **62** (2012), 989–1013.

[De] C. Demarche, *Groupe de Brauer non ramifié d'espaces homogènes à stabilisateurs finis*, Math. Ann. **346** (2010), 949–968.

[DZ1] R. Dvornicich, U. Zannier, *Local-global divisibility of rational points in some commutative algebraic groups*, Bull. Soc. Math. France **129** (2001), 317–338.

[DZ2] R. Dvornicich, U. Zannier, *An analogue for elliptic curves of the Grunwald–Wang example*, C. R. Acad. Sci. Paris, Ser. I **338** (2004), 47–50.

[DZ3] R. Dvornicich, U. Zannier, *On a local–global principle for the divisibility of a rational point by a positive integer*, Bull. Lond. Math. Soc. **39** (2007), 27–34.

[GS] N. Gordeev, J. Saxl, *Products of conjugacy classes in Chevalley groups over local rings*, Algebra i Analiz **17** (2005), no. 2, 96–107; English transl. in St. Petersburg Math. J. **17** (2006), 285–293.

[GKP] F. Grunewald, B. Kunyavskiĭ, E. Plotkin, *Characterization of solvable groups and solvable radical*, Intern. J. Algebra Computation **23** (2013), 1011–1062.

[GKKL] R. N. Gupta, A. Khurana, D. Khurana, T. Y. Lam, *Rings over which the transpose of every invertible matrix is invertible*, J. Algebra **322** (2009), 1627–1636.

[Ha1] D. Harari, *Flèches de spécialisation en cohomologie étale et applications arithmétiques*, Bull. Soc. Math. France **125** (1997), 143–166.

[Ha2] D. Harari, *Quelques propriétés d'approximation reliées à la cohomologie galoisienne d'un groupe algébrique fini*, Bull. Soc. Math. France. **135** (2007), 549–564.

[HSVZ] R. Hazrat, A. Stepanov, N. Vavilov, Z. Zhang, *Commutator width in Chevalley groups*, Note Mat. **33** (2013), 139–170.

[Ho] R. D. Horowitz, *Characters of free groups represented in the two-dimensional special linear group*, Comm. Pure Appl. Math. **25** (1972), 635–649.

[Il] M. Illengo, *Cohomology of integer matrices and local–global divisibility on the torus*, J. Théor. Nombres Bordeaux **20** (2008), 327–334.

[JLO] S. Jambor, M. W. Liebeck, E. A. O'Brien, *Some word maps that are non-surjective on infinitely many finite simple groups*, Bull. London Math. Soc. **45** (2013), 907–910.

[KBKP] A. Kanel-Belov, B. Kunyavskiĭ, E. Plotkin, *Word equations in simple groups and polynomial equations in simple algebras*, Vestnik St. Petersburg Univ. Math. **46** (2013), no. 1, 3–13.

[KBMR1] A. Kanel-Belov, S. Malev, L. Rowen, *The images of non-commutative polynomials evaluated on 2×2 matrices*, Proc. Amer. Math. Soc. **140** (2012), 465–478.

[KBMR2] A. Kanel-Belov, S. Malev, L. Rowen, *The images of non-commutative polynomials evaluated on 3×3 matrices*, `arXiv:1306.4389`, to appear in Proc. Amer. Math. Soc.

[KBMR3] A. Kanel-Belov, S. Malev, L. Rowen, *Power-central polynomials on matrices*, `arXiv:1310.1598`.

[KG] S. A. Katre, A. S. Garge, *Matrices over commutative rings as sums of k-th powers*, Proc. Amer. Math. Soc. **141** (2013), 103–113.

[KL] D. Khurana, T. Y. Lam, *Generalized commutators in matrix rings*, Linear Multilin. Algebra **60** (2012), 797–827.

[LR] T. J. Laffey, R. Reams, *Integral similarity and commutators of integral matrices*, Linear Algebra Appl. **197/198** (1994), 671–689.

[LOST] M. W. Liebeck, E. A. O'Brien, A. Shalev, P. H. Tiep, *The Ore conjecture*, J. Europ. Math. Soc. **12** (2010), 939–1008.

[Li] D. Lissner, *Matrices over polynomial rings*, Trans. Amer. Math. Soc. **98** (1961), 285–305.

[LA] G. Lucchini Arteche, *Groupe de Brauer non ramifié des espaces homogènes à stabilisateur fini*, J. of Algebra **411** (2014), 129–181.

[Male] S. Malev, *The images of non-commutative polynomials evaluated on* 2×2 *matrices over an arbitrary field*, J. Alg. Appl. **13** (2014), no. 6, 1450004, 12pp.

[Mall] G. Malle, *The proof of Ore's conjecture (after Ellers–Gordeev and Liebeck–O'Brien–Shalev–Tiep).* Astérisque **361** (2014), exp. 1069, 325–348.

[Man] Yu. I. Manin, *Cubic Forms: Algebra, Geometry, Arithmetic*, Nauka, Moscow, 1972; English transl., 2nd ed., North-Holland, Amsterdam, 1986.

[Ne] D. Neftin, *The tame and wild Grunwald problems*, preprint available at `http://www-personal.umich.edu/~neftin/manu.htm`.

[NSW] J. Neukirch, A. Schmidt, K. Wingberg, *Cohomology of Number Fields*, Grundlehren Math. Wiss., vol. 323, Springer-Verlag, Berlin, 2000.

[Pa1] L. Paladino, *Local–global divisibility by* 4 *in elliptic curves defined over* $\mathbb{Q}$, Ann. Mat. Pura Appl. (4) **189** (2010), 17–23.

[Pa2] L. Paladino, *Elliptic curves with* $\mathbb{Q}(\mathcal{E}[3]) = \mathbb{Q}(\zeta_3)$ *and counterexamples to local–global divisibility by* 9, J. Théor. Nombres Bordeaux **22** (2010), 139–160.

[Pa3] L. Paladino, *On counterexamples to local–global divisibility in commutative algebraic groups*, Acta Arith. **148** (2011), 21–29.

[PRV] L. Paladino, G. Ranieri, E. Viada, *On local–global divisibility by* p^n *in elliptic curves*, Bull. Lond. Math. Soc. **44** (2012), 789–802.

[Pr] G. Prasad, *Elementary proof of a theorem of Bruhat–Tits–Rousseau and of a theorem of Tits*, Bull. Soc. Math. France **110** (1982), 197–202.

[Pu] S. Pumplün, *Sums of d-th powers in non-commutative rings*, Beiträge Algebra Geom. **48** (2007), 291–301.

[Roq] P. Roquette, *The Brauer–Hasse–Noether theorem in historical perspective*, Schriften der Mathematisch-Naturwissenschaftlichen Klasse der Heidelberger Akademie der Wissenschaften, vol. 15, Springer-Verlag, Berlin–New York, 2005.

[Ros] M. Rosset, *Elements of trace zero and commutators*, Ph.D. Thesis, Bar-Ilan Univ., 1997.

[RR] M. Rosset, S. Rosset, *Elements of trace zero that are not commutators*, Comm. Algebra **28** (2000), 3059–3072.

[Sal1] D. J. Saltman, *Generic Galois extensions and problems in field theory*, Adv. Math. **43** (1982), 250–283.

[Sal2] D. J. Saltman, *Retract rational fields and cyclic Galois extensions*, Israel J. Math. **47** (1984), 165–215.

[San] J.-J. Sansuc, *Groupe de Brauer et arithmétique des groupes algébriques linéaires sur un corps de nombres*, J. reine angew. Math. **327** (1981), 12–80.

[Sh1] A. Shalev, *Commutators, words, conjugacy classes and character methods*, Turkish J. Math. **31** (2007), 131–148.

[Sh2] A. Shalev, *Applications of some zeta functions in group theory*, in: "Zeta Functions in Algebra and Geometry" (A. Campillo *et al.*, Eds.), Contemp. Math., vol. 566, Amer. Math. Soc., Providence, RI, 2012, pp. 331–344.

[Sh3] A. Shalev, *Some results and problems in the theory of word maps*, in: "Erdös Centennial" (L. Lovász, I. Ruzsa, V. T. Sós, Eds.), Bolyai Soc. Math. Studies, vol. 25, Springer, 2013, pp. 611–649.

[So] C. M. Sorensen, *A proof of the Breuil–Schneider conjecture in the indecomposable case*, Ann. Math. **177** (2013), 367–382.

[Sta] A. Stasinski, *Similarity and commutators of matrices over principal ideal rings*, `arXiv:1211.6872`, to appear in Trans. Amer. Math. Soc.

[Ste] R. Steinberg, *On power maps in algebraic groups*, Math. Res. Lett. **10** (2003), 621–624.

[Sw] R. G. Swan, *Noether's problem in Galois theory*, in: "Emmy Noether in Bryn Mawr" (B. Srinivasan, J. Sally, Eds.), Springer, New York–Berlin, 1983, pp. 21–40.

[Sp] Š. Špenko, *On the image of a noncommutative polynomial*, J. Algebra **377** (2013), 298–311.

[Th] A. Thom, *Convergent sequences in discrete groups*, Canad. Math. Bull. **56** (2013), 424–433.

[Va1] L. N. Vaserstein, *On the sum of powers of matrices*, Linear Multilin. Algebra **21** (1987), 261–270.

[Va2] L. N. Vaserstein, *Noncommutative number theory*, in: "Algebraic K-theory and Algebraic Number Theory (Honolulu, HI, 1987)" (M. R. Stein, R. K. Dennis, Eds.), Contemp. Math., vol. 83, Amer. Math. Soc., Providence, RI, 1989, pp. 445–449.

[Vo1] V. E. Voskresenskiĭ, *On the question of the structure of the subfield of invariants of a cyclic group of automorphisms of the field* $\mathbf{Q}(x_1, \ldots x_n)$, Izv. Akad. Nauk SSSR Ser. Mat. **34** (1970), 366–375; English transl. in Math. USSR Izv. **4** (1970), 371–380.

[Vo2] V. E. Voskresenskiĭ, *Algebraic Tori*, Nauka, Moscow, 1977. (Russian.)

[Wo] S. Wong, *Power residues on abelian varieties*, Manuscripta Math. **102** (2000), 129–138.

12

On the ℓ-adic regulator as an ingredient of Iwasawa theory

L. V. Kuz'min
NRC "Kurchatov Institute"
E-mail address: helltiapa@mail.ru

Contents

12.1 Introduction

A starting-point of the Iwasawa theory is an analogy between algebraic number fields and fields of algebraic functions in one indeterminate over a finite field of constants [3]. Let X be a curve over a finite field $\mathbb{F}_q$ of $q = p^n$ elements, $K = k(X)$ the function field of X. For a given prime $\ell \neq p$ let $T_\ell(K)$ be the Galois group of $L/(\bar{\mathbb{F}}_q K)$, where L is the maximal Abelian unramified ℓ-extension of $\bar{\mathbb{F}}_q K$. The Frobenius automorphism $\varphi \in G(\bar{\mathbb{F}}_q/\mathbb{F}_q)$ acts on $T_\ell(K) \otimes_{\mathbb{Z}_\ell} \mathbb{Q}_\ell$ hence we have a characteristic polynomial $P(X) \in \mathbb{Z}[X]$ of this action and the system of its eigenvalues $\{\alpha_i\}$, $1 \leqslant i \leqslant 2g$, where g is the genus of X. This polynomial is equal to the numerator of the zeta-function of X. The eigenvalues α_i satisfy the Riemann conjecture, that is, $|\alpha_i| = \sqrt{q}$. Moreover, there is a skew-symmetric Weil scalar product on $T_\ell(K)$.

Arithmetic and Geometry, ed. Luis Dieulefait *et al.* Published by Cambridge University Press.

The Iwasawa theory tries to generalize all these results to algebraic number fields. For such a field K one defines K_∞ to be the cyclotomic $\mathbb{Z}_\ell$-extension of K. Then, as in the functional case, one defines L to be the maximal Abelian unramified ℓ-extension of K_∞ and $\bar{T}_\ell(K) = \bar{T}_\ell(K_\infty) = G(L/K_\infty)$. So $\bar{T}_\ell(K)$ is a Λ-module, where $\Lambda = \mathbb{Z}_\ell[[\Gamma]]$ is a completed group ring of Γ and $\Gamma = G(K_\infty/K)$. This module is known as the Tate module of K or the Iwasawa module. Consider the simplest case when K is a cyclotomic field. In this case one has $\bar{T}_\ell(K) = \bar{T}^+(K) \oplus \bar{T}_\ell^-(K)$ corresponding to the action of the automorphism of complex conjugation j. Then there is the so called "the main conjecture of Iwasawa theory" [5, 20] (which is stated for cyclotomic fields and in some other cases) that expresses an action of Γ on $\bar{T}_\ell^-(K)$ in terms of some functions (Iwasawa series), which may be defined in terms of the ℓ-adic L-functions.

By the analogy with the functional case one can expect that, under assumption that K contains an ℓ-th primitive root of unity, there exists an analog of Weil scalar product on $\bar{T}_\ell(K)$. But one can see easily that it is impossible. Indeed, this product must be compatible with the action of the Galois group on $\bar{T}_\ell(K)$ and hence it must induce a non-degenerate product $\bar{T}^-(K) \times \bar{T}^+(K) \to V$, where V is the projective limit of ℓ-primary roots of unity in K_∞. But the group $\bar{T}_\ell^-(K)$ is infinite (if it is non-zero), whereas $\bar{T}_\ell^+(K)$ is supposed to be finite by the Greenberg conjecture [2].

We can resume the situation as follows. The group $\bar{T}_\ell^-(K)$ is a right object but it is only one half of the right Iwasawa module for an algebraic number field. In [8] the author undertook an attempt to define a good Iwasawa module for a CM-field K. (It was denoted by $A_\ell(K)$ in that paper). In brief, the definition is as follows. Let K_n be an n-layer of K_∞/K, that is, $[K_n : K] = \ell^n$, and M_n the maximal Abelian ℓ-extension of K_n unramified outside ℓ. By class field theory the Galois group $X_n = G(M_m/K_n)$ enters the exact sequence

$$U_S(K_n)[\ell] \xrightarrow{\alpha} \prod_{v|\ell} K_{n,v}^\times[\ell] \longrightarrow X_n \longrightarrow \tilde{Cl}_\ell(K_n) \longrightarrow 1, \tag{12.1}$$

where $K_{n,v}$ is the completion of K_n relative to the place v, $U_S(K_n)$ the group of S-units in K_n (S is the set of all $v|\ell$), symbol $[\ell]$ means pro-ℓ-completion and $\tilde{Cl}_\ell(K_n)$ is a factor of ℓ-class group of K_n by the subgroup generated by all prime divisors in S.

For a local field $K_{n,v}$ we put $\tilde{H}(K_v) = (K_{n,v}^\times/\mu(K_{n,v})[\ell])$ and $\tilde{H}(K_n) = \prod_{v|\ell} \tilde{H}(K_{n,v})$. Let $\tilde{U}_S(K_n)$ be the image of the natural map $U_S(K_n)[\ell] \to \tilde{H}(K_n)$ induced by α in (12.1). Then (12.1) induces an exact sequence (in additive notations)

$$0 \longrightarrow \tilde{U}_S(K_n) + \tilde{H}(K_n)^- \longrightarrow \tilde{H}(K_n) \longrightarrow Y_n \longrightarrow \tilde{Cl}_\ell(K_n) \longrightarrow 0.$$

Thus Y_n is a factor module of X_n, and by Galois theory there is the unique subfield $L_n \subseteq M_n$ such that the Galois group $G(L_n/K_n)$ equals Y_n. We put $L = \cup_n L_n$ and $A_\ell(K) = G(L/K_\infty) = \varprojlim Y_n$.

In [8] we had studied the Λ-module $A_\ell(K)$ for a CM-field K. The main result of that paper was a construction of an inner skew-symmetric product

$$A_\ell(K) \times A_\ell(K) \longrightarrow V, \quad V = \varprojlim \mu_\ell(K_n),$$

that is completely analogous to Weil scalar product.

The module $A_\ell(K)$ is Λ-torsion and, assuming that the Iwasawa μ-invariant $\mu_\ell(K)$ vanishes, we see that $A_\ell(K)$ is a free $\mathbb{Z}_\ell$-module of finite rank, which we denote by $\lambda(A_\ell(K))$. In this case we have the following theorem, which is an analog of Riemann-Hurwitz formula.

Theorem 12.1. *[8, Prop. 5] Let K be a field of CM-type containing a primitive ℓ-th root of unity and let K' be a field of CM-type, which is an ℓ-extension of K. Suppose that $\mu_\ell(K) = 0$. Then one has*

$$\lambda(A_\ell(K')) - 2 = [K'_\infty : K_\infty](\lambda(A_\ell(K)) - 2) + \sum_{\mathfrak{p} \in S_1} (e_\mathfrak{p} - 1),$$

where S_1 consists of all places of the field K'_∞ which split completely in the extension K/K^+ and either are divisors of ℓ or ramify in the extension K'/K. If $\mathfrak{p}$ is a place, $e_\mathfrak{p}$ denotes the local degree of the extension K'_∞/K_∞ at $\mathfrak{p}$, and for $\mathfrak{p} \nmid \ell$, $e_\mathfrak{p}$ is also equal to the ramification index of $\mathfrak{p}$ in K'_∞/K_∞.

So there is a problem to generalize the construction of $A_\ell(K)$ to the case of an arbitrary algebraic number field K. This problem yields a number of interesting conjectures and results, which will be the theme of this paper. Because of lack of the space we will omit full proofs of presented results. Instead of this we will explain our goals and the problems we meet with. Note that another approach to this problem was suggested in [22]. We shall not discuss it here, since it is based on absolutely different ideas.

Our paper is organized as follows.

In Section 12.2 we define the ℓ-adic regulator $R_\ell(K)$ of an algebraic number field K and the relative ℓ-adic regulator $R_\ell(K/k)$ of a finite extension of algebraic number fields and formulate our main conjectures: the conjecture on the ℓ-adic regulator (Conjectures 12.3) and the feeble conjecture on the ℓ-adic regulator (Conjecture 12.6). Then we state the first results on this conjectures. These conjectures follow from the ℓ-adic Schanuel conjecture (see Conjecture 12.2 at the end of this section) and hence are very probable. Nevertheless, it is interesting to give an unconditional proof of them in some particular cases.

In brief, an analog of $H^+(K_n)$ is well defined in any case, and we are going to define $\tilde{H}^-(K_n)$ for an arbitrary K as an orthogonal complement to $\tilde{H}^+(K_n)$ relative to the scalar product induced by local scalar products

$$U(K_{n,v}) \times U(K_{n,v}) \to \mathbb{Q}_\ell, \quad (x, y) \to \mathrm{Sp}_{K_n/\mathbb{Q}_\ell}(\log x \cdot \log y), \tag{12.2}$$

where $U(K_{n,v})$ is the group of units of $K_{n,v}$ and log means the ℓ-adic logarithm.

In Section 12.3 we introduce some tools that will be used in sequel. Namely, we define a Galois module $V(K_{n,v})$ that is close to $U(K_{n,v})$ and, assuming that $K_{n,v}$ is Abelian, define an inner product of the form $V(K_{\infty,v}) \times V(K_{\infty,v}) \to \Lambda_p$, where $\Lambda_p = \mathbb{Z}_\ell[[G(K_\infty/K_p)]]$, $V(K_{\infty,v}) = \varprojlim V(K_{n,v})$ and the limit is taken with respect to the norm maps (12.23). Let K be an algebraic number field Abelian over ℓ. This means that the completions K_v are Abelian for all $v|\ell$. Then we put $V(K) = \prod_{v|\ell} V(K_v)$ and $V(K_\infty) = \prod_{v|\ell} V(K_{\infty,v})$. If p is sufficiently large then the products (12.23) for all $v|\ell$ yields a semi-local product $T_\infty\colon V(K_\infty) \times V(K_\infty) \to \Lambda_p$ (see (12.25)). Thus we see immediately that there exists an orthogonal complement $V^-(K_\infty)$ of the Λ_p-module $V^+(K_\infty) = V(K_\infty) \cap H^+(K_\infty)$. We show (Theor. 12.7) that Conjecture 12.6 is equivalent to the assertion that $V^+(K_\infty) \cap V^-(K_\infty) = 0$.

In Section 12.4 we generalize the results of preceding section to the case of non-Abelian local fields. Using some complicated techniques (Laurent series with infinite principal part), we state the results (Theor. 12.10) analogous but more feeble than those of Section 12.3. Nevertheless these results are sufficient for application to Conjecture 12.6 that we give in Section 12.6.

In Section 12.5 we consider a relation between the product (12.25) and the norm residue symbol. Using this last symbol man can define an inner product $\langle\, , \rangle\colon V(K_\infty) \times V(K_\infty) \to \Lambda_p$, which looks as (12.25). So, if K is Abelian over ℓ, then there is an isomorphism $\varphi\colon V(K_\infty) \to V(K_\infty)$ such that $\langle x, y\rangle = T_\infty(x, \varphi(y))$ for any $x, y \in V(K_\infty)$. Since $V^+(K_\infty)$ is a maximal isotropic submodule of $V(K_\infty)$, we obtain that $V^+(K_\infty) \cong V^-(K_\infty)$ as H-modules for any finite group of automorphisms H (Prop. 5.1).

In Section 12.6 we use the obtained results to prove Conjecture 12.3 in some cases. We consider a finite ℓ-extension of algebraic number fields K/k. Assuming that Conjecture 12.6 holds for k and k satisfies some other natural conditions, we prove that Conjecture 12.6 holds for K as well.For example, if k is Abelian then Conjecture 12.6 holds for any ℓ-extension of k. Moreover, if K is Abelian over ℓ then we obtain a bit more (see Theorems 12.15, 12.16).

In Section 12.7 we study some other aspects of the situation. Firstly, we prove that Theor. 12.1 can be generalized to the case when K/k is a finite ℓ-extension of algebraic number fields, k contains a primitive ℓ-th root of unity if K is Abelian over ℓ and k satisfies the conditions of Theorems 12.15, 12.16 (see Theor. 12.21, 12.22). Note that there appears some new module $D(K_\infty)$ ($D'(K_\infty)$ if $\ell = 2$), which is zero in CM-case.

In Section 12.8 we introduce a new notion of the ℓ-adic regulator $\mathfrak{R}_\ell(K)$, assuming that ℓ splits completely in K. In distinction with $R_\ell(K)$, this last regulator is defined not in terms of the ℓ-adic logarithms but in terms of divisors. Conjecture 12.2 yields that $\mathfrak{R}_\ell(K) \neq 0$ for any K. Moreover, we give an unconditional proof of this last inequality for some Abelian extensions of imaginary quadratic field.

In conclusion we will state a particular case of the ℓ-adic Schanuel conjecture that we shall use in this paper.

Conjecture 12.2. *Let $x_1, \ldots, x_n \in \bar{\mathbb{Q}}_\ell^\times$ be algebraic over $\mathbb{Q}$ and $\ell, x_1, \ldots, x_n$ are multiplicatively independent. Then the ℓ-adic logarithms* $\log x_1, \ldots, \log x_n$ *are algebraically independent over $\mathbb{Q}$.*

A part of results of this last section was obtained while the author was a guest of the Trimester Program "Arithmetic and Geometry", Hausdorff research institute for mathematics, Bonn. The author is pleased to thank the collaborators of the institute for hospitality and the participants of the Trimester for stimulating communion.

12.2 The main definitions and conjectures

Now let K be any algebraic number field. In this case there is the cyclotomic $\mathbb{Z}_\ell$-extension K_∞/K and a Galois module $H(K_\infty)$. This last module may be defined in the same manner as in the case of CM-fields. Namely, if $K_\infty = \cup_n K_n$, where $[K_n : K] = \ell^n$ then we put $H(K_\infty) = \varprojlim \tilde{H}(K_n)$, where $\tilde{H}(K_n) = \prod_{v|\ell} K_{n,v}^\times[\ell]/\mu(K_{n,v})$ and the limit is taken with respect to the norm maps. Since there are only finitely many places $v|\ell$ in K_∞, we can also write $H(K_\infty) = \prod_{v|\ell} H(K_{\infty,v})$, where v runs through all places over ℓ in K_∞ and $H(K_{\infty,v}) = \varprojlim K_{n,v}^\times[\ell]/\mu(K_{n,v})$. If K is a CM-field then there is the automorphism of complex conjugation j and for $\ell \neq 2$ and any $\mathbb{Z}_\ell$-module A one has a decomposition $A = A^+ \oplus A^-$, where $A^+ = e^+A$ and $A^- = e^-A$, and $e^+ = (1+j)/2$, $e^- = (1-j)/2$. If $\ell = 2$ then we put $A^+ = (1+j)A$, and $A^- = (1-j)A$. Thus, if $\ell = 2$ and A has no 2-torsion then $A/(A^+ \oplus A^-)$ has period 2.

Our goal here is to generalize this decomposition to the case of any field, that is, to define Galois submodules $H^+(K_\infty), H^-(K_\infty) \subseteq H(K_\infty)$, whose properties are analogous to those of $H^+(K_\infty)$ and $H^-(K_\infty)$ for CM-fields.

There is quite natural way to define $H^+(K_\infty)$. Indeed, put $\bar{U}(K_\infty) = \varprojlim(U(K_n)[\ell]/\mu(K_n))$. Then $\bar{U}(K_\infty) \subseteq H(K_\infty)$ and $\bar{U}(K_\infty)$ is a Γ-module of rank $r_1 + r_2$ (see [7, Theorem 7.2]), where r_1 and r_2 is the number of real and complex places of K, respectively. Then we define $H^+(K_\infty)$ as the minimal Γ-submodule of $H(K_\infty)$ such that $\bar{U}(K_\infty) \subseteq H^+(K_\infty)$ and $H(K_\infty)/H^+(K_\infty)$ has no torsion.

The problem of definition of $H^-(K_\infty)$ is much more complicated. It seems convenient to introduce some product on $H(K_\infty)$, which is non-degenerate, and such that its restriction to $H^+(K_\infty)$ is non-degenerate as well. Then man can define $H^-(K_\infty)$ as an orthogonal complement of $H^+(K_\infty)$ with respect to this product. This is a general idea, and to realize it we have to overcome many difficulties. Moreover, we can realize this idea completely only under some additional assumptions. Nevertheless, this idea leads us to many interesting conjectures, phenomena and results.

To start with, for a Γ-extension K_∞/K of algebraic number fields we put $\tilde{\mathcal{A}}(K_n) = \prod_{v|\ell} U(K_{n,v})/\mu(K_{n,v})$ and $\bar{\mathcal{A}}(K_\infty) = \varprojlim \tilde{\mathcal{A}}(K_n)$, where the limit is taken with respect to the norm maps. We wish to define our product on $\bar{\mathcal{A}}(K_\infty)$, but at first we have to settle the situation for a finite level. Note that we have a product on the Galois module $\tilde{\mathcal{A}}(K)$. Indeed, for a local field K_v we have a product

$$(\, , \,)_v \colon U(K_v)/\mu(K_v) \times U(K_v)/\mu(K_v) \longrightarrow \mathbb{Q}_\ell, \tag{12.3}$$

where $(x, y)_v = \mathrm{Sp}_{K_v/\mathbb{Q}_\ell}(\log x \log y)$ and $\log \colon U(K_v) \to K_v$ is the ℓ-adic logarithm in the sense of Iwasawa. At first glance the product (12.3) is far from being good, though it is non-degenerate. We shall study its properties more carefully in the next section.

If $x, y \in \tilde{\mathcal{A}}(K)$ then $x = \prod_{v|\ell} x_v$, $\quad y = \prod_{v|\ell} y_\ell$, where we have $x_v, y_v \in U(K_v)/\mu(K_v)$, and we can put

$$S(x, y) = \sum_{v|\ell} (x_v, y_v)_v \in \mathbb{Q}_\ell. \tag{12.4}$$

Note that the sum of local traces is a global one, thus, one has $S(x, y) = \mathrm{Sp}_{K/\mathbb{Q}}(\log_\ell x \log_\ell y)$, where $\log_\ell$ means a homomorphism $\tilde{\mathcal{A}}(K) \to \prod_{v|\ell} K_v$, which coincides with the ℓ-adic logarithm log on each v-component, and $\mathrm{Sp}_{K/\mathbb{Q}}$ denotes the map $\prod_{v|\ell} K_v \cong K \otimes_{\mathbb{Q}} \mathbb{Q}_\ell \to \mathbb{Q}_\ell$ induced by the trace map from K to $\mathbb{Q}$. So we obtain a non-degenerate product of the form $S \colon \tilde{\mathcal{A}}(K) \times \tilde{\mathcal{A}}(K) \to \mathbb{Q}_\ell$.

There is a natural map $\alpha \colon U(K)[\ell] \to \tilde{\mathcal{A}}(K)$, and it is natural to assume that S rests non-degenerate after restriction to its image. Thus we come to the following conjecture.

Conjecture 12.3. (Conjecture on the ℓ-adic regulator.)*[9] Let (K, ℓ) be an algebraic number field and a rational prime number respectively. Let $\varepsilon_1, \dots, \varepsilon_r$ be a system of fundamental units of K and $\varepsilon_0 = 1 + \ell$ ($\varepsilon_0 = 5$ if $\ell = 2$). Define the ℓ-adic regulator of K by*

$$R_\ell(K) = \det(S(\varepsilon_i, \varepsilon_j)) \in \mathbb{Q}_\ell, \quad 0 \leqslant i, j \leqslant r. \tag{12.5}$$

Then for any (K, ℓ) one has $R_\ell(K) \neq 0$.

Note that $R_\ell(K)$ does not depend on the choice of the basis in $U(K)$, since it is defined up to the square of the determinant of transition matrix, which is unimodular. We note also that for a totally real K the regulator $R_\ell(K)$ in (12.5) is almost a square of the Leopoldt ℓ-adic regulator $\mathbf{R}_\ell(K)$. More precisely, man has a relation [17, Proposition 2.1]

$$R_\ell(K) = (\log \varepsilon_0)^2 [K : \mathbb{Q}]^2 \mathbf{R}_\ell(K)^2.$$

Thus non-vanishing of $R_\ell(K)$ for a totally real field is equivalent to the validity of Leopoldt conjecture. In general, Conjecture 12.3 is essentially stronger. It asserts that $\varepsilon_0, \dots, \varepsilon_r$ generate a submodule in $\tilde{A}(K)$ of $\mathbb{Z}_\ell$-rank $r + 1$ and, moreover, the product (12.4) is non-degenerate being restricted to this submodule.

Theorem 12.4. *Conjecture 12.3 is a consequence of Conjecture 12.2.*

We can assume that K is Galois over $\mathbb{Q}$ and contains an imaginary quadratic field k. Then there exists an Artin unit ε in $U(K)$. This means that $\tau(\varepsilon) = \varepsilon$, where τ is an automorphism of complex conjugation in K, and the elements $h(\varepsilon)$, $h \in H = G(K/k)$, generate a subgroup of finite index in $U(K)$. Then, putting $\varepsilon' = \varepsilon_0 \varepsilon$ and $T_h = \log(h(\varepsilon'))$ for $h \in H$, we see that $R_\ell(K)$ is a non-zero polynomial in T_h for $h \in H$ of degree $2r$ with coefficients in $\mathbb{Z}$. Since the elements $h(\varepsilon')$, $h \in H$, are multiplicatively independent and hence their logarithms are algebraically independent over $\mathbb{Q}$ by Conjecture 12.2, we get $R_\ell(K) \neq 0$. This proves the theorem.

There is one particular case, when Conjecture 12.3 can be proved unconditionally.

Theorem 12.5. *[9, Theorem 3.1] Let K be an Abelian extension of an imaginary quadratic field k. Suppose that K is Galois over $\mathbb{Q}$ and the only non-trivial automorphism $\tau \in G(k/\mathbb{Q})$ acts on the Galois group $H = G(K/k)$ by the rule $\tau(h) = \tau h \tau^{-1} = h^{-1}$ for any $h \in H$. Then $R_\ell(K) \neq 0$ for any prime ℓ.*

Proof. For any $h_i, h_j \in H$ one has

$$\tau(\mathrm{Sp}_{K/k}(\log_\ell h_i(\varepsilon')\log_\ell h_j(\varepsilon')) = \mathrm{Sp}_{K/k}(\log_\ell \tau h_i(\varepsilon')\log_\ell \tau h_j(\varepsilon'))$$
$$= h_ih_j\mathrm{Sp}_{K/k}(h_ih_j(\log_\ell h_i^{-1}(\varepsilon')\log_\ell h_j^{-1}(\varepsilon'))) = \mathrm{Sp}_{K/k}(\log_\ell h_j(\varepsilon')\log_\ell h_i(\varepsilon')).$$

Thus, one has $\mathrm{Sp}_{K/\mathbb{Q}}(\log_\ell u_1 \log_\ell u_2) = 2\mathrm{Sp}_{K/k}(\log_\ell u_1 \log_\ell u_2)$ for any $u_1, u_2 \in U(K)$, and we can finish the proof by the same arguments, which used Brumer in his proof of Leopoldt conjecture for Abelian fields [1].

Our definition of the ℓ-adic regulator allows us to introduce an important notion of the relative ℓ-adic regulator. Let K/k be a finite extension of algebraic number fields. Let $U(K/k)$ be the relative group of units of K/k, that is,

$$U(K/k) = \{x \in U(K) \mid N_{K/k}(x) \in \mu(k)\}, \tag{12.6}$$

where $N_{K/k}$ is the norm map from K to k. □

Definition. Let $u_1, \ldots, u_t$ be some basis of $U(K/k)$. Then the ℓ-adic regulator $R_\ell(K/k)$ is defined by

$$R_\ell(K/k) = \det(\mathrm{Sp}_{K/\mathbb{Q}}(\log_\ell u_i \log_\ell u_j)) \in \mathbb{Q}_\ell, \quad 1 \leqslant i, j \leqslant t.$$

For a finite extension of algebraic number fields K/k and prime ℓ the condition $R_\ell(K/k) \neq 0$ means that α induces an injection $(U(K/k)/\mu_\ell)[\ell] \to \tilde{A}(K)$ an the product (12.4) is non-degenerate on the image of this injection.

Let K/k be an extension of algebraic number fields. Then $R_\ell(K) \neq 0$ if and only if $R_\ell(k) \neq 0$ and $R_\ell(K/k) \neq 0$. Moreover, there is an explicitly computable non-zero constant $c(K/k) \in \mathbb{Q}$ such that $R_\ell(K) = c(K/k)R_\ell(k)R_\ell(K/k)$ [17, Prop. 2.3](see also Sect. 7).

Conjecture 12.6. Feeble conjecture on the ℓ-adic regulator. *Let K be an algebraic number field and K_∞ the cyclotomic $\mathbb{Z}_\ell$-extension of K. Then there is an index n_0, which depends only on K and ℓ, such that for any $m > n \geqslant n_0$ man has $R_\ell(K_m/K_n) \neq 0$.*

This conjecture is in the same position to Conjecture 12.3 as the feeble Leopoldt conjecture to Leopoldt conjecture. The feeble Leopoldt conjecture is, in fact, a theorem, whereas Conjecture 12.6 is stated only for some particular cases (see Sections 12.6 and 12.8). Conjecture 12.6 is more tractable than Conjecture 12.3, and in the next section we discuss some approach to it.

12.3 Local and semi-local duality

Let k be a finite cyclotomic extension of $\mathbb{Q}_\ell$ that contains a primitive root of unity of degree ℓ (of degree 4 if $\ell = 2$). By ζ_n we denote a primitive root of unity of degree ℓ^{n+1} (of degree 2^{n+2} for $\ell = 2$). By ξ_r we denote a primitive root of unity of degree r for some r prime to ℓ. Thus, k is of the form $k = \mathbb{Q}_\ell(\zeta_{n_0}, \xi_r)$ for $n_0 \geqslant 0$, and $H := \mathbb{Q}_\ell(\xi_r)$ is the maximal unramified subfield of k. Put $k_n = k(\zeta_n)$. Then $k_n = k$ for $n \leqslant n_0$, $[k_n : k] = \ell^{n-n_0}$ for $n > n_0$, and $k_\infty := \cup_n k_n$ is the cyclotomic $\mathbb{Z}_\ell$-extension of k. If A is a full lattice in k_n then we put

$$A^\perp = \{ y \in k_n \mid \mathrm{Sp}_{k_n/\mathbb{Q}_\ell}(xy) \in \mathbb{Z}_\ell \text{ for all } x \in A \}.$$

Our first goal here is to determine the dual lattice $B_n := (\log U(k_n))^\perp$, where $U(k_n)$ is the group of units of k.

Note that this group was determined by Iwasawa in [4], and this remarkable result was a starting point for our paper [10].

Let $\mathcal{O}_H$ be the ring of integers of H, $[H : \mathbb{Q}_\ell] = f$, and φ the Frobenius automorphism of H. Let $\alpha, \beta \in \mathcal{O}_H$ form integer normal dual bases of H. It means that

$$\mathrm{Sp}_{H/\mathbb{Q}_\ell}(\varphi^i(\alpha) \cdot \beta) = \begin{cases} 1 & \text{if } i = 0, \\ 0 & \text{if } i = 1, \dots, f-1. \end{cases} \tag{12.7}$$

Put

$$\Theta_n(\beta, i) - q_n^{-1} \sum_{s=0}^{m-1} \zeta_n^{-i\ell^s} \varphi^{s+1-m}(\beta), \tag{12.8}$$

where $m = n+1$ if $\ell \neq 2$ ($m = n+2$ if $\ell = 2$) and $q_n = \ell^m$.

Generalizing the results of Iwasawa, we had proved in [10, Prop. 1.2] that for any indices i, i_1, i_2 prime to ℓ we have

$$\Omega_n(\beta, i_1, i_2) : = \Theta_n(\beta, i_1) - \Theta_n(\beta, i_2) \in B_n, \tag{12.9}$$

$$\Lambda_n(\beta, i) : = \Theta_n(\varphi(\beta), i) - \Theta_n(\beta, i) - q_n^{-1}\varphi(\beta) \in B_n. \tag{12.10}$$

By $\mathfrak{M}(k_n)$ (resp. by $\mathfrak{X}(k_n)$) we denote an additive Galois $\mathbb{Z}_\ell$-submodule of k_n generated by $\Theta_n(\beta, 1)$ (resp. by all $\Omega_n(\beta, i_1, i_2)$ and $\Lambda_n(\beta, i)$).

Thus, $\mathfrak{M}(k_n)$ is a cyclic free $\mathbb{Z}_\ell[G(k_n/\mathbb{Q}_\ell)]$-module, which is generated by any of two elements $\Theta_n(\alpha, 1)$ and $\Theta_n(\beta, -1)$, where α and β are as in (12.7).

We wish to introduce a construction that unites the products (17.10) for all n. To do this, we at first define this construction for $\mathfrak{M}(k_n)$, then for $\mathfrak{M}(k_n)^\perp$, and at last for $\mathfrak{X}(k_n)^\perp$. Note that $\mathfrak{X}(k_n)^\perp$ is close to the group $\log U(k_n)$. Indeed, let $u_n \in U(k_n)$ be an element such that $k_n(\sqrt[q_n]{u_n})$ is the unramified extension of k_n of degree q_n. We denote by $V(k_n)$ a multiplicative group generated by $U(k_n)/\mu(k_n)$ and $\sqrt[q_n]{u_n}$. Then $V(k_n)$ is a $G(k_n/\mathbb{Q}_\ell)$-module and $V(k_n)/(U(k_n)/\mu(k_n)) \cong \mu_\ell(k_n)$.

To construct our product, we put for any $x, y \in k_n$

$$\mathfrak{T}_n(x, y) = \sum_{\sigma \in G(k_n/\mathbb{Q}_\ell)} \mathrm{Sp}_{k_n/\mathbb{Q}_\ell}(\sigma^{-1}(x) \cdot y)\sigma \in \mathbb{Q}_\ell[G(k_n/\mathbb{Q}_\ell)]. \qquad (12.11)$$

One can easily check that (12.11) is compatible with the trace maps, that is, for any $m > n$ there is a commutative diagram

$$\begin{array}{ccccc} k_m & \times & k_m & \xrightarrow{\mathfrak{T}_m} & \mathbb{Q}_\ell[G(k_m/\mathbb{Q}_\ell)] \\ \downarrow \mathrm{Sp}_{m,n} & & \downarrow \mathrm{Sp}_{m,n} & & \downarrow \pi_{m,n} \\ k_n & \times & k_n & \xrightarrow{\mathfrak{T}_n} & \mathbb{Q}_\ell[G(k_n/\mathbb{Q}_\ell)] \end{array} \qquad (12.12)$$

where $\mathrm{Sp}_{m,n}$ is the trace from k_m to k_n and $\pi_{m,n}$ is the natural projection of the group rings.

Now we assume that the roots of unity $\zeta_n \in k_n$ are coherent for all n, that is, $\zeta_{n+1}^\ell = \zeta_n$. Then, passing to projective limits in (12.12), we obtain a product

$$\mathfrak{T}_\infty : \tilde{k}_\infty \times \tilde{k}_\infty \longrightarrow \tilde{\Lambda} := \mathbb{Q}_\ell[[G(k_\infty/\mathbb{Q}_\ell)]], \qquad (12.13)$$

where $\tilde{k}_\infty = \varprojlim k_n$ and $\tilde{\Lambda} = \varprojlim \mathbb{Q}_\ell[G(k_n/\mathbb{Q}_\ell)]$. This product is non-degenerate and sesquilinear in the sense that

$$\mathfrak{T}_\infty(\lambda x, \mu y) = \lambda \mathfrak{T}_\infty(x, y)\mu^* \text{ for any } \lambda, \mu \in \tilde{\Lambda}, \qquad (12.14)$$

where $*$ means an anti-automorphism of $\tilde{\Lambda}$ induced by the map $\sigma \to \sigma^{-1}$ for any $\sigma \in G(k_\infty/\mathbb{Q}_\ell)$.

Now, restricting (12.13) to the Galois module $\mathfrak{M}(k_\infty) := \varprojlim \mathfrak{M}(k_n)$, we obtain a product of the form

$$\mathfrak{M}(k_\infty) \times \mathfrak{M}(k_\infty) \longrightarrow \tilde{\Lambda}. \qquad (12.15)$$

Note that, under our choice of the roots ζ_n, the generators $\Theta_n(\alpha, 1)$ and $\Theta_n(\beta, -1)$ behave well with respect to the trace maps. To be precise, the sequences $\{\Theta_n(\varphi^{-n}(\alpha), 1)\}$ and $\{\Theta_n(\varphi^{-n}(\beta), -1)\}$ are compatible with respect to the trace maps, and we can put $\Theta_\infty = \varprojlim \Theta_n(\varphi^{-n}(\alpha), 1)$, $\Theta'_\infty = \varprojlim \Theta_n(\varphi^{-n}(\beta), -1)$.

By next step we can compute explicitly the element $\Phi_\infty := \mathfrak{T}_\infty(\Theta_\infty, \Theta'_\infty)$. Indeed, we have $\Phi_\infty = \varprojlim \Phi_n$, where $\Phi_n = \mathfrak{T}_n(\Theta_n(\varphi^{-n}(\alpha), 1), \Theta_n(\beta, -1)) = (\Theta_n(\alpha, 1), \Theta_n(\beta, -1))$. The calculation of this last element reduces to calculation of the traces of different powers of ζ_n. These calculations were carried out in [10, Prop. 3.1]. It was stated there that $\Phi_n = -\ell^{-(n+2)}\mathrm{Sp}_{G_n} + \Psi_n$ if $\ell \neq 2$ and $\Phi_n = \ell^{-1}\Psi_n$ if $\ell = 2$, where $G_n = G(k_n/H)$ and

$$\Psi_n = \ell^{-(n+2)} \left[\ell\, \mathrm{Sp}_{n,0} + \sum_{r=0}^{n-1} \left(\ell\, \mathrm{Sp}_{n,r+1} - \mathrm{Sp}_{n,r}\right) \right]. \tag{12.16}$$

In this formula $\mathrm{Sp}_{n,i}$ means $\mathrm{Sp}_{k_m/k_i} = \sum_{\sigma \in G(k_n/k_i)} \sigma$, and we put $k_i = H(\zeta_i)$ for any $i \geqslant 0$.

Our next goal is to pass from $\mathfrak{M}(k_n)$ to $\mathfrak{M}(k_n)^{\perp}$. To do this, we note that the element Φ_n is invertible in $\mathbb{Q}_\ell[G(k_n/\mathbb{Q}_\ell)]$ for any n, whereas the elements Φ_n^{-1} can be given explicitly. Namely, we have [10, Prop. 3.2]

$$\Psi_n^{-1} = \ell^{n+1} - \sum_{r=1}^{n} (\ell - 1)\ell^{n+1-2r}\mathrm{Sp}_{n,n-r} \tag{12.17}$$

and for $\ell \neq 2$

$$\Phi_n^{-1} = \Psi_n^{-1} + \ell^{1-n}\mathrm{Sp}_{G_n}.$$

Then, applying an analog of (12.14) to the product $\mathfrak{T}_n$, we obtain immediately $\mathfrak{T}_n(\Phi_n^{-1}(\Theta(k_n)), \Theta'(k_n)) = \Phi_n^{-1}\Phi_n = 1$. In other words, this means that $\mathfrak{M}(k_n)^{\perp} = \Phi_n^{-1}\mathfrak{M}(k_n)$ and $\mathfrak{M}(k_n)^{\perp}$ is a free cyclic Galois module over the group ring $\mathbb{Z}_\ell[G(k_n/\mathbb{Q}_\ell)]$, which may be generated by any of two elements $\xi(k_n) = \Phi_n^{-1}(\Theta(k_n))$ and $\xi'(k_n) = \Phi_n^{-1}(\Theta'(k_n))$ [10, Prop.3.3]. Using explicit representations (12.8) and (12.17), we obtain

$$\xi(k_n) = \delta\alpha + \sum_{s=0}^{m-1} \ell^{-s}\zeta_n^{-\ell^s}\varphi^{s-m+1}(\alpha), \tag{12.18}$$

$$\xi'(k_n) = \delta\beta + \sum_{s=0}^{m-1} \ell^{-s}\zeta_n^{\ell^s}\varphi^{s-m+1}(\beta), \tag{12.19}$$

where $\delta = -\ell^{1-m}$ if $\ell \neq 2$ and $\delta = 0$ if $\ell = 2$. The elements $\xi(k_n)$ and $\xi'(k_n)$ are compatible for different n with respect to the trace maps and, passing to inverse limits, we obtain two elements $\xi(k_\infty) := \varprojlim \xi(k_n)$ and $\xi'(k_\infty) := \varprojlim \xi'(k_n)$, each of which generates a free $\mathbb{Z}_\ell[[(G(k_\infty/\mathbb{Q}_\ell)]]$-module $\mathfrak{M}(k_\infty)^{\perp} = \varprojlim \mathfrak{M}(k_n)^{\perp}$. Moreover, we have $\mathfrak{T}_\infty(\xi(k_\infty, \xi'(k_\infty)) = \Phi_\infty^{-1}$.

If $x, y \in \mathfrak{M}(k_\infty)^{\perp}$ then there are the unique presentations $x = \lambda\xi(k_\infty)$, $y = \mu\xi'(k_\infty)$ for some $\lambda, \mu \in \Lambda := \mathbb{Z}_\ell[[G(k_\infty/\mathbb{Q}_\ell)]]$. Thus, $\mathfrak{T}_\infty(x, y) = \lambda\Phi^{-1}\mu* = \lambda\mu * \Phi^{-1}$, since the Galois group $G(k_\infty/\mathbb{Q}_\ell)$ is Abelian. So, by definition the product $\mathfrak{T}_\infty$ takes its values in the very large ring

$\mathbb{Q}_\ell[[G(k_\infty/\mathbb{Q}_\ell)]]$. But in fact these values belong to a free Λ-module of rank 1, namely to $\tilde{\Lambda}\Phi^{-1}$.

Fix some index $p > 0$. For any $n \geqslant p$ there is a natural restriction map $\mathrm{res}_n\colon \mathbb{Q}_\ell[G(k_n/\mathbb{Q}_\ell)] \to \mathbb{Q}_\ell[G(k_n/k_p)]$, where we put $\mathrm{res}_n(\sum_{\sigma\in G(k_n/\mathbb{Q}_\ell)} a_\sigma\sigma) = \sum_{\sigma\in G(k_n/k_p)} a_\sigma\sigma$. Passing to inverse limit, we get a map $\mathrm{res}_\infty\colon \mathbb{Q}_\ell[[G(k_\infty)]] \to \mathbb{Q}_\ell[[G(k_\infty/k_p)]]$. Then, taking into account that $(\gamma_p-1)\Phi_n^{-1} \in \mathbb{Q}_\ell[G(k_n/k_p)]$, hence $(\gamma_p-1)\Phi_\infty^{-1} \in \mathbb{Q}_\ell[[G(k_\infty/k_p)]]$, and putting $\bar{\mathfrak{T}}_\infty(x,y) = \mathrm{res}_\infty(\mathfrak{T}_\infty(x,y))$, we get an inner product [10, Prop. 3.5]

$$\bar{\mathfrak{T}}_\infty\colon (\gamma_p-1)\mathfrak{M}(k_\infty)^\perp \times \mathfrak{M}(k_\infty)^\perp \longrightarrow (\gamma_p-1)B_1, \qquad (12.20)$$

where $(\gamma_p-1)B_1 = (\gamma_p-1)\Lambda_p\Phi_\infty^{-1}$ and $\Lambda_p := \mathbb{Z}_\ell[[G(k_\infty/k_p)]]$ is the ring of formal power series in $T_p = \gamma_p - 1$.

Since Λ_p is an integral domain, the multiplication by (γ_p-1) induces an isomorphism $(\gamma_p-1)\colon \Lambda_p \to (\gamma_p-1)\Lambda_p$. So, we can modify (12.20) once more, putting $\tilde{\mathfrak{T}}_\infty(x,y) = h \in \Lambda_p$ if $\bar{\mathfrak{T}}(x,y) = (\gamma_p-1)h\Phi_\infty^{-1}$. In such a way we obtain an inner product

$$\tilde{\mathfrak{T}}_\infty\colon \mathfrak{M}(k_\infty)^\perp \times \mathfrak{M}(k_\infty)^\perp \longrightarrow \Lambda_p. \qquad (12.21)$$

Next step is passing from $\mathfrak{M}(k_n)^\perp$ to $\mathfrak{X}(k_n)^\perp = \log V(k_n)$. We prove [10, Theor. 2.1] that $(\sigma-1)\mathfrak{X}(k_n)^\perp = (\sigma-1)\mathfrak{M}(k_n)^\perp$ for any $\sigma \in G_n = G(k_n/H)$. So, passing to the limit, we get an isomorphism $(\gamma_p-1)\mathfrak{X}(k_\infty)^\perp = (\gamma_p-1)\mathfrak{M}(k_\infty)^\perp$. Since (12.21), in fact, depends only on the elements $(\gamma_p-1)x$, $(\gamma_p-1)y$, which are the elements of $(\gamma_p-1)\mathfrak{X}(k_\infty)^\perp$, this product defines an inner product of the form

$$\hat{\mathfrak{T}}_\infty\colon \mathfrak{X}(k_\infty)^\perp \times \mathfrak{X}(k_\infty)^\perp \longrightarrow \Lambda_p. \qquad (12.22)$$

Since the ℓ-adic logarithm induces an isomorphism $\log\colon V(k_\infty) \to \mathfrak{X}(k_\infty)^\perp$, we obtain an inner product

$$T_\infty\colon V(k_\infty) \times V(k_\infty) \longrightarrow \Lambda_p, \qquad (12.23)$$

where for $x, y \in V(k_\infty)$ we put $T_\infty(x,y) = \hat{\mathfrak{T}}(\log x, \log y)$.

It is easy to generalize our results to the case when k is an arbitrary finite Abelian extension of $\mathbb{Q}_\ell$. Indeed, put $F = k(\zeta_0)$. Then $\Delta := G(F_\infty/k_\infty)$ is a cyclic group and $|\Delta|\,|(\ell-1)$ ($|\Delta| = 2$ if $\ell = 2$). If we put $V(k_\infty) = N_\Delta V(F_\infty)$, then we have a non-degenerate inner product $T_{k_\infty}\colon V(k_\infty) \times V(k_\infty) \to \Lambda_p$, which is defined as follows. Let $x, y \in V(k_\infty)$. We have two natural maps: the inclusion $i\colon V(k_\infty) \hookrightarrow V(F_\infty)$ and the norm map $N_\Delta\colon V(F_\infty) \to V(k_\infty)$. Let $z \in V(F_\infty)$ be such an element that $N_\Delta(z) = y$. Then we put $\mathfrak{T}_{k_\infty}(x,y) = T_{F_\infty}(i(x), z)$, where T_{F_∞} is the product (12.23) defined for F_∞.

Now we consider the semi-local situation. Let K be an algebraic number field such that the completions K_v are Abelian for all $v|\ell$. Let p be such an index that any place $v|\ell$ is purely ramified in K_∞/K_p. Thus, for any $v|\ell$ we have a canonical isomorphism $G(K_\infty/K_p) \cong G(K_{n,v}/K_{p,v})$. So, we have for any $v|\ell$ a product of the form (12.23)

$$T_{v,\infty}\colon V(K_{\infty,v}) \times V(K_{\infty,v}) \longrightarrow \Lambda_{p,v}, \tag{12.24}$$

where we can identify $\Lambda_{p,v}$ with Λ_p. Then, putting $V(K_\infty) = \prod_{v|\ell} V(K_{\infty,v})$, we obtain an inner product

$$T_\infty\colon V(K_\infty) \times V(K_\infty) \longrightarrow \Lambda_p, \tag{12.25}$$

where for $x, y \in V(K_\infty)$, $x = \prod_{v|\ell} x_v$, $y = \prod_{v|\ell} y_v$ we put $T_\infty(x, y) = \sum_{v|\ell} T_{\infty,v}(x_v, y_v)$. Note that (12.25) is defined via the sum of local traces that coincides with global one.

Let be given a $G(K_\infty/K_p)$-module $A \subset V(K_\infty)$ and $B = A^\wedge$ is an orthogonal complement of A with respect to (12.25). Since (12.25) is sesquilinear, B is also a $G(K_\infty/K_p)$-module. One has also $\operatorname{rk} A + \operatorname{rk} B = \operatorname{rk} V(K_\infty) = [K_p : \mathbb{Q}]$, where rk means the rank over Λ_p. Let $a \in A, b \in B$, where $a = \varprojlim a_n, b = \varprojlim b_n$, $a_n, b_n \in V(K_n) = \prod_{v|\ell} V(K_{n,v})$. Then it follows from the definition of T_∞ that $\mathrm{Sp}_{K_n/\mathbb{Q}}((\gamma_p - 1)\log_\ell(a_n)\log_\ell(b_n)) = 0$ for all $n > p$. Applying this to the case $A = V^+(K_\infty)$ and denoting $(V^+(K_\infty))^\wedge$ by $V^-(K_\infty)$, we get the following statement [10, Theor. 5.1].

Theorem 12.7. *Let K be Abelian over ℓ, that is the completions K_v are Abelian for all $v|\ell$. Put $C(K_\infty) = V^+(K_\infty) \cap V^-(K_\infty)$. Then $C(K_\infty) = 0$ if and only if Conjecture 12.6 holds for (K, ℓ).*

Problem 12.1. *Let $a, b \in V(K_\infty$, $a \in V^+(K_\infty)$, $b \in V^-(K_\infty)$ and, for a given n, let a_n and b_n be the images of a and b relative to the natural projection $V(K_\infty) \to V(K_n)$. Is it true that $S(a_n, b_n) = 0$, where S is the product* (12.4)*?*

12.4 Duality results for non-Abelian local fields

Now we consider the case, when k is any (non-Abelian) finite extension of $\mathbb{Q}_\ell$. We wish to generalize the results of the preceding section to the case of the cyclotomic $\mathbb{Z}_\ell$-extension k_∞/k. To do this, we have firstly to formulate the results of the preceding section in the form, which is maximally independent of the special properties of cyclotomic fields. For example, such elements as $\xi(k_n)$ and $\Theta(k_n)$ cannot be defined in the general setting. So our first goal is to formulate the results of Section 12.3 in the form, which is suitable for generalization. This was done in [12].

Thus for a while we suppose that k is a cyclotomic field of the form $k = H(\zeta_0)$, $k_n = H(\zeta_n)$ and $k_\infty = \cup_n k_n$, where H is an unramified extension of $\mathbb{Q}_\ell$. Put $\Gamma = G(k_\infty/k)$ and $\Gamma_n = G(k_\infty/k_n)$. Let γ be a fixed topological generator of Γ and $\gamma_n = \gamma^{\ell^n}$. We define the groups $V(k_n)$ and $V(k_n)^\perp$ as in Section 12.3. We put for $n > 0$, as in Section 12.3,

$$\Phi_n = q_n^{-1}\ell^{-1}\left[\delta \mathrm{Sp}_{G_n} + \ell \mathrm{Sp}_{\Gamma/\Gamma_n} + \sum_{i=0}^{n-1} \mathcal{F}_n^{(i)}\right] \in \mathbb{Q}_\ell[G_n],$$

where $\delta = -1$ if $\ell \neq 2$, $\delta = 0$ if $\ell = 2$, $G_n = G(k_n/H)$ and

$$\mathcal{F}_n^{(i)} = \ell \mathrm{Sp}_{\Gamma_{n-i}/\Gamma_n} - \mathrm{Sp}_{\Gamma_{n-i-1}/\Gamma_n} \in \mathbb{Q}_\ell[G_n].$$

Theorem 12.8. *Let k_n be as above. Then for a fixed index $p \geqslant 0$ and any $n > p$ we have*

$$\operatorname{Ker} \mathrm{Sp}_{k_n/k_p}(\log V(k_n)^\perp) = (\gamma_p - 1)\Phi_n(\log V(k_n)),$$

where $\operatorname{Ker} \mathrm{Sp}_{k_n/k_p}$ *is the kernel of the corresponding trace map.*

This theorem was stated in [12, Theor. A] for $p = 0$ and then generalized in [15] for any p.

We shall say a few words on the proof given in [12], since this proof can be generalize to the non-Abelian case. The proof breaks in two parts. At first, we prove that

$$(\gamma_p - 1)\Phi_n(\log U(k_n)) \subseteq (\log U(k_n))^\perp, \tag{12.26}$$

where $U(k_n)$ is the group of units of k_n. In the second part we prove that (12.26) implies an inclusion

$$(\gamma_p - 1)\Phi_n(\log V(k_n)) \subseteq \operatorname{Ker} \mathrm{Sp}_{k_n/k_p}(\log V(k_n))^\perp, \tag{12.27}$$

which in fact is an equality.

While the second part of the proof is rather routine, the first part is more interesting, and we will to discuss it in some details. Obviously, we have to prove that

$$A := \mathrm{Sp}_{k_n/\mathbb{Q}_\ell}((\gamma_p - 1)\Phi_n x \cdot y) \in \mathbb{Z}_\ell \tag{12.28}$$

for any $x, y \in \log U(k_n)$. The group $\log U(k_n)$ is generated by the Artin-Hasse logarithms $\eta_n(\alpha, a) := \sum_{r=0}^{\infty} \varphi^{-r}(\alpha)\ell^{-r}\pi_n^{a\ell^r}$, where α belongs to the ring of integers $\mathcal{O}_H$ of H, φ is the Frobenius automorphism of H, $a = 1, 2, \ldots$ and

$\pi_n = 1 - \zeta_n$ is a local parameter of k_n. So it is enough to prove (12.28) for $x = \eta_n(\beta, b)$ and $y = \eta_n(\alpha, a)$.

So we put $C_n^i(\alpha, a, r) = \mathcal{F}_n^{(i)}(\varphi^r(\alpha)\ell^{-r}\pi_n^{a\ell^r})$. For $r < 0$ we put $C_n^i(\alpha, a, r) = 0$ by definition. Taking into account that $1 = \sum_{i=0}^{n-1} \ell^{-(i+1)}\mathcal{F}_n^{(i)} + \ell^{-n}\mathrm{Sp}_{\Gamma/\Gamma_n}$ and the elements $\mathcal{F}_n^{(i)}$, $\mathcal{F}_n^{(j)}$ are orthogonal for $i \neq j$, we may rewrite (12.28) for the above-mentioned x and y in the form

$$A = \sum_{-\infty<r,s<\infty} \sum_{i=0}^{n-1} q_n^{-1}\ell^{-(i+2)}\mathrm{Sp}_{k_n/\mathbb{Q}_\ell}\Big[C_n^i(\alpha, a, r+i) \times (\gamma - 1)\big(C_n^i(\beta, b, s+i)\big)\Big].$$

To prove the theorem, it is enough to show that for any integers r, s we have

$$B(r,s) := \sum_{i=0}^{n-1} q_n^{-1}\ell^{-(i+2)}\mathrm{Sp}_{k_n/\mathbb{Q}_\ell}\Big[C_n^i(\alpha, a, r+i) \times (\gamma - 1)\big(C_n^i(\beta, b, s+i)\big)\Big] \equiv 0 \pmod{\mathbb{Z}_\ell}. \quad (12.29)$$

Since one can easily prove by induction on r a congruence $\pi_{n+1}^{a\ell^{r+1}} \equiv \pi_n^{a\ell^r} \pmod{\ell^{r+1}}$, it follows that (see [12, Lemma 3])

$$B(r,s) \equiv q_n^{-1}\ell^{-(n+1)} \sum_{j=n}^{\infty} \mathrm{Sp}_{k_j/\mathbb{Q}_\ell}\Big[C_j^{n-1}(\varphi^{n-j}(\alpha), a, n+r-1) \times (\gamma - 1)\big(C_j^{n-1}(\varphi^{n-j}(\beta), b, n+s-1)\big)\Big] \pmod{\mathbb{Z}_\ell}. \quad (12.30)$$

Namely, we prove that the i-th summand in (12.29) is congruent to the $(n+i)$-th summand in the last sum modulo $\mathbb{Z}_\ell$ and all summands that correspond $j \geqslant 2n - 1$ are congruent 0 modulo $\mathbb{Z}_\ell$.

Thus instead of sum (12.29), which include summands of different form but all belonging to k_n, we obtain another sum, all terms of which look uniformly, but belong to different fields. Note that we can also add to the sum (12.30) all terms with $j < n$, since all these terms are zero because of the factor $(\gamma - 1)$.

The next step is to introduce some universal power series that presents the terms of (12.30). This is done in [12, Sect. 3]. Namely, we prove that there is a formal power series $F(T) \in \mathcal{O}_H[[T]]$ such that the j-th term of (12.30) equals $\mathrm{Sp}_{k_j/\mathbb{Q}_\ell}(F^{(-j)}(\pi_j))$ times some power of ℓ independent of j. Here $F^{-j}(T)$ means the result of action of φ^{-j} on the coefficients of $F(T)$. Obviously, this twisting by a power of Frobenius does not change the trace. The final almost obvious fact we need to state the theorem is the following [12, Lemma 6]. For any formal power series $F(T) \in \mathcal{O}_H[[T]]$ we have $\sum_{j=-1}^{\infty} \mathrm{Sp}_{k_j/\mathbb{Q}_\ell} F^{(-j)}(\pi_j) = 0$, where $\pi_{-1} = 0$. (A slight modification is necessary if $\ell = 2$.) This proves the theorem.

Now we have to repeat this proof for arbitrary finite (non-Abelian) extension of $\mathbb{Q}_\ell$. This was done in [14]. The first problem is to define a coherent system of local parameters. So let K be any finite extension of $\mathbb{Q}_\ell$ and K_∞ the cyclotomic $\mathbb{Z}_\ell$-extension of K. Assume that K contains a primitive ℓ-th root of unity ($\sqrt{-1} \in K$ if $\ell = 2$). Let H be the maximal unramified subfield of K_∞, $k_n = H(\zeta_n)$ and $K_n = K \cdot k_n$. Then for all sufficiently large n the extension K_n/k_n is purely ramified of some degree e, which does not depend on n.

Proposition 12.9. *[14, prop. 1.2] Put $R = \mathcal{O}_H[[T]]$. Then there is an irreducible (over R) monic polynomial $F(X, T) \in R[X]$ of degree e such that $F(X, \pi_n)$ is irreducible in $\mathcal{O}(k_n)[X]$ for any $n \geqslant 0$. For all sufficiently large n there is a local parameter λ_n in K_n that satisfies $F^{(-n)}(\lambda_n, \pi_n) = 0$. There is a positive integer δ (independent of n) such that the different $\mathcal{D}(K_n/k_n)$ satisfies the condition $\mathcal{D}(K_n/k_n) = (\lambda_n^\delta)$ for all sufficiently large n. Moreover, one can choose the elements λ_n so that for all sufficiently large n*

$$\lambda_{n+1}^\ell \equiv \lambda_n \pmod{\ell\lambda^{-\delta}}. \tag{12.31}$$

Remark. The equality $\mathcal{D}(K_n/k_n) = (\lambda_n^\delta)$ was originally stated in [23].

Sketch of the proof. Let q be such an index that $\nu_\ell(\mathcal{D}(K_q/k_q)) \leqslant 1/2$, where ν_ℓ is the ℓ-adic exponent. Let λ_q be any local parameter of K_q and $f(X)$ the minimal polynomial of λ_q over k_q. Thus one can write $f(X)$ in the form $f(X) = X^e + a_{e-1}x^{e-1} + \cdots + a_0$, where $a_i = a_i(\pi_q)$ are formal power series in π_q. We put $f(X, T) = X^e + a_{e-1}(T)X^{e-1} + \cdots + a_0(T)$ and $F(X, T) = f^{(q)}(X, T)$. So we have $F^{(-q)}(\lambda_q, \pi_q) = 0$. Now let λ'_{q+1} be a root of $F^{-(q+1)}(X, \pi_{q+1})$. Since $(a+b)^\ell \equiv a^\ell + b^\ell \pmod{\ell}$ for any integer elements a, b, we obtain that $0 = (F^{-(q+1)}(\lambda_{q+1}, \pi_{q+1}))^\ell \equiv F^{(-q)}(\lambda'^{\ell}_{q+1}, \pi_q) \pmod{\ell}$. Therefore λ'^{ℓ}_{q+1} is a good approximation to a root of $F^{(-q)}(X, \pi_q)$. Moreover, in this way we obtain one-to-one correspondence between all roots of $F^{(-(q+1))}(X, \pi_{q+1})$ and $F^{(-q)}(X, \pi_q)$. We take for λ_{q+1} such a root that λ_{q+1}^ℓ is close to λ_q.

To show that $k_{q+1}(\lambda_{q+1}) = K_{q+1}$ it is enough to prove that $\lambda_q \in F := k_{q+1}(\lambda_{q+1})$. But it follows from the Hensel lemma, since λ_{q+1}^ℓ is an approximation in F of the root λ_q. Then we define λ_{q+2} in the same way and so on. This proves the proposition.

Now we put $\eta_n(\alpha, a) = \sum_{r=0}^{\infty} \varphi^{-r}(\alpha)\ell^{-r}\lambda_n^{a\ell^r}$. We would like to show that (12.28) holds for $y = \eta_n(\beta, b)$ and $x = \eta_n(\alpha, a)$. But (12.28) fails to be true in this case. So we can state only some weak version of this result. Thus we shall proceed as follows. Reasoning as in the proof of Theorem 12.8, where we have proved that the right sides of (12.29) and (12.30) are congruent modulo $\mathbb{Z}_\ell$, we prove, using (12.31), (see [14, Theor. 2.1] that for some p that depends only on K and any $n > p$ one has a congruence

$$A = A(\alpha, a, \beta, b) := \mathrm{Sp}_{K_n/\mathbb{Q}_\ell}((\gamma_p - 1)\Phi_n(x) \cdot y) \tag{12.32}$$
$$\equiv \ell^{-m-n+p-1} \sum_{j=n}^{\infty} \mathrm{Sp}_{K_j/\mathbb{Q}_\ell} \Big[(\gamma_p - 1)\big(\mathcal{F}_j^{(n-p-1)}(\eta_j(\varphi^{n-j}(\alpha), a))\big)$$
$$\times \mathcal{F}_j^{(n-p-1)}(\eta_j(\varphi^{n-j}(\beta), b))\Big] \pmod{\mathbb{Z}_\ell}.$$

Again, as in the proof of Theorem 12.8, we wish to define some universal series $G(T)$ such that $G^{-j}(\lambda_j) = \mathcal{F}_j^{n-p-1}(\eta_j(\varphi^{n-j}(\alpha, a)))$. So we consider the ring of formal power series $R := \mathcal{O}_H[[T]](\zeta_\sigma)$ and an automorphism σ of this ring defined by $\sigma(T) = T - (1-T)(1-\zeta_\sigma)$, where ζ_σ is some root of unity (maybe non-primitive) of degree ℓ^{n-p}. All such σ form a cyclic group of order ℓ^{n-p}. Note that after specialization $Y = \pi_j$ we obtain an automorphism $\sigma[j]$ of $\mathcal{O}_{K_j}$ of the form $\sigma[j](\zeta_j) = \zeta_j\zeta_\sigma$. Now let λ be a root of the polynomial $F(X, T)$ defined in Prop. 12.9. Then one can present T as a formal power series in λ, hence $R[\lambda] \cong \mathcal{O}_H[[\lambda]]$. Our goal is to extend the action of the automorphisms σ on $R(\lambda, \mu)$, where μ is the group of all roots of unity of degree ℓ^{n-p}, but to do this we have to consider instead of $\mathcal{O}_H(\mu)[[\lambda]]$ the ring of formal Laurent power series with infinite principal part. So let $\tilde{R} = \mathcal{O}_H(\mu)\{\{\lambda\}\}$ be the ring of Laurent power series of the form $\sum_{i=-\infty}^{\infty} a_i\lambda^i$, where $a_i \in \mathcal{O}_H(\mu)$ and a_i tends to zero if i tends to $-\infty$. Let $F_\sigma(X, T) \in R(\mu)[X]$ be a result of the action of σ on the coefficients $a_i(T)$ of $F(X, T)$. Then $F_\sigma(\lambda, T) \equiv 0 \pmod{1 - \zeta_\sigma}$. So λ is an approximation of the root of F_σ, and we can, using the Hensel lemma, extend λ uniquely to the root $\sigma(\lambda) \in \tilde{R}$ of F_σ. (We have $\sigma(\lambda)_0 = \lambda$, $\sigma(\lambda)_1 = \lambda - F_\sigma(\lambda)/F'_\sigma(\lambda)$ and so on.) Using this expressions for $\sigma(\lambda)$ for any $\zeta_\sigma \in \mu$, one can define the above-mentioned series $G(T)$. Note that in our case $G(T)$ is not a formal power series, but a formal Laurent series with coefficients in $\mathcal{O}_H$ and infinite principal part. Moreover, we obtain an explicit estimate for the ℓ-adic exponents of its coefficients at negative powers of T.

The next step is as follows. Though $\mathcal{F}_j^{n-p-1}(f(\lambda_j))$ is a function in λ_j, in fact it is a function in λ_{j-s} because $\mathcal{F}_j^s = \mathcal{F}_{j-s}^{(0)}\mathrm{Sp}_{K_j/K_{j-s}}$, where $s = n-p-1$. We can translate all this into the language of formal Laurent series. That is, we can define a new Laurent series $D(X) = \sum_{i=-\infty}^{\infty} d_iX^i \in \mathcal{O}_H\{\{X\}\}$ such that $D^{n-j}(\lambda_{j-n+p+1}) = \mathcal{F}_j^{n-p-1}(\eta_j(\varphi^{n-j}(\alpha), a))$ for all sufficiently large j. The advantage of using $D(X)$ instead of $G(T)$ is that its coefficients at the negative powers of X tends to zero very fastly. Namely, we prove [14, Theor. 5.2] that one has

$$\nu_\ell(d_i) > -\frac{i}{2\delta\ell} \qquad (12.33)$$

for any $i < 0$, where δ is as in Prop. 12.9.

Another sort of automorphisms we have to deal with is the automorphism γ_p defined by $\gamma_p(\zeta_j) = \zeta_j^{\varkappa}$, where $\varkappa = \varkappa(\gamma_p) \in \mathbb{Z}_\ell^\times$. Again, we can consider γ_p as an automorphism of R defined by $\gamma_p(T) = 1 - (1-T)^{\varkappa}$. Using the Hensel lemma one can uniquely extend the action of γ_p to $\tilde{R}$. We prove [14, Theor. 6.3] that there is a formal Laurent series $\mathcal{E}(X)$ such that for any $j \geqslant n$ and $\ell \neq 2$ one has

$$(\gamma_p - 1)D^{(-j)}(\lambda_{j-s}) = \pi_{j-n}\mathcal{E}^{-j}(\lambda_{j-s}), \qquad (12.34)$$

where $s = n - p - 1$. (For $\ell = 2$ (12.34) needs a slight modification.) Note that the coefficients of $\mathcal{E}(X)$ at the negative powers of X satisfies the condition (12.33). Now we can state and prove the main theorem of [14].

Theorem 12.10. *Let K be an arbitrary finite extension of $\mathbb{Q}_\ell$, $K_n = K \cdot k_n$ and $K_\infty = \cup K_n$. Then there are an index p and a non-negative constant C, depending only on K_∞/K, such that for any $n > p$ there exists a $\mathbb{Z}_\ell$-submodule M_n of the $\mathbb{Z}_\ell$-module* $\log(U(K_n))$ *with the property that the group $(log(U(K_n))/M_n$ can be generated by at most Cn generators and*

$$(\gamma_p - 1)\Phi_n(M_n) \subseteq (\log U(K_n))^{\perp}.$$

Moreover, one can effectively determine p and C.

Sketch of the proof. Let the elements $\eta_{n,1} = \eta_n(\alpha_1, a_1), \ldots, \eta_{n,t} = \eta_n(\alpha_t, a_t)$ generate the $\mathbb{Z}_\ell$-module $\log(U(K_n))$ for a given n. Then any element $\eta_n \in \log(U(K_n))$ can be written as

$$\eta_n = \sum_{i=1}^{t} c_i \eta_{n,i}$$

with some $c_i \in \mathbb{Z}_\ell$.

For each $i = 1, \ldots, t$ there is a Laurent series $\mathcal{E}_i(X)$ such that

$$(\gamma_p - 1)\mathcal{F}_j^{(s)}(\eta_{j,i}) = \pi_{j-n}\mathcal{E}_i^{(-j)}(\lambda_{j-s}), \qquad s = n - p - 1,$$

where $\eta_{j,i} = \eta_j(\varphi^{n-j}(\alpha_i), a_i)$. Putting

$$\eta_j = \sum_{i=1}^{t} c_i \eta_{j,i}, \qquad \mathcal{E}(X) = \sum_{i=1}^{t} c_i \mathcal{E}_i(X),$$

for any $j \geqslant n$, we see that

$$(\gamma_p - 1)\mathcal{F}_j^{(s)}(\eta_j) = \pi_{j-n}\mathcal{E}^{(-j)}(\lambda_{j-s}), \tag{12.35}$$

for every $j \geqslant n$.

The coefficient e_ν of X^ν in $\mathcal{E}(X)$ has the form $e_\nu = \sum_t c_i e_{i,\nu}$, where $e_{i,\nu}$ is the coefficient of X^ν in $\mathcal{E}_i(X)$. Thus e_ν is a linear function in c_i with coefficients $e_{i,\nu}$, and we can consider a subgroup M_n of $\log(U(K_n))$ consisting of those η_n that the corresponding series $\mathcal{E}(X)$ has zero coefficients e_ν whenever ν satisfies $|\nu| \leqslant (n-p+2)\ell\delta$. Thus there are $r_n := 8(n-p+2)\ell\delta + 1$ coefficients e_ν in $\mathcal{E}(X)$ that must vanish, that is, M_n is defined in $\log(U(K_n))$ by r_n linear equations with coefficients in $\mathcal{O}_H$ or by $r_n f$ equations with coefficients in $\mathbb{Z}_\ell$, where $f = [H : \mathbb{Q}_\ell]$.

We claim that for any $\eta_n \in M_n$ and any $\eta_{n,i}$ as above one has

$$A' := \mathrm{Sp}_{K_n/\mathbb{Q}_\ell}(\Phi_n \eta_n \cdot \eta_{n,i}) \in \mathbb{Z}_\ell.$$

Let $D(X)$ be the series corresponding to $\eta_{n,i}$, that is, for any $j \geqslant n$ one has

$$\mathcal{F}_j^{(s)}(\eta_{j,i}) = D^{(-j)}(\lambda_{j-s}). \tag{12.36}$$

Then by (12.32), (12.35) and (12.36) we have

$$A' \equiv \ell^{-m-n+p-1} \sum_{j=n}^{\infty} \mathrm{Sp}_{K_j/\mathbb{Q}_\ell}(\pi_{j-n}\mathcal{E}^{(-j)}(\lambda_{j-s})D^{(-j)}(\lambda_{j-s})) \pmod{\mathbb{Z}_\ell},$$

where π_{j-n} must be replaced by π_{j-n-1} if $\ell = 2$.

If $\mathcal{E}(X) = \sum_{\nu=-\infty}^{\infty} e_\nu X^\nu$, then we put

$$\mathcal{E}(X) = \mathcal{E}_1(X) + \mathcal{E}_2(X),$$

where

$$\mathcal{E}_1(X) = \sum_{\nu=-\infty}^{-4(n-p+2)-1} e_\nu X^\nu, \quad \mathcal{E}_2(X) = \sum_{\nu=4(n-p+2)+1}^{\infty} e_\nu X^\nu.$$

We define $D_1(X)$ and $D_2(X)$ analogously. Then

$$A' \equiv \sum_{1 \leqslant \alpha,\beta \leqslant 2} A'_{\alpha,\beta} \pmod{\mathbb{Z}_\ell},$$

where

$$A'_{\alpha,\beta} = \ell^{-m-n+p-1} \sum_{j=n}^{\infty} \mathrm{Sp}_{K_j/\mathbb{Q}_\ell}(\pi_{j-n}\mathcal{E}_\alpha^{(-j)}(\lambda_{j-s}) D_\beta^{(-j)}(\lambda_{j-s})).$$

We can prove that $A'_{1,1} \equiv A'_{1,2} \equiv A'_{2,1} \equiv 0 \pmod{\mathbb{Z}_\ell}$, using the estimates of coefficients of these series of the form (12.33). The series $A'_{2,2}$ is a formal power series, and we can treat it as in the proof of Theorem 12.8. This finishes the proof of the theorem.

This theorem means that the situation for non-Abelian local fields is in some sense close to that for Abelian fields. We shall see in Section 12.6 that this fact yields interesting consequences concerning Conjecture 12.6 (see Theorem 12.19).

To analyze the non-Abelian situation, we introduced in [15] two sequences of finite Abelian ℓ-groups $\mathcal{A}_n^{(p)}$ and $\mathcal{B}_n^{(p)}$. Namely, if K is a non-Abelian finite extension of $\mathbb{Q}_\ell$ and K_n are defined as above, we put $V_n = V(K_n)$ and

$$\mathcal{A}_n^{(p)} = (\gamma_p - 1)\Phi_n(\log V_n)/\mathcal{M}_n^{(p)},$$

$$\mathcal{B}_n^{(p)} = \mathrm{Ker}\,\mathrm{Sp}_{K_n/K_p}((\log V_n)^\perp)/\mathcal{M}_n^{(p)},$$

where $\mathcal{M}_n^{(p)} = (\gamma_p - 1)\Phi_n(\log V_n) \cap \mathrm{Ker}\,\mathrm{Sp}_{K_n/K_p}((\log V_n)^\perp)$.

Proposition 12.11. *[15, Theor. 2.1] For all sufficiently large n we have*

$$d(\mathcal{A}_n^{(p)}) = d(\mathcal{B}_n^{(p)}), \quad \mathcal{A}_n^{(p)} = \ell\mathcal{A}_{n+1}^{(p)}, \quad \mathcal{B}_n^{(p)} = \ell\mathcal{B}_{n+1}^{(p)},$$

where $d(A)$ means the minimal number of generators of an Abelian ℓ-group A.

In [16] it was stated by direct calculation that for any non-Abelian K and $\ell > 3$ one has $\mathcal{A}_{p+1}^{(p)} \neq 0$ and $\mathcal{B}_{p+1}^{(p)} \neq 0$ for all sufficiently large p. This result holds also for $\ell = 3$ with one exceptional case. It was also proved there that $\mathcal{A}_n^{(p)} \cong \mathcal{B}_n^{(p)}$ for all sufficiently large n and p [16, Theor. 3.3].

Moreover, recently the author have proved that for any non-Abelian K the groups $\mathcal{A}_n^{(p)}$ and $\mathcal{B}_n^{(p)}$ are non-zero for any fixed p and all sufficiently large n and $d(\mathcal{A}_n^{(p)}) \to \infty, \quad d(\mathcal{B}_n^{(p)}) \to \infty$ if $n \to \infty$.

To say the truth, our understanding of pairing (17.10) for non-Abelian local fields is far from being good. We shall return to this question at the end of Section 12.6. Here we should like to note the following problem.

Problem 12.2. *Let K be a finite non-Abelian extension of $\mathbb{Q}_\ell$ and the groups $\mathcal{A}_n^{(p)}$, $\mathcal{B}_n^{(p)}$ be as above. Is there any connection between these groups as Galois modules?*

12.5 Explicit formulae for the norm residue symbol

In this section we consider the norm residue symbol (the Hilbert symbol) for local number fields. It appears that in Abelian case this symbol is very close to the product T_∞ from the preceding section. It explains why these formulae are applicable to the ℓ-adic regulator. Though our main results are valid only for Abelian local fields, we start with considering a general case.

Let k be a finite extension of $\mathbb{Q}_\ell$. Suppose that k contains a primitive ℓ-th root of unity if $\ell \neq 2$ (k contains $\sqrt{-1}$ if $\ell = 2$). Let $k_\infty = \cup_n k_n$ be the cyclotomic $\mathbb{Z}_\ell$-extension of k and $k_n = k(\zeta_n)$. Let $\mu_\ell(k_n)$ be the ℓ-component of $\mu(k_n)$. Hence $\mu_\ell(k_n) = \langle \zeta_n \rangle$ for all sufficiently large n.

Then for any n one has the non-degenerate Hilbert symbol

$$(\ ,\ )_n : k_n^\times \times k_n^\times \longrightarrow \mu_\ell(k_n). \tag{12.37}$$

If $a, b \in k_n^\times$ and $q_n = |\mu_\ell(k_n)|$ then $(a, b) = \vartheta(a)(\sqrt[q_n]{b})/\sqrt[q_n]{b}$, where $\vartheta(a)$ is the element of the Galois group G^{ab} of the maximal Abelian extension of k_n that corresponds to a in the sense of local class field theory. Let $k_n^\times[\ell]$ be the pro-ℓ-completion of $k_n^\times$ and put $H(k_n) = \hat{H}(k_n)/\mu_\ell(k_n)$, where $\hat{H}(k_n)$ is the group of universal norms in $k_n^\times[\ell]$ from k_∞. It follows from local class field theory that (12.37) induces a non-degenerate product

$$H(k_n) \times H(k_n) \longrightarrow \mu_\ell(k_n). \tag{12.38}$$

We choose the roots $\zeta_n \in k_n$ in such a way that $\zeta_{n+1}^\ell = \zeta_n$ for all n and identify $\mu_\ell(k_n)$ with $\mathbb{Z}(m) = \mathbb{Z}/\ell^m\mathbb{Z}$ by putting $1 \leftrightarrow \zeta_n$, where $\ell^m = q_n$. We put $H(k_n)(m) = H(k_n)/(H(k_n))^{\ell^m}$ and define $H(k_n)(m)[-1]$ as $H(k_n)(m)$ with twisted Galois action, that is, $\sigma[-1](y) = \varkappa(\sigma)\sigma(y)$, where $y \in H(k_n)(m)$ and $\varkappa(\sigma)$ is the cyclotomic character defined by $\sigma(\zeta_n) = \zeta_n^{\varkappa(\sigma)}$. Then (12.38) turns into the product

$$H(k_n)(m) \times H(k_n)(m)[-1] \longrightarrow \mathbb{Z}(m),$$

satisfying the relation $(x, y)_n = (\sigma(x), \sigma[-1](y))_n$ for any $x, y \in H(k_n)(m)$.

Now, as in Section 12.3, we choose some index p and for all $n > p$ put

$$\langle x, y\rangle_n = \sum_{\sigma \in G(k_n/k_p)} (\sigma^{-1}(x), y)_n \sigma \in \mathbb{Z}(m)[G(k_n/k_p)]. \tag{12.39}$$

The products (12.39) are compatible relative to the norm maps and, after passing to inverse limit, we obtain an inner sesquilinear product [11, Prop. 2.2]

$$\langle\, ,\, \rangle_\infty \colon H(k_\infty) \times H(k_\infty)[-1] \longrightarrow \Lambda_p, \tag{12.40}$$

where Λ_p was defined in Section 12.3.

Since $(\gamma_p - 1)H(k_\infty) \subseteq V(k_\infty)$, where γ_p and $V(k_\infty)$ are as in Section 12.3, we can extend (12.40) to the product on $V(k_\infty) \times V(k_\infty)$. This last product has its values, generally speaking, in the quotient field L_p of Λ_p, but in fact all these values lie in Λ_p, and (12.40) induces an inner sesquilinear product [11, Theor. 3.1]

$$\langle\, ,\, \rangle_\infty \colon V(k_\infty) \times V(k_\infty)[-1] \longrightarrow \Lambda_p. \tag{12.41}$$

In what follows, we assume that k is an Abelian field. Since k contains also a primitive ℓ-th root of unity, we may assume k to be a cyclotomic field. So we have also the product T_∞ defined in Section 12.3. Thus we have two isomorphisms $f_1 \colon V(k_\infty) \cong \mathrm{Hom}_{\Lambda_p}(V(k_\infty), \Lambda_p)$ and $f_2 \colon V(k_\infty)[-1] \cong \mathrm{Hom}_{\Lambda_p}(V(k_p), \Lambda_p)$, where f_1 and f_2 are induced by T_∞ and by (12.41) respectively. In other words, there is an isomorphism $\psi = f_1^{-1} \circ f_2 \colon V(k_\infty)[-1] \to V(k_\infty)$ such that $\langle x, y\rangle_\infty = T_\infty(x, \psi(y))$ and the problem is to describe ψ explicitly.

At first we prove that ψ does not depend on p [11, Prop. 4.4] and then prove the main theorem of [11].

Theorem 12.12. *[11, Theor. 4.1] Let k be a finite Abelian extension of $\mathbb{Q}_\ell$ and k_∞ the cyclotomic $\mathbb{Z}_\ell$-extension of k. Then the twisted automorphism ψ satisfies the following two equivalent conditions:*

$$\psi(\eta(k_\infty)) = \eta(k_\infty), \qquad \psi(\eta'(k_\infty)) = -\eta'(k_\infty) \tag{12.42}$$

(we use an additive notation for operations in $V(k_\infty)$). Each of these conditions defines uniquely the automorphism ψ.

The proof of the theorem given in [11] is rather complicated, but its idea is simple. We use the explicit formulae for the norm residue symbol of Sen [21], which express this symbol in terms of the logarithmic derivative. To be precise, one has

$$(\beta, \alpha)_n = \zeta_n^{-q_n S_n(\zeta_n (f'(\pi_n)/f(\pi_n)) \log \alpha)}, \tag{12.43}$$

where $\alpha, \beta \in k_n^\times$, $\pi_n = 1 - \zeta_n$ is a local parameter of k_n, $\beta = f(\pi_n)$, $f(X) \in \mathcal{O}_H[X]$ and $\mathcal{O}_H$ is the ring of integers of the maximal unramified subfield H

of k_n. Note that the element α in (12.43) satisfies rather restrictive condition $\alpha \equiv 1 \pmod{\lambda^2}$, where $\lambda = 1 - \zeta_0$ ($\lambda = 2$ if $\ell = 2$).

We prove that the setting $\beta \to -q_n \zeta_n f'(\pi_n)/f(\pi_n)$, where $f(X)$ is as in (12.43), defines a map

$$\delta_n' \colon U^{(1)}(k_n) \longrightarrow \ell^{-q_n}\mathcal{O}(k_n)/\ell^{-q_n}\mathcal{D}(k_n), \tag{12.44}$$

where $\mathcal{O}(k_n)$ and $\mathcal{D}(k_n)$ are the ring of integers and the different of k_n respectively.

Passing in (12.44) to inverse limit (relative to the norm maps on the left and to the trace maps on the right side), we get a map $\delta_\infty' \colon U(k_\infty) \to \tilde{k}_\infty := \varprojlim k_n$. Using the behaviour of the symbol (12.37) under variation of the ground field, we can get rid of restrictions on α. As a result we obtain a formula $\langle u, v\rangle_\infty = T_\infty(\log u, \delta_\infty'(v))$ that is valid for any $u, v \in U(k_\infty)$ [11, Cor. to Lemma 5.2]. So, to prove the theorem we have to calculate $\delta_\infty'(v)$ for a certain special element $v \in U(k_\infty)$.

Comparing the explicit formulae for $\Theta'(k_n)$ and $\xi'(k_n)$ and reasoning quite naively, we see that $\Theta'(k_n) = \zeta_n d(\xi'(k_n))$, where d means differentiation relative to ζ_n. To make this approach correct, we define a special power series $f(X) \in \mathcal{O}_H[X]$ and prove that the elements $\exp(f_n(\pi_n)) \in U(k_n)$ form a coherent sequence relative to the norm maps. Here $f_n(X)$ means a result of the action of the power φ^{-n} of Frobenius on the coefficients of $f(X)$. Let γ be a topological generator of $G(k_\infty/k_p)$ and $\varkappa = \varkappa(\gamma) \in \mathbb{Z}_\ell$ is defined by the condition $\gamma(\zeta_n) = \zeta_n^{\varkappa}$ for any n. Then we prove [11, Lemma 5.5] that $f_n(\pi_n) = (\gamma - \varkappa)(\gamma - 1)\xi'(k_n)$ and $\zeta_n f_n'(\pi_n) = \varkappa(\varkappa\gamma - 1)(\gamma - 1)\Theta'(k_n)$. This conclude the proof of the theorem.

Now let K be an algebraic number field containing a primitive ℓ-th root of unity ζ_0 ($\sqrt{-1} \in K$ if $\ell = 2$). Let $K_\infty = \cup_n K_n$ be the cyclotomic $\mathbb{Z}_\ell$-extension of K. Then $K_n = K(\zeta_n)$ for all sufficiently large n. There are only finitely many places $v|\ell$ in K_∞, and we put

$$H(K_\infty) = \prod_{v|\ell} H(K_{\infty,v}), \qquad V(K_\infty) = \prod_{v|\ell} V(K_{\infty,v}),$$

where the products are taken over all places over ℓ in K_∞. We choose a coherent family of roots of unity $\zeta_n \in K_n$ such that $\zeta_{n+1}^\ell = \zeta_n$ for any n and, since $K_n \subset K_{n,v}$, we may consider $\{\zeta_n\}_{n\geqslant 0}$ as a coherent sequence of roots of unity in the sequence of local fields $\{K_{n,v}\}_{n\geqslant 0}$ for any $v|\ell$. Thus, putting $1 \leftrightarrow \zeta_n$, we get isomorphisms $\mu_\ell(K_n) \cong \mathbb{Z}(m)$ for any n. So, passing to the limit, we obtain an isomorphism $\mathbb{Z}_\ell \cong T[-1]$, where $T = \varprojlim \mu_\ell(K_n)$ and $[-1]$ denotes twisted Galois action as above.

We choose some index p such that for any $n > p$ we have $[K_n : K_p] = [K_{n,v} : K_{n,p}] = \ell^{n-p}$ for any $v|\ell$. Then, denoting the local product (12.39) by $\langle\,,\,\rangle_{n,v}$, we define a product

$$\langle x, y\rangle_n = \sum_{v|\ell} \langle x_v, y_v\rangle_{n,v} \in \mathbb{Z}(m)[G(K_n/K_p)], \tag{12.45}$$

where $x = \prod_{v|\ell} x_v$, $y = \prod_{v|\ell} y_v \in H(K_\infty)$.

Just as in the local case, passing to inverse limit in (12.45), we get an inner sesquilinear product

$$\langle\,,\,\rangle_\infty \colon H(K_\infty) \times H(K_\infty)[-1] \longrightarrow \Lambda_p,$$

which, in turn, induces an inner sesquilinear product

$$\langle\,,\,\rangle_\infty \colon V(K_\infty) \times V(K_\infty) \longrightarrow \Lambda_p.$$

The latter product is connected with (12.41) by a formula analogous to (12.45).

Since we have also the product T_∞ of Section 12.3, we again obtain a twisted isomorphism $\psi \colon V(K_\infty) \cong V(K_\infty)$. Obviously, ψ coincides with local ψ_v of Theorem 12.8 on each local component $V(K_{\infty,v})$. So we can write $\psi = \prod_{v|\ell} \psi_v$. For any $x, y \in V(K_\infty)$ one has $\langle x, y\rangle_\infty = T_\infty(x, \psi(y))$. The group $V^+(K_\infty)$ is the maximal isotropic subgroup of $V(K_\infty)$. This implies the following characterization of the complement $V^-(K_\infty)$ defined at the end of Section 12.3.

Theorem 12.13. *[11, Prop. 7.2] Let $V^-(K_\infty)$ be the orthogonal complement to $V^+(K_\infty)$ relative to the product T_∞. Then $V^-(K_\infty) = \psi(V^+(K_\infty))$. More generally, for any integer i the module $\psi^i(V^+(K_\infty))$ coincides with the orthogonal complement to $\psi^{1-i}(V^+(K_\infty))$ relative to the product T_∞ and $\psi^i(V^+(K_\infty))$ coincides with the orthogonal complement to $\psi^{-i}(V^+(K_\infty))$ relative to the product $\langle\,,\,\rangle_\infty$.*

As an immediate consequence of Theorem 12.10 we get the following statement.

Proposition 12.14. *Let H be a finite ℓ-group of automorphisms of K_∞ Then $V^+(K_\infty)$ and $V^-(K_\infty)$ are isomorphic as H-modules.*

Proof. Since H is finite, the cyclotomic character $\varkappa$ is trivial on H. Hence ψ is an H-isomorphism.

12.6 Applications to the feeble conjecture on the ℓ-adic regulator

The results of preceding sections enable us to prove Conjecture 12.6 for some particular cases.

Theorem 12.15. *[10, Theor. 6.1] Let $\ell \neq 2$, K be a finite ℓ-extension of an algebraic number field k, k_∞ the cyclotomic $\mathbb{Z}_\ell$-extension of k and $K_\infty = K \cdot k_\infty$ the cyclotomic $\mathbb{Z}_\ell$-extension of K. Let K be Abelian over ℓ, that is, the completions K_v are Abelian for any $v|\ell$. Suppose that the Iwasawa μ-invariant $\mu_\ell(k)$ vanishes. Suppose also that k contains a primitive ℓ-th root of unity, Conjecture 12.6 holds for k_∞/k, (that is, $V^+(k_\infty) \cap V^-(k_\infty) = 0$ by Theorem 12.4) and $D(k_\infty) := V(k_\infty)/(V^+(k_\infty) \oplus V^-(k_\infty))$ is finitely generated over $\mathbb{Z}_\ell$. Then $V^+(K_\infty) \cap V^-(K_\infty) = 0$, in other words, Conjecture 12.6 holds for K_∞/K and, moreover, $D(K_\infty) := V(K_\infty)/(V^+(K_\infty) \oplus V^-(K_\infty))$ is finitely generated over $\mathbb{Z}_\ell$.*

Sketch of the proof. We may assume that K_∞/k_∞ is Galois with a Galois group G. By Iwasawa theorem on the μ-invariant we have $\mu_\ell(K) = 0$. This means that the ℓ-ranks of the ℓ-class groups $Cl_\ell(K_n)$ are restricted by some λ that does not depend on n. We can deduce from this fact that the Tate cohomology groups $H^i(G, U(K_\infty))$ are finite.

The equality $\mu_\ell = 0$ combined with the fact that $\zeta_0 \in k$ yield that $V^+(K_\infty)/U(K_\infty)$ is a finitely generated $\mathbb{Z}_\ell$-module. Hence the cohomologies $H^i(G, V^+(K_\infty))$ are also finite. By Proposition 12.14 we obtain that the groups $H^i(G, V^-(K_\infty))$ are finite as well.

Put $A_\ell = A/\ell A$ for an Abelian group A. Then $V(K_\infty)_\ell$, $V^+(K_\infty)_\ell$ and $V^-(K_\infty)_\ell$ are $\mathbb{Z}/\ell\mathbb{Z}[G]$-modules with finite cohomologies, and the natural injections $V^+(K_\infty) \hookrightarrow V(K_\infty)$ and $V^-(K_\infty) \hookrightarrow V(K_\infty)$ induce injections $V^+(K_\infty)_\ell \hookrightarrow V(K_\infty)_\ell$ and $V^-(K_\infty)_\ell \hookrightarrow V(K_\infty)_\ell$. If Conjecture 12.6 fails to be true for K_∞/K then by Theorem 12.4 $C(K_\infty) \neq 0$. Hence $C(K_\infty)_\ell$ is infinite, but $C(K_\infty)_\ell \subset V^+(K_\infty)_\ell \cap V^-(K_\infty)_\ell$. Therefore this last group is infinite and so is the group $(V^+(K_\infty)_\ell \cap V^-(K_\infty)_\ell)^G$. On the other hand, the natural inclusion $V(k_\infty) \hookrightarrow V(K_\infty)$ implies inclusions $V^+(k_\infty)_\ell \hookrightarrow (V^+(K_\infty)_\ell)^G$ and $V^-(k_\infty)_\ell \hookrightarrow (V^-(K_\infty)_\ell)^G$, both with finite co-kernels. If $D(k_\infty)$ is finitely generated over $\mathbb{Z}_\ell$ then $V^+(k_\infty)_\ell \cap V^-(k_\infty)_\ell$ is also finite. (In fact it is a zero group, since $V^+(k_\infty)_\ell$ and $V^-(k_\infty)_\ell$ contain no non-trivial finite submodules.) This contradicts to infiniteness of the group $(V^+(K_\infty)_\ell \cap V^-(K_\infty)_\ell)^G$. The theorem is proved.

Note that among other things we have proved in Theorem 12.12 that $D(K_\infty)$ is finitely generated over $\mathbb{Z}_\ell$. We may treat this result as a complement to Iwasawa theorem on the μ-invariant.

If $\ell = 2$ then the situation is a bit more complicated. In this case $D(k_\infty)$ always has non-zero μ-invariant. Indeed, Theorem 12.8 shows that the map $\psi : V(k_\infty) \to V(k_\infty)$ induces the identity map on $V(k_\infty)_2$, hence the groups $V^+(k_\infty)_2$ and $V^-(k_\infty)_2$ coincide as subgroups of $V(k_\infty)_2$. Denote this last subgroup by $\tilde{P}$, and by P we denote the maximal subgroup of $V(k_\infty)_2$ that contains $\tilde{P}$ as a subgroup of finite index. Let $V'(k_\infty)$ be the pre-image of P under the natural projection $V(k_\infty) \to V(k_\infty)_2$. At last, we put $D'(k_\infty) = V'(k_\infty)/(V^+(k_\infty) \oplus V^-(k_\infty))$. We define the groups $V'(K_\infty)$ and $D'(K_\infty)$ analogously. Then for $\ell = 2$ we have the following version of Theorem 12.15.

Theorem 12.16. *[11, Theor. 7.1] Let $\ell = 2$, let k be any extension of $\mathbb{Q}$ containing $\sqrt{-1}$, and let K be a finite 2-extension of the field k, Abelian over 2. Let k_∞ and K_∞ be the cyclotomic $\mathbb{Z}_\ell$-extensions of k and K respectively. Suppose that $V^+(k_\infty) \cap V^-(k_\infty) = 0$, the Iwasawa μ-invariant $\mu_2(k)$ vanishes and $D'(k_\infty)$ is finitely generated over $\mathbb{Z}_\ell$. Then $V^+(K_\infty) \cap V^-(K_\infty) = 0$, that is Conjecture 12.6 holds for K_∞/K, the group $D'(K_\infty)$ is finitely generated over $\mathbb{Z}_\ell$ and $\mu(D(K_\infty)) = [K : \mathbb{Q}]/2$.*

Suppose that k is a CM-field. Then $V^+(k_\infty) \cap V^-(k_\infty) = 0$, since the complex conjugation j acts on $V^+(k_\infty)$ and $V^-(k_\infty)$ by multiplication by +1 and -1 respectively. So we have $D(k_\infty) = 0$ if $\ell \neq 2$, and $D(k_\infty)$ has exponent two and Iwasawa μ-invariant $[k : \mathbb{Q}]/2$ if $\ell = 2$. Hence Theorems 12.15 and 12.16 yield the following result.

Theorem 12.17. *In the notation of Theorem 12.15 let k be a CM-field containing ζ_0 (containing $\sqrt{-1}$ if $\ell = 2$) and such that $\mu_\ell(k) = 0$. Then $C(K_\infty) = 0$, that is, Conjecture 12.6 holds for K_∞/K. If $\ell \neq 2$ then $D(K_\infty)$ is a finitely generated $\mathbb{Z}_\ell$-module. If $\ell = 2$ then $D'(K_\infty)$ is a finitely generated $\mathbb{Z}_2$-module and $D(K_\infty)/D'(K_\infty)$ is of exponent two and has μ-invariant $[K : \mathbb{Q}]/2$.*

Note that the condition $\mu_\ell(k) = 0$ holds if k is Abelian by the Ferrero-Washington theorem. Therefore we have the following unconditional result.

Theorem 12.18. *Let k be an Abelian algebraic number field and K an ℓ-extension of k Abelian over ℓ. Then Conjecture 12.6 holds for the cyclotomic $\mathbb{Z}_\ell$-extension K_∞/K. The module $D(K_\infty)$ defined above (the module*

$D'(K_\infty)$ if $\ell = 2$) is finitely generated over $\mathbb{Z}_\ell$. If $\ell = 2$ and $\sqrt{-1} \in K$ then $D(K_\infty)/D'(K_\infty)$ is of exponent two and its μ-invariant is equal to $[K : \mathbb{Q}]/2$.

The results of Section 12.4 enable us to generalize all these results to the case, when K has an arbitrary ramification over ℓ. Note that in non-Abelian case we don't know if there exists the complement $V^-(K_\infty)$. So we define $V^+(K_n)$ as the image of the natural projection $V^+(K_\infty) \to V(K_n)$ and put for any intermediate field K_n in the cyclotomic $\mathbb{Z}_\ell$-extension K_∞/K

$$V^-(K_n) = \{x \in V(K_n) \mid \mathrm{Sp}_{K_n/\mathbb{Q}_\ell}(\log x \cdot \log y) = 0 \text{ for any } y \in V^+(K_n)\}.$$

Theorem 12.19. *[15, Theor. 6.1] Let ℓ be an odd prime number and let K be an algebraic number field that satisfies the following conditions:*

(i) K contains a primitive ℓ-th root of unity ζ_0;
(ii) $\mu_\ell(K) = 0$;
(iii) the field K and its cyclotomic $\mathbb{Z}_\ell$-extension satisfy the feeble conjecture on the ℓ-adic regulator (Conjecture 12.6);
(iv) There is a constant c (independent of n) such that for all $n > 0$

$$d\big(V(K_n)/(V^+(K_n) + V^-(K_n))\big) < cn.$$

Then the following conditions are satisfied by any field L that is a finite ℓ-extension of K:

(ii') $\mu_\ell(L) = 0$;
(iii') the field L and its cyclotomic $\mathbb{Z}_\ell$-extension L_∞/L satisfy the feeble conjecture on the ℓ-adic regulator;
(iv') there is a constant c' (independent of n) such that for all $n > 0$

$$d\big(V(L_n)/(V^+(L_n) + V^-(L_n))\big) < c'n.$$

The proof is similar to that of Theorem 12.15. We may suppose that L is a Galois extension of K with Galois group G. Let $\{A_n\}$ be a sequence of G-modules. We say that this sequence has bounded cohomologies in dimension i if there is $c = c(i)$ (independent of n) such that $d(\hat{H}^i(G, A_n)) < cn$. Obviously, two sequences $\{V(L_n)\}$ and $\{V^+(K_n)\}$ has bounded cohomologies. Using Theorem 12.10, one can prove that $\{V^-(K_n)\}$ also has bounded cohomologies. Then, reasoning as in the proof of Theorem 12.15 and putting $C(L_n) = V^+(L_n) \cap V^-(L_n)$, we obtain that $d(C(L_n)) < c_1 n$ for some c_1 (independent of n). But $C(L_n)$ are $G(L_n/L)$-modules without $\mathbb{Z}_\ell$-torsion. Hence if their ranks grow, this growth cannot be so slow. This contradiction proves the theorem.

Again, one has analogs of Theorem 12.19 for the cases of CM-field K or of Abelian K, which are similar to Theorems 12.17 and 12.18 respectively. Note that in [18] we have generalized Theorem 12.19 to the case $\ell = 2$. This generalization yields also generalization of Theorems 12.17 and 12.18. In particular, Conjecture 12.6 holds for $\ell = 2$ and for any L that is a 2-extension of an Abelian field K.

Problem 12.3. *Let $V^-(K_n)$ be as in Theor. 12.19 and $V^-(K_\infty) = \varprojlim V^-(K_n)$. Is it true that* $\mathrm{rk}(V(K_\infty)) = \mathrm{rk}(V^+(K_\infty)) + \mathrm{rk}(V^-(K_\infty))$*?*

In [8] in construction of an analog of the Weil product we used the fact that $H^-(K_\infty)$ is isotropic relative to the norm residue symbol. The module $V^-(K_\infty)$, that we have constructed, as a rule has not this property. We may improve $V^-(K_\infty)$, using the following simple statement.

Proposition 12.20. *Let A be a finite free $\mathbb{Z}/\ell^n\mathbb{Z}$-module and $A \times A \to \mathbb{Z}/\ell^n\mathbb{Z}$ a skew-symmetric non-degenerate product on A. Let $B \subset A$ be a maximal isotropic submodule of A such that $A \cong B \oplus C$ for some C. Then there is a canonically defined $C' \subset A$ (C' depends on C) such that $A \cong B \oplus C'$ and C' is isotropic.*

At last, note that in Section 12.8 we shall discuss another approach to Conjecture 12.6, which is based on completely different ideas.

12.7 An Analog of the Riemann-Hurwitz formula for some ℓ-extensions of algebraic number fields

The theorem 12.1 of introduction can be generalized to some ℓ-extensions K/k, where K are not of CM-type. Let k_∞/k be the cyclotomic $\mathbb{Z}_\ell$-extension of k and N the maximal unramified Abelian ℓ extension of k_∞ such that in N/k_∞ all places $v|\ell$ completely split. Let M be the maximal Abelian ℓ-extension of the field k_∞ not ramified outside ℓ, and M' the maximal subfield of M such that the Galois group $G(M'/k_\infty)$ is a torsion free Λ-module, where $\Lambda = \mathbb{Z}_\ell[[\Gamma]]$ and $\Gamma = G(k_\infty/k)$. Let $\bar{N}$ be the maximal Abelian unramified ℓ-extension of k_∞. Let $N' = N \cap M'$ and $\bar{N}' = \bar{N} \cap M'$. We denote the Galois groups $G(N/N')$, $G(N'/k_\infty)$ and $G(\bar{N}'/N')$ by $T'_\ell(k_\infty)$, $T''_\ell(k_\infty)$ and $R''_\ell(k_\infty)$ respectively. These groups are torsion Λ-modules, and we will denote their λ-invariants by $\lambda'(k_\infty)$, $\lambda''(k_\infty)$ and $r''(k_\infty)$.

Theorem 12.21. *[13, Theor. 1] (For the case $\ell = 2$ see the remark at the end of [11].) Let k contains a primitive ℓ-th root of unity ($\sqrt{-1} \in k$ if $\ell = 2$), K a finite ℓ-extension of k, Abelian over ℓ. (This means that K_v are Abelian for all $v|\ell$.) Suppose that $D(k_\infty)$ defined in Section 12.6 is a torsion Λ-module, and that the torsion Λ-modules $T_\ell(k_\infty)$ and $D(k_\infty)$ ($D'(k_\infty)$ if $\ell = 2$) have zero Iwasawa μ-invariants.*

Then $D(K_\infty)$ is a torsion Λ-module, and the following relation holds:

$$d(K_\infty) + 2\lambda'(K_\infty) + 4\lambda''(K_\infty) + 4r''(K_\infty) - 4$$
$$= [K_\infty : k_\infty][d(k_\infty) + 2\lambda'(k_\infty) + 4\lambda''(k_\infty) + 4r''(k_\infty) - 4] + 2\sum_{v\nmid\ell}(e_v - 1), \tag{12.46}$$

where v runs through all places $v \nmid \ell$ of the field K_∞, e_v is the ramification index of the place v in the extension K_∞/k_∞, and $d(k_\infty) = \lambda(D(k_\infty))$.

The idea of the proof is as follows. It is enough to prove the theorem in the case when K_∞/k_∞ is a cyclic extension of degree ℓ. Let G denotes its Galois group. For a G-module A such that the Tate cohomology groups $\hat{H}^i(G, A)$ are finite for all i, we define the Euler characteristic by the formula $\chi(A) = \dim_{\mathbb{F}_\ell} H^2(G, A) - \dim_{\mathbb{F}_\ell} H^1(G, A)$. If the G-module A is finitely generated over $\mathbb{Z}_\ell$ then $\mathrm{rk} A = \ell \mathrm{rk} A^G - (\ell - 1)\chi(A)$. So we compute $\chi(A)$ for all modules entering the formula of the theorem and it yields the proof.

As an application of this theorem we take $k = \mathbb{Q}(\sqrt{-3})$ and $K = k(\sqrt[3]{a})$, where $a \in \mathbb{Z}$ and K_∞/k_∞ ramifies exactly in two places. Then all invariants in (12.46) vanish, and we get a decomposition $V(K_\infty) = V^+(K_\infty) \oplus V^-(K_\infty)$. Suppose that K_∞/k_∞ ramifies in three places. Then Theorem 12.21 yields that either $\lambda''(K_\infty) = 2$ and $\lambda'(K_\infty) = d(K_\infty) = r''(K_\infty) = 0$ or $d(K_\infty) = 4$ and all other invariants are zero.

The second case takes place for $K = k(\sqrt[3]{550})$ and $K = k(\sqrt[3]{460})$ (see [13, examples 6.1 and 6.2]. In these two examples $T_\ell(K_\infty)$ is finite but non-zero.

Problem 12.4. *Are there algebraic number fields K such that $\lambda''(K_\infty) > 0$?*

If K is of CM-type then the condition $\lambda''(K_\infty) = 0$ is equivalent to the Greenberg conjecture [2]. So we can consider the claim $\lambda''(K_\infty) = 0$ as a generalization of Greenberg conjecture that has sense for any algebraic number field. The problem is if it holds true or not.

12.8 On a new type of the ℓ-adic regulator

The regulator $R_\ell(K)$ that we had discussed in the preceding sections isn't quite perfect from the viewpoint of Iwasawa theory.

For example, suppose that prime ℓ splits completely in an algebraic number field K. In this case for any intermediate subfield K_n of the cyclotomic $\mathbb{Z}_\ell$-extension K_∞/K one has a regulator $R_\ell(K_n)$ that characterize the ℓ-adic behaviour of the group of units $U(K_n)$ of K_n. As it was shown in [17], the behaviour of the ℓ-adic exponents $\nu_\ell(R_\ell(K_n))$ for increasing n shows some interesting features. This is connected with the fact that the properties of the sequence of the group of units $\{U(K_n)\}_{n\geqslant 0}$ reflect in a single object $\bar{U}(K_\infty) := \varprojlim(U(K_n)[\ell]/\mu(K_n))$, which is a Galois module (relative to the action of $\Gamma = G(K_\infty/K)$). We have a natural projection $\pi\colon \bar{U}(K_\infty) \to U(K)[\ell]/\mu_\ell(K)$, but in our case this projection is a zero map. (At any rate, if the Leopoldt conjecture holds for (K, ℓ).) So it is natural to consider instead of $U(K)$ another object that behaves better from the viewpoint of Iwasawa theory.

Namely, let S denotes the set of all places over ℓ in a given field and $U_S(K_n)$ the group of S-units in K_n. Let $U_{S,1}(K_n)$ be the subgroup of universal norms in $U_S(K_n)[\ell]/\mu_\ell(K_n)$, that is

$$U_{S,1}(K_n) = \cap_{m>n} N_{K_m/K_n}(U_S(K_m)[\ell]/\mu_\ell(K_m)). \tag{12.47}$$

The new ℓ-adic regulator that we wish to define should characterize the ℓ-adic behaviour of $U_{S,1}(K_n)$. But now we shall treat only a particular case $n = 0$, that is, ℓ splits completely in K. To give the definition of our new regulator we have to state some properties of the group $U_{S,1}(K)$.

For any S-unit u there is the principal divisor (u) of u. We can extend this map to the pro-ℓ-completion of the group of S-units and obtain a well-defined map $\mathrm{Div}\colon U_S(K)[\ell] \to D(K)$, where $D(K)$ is a free $\mathbb{Z}_\ell$-module (written additively) generated by the set S of all places $v|\ell$ in K. Let $\mathfrak{l}_1, \cdots, \mathfrak{l}_n$ be all prime divisors of ℓ in K. If $x, y \in D(K)$ and $x = \sum_{i=1}^n a_i \mathfrak{l}_i$, $y = \sum_{i=1}^n b_i \mathfrak{l}_i$, we put

$$\langle x, y\rangle = \sum_{i=1}^n a_i b_i \in \mathbb{Z}_\ell. \tag{12.48}$$

Thus (12.48) defines a non-degenerate bilinear product of the form $D(K) \times D(K) \to \mathbb{Z}_\ell$. If K is Galois over some subfield with a Galois group G then the natural action G on S extends to the action on $D(K)$, and the product (12.48) satisfies the condition $\langle \sigma(x), \sigma(y)\rangle = \langle x, y\rangle$ for any $x, y \in D(K)$, $\sigma \in G$. Under assumption that the Leopoldt conjecture holds for (K, ℓ), one can state easily that Div defines an injection Div: $U_{S,1}(K) \hookrightarrow D(K)$. The next theorem [19, Theor. 1] characterize the group $U_{S,1}(K)$ as Galois module (as $\mathbb{Z}_\ell$-module if K isn't Galois). Note that in this theorem we put no restrictions on decomposition type of prime l in K.

Theorem 12.22. *Let K be a finite Galois extension of $\mathbb{Q}$ with a Galois group G and $R = \mathbb{Q}_\ell[G]$. Suppose that the Leopoldt conjecture holds for (K, ℓ). Then there an isomorphism of R-modules*

$$\varphi\colon U_{S,1}(K) \otimes_{\mathbb{Z}_\ell} \mathbb{Q}_\ell \cong (U(K) \otimes_{\mathbb{Z}} \mathbb{Q}_\ell) \oplus \mathbb{Q}_\ell$$

(here we write the group operation in $U_{S,1}(K)$ and $U(K)$ additively). Without assumption that K is Galois over $\mathbb{Q}$, the module $U_{S,1}(K)$ has $\mathbb{Z}_\ell$-rank $r_1 + r_2$, where r_1 and r_2 is the number of real and complex places of K respectively.

The idea of proof is as follows. Let M be the maximal Abelian ℓ-extension of K_∞ unramified outside S and $X = G(M/K_\infty)$. Let M_0 be the maximal subfield of M Abelian over K. Then we can characterize the Galois group $G(M_0/K)$ in two ways. Firstly, M_0 is an Abelian extension of K hence its Galois group over K can be characterized in terms of class field theory. This characterization depends on $U(K)$. On the other hand, denoting by M_n the maximal subfield of M Abelian over K_n and putting $X'_n = G(M_n/K_n)$ we can characterize any X'_n via class field theory, as above, and put $X = \varprojlim X'_n$, where the limit is taken relative to the norm maps. Thus, factoring X by the action of γ, where γ is a topological generator of $\Gamma = G(K_\infty/K)$, we obtain a characterization of $X_0 = X/(\gamma - 1)X \cong G(M_0/K_\infty)$ in terms of $U_{S,1}(K)$. Comparing these two characterizations, we obtain the theorem.

Using this theorem, we can define our new regulator as follows.

Definition. Let $e_1, \cdots, e_t$ be a $\mathbb{Z}_\ell$-basis of $U_{S,1}(K)$, where $t = r_1 + r_2$. Then we put

$$\mathfrak{R}_\ell(K) = \det(\langle \mathrm{Div}(e_i), \mathrm{Div}(e_j) \rangle), \qquad 1 \leqslant i, j \leqslant t,$$

where $\langle\ ,\ \rangle$ is the product (12.48).

Conjecture 12.23. *Let K be an algebraic number field and ℓ a prime that splits completely in K. Then $\mathfrak{R}_\ell(K) \neq 0$.*

Theorem 12.24. *[19, Theor. 2] Let K be an algebraic number field and prime ℓ splits completely in K. Then Conjecture 12.2 yields that $\mathfrak{R}_\ell(K) \neq 0$.*

Sketch of the proof. We may assume that K is Galois over $\mathbb{Q}$ and K contains an imaginary quadratic field k. Put $G = G(K/\mathbb{Q})$ and $H = G(K/k)$. Let $\alpha \in U_S(K)$ be such an element that $(\alpha) = \mathfrak{l}_1^{\mathbf{h}}$, where $\mathfrak{l}_1$ is a prime divisor of ℓ in K and $\mathbf{h}$ any number such that $\mathfrak{l}_1^{\mathbf{h}}$ is principal. It follows from Conjecture 12.2 that the elements $\log_\ell(\sigma(\alpha))$, $\sigma \in G$ generate in $K \otimes_{\mathbb{Q}} \mathbb{Q}_\ell$ a subspace of co dimension one. (The only relation is $\sum_\sigma \log_\ell(\sigma(\alpha)) = 0$.) The proof of this fact is similar to the proof that the Leopoldt conjecture follows from Conjecture 12.2.

Let $\varepsilon \in U(K)$ be an Artin unit, that is, $\tau(\varepsilon) = \varepsilon$ for an automorphism of complex conjugation $\tau \in G$ and the elements $h(\varepsilon)$, $h \in H$, generate a subgroup of finite index in $U(K)$. Then there exist coefficients c_h, $h \in H$, defined up to a constant summand and such that

$$\log_\ell \varepsilon = (1+\tau)\sum_{h\in H} c_h h(\alpha).$$

Using Conjecture 12.2, one can show that the set of n elements c_h, $h \in H$, where $|H| = n$, has transcendence degree $n-1$ over $\mathbb{Q}$. The element

$$z = \varepsilon^{-1}\prod_{h\in H}(h(\alpha^{c_h})^{1+\tau}$$

has the property $\log_\ell(z) = 0$. One can deduce from this fact that $y = z^{\ell^m} \in U_{S,1}(K)$ for some sufficiently large m. Then the elements $h(y)$, $h \in H$ generate a subgroup of finite index in $U_{S,1}(K)$, and it is enough to check that this system of elements has non-zero regulator $\mathfrak{R}$. We show that this last regulator is a non-zero polynomial with rational integer coefficients in indeterminates $c_1, \ldots, c_{n-1}$. Hence $\mathfrak{R} \neq 0$ and $\mathfrak{R}_\ell(K) \neq 0$. This proves the theorem.

It is desirable to give a non-conditional proof of conjecture 12.23 in some cases. In this direction we have proved the following result.

Theorem 12.25. *[19, Theor. 3] Let K be an Abelian extension of an imaginary quadratic field k with Galois group $H = G(K/k)$. Suppose that K is Galois over $\mathbb{Q}$ and prime ℓ splits completely in K. Let the complex conjugation $\tau \in G = G(K/\mathbb{Q})$ acts on H by the rule $\tau(h) = \tau h \tau^{-1} = h^{-1}$ for any $h \in H$. Let the exponent m of H be such that the primitive m-th root of unity $\zeta_m \in \bar{\mathbb{Q}}_\ell$ is conjugated with ζ_m^{-1} over $\mathbb{Q}_\ell$. Then one has $\mathfrak{R}_\ell(K) \neq 0$.*

Sketch of the proof. Let $\mathfrak{p}_1$, $\mathfrak{p}_2$ be two divisors of ℓ in k. Then $S = S_1 \cup S_2$, where S_i is the set of all places in K over $\mathfrak{p}_i$ for $i = 1, 2$. We put $D_i(K) = \mathbb{Z}_\ell[S_i]$ and define two products as in (12.48)

$$\langle\,,\rangle_1\colon D_1(K)\times D_1(K)\longrightarrow \mathbb{Z}_\ell,\quad \langle\,,\rangle_2\colon D_2(K)\times D_2(K)\longrightarrow \mathbb{Z}_\ell.$$

If $x, y \in D(K)$ then $x = x_1+x_2$ and $y = y_1+y_2$, where $x_i, y_i \in D_i$, $i = 1, 2$, and we can define a new product $\langle x, y\rangle_k = \langle x_1, y_1\rangle \mathfrak{p}_1 + \langle x_2, y_2\rangle_2 \mathfrak{p}_2$.

The first step is as in the proof of Theor. 12.5. We prove, using Theor. 12.22, that one has $\langle \tau(x), \tau(y)\rangle_k = \langle x, y\rangle_k$. It means that $\langle x, y\rangle = 2\langle x_1, y_1\rangle_1 = 2\langle x_2, y_2\rangle_2$.

By Theor. 12.22 we have $U_{S,1}(K) \otimes_{\mathbb{Z}_\ell} \mathbb{Q}_\ell \cong \mathbb{Q}_\ell[H]$. Thus, if φ is a $\mathbb{Q}_\ell$-irreducible character of H then at least one of two maps $\pi_i : U_{S,1}(K)_\varphi \to D_i(K)$, $i = 1, 2$ must be an injection. Suppose that for a given φ the map π_1 is an injection hence $\mathrm{Im}(\pi_1(U_{S,1}(K)_\varphi))$ is a subgroup of finite index in $D(K)_{1,\varphi}$. Any φ is of the form $\varphi = \sum_{\chi|\varphi} \chi$, where χ runs through all absolutely irreducible characters of H that are conjugated over $\mathbb{Q}_\ell$. By $\bar{\varphi}$ we denote the $\mathbb{Q}_\ell$-irreducible character $\sum_{\chi|\varphi} \chi^{-1}$. Since the product $\langle\, , \rangle_1$ is non-degenerate, we obtain that $\langle\, , \rangle_1$ induces a non-degenerate product $D(K)_{1,\varphi} \times D(K)_{1,\bar{\varphi}} \to \mathbb{Z}_\ell$. But in our case one has $\varphi = \bar{\varphi}$. Hence the product $U_{S,1}(K)_\varphi \times U_{S,1}(K)_\varphi \to \mathbb{Z}_\ell$ induced by $\langle\, , \rangle$ is non-degenerate. This proves the theorem, since φ is an arbitrary $\mathbb{Q}_\ell$-irreducible character of H.

The next proposition characterize those ℓ that satisfy the conditions of Theorem 12.25 for a given K. □

Proposition 12.26. *Let K be Galois over $\mathbb{Q}$ and Abelian over an imaginary quadratic field k, as in Theor. 12.25. Suppose that $\tau \in G(k/\mathbb{Q})$, $\tau \neq 1$, acts on $H = G(K/k)$ by inversion. Let m be an exponent of H, ξ_m a primitive m-th root of unity and T the set of all primes that do not divide m.*

If $K \cap \mathbb{Q}(\xi_m)$ is totally real then there are infinitely many $\ell \in T$ such that (K, ℓ) satisfy all conditions of Theor. 12.25. Therefore $\mathfrak{R}_\ell(K) \neq 0$ for these ℓ.

If $K \cap \mathbb{Q}(\xi_m)$ is an imaginary field then there is no $\ell \in T$ that satisfies the conditions of Theor. 12.25.

In conclusion we return to Conjecture 12.6.

Theorem 12.27. *Let K be an algebraic number field and prime ℓ splits completely in K. Suppose that $\mathfrak{R}_\ell(K) \neq 0$. then the feeble conjecture on the ℓ-adic regulator (Conjecture 12.6) is valid for (K, ℓ).*

Sketch of the proof. Let $H(K_\infty)$ be as in Section 12.5. One has $(\gamma - 1)H(K_\infty) \subset V(K_\infty)$, where γ is a topological generator of $G(K_\infty/K)$. Therefore one can extend the product (12.25) by Λ-sesquilinearity to the product

$$H(K_\infty) \times H(K_\infty) \longrightarrow (\gamma - 1)^{-2}\Lambda.$$

In turn, this product leads to the product

$$H(K_\infty) \times H(K_\infty) \longrightarrow (\gamma - 1)^{-2}\Lambda/(\gamma - 1)^{-1}\Lambda \cong \mathbb{Z}_\ell. \qquad (12.49)$$

Since $H(K_\infty)/(\gamma - 1)H(K_\infty) \cong D(K)$ and the product (12.49) vanishes on $(\gamma - 1)H(K_\infty)$, the product (12.49) induces a non-degenerate product

$$D(K) \times D(K) \longrightarrow \mathbb{Z}_\ell, \qquad (12.50)$$

which up to a constant factor coincides with (12.48).

Suppose that Conjecture 12.6 fails to hold for K_∞/K. Then by Theor. 12.7 one has $C(K_\infty) = V^+(K_\infty) \cap V^-(K_\infty) \neq 0$. We define $H^+(K_\infty) \subseteq H(K_\infty)$ as the minimal Λ submodule of $H(K_\infty)$ such that $V^+(K_\infty) \cap H(K_\infty) \subseteq H^+(K_\infty)$ and $H(K_\infty)/H^+(K_\infty)$ has no Λ-torsion. We define $H^-(K_\infty)$ analogously. Let $D^+(K)$ and $D^-(K)$ be the image of the natural projection $\pi : H^+(K_\infty) \to D(K)$ and $H^-(K_\infty) \to D(K)$ respectively. If $C(K_\infty) \neq 0$ then $F(K_\infty) := H^+(K_\infty) \cap H^-(K_\infty) \neq 0$ and $H(K_\infty)/F(K_\infty)$ has no Λ-torsion. Let F_1 be the image in $D(K)$ of $F(K_\infty)$ relative to π.

Since the product (12.49) is zero on $F(K_\infty) \times H^+(K_\infty)$, the product (12.50) is zero on $F_1 \times D^+(K)$, hence the product (12.50) is degenerate on $D^+(K) \times D^+(K)$. But $D^+(K)$ contains $\mathrm{Div}U_{S,1}(K)$ as a subgroup of finite index. Therefore the product (12.48) is degenerate on $\mathrm{Div}(U_{S,1})$ and $\mathfrak{R}_\ell(K) = 0$. This finishes the proof. □

Problem 12.5. *We started from an analogy between functional and number cases. One can ask if there are any analogs of the considered results in the functional case?*

References

[1] Brumer A. "On the units of algebraic number fields". Math. **14** (1967), 121-124.

[2] Greenberg R. "On the Iwasawa invariants of totally real number fields". Amer. J. Math. **98** (1976), 263-284.

[3] Hartshorne R. Algebraic geometry, Springer-Verlag, New York, Heidelberg, Berlin, 1977.

[4] Iwasawa K. "On some modules in the theory of cyclotomic fields". J. Math. Soc. Japan. **20** (1964), 42-82.

[5] "Lectures on p-adic L-functions", Princeton Univ. Press and Univ. of Tokyo Press, Princeton 1972.

[6] Iwasawa K. "On the μ-invariants of $\mathbb{Z}_\ell$-extensions". Number theory, algebraic geometry and commutative algebra in honor of Akizuki, Tokyo 1979, 1-11.

[7] Kuz'min L. V. "The Tate module for algebraic number fields". Math. USSR-Izv., **6**:2 (1972), 263-321.

[8] Kuz'min L. V. "Some duality theorems for cyclotomic Γ-extensions of algebraic number fields of CM-type". Math. USSR-Izv., **14**:3 (1980), 441-498.

[9] Kuz'min L. V. "Some remarks on the ℓ-adic Dirichlet theorem and the ℓ-adic regulator". Math. USSR-Izv., **19**:3 (1982), 445-478.

[10] Kuz'min L. V. "Some remarks on the ℓ-adic regulator. II". Math. USSR-Izv., **35** (1980), 113-144.

[11] Kuz'min L. V. "New explicit formulas for the norm residue symbol, and their applications". Math. USSR-Izv., **37**:3 (1991), 555-586.

[12] Kuz'min L. V. "A new proof of a duality theorem on the ℓ-adic logarithms of local units". J.Math Sci. **95**:1 (1999), 1996-2005.

[13] Kuz'min L. V. "An analog of the Riemann-Hurwitz formula for one type of ℓ-extensions of algebraic number fields". Math. USSR-Izv. **36** (1991):2, 325-347.

[14] Kuz'min L. V. "Some remarks on the ℓ-adic regulator. III". Izv. Math. **63**:6 (1999), 1089-1138.

[15] Kuz'min L. V. "Some remarks on the ℓ-adic regulator. IV". Izv. Math. **64**:2 (2000), 265-310.

[16] Kuz'min L. V. "On some property of the ℓ-adic logarithms of units of non-Abelian local fields". Izv. Math. **70**:5, (2006), 949-974.

[17] Kuz'min L. V. "Some remarks on the ℓ-adic regulator.V. Growth of the ℓ-adic regulator in the cyclotomic $\mathbb{Z}_\ell$-extension of an algebraic number field". Izv. Math. **73**:5, (2009), 959-1021.

[18] Kuz'min L. V. "The feeble conjecture on the 2-adic regulator for some 2-extensions". Izv. Math. **76**:2 (2012), 346-355.

[19] Kuz'min L. V. "On a new type of the ℓ-adic regulator for algebraic number fields (The ℓ-adic regulator without logarithms)". Izv. Math., **79**.1 (2015), 109–144.

[20] Lang S. "Cyclotomic fields" I, II Grad. Texts in Math., **121** Springer-Verlag, New York, 1990.

[21] Sen Sh. "On explicit reciprocity laws". J. Reine Angew. Math. **313** 91980), 1-26.

[22] Winberg K. "Duality theorems for Γ-extensions of algebraic number fields". Comp. Math. **55** (1985), 333-381.

[23] Wintenberger J.-P. "Le corps des norms de certaines extensions infinies de corps locaux; applications". Ann. Sci. Ecole Norm. Sup. (4) **16**:1 (1983), 59-89.

13

On a counting problem for G-shtukas

Ngo Dac Tuan

CNRS - Université de Paris Nord (Paris 13), LAGA - Département de Mathématiques, 99 avenue Jean-Baptiste Clément, 93430 Villetaneuse, France

E-mail address: ngodac@math.univ-paris13.fr

ABSTRACT. In this paper, we give a fixed point formula of stacks of G-shtukas with arbitrary modifications, which generalizes the works of Drinfeld and Lafforgue on stacks of Drinfeld's shtukas.

Moduli spaces (or stacks) of shtukas were introduced by Drinfeld [4] and they are associated to the data consisting of a reductive group G over a function field F and a collection of dominant coweights of this group which are also called modifications. These algebraic stacks are considered to be an analog version of Shimura varieties over function field and it is expected that they would play an important role in the Langlands program over function fields.

In the seventies, Drinfeld studied the so-called moduli space of Drinfeld's shtukas for GL_2 and proved that the Langlands' correspondence for GL_2 could be realized in the cohomology of this moduli space [4, 5]. Later, following the same strategy, Laurent Lafforgue proved the Langlands' correspondence for GL_n by realizing it in the cohomology of the stack of Drinfeld's shtukas for GL_n [13, 14, 15]. When the reductive group G is an interior form of GL_n, different stacks associated to this group have been studied in the literature, see Laumon [19, 20], Laumon-Rapoport-Stuhler [21], Lau [17, 18] and Ngô Bao-Châu [24]. In the direction of a general reductive group G, the first result is due to Varshavsky and Kazhdan [33, 7]. Note that, by a completely new and different method, Vincent Lafforgue [16] has proved a canonical decomposition of the space of cuspidal automorphic forms for G by studying all the stacks of G-shtukas.

2010 Mathematics Subject Classification: 22E55, 11G09, 14D20.
Keywords: Drinfeld shtukas, moduli spaces, function fields, fixed point formula, Arthur-Selberg trace formula.

Arithmetic and Geometry, ed. Luis Dieulefait *et al.* Published by Cambridge University Press.

This paper is a continuation of our previous articles [26, 28] in which we are interested in counting the number of fixed points for stacks of G-shtukas under a power of the Frobenius and a Hecke correspondence. We hope that it would constitute a step towards the study of the cohomology of these stacks in the same spirit of the works of Drinfeld and Lafforgue. In all the works cited above, the classification of φ-spaces (for GL_n) due to Drinfeld plays a primordial role in the counting problem. Roughly speaking, it could be considered as an analog version of the Honda-Tate theory over function fields. To our knowledge, we do not know any generalization of this classification for a general reductive group G. In the articles [26, 28], we have suggested a way to by-pass this obstacle and have given a formula for the number of fixed points for the elliptic part. Essentially, the number of fixed points for *the elliptic part* for G-shtukas is a sum indexed by *elliptic semisimple elements of* $G(F)$ of product of twisted orbital integrals and orbital integrals for some test function. The main goal of this paper is to extend this formula by dealing with all (not necessarily semisimple) elements of $G(F)$. Since we are working in positive characteristic, we do not have the canonical Jordan decomposition for elements of $G(F)$. What saves the day is the observation that, for any element γ in $G(F)$, there always exists a power of γ which is semisimple.

The article is organized as follows. In Section 13.1 and Section 13.2, we recall the definition of stacks classifying G-shtukas, its basic properties, the morphisms of partial Frobenius and the Hecke correspondences. In Sections 13.3-13.6, we present a formula for the number of fixed points of a power of the Frobenius and a Hecke correspondence on a fiber of these stacks, see Theorem 13.21. The formula is given in terms of twisted orbital integrals. In Section 13.7, we give a slight generalization of this formula by counting the number of fixed points of a power of the partial Frobenius and a Hecke correspondence on a fiber of stacks of G-shtukas, see Theorem 13.25. We also illustrate our calculation in the case of the general linear group $G = \mathrm{GL}_n$, see 13.28.

Acknowledgements. The author thanks Ngô Bao-Châu for his interest and helpful conversations on this work. Different parts of the work were done while the author visited the Hausdorff Institute for Mathematics (HIM) and the Vietnam Institute for Advanced Study in Mathematics (VIASM), which I thank for stimulating atmosphere and financial support. My research is partly supported by the grant ANR-10-BLAN 0114 "Arithmétique des Variétés de Shimura et des formes automorphes et Applications", the grant LIA Formath Vietnam and the grant ARCUS via the Université de Paris-Nord (Paris 13).

Notations

In this article, let X be a smooth projective and geometrically connected curve over a finite field $\mathbb{F}_q$ of positive characteristic p. Denote by F the function field of X. Let x be a place of F, denote by F_x the x-adic completion of F and $\mathcal{O}_x$ its ring of integers. We fix an algebraic closure k of $\mathbb{F}_q$ and denote by $\bar{X}$, $\bar{F}$, $\bar{F}_x$ and $\bar{\mathcal{O}}_x$ the corresponding objects obtained by base change from $\mathbb{F}_q$ to k, for example $\bar{X} = X \times_{\mathbb{F}_q} k$, $\bar{F} = F \otimes_{\mathbb{F}_q} k$, etc.

Let G be a (connected) split reductive group over $\mathbb{F}_q$, hence over F, whose derived group G_{der} is always supposed to be simply connected. This hypothesis is not necessary. However, it allows us to use several results of Steinberg [32] and Kottwitz [8], hence simplifies lots of arguments. Denote by B and T a Borel subgroup and a maximal torus of G defined over $\mathbb{F}_q$, with $T \subset B$. Denote by B^{der} et T^{der} (resp. B^{ad} and T^{ad}) the induced Borel subgroup and the induced maximal torus of G^{der} (resp. G^{ad}). With these notations, we define the set of simple roots, dominant (rational) weights and coweights of G with respect to (T, B).

In this article, every scheme or stack will be defined over $\mathbb{F}_q$. Let $\mathcal{Y}$ and $\mathcal{Z}$ be two such stacks, then by $\mathcal{Y} \times \mathcal{Z}$, we always mean the fiber product $\mathcal{Y} \times_{\mathbb{F}_q} \mathcal{Z}$.

13.1 Stack of G-shtukas

We begin by recalling the definition of G-shtukas, the stack classifying such objects and its properties. The principal references of this section are Lafforgue's book [13], Varshavsky's article [33] and that of Ngô Bao-Châu [24].

13.1.1 Stack of G-torsors. Modifications

We consider the stack Bun_G classifying G-torsors over the curve X. Let $\overline{T}$ be a finite subscheme of $\overline{X}$, and let $\mathcal{V}$, $\mathcal{V}'$ be two G-torsors over $\overline{X}$. By definition, *a $\overline{T}$-modification of $\mathcal{V}$ in $\mathcal{V}'$* consists of an isomorphism between $\mathcal{V}$ and $\mathcal{V}'$ outside $\overline{T}$, i.e.

$$\phi : \mathcal{V} \,|_{\overline{X}-\overline{T}} \xrightarrow{\sim} \mathcal{V}' \,|_{\overline{X}-\overline{T}} .$$

Let $\overline{x}$ be a geometric point in $\overline{T}$. Denote by $\mathcal{O}_{\overline{x}}$ the completion of $\mathcal{O}_{\overline{X}}$ at the place $\overline{x}$, and by $F_{\overline{x}}$ its fraction field. Given a $\overline{T}$-modification

$$\phi : \mathcal{V} \,|_{\overline{X}-\overline{T}} \xrightarrow{\sim} \mathcal{V}' \,|_{\overline{X}-\overline{T}}$$

between two G-torsors $\mathcal{V}$ et $\mathcal{V}'$, we get an isomorphism $V \xrightarrow{\sim} V'$ between the generic fibers of $\mathcal{V}$ and $\mathcal{V}'$ which induces an isomorphism $\phi_{\overline{x}} : V_{\overline{x}} \xrightarrow{\sim} V'_{\overline{x}}$ between the completion of V et V' at $\overline{x}$. If we choose a trivialization of $V_{\overline{x}}$, then the completion of $\mathcal{V}$ at $\overline{x}$ defines an element of the quotient $G(F_{\overline{x}})/G(\mathcal{O}_{\overline{x}})$. Similarly, if we choose a trivialization of $V'_{\overline{x}}$, then the completion of $\mathcal{V}'$ at $\overline{x}$ defines an element of the quotient $G(F_{\overline{x}})/G(\mathcal{O}_{\overline{x}})$. Using the above isomorphism $\phi_{\overline{x}}$, $V_{\overline{x}}$ and $V'_{\overline{x}}$ can be identified, hence we get a double class in the double quotient $G(\mathcal{O}_{\overline{x}})\backslash G(F_{\overline{x}})/G(\mathcal{O}_{\overline{x}})$. We define the relative position of ϕ to be this double class.

Since G is a split reductive group over $\mathbb{F}_q$, the Cartan decomposition gives us:

$$G(F_{\overline{x}}) = \bigsqcup_{\lambda} G(\mathcal{O}_{\overline{x}})\varpi_{\overline{x}}^{\lambda} G(\mathcal{O}_{\overline{x}})$$

where λ runs through the set of dominant coweights of G, and $\varpi_{\overline{x}}$ denotes a uniformizing element of X at $\overline{x}$. Each double class contains a unique element of the form $\varpi_{\overline{x}}^{\lambda}$ where λ is a dominant coweight. We define this dominant coweight as *the invariant of ϕ at $\overline{x}$*.

13.1.2 Hecke stacks

Recall that the set of dominant coweights of G is equipped with an order relation. Let λ and λ' be two dominant coweights of G, we say that $\lambda \leq \lambda'$ if and only if $\lambda' - \lambda$ is a positive integral linear combination of simple coroots of G.

Following Beilinson and Drinfeld, for each dominant coweight λ of G, we consider the stack $\mathrm{Hecke}_{\leq\lambda}$ classifying simple modifications bounded by λ as follows. For each scheme S (over $\mathbb{F}_q$), $\mathrm{Hecke}_{\leq\lambda}(S)$ is the groupoid consisting of the following data:

i) two G-torsors $\mathcal{V}$ and $\mathcal{V}'$ over $X \times S$,
ii) a morphism $x : S \longrightarrow X$, and we denote by $\Gamma(x)$ the graph of x,
iii) an isomorphism

$$\phi : \mathcal{V}\,|_{X\times S-\Gamma(x)} \xrightarrow{\sim} \mathcal{V}'\,|_{X\times S-\Gamma(x)}$$

such that, for each geometric point $\overline{s}$ of S, the $\overline{s}$-modification $\phi_{\overline{s}}$ admits an invariant (at $x(\overline{s})$) bounded by λ.

Denote by Hecke_{λ} the substack of $\mathrm{Hecke}_{\leq\lambda}$ by replacing the condition iii) by the following condition:

iii') an isomorphism

$$\phi : \mathcal{V} \mid_{X\times S-\Gamma(x)} \xrightarrow{\sim} \mathcal{V}' \mid_{X\times S-\Gamma(x)}$$

such that, for each geometric point $\overline{s}$ of S, the $\overline{s}$-modification $\phi_{\overline{s}}$ admits λ as the invariant (at $x(\overline{s})$).

It is proved that Hecke_{λ} is an open substack of $\mathrm{Hecke}_{\leq\lambda}$. Furthermore, we get the following proposition and we send the reader to [2] for the proof:

Proposition 13.1. *The morphism*

$$\begin{aligned} \mathrm{Hecke}_{\leq\lambda} &\longrightarrow X \times \mathrm{Bun}_G \\ (\mathcal{V}, \mathcal{V}', x) &\mapsto (x, \mathcal{V}) \end{aligned}$$

is representable and projective. And the restriction of this morphism to the open substack Hecke_{λ} *is smooth.*

More generally, to each collection of dominant coweights $\underline{\lambda} = (\lambda_1, \ldots, \lambda_n)$ of G, we introduce the Hecke stack $\mathrm{Hecke}_{\leq\underline{\lambda}}$ (resp. the stack $\mathrm{Hecke}_{\underline{\lambda}}$) classifying successive modifications, see [33, 24]. For each scheme S, it classifies the data consisting of

i) $(n+1)$ G-torsors $\mathcal{V}_0, \ldots, \mathcal{V}_n$ over $X \times S$,
ii) n morphisms $x_i : S \longrightarrow X$ $(1 \leq i \leq n)$ and we denote by $\Gamma(x_i)$ the graph of x_i,
iii) a collection of isomorphisms

$$\phi_i : \mathcal{V}_{i-1} \mid_{X\times S-\Gamma(x_i)} \xrightarrow{\sim} \mathcal{V}_i \mid_{X\times S-\Gamma(x_i)}, \quad 1 \leq i \leq n,$$

such that, for each geometric point $\overline{s}$ of S, the $\overline{s}$-modification $\phi_{i,\overline{s}}$ admits an invariant (at $x_i(\overline{s})$) bounded by λ_i (resp. the $\overline{s}$-modification $\phi_{i,\overline{s}}$ admits λ_i as the invariant (en $x_i(\overline{s})$)).

We have immediately the following proposition:

Proposition 13.2. *The morphism*

$$\begin{aligned} \mathrm{Hecke}_{\leq\underline{\lambda}} &\longrightarrow X^n \times \mathrm{Bun}_G \\ (\mathcal{V}_0, \mathcal{V}_1, \ldots, \mathcal{V}_n, x_1, x_2, \ldots, x_n) &\mapsto (x_1, x_2, \ldots, x_n, \mathcal{V}_0) \end{aligned}$$

is representable and projective. And the restriction of this morphism to the open substack $\mathrm{Hecke}_{\underline{\lambda}}$ *is smooth.*

Over the complementary $X^n \setminus \Delta$ of the union of the diagonals of X^n, we get *a factorization property* which will be useful later.

Proposition 13.3. *Over the complementary $X^n \setminus \Delta$ of the union of the diagonals of X^n, the Hecke stack of successive modifications* $\mathrm{Hecke}_{\leq\underline{\lambda}}$ *is canonically isomorphic to the fiber product of the morphisms* $\mathrm{Hecke}_{\leq\lambda_j} \longrightarrow \mathrm{Fib}_G$ *for* $j = 1, 2, \dots, n$.

Proof. See [24]. □

13.1.3 Stack of G-shtukas

We keep the notations of the previous section. We are ready to introduce the stack of G-shtukas $\mathrm{Cht}_{\leq\underline{\lambda}}$ associated to the collection of dominant coweights $\underline{\lambda}$ of G. For each scheme S, $\mathrm{Cht}_{\leq\underline{\lambda}}(S)$ classifies the data consisting of

i) $(n+1)$ G-torsors $\mathcal{V}_0, \dots, \mathcal{V}_n$ over $X \times S$,
ii) n morphisms $x_i : S \longrightarrow X$ $(1 \leq i \leq n)$ and we denote by $\Gamma(x_i)$ the graph of x_i,
iii) a collection of isomorphisms

$$\phi_i : \mathcal{V}_{i-1} \mid_{X\times S-\Gamma(x_i)} \xrightarrow{\sim} \mathcal{V}_i \mid_{X\times S-\Gamma(x_i)}, \quad 1 \leq i \leq n,$$

such that, for each geometric point $\overline{s}$ of S, the $\overline{s}$-modification $\phi_{i,\overline{s}}$ admits an invariant (at $x_i(\overline{s})$) bounded by λ_i,
iv) an isomorphism $\mathcal{V}_0^\sigma := (\mathrm{Id}_X \times \mathrm{Frob}_S)^*\mathcal{V}_0 \xrightarrow{\sim} \mathcal{V}_n$.

By composition, we get an isomorphism

$$t : \mathcal{V}_0^\sigma \mid_{X\times S-\bigcup_{i=1}^n \Gamma(x_i)} \xrightarrow{\sim} \mathcal{V}_0 \mid_{X\times S-\bigcup_{i=1}^n \Gamma(x_i)}.$$

If the graphs $\Gamma(x_i)$ are disjoint, the isomorphisms ϕ_i can be determined by t so that we have a simpler description of the stack of G-shtukas over the complementary $X^n \setminus \Delta$ of the union of the diagonals of X^n.

The stack of G-shtukas fits into the following cartesian diagram:

$$\begin{array}{ccc} \mathrm{Cht}_{\leq\underline{\lambda}} & \longrightarrow & \mathrm{Fib}_G \\ \downarrow & & \downarrow \\ \mathrm{Hecke}_{\leq\underline{\lambda}} & \longrightarrow & \mathrm{Fib}_G \times \mathrm{Fib}_G. \end{array}$$

The horizontal maps are given as follows:

$$\left(\mathcal{V}_0, \dots, \mathcal{V}_n; x_1, \dots, x_n; \mathcal{V}_0^\sigma \xrightarrow{\sim} \mathcal{V}_n\right) \mapsto \mathcal{V}_0,$$
$$(\mathcal{V}_0, \dots, \mathcal{V}_n; x_1, \dots, x_n) \mapsto (\mathcal{V}_0, \mathcal{V}_n).$$

The vertical maps are given as follows:

$$(\mathcal{V}_0, \ldots, \mathcal{V}_n; x_1, \ldots, x_n; \mathcal{V}_0^\sigma \xrightarrow{\sim} \mathcal{V}_n) \mapsto (\mathcal{V}_0, \ldots, \mathcal{V}_n; x_1, \ldots, x_n)$$
$$\mathcal{V}_0 \mapsto (\mathcal{V}_0, \mathcal{V}_0^\sigma).$$

As before, we define the open substack $\mathrm{Cht}_{\underline{\lambda}}$ of $\mathrm{Cht}_{\leq\underline{\lambda}}$ by replacing the inequalities $\leq \lambda_i$ in the condition (iii) by the equalities.

13.1.4 Level structures

By a level of X, we always mean a finite subscheme of X.

Definition 13.4. Let I be a level of X and S be a scheme. We consider a G-shtuka $\widetilde{\mathcal{V}} = (\mathcal{V}_0, \ldots, \mathcal{V}_n; x_1, \ldots, x_n; \mathcal{V}_0^\sigma \xrightarrow{\sim} \mathcal{V}_n)$ over S such that the associated points $x_1, x_2, \ldots, x_n$ do not meet I. By definition, an I-level structure of $\widetilde{\mathcal{V}}$ consists of an isomorphism

$$u : \mathcal{V}_0 |_{I \times S} \xrightarrow{\sim} G(\mathcal{O}_I) \times S$$

such that the following diagram is commutative:

$$\begin{array}{ccccc} \mathcal{V}_0 |_{I\times S} & \xrightarrow{\sim} \cdots \xrightarrow{\sim} & \mathcal{V}_n |_{I\times S} & \xrightarrow{\sim} & \mathcal{V}_0^\sigma |_{I\times S} \\ {\scriptstyle u}\downarrow & & & & \downarrow{\scriptstyle u^\sigma} \\ G(\mathcal{O}_I) \times S & & = & & G(\mathcal{O}_I) \times S. \end{array}$$

Denote by $\mathrm{Cht}_{\leq\underline{\lambda},I}$ the stack classifying G-shtukas equipped with an I-level structure. We call *the characteristic morphism* the natural morphism

$$\mathrm{Cht}_{\leq\underline{\lambda},I} \longrightarrow (X - I)^n.$$

We are now presenting an equivalent definition of level structures [26]. We consider G as the trivial G-torsor over X with the action of G on itself on the left and G_I the restriction of G on the finite subscheme I of X. We consider the functor $\mathcal{G}'^I$ which, for each X-scheme S, classifies the subgroup of automorphisms of $G \times_X S$ which commute with the action of G on the left and induce the identity action on $G_I \times_X S$. This functor is representable by a group scheme $\mathcal{G}'^I$ on X. It is equipped with a map to G which is an isomorphism outside I. In general, it is not flat, hence not smooth, we will consider the smooth group scheme associated to it as explained in Raynaud [3, Théorème 3, page 61]. We denote it by $\mathcal{G}^I$ equipped with a canonical morphism $\mathcal{G}^I \to \mathcal{G}'^I$.

It is proved [26] that a shtuka $\widetilde{\mathcal{V}}$ equipped with an I-level structure is equivalent to the data consisting of

i) $(n+1)$ $\mathcal{G}^I$-torsors $\mathcal{V}_0, \ldots, \mathcal{V}_n$ over $X \times S$,
ii) n morphisms $x_i : S \longrightarrow X$ $(1 \leq i \leq n)$ and we denote by $\Gamma(x_i)$ the graph of x_i,
iii) a collection of isomorphisms

$$\phi_i : \mathcal{V}_{i-1} \mid_{X\times S-\Gamma(x_i)} \xrightarrow{\sim} \mathcal{V}_i \mid_{X\times S-\Gamma(x_i)}, \quad 1 \leq i \leq n,$$

such that, for each geometric point $\overline{s}$ of S, the $\overline{s}$-modification $\phi_{i,\overline{s}}$ admits an invariant (at $x_i(\overline{s})$) bounded by λ_i (resp. the $\overline{s}$-modification $\phi_{i,\overline{s}}$ admits λ_i as the invariant (en $x_i(\overline{s})$)).
iv) an isomorphism $\mathcal{V}_0^\sigma := (\mathrm{Id}_X \times \mathrm{Frob}_S)^*\mathcal{V}_0 \xrightarrow{\sim} \mathcal{V}_n$.

In other words, we have defined a shtuka associated to the group scheme $\mathcal{G}^I$ instead of G.

13.1.5 Affine grassmannian and local model

We keep the notations of the previous section. We recall now the notion of the local and global affine grassmannians. Let λ be a dominant coweight of G. We define the global affine grassmannian $\mathcal{G}r_{\leq\lambda}$ as follows. For each scheme S, $\mathcal{G}r_{\leq\lambda}(S)$ classifies the data consisting of

i) two G-torsors $\mathcal{V}$ and $\mathcal{V}'$ over $X \times S$,
ii) a morphism $x : S \longrightarrow X$, and denote by $\Gamma(x)$ the graph of x,
iii) an isomorphism

$$\phi : \mathcal{V} \mid_{X\times S-\Gamma(x)} \xrightarrow{\sim} \mathcal{V}' \mid_{X\times S-\Gamma(x)}$$

such that, for each geometric point $\overline{s}$ of S, the $\overline{s}$-modification $\phi_{\overline{s}}$ admits an invariant (at $x(\overline{s})$) bounded by λ,
iv) and a trivialization of $\mathcal{V}'$, i.e. an isomorphism between $\mathcal{V}'$ and the trivial G-torsor on $X \times S$.

Recall now the definition of the local affine grassmannian Gr. It is an ind-scheme which classifies G-torsors over the formal local disc equipped with a trivialization on the punctured disc. The closed sub-ind-scheme $Gr_{\leq\lambda}$ (resp. the locally closed sub-ind-scheme Gr_λ) of Gr are defined by requiring that the Hodge invariant is bounded by λ (resp. equal to λ).

Proposition 13.5. *Locally for the étale topology, the natural morphism*

$$\mathcal{G}r_{\leq\lambda} \longrightarrow X$$

is a fibration with fiber $Gr_{\leq\lambda}$.

Proof. See [33, Lemma A.8]. □

More generally, let $\underline{\lambda} = (\lambda_1, \ldots, \lambda_n)$ be a collection of dominant coweights of G. Then, the global affine grassmannian $\mathcal{G}r_{\leq\underline{\lambda}}$ is defined as the stack which, for each scheme S, classifies the data consisting of

i) G-torsors $\mathcal{V}_0, \ldots, \mathcal{V}_n$ sur $X \times S$,
ii) morphisms $x_i : S \longrightarrow X$ $(1 \leq i \leq n)$, and denote by $\Gamma(x_i)$ the graph of x_i,
iii) isomorphisms

$$\phi_i : \mathcal{V}_{i-1} \mid_{X\times S-\Gamma(x_i)} \xrightarrow{\sim} \mathcal{V}_i \mid_{X\times S-\Gamma(x_i)}, \quad 1 \leq i \leq n,$$

such that, for each geometric point $\overline{s}$ of S, the $\overline{s}$-modification $\phi_{i,\overline{s}}$ admits an invariant (at $x_i(\overline{s})$) $\leq \lambda_i$ (resp. the $\overline{s}$-modification $\phi_{i,\overline{s}}$ admits an invariant (at $x_i(\overline{s})$) λ_i),
iv) and a trivialization of $\mathcal{V}_n$, i.e. an isomorphism between $\mathcal{V}_n$ and the trivial G-torsor over $X \times S$.

It is shown that:

Proposition 13.6. *i) Over the complementary $X^n \setminus \Delta$ of the union of the diagonals of X^n, $\mathcal{G}r_{\leq\underline{\lambda}}$ is cannonically isomorphic to the product $\prod_i \mathcal{G}r_{\leq\lambda_i}$.*
ii) $\mathcal{G}r_{\leq\underline{\lambda}}$ (resp. $\mathcal{G}r_{\underline{\lambda}}$) is a local model of $\mathrm{Cht}_{\leq\underline{\lambda}}$ *(resp.* $\mathrm{Cht}_{\underline{\lambda}}$*). In other words, for every point u of* $\mathrm{Cht}_{\leq\underline{\lambda}}$*, there exists an étale neighborhood $U_u \longrightarrow$* $\mathrm{Cht}_{\leq\underline{\lambda}}$ *of u and an étale morphism $U_u \longrightarrow \mathcal{G}r_{\leq\underline{\lambda}}$.*

Proof. See [33, Lemma A.8] and [33, Theorem 2.20]. □

13.1.6 Morphisms of partial Frobenius

Let $\underline{\lambda} = (\lambda_1, \lambda_2, \ldots, \lambda_n)$ be a collection of dominant coweights of G and let I be a finite subscheme of X. We have introduced the stack $\mathrm{Cht}_{I,\leq\underline{\lambda}}$ classifying G-shtukas with I-level structures. It is equipped with the characteristic morphism

$$\mathrm{Cht}_{I,\leq\underline{\lambda}} \longrightarrow (X - I)^n.$$

For each integer i such that $1 \leq i \leq n$, we consider the morphism of partial Frobenius of $(X - I)^n$ given by

$$\sigma_i(x_1, \ldots, x_n) = (x_1, \ldots, x_{i-1}, \mathrm{Frob}(x_i), x_{i+1}, \ldots, x_n).$$

Denote by Λ the set of points $(x_1, \dots, x_n) \in (X - I)^n$ such that $\mathrm{Frob}^m(x_i) \neq \mathrm{Frob}^n(x_j)$ for all couples of positive integers (m, n) and all couples of indexes $i \neq j$; then Λ is open in $(X - I)^n$.

Proposition 13.7. *Over the open subscheme Λ, we have an isomorphism of partial Frobenius:*

$$\sigma_i : \mathrm{Cht}_{I,\leq\underline{\lambda}} \times_{(X-I)^n} \Lambda \longrightarrow \mathrm{Cht}_{I,\leq\underline{\lambda}} \times_{(X-I)^n} \Lambda$$

which suits into the commutative diagram:

$$\begin{array}{ccc} \mathrm{Cht}_{I,\leq\underline{\lambda}} \times_{(X-I)^n} \Lambda & \xrightarrow{\sigma_i} & \mathrm{Cht}_{I,\leq\underline{\lambda}} \times_{(X-I)^n} \Lambda \\ \downarrow & & \downarrow \\ \Lambda & \xrightarrow{\sigma_i} & \Lambda. \end{array}$$

Moreover, the open substack $\mathrm{Cht}_{I,\underline{\lambda}} \times_{(X-I)^n} \Lambda$ is stable under the isomorphisms of partial Frobenius.

Proof. To simplify the exposition, we can suppose that $i = 1$. Let S be a scheme and let

$$\widetilde{\mathcal{V}} = (\mathcal{V}_0, \mathcal{V}_1, \dots, \mathcal{V}_n, x_1, \dots, x_n, \mathcal{V}_0^\sigma \xrightarrow{\sim} \mathcal{V}_n, \iota)$$

be an object of the groupoid $\mathrm{Cht}_{I,\leq\underline{\lambda}} \times_{(X-I)^n} \Lambda(S)$. We will delete the G-torsor $\mathcal{V}_0$ from the beginning of the chain and add $\mathcal{V}_0^\sigma$ at the end to obtain a new object

$$(\mathcal{V}'_0, \mathcal{V}'_1, \dots, \mathcal{V}'_n, x_2, \dots, x_n, \sigma_1(x_1), \mathcal{V}_0'^\sigma \xrightarrow{\sim} \mathcal{V}'_n, \iota')$$

of $\mathrm{Cht}_{I,\leq(\lambda_2,\dots,\lambda_n,\lambda_1)}$. The associated points in $(X-I)^n$ is $(x_2, \dots, , x_n, \sigma_1(x_1))$. Since we are working over the open subscheme Λ, the factorization property 13.3 implies that this object induces an object $\sigma_1(\widetilde{\mathcal{V}})$ of $\mathrm{Cht}_{I,\leq\underline{\lambda}}$ over $(\sigma_1(x_1), x_2, \dots, x_n)$. It is obvious that σ_1 verifies all the required properties. The proof is finished. □

The morphisms of partial Frobenius commute with each other and their composition gives the usual Frobenius.

13.1.7 Truncating parameters

Definition 13.8. Let p be a rational dominant coweight in $X_*T^{\mathrm{der}} \otimes \mathbb{Q}$. We will say that p is sufficiently convex if, for each simple root α of G, $\langle\alpha, p\rangle$ is sufficiently big.

Let p be a rational dominant coweight in $X_*T^{\mathrm{der}} \otimes \mathbb{Q}$. Let S be a scheme and $\mathcal{V}$ be a G-torsor over the fiber product $X \times S$. We will say that $p(\mathcal{V}) \leq p$ if, for each geometric point $s \in S$, each B-reduction $\mathcal{V}_s^B$ of $\mathcal{V}_s$ and each dominant weight $\lambda \in X^*T^{\mathrm{ad}}$, we have the inequality:

$$\deg(\mathcal{V}_{s,\lambda}^B) \leq \langle \lambda, p \rangle,$$

where $\mathcal{V}_{s,\lambda}^B$ is the line bundle induced by $\mathcal{V}_s^B$ via the character λ.

Then, we define the substack $\mathrm{Bun}_G^{\leq p}$ of Bun_G which, for each scheme S, classifies G-torsors $\mathcal{V}$ over the product $X \times S$ such that $p(\mathcal{V}) \leq p$. It is proved that the substack $\mathrm{Bun}_G^{\leq p}$ is an open substack of the stack Bun_G, see [33, Lemma A.3].

Then, let $\underline{\lambda} = (\lambda_1, \lambda_2, \ldots, \lambda_n)$ be a collection of dominant coweights of G and let I be a finite subscheme of X. Recall that we have defined the stack $\mathrm{Cht}_{I,\leq\underline{\lambda}}$ classifying G-shtukas with I-level structures. We define the open substack $\mathrm{Cht}_{I,\leq\underline{\lambda}}^{\leq p}$ of $\mathrm{Cht}_{I,\leq\underline{\lambda}}$ by requiring that the associated G-torsor $\mathcal{V}_0$ satisfies the condition that $p(\mathcal{V}_0) \leq p$.

Remark 13.9. If the reductive group G is the general linear group GL_n, we rediscover the notion of truncation by the polygon of Harder-Narasimhan for vector bundles of rank n over the curve X. It gives a definition of truncating parameters for GL_n-shtukas (or shtukas of rank n) over X. Lafforgue introduced a slightly different definition of truncating parameters for shtukas of rank n over X, see [13, Chapter II, section 2]. However, he observed that if the parameter p is sufficiently convex, then the two truncating parameters give the same open substack. We refer the reader to [13] for more details.

13.2 Hecke operators

13.2.1 Notations

We denote by $\mathbb{A}$ the ring of adèles of F, $\mathcal{O}_{\mathbb{A}} = \prod_{x \in |X|} \mathcal{O}_x$ the ring of integers. Let $K = \prod_{x \in |X|} K_x$ be the maximal compact subgroup of $G(\mathbb{A})$ defined by $K_x = G(\mathcal{O}_x)$ for each place x of F. For each finite set of places T of X, denote by

$$\mathbb{A}_T = \prod_{x \in T} F_x, \quad \mathcal{O}_T = \prod_{x \in T} \mathcal{O}_x,$$
$$\mathbb{A}^T = \mathbb{A}/\mathbb{A}_T, \quad \mathcal{O}^T = \mathcal{O}_{\mathbb{A}}/\mathcal{O}_T,$$

and

$$K_T = \prod_{x \in T} K_x, \quad K^T = \prod_{x \notin T} K_x.$$

We consider a level I of X which does not meet T. Denote by K_I (resp. K_I^T) the kernel of the surjective map $K \longrightarrow G(\mathcal{O}_I)$ (resp. $K^T \longrightarrow G(\mathcal{O}_I)$). Then we get $K_I = \prod_{x \in |X|} K_{I,x}$ where $K_{I,x}$ is the open compact subgroup of K_x defined as the kernel of the map $K_x \to G(\mathcal{O}_I)$.

The group $G(\mathbb{A})$ (resp. $G(\mathbb{A}^T)$) is equipped with a Haar measure dg (resp. dg^T). It is normalized such that $dg(K) = 1$ (resp. $dg^T(K^T) = 1$). We denote by dg_x the normalized Haar measure of $G(F_x)$ such that $dg_x(K_x) = 1$. Hence, we obtain $dg = dg^T \times \prod_{x \in T} dg_x$.

For each place x, we denote by $\mathcal{H}_{I,x}$ the vector space of locally constant functions with compact support $f : G(F_x) \to \mathbb{Q}$, invariant on the left and on the right by the action of the open compact subgroup $K_{I,x}$. The characteristic functions ϕ_{β_x} of double-classe $\beta_x \in K_{I,x} \backslash G(F_x)/K_{I,x}$ form a basis of $\mathcal{H}_{I,x}$. The convolution product with respect to the Haar measure dg_x on the vector space $\mathcal{H}_{I,x}$ equips it with an algebra structure. For each finite set of places T' of X, we denote by $\mathcal{H}_{I,T'} = \bigotimes_{x \in T'} \mathcal{H}_{I,x}$ and define $\mathcal{H}_I$ (resp. $\mathcal{H}_I^T$) as the inductive limit of $\mathcal{H}_{I,T'}$ where T' runs through the set of all the finite sets of places of X (resp. of $X - T$): if $T' \subset T''$, we have an injective map

$$\bigotimes_{x \in T'} \mathcal{H}_{I,x} \to \bigotimes_{x \in T''} \mathcal{H}_{I,x}$$

defined by

$$\phi \mapsto \phi \otimes \bigotimes_{x \in T'' - T'} \mathbf{1}_{K_{I,x}}$$

where $\mathbf{1}_{K_{I,x}}$ is the characteristic function of $K_{I,x}$. The algebra $\mathcal{H}_I$ has a basis $\bigotimes_{x \in T'} \phi_{\beta_x}$ indexed by the set of all the couples $(T', (\beta_x)_{x \in T'})$ where T' is a finite set of places of X and $\beta_x \in K_{I,x} \backslash G(F_x)/K_{I,x}$ is a double class modulo by the obvious equivalent relation.

13.2.2 Hecke operators

Let I be a level of X and T' be a finite set of places of X which can eventually meet the level I. For each place $x \in T'$, we are given a double class $\beta_x \in K_{I,x} \backslash G(F_x)/K_{I,x}$. The Hecke operator $\bigotimes_{x \in T'} \beta_x$ acts on $\mathrm{Cht}_{\underline{\lambda},I} \otimes_{(X-I)^n} (X - I - T')^n$ and $\mathrm{Cht}_{\leq \underline{\lambda},I} \otimes_{(X-I)^n} (X - I - T')^n$ by correspondence which will be introduced in this section.

Let x be a closed point of T'. Let Y be a flattening neighborhood of x in X. A shtuka $\widetilde{\mathcal{V}} \in \mathrm{Cht}_{\underline{\lambda},I} \otimes_{(X-I)^n} (X - I - T')^n(S)$ induces by restriction to Y a $\mathcal{G}^I$-torsor $\mathcal{V}_Y$ over the fiber product $Y \times S$ equipped with an isomorphism $\mathcal{V}_Y^{\sigma} \xrightarrow{\sim} \mathcal{V}_Y$. Since $\mathcal{G}^I$ has connected fibers, it is equivalent to give a S-point of the classifying stack of the finite group $\mathcal{G}^I(\mathcal{O}_Y)$, i.e. a $\mathcal{G}^I(\mathcal{O}_Y)$-torsor $\mathcal{V}_Y^{\natural}$ over S. If Y runs through the set of all flattening neighborhoods of x in X, we obtain a $K_{I,x}$-torsor $\mathcal{V}_x^{\natural}$ over S where $K_{I,x} = \mathcal{G}^I(\mathcal{O}_x)$.

A correspondence of type $(T', (\beta_x)_{x\in T'})$ between two shtukas $\widetilde{\mathcal{V}}, \widetilde{\mathcal{V}}' \in \mathrm{Cht}_{\underline{\lambda},I} \otimes_{(X-I)^n} (X - I - T')^n(S)$ with the same associated points $x_1, \ldots, x_n$ consists of an isomorphism

$$t' : \widetilde{\mathcal{V}}|_{(X-T')\times S} \xrightarrow{\sim} \widetilde{\mathcal{V}}'|_{(X-T')\times S}$$

between the restrictions of $\widetilde{\mathcal{V}}$ and $\widetilde{\mathcal{V}}'$ over the open complementary of T', verifying the following condition. For each closed point $x \in T'$, let $\mathcal{V}_x^{\natural}$ and $\mathcal{V}'^{\natural}_x$ be the $K_{I,x}$-torsors over S associated to these shtukas $\widetilde{\mathcal{V}}$ and $\widetilde{\mathcal{V}}'$. The map t' gives an isomorphism between the $\mathcal{G}^I(F_x)$-torsors induced by $\mathcal{V}_x^{\natural}$ et $\mathcal{V}'^{\natural}_x$. This isomorphism gives in turn a function locally constant on S in the set of double-classe $K_{I,x}\backslash\mathcal{G}^I(F_x)/K_{I,x}$. We require that this function is constant and its value is β_x. Moreover, we have the following result:

Proposition 13.10. *For each* $(T', (\beta_x)_{x\in T'})$ *as above, let consider the functor* $\Phi(T', (\beta_x)_{x\in T'})$ *over*

$$\mathrm{Cht}_{\underline{\lambda},I} \times_{(X-I)^n} (X - I - T')^n$$

which associates to any G-shtuka $\widetilde{\mathcal{V}}$ *with an I-level structure such that its associated points do not meet* T' *the category of couples consisting of a G-shtuka* $\widetilde{\mathcal{V}}'$ *with an I-level structure and the same associated points as those of* $\widetilde{\mathcal{V}}$ *and a correspondence between* $\widetilde{\mathcal{V}}$ *and* $\widetilde{\mathcal{V}}'$ *of type* $(T', (\beta_x)_{x\in T'})$*. Then, the functor* $\Phi(T', (\beta_x)_{x\in T'})$ *is representable by a finite and étale morphism over* $\mathrm{Cht}_{\underline{\lambda},I} \times_{(X-I)^n} (X - I - T')^n$*.*

Remark 13.11. A similar argument to that given by Lafforgue [13, Chapter I, section 4] in the case $G = \mathrm{GL}_n$ implies that we could extend the above construction and define a map from the Hecke algebra $\mathcal{H}_I$ to the algebra of étale correspondences in the generic fiber of $\mathrm{Cht}_{\underline{\lambda},I}$ over $(X - I)^n$. Moreover, this action commutes with the morphisms of partial Frobenius, hence with the Frobenius.

Remark 13.12. Let p be a truncating parameter sufficiently convex with respect to $\underline{\lambda}$. We consider the correspondence $\Phi(T', (\beta_x)_{x\in T'})$ and we suppose that, for each $x \in T'$, β_x lies in $K_{I,x}\backslash G(\mathcal{O}_x)/K_{I,x}$. Then, the open substack $\mathrm{Cht}^{\leq p}_{\underline{\lambda},I} \times_{(X-I)^n} (X - I - T')^n$ is stable by this correspondence.

13.3 Counting the number of fixed points by a power of the Frobenius

13.3.1 Notations

Let $s \geq 1$ be a positive integer, $x_1, \ldots, x_n \in X(\mathbb{F}_{q^s})$ be distinct points of $(X - I)$ with value in $\mathbb{F}_{q^s}$. Let $\overline{T} = \{\overline{x}_1, \ldots, \overline{x}_n\}$ where $\overline{x}_i$ is a geometric point over x_i. We will fix a finite set of places T' of X which does not meet the set $\{x_1, \ldots, x_n\}$ and for each place $x \in T'$, we fix a double-class $\beta_x \in K_{I,x}\backslash G(F_x)/K_{I,x}$. Let $\Phi(T', (\beta_x)_{x \in T'})$ be the associated Hecke correspondence defined in the previous section.

The main goal of this article is to give an expression of the number of fixed points $\sigma^s \circ \Phi(T', (\beta_x)_{x\in T'})$ in the fiber $\mathrm{Cht}_{\underline{\lambda},I}(\overline{x}_1, \ldots, \overline{x}_n)$ de $\mathrm{Cht}_{\underline{\lambda},I}$ over $(\overline{x}_1, \ldots, \overline{x}_n)$ by a formula which could be compared to the geometric side of the Arthur-Selberg trace formula of G for a suitable test function.

13.3.2 Fixed points by a power of the Frobenius

We will give the first description of the category of fixed points $\sigma^s \circ \Phi(T', (\beta_x)_{x\in T'})$ in $\mathrm{Cht}_{\underline{\lambda},I}(\overline{x}_1, \ldots, \overline{x}_n)$. Let $\mathcal{C}^I(T, T', s)$ be the groupoid of triples $(\mathcal{V}, t, t')$ consisting of:

i) $\mathcal{V}$ is a $\mathcal{G}^I$-torsor over the geometric curve $\overline{X}$,
ii) a $\overline{T}$-modification $t : \mathcal{V}^{\sigma} |_{\overline{X}-\overline{T}} \xrightarrow{\sim} \mathcal{V} |_{\overline{X}-\overline{T}}$,
iii) a $\overline{T}'$-modification $t' : \mathcal{V}^{\sigma^s} |_{\overline{X}-\overline{T}'} \xrightarrow{\sim} \mathcal{V} |_{\overline{X}-\overline{T}'}$,

such that the maps t et t' commute, i.e. the following diagram is commutative:

$$\begin{array}{ccc} \mathcal{V}^{\sigma^{s+1}} |_{\overline{X}-\overline{T}\cup\overline{T}'} & \xrightarrow{\sigma^s(t)} & \mathcal{V}^{\sigma^s} |_{\overline{X}-\overline{T}\cup\overline{T}'} \\ {\scriptstyle \sigma(t')}\downarrow & & \downarrow{\scriptstyle t'} \\ \mathcal{V}^{\sigma} |_{\overline{X}-\overline{T}\cup\overline{T}'} & \xrightarrow{t} & \mathcal{V} |_{\overline{X}-\overline{T}\cup\overline{T}'} \end{array} .$$

The definition of an isomorphism between such objects is obvious.

We introduce the sub-category $\mathcal{C}^I_{\lambda_{\overline{T}}, \beta_{T'}}(T, T', s)$ whose objects are triples $(\mathcal{V}, t, t') \in \mathcal{C}^I(T, T', s)$ which verify the following conditions:

iv) the invariant $\mathrm{inv}_{\overline{x}_i}(t)$ is equal to λ_i,
v) for each place $x \in T'$, the correspondence t' between the shtuka $(\mathcal{V}, t)$ and the shtuka $\sigma^s(\mathcal{V}, t)$ is of type β_x, see Section 13.2.2.

In the triple $(\mathcal{V}, t, t') \in \mathcal{C}^I_{\lambda_{\overline{T}}, \beta_{T'}}(T, T', s)$, the couple $\widetilde{\mathcal{V}} = (\mathcal{V}, t)$ defines a G-shtuka with an I-level structure and t' defines a correspondence between $\widetilde{\mathcal{V}}^{\sigma^s}$ and $\widetilde{\mathcal{V}}$ of type $(T', (\beta_x)_{x \in T'})$. Hence, we obtain:

Proposition 13.13. *The groupoid $\mathcal{C}^I_{\lambda_{\overline{T}}, \beta_{T'}}(T, T', s)$ is the groupoid of fixed points of $\sigma^s \circ \Phi(T', (\beta_x)_{x \in T'})$ in $\mathrm{Cht}_{\underline{\lambda}, I}(\overline{x}_1, \ldots, \overline{x}_n)$.*

Forgetting the level structures, we will denote by

$$\mathcal{C}(T, T', s) \text{ et } \mathcal{C}_{\lambda_{\overline{T}}, \beta_{T'}}(T, T', s)$$

the resulting groupoids. There are the forgetting functors

$$\mathcal{C}^I(T, T', s) \to \mathcal{C}(T, T', s)$$

et

$$\mathcal{C}^I_{\lambda_{\overline{T}}, \beta_{T'}}(T, T', s) \to \mathcal{C}_{\lambda_{\overline{T}}, \beta_{T'}}(T, T', s).$$

13.3.3

We consider the counting number $\mathrm{Lef}^I_{\lambda_{\overline{T}}, \beta_{T'}}(T, T', s)$ defined by the formula

$$\begin{aligned} \mathrm{Lef}^I_{\lambda_{\overline{T}}, \beta_{T'}}(T, T', s) &:= dg(K_I) \cdot \#\mathcal{C}^I_{\lambda_{\overline{T}}, \beta_{T'}}(T, T', s) \\ &= dg(K_I) \cdot \sum \frac{1}{\#\,\mathrm{Isom}(\mathcal{V}, t, t')} \end{aligned}$$

where the sum runs through the set of representatives of isomorphism classes of objects $(\mathcal{V}, t, t')$ of $\mathcal{C}^I_{\lambda_{\overline{T}}, \beta_{T'}}(T, T', s)$. This number is in general infinite because the connected components of stack of G-shtukas are not always of finite type.

Therefore we suggest to modify the counting problem as follows. Let Z be the center of G. We fix a discrete sub-group $J \subset Z(\mathbb{A}^{T \cup I}) \subset Z(\mathbb{A})$ which is cocompact in $Z(\mathbb{A})/Z(F)$. The group J acts on $\mathrm{Cht}_{\underline{\lambda}, I}(\overline{x}_1, \ldots, \overline{x}_n)$ by Hecke correspondences. We will consider the quotient of the stack $\mathrm{Cht}_{\underline{\lambda}, I}(\overline{x}_1, \ldots, \overline{x}_n)$ by the action of J. It is equivalent to add formally isomorphisms between an object $\widetilde{\mathcal{V}} \in \mathrm{Cht}_{\underline{\lambda}, I}(\overline{x}_1, \ldots, \overline{x}_n)$ and its image under the action of an element $j \in J$. By applying the same procedure to the groupoids $\mathcal{C}^I(T, T', s)$ and $\mathcal{C}^I_{\lambda_{\overline{T}}, \beta'_T}(T, T', s)$ we obtain the groupoids $\mathcal{C}^I(T, T', s)_J$ and $\mathcal{C}^I_{\lambda_{\overline{T}}, \beta'_T}(T, T', s)_J$

We define the number

$$\mathrm{Lef}^{I,\leq p}_{\lambda_{\overline{T}},\beta_{T'}}(T,T',s)_J := dg(K_I)\cdot \#\mathcal{C}^{I,\leq p}_{\lambda_{\overline{T}},\beta_{T'}}(T,T',s)_J$$
$$= dg(K_I)\cdot \sum \frac{1}{\#\,\mathrm{Isom}(\mathcal{V},t,t')}$$

where the sum runs through the set of representatives of isomorphism classes of objects $(\mathcal{V},t,t')$ of $\mathcal{C}^{I}_{\lambda_{\overline{T}},\beta_{T'}}(T,T',s)_J$ with $p(\mathcal{V})\leq p$.

If we restrict the formula to a so-called elliptic part, the truncating function will be always 1 and we will get the equality:

$$\mathrm{Lef}^{I,\mathrm{ell}}_{\lambda_{\overline{T}},\beta_{T'}}(T,T',s)_J = dg(K_I)\cdot \#\mathcal{C}^{I,\mathrm{ell}}_{\lambda_{\overline{T}},\beta_{T'}}(T,T',s)_J$$
$$= dg(K_I)\cdot \sum \frac{1}{\#\,\mathrm{Isom}(\mathcal{V},t,t')}$$

where the sum runs through the set of representatives of isomorphism classes of so-called *elliptic objects* $(\mathcal{V},t,t')$ of $\mathcal{C}^{I}_{\underline{\lambda},\beta_{T'}}(T,T',s)_J$. In our previous work [26, 28], we have found an expression of this number which is a sum indexed by the elliptic elements of $G(F)$ of products of twisted orbital integrals and orbital integral. In this article, we extend this result to get an integral expression of $\mathrm{Lef}^{I,\leq p}_{\lambda_{\overline{T}},\beta_{T'}}(T,T',s)_J$.

13.4 Description of isogeny classes

13.4.1 Functor of generic fiber

We consider the groupoid $C(T,s)$ of triples (V,τ,τ') consisting of

i) V is a G-torsor over $\bar{F}=F\otimes_{\mathbb{F}_q}k$,
ii) $\tau: V^{\sigma}\longrightarrow V$ is a σ-linear application,
iii) $\tau': V^{\sigma^s}\longrightarrow V$ is a σ^s-linear application,

which verify the following conditions:

a) τ et τ' commute with each other,
b) for each place $x\notin T$, (V_x,τ_x) is isomorphic to

$$(G(F_x\widehat{\otimes}_{\mathbb{F}_q}k),\ \mathrm{id}\,\widehat{\otimes}_{\mathbb{F}_q}\sigma),$$

c) for each place $x\in T$, (V_x,τ'_x) is isomorphic to

$$(G(F_x\widehat{\otimes}_{\mathbb{F}_q}\mathbb{F}_{q^s}\widehat{\otimes}_{\mathbb{F}_{q^s}}k),\ \mathrm{id}\,\widehat{\otimes}_{\mathbb{F}_{q^s}}\sigma^s).$$

We will define a functor called the functor of generic fiber from the category $\mathcal{C}(T, T', s)$ to the category $C(T, s)$. To each triple $(\mathcal{V}, t, t')$ of $\mathcal{C}(T, T', s)$, we associate the object (V, τ, τ') in $C(T, s)$ which is called the generic fiber. This construction can be extended without difficulties to triples $(\mathcal{V}, t, t')$ de $\mathcal{C}^I(T, T', s)$ using the forgetting functor.

Denote by V the generic fiber of $\mathcal{V}$. It is a G-torsor over $F \otimes_{\mathbb{F}_q} k$. The modifications t and t' induce the isomorphisms $\tau : V^{\sigma} \longrightarrow V$ and $\tau' : V^{\sigma^s} \longrightarrow V$. It is proved [26] that this triple (V, τ, τ') is an object in the groupoid $C(T, s)$. Since the generic fiber is stable by the Hecke action of elements in J, we obtain a well-defined functor $\mathcal{C}(T, T', s)_J \rightarrow C(T, s)$.

13.4.2 Stable conjugacy classes and Kottwitz's triples

Following [26], to each triple $(V, \tau, \tau') \in C(T, s)$, we will associate a Kottwitz's triple $(\gamma_0; (\gamma_x)_{x \notin T}, (\delta_x)_{x \in T})$ where

a) γ_0 is a stable conjugacy class of $G(F)$,
b) for each place $x \notin T$, γ_x is a conjugacy class of $G(F_x)$ which is stably conjugate to γ_0. Moreover, for almost every place $x \notin T$, γ_x is conjugate to γ_0,
c) for each place $x \in T$, δ_x is a σ-conjugacy class of $G(F_x \hat{\otimes}_{\mathbb{F}_q} \mathbb{F}_{q^s})$ whose norm is stably conjugate to γ_0.

We will say that $(\mathcal{V}, t, t')$ or (V, τ, τ') is *semisimple* (resp. *elliptic*) if the associated conjugacy class γ_0 of $G(F)$ is semisimple (resp. elliptic). Note that the notion of Kottwitz's triples is borrowed from the work of Kottwitz on a similar counting problem on Shimura varieties [11, 12].

13.4.3 Basic elements after Kottwitz

In this section, we recall important results of Kottwitz on the set BH of σ-conjugacy classes associated to a connected reductive group H over a non-archimedian local field. We refer the reader to the excellent articles [9, 30] for proofs and details.

Let x be a place of F and H be a connected reductive group over F_x. Denote by Γ_x the Galois group of F_x and by $\widehat{H}$ the dual group of H. We consider the set $H(F_x \widehat{\otimes}_{\mathbb{F}_q} k)$ equipped with the Frobenius action σ which acts trivially on F_x and acts by the usual Frobenius on k. We define $BH(F_x)$ the set of σ-conjugacy classes of $H(F_x \widehat{\otimes}_{\mathbb{F}_q} k)$.

Let T be a maximal torus of H defined over F_x, we denote by W the corresponding Weyl group. Kottwitz has introduced the Newton map:

$$\nu : BH(F_x) \longrightarrow (X_*T/W)^{\Gamma_x}.$$

Then, Kottwitz has defined the set $BH_{\text{basic}}(F_x)$ of so-called *basic elements* characterized by the following property: an element b in $BH(F_x)$ is basic if and only if its image $\nu(b)$ by the Newton map factorizes through the center $Z(H)$ of H. Kottwitz proved:

Theorem 13.14 (Kottwitz). *i) An element b in $BH(F_x)$ is basic if and only if the group of twisted centralizers*

$$J_b(F) = \{g \in H(F_x \widehat{\otimes}_{\mathbb{F}_q} k) : g^{-1} b\sigma(b) = b\}$$

is the group of F_x-points of an interior form of H.

ii) There exists an isomorphism of functors from the category of connected reductive groups over F_x to the category of the set

$$BH_{\text{basic}}(F_x) \longrightarrow (Z(\widehat{H})^{\Gamma_x})^*$$

of the set $BH_{\text{basic}}(F_x)$ to the group of algebraic characters of $Z(\widehat{H})^{\Gamma_x}$. This isomorphism of functors gives an isomorphism of the subgroup $H^1(F_x, H)$ to the subgroup of characters of $Z(\widehat{H})^{\Gamma_x}$ which are trivial on the neutral component.

In particular, with the above structure, the set $BH_{\text{basic}}(F_x)$ is an abelian group.

iii) The isomorphism of functors $BH_{\text{basic}}(F_x) \longrightarrow (Z(\widehat{H})^{\Gamma_x})^*$ can be extended to a morphism of functors

$$BH(F_x) \longrightarrow (Z(\widehat{H})^{\Gamma_x})^*.$$

13.4.4 Existence of a global element.

Let γ be a semisimple element of $G(F)$. Denote by H the centralizer G_γ of γ in G; it is a connected reductive group over F since G_{der} is simply connected. We get a natural map

$$BH(F) \longrightarrow \prod_x BH(F_x)$$

that, for each place x, sends a global σ-conjugacy class $b \in BH(F)$ to its local σ-conjugacy class $b_x \in BH(F_x)$. Following [26], we define the global invariant

$$\text{inv}(b) : Z(\hat{H})^\Gamma \longrightarrow \mathbb{C}^*$$

to be the sum of the local invariants

$$\mathrm{inv}(b_x)_{|Z(\hat{H})^\Gamma} : Z(\hat{H})^\Gamma \longrightarrow Z(\hat{H})^{\Gamma_x} \longrightarrow \mathbb{C}^*.$$

and it is proved that $\mathrm{inv}(b) = 0$, i.e. the global invariant associated to b is the trivial character of $Z(\hat{H})^\Gamma$. Under certain hypothesis, this condition is sufficient to prove the existence of a global element, see [28]. Remember that the case of tori is already proved in [26] and also known by Varshavsky [34].

Proposition 13.15. *Let H be a connected reductive group over F. Given the elements $b_x \in BH(F_x)$ which are supposed to be basic and trivial for almost every place x and which verify the condition $\sum_x \mathrm{inv}(b_x) = 0$. Then, there exists an element b of the form $b = hz$ with $h \in H^1(F, H)$ and $z \in Z(H)(\bar{F})$ such that $b = b_x$ in $BH(F_x)$.*

Moreover, the number of such elements is equal to

$$\ker^1(F, H) = \ker(H^1(F, H) \longrightarrow \prod_x H^1(F_x, H)).$$

13.4.5 A necessary and sufficient condition for stable conjugacy classes

Definition 13.16. Let γ_0 be a stable conjugacy class of $G(F)$. We say that γ_0 is a (T, s)-norm if, for each place $x \in T$, there exists a σ-conjugacy class δ_x of $G(F_x \hat{\otimes}_{\mathbb{F}_q} \mathbb{F}_{q^s})$ whose norm is stably conjugate to γ_0.

Let γ_0 be a stable conjugacy class of $G(F)$ which is also a (T, s)-norm. There exists a positive integer k big enough such that the element $\gamma_0^{p^k}$ is semisimple. We choose a such element and denote it by γ, i.e. $\gamma = \gamma_0^{p^k}$. For each place $x \in T$, we fix a σ-conjugacy class δ_x de $G(F_x \hat{\otimes}_{\mathbb{F}_q} \mathbb{F}_{q^s})$ whose norm is stably conjugate to γ_0. There is an element $g_x \in G(\bar{F}_x)$ such that $\gamma_0 = g_x^{-1} N\delta_x g_x$. Then the element $\tau_x := g_x^{-1}\delta_x\sigma(g_x)$ commutes with γ_0, hence, it belongs to $G_{\gamma_0}(\bar{F}_x)$. Since G_{γ_0} is a subgroup of G_γ, τ_x defines an element, says $\mathrm{inv}_x(\gamma_0, s)$, in $BG_\gamma(F_x)$.

Lemma 13.17. *The class $\mathrm{inv}_x(\gamma_0, s)$ is basic in $BG_\gamma(F_x)$. Moreover, we have:*

$$\nu(\mathrm{inv}_x(\gamma_0, s)) = \frac{\nu(\gamma_0)}{s} = \frac{\nu(\gamma)}{p^k s}$$

where ν is the Newton map constructed by Kottwitz.

Proof. Put $\tau_x' = g_x^{-1}\sigma^s(g_x)$, then $\gamma_0 = \tau_x^s\tau_x'^{-1}$. Since γ_0 and τ_x commute with each other, τ_x and τ_x' commute with each other as well. Since τ_x' is of torsion, it implies that $\gamma_0^m = \tau_x^{ms}$ for a certain positive integer m. So we get an equality:

$$\gamma^m = \gamma_0^{mp^k} = \tau_x^{mp^k s}.$$

It is obvious that γ is always basic in $BG_\gamma(F_x)$, we deduce that $\mathrm{inv}_x(\gamma_0, s)$ is basic in $BG_\gamma(F_x)$ and that

$$\nu(\mathrm{inv}_x(\gamma_0, s)) = \frac{\nu(\gamma_0)}{s} = \frac{\nu(\gamma)}{p^k s}.$$

The proof is finished. □

The above lemma implies that the class $\mathrm{inv}_x(\gamma_0, s) \in BG_{\gamma,\mathrm{basic}}(F_x)$ whose definition depends on the choices of δ_x and g_x is well defined in $BG_{\gamma,\mathrm{basic}}(F_x)$ modulo the action of the torsion group $H^1(F_x, G_\gamma)$.

Given the local invariants $(\mathrm{inv}_x(\gamma_0, s))_{x\in T}$, we define the global invariant

$$\mathrm{inv}_T(\gamma_0, s) : Z(\hat{G})^\Gamma \longrightarrow \mathbb{C}^*$$

as follows. For each place $x \in T$, the local invariant $\mathrm{inv}_x(\gamma_0, s)$ is a basic element in $BG_{\gamma,\mathrm{basic}}(F_x)$. Following Kottwitz, it induces a character $\mathrm{inv}_x(\gamma_0, s) : Z(\hat{G})^{\Gamma_x} \longrightarrow \mathbb{C}^*$, then by restriction, we get a character $\mathrm{inv}_x(\gamma_0, s)_{|Z(\hat{G})^\Gamma} : Z(\hat{G})^\Gamma \longrightarrow \mathbb{C}^*$. The global invariant $\mathrm{inv}_T(\gamma_0, s)$ is defined as the sum of local characters:

$$\mathrm{inv}_T(\gamma_0, s) = \sum_{x\in T} \mathrm{inv}_x(\gamma_0, s).$$

Definition 13.18. With the above notations, we say that a stable conjugacy class γ_0 of $G(F)$ is (T, s)-admissible if it is a (T, s)-norm and that its global invariant $\mathrm{inv}_T(\gamma_0, s)$ lies in $\mathrm{Im}(\bigoplus_{x\in|X|} H^1(F_x, G_\gamma) \longrightarrow H^1(F, G_\gamma))$.

The rest of this section is devoted to the proof of the following proposition:

Proposition 13.19. *i) Let (V, τ, τ') be an element of $C(T, s)$; denote by γ_0 its associated stable conjugacy class in $G(F)$. Then, γ_0 is (T, s)-admissible.*

ii) Conversely, let γ_0 be a stable conjugacy class which is (T, s)-admissible. Then, there exists an element (V, τ, τ') in $C(T, s)$ whose associated stable conjugacy class is γ_0 and that $\tau \in Z(G_{\gamma_0})(\bar{F})$. Moreover, any triple (V, τ_0, τ_0') in $C(T, s)$ whose associated stable conjugacy class is γ_0 can be written in the form: $\tau_0 = h\tau$ with $h \in H^1(F, G_{\gamma_0})$.

In particular, the set $V(\gamma_0)$ of such triples is in bijection with $H^1(F, G_{\gamma_0})$.

Proof. i) Let (V, τ, τ') be an element in $C(T, s)$ and let $(\gamma_0; (\gamma_x)_{x\notin T}, (\delta_x)_{x\in T})$ be its associated Kottwitz's triple. By definition of a Kottwitz's triple, it is clear that γ_0 is always a (T, s)-norm. Moreover, for each place $x \in T$, the local invariants $\mathrm{inv}_x(\gamma_0, s)$ and $\mathrm{inv}_x(\gamma_0, \gamma_x, \delta_x)$ are the same. So we get the equality:

$$\mathrm{inv}_T(\gamma_0, s) = \sum_{x\in T} \mathrm{inv}_x(\gamma_0, s) = \sum_{x\in T} \mathrm{inv}_x(\gamma_0, \gamma_x, \delta_x).$$

We know that

a) For each place $x \notin T$, the local invariant $\mathrm{inv}_x(\gamma_0, \gamma_x, \delta_x)$ lies in the subgroup $H^1(F_x, G_\gamma)$ of $BG_{\gamma,\mathrm{basic}}(F_x)$.
b) The global invariant $\mathrm{inv}(\gamma_0, \gamma_x, \delta_x) = \sum_{x\in|X|} \mathrm{inv}_x(\gamma_0, \gamma_x, \delta_x)$ vanishes.

It implies that $\mathrm{inv}_T(\gamma_0, s)$ lies in $\mathrm{Im}(\bigoplus_{x\in|X|} H^1(F_x, G_\gamma) \longrightarrow H^1(F, G_\gamma))$, i.e. γ_0 is (T, s)-admissible.

ii) Let γ_0 a stable conjugacy class of $G(F)$ which is (T, s)-admissible. Since γ_0 is (T, s)-admissible, we can add to the collection $(\mathrm{inv}_x(\gamma_0, s))_{x\in T}$ the elements $(\mathrm{inv}_x(\gamma_0, s))_{x\notin T} \in \bigoplus_{x\notin T} H^1(F_x, G_\gamma)$ such that the global invariant

$$\sum_{x\in|X|} \mathrm{inv}_x(\gamma_0, s)$$

vanishes. Following [28], there exists an element $\tau_0 \in G_\gamma(\bar{F})$ such that, for every place x, we have the equality $\mathrm{inv}_x(\tau_0) = \mathrm{inv}_x(\gamma_0, s)$. Further, this element is of the form $h\tau$ with $h \in H^1(F, G_\gamma)$ and $\tau \in Z(G_\gamma)(\bar{F})$. In particular, τ lies in $Z(G_{\gamma_0})(\bar{F})$.

Put $\tau' = \gamma_0^{-1}\tau^s$, then we claim that (V, τ, τ') is an element of $C(T, s)$ whose associated stable conjugacy class is γ_0. In fact, we have to show that this triple satisfies the three properties required in the definition of $C(T, s)$, see Section 13.4.1. Since τ et γ_0 commute with each other, by construction, τ and τ' commute too. So the first property is shown. Next, for each place $x \notin T$, since $\mathrm{inv}_x(\tau_0)$ lies in $H^1(F_x, G_\gamma)$, so does $\mathrm{inv}_x(\tau)$. Since the derived group G_{der} of G is simply connected, the group $H^1(F_x, G)$ vanishes, hence we get the second property. Finally, for each place $x \in T$, since γ_0 and τ^s have the same image by the Newton map, it implies that τ' is of torsion. The same argument implies the third property.

To conclude, we remark that by construction, the stable conjugacy class associated to the triple (V, τ, τ') is γ_0. The proof is finished □

13.4.6 The automorphism group of (V, τ, τ')

Let (V, τ, τ') be an element of $C(T, s)$ and let $(\gamma_0, (\gamma_x)_{x\in T}, (\delta_x)_{x\notin T})$ be the associated Kottwitz's triple. In this section, we will show that the automorphism group $\mathrm{Aut}(V, \tau, \tau')$ of (V, τ, τ') is an interior form J_{γ_0} of G_{γ_0} defined over F.

Choose a representative of the conjugacy class γ_0 in $G(F)$, says γ_0. Suppose that $\gamma_0 = \tau^s\tau'^{-1}$. Since γ_0 and τ commute, τ acts on the group of centralizers $G_{\gamma_0}(F \otimes_{\mathbb{F}_q} k)$. In the one hand, the automorphism group $\mathrm{Aut}(V, \tau, \tau')$ is the subgroup of fixed points by τ of the semi-direct product $G_{\gamma_0}(F \otimes_{\mathbb{F}_q} k) \rtimes \langle\tau\rangle$. On the other hand, G_{γ_0} is the subgroup of fixed points by σ of the semi-direct product $G_{\gamma_0}(F \otimes_{\mathbb{F}_q} k) \rtimes \langle\sigma\rangle$. It suffices to verify that for some positive integer n, we have an isomorphism of semi-direct products:

$$G_{\gamma_0}(F \otimes_{\mathbb{F}_q} k) \rtimes \langle\tau^n\rangle \xrightarrow{\sim} G_{\gamma_0}(F \otimes_{\mathbb{F}_q} k) \rtimes \langle\sigma^n\rangle.$$

In fact, there exists a positive integer m such that τ^{ms} is central in $G_\gamma(F)$. We write $n = ms$ and obtain the desired isomorphism

$$G_{\gamma_0}(F \otimes_{\mathbb{F}_q} k) \rtimes \langle\tau^n\rangle \xrightarrow{\sim} G_{\gamma_0}(F \otimes_{\mathbb{F}_q} k) \rtimes \langle\sigma^n\rangle.$$

By the Hasse's principle for adjoint groups, the automorphism group $J_{\gamma_0} = \mathrm{Aut}(V, \tau, \tau')$ which is an interior form of G_{γ_0} is completely determined by its local components $J_{\gamma_0,x}$:

- For each place $x \notin T$, the local component $J_{\gamma_0,x}$ is the centralizer of γ_x in $G(F_x)$.
- For each place $x \in T$, the local component $J_{\gamma_0,x}$ is the twisted centralizer of δ_x in $G(F_x\hat{\otimes}_{\mathbb{F}_q}\mathbb{F}_{q^s})$.

13.5 Description of an isogeny class

13.5.1 Adelic description of fixed points

Let $\mathcal{V}$ be a $\mathcal{G}^I$-torsor over k. The generic fiber V of $\mathcal{V}$ is a G-torsor over $\bar{F} = F \otimes_{\mathbb{F}_q} k$, hence it is trivial. We will fix an isomorphism between V and the trivial G-torsor over $\bar{F}$. Then, to give a $\mathcal{G}^I$-torsor $\mathcal{V}$ over k equipped with a trivialization of the generic fiber is equivalent to give, for each place $x \in |X|$, an element g_x in $G(F_x\widehat{\otimes}_{\mathbb{F}_q}k)/K_{I,x}$ which is trivial almost everywhere. If we forget the choice of the trivialization, we will say that two such data $(g_x)_{x\in|X|}$ and $(g'_x)_{x\in|X|}$ are equivalent if there exists an element $g \in G(\bar{F})$ such that $g'_x = gg_x$.

Let $(\mathcal{V}, \tau, \tau')$ be an element of $\mathcal{C}^I(T, T', s)$. Then, to give a such element is equivalent to give the data consisting of

i) two elements $\tau, \tau' \in G(\bar{F})$,
ii) for each place $x \in |X|$, an element g_x of $G(F_x \hat{\otimes}_{\mathbb{F}_q} k)/K_{I,x}$ which is trivial almost everywhere,

which verify the following conditions:

1) τ and τ' commute, i.e. $\tau\tau' = \tau'\tau$.
2) For each place $x \in T$, $g_x^{-1}\tau\sigma(g_x) \in K_x \varpi_x^{\lambda_x} K_x$.
3) For each place $x \notin T$, $g_x^{-1}\tau\sigma(g_x) \in K_x$.
4) For each place $x \in T$, $g_x^{-1}\tau'\sigma^s(g_x) \in K_{I,x}$.
5) For each place $x \notin T$, $g_x^{-1}\tau'\sigma^s(g_x) \in K_{I,x}\beta_x K_{I,x}$.

Two such data $(\tau_1, \tau_1', (g_{1,x})_{x\in|X|})$ et $(\tau_2, \tau_2', (g_{2,x})_{x\in|X|})$ are equivalent if there exists an element $g \in G(\bar{F})$ such that $\tau_2 = g^{-1}\tau_1\sigma(g)$, $\tau_2' = g^{-1}\tau_1'\sigma^s(g)$ and for each place x, $g_{2,x} = gg_{1,x}$.

13.5.2 Adelic description of the functor of generic fiber

Let $(\mathcal{V}, \tau, \tau')$ be an element of $\mathcal{C}^I(T, T', s)$ and let $(\tau, \tau', (g_x)_{x\in|X|})$ be the associated adelic data as above. The functor of generic fiber will associate to it the data consisting of two elements $\tau, \tau' \in G(\bar{F})$ which verify the following conditions:

1) τ and τ' commute, i.e. $\tau\tau' = \tau'\tau$.
2b) For each place $x \notin T$, there exists an element $g_{0,x} \in G(F_x \hat{\otimes}_{\mathbb{F}_q} k)$ such that $g_{0,x}^{-1}\tau\sigma(g_{0,x}) = 1$.
3b) For each place $x \in T$, there exists an element $g_{0,x} \in G(F_x \hat{\otimes}_{\mathbb{F}_q} k)$ such that $g_{0,x}^{-1}\tau'\sigma^s(g_{0,x}) = 1$.

In fact, the functor of generic fiber consists of forgetting the local data $((g_x)_{x\in|X|})$. By a theorem of Lang, every element of K_x can be written in the form $g_{0,x}^{-1}\sigma(g_{0,x})$ for some $g_{0,x} \in G(F_x \hat{\otimes}_{\mathbb{F}_q} k)$. It implies that the last two conditions 2b) and 3b) are satisfied.

Let (τ, τ') be an element of $C(T, s)$. We fix a choice of elements $(g_{0,x})_{x\in|X|}$ as above. We will give an adelic description of the Kottwitz's triple associated to (τ, τ'). Put $\gamma_0 = \tau^s\tau'^{-1}$; it is an element of $G(\bar{F})$. We have already seen that the stable conjugacy class of γ_0 is fixed by the Frobenius, hence it contains an element of $G(F)$. If necessary replacing (τ, τ') by an equivalent datum, we can suppose that γ_0 is in $G(F)$. For each place $x \notin T$, since $g_{0,x}^{-1}\tau\sigma(g_{0,x}) = 1$, the element $\gamma_x := g_{0,x}^{-1}\gamma_0 g_{0,x}$ is fixed by the Frobenius, hence it is an element in $G(F_x)$. For each place $x \in T$, since $g_{0,x}^{-1}\tau'\sigma^s(g_{0,x}) = 1$, the element $\delta_x := g_{0,x}^{-1}\tau\sigma(g_{0,x})$ is fixed by σ^s, it is an element in $G(F_x\hat{\otimes}_{\mathbb{F}_q}\mathbb{F}_{q^s})$. To conclude, note that the triple $(\gamma_0, (\gamma_x)_{x\notin T}, (\delta_x)_{x\in T})$ is exactly the Kottwitz's triple associated to (τ, τ').

13.5.3 Adelic description of an isogeny class. Choice of base points

Let (τ, τ') be an element of $C(T, s)$. We will give an adelic description of fixed points $(\mathcal{V}, \tau, \tau')$ over (τ, τ'). As before, we always suppose that the element $\gamma_0 = \tau^s\tau'^{-1}$ is an element in $G(F)$. We fix a choice of base points $(g_{0,x})_{x\in|X|}$. To give a fixed point $(\mathcal{V}, \tau, \tau')$ over (τ, τ') is equivalent to give a collection of elements $(g_x \in G(F_x\hat{\otimes}_{\mathbb{F}_q}k)/K_{I,x})_{x\in|X|}$ which verify the following conditions:

1) They are trivial almost everywhere.
2) For each place $x \in T$, $g_x^{-1}\tau\sigma(g_x) \in K_x\varpi_x^{\lambda_x}K_x$.
3) For each place $x \notin T$, $g_x^{-1}\tau\sigma(g_x) \in K_x$.
4) For each place $x \in T$, $g_x^{-1}\tau'\sigma^s(g_x) \in K_{I,x}$.
5) For each place $x \notin T$, $g_x^{-1}\tau'\sigma^s(g_x) \in K_{I,x}\beta_x K_{I,x}$.

For each place $x \in |X|$, put $h_x = g_{0,x}^{-1}g_x$. The condition 3) requires that, for each place $x \notin |T|$, $h_x^{-1}\sigma(h_x) \in K_x$, it is equivalent to ask that $h_x \in G(F_x)/K_{I,x}$. Similarly, the condition 4) requires that, for each place $x \in T$, $h_x^{-1}\sigma^s(h_x) \in K_x$, which is equivalent to say that $h_x \in G(F_x\hat{\otimes}_{\mathbb{F}_q}\mathbb{F}_{q^s})/G(\mathcal{O}_x\hat{\otimes}_{\mathbb{F}_q}\mathbb{F}_{q^s})$.

Recall that we have already defined $\gamma_x := g_{0,x}^{-1}\gamma_0 g_{0,x} \in G(F_x)$ and $\delta_x := g_{0,x}^{-1}\tau\sigma(g_{0,x}) \in G(F_x\hat{\otimes}_{\mathbb{F}_q}\mathbb{F}_{q^s})$. With these notations, the conditions 2) and 5) are equivalent to the following:

2b) For each place $x \in T$, $h_x^{-1}\delta_x\sigma(h_x) \in K_x\varpi_x^{\lambda_x}K_x$.
5b) For each place $x \notin T$, $h_x^{-1}\gamma_x\sigma^s(h_x) \in K_{I,x}\beta_x K_{I,x}$.

To summarize, we have proved:

Proposition 13.20. *We keep all the above notations. Then, to give a fixed point $(\mathcal{V}, \tau, \tau')$ over (τ, τ') is equivalent to give a collection of elements $(g_x = g_{0,x}h_x)_{x\in|X|}$ where the elements*

$$(h_x)_{x\in|X|} \in \prod_{x\in T} G(F_x\hat{\otimes}_{\mathbb{F}_q}\mathbb{F}_{q^s})/G(\mathcal{O}_x\hat{\otimes}_{\mathbb{F}_q}\mathbb{F}_{q^s}) \times \prod_{x\notin T} G(F_x)/K_{I,x}$$

verify the following conditions:

2b) *For each place* $x \in T$, $h_x^{-1}\delta_x\sigma(h_x) \in K_x\varpi_x^{\lambda_x}K_x$.
5b) *For each place* $x \notin T$, $h_x^{-1}\gamma_x h_x \in K_{I,x}\beta_x K_{I,x}$.

13.5.4 Adelic description of the truncating function

We keep the notations of the previous section. Following Section 13.1.7, for each truncating parameter p, we have defined a characteristic function of the set of G-torsors over k bounded by p. In adelic terms, it is just a characteristic function over $G(\bar{\mathbb{A}})$ which is invariant on the left by $G(\bar{F})$ and invariant on the right by the sub-group $\prod_{x\in|X|} K_x$. Let $(\mathcal{V}, \tau, \tau')$ be a fixed point and let $(g_x = g_{0,x}h_x)_{x\in|X|}$ be the corresponding adèle. Then, we have the equality:

$$\mathbf{1}(p(\mathcal{V}) \leq p)) = \mathbf{1}(p((g_x)_{x\in|X|}) \leq p)).$$

13.6 Integral expression of the number of fixed points for G-shtukas

13.6.1 Integral expression

In this section, we will give a formula for the number of fixed points $\mathrm{Lef}^{I,\leq p}_{\lambda_{\bar{T}},\beta_{T'}}(T, T', s)_J$. Let (V, τ, τ') be an object of $C(T, s)$, denote by $(\gamma_0; (\gamma_x)_{x\notin T}, (\delta_x)_{x\in T})$ its associated Kottwitz's triple. Then, the number of isomorphism classes

$$dg(K_I) \cdot \sum_{(\mathcal{V},t,t')} \frac{1}{\#\mathrm{Isom}(\mathcal{V}, t, t')}$$

where the sum runs through the set of representatives of isomorphism classes of triples $(\mathcal{V}, t, t')$ of $C^I_{\lambda_{\bar{T}},\beta_{T'}}(T, T', s)_J$ satisfying the condition that the generic fiber is isomorphic to (V, τ, τ') and that the shtuka $\mathcal{V}$ is bounded by p is finite and is equal to

$$\int_{J J_{\gamma_0}(F)\backslash \prod_{x\in T} G(F_x\hat{\otimes}_{\mathbb{F}_q}\mathbb{F}_{q^s})\times G(\mathbb{A}^T)} \prod_{x\in T} \phi_{\lambda_x}(h_x^{-1}\delta_x\sigma(h_x))$$
$$\times \prod_{x\notin T} \phi_{\beta_x}(h_x^{-1}\gamma_x h_x) \times \mathbf{1}(p((g_{0,x}h_x)_{x\in|X|}) \leq p)) \cdot dh$$

where

- for each place $x \in T$, ϕ_{λ_x} is the spherical function

$$\phi_{\lambda_x} = \bigotimes_{\substack{y\in T\otimes_{\mathbb{F}_q}\mathbb{F}_{q^s}\\ y|x}} \phi_{\lambda_y} \in \mathcal{H}_x \otimes_{\mathbb{F}_q} \mathbb{F}_{q^s} := \bigotimes_{\substack{y\in T\otimes_{\mathbb{F}_q}\mathbb{F}_{q^s}\\ y|x}} \mathcal{H}_y,$$

- for each place $x \in T'$, ϕ_{β_x} is the characteristic function of the double class $\beta_x \in K_{I,x}\backslash G(F_x)/K_{I,x}$,
- for each place $x \notin T \cup T'$, ϕ_{β_x} is the characteristic function of the unit double class of $G(\mathcal{O}_x)\backslash G(F_x)/G(\mathcal{O}_x)$.

We have proved that there exists a triple (V, τ, τ') whose stable conjugacy class is γ_0 if and only if γ_0 is (T, s)-admissible. Moreover, if γ_0 is (T, s)-admissible, the set $V(\gamma_0)$ of triples (V, τ, τ') whose stable conjugacy class is γ_0 is in bijection with $H^1(F, G_{\gamma_0})$.

We obtain then an integral expression of the number of fixed points in the stack of G-shtukas:

Theorem 13.21. *With the notations as before, we have:*

$$\mathrm{Lef}^{I,\leq p}_{\lambda_{\overline{T}},\beta_{T'}}(T, T', s)_J = \sum_{\gamma_0}\sum_{V(\gamma_0)} \int_{JJ_{\gamma_0}(F)\backslash\prod_{x\in T} G(F_x\hat{\otimes}_{\mathbb{F}_q}\mathbb{F}_{q^s})\times G(\mathbb{A}^T)}$$

$$\prod_{x\in T}\phi_{\lambda_x}(h_x^{-1}\delta_x\sigma(h_x))\prod_{x\notin T}\phi_{\beta_x}(h_x^{-1}\gamma_x h_x)\cdot \boldsymbol{1}(p((g_{0,x}h_x)_{x\in|X|})\leq p))\cdot dh$$

where the first sum runs through all (T, s)-admissible stable conjugacy classes γ_0 of $G(F)$, and the second sum runs through the set $V(\gamma_0)$ of triples (V, τ, τ') whose stable conjugacy class is γ_0.

13.6.2 The elliptic terms

Suppose that γ_0 is elliptic. Since γ_0 is semisimple, (V, τ, τ') is determined by its associated Kottwitz's triple $(\gamma_0; (\gamma_x)_{x\notin T}, (\delta_x)_{x\in T})$. Moreover, the characteristic truncating function is always 1 so that the choice of base points does not appear in this formula. The formula was obtained in the article [28]:

$$\mathrm{Lef}^{I,\mathrm{ell}}_{\lambda_{\overline{T}},\beta_{T'}}(T, T', s)_J = \sum_{(\gamma_0;\gamma_x,\delta_x)} \ker\left(H^1(F, G_{\gamma_0}) \longrightarrow \prod_{x\in|X|} H^1(F_x, G_{\gamma_0})\right)\cdot$$

$$\cdot\,\mathrm{vol}(JJ_{\gamma_0}(F)\backslash J_{\gamma_0}(\mathbb{A}))\prod_{x\notin T}\mathbf{O}_{\gamma_x}(\phi_{\beta_x})\prod_{x\in T}\mathbf{TO}_{\delta_x}(\phi_{\lambda_x})$$

where the sum runs through all Kottwitz's triples $(\gamma_0; (\gamma_x)_{x\notin T}, (\delta_x)_{x\in T})$ such that γ_0 is elliptic and that the global invariant $\mathrm{inv}(\gamma_0; \gamma_x, \delta_x)$ vanishes.

13.7 The number of fixed points under the actions of partial Frobenius

In this section, we give a slight generalization of the previous results and obtain a formula for the number of fixed points under the actions of partial Frobenius. This result generalizes a unpublished result of de Ngô Bao Châu [23]. The proofs are almost straightforward. So we will only state the results and we let the reader to fill the details of the proofs. We will recall the definition of the partial Frobenius, then introduce the groupoid of fixed points by the composition of a power of partial Frobenius with a Hecke operator. Finally, we present an integral expression for the number of fixed points.

13.7.1 Fixed points by the actions of partial Frobenius

Let $s \geq 1$ be a positive integer and $x_1, \ldots, x_n$ be distinct places of $X - I$. Put $\overline{T} = \{\overline{x}_1, \ldots, \overline{x}_n\}$ where $\overline{x}_i$ is a geometric point over x_i. Let $\underline{s} = (s_1, \ldots, s_n)$ be a collection of positive integers verifying that, for every $1 \leq i \leq n$, s_i is a multiple of $\deg(x_i)$: we put $s_i = \deg(x_i)s_i'$ with some positive integer s_i'. We fix a finite set of places T' disjoint from $\{x_1, \ldots, x_n\}$ and for each place $x \in T'$, we fix a double class $\beta_x \in K_{I,x}\backslash G(F_x)/K_{I,x}$. Let $\partial^{\underline{s}} = (\sigma_1^{s_1}, \ldots, \sigma_n^{s_n})$ be a composition of partial Frobenius and $\Phi(T', (\beta_x)_{x\in T'})$ be the associated Hecke correspondence over $\mathrm{Cht}_{\underline{\lambda},I}(\overline{x}_1, \ldots, \overline{x}_n)$.

As before, we fix a discrete subgroup $J \subset Z(\mathbb{A}^{T\cup I}) \subset Z(\mathbb{A})$ which is cocompact in $Z(\mathbb{A})/Z(F)$. The group J acts on $\mathrm{Cht}_{\underline{\lambda},I}(\overline{x}_1, \ldots, \overline{x}_n)$ by Hecke correspondences. We will consider the quotient of the stack $\mathrm{Cht}_{\underline{\lambda},I}(\overline{x}_1, \ldots, \overline{x}_n)$ by the action of J and the groupoid of the fixed points $\mathrm{Fixe}(\partial^{\underline{s}} \circ \Phi(T', (\beta_x)_{x\in T'}), \mathrm{Cht}_{\underline{\lambda},I}(\overline{x}_1, \ldots, \overline{x}_n)_J)$ by the correspondence $\partial^{\underline{s}} \circ \Phi(T', (\beta_x)_{x\in T'})$ in $\mathrm{Cht}_{\underline{\lambda},I}(\overline{x}_1, \ldots, \overline{x}_n)_J$. We refer the reader to the article [23] for a detailed description of this groupoid for $\mathcal{D}$-shtukas. The generalization to G-shtukas is straightforward.

We define the number

$$\begin{aligned}\mathrm{Lef}^{I,\leq p}_{\lambda_{\overline{T}},\beta_{T'}}(T, T', \underline{s})_J &= dg(K_I) \cdot \#\mathcal{C}^{I,\leq p}_{\lambda_{\overline{T}},\beta_{T'}}(T, T', s)_J \\ &= dg(K_I) \cdot \sum \frac{1}{\#\,\mathrm{Isom}(\mathcal{V}, t, t')}\end{aligned}$$

where the sum runs through the set of representatives of isomorphism classes of objects $(\mathcal{V}, t, t')$ of $\mathrm{Fixe}(\partial^{\underline{s}} \circ \Phi(T', (\beta_x)_{x\in T'}), \mathrm{Cht}_{\underline{\lambda},I}(\overline{x}_1, \ldots, \overline{x}_n)_J)$ with $p(\mathcal{V}) \leq p$.

13.7.2 Integral expression of the number of fixed points

We introduce a generalization of the category $C(T, s)$, denoted by $C(T, \underline{s})$ classifying triples (V, τ, γ_0) consisting of

i) V is a G-torsor over $\bar{F} = F \otimes_{\mathbb{F}_q} k$,
ii) $\tau : V^\sigma \longrightarrow V$ is a σ-linear map,
iii) $\gamma_0 : V \longrightarrow V$ is a linear map,

which verify the following properties:

1) τ and γ_0 commute with each other, i.e. $\gamma_0^{-1}\tau\sigma(\gamma_0) = \tau$,
2) for each place $x \notin T$, (V_x, τ_x) is isomorphic to

$$(G(F_x \widehat{\otimes}_{\mathbb{F}_q} k), \mathrm{id}\,\widehat{\otimes}_{\mathbb{F}_q} \sigma),$$

3) for each place $x_i \in T$, $(V_{x_i}, \gamma_0^{-1}\tau^{s_i})$ is isomorphic to

$$(G(F_{x_i} \widehat{\otimes}_{\mathbb{F}_q} \mathbb{F}_{q^{s_i}} \widehat{\otimes}_{\mathbb{F}_{q^{s_i}}} k), \mathrm{id}\,\widehat{\otimes}_{\mathbb{F}_{q^{s_i}}} \sigma^{s_i}).$$

The definition of isomorphisms between two such triples is immediate.

Let (V, τ, γ_0) be an element in $C(T, \underline{s})$. We will choose a trivialization of V, so τ and γ_0 can be considered as elements of $G(\bar{F})$. The condition 1) implies that the stable conjugacy class γ_0 is stable under the action of the Frobenius, hence contains an element of $G(F)$. *From now on, we always suppose that γ_0 is in $G(F)$.*

Definition 13.22. Let γ_0 be a stable conjugacy class of $G(F)$. We will say that γ_0 is a $(T, \underline{s})$-norm if, for each place $x_i \in T$, there exists a σ-conjugacy class δ_{x_i} of $G(F_{x_i} \hat{\otimes}_{\mathbb{F}_q} \mathbb{F}_{q^{s_i}})$ whose norm is stably conjugate to γ_0.

Let γ_0 be a stable conjugacy class of $G(F)$ which is *semisimple* and is a $(T, \underline{s})$-norm. By a similar manner as before, we define the global invariant $\mathrm{inv}_T(\gamma_0, \underline{s})$ which lies in $(Z(\hat{G})^\Gamma)^*$ modulo $\bigoplus_{x \in |T|} H^1(F_x, G_\gamma)$.

Definition 13.23. With the previous notations, we will say that a stable *semisimple* conjugacy class γ_0 of $G(F)$ is $(T, \underline{s})$-admissible if it is a $(T, \underline{s})$-norm and that its global invariant $\mathrm{inv}_T(\gamma_0, \underline{s})$ lies in $\mathrm{Im}(\bigoplus_{x \in |X|} H^1(F_x, G_{\gamma_0}) \longrightarrow H^1(F, G_{\gamma_0}))$.

We can extend this definition to all the stable conjugacy classes by following the same procedure as described in Section 13.4.5.

Proposition 13.24. *i) Let (V, τ, γ_0) be an element in $C(T, \underline{s})$ with $\gamma_0 \in G(F)$. Then, γ_0 is $(T, \underline{s})$-admissible.*

ii) Conversely, let γ_0 be a stable conjugacy class which is $(T, \underline{s})$-admissible. Then, there exists an element (V, τ_0, γ_0) in $C(T, \underline{s})$ with $\tau_0 \in Z(G_{\gamma_0})(\bar{F})$. Moreover, a triple (V, τ, γ_0) in $C(T, \underline{s})$ can be written in the form: $\tau = h\tau_0$ with $h \in H^1(F, G_{\gamma_0})$.

In particular, the set $V(\gamma_0)$ of such triples $(V, \tau, \gamma_0) \in C(T, \underline{s})$ is in bijection with $H^1(F, G_{\gamma_0})$.

Let γ_0 be a stable conjugacy class of $G(F)$ which is $(T, \underline{s})$-admissible and let (V, τ, γ_0) be an element in $C(T, \underline{s})$. There exist elements $(g_{0,x} \in G(\bar{F}_x))_{x \in |X|}$ such that

- For each place $x \notin T$, we have: $\tau = g_{0,x}^{-1}\sigma(g_{0,x})$.
- For each place $x_i \in T$, we have: $\gamma_0^{-1}\tau^{s_i} = g_{0,x_i}^{-1}\sigma^{s_i}(g_{0,x_i})$.

We fix a choice of these base points. We can see that:

- For each place $x \notin T$, the element $\gamma_x := g_{0,x}^{-1}\gamma_0 g_{0,x}$ is in $G(F_x)$.
- For each place $x_i \in T$, the element $\delta_{x_i} := g_{0,x_i}^{-1}\tau\sigma(g_{0,x_i})$ is in $G(F_{x_i}\widehat{\otimes}_{\mathbb{F}_q}\mathbb{F}_{q^{s_i}})$.

We show that the automorphism group J_{γ_0} of (V, τ, γ_0) is an interior form of G_{γ_0} defined over F whose local components are completely determined by the elements $(\gamma_x)_{x \notin T}$ and $(\delta_{x_i})_{x_i \in T}$ defined above.

Finally, we obtain the integral expression of the number of fixed points:

Theorem 13.25. *With the above notation, we get:*

$$\mathrm{Lef}^{I,\leq p}_{\lambda_{\overline{T}},\beta_{T'}}(T, T', \underline{s})_J = \sum_{\gamma_0}\sum_{V(\gamma_0)}\int_{J J_{\gamma_0}(F)\backslash \prod_{x_i \in T} G(F_{x_i}\hat{\otimes}_{\mathbb{F}_q}\mathbb{F}_{q^{s_i}})\times G(\mathbb{A}^T)}$$

$$\prod_{x_i \in T}\phi_{\lambda_{x_i}}(h_{x_i}^{-1}\delta_{x_i}\sigma(h_{x_i}))\prod_{x \notin T}\phi_{\beta_x}(h_x^{-1}\gamma_x h_x)\cdot \mathbf{1}(p((g_{0,x}h_x)_{x\in|X|}) \leq p))\cdot dh$$

where

- *the first sum runs through all the stable conjugacy classes γ_0 of $G(F)$ which are $(T, \underline{s})$-admissible,*
- *the second sum runs through the set $V(\gamma_0)$ of the elements (V, τ, γ_0) in $C(T, \underline{s})$,*

- *for each place $x_i \in T$, $\phi_{\lambda_{x_i}}$ is the spherical function*

$$\phi_{\lambda_{x_i}} = \bigotimes_{\substack{y \in T \otimes_{\mathbb{F}_q} \mathbb{F}_{q^{s_i}} \\ y|x_i}} \phi_{\lambda_y} \in \mathcal{H}_{x_i} \otimes_{\mathbb{F}_q} \mathbb{F}_{q^{s_i}} := \bigotimes_{\substack{y \in T \otimes_{\mathbb{F}_q} \mathbb{F}_{q^{s_i}} \\ y|x_i}} \mathcal{H}_y,$$

- *for each place $x \in T'$, ϕ_{β_x} is the characteristic function of the double class $\beta_x \in K_{I,x} \backslash G(F_x)/K_{I,x}$,*
- *for each place $x \notin T \cup T'$, ϕ_{β_x} is the characteristic function of the unit double class $G(\mathcal{O}_x)\backslash G(F_x)/G(\mathcal{O}_x)$.*

13.7.3 Example: the case of the general linear group $\mathbf{G = GL}_n$

In this section, we consider the case of the general linear group $G = \mathrm{GL}_n$. It is well known that, for any field F, if two elements in $\mathrm{GL}_n(F)$ are stably conjugate, then they are in fact conjugate. In other words, every stable conjugacy class contains only one conjugacy class. This assertion allows us to obtain a simpler expression of the number of fixed points for GL_n-chtoucas.

As stated in the introduction, different counting problems for GL_n-chtoucas have been treated in the literature. When the modifications are elementary, it was done by Drinfeld [4, 5] and Lafforgue [13]. For elliptic $\mathcal{D}$-modules, it was done by Laumon, Rapoport and Stuhler [21]. Contrary to our result, their formulas contain only orbital integrals. The first formula containing twisted orbital integrals was appeared in the work of Laumon [19, 20] using a strategy similar to that of Kottwitz on certain Shimura varieties. Then, Ngô Bao Châu [24] and Lau [17] have given similar expressions for the number of fixed points for $\mathcal{D}$-chtoucas. It is important to note that all these works are based on the classification of ϕ-spaces of Drinfeld, which is different from our approach.

Let γ_0 be an element of $\mathrm{GL}_n(F)$ and let γ be a semisimple element of $\mathrm{GL}_n(F)$ which has the same characteristic polynomial as γ_0. The F-algebra $F[\gamma]$ can be written in the form

$$F[\gamma] = E_1 \times E_2 \times \ldots \times E_k$$

where E_i are the finite field extensions of F, with $\sum_{j=1}^{k}[E_j : F] = n$. For each index $1 \leq j \leq k$, denote by γ_j the projection of $\gamma \in F[\gamma]$ in E_j. Let X_{E_j} be the normalized curve over X whose function field is E_j. Then, γ_j induces a divisor $\mathrm{div}(\gamma_j)$ over X_{E_j} and for each point $x_i \in T$, denote by $\mathrm{div}_{x_i}(\gamma_j)$ the part of $\mathrm{div}(\gamma_i)$ supported by x_i.

Proposition 13.26. *The element γ_0 is $(T, \underline{s})$-admissible if and only if it satisfies the following conditions:*

i) *For each place* $x_i \in T$, γ_0 *is a* s_i*-norm in* $\mathrm{GL}_n(F_{x_i})$.
ii) *For each index* $1 \le j \le k$, *we have the equality:* $\sum_{x_i \in T} \deg(\mathrm{div}_{x_i}(\gamma_j)/s_i) = 0$.

Proof. The proposition follows from the previous observation that, for each place x, we have $H^1(F_x, \mathrm{GL}_{n,\gamma}) = 1$. □

Proposition 13.27. *i) Let* (V, τ, γ_0) *be an element of* $C(T, \underline{s})$*. Then,* γ_0 *is always* $(T, \underline{s})$*-admissible.*
ii) Conversely, if γ_0 *is a* $(T, \underline{s})$*-admissible element of* $\mathrm{GL}_n(F)$*, then there exists a unique element* (V, τ, γ_0) *in* $C(T, s)$*. Moreover,* τ *lies in* $Z(\mathrm{GL}_{n,\gamma_0})(\bar{F})$ *and the automorphism group of* (V, τ, γ_0) *is exactly* G_{γ_0}.

Proof. It follows from the fact that $H^1(F, \mathrm{GL}_{n,\gamma_0}) = 1$. □

Now, we choose the base points $(g_{0,x})_{x \in |X|}$ such that $g_{0,x} \in \mathrm{GL}_{n,\gamma_0}(\bar{F}_x)$. In particular, for each place $x \notin T$, we can take $\gamma_x = \gamma_0$.

The expression for the number of fixed points for GL_n-chtoucas is given as follows:

Theorem 13.28. *We keep the previous notations. Then, we have:*

$$\mathrm{Lef}^{I,\le p}_{\lambda_{\overline{T}},\beta_{T'}}(T, T', \underline{s})_J = \sum_{\gamma_0} \int_{J G_{\gamma_0}(F)\backslash \prod_{x_i \in T} G(F_{x_i} \hat{\otimes}_{\mathbb{F}_q} \mathbb{F}_{q^{s_i}}) \times G(\mathbb{A}^T)}$$

$$\prod_{x_i \in T} \phi_{\lambda_{x_i}}(h_{x_i}^{-1} \delta_{x_i} \sigma(h_{x_i})) \times \prod_{x \notin T} \phi_{\beta_x}(h_x^{-1} \gamma h_x) \times \mathbf{1}(p((g_{0,x} h_x)_{x \in |X|}) \le p)) \cdot dh_x$$

where the sum runs through all the conjugacy class γ_0 *of* $\mathrm{GL}_n(F)$ *which are* $(T, \underline{s})$*-admissible.*

References

[1] A. Beauville, Y. Laszlo, *Un lemme de descente*, C. R. Acad. Sci. Paris Sér. I Math., **320** (1995), 335-340.

[2] A. Beilinson, V. Drinfeld, *Quantization of Hitchin's Integrable System and Hecke Eigensheaves*, preprint.

[3] S. Bosch, W. Lutkebohmert, M. Raynaud, *Neron models*, Ergeb. der Math. 21. Springer Verlag 1990.

[4] V. Drinfeld, *Varieties of modules of F-sheaves*, Functional Analysis and its Applications, **21** (1987), 107-122.

[5] V. Drinfeld, *Cohomology of compactified moduli varieties of F-sheaves of rank* 2, J. Soviet Math., **46**(2) (1989), 1789-1821.

[6] M. GREENBERG, *Schemata over local rings: II*, Annals of Mathematics, **78**(2) (1963), 256-266.

[7] D. KAZHDAN, Y. VARSHAVSKY, article in preparation.

[8] R. KOTTWITZ, *Rational conjugacy classes in reductive groups*, Duke Math. J., **49** (1982), no. 4, 785-806.

[9] R. KOTTWITZ, *Isocrystals with additional structure*, Compositio Math., **56**(2) (1985), 201-220.

[10] R. KOTTWITZ, *Stable trace formula : elliptic singular terms*, Math. Ann., **275** (1986) 365-399.

[11] R. KOTTWITZ, *Shimura varieties and λ-adic representations*, Automorphic forms, Shimura varieties, and L-functions, Vol. I (Ann Arbor, MI, 1988), 161-209, Perspect. Math., 10, Academic Press, Boston, MA, 1990.

[12] R. KOTTWITZ, *Points on some Shimura varieties over finite fields*, J. Amer. Math. Soc., **5** (1992), 373-444.

[13] L. LAFFORGUE, *Chtoucas de Drinfeld et conjecture de Ramanujan-Petersson*, Astérisque, **243** (1997), +329pp.

[14] L. LAFFORGUE, *Une compactification des champs classifiant les chtoucas de Drinfeld*, J. Amer. Math. Soc., **11**(4) (1998), 1001-1036.

[15] L. LAFFORGUE, *Chtoucas de Drinfeld et correspondance de Langlands*, Invent. Math., **147** (2002), 1-241.

[16] V. LAFFORGUE, *Chtoucas pour les groupes réductifs et paramétrisation de Langlands globale*, arXiv:1209.5352v4.

[17] E. LAU, *On generalised $\mathcal{D}$-shtukas*, Dissertation, Bonner Mathematische Schriften [Bonn Mathematical Publications], Universitat Bonn, Mathematisches Institut, Bonn, **369** (2004), +110 pp.

[18] E. LAU, *On degenerations of $\mathcal{D}$-shtukas*, Duke Math. J. **140**(2) (2007), 351-389.

[19] G. LAUMON, *Cohomology of Drinfeld modular varieties. Part I. Geometry, counting of points and local harmonic analysis*, Cambridge Studies in Advanced Mathematics, Cambridge University Press, Cambridge, **41** (1996), +344 pp.

[20] G. LAUMON, *Cohomology of Drinfeld modular varieties. Part II. Automorphic forms, trace formulas and Langlands correspondence*, with an appendix by Jean-Loup Waldspurger, Cambridge Studies in Advanced Mathematics, Cambridge University Press, Cambridge, **56** (1997), +366 pp.

[21] G. LAUMON, M. RAPOPORT, U. STUHLER, *$\mathcal{D}$-elliptic sheaves and the Langlands correspondence*, Invent. Math. **113**(2) (1993), 217-338.

[22] B.-C. NGÔ, private communication.

[23] B.-C. NGÔ, *$\mathcal{D}$-chtoucas de Drinfeld à modifications symétriques et identité de changement de base*, long version, arXiv 2003.

[24] B.-C. NGÔ, *$\mathcal{D}$-chtoucas de Drinfeld à modifications symétriques et identité de changement de base*, Annales Scientifiques de l'ENS, **39** (2006), 197-243.

[25] B.-C. NGÔ, *Fibration de Hitchin et endoscopie*, Invent. Math., **164** (2006), 399-453.

[26] B.-C. NGÔ, T. NGÔ DAC, *Comptage de G-chtoucas: la partie régulière ellitique*, Journal de l'Institut Mathématique de Jussieu, **7**(1) (2008), 181-203.

[27] T. NGÔ DAC, *Compactification des champs de chtoucas et théorie géométrique des invariants*, Astérisque, **313** (2007), +124pp.

[28] T. NGÔ DAC, *Comptage de G-chtoucas: la partie ellitique*, to appear, Compositio Math.

[29] M. RAPOPORT, *A guide to the reduction of Shimura varieties*, Astérisque, **298** (2005), 271-318.

[30] M. RAPOPORT, M. RICHARTZ, *On the classification and specialization of F-isocrystals with additional structure*, Compositio Math., **103** (1996), 153-181.

[31] J-P. SERRE, *Cohomologie galoisienne*, 5e édition, Lecture Notes in Mathematics, **5** (1994), +181pp.

[32] R. STEINBERG, *Regular elements of semisimple algebraic groups*, Publications de l'IHES, **25** (1965), 49-80.

[33] Y. VARSHAVSKY, *Moduli spaces of principal F-bundles*, Selecta Math., **10** (2004), 131-166.

[34] Y. VARSHAVSKY, private communication.

14

Modular forms and Calabi-Yau varieties

Kapil Paranjape[†,1] *and Dinakar Ramakrishnan*[‡,2]

[†] Department of Mathematical Sciences, IISER, Mohali, India
E-mail address: kapil@iisermohali.ac.in

[‡] Department of Mathematicss, California Institute of Technology
Pasadena, CA 91125
E-mail address: dinakar@caltech.edu

Introduction

Let $f(z) = \sum_{n=1}^{\infty} a_n q^n$ be a holomorphic newform of weight $k \geq 2$ relative to $\Gamma_1(N)$ acting on the upper half plane $\mathcal{H}$. Suppose the coefficients a_n are all rational. When $k = 2$, a celebrated theorem of Shimura asserts that there corresponds an elliptic curve E over $\mathbb{Q}$ such that for all primes $p \nmid N, a_p = p + 1 - |E(\mathbb{F}_p)|$. Equivalently, there is, for every prime ℓ, an ℓ-adic representation ρ_ℓ of the absolute Galois group $\mathfrak{G}_\mathbb{Q}$ of $\mathbb{Q}$, given by its action on the ℓ-adic Tate module of E, such that a_p is, for any $p \nmid \ell N$, the trace of the Frobenius Fr_p at p on ρ_ℓ.

The primary aim of this article is to provide some positive evidence for the expectation of Mazur and van Straten that for every $k \geq 2$, any (normalized) newform f of weight k and level N should, if it has rational coefficients, have an associated Calabi-Yau variety $X/\mathbb{Q}$ of dimension $k - 1$ such that

- (Ai) The $\{(k-1, 0), (0, k-1)\}$-piece of $H^{k-1}(X)$ splits off as a submotive M_f over $\mathbb{Q}$,
- (Aii) $a_p = \mathrm{tr}(Fr_p \mid M_{f,\ell})$, for almost all p, and
- (Aiii) $\det(M_{f,\ell}) = \chi_\ell^{k-1}$,

where χ_ℓ is the ℓ-adic cyclotomic character.

In fact we expect (Aiii) to hold for every p not dividing ℓN.

[1] Partly supported by the the JCBose Fellowship of the DST (SR/S2/JCB-16/2010) and IMSc, Chennai

[2] Partly supported by the NSF grants DMS-0701089 and DMS-1001916

Arithmetic and Geometry, ed. Luis Dieulefait *et al.* Published by Cambridge University Press.

Furthermore, we even hope that in addition the following holds:

(Aiv) X admits an involution τ which acts by -1 on $H^0(X, \Omega^{k-1})$.

We anticipate that the involution τ can be chosen in such a way to make the quotient X/τ a rational variety.

This typically holds in our examples below. This extra structure is natural to want and is needed for understanding a variety of operations like products and twists. Such a τ obviously exists for $k = 2$, in which case X is (by Shimura) an elliptic curve over $\mathbb{Q}$, given by an equation $y^2 = f(x)$, and the involution τ sends (x, y) to $(-x, y)$, thus acting by -1 on the holomorphic differential $\omega = dx/y$ which spans $H^0(X, \Omega^1)$. (Of course X/τ is in this case $\mathbb{P}^1$.)

To fix ideas, one could think of a motive as a semisimple motive M relative to absolute Hodge cycles ([DMOS]), with avatars $(M_B, M_{\mathrm{dR}}, M_\ell)$ of Betti, de Rham and ℓ-adic realizations. Since every Calabi-Yau manifold of dimension 1 is an elliptic curve, (Ai) through (Aiv) provide a natural extension of what one has for $k = 2$. Of course for $k > 2$, one knows by Deligne ([Del]) that there is an irreducible, two-dimensional ℓ-adic representation ρ_ℓ of $\mathrm{Gal}(\overline{\mathbb{Q}}/\mathbb{Q})$ so that (Aii), $(Aiii)$ hold with $M_{f,\ell}$ replaced by ρ_ℓ.

In the first part of this article we will focus on the forms f of even weight for small levels, before moving on in the second part to formulate an analogue for regular self-dual cusp forms on $\mathrm{GL}(n)/\mathbb{Q}$ and see how the framework is compatible with the principle of functoriality.

When the weight is odd, the $\mathbb{Q}$-rationality forces f to be of CM type, and for weight 3, we refer to the paper of Elkies and Schuett ([ES]) for a beautiful result.

For non-CM newforms f of weight $k > 2$ with $\mathbb{Q}$-coefficients, we in fact hope for more, namely that the cohomology ring of X will be spanned by M and the Hodge/Tate classes of various degrees; in particular, X should be rigid in this case.

It is a difficult problem in dimensions > 3 to find smooth models of varieties with trivial canonical bundles, and for this reason we formulate our questions for such varieties with mild singularities. By a *Calabi-Yau variety* over a field k, we will mean an n-dimensional projective variety X/k on which the canonical bundle $\mathcal{K}_X$ is defined such that

(CY1) $\mathcal{K}_X$ is trivial; and
(CY2) $H^m(X, \mathcal{O}_X) = 0$ for all (strictly) positive $m < n$.

More precisely, we will want such an X to be normal and Cohen-Macaulay, so that the dualizing sheaf $\mathcal{K}_X$ is defined, with the singular locus in codimension at least 2, so that $\mathcal{K}_X$ defines a Weil divisor; finally, X should be Gorenstein, so that $\mathcal{K}_X$ will represent a Cartier divisor. Ideally we would like to singular locus X_{sing} to be of dimension $\leq \left[\frac{n-1}{2}\right]$.

In addition to these properties, we would ideally also like X to be realized as a double cover $\pi : X \to Y$, where Y is a projective smooth rational variety with negative ample canonical divisor. (Such an X will be automatically Gorenstein so that $\mathcal{K}_X$ is defined, and $\pi_*(\mathcal{O}_X) = \mathcal{O}_Y \oplus L$, for a line bundle L on Y with L^2 giving the branch locus; moreover, $\pi_*(\mathcal{K}_X) = \mathcal{K}_Y \oplus \mathcal{K}_Y \otimes L^{-1}$, forcing $L = \mathcal{K}_Y$.)

Here is our first result:

Theorem 14.1. *Fix* $\Gamma = \Gamma_1(N)$, $N \leq 5$. *Let* k *be the first even weight s.t. dim*$(S_k(\Gamma)) = 1$. *Then* $\exists$ *a Calabi-Yau variety* $V(f)/\mathbb{Q}$ *with an involution* τ *associated to the new generator* f *of* $S_k(\Gamma)$ *satisfying* (Ai) *through* (Aiv). *In fact, when* $N \leq 5$, $V = V(f)$ *is birational over* $\mathbb{Q}$ *to the Kuga-Sato variety* $\tilde{\mathcal{E}}_N^{(k-2)}$.

In particular, this applies to the Delta function $\Delta(z) = \sum_{n=1}^{\infty} \tau(n) q^n = \prod_{m\geq 1}(1-q^m)^{24}$, for $N = 1, k = 12$. By the Kuga-Sato variety, we mean a smooth compactification (over $\mathbb{Q}$) of the fibre product E_N^{k-2} of the universal elliptic curve E_N over the model over $\mathbb{Q}$ of the modular curve $\Gamma_1(N)\backslash\mathcal{H}$.

It is well known that $S_k(\Gamma_1(N))$ is **one-dimensional** when (N, k) equals $(\mathbf{1}, \mathbf{12})$, $(1, 16)$, $(\mathbf{2}, \mathbf{8})$, $(2, 10)$, $(\mathbf{3}, \mathbf{6})$, $(3, 8)$, $(\mathbf{4}, \mathbf{6})$, $(\mathbf{5}, \mathbf{4})$, $(5, 6)$, $(\mathbf{6}, \mathbf{4})$, $(7, 4)$. (The values in bold are the cases to which the Theorem applies.) For example, for the case $(\mathbf{2}, \mathbf{8})$, the generator is

$$f(z) = q - 8q^2 + 12q^3 - 210q^4 + 1016q^5 + \ldots$$

Recall that a newform $f(z) = \sum_{n=1}^{\infty} a_n q^n$ is of *CM-type* iff there is an odd, quadratic Dirichlet character δ such that $a_p = a_p\delta(p)$ for almost all primes p. Equivalently, if K is the imaginary quadratic field cut out by δ, $a_p = 0$ for all p which are inert in K.

As a first step, for general N, we may ask for a *potential statement*, i.e. the association, over a finite extension F of $\mathbb{Q}$, of a Calabi-Yau variety V/F to a newform f with $\mathbb{Q}$-coefficients. Here we state a modest result in this direction, already known in different ways, just to show that it fits into our framework:

Proposition 14.2. Let f be a newform of weight $k \geq 3$ of CM type with rational coefficients. Then $\exists$ a Calabi-Yau $(k-1)$-fold X defined over a number field F such that (Ai), (Aii) and (Aiii) hold over F. This X arises as a Kummer variety associated to an elliptic curve E with complex multiplication. When $k \leq 4$, X can be taken to be a smooth model.

For $k = 4$, all but one form f have been treated in Cynk-Schuett ([CS]).

Again, when $k = 3$, there is a much more precise and satisfactory result over $\mathbb{Q}$ in the work of Elkies and Schuett [ES].

In the converse direction, if M is a simple motive over $\mathbb{Q}$ of rank 2, with coefficients in $\mathbb{Q}$, of Hodge type $\{(w, 0), (0, w)\}$ with $w > 0$, then the general philosophy of Langlands, and also a conjecture of Serre, predicts that M should be modular and be associated to a newform f of weight $w + 1$ with rational coefficients. This is part of a very general phenomenon, and applies to motives occurring in the cohomology of smooth projective varieties over $\mathbb{Q}$. In any case, it applies in particular to Calabi-Yau threefolds over $\mathbb{Q}$ whose $\{(3, 0), (0, 3)\}$-part splits off as a submotive. In this context, there have been a number of beautiful results, some of which have been described in the monographs [YL] and [YYL]. See also [Mey], [GKY] and [GY]. They are entirely consistent with what we are trying to do in the opposite direction, and also provide supporting examples. It should perhaps be noted that these results (relating to the modularity of rigid Calabi-Yau threefolds) can now be deduced *en masse* from the proof of Serre's conjecture due to Khare and Wintenberger ([KW]), with a key input from Kisin ([Kis]). See also the article of Dieulefait ([Die]).

Let us now move to a more general situation. Fix any positive integer n and suppose that f is a (new) Hecke eigen-cuspform on the symmetric space

$$\mathcal{D}_n := \mathrm{SL}(n, \mathbb{R})/\mathrm{SO}(n),$$

relative to a congruence subgroup Γ of $\mathrm{SL}(n, \mathbb{Z})$, which is *algebraic* and *regular*. For $n = 2$, f is algebraic and regular iff it is holomorphic of weight ≥ 2. In general, one considers the cuspidal automorphic representation π of $\mathrm{GL}(n, \mathbb{A})$ which is generated by f, and by Langlands the archimedean component π_∞ corresponds to an n-dimensional representation σ_∞ of the real Weil group $W_\mathbb{R}$, which contains $\mathbb{C}^*$ as a subgroup of index 2. One says ([Clo1]) that π is algebraic if the restriction of σ_∞ to $\mathbb{C}^*$ is a sum of characters χ_j of the form $z \to z^{p_j}\bar{z}^{q_j}$, with $p_j, q_j \in \mathbb{Z}$, and it is regular iff $\chi_i \neq \chi_j$ when $i \neq j$. Such an f contributes to the *cuspidal cohomology* $H^*_{\mathrm{cusp}}(\Gamma\backslash\mathcal{D}_n, V)$ relative to a *local coefficient system* V in a specific degree $w = w(f)$. Moreover, f is rational over a number field $\mathbb{Q}(f)$, defined by the Hecke action on cohomology, which preserves the cuspidal part (*loc. cit.*). There is conjecturally a motive $M(f)$ over $\mathbb{Q}$ of rank n, with coefficients in $\mathbb{Q}(f)$, and weight w. By Clozel [Clo2], [CHL], $M_\ell(f)$ exists for suitable f which are in addition *essentially self-dual*.

Now let f be an algebraic, regular, essentially self-dual newform of weight w relative to $\Gamma \subset \mathrm{SL}(n, \mathbb{Z})$, with L-function $L(s, f) = \prod_p L_p(s, f)$, such that $\mathbb{Q}(f) = \mathbb{Q}$. Then our question is if there exists a Calabi-Yau variety $X/\mathbb{Q}$ of dimension w with an involution τ such that

(Ci) There is a submotive $M(f)$ of $H^w(X)$ of rank n such that $M(f)^{(w,0)} = H^{w,0}(X)$,
(Cii) $L_p(s, f) = L_p(s, M_\ell(f))$ for almost all p, and
(Ciii) the quotient of X bt τ is a rational variety.

where $L_p(s, M_\ell(f))$ equals, at any prime $p \neq \ell$ where the ℓ-adic realization $M_\ell(f)$ is unramified, $\det(I - Fr_p p^{-s} \mid M_\ell(f))^{-1}$. Again we would like to be able to find an X having good reduction outside the primes dividing tN, where n is the level of f and t the order of torsion in Γ, such that (Cii) holds for any such $p \neq \ell$.

Thanks to the *principle of functionality*, one should be able to obtain a certain class of $\mathbb{Q}$-rational, regular, algebraic, essentially self-dual newforms f by transferring forms on (the symmetric domains of) smaller reductive $\mathbb{Q}$-subgroups G of $\mathrm{GL}(n)$. The simplest instance of this phenomenon is given by the symmetric powers $\mathrm{sym}^m(g)$ of classical $\mathbb{Q}$-rational, non-CM newforms g of weight k. One knows by Kim and Shahidi ([KS], [Kim]) that for $m \leq 4$, $f = \mathrm{sym}^m(g)$ is a cusp form on $\mathrm{GL}(m+1)$. (Recently, this has been extended to $m = 5$ (and further) in the works of Clozel and Thorne, and of Dieulefait.) Here is our third result:

Theorem 14.3. *Let g be a non-CM, elliptic modular newform of weight 2, level N and trivial character, whose coefficients a_n lie in $\mathbb{Q}$. Then for for any $m > 0$, there is a Calabi-Yau variety X_m with an involution τ over $\mathbb{Q}$ of dimension m associated to (g, sym^m) such that (Ci), (Cii), $(Ciii)$ hold relative to $M_\ell = \mathrm{sym}^m(\rho_\ell(g))$. Moreover, for $m \leq 3$, X_m can be taken to be non-singular, with good reduction outside N.*

Here X_2 is just the familiar Kummer surface attached to $E \times E$, where $E = X_1$ is the elliptic curve/$\mathbb{Q}$ defined by g. But the case $m = 3$ is interesting, especially since it is not rigid, thanks to the Hodge type being $\{(3,0), (2,1), (1,2), (0,3)\}$, with each Hodge piece being one-dimensional. In fact, in that case, $\mathrm{sym}^3(g)$ corresponds (by [RS]) to a holomorphic Siegel modular cusp form F of genus 2 and (Siegel) weight 3. Such an F contributes to the cohomology in degree 3 of the Siegel modular threefold V of level N^3. Since the geometric genus of V is typically > 1, it cannot be Calabi-Yau. However, there should be, as predicted by the Hodge and Tate conjectures, an algebraic correspondence between X_3 and V (for any N).

One also knows (cf. [Ram]) that given two non-CM newforms g, h of weights $k, r \geq 2$ respectively, then there is an algebraic automorphic form $f = g \boxtimes h$ on $\mathrm{GL}(4)/\mathbb{Q}$, which will be cuspidal and regular if $k \neq r$. If g, h are $\mathbb{Q}$-rational, then so is f. Moreover, f is essentially self-dual because g and h are.

Theorem 14.4. *Let g, h be $\mathbb{Q}$-rational, non-CM newforms as above of respective weights $k, r > 1$, with $k \neq r$. Suppose we have Calabi-Yau varieties with involutions $(X(g), \tau_g)$, $(X(h), \tau_h)$ over $\mathbb{Q}$ attached to g, h respectively, satisfying (Ai) through $A(iv)$. Put $f = g \boxtimes h$, so that $w(f) = (k-1)(r-1)$. Then there is a Calabi-Yau variety with involution $(X(f), \tau_f)$ over $\mathbb{Q}$ of dimension $w(f)$ such that (Ci) through $(Ciii)$ hold.*

The point is that the product $Z := X(g) \times X(h)$ has the desired submotive in degree w, but it also has global holomorphic m-forms for $m = k - 1$ and $m = r - 1$. We exhibit an involution τ on Z such that when we take the quotient by τ, these forms get killed and we get a Calabi-Yau variety with reasonable singularities. To get unconditional examples of this Theorem, take $k = 2$ and choose h to be one of the examples of Theorem A of weight $r > 2$. To be specific, we may take g to be the newform of weight 2 and level 11, and h to be the newform of weight 4 and level 5, in which case $f = g \boxtimes h$ has level 55^2, and $X(f)$ is a Calabi-Yau fourfold. A similar inductive construction of Calabi-Yau varieties is also found in a paper of Cynk and Hulek ([CH]).

In sum, it is a natural question, given Shimura's work on forms of weight 2, if there are Calabi-Yau varieties with an involution associated to forms of higher weight with rational coefficients, and a preliminary version of this circle of questions was raised by the second author in a talk at the Borel memorial conference at Zhejiang University in Hangzhou, China, in 2004, and quite appropriately, Dick Gross, who was in the audience, cautioned against hoping for too much without sufficient evidence. Over the past years, there has been some positive evidence, though small, and even if there is no V in general, especially for non-CM forms f of even weight, the examples where one has nice Calabi-Yau varieties V enriches them considerably, and one of our aims is to understand the Hecke eigenvalues a_p in such cases a bit better in terms of counting points of V mod p. Since then we have learnt from [ES] (and Noriko Yui) that the question of existence of a Calabi-Yau variety V associated to f were earlier raised by Mazur and van Stratten. What we truly hope for is that in addition, V will be equipped with an involution τ acting by -1 on the unique global holomorphic form of maximal degree, so one can form products, etc., and also deal with quadratic twists. We have some interesting examples in the three-dimensional case (where $k = 4$), and a sequel to this paper will also contain a discussion of these matters, as well as a way to get nicer models in certain higher dimensional examples.

The Ramanujan coefficients $\tau(p)$ of the Delta function have been a source of much research. Our own work was originally motivated by the desire to express them in terms of the zeta function of a Calabi-Yau variety, though our path has diverged somewhat. In a different direction, a very interesting monograph of Edixhoven, et al. ([Edi]) yields a deterministic algorithm for computing $\tau(p)$ with expected running time which is polynomial in $\log p$. They do this by relating the associated Galois representation mod ℓ to the geometry of certain effective divisors on the modular curve $X_1(5\ell)$.

We thank all the people who have shown interest in this work, and a special thanks must go to Matthias Schuett who made a number of useful comments on the earlier version (2008) as well as a recent version (of two months ago). One of us (K.P.) would like to thank the JCBose Fellowship of the DST (SR/S2/JCB-16/2010) as well as Caltech and the IMSc, Chennai, where some of the work was done, and the second author (D.R.) would like to thank the NSF for continued support through the grants DMS-0701089 and DMS-1001916, and also IMSc, Chennai, for having him visit at various times. In addition, the second author (D.R.) would like to thank the Hausdorff Institute for their hospitality during his visit there for a week in February to attend and speak at the Session organized by Dieulefait and Wintenberger.

14.1 An intuitive picture of the geometric construction

In this section we will give an idea behind our construction of varieties with trivial canonical bundles arising as birational models of elliptic modular varieties V with non-positive canonical bundles. (S.T. Yau has informed us that he earlier had a similar construction, albeit in a different context.) We will use the intuitive language of divisors and linear systems. The content here will not be used in the succeeding sections, where we will use sheaves and do everything precisely in the different cases at hand.

Recall that V arise as fibre products of the universal family of elliptic curves E with additional structures over the modular curve associated to a congruence subgroup Γ of $\mathrm{SL}(2, \mathbb{Z})$. When Γ is $\Gamma_1(N)$ (resp. $\Gamma_0(N)$), the additional structure is a point (resp. subgroup) of order N. We have a slight preference for $\Gamma_1(N)$ over $\Gamma_0(N)$, because in the latter case, one gets, due to the existence of $-I$, only a coarse moduli space.

Suppose we want to parametrize triples of the form $(E, S; R)$ where E is a curve of genus 1, S a finite set of points on E, and R a finite set of linear equivalence relations on the points S. We will think of E as a curve of genus 1 without a specified origin (by making an identification with line bundles of degree 1 on it), hence with no group structure, and the relations in R are taken to hold in the divisor class group. (Once we pick a point o of E, we can of course identify this class group with $E \times \mathbb{Z}$ by sending (e, n) to $((e - o) + n.o)$.) We will take at least a portion of S to consist of general points, and we assume that R contains all the relations between the chosen points. Additional relations may hold for special triples (E,S,R) but generically, only those in R will hold.

Suppose also that there is a surface X, and a linear system P of divisors linear equivalent to $-K_X$, where K_X is the canonical divisor of X. Assume that for a "general" datum $(E, S; R)$ we have a uniquely determined element of P, so that the projective space P parametrizes triples as above up to birational isomorphism. Here by "general datum" we mean a point in an open subset of the parameter space (or moduli stack to be precise).

Let $W \subset \Gamma(X, \mathcal{O}(-K_X))$ be the subspace so that P is the associated projective space and $n = \operatorname{rank} W$. We have a natural homomorphism

$$\phi : V \otimes \mathcal{O}_{X^n} \to \oplus_{i=1}^n pr_i^* \mathcal{O}(-K_X),$$

where pr_i denotes the projection onto the i-th factor. The divisor D where the determinant $\det(\phi)$ vanishes parametrizes, birationally, tuples of the form $(E, S; R; p_1, \ldots, p_n)$ where the p_i are n additional points which are not subject to any additional relation. Moreover, D has trivial canonical bundle.

If n is at most the dimension of the linear system, then the parameter space is rational. But new things happen when n is larger than that, when one gets a divisor on a product of rational surfaces, in fact given by the vanishing of $\det(\phi)$. This moduli space V_n, say, fibers over the rational variety V_{n-1}, with the general fibre being an elliptic curve. The involution $x \mapsto -x$ on the general fibre gives rise to an involution τ on V_n. So V_n/τ fibers over a rational variety with $\mathbb{P}^1$ fibres. Hence V_n/τ is unirational.

14.2 The modular varieties of interest

In the context of the congruence subgroup $\Gamma_1(N)$ of $\mathrm{SL}(2, \mathbb{Z})$, the elliptic modular varieties V with non-positive canonical bundles are associated with pairs (N, k) such that there is at most one modular form of level N and weight k. The complete list of such pairs is given in the following table:

N	k
1	$\leq 23, 25, 26$, odd
2	≤ 11, odd
3	≤ 8
4	≤ 6
5, 6	≤ 4
7, 8	≤ 3
9, 10, 11, 12, 14, 15	2

One can similarly make the corresponding tables for the groups $\Gamma_0(N)$ and $\Gamma_0(N^2) \cap \Gamma_1(N)$.

14.3 The Calabi-Yau 11-fold associated to Δ

Δ is a generator of $S_{12}(\mathrm{SL}(2, \mathbb{Z}))$. The object is to show that the Kuga-Sato variety $\mathcal{E}^{(10)}$ is birational to an eleven-dimensional Calabi-Yau variety V. We will use $\equiv$ to denote birational equivalence. It is easy to see that for any $r \geq 0$,

$$\mathcal{E}^{(r)} \equiv \mathcal{M}_1(r+1),$$

where $\mathcal{M}_g(k)$ is the *moduli space of genus g curves with k marked points*, with compactification $\overline{\mathcal{M}}_g(k)$.

So we need to find a *birational model* V of $\overline{\mathcal{M}}_1(11)$ such that V is Calabi-Yau. Let S be the surface obtained by *blowing up* 4 *general points* P_1, P_2, P_3, P_4 in $\mathbb{P}^2$. Let E be an elliptic curve with $n+5$ general points $Q_0, Q_1, \ldots, Q_{n+4}$, and use $|3Q_0|$ to define a morphism $E \to \mathbb{P}^2$.

Using an automorphism of $\mathbb{P}^2$ we may assume: $Q_i = P_i$ for $1 \leq i \leq 4$. The embedding $E \to \mathbb{P}^2$ lifts to a morphism $\varphi : E \to S$, and the *adjunction formula* gives $\varphi(E) \in |\mathcal{K}_S^{-1}|$.

We get a *rational map* $\mathcal{M}_1(n+5) \to S^n$,

$$(E, \{Q_0, \ldots, Q_{n+4}\}) \to (P_5, \ldots, P_{n+4}).$$

$W := \Gamma(S, \mathcal{K}_S^{-1})$ has dimension 6, and there exists a hom (of sheaves on S^n):

$$f_n : W \otimes \mathcal{O}_{S^n} \to \mathcal{K}_S^{-1} \boxtimes \cdots \boxtimes \mathcal{K}_S^{-1}.$$

$\mathrm{Ker}(f_n)$ is the vector space of *sections in W vanishing at* $= P_5, \ldots, P_{n+4}$. The associated projective space then identifies with the collection of all (general) points in $\mathcal{M}_1(11)$ giving rise to this point on S^n.

Put

$$V_n := \mathrm{Proj}_{S^n}(\mathrm{coker}({}^t f_n))$$

where

$$^t f_n : \mathcal{K}_S \boxtimes \cdots \boxtimes \mathcal{K}_S \to W^\vee \otimes \mathcal{O}_S^\vee.$$

Then V_n is birational to $\mathcal{M}_1(n+5)$. We have corank$({}^t f_n) = 6-n$ at a general point of S^n. And there is a natural map

$$\pi : V_n \to S^n$$

$\mathbf{n \leq 5}$ π is surjective with fibres $\mathbb{P}^{5-n}$. Hence V_n, which is $\equiv \overline{\mathcal{M}}_1(n+5)$, is a *rational variety* in this case. Note that V_5 is just S^5.

$\mathbf{n = 6}$ $V = V_6$ is a (reduced) *divisor in* S^6, hence Gorenstein. It is defined by the vanishing of $\det(f_n)$, which is a section of $\mathcal{K}_{S^n}^{-1}$. So $\mathcal{K}_V$ **is trivial.** We already know that $h^{(11,0)} = 1$ and $h^{(p,0)} = 0$ for $0 < p < 11$ for $\tilde{\mathcal{E}}^{10}$. These also hold for V. So V is Calabi-Yau. Also, the whole construction is rationally defined.

$V = V_6$ fibers over V_5 with fibres of dimension 1, corresponding to cubics passing through 10 points. The natural involution on the general fibre, which is an elliptic curve, gives rise to one, call it τ, on V. The quotient V/τ is unirational because it is a family of rational curves on a smooth rational surface. Clearly, τ must act by -1 on the one dimensional space $H^0(V, \Omega^{11})$. □

14.4 A C-Y 7-fold occurring in level 2

Let E be an elliptic curve with origin $o \in E$, $x \in E$ a point of order 2 and $y, z \in E$ some other (general) points. Under the morphism $a : E \to |2[o] + [y]|$, the divisors $2[o] + [y]$ and $2[x] + [y]$ are linear sections. Of the four points o, x, y and z we may assume (under the hypothesis of generality on y and z) that no three become collinear under the morphism a. Thus, we can identify these with the points $(0 : 0 : 1)$, $(1 : 0 : 0)$, $(0 : 1 : 0)$ and $(1 : 1 : 1)$ respectively in order to identify $|2[o] + [y]|$ with $\mathbb{P}^2$.

Conversely, let E be a cubic curve in $\mathbb{P}^2$ which has the following properties:

(1) E passes through the points $(0 : 0 : 1)$, $(0 : 1 : 0)$, $(1 : 0 : 0)$ and $(1 : 1 : 1)$.
(2) The line $Y = 0$ is tangent to E at the point $(0 : 0 : 1)$.
(3) The line $Z = 0$ is tangent to E at the point $(0 : 1 : 0)$.

Then E is a curve of genus 1 for which we take $o = (0 : 0 : 1)$ as the origin of a group law. Let $x = (0 : 1 : 0)$, $y = (1 : 0 : 0)$ and $z = (1 : 1 : 1)$. Then we obtain the relation

$$2[o] + [y] \simeq 2[x] + [y]$$

It follows that $2x = o$. Thus we have obtained (E, o, x, y, z) of the type we started with.

Direct calculation shows that the linear system of cubics in $\mathbb{P}^2$ that satisfy the conditions above is the linear span of $X^2Y - XYZ$, $X^2Z - XYZ$, $Y^2Z - XYZ$ and $YZ^2 - XYZ$.

Here is an **alternate construction in level** 2:

Let E be an elliptic curve with origin $o \in E$, $x \in E$ a point of order 2 and $y, z \in E$ some other points. The morphism $a : E \to |2[o]|$ has fibres $2[o]$, $[y]+[-y]$ and $[z]+[-z]$ which we map to 0, 1 and ∞ respectively in order to identify $|2[o]|$ with $\mathbb{P}^1$. The morphism $b : E \to |[o]+[x]|$ has fibres $[o]+[x]$, $[y]+[x-y]$ and $[z]+[x-z]$ which we map to 0, 1 and ∞ in order to identify $|[o]+[x]|$ with $\mathbb{P}^1$. Thus we obtain a morphism $a \times b : E \to \mathbb{P}^1 \times \mathbb{P}^1$ which is constructed canonically from the data (E, o, x, y, z).

Conversely let E be a curve of type (2, 2) in $\mathbb{P}^1 \times \mathbb{P}^1$ which has the following properties:

(1) E passes through the points $(0, 0)$, $(1, 1)$ and (∞, ∞).
(2) The line $\{0\} \times \mathbb{P}^1$ is tangent to E at the point $(0, 0)$.
(3) If $(u, 0)$ is the residual point of intersection of E with $\mathbb{P}^1 \times \{0\}$, then $\{u\} \times \mathbb{P}^1$ is tangent to E at this point.

Then E is a curve of genus 1 for which we take $o = (0, 0)$ as the origin in a group law. Let $x = (u, 0)$, $y = (1, 1)$ and $z = (\infty, \infty)$. We obtain the identity $2[o] \simeq 2[x]$, from which it follows that $2x = o$. Thus we have recovered the data (E, o, x, y, z).

14.5 A remark on elliptic curves with level 3 structure

Let E be an elliptic curve with origin $o \in E$, $x \in E$ a point of order 3 and $y \in E$ some other point. The morphism $a : E \to |2[o]|$ has fibres $2[o]$, $[x]+[2x]$ and $[y]+[-y]$ which we map to 0, 1 and ∞ respectively in order to identify $|2[o]|$ with $\mathbb{P}^1$. The morphism $b : E \to |[o]+[x]|$ has fibres $[o]+[x]$, $2[2x]$ and $[y]+[x-y]$ which we map to 0, 1 and ∞ respectively in order to identify $|[o]+[x]|$ with $\mathbb{P}^1$. Thus we obtain a morphism $a \times b : E \to \mathbb{P}^1 \times \mathbb{P}^1$ which is constructed canonically from the data (E, o, x, y).

Conversely, let E be a curve of type (2, 2) in $\mathbb{P}^1 \times \mathbb{P}^1$ which has the following properties:

(1) E is tangent to the line $\{0\} \times \mathbb{P}^1$ at the point $(0, 0)$.
(2) E is tangent to the line $\mathbb{P}^1 \times \{1\}$ at the point $(1, 1)$.
(3) E passes through the points $(1, 0)$ and $\infty, \infty)$.

Then E is a curve of genus 1 and we use $o = (0, 0)$ as the origin of a group law on E. Let $x = (1, 0)$, $y = (\infty, \infty)$ and $p = (1, 1)$. We obtain the identities,

$$2[o] \simeq [x] + [p] \qquad\qquad [o] + [x] \simeq 2[p]$$

It follows that $p = 2x$ and $3x = o$. We have thus recovered the data (E, o, x, y) that we started with.

14.6 Level 3 and a CY 5-fold

We will construct a 5-fold with trivial canonical bundle and singularities only in dimension 2 or less such that its middle cohomology represents the motive of the (unique) modular form of level 3 and weight 6.

Consider the linear system P of cubics in $\mathbb{P}^2$ that is spanned by the curves $X^2Y - XYZ$, $Y^2Z - XYZ$ and $Z^2X - XYZ$; this system is stable under the cyclic automorphism $X \to Y \to Z \to X$ of $\mathbb{P}^2$. Each curve in the linear system P is tangent to the line $Z = 0$ at the point $p_Y = (0 : 1 : 0)$; similarly, the curve is tangent to $X = 0$ at the point $p_Z = (0 : 0 : 1)$ and to $Y = 0$ at the point $p_X = (1 : 0 : 0)$. Moreover, each curve passes through the point $p_0 = (1 : 1 : 1)$. We note that the linear system P is precisely the collection of cubic curves in $\mathbb{P}^2$ that satisfy these conditions.

In the divisor class group of a smooth curve in this linear system we obtain the identities

$$2p_Y + p_X = 2p_Z + p_Y = 2p_X + p_Z = p_X + p_0 + r$$

where r denotes the remaining point of intersection of the curve with the line $Y = Z$ that joins p_X and p_0. In particular, we note that $p_Y - p_X$ is of order 3 in this class group and $p_Z - p_X = 2(p_Y - p_X)$.

Conversely, suppose we are given a smooth curve E of genus 1 and a line bundle ξ of order 3 on E; moreover, suppose that three distinct points p, q and r are marked on E. We then obtain two additional points a and b on E such that $a - p = \xi$ and $b - p = 2\xi$ in the divisor class group of E. Consider the morphism $E \to \mathbb{P}^2$ that is given by the linear system of the divisor $p + q + r$. Moreover, we choose co-ordinates on $\mathbb{P}^2$ so that the point p goes to p_X, q goes to p_0, a goes to p_Y and b goes to p_Z. This gives an embedding of E as a curve in $\mathbb{P}^2$ that belongs to the linear system P.

Let S denote the surface obtained by blowing up $\mathbb{P}^2$ at the four points p_X, p_Y, p_Z and p_0, and then further blowing up the resulting surface at the "infinitely near points" that correspond to $Z = 0$ at p_Y, to $X = 0$ at p_Z and to $Y = 0$ at p_X. Let H denote the inverse image in S of a general line in $\mathbb{P}^2$; let E_X, E_Y, E_Z and E_0 denote the strict transforms of the exceptional loci of the first blow-up over the points p_X, p_Y, p_Z and p_0 respectively; let F_X, F_Y and F_Z denote the exceptional divisors of the second blow-up. The anti-canonical divisor $-K_S = 3H - E_X - E_Y - E_Z - E_0 - 2(F_X - F_Y - F_Z)$ has a base-point free complete linear system $|-K_S|$ which can be identified with P. Let T denote the natural incidence locus in $S \times P$. The variety

$$X = T \times_P T \times_P T$$

is a singular 5-fold which is Gorenstein and has trivial canonical bundle. Moreover, an open subset of X_0 parametrizes tuples of the form $(E, \xi, p, q, r, s, t, u)$ where E is a curve of genus 1, ξ is a line bundle of order 3 on E and p, q, r, s, t and u are six distinct points on E.

Let L_X, L_Y, L_Z denote the strict transforms in S of the lines in $\mathbb{P}^2$ defined by $X = 0$, $Y = 0$, $Z = 0$ respectively. Further, let R be the strict transform in S of the curve in $\mathbb{P}^2$ defined by

$$X^2Z + Y^2X + Z^2Y - 3XYZ = 0$$

This is the unique cubic in $\mathbb{P}^2$ that has a node at p_0 and is tangent to $X = 0$ at p_Y, to $Y = 0$ at p_Z and to $Z = 0$ at p_X. It follows that R is a smooth rational curve that meets E_0 in a pair of distinct points and the triple (R, L_X, E_Y) (respectively (R, L_Y, E_Z) and (R, L_Z, E_X)) consists of smooth curves that meet pairwise transversally.

The morphism $S \to P^*$ induced by the linear system P can be factorized via a double cover $S \to W$ which is ramified along R. Each of the curves L_X and E_Y (respectively L_Y and E_Z; L_Z and E_X) is mapped isomorphically onto the same smooth irreducible curve G_Z (respectively G_X; G_Y) in W; the curve R is mapped isomorphically onto the branch locus Q in W. The morphism $W \to P^*$ collapses the curves G_X (respectively G_Y and G_Z) to a point q_X (respectively q_Y and q_Z) in P^*; in fact $W \to P^*$ is identified with the blow-up of P^* at these points. Moreover, Q is mapped to a plane quartic $\overline{Q}$ which has cusps at these three points.

Let T denote the incidence locus in $P \times S$ as above. It is the pull-back via $S \to P^*$ of the natural incidence locus $I \subset P \times P^*$. The latter can be identified (via the projection $I \to P^*$) with the projective bundle of one-dimensional linear subspaces of the tangent bundle of P^*. Hence, the exceptional curve G_X (respectively G_Y and G_Z) of the blow-up $W \to P^*$ can be identified with the fibre I_X of $I \to P^*$ over the point q_X (respectively q_Y and q_Z). Thus we obtain natural maps $E_\alpha \to T$ and $L_\alpha \to T$ that are sections of the $\mathbb{P}^1$-bundle $T \to S$ over the curves E_α and L_α respectively; let $\tilde{E_\alpha}$ and $\tilde{L_\alpha}$ denote the images. Let T_X (respectively T_Y; T_Z) denote the fibre of T over the point of intersection of L_Y and E_Z (respectively L_Z and E_X; L_X and E_Y).

The tangent direction along $\overline{Q}$ gives a rational morphism (defined outside the cusps) from $\overline{Q}$ to I. It follows that this extends to a section $R \to T$ of $T \to S$ over R and gives a curve $\tilde{R}$ in T. The quadruple of curves $\tilde{R}$, T_X, $\tilde{L_Y}$, $\tilde{E_Z}$ (respectively, $\tilde{R}$, T_Y, $\tilde{L_Z}$, $\tilde{E_X}$; $\tilde{R}$, T_Z, $\tilde{L_X}$, $\tilde{E_Y}$) meet pairwise transversally in a single point r_X (respectively r_Y; r_Z) in T. The curve $\tilde{R}$ in T is mapped to a nodal cubic $\overline{R}$ in P for which I_X, I_Y and I_Z are inflectional tangents. The curves T_X, $\tilde{L_Y}$, $\tilde{E_Z}$ (respectively T_Y, $\tilde{L_Z}$, $\tilde{E_X}$; T_Z, $\tilde{L_X}$, $\tilde{E_Y}$) in T lie over I_X (respectively I_Y; I_Z) in P.

The singular locus of the morphism $T \to P$ consists of the curves $\tilde{R}$, T_α, $\tilde{L_\alpha}$ and $\tilde{E_\alpha}$ for $\alpha = X, Y, Z$ as described above.

The singular fibres of $T \to P$ then have the following description:

(1) If a is a smooth point of $\overline{R}$ which is not a point of inflection then the fibre C_a is a rational curve in S with a single ordinary node.
(2) If b which is on an inflectional tangent (i. e. one one of the lines I_X, I_Y, I_Z) of $\overline{R}$ but is *not* a point of inflection of $\overline{R}$ then the fibre C_b is a curve with three components and three nodes (i. e. a "triangle" of $\mathbb{P}^1$'s).
(3) If c is a point of inflection of the curve $\overline{R}$, then C_c consists of three $\mathbb{P}^1$'s that pass through a point and (since C_c lies on a smooth surface S) is locally a complete intersection.
(4) If d is the node of $\overline{R}$ then the fibre C_d consists of a pair of smooth $\mathbb{P}^1$'s in S that meet in a pair of points. In fact the curves are E_0 and the strict transform in S of the curve in $\mathbb{P}^2$ defined by the equation

$$X^2Y + Y^2Z + Z^2X - 3XYZ = 0$$

In particular, the elliptic fibration $T \to P$ is semi-stable but for the three fibres over the points of inflection of $\overline{R}$.

Now consider the variety $X = T \times_P T \times_P T$. The singular points of X_0 consist of triples (x, y, z) of points of T, where at least two of these points are critical points for the morphism $T \to P$. In particular, these points lie over the union of R and I_X, I_Y and I_Z. Since the singular points of each of the fibres described above are isolated, it follows that the singular locus of X has components of dimension at most 2.

14.7 Level 4 and a C-Y 5-fold

Let E be an elliptic curve with origin $o \in E$, $x \in E$ a point of order 4. Under the morphism $a : E \to |3[o]|$, the divisors $3[o]$, $2[2x] + [o]$, $[2x] + 2[x]$ and $[o] + [x] + [3x]$ are linear sections of the image curve. Let p denote the point of intersection of the lines corresponding to $3[o]$ and $[2x] + 2[x]$. No three of the points o, $2x$, $3x$ and p are collinear. Thus we can identify $|3[o]|$ with $\mathbb{P}^2$ in such a way that o is identified with $(0 : 1 : 0)$, $3x$ is identified with $(0 : 0 : 1)$, p is identified with $(1 : 0 : 0)$ and $3x$ is identified with $(1 : 1 : 1)$. Thus we obtain a morphism $a : E \to \mathbb{P}^2$ which is constructed canonically from the data (E, o, x).

Conversely, let E be a cubic curve in $\mathbb{P}^2$ which has the following properties:

(1) E passes through the points $(0 : 1 : 0)$, $(0 : 0 : 1)$, $(1 : 1 : 1)$ and $(1 : 0 : 1)$.
(2) The line $Z = 0$ is an inflectional tangent to E (at the point $(0 : 1 : 0)$).
(3) The line $X = 0$ is tangential to the curve E at the point $(0 : 0 : 1)$.
(4) the line $Y = 0$ is tangential to the curve E at the point $(1 : 0 : 1)$.

Then E is a curve of genus 1 and we use $o = (0 : 1 : 0)$ as the origin of a group law on E. Let $x = (1 : 0 : 1)$, $p = (0 : 0 : 1)$ and $q = (1 : 1 : 1)$. We obtain the identities

$$3[0] \simeq [o] + 2[p] \qquad \simeq [p] + 2[x] \simeq [o] + [q] + [x]$$

It follows that $2p = o$, $2x = p$ and $q = -x = 3x$. Thus we have recovered the data (E, o, x) that we started with.

Direct calculation shows us that the linear system of cubic curves in $\mathbb{P}^2$ that satisfy the above conditions is the linear span of $YZ(Y - Z)$ and $X(X - Z)^2$.

Here is an **alternate construction in level** 4:

Let E be an elliptic curve with origin $o \in E$, $x \in E$ a point of order 4 and $y \in E$ some other point. The morphism $a : E \to |2[o]|$ has fibres $2[o]$, $2[2x]$ and $[y] + [-y]$ which we map to 0, 1 and ∞ respectively in order to identify $|2[o]|$ with $\mathbb{P}^1$. The morphism $b : E \to |[o] + [x]|$ has fibres $[o]+[x]$, $[2x]+[3x]$ and $[y]+[x-y]$ which we map to 0, 1 and ∞ respectively in order to identify $|[o] + [x]|$ with $\mathbb{P}^1$. Thus we obtain a morphism $a \times b : E \to \mathbb{P}^1 \times \mathbb{P}^1$ which is constructed canonically from the data (E, o, x, y).

Conversely, let E be a curve of type $(2, 2)$ in $\mathbb{P}^1 \times \mathbb{P}^1$ which has the following properties:

(1) E is tangent to the line $\{0\} \times \mathbb{P}^1$ at the point $(0, 0)$.
(2) E is tangent to the line $\{1\} \times \mathbb{P}^1$ at the point $(1, 1)$.

(3) If E meets $\mathbb{P}^1 \times \{0\}$ at $(0, 0)$ and $(u, 0)$ and E meets $\mathbb{P}^1 \times \{1\}$ at $(1, 1)$ and $(v, 1)$; then $u = v$.
(4) E passes through the point (∞, ∞).

Then E is a curve of genus 0 and we use $o = (0, 0)$ as the origin of a group law on E. Let $x = (u, 0)$, $p = (v, 1)$ and $q = (1, 1)$. We obtain the identities,

$$2[o] \simeq 2[q] \qquad [o] + [x] \simeq [p] + [q]$$
$$2[0] \simeq [x] + [p] \qquad \text{(from condition 3 above)}$$

It follows that $q = 2x$, $p = 3x$ and $4x = o$. Let $y = (\infty, infty)$. We have thus recovered the data (E, o, x, y) that we started with. Note that in this construction, E is in $S = \mathbb{P}^1 \times \mathbb{P}^1$, and the Calabi-Yau variety V is a divisor on $S^4 = (\mathbb{P}^1)^8$.

14.8 Forms of weight 4 and Calabi-Yau 3-folds

The construction of this section is adapted from that of Schoen's article [Sch1], where he has associated a modular form f of weight 4 in this case to a special quintic threefold X. However, we need to modify his construction to obtain an involution τ such that the quotient X/τ is a rational threefold whose homology is the same as its Chow group. (Hence, the rational variety has no "interesting" motives other than powers of the Tate motive.) Moreover, the "additional" motive on the C-Y double cover is exactly the one associated (in [Sch2]) to the modular form f.

Let E be an elliptic curve with $o \in E$ as its origin and $x \in E$ a point of order 5. Under the morphism $a : E \to |3[o]|$, the divisors $3[o]$, $[o] + [x] + [4x]$, $2[x]+[3x]$ and $2[3x]+[4x]$ are linear sections. There is a unique identification of $|3[o]|$ with $\mathbb{P}^2$ under which these sections are identified with $Z = 0$, $X = 0$, $X + Y + Z = 0$ and $Y = 0$ respectively.

Conversely, let E be a cubic curve in $\mathbb{P}^2$ which has the following properties:

(1) E passes through the points $(0 : 1 : 0)$, $(0 : 1 : -1)$, $(1 : 0 : -1)$ and $(0 : 0 : 1)$.
(2) The line $Z = 0$ is an inflectional tangent to E (at the point $(0 : 1 : 0)$).
(3) The line $X + Y + Z = 0$ is tangent to E at the point $(1 : 0 : -1)$.
(4) The line $Y = 0$ is tangent to E at the point $(0 : 1 : -1)$.

The E is a curve of genus 1. Let $o = (0 : 1 : 0)$, $x = (0 : 1 : -1)$, $p = (0 : 0 : 1)$ and $q = (1 : 0 : -1)$. We use o as the origin of the group law on E. We obtain the identities,

$$3[o] \simeq [o] + [x] + [p] \simeq 2[x] + [q] \simeq 2[q] + [p].$$

It follows that $q = 2x$, $p = 4x$ and $5x = o$. Thus we have obtained the data (E, o, x) that we started with.

Direct calculation shows us that the linear system of cubics in $\mathbb{P}^2$ satisfying the above conditions is the linear span of $YZ(X+Y+Z)$ and $YZ(Y+Z) - X(X+Z)^2$.

Let E be an elliptic curve with $o \in E$ as its origin and $x \in E$ a point of order 5. The morphism $a : E \to |2[o]|$ has fibres $2[o]$, $[x]+[4x]$ and $[2x]+[3x]$, which we map to 0, 1 and ∞ to identify $|2[o]|$ with $\mathbb{P}^1$. Similarly, the morphism $b : E \to |2[x]|$ has fibres $2[x]$, $[o]+[2x]$ and $[3x]+[4x]$, which we map to 0, 1 and ∞ to identify $|2[x]|$ with $\mathbb{P}^1$. Thus we obtain a morphism $a \times b : E \to \mathbb{P}^1 \times \mathbb{P}^1$, which is constructed canonically from the data (E, o, x).

Conversely, let E be a curve of type (2, 2) in $\mathbb{P}^1 \times \mathbb{P}^1$ which is

(1) tangent to $\mathbb{P}^1 \times \{0\}$ at the point $(1, 0)$,
(2) tangent to $\{0\} \times \mathbb{P}^1$ at the point $(0, 1)$,
(3) passes through the points $(\infty, 1)$, (∞, ∞), and $(1, \infty)$.

Then E is a curve of genus 1. Let $o = (0, 1)$ and $x = (1, 0)$, which are points on E. We use o as the origin of the group law on E. Let p, q, r denote the points $(\infty, 1)$, (∞, ∞) and $(1, \infty)$ respectively. We obtain the identities,

$$2[o] \simeq [p] + [q] \qquad 2[x] \simeq [q] + [r]$$
$$2[o] \simeq [x] + [r] \qquad 2[x] \simeq [o] + [p]$$

We solve these to show that $a = 2x$, $b = 3x$, $c = 4x$, and $5x = o$. We have thus recovered the data (E, o, x) that we started with.

Now, appealing to the fibre product paper of Schoen ([Sch1]), we can deduce that our canonical object is Calabi-Yau.

14.9 CM forms and C-Y varieties over suitable extensions

Let E be an elliptic curve and $n \geq 1$. Put

$$B := \{x \in E^{n+1} \mid \sum_{j=1}^{n+1} x_j = 0\},$$

which admits an action by the alternating group A_{n+1}. Consider the quotient

$$X := B/A_{n+1}.$$

The following result is proved in [PR], where the smoothness of the model for $n = 3$ appeals to ideas of Cynk and Hulek.

Theorem. *X has trivial canonical bundle, with $H^0(X, \Omega_X^p) = 0$ if $0 < p < n$. If $n \leq 3$, there is a smooth model $\tilde{X}$ which is Calabi-Yau.*

A submotive M of rank 4 splits off of $H^3(\tilde{X})$ corresponding to $\mathrm{sym}^3(H^1(E))$, of Hodge type
$\{(3,0), (2,1), (1,2), (0,3)\}$. It is simple iff E is not of CM type, and in this case $\tilde{X}$ is not rigid.

Here is a *sketch of proof of Proposition 2*. Let Ψ be the Hecke character (of weight $k-1$) of an imaginary quadratic field K attached to f, so that $L(s, f) = L(s, \Psi)$. Pick an algebraic Hecke character λ of K of weight 1, with an associated elliptic curve E over a finite extension F_0. Then Ψ/λ^{k-1} is a finite order character ν. One attaches (by the Theorem above) a Calabi-Yau variety V, smooth for $k \leq 4$, to E^{k-1}. Then it has the requisite properties relative to f over the number field $F = F_0(\nu)$, having the correct traces of Frobenius elements at the primes outside a finite set S. There are a lot of choices for λ, which can be used to make F and S more precise. For example, when the discriminant $-D$ of K is either odd or divisible by 8, with $D > 4$, we may choose λ to be a *canonical* Hecke character of weight one ([Roh]), which is K-valued, equivariant relative to the complex conjugation of K, and is ramified only at the primes dividing D. There is an associated CM elliptic curve E over an explicit number field, which is a quadratic subextension of the Hilbert class field of K when D is odd), studied deeply by Gross in [Gro], which is isogenous to all of its Galois conjugates. We start with the C-Y variety V associated to E^{k-1}. In this case S involves only the primes dividing the level of f and D. □

We will discuss Theorem 3 elsewhere in detail, where for the key case $m = 3$ is partly understood via the descent to $\mathrm{GSp}(4)/\mathbb{Q}$ ([RS]).

14.10 Behavior under taking products

In this section we exploit an idea of Claire Voisin. M. Schuett has remarked that Cynk and Hulek have also used such an argument.

Suppose X_i is a double cover of smooth variety Y_i branched along a smooth divisor D_i for $i = 1, 2$; let ι_i denote the associated involutions on X_i.

The product variety $X_1 \times X_2$ carries an action of the involution $\iota = \iota_1 \times \iota_2$ which has $D_1 \times D_2$ as its fixed locus.This is a codimension two transverse intersection of the divisors $D_1 \times X_2$ and $X_1 \times D_2$.

Let X_{12} be the blow-up of $X_1 \times X_2$ along $D_1 \times D_2$ and E_{12} be the exceptional divisor. Then E_{12} is isomorphic to the projective bundle over $D_1 \times D_2$

associated with the rank two-vector bundle $L_1 \oplus L_2$, where $L_i = p_i^* N_{D_i/X_i}$. The strict transform E_1 of $D_1 \times X_2$ (respectively E_2 of $X_1 \times D_2$) in X_{12} is a divisor that meets E_{12} in a section of this projective bundle; the two intersections $E_1 \cap E_{12}$ and $E_2 \cap E_{12}$ are disjoint.

Since $D_1 \times D_2$ is the fixed locus of ι, this involution lifts to X_{12}; we denote this lift also by ι by abuse of notation. The (scheme-theoretic) fixed locus of ι on X_{12} is the smooth divisor E_{12} and hence the quotient Z_{12} of X_{12} by this involution is a smooth variety. In other words, X_{12} is a double cover of Z_{12} branched along the smooth divisor D_{12} which is the image of E_{12}.

The involution $\iota_1 \times id_{X_2}$ also lifts to X_{12} since the base of the blow-up is contained in its fixed locus. Moreover, it commutes with ι and hence descends to an involution τ on Z_{12}. The (scheme-theoretic) fixed locus of this involution on Z_{12} is the *disjoint* union of the images of E_1 and E_2 and is thus again a smooth divisor. Thus Z_{12} is the double cover of a variety Y_{12} branched along a smooth divisor. Moreover, one checks that Y_{12} is just the blow-up of $Y_1 \times Y_2$ along $D_1 \times D_2$ (where by abuse of notation we are using the same notation for the divisors D_i in X_i and for their images in Y_i).

Finally, let us calculate the canonical bundle of Z_{12} and check if it is trivial. First note that the double cover X of a smooth variety Y branched along a smooth divisor D is obtained by choosing an isomorphism of $\mathcal{O}_Y(D)$ with the square of some line bundle L on Y. In this case, the canonical bundle of X is the pull back of $K_Y \otimes L$.

Next let L_i denote the square root of $\mathcal{O}_{Y_i}(D_i)$. The canonical bundle of K_{X_i} is the pull-back of $K_{Y_i} \otimes L_i$; so if X_i is a Calabi-Yau variety, then K_{Y_i} is the dual of L_i. If E denotes the exceptional locus of the morphism $e : Y_{12} \to Y_1 \times Y_2$, then we see that if $\hat{D}_1$ denotes the strict transform of $D_1 \times Y_2$ in Y_{12} we have

$$\mathcal{O}_{Y_{12}}(\hat{D}_1 + E) = e^* \mathcal{O}_{Y_1 \times Y_2}(D_1 \times Y_2)$$

Similarly, we have

$$\mathcal{O}_{Y_{12}}(\hat{D}_2 + E) = e^* \mathcal{O}_{Y_1 \times Y_2}(Y_1 \times D_2)$$

It follows that

$$\mathcal{O}_{Y_{12}}(\hat{D}_1 + \hat{D}_2) = e^*(L_1 \otimes L_2 \otimes \mathcal{O}_{Y_{12}}(-E))^2$$

Since the canonical bundle of Y_{12} is

$$K_{Y_{12}} = e^* K_{Y_1 \times Y_2} \otimes \mathcal{O}_{Y_{12}}(E) = e^*(L_1 \boxtimes L_2)^{-1} \otimes \mathcal{O}_{Y_{12}}(E)$$

it follows that the canonical bundle of the double cover Z_{12} is trivial.

Theorem 4 now follows by applying this construction. □

Appendix: A consequence of the Hirzebruch Riemann-Roch theorem

The object of this appendix to the paper is to say that there is no obstruction from the Hirzebruch Riemann-Roch theorem for the cohomology of an odd dimensional Calabi-Yau manifold to be of the structure that we prescribe in the main conjecture we pursue in this paper. Specifically, there can be C-Y manifolds X whose Hodge structure looks like a sum of $H^{p,p}$'s (for various p) with $H^{n,0} \oplus H^{0,n}$, where n is the dimension of X.

Theorem. *For any odd dimensional smooth projective variety X of dimension $m \geq 3$, the Chern class $c_m(T_X)$ does not enter into the expression for the Euler characteristic $\chi(\mathcal{O}_X)$ when it is written as a sum of monomials in Chern classes.*

Proof. Let X be any smooth projective variety. Recall that the Todd classes of a variety are the multiplicative classes defined by the generating function

$$\mathrm{td}(t) = \frac{t}{1-\exp(-t)} = 1 + \frac{t}{2} + \frac{t^2}{12} - \frac{t^4}{720} + \frac{t^6}{30240} + O(t^8)$$

Fix an integer m and for $i = 1, \ldots, m$, let β_i be algebraic numbers (depending on m) such that

$$\mathrm{td}(t) \equiv \prod_{i=1}^{m} (1 + \beta_i t) \mod t^{m+1}$$

For X of dimension m, let c_i for $i = 1, \ldots, m$ be the Chern classes of its tangent bundle. Let γ_i for $i = 1, \ldots, m$ be the Chern roots so that

$$1 + c_1 t + c_2 t^2 + \cdots + c_m t^m = \prod_{i=1}^{m} (1 + \gamma_i t)$$

The Todd polynomial of the variety is then given by

$$\mathrm{Todd}(t) = \prod_{i=1}^{m} \mathrm{td}(\gamma_i t)$$

The coefficient of t^m in Todd(t) is the m-th Todd class Todd$_m$ of the variety. Making use of the above expression for $\mathrm{td}(t) \mod t^{m+1}$,

$$\mathrm{Todd}(t) \equiv \prod_{i,j=1}^{m} (1+\beta_j \gamma_i t) \equiv \prod_{j=1}^{m} \left(1 + c_1(\beta_j t) + c_2(\beta_j t)^2 + \cdots + c_m(\beta_j t)^m\right),$$

where $\equiv$ denotes congruence modulo t^{m+1}. Consequently, the coefficient of c_m in the m-th Todd class Todd$_m$ is $\sum_{j=1}^{m} \beta_j^m$.

We can compute this as follows. Consider the function $f(t) = \log(\mathrm{td}(t))$. It has an expression modulo t^{m+1} as

$$f(t) = \sum_{j=1}^{m} \log(1 + \beta_j t) = \sum_{j=1}^{m} \sum_{k=1}^{\infty} \frac{(-1)^{k-1}}{k} (\beta_j t)^k \mod t^{m+1}$$

On the other hand we have $f(-t) = f(t) - t$ so that in the expression of $f(t)$ as a power series in t, all the odd degree terms except t have coefficient 0. In particular, it follows that $\sum_{j=1}^{m} \beta_j^m$ is 0 whenever m is odd and $m > 1$.

To summarize, we have proved that the coefficient of the top Chern class in the Todd class of an m-dimensional (smooth projective) variety X is 0 if m is odd and > 1. Since this top Chern class can be identified with the Euler characteristic of X, we have the result that we claimed.

References

[Clo1] L. Clozel. Motifs et formes automorphes: applications du principe de fonctorialité. In *Automorphic forms, Shimura varieties, and L-functions, Vol. I (Ann Arbor, MI, 1988)*, volume 10 of *Perspect. Math.*, pages 77–159. Academic Press, Boston, MA, 1990.

[Clo2] L. Clozel. Représentations galoisiennes associées aux représentations automorphes autoduales de GL(n). *Inst. Hautes Études Sci. Publ. Math.* (1991), 97–145.

[CHL] L. Clozel, M. Harris, and J.-P. Labesse, Construction of automorphic Galois representations, I. On the stabilization of the trace formula. *Stab. Trace Formula Shimura Var. Arithm. Appl.* **1**, Int. Press, Somerville, MA, 2011, pages 497–527.

[CH] S. Cynk and K. Hulek. Higher-dimensional modular Calabi-Yau manifolds. *Canad. Math. Bull.* **50**(2007), 486–503. http://dx.doi.org/10.4153/CMB-2007-049-9

[CS] S. Cynk and M. Schütt. Generalised Kummer constructions and Weil restrictions. *J. Number Theory* **129**(2009), 1965–1975. http://dx.doi.org/10.1016/j.jnt.2008.09.010

[Del] P. Deligne. Formes modulaires et reprsentations ℓ-adiques. *Séminaire Bourbaki 1968/69* **355**(1971), 139–172.

[DMOS] P. Deligne, J. S. Milne, Arthur Ogus, and Kuang-yen Shih. *Hodge cycles, motives, and Shimura varieties*, volume 900 of *Lecture Notes in Mathematics*. Springer-Verlag, Berlin, 1982.

[Die] L. Dieulefait. On the modularity of rigid Calabi-Yau threefolds: Epilogue. *Proceedings of the trimester on Diophantine Equations at the Hausdorff Institute* (2010), 1–7.

[Edi] B. Edixhoven. Introduction, main results, context. In *Computational aspects of modular forms and Galois representations*, volume 176 of *Ann. of Math. Stud.*, pages 1–27. Princeton Univ. Press, Princeton, NJ, 2011.

[ES] N. D. Elkies and M. Schütt. Modular forms and K3 surfaces. *Adv. Math.* **240**(2013), 106–131. http://dx.doi.org/10.1016/j.aim.2013.03.008

[GKY] F. Q. Gouvêa, I. Kiming, and N. Yui. Quadratic twists of rigid Calabi-Yau threefolds over $\mathbb{Q}$. In *Arithmetic and geometry of K3 surfaces and Calabi-Yau threefolds*, volume 67 of *Fields Inst. Commun.*, pages 517–533. Springer, New York, 2013. http://dx.doi.org/10.1007/978-1-4614-6403-7_20

[GY] F. Q. Gouvêa and N. Yui. Rigid Calabi-Yau threefolds over $\mathbb{Q}$ are modular. *Expo. Math.* **29**(2011), 142–149. http://dx.doi.org/10.1016/j.exmath.2010.09.001

[Gro] B. H. Gross. *Arithmetic on elliptic curves with complex multiplication*, volume 776 of *Lecture Notes in Mathematics*. Springer, Berlin, 1980. With an appendix by B. Mazur.

[KW] C. Khare and J.-P. Wintenberger. Serre's modularity conjecture. I. *Invent. Math.* **178**(2009), 485–504. http://dx.doi.org/10.1007/s00222-009-0205-7

[Kim] H. H. Kim. Functoriality for the exterior square of GL_4 and the symmetric fourth of GL_2. *J. Amer. Math. Soc.* **16**(2003), 139–183 (electronic). With Appendix 1 by Dinakar Ramakrishnan and Appendix 2 by Kim and Peter Sarnak.

[KS] H. H. Kim and F. Shahidi. Functorial products for $\mathrm{GL}_2 \times \mathrm{GL}_3$ and the symmetric cube for GL_2. *Ann. of Math. (2)* **155**(2002), 837–893. With an appendix by Colin J. Bushnell and Guy Henniart.

[Kis] M. Kisin. Modularity of 2-adic Barsotti-Tate representations. *Invent. Math.* **178**(2009), 587–634. http://dx.doi.org/10.1007/s00222-009-0207-5

[Mey] C. Meyer. *Modular Calabi–Yau threefolds*. Fields Institute Monographs, 22. American Mathematical Society, Providence, RI, 2005.

[PR] K. Paranjape and D. Ramakrishnan. Quotients of E^n by $\mathfrak{A}_{n+1}$ and Calabi-Yau manifolds. In *Algebra and number theory*, pages 90–98. Hindustan Book Agency, Delhi, 2005.

[Ram] D. Ramakrishnan. Modularity of the Rankin-Selberg L-series, and multiplicity one for SL(2). *Ann. of Math. (2)* **152**(2000), 45–111.

[RS] D. Ramakrishnan and F. Shahidi. Siegel modular forms of genus 2 attached to elliptic curves. *Math. Res. Lett.* **14**(2007), 315–332.

[Roh] D. E. Rohrlich. On the L-functions of canonical Hecke characters of imaginary quadratic fields. *Duke Math. J.* **47**(1980), 547–557. http://projecteuclid.org/euclid.dmj/1077314180

[Sch1] C. Schoen. On fiber products of rational elliptic surfaces with section. *Math. Z.* **197**(1988), 177–199. http://dx.doi.org/10.1007/BF01215188

[Sch2] A. J. Scholl. Motives for modular forms. *Invent. Math.* **100**(1990), 419–430. http://dx.doi.org/10.1007/BF01231194

[YL] N. Yui and J. D. Lewis, editors. *Calabi-Yau varieties and mirror symmetry*, volume 38 of *Fields Institute Communications*, Providence, RI, 2003. American Mathematical Society.

[YYL] N. Yui, S.-T. Yau, and J. D. Lewis, editors. *Mirror symmetry. V*, volume 38 of *AMS/IP Studies in Advanced Mathematics*, Providence, RI, 2006. American Mathematical Society.

15

Derivative of symmetric square p-adic L-functions via pull-back formula

*Giovanni Rosso**

In this paper we recall the method of Greenberg and Stevens to calculate derivatives of p-adic L-functions using deformations of Galois representation and we apply it to the symmetric square of a modular form Steinberg at p. Under certain hypotheses on the conductor and the Nebentypus, this proves a conjecture of Greenberg and Benois on trivial zeros.

Contents

15.1 Introduction

Let M be a motive over $\mathbb{Q}$ and suppose that it is pure of weight zero and irreducible. We suppose also that $s = 0$ is a critical integer à la Deligne.

Fix a prime number p and let V be the p-adic representation associated to M. We fix once and for all an isomorphism $\mathbb{C} \cong \mathbb{C}_p$. If V is semistable, it

*PhD Fellowship of Fund for Scientific Research - Flanders, partially supported by a JUMO grant from KU Leuven (Jumo/12/032), a ANR grant (ANR-10-BLANC 0114 ArShiFo) and a NSF grant (FRG DMS 0854964).

Arithmetic and Geometry, ed. Luis Dieulefait *et al.* Published by Cambridge University Press.

is conjectured that for each regular submodule D [Ben11, §0.2] there exists a p-adic L-function $L_p(V, D, s)$. It is supposed to interpolate the special values of the L-function of M twisted by finite-order characters of $1 + p\mathbb{Z}_p$ [PR95], multiplied by a corrective factor (to be thought of as a part of the local epsilon factor at p) which depends on D. In particular, we expect the following interpolation formula at $s = 0$;

$$L_p(V, D, s) = E(V, D)\frac{L(V, 0)}{\Omega(V)},$$

for $\Omega(V)$ a complex period and $E(V, D)$ some Euler type factors which conjecturally have to be removed in order to permit p-adic interpolation (see [Ben11, §2.3.2] for the case when V is crystalline). It may happen that certain of these Euler factors vanish. In this case the connection with what we are interested in, the special values of the L-function, is lost. Motivated by the seminal work of Mazur-Tate-Teitelbaum [MTT86], Greenberg in the ordinary case [Gre94] and Benois [Ben11] have conjectured the following;

Conjecture 15.1. *[Trivial zeros conjecture] Let e be the number of Euler-type factors of $E(V, D)$ which vanish. Then the order of zeros at $s = 0$ of $L_p(V, D, s)$ is e and*

$$\lim_{s\to 0} \frac{L_p(V, D, s)}{s^e} = \mathcal{L}(V, D)E^*(V, D)\frac{L(V, 0)}{\Omega(V)} \tag{15.1}$$

for $E^*(V, D)$ the non-vanishing factors of $E(V, D)$ and $\mathcal{L}(V, D)$ a non-zero number called the $\mathcal{L}$-invariant.

There are many different ways in which the $\mathcal{L}$-invariant can be defined. A first attempt at such a definition could that of an *analytic* $\mathcal{L}$-invariant

$$\mathcal{L}^{\mathrm{an}}(V, D) = \frac{\lim_{s\to 0} \frac{L_p(V,D,s)}{s^e}}{E^*(V, D)\frac{L(V,0)}{\Omega(V)}}.$$

Clearly, with this definition, the above conjecture reduces to the statement on the order of $L_p(V, D, s)$ at $s = 0$ and the non-vanishing of the $\mathcal{L}$-invariant.

In [MTT86] the authors give a more arithmetic definition of the $\mathcal{L}$-invariant for an elliptic curve, in terms of an *extended* regulator on the *extended* Mordell-Weil group. The search for an intrinsic, Galois theoretic interpretation of this error factor led to the definition of the *arithmetic* $\mathcal{L}$-invariant $\mathcal{L}^{\mathrm{ar}}(V, D)$ given by Greenberg [Gre94] (resp. Benois [Ben11]) in the ordinary case (resp. semistable case) using Galois cohomology (resp. cohomology of (φ, Γ)-module).

For two-dimensional Galois representations many more definitions have been proposed and we refer to [Col05] for a detailed exposition.

When the p-adic L-function can be constructed using an Iwasawa cohomology class and the big exponential [PR95], one can use the machinery developed in [Ben12, §2.2] to prove formula (15.1) with $\mathcal{L} = \mathcal{L}^{\mathrm{al}}$. Unluckily, it is a very hard problem to construct classes in cohomology which are related to special values. Kato's Euler system has been used in this way in [Ben12] to prove many instances of Conjecture 15.1 for modular forms. It might be possible that the construction of Lei-Loeffler-Zerbes [LLZ12] of an Euler system for the Rankin product of two modular forms could produce such Iwasawa classes for other Galois representations; in particular, for $V = \mathrm{Sym}^2(V_f)(1)$, where V_f is the Galois representation associated to a weight two modular form (see also [PR98] and the upcoming work of Dasgupta on Greenberg's conjecture for the symmetric square p-adic L-functions of ordinary forms). We also refer the reader to [BDR14].

We present in this paper a different method which has already been used extensively in many cases and which we think to be more easy to apply at the current state: the method of Greenberg and Stevens [Gre94]. Under certain hypotheses which we shall state in the next section, it allows us to calculate the derivative of $L_p(V, D, s)$.

The main ingredient of their method is the fact that V can be p-adically deformed in a one-dimensional family. For example, modular forms can be deformed in a Hida-Coleman family. We have decided to present this method because of the recent developments on families of automorphic forms [AIP12, Bra13] have opened the door to the construction of families of p-adic L-function in many different settings. For example, we refer to the ongoing PhD thesis of Z. Liu on p-adic L-functions for Siegel modular forms.

Consequently, we expect that one could prove many new instances of Conjecture 15.1.

In Section 15.4 we shall apply this method to the case of the symmetric square of a modular form which is Steinberg at p. The theorem which will be proved is the following:

Theorem 15.2. *Let f be a modular form of trivial Nebentypus, weight k_0 and conductor Np, N squarefree and prime to p. If $p = 2$, suppose then $k = 2$. Then Conjecture 15.1 (up to the non-vanishing of the $\mathcal{L}$-invariant) is true for $L_p(s, \mathrm{Sym}^2(f))$.*

In this case there is only one choice for the regular submodule D and the trivial zero appears at $s = k_0 - 1$. This theorem generalizes [Ros13b, Theorem 1.3] in two ways: we allow $p = 2$ in the ordinary case and when $p \neq 2$ we do not require N to be even. In particular, we cover the case of the symmetric square 11-adic L-function for the elliptic curve $X_0(11)$ and of the 2-adic L-function for the symmetric square of $X_0(14)$.

In the ordinary case (i.e. for $k = 2$), the proof is completly independent of [Ros13b] as we can construct directly a two-variable p-adic L-function. In the finite slope case we can not construct a two-variable function with the method described below and consequently we need the two-variable p-adic L-function of [Ros13b, Theorem 4.14], which has been constructed only for $p \neq 2$.

In the ordinary setting, the same theorem (but with the hypothesis at 2) has been proved by Greenberg and Tilouine (unpublished). The importance of this formula for the proof of the Greenberg-Iwasawa Main Conjecture [Urb06] has been put in evidence in [HTU97].

We improve on the result of [Ros13b] making use of a different construction of the p-adic L-function, namely that of [BS00]. We express the complex L-function using Eisenstein series for GSp_4, a pullback formula to the Igusa divisor and a double Petersson product. We are grateful to É. Urban for having suggested this approach to us. In Section 15.3 we briefly recall the theory of Siegel modular forms and develop a theory of p-adic modular forms for $\mathrm{GL}_2 \times \mathrm{GL}_2$ necessary for the construction of p-adic families of Eisenstein series.

Acknowledgements. This paper is part of the author's PhD thesis and the author is very grateful to his director J. Tilouine for the constant guidance and the attention given. Thanks go to É. Urban for inviting the author to Columbia University (where this work has seen the day) and for the generosity with which he shared with his ideas and insights. R. Casalis is thanked for interesting discussions on Kähler differentials and J. Welliaveetil for useful remarks.

15.2 The method of Greenberg and Stevens

The aim of this section is to recall the method of Greenberg and Stevens [GS93] to calculate analytic $\mathcal{L}$-invariant. This method has been used successfully many other times [Mok09, Ros13a, Ros13b]. It is very robust and easily adaptable to many situations in which the expected order of the trivial zero is one. We also describe certain obstacles which occur while trying to apply this method to higher order zeros.

We let K be a p-adic local field, $\mathcal{O}$ its valuation ring and Λ the Iwasawa algebra $\mathcal{O}[[T]]$. Let V be a p-adic Galois representation as before.

We denote by $\mathcal{W}$ the rigid analytic space whose $\mathbb{C}_p$-points are $\mathrm{Hom}_{\mathrm{cont}}(\mathbb{Z}_p^*, \mathbb{C}_p^*)$. We have a map $\mathbb{Z} \to \mathcal{W}$ defined by $k \mapsto [k] : z \to z^k$.

Let us fix once and for all, if $p \neq 2$ (resp. $p = 2$) a generator u of $1 + p\mathbb{Z}_p$ (resp. $1+4\mathbb{Z}_2$) and a decomposition $\mathbb{Z}_p^* = \mu \times 1+p\mathbb{Z}_p$ (resp $\mathbb{Z}_2^* = \mu \times 1+4\mathbb{Z}_2$). Here $\mu = \mathbb{G}_m^{\text{tors}}(\mathbb{Z}_p)$. We have the following isomorphism of rigid spaces:

$$\mathcal{W} \cong \mathbb{G}_m^{\text{tors}}(\mathbb{Z}_p)^{\wedge} \times B(1, 1^-)$$
$$\kappa \mapsto (\kappa_{|\mu}, \kappa(u)),$$

where the first set has the discrete topology and the second is the rigid open unit ball around 1. Let $0 < r < \infty$, $r \in \mathbb{R}$, we define

$$\mathcal{W}(r) = \left\{ (\zeta, z) | \zeta \in \mathbb{G}_m^{\text{tors}}(\mathbb{Z}_p), |z-1| \leq p^{-r} \right\}.$$

We fix an integer h and we denote by $\mathcal{H}_h$ the algebra of h-admissible distributions over $1 + p\mathbb{Z}_p$ (or $1 + 4\mathbb{Z}_2$ if $p = 2$) with values in K. Here we take the definition of admissibility as in [Pan03, §3], so measures are one-admissible, i.e. $\Lambda \otimes \mathbb{Q}_p \cong \mathcal{H}_1$. The Mellin transform gives us a map

$$\mathcal{H}_h \to \text{An}(0),$$

where An(0) stands for the algebra of $\mathbb{Q}_p$-analytic, locally convergent functions around 0. If we see $\mathcal{H}_h$ as a subalgebra of the ring of formal series, this amounts to $T \mapsto u^s - 1$.

We suppose that we can construct a p-adic L-function for $L_p(s, V, D)$ and that it presents a single trivial zero.

We suppose also that V can be deformed in a p-adic family $V(\kappa)$. Precisely, we suppose that we are given an affinoid $\mathcal{U}$, finite over $\mathcal{W}(r)$. Let us write $\pi : \mathcal{U} \to \mathcal{W}(r)$. Let us denote by $\mathcal{I}$ the Tate algebra corresponding to $\mathcal{U}$. We suppose that $\mathcal{I}$ is integrally closed and that there exists a *big* Galois representation $V(\kappa)$ with values in $\mathcal{I}$ and a point $\kappa_0 \in \mathcal{U}$ such that $V = V(\kappa_0)$.

We define $\mathcal{U}^{\text{cl}}$ to be the set of $\kappa \in \mathcal{U}$ satisfying the following conditions:

- $\pi(\kappa) = [k]$, with $k \in \mathbb{Z}$,
- $V(\kappa)$ is motivic,
- $V(\kappa)$ is semistable as $G_{\mathbb{Q}_p}$-representation,
- $s = 0$ is a critical integer for $V(\kappa)$.

We make the following assumption on $\mathcal{U}^{\text{cl}}$.

(CI) For every $n > 0$, there are infinitely many κ in $\mathcal{U}^{\text{cl}}$ and $s \in \mathbb{Z}$ such that:

- $|\kappa - \kappa_0|_{\mathcal{U}} < p^{-n}$
- s critical for $V(\kappa)$,
- $s \equiv 0 \,(\text{mod}\, p)^n$.

This amounts to asking that the couples $(\kappa, [s])$ in $\mathcal{U} \times \mathcal{W}$ with s critical for $V(\kappa)$ accumulate at $(\kappa_0, 0)$.

We suppose that there is a global triangulation $D(\kappa)$ of the (φ, Γ)-module associated to $V(\kappa)$ [Liu13] and that this induces the regular submodule used to construct $L_p(s, V, D)$.

Under these hypotheses, it is natural to conjecture the existence of a two-variable p-adic L-function (depending on $D(\kappa)$) $L_p(\kappa, s) \in \mathcal{I}\hat{\otimes}\mathcal{H}_h$ interpolating the p-adic L-functions of $V(\kappa)$, for all κ in $\mathcal{U}^{\mathrm{cl}}$. Conjecturally [Pan94, Pot13], h should be defined solely in terms of the p-adic Hodge theory of $V(\kappa)$ and $D(\kappa)$.

We make two hypotheses on this p-adic L-function.

i) There exists a subspace of dimension one $(\kappa, s(\kappa))$ containing $(\kappa_0, 0)$ over which $L_p(\kappa, s(\kappa))$ vanishes identically.

ii) There exists an *improved* p-adic L-function $L_p^*(\kappa)$ in $\mathcal{I}$ such that $L_p(\kappa, 0) = E(\kappa)L_p^*(\kappa)$, for $E(\kappa)$ a non-zero element which vanishes at κ_0.

The idea is that *i)* allows us to express the derivative we are interested in in terms of the "derivative with respect to κ". The latter can be calculated using *ii)*. In general, we expect that $s(\kappa)$ is a simple function of $\pi(\kappa)$.

Let $\log_p(z) = -\sum_{n=1}^{\infty} \frac{(1-z)^n}{n}$, for $|z-1|_p < 1$ and

$$\mathrm{Log}_p(\kappa) = \frac{\log_p(\kappa(u^r))}{\log_p(u^r)},$$

for r any integer big enough.

For example, in [Gre94, Mok09] we have $s(\kappa) = \frac{1}{2}\mathrm{Log}_p(\pi(\kappa))$ and in [Ros13a, Ros13b] we have $s(\kappa) = \mathrm{Log}_p(\pi(\kappa)) - 1$. In the first case the line corresponds to the vanishing on the central critical line which is a consequence of the fact that the ε-factor is constant in the family. In the second case, the vanishing is due to a line of trivial zeros, as all the motivic specializations present a trivial zero.

The idea behind *ii)* is that the Euler factor which brings the trivial zero for V varies analytically along $V(\kappa)$ once one fixes the cyclotomic variable. This is often the case with one-dimensional deformations. If we allow deformations of V in more than one variable, it is unlikely that the removed Euler factors define p-analytic functions, due to the fact that *eigenvalues of the crystalline Frobenius do not vary p-adically* or equivalently, that the Hodge-Tate weights are not constant.

We now give the example of families of Hilbert modular forms. For simplicity of notation, we consider a totally real field F of degree d where p is split. Let $\mathbf{f}$ be a Hilbert modular form of weight $(k_1, \ldots, k_d)$ such that the parity of

k_i does not depend on i. We define $m = \max(k_i - 1)$ and $v_i = \frac{m+1-k_i}{2}$. We suppose that $\mathbf{f}$ is nearly-ordinary [Hid89] and let $\mathbf{F}$ be the only Hida family to which $\mathbf{f}$ belongs. For each p-adic place $\mathfrak{p}_i$ of F, the corresponding Hodge-Tate weights are $(v_i, m - v_i)$. This implies that the Fourier coefficient $a_{\mathfrak{p}_i}(\mathbf{F})$ is a p-adic analytic function only if it is divided by p^{v_i}. Unluckily, $a_{\mathfrak{p}_i}(\mathbf{F})$ is the number which appears in the Euler type factor of the evaluation formula for the p-adic L-function of $\mathbf{F}$ or $\mathrm{Sym}^2(\mathbf{F})$. This is why in [Mok09, Ros13a] the authors deal only with forms of parallel weight. It seems very hard to generalize the method of Greenberg and Stevens to higher order derivatives without new ideas.

It may happen that the Euler factor which brings the trivial zero for V is (locally) identically zero on the whole family; this is the case for the symmetric square of a modular form of prime-to-p conductor and more generally for the standard L-function of parallel weight Siegel modular forms of prime-to-p level. That's why in [Ros13a, Ros13b] and in this article we can deal only with forms which are Steinberg at p.

We have seen in the examples above that $s(\kappa)$ is a linear function of the weight. Consequently, one needs to evaluate the p-adic L-function $L_p(V(\kappa), s)$ at s which are big for the archimedean norm. When s is not a critical integer it is quite a hard problem to evaluate the p-adic L-function. This is why we have supposed (**CI**). It is not a hypothesis to be taken for granted. One example is the spinor L-function for genus two Siegel modular forms of any weight, which has only one critical integer.

The improved p-adic L-function is said so because $L_p^*(\kappa_0)$ is supposed to be exactly the special value we are interested in.

The rest of the section is devoted to make precise the expression "derive with respect to κ".

We recall some facts about differentials. We fix a Λ-algebra $\mathcal{I}_1$. We suppose that $\mathcal{I}_1$ is a DVR and a K-algebra. Let $\mathcal{I}_2$ be an integral domain and a local ring which is finite, flat and integrally closed over $\mathcal{I}_1$. We have the first fundamental sequence of Kähler differentials

$$\Omega_{\mathcal{I}_1/K} \otimes_{\mathcal{I}_1} \mathcal{I}_2 \to \Omega_{\mathcal{I}_2/K} \to \Omega_{\mathcal{I}_2/\mathcal{I}_1} \to 0.$$

Under the hypotheses above, we can write $I_2 = \frac{I_1[X]}{P(X)}$. Then, every K-linear derivation of $\mathcal{I}_1$ can be extended to a derivation of $\mathcal{I}_2$ and [Mat80, Theorem 57 ii)] ensures us that the first arrow is injective.

Let P_0 be the prime ideal of $\mathcal{I}$ corresponding to the point κ_0 and P the corresponding ideal in $\mathcal{O}(\mathcal{W}(r))$. We take $\mathcal{I}_1 = \mathcal{O}(\mathcal{W}(r))_P$ and $\mathcal{I}_2 = \mathcal{I}_{P_0}$. The assumption that $\mathcal{I}_2$ is integrally closed is equivalent to ask that $\mathcal{U}$ is smooth at κ_0. In many cases, we expect that $\mathcal{I}_1 \to \mathcal{I}_2$ is étale; under this hypothesis, we can appeal to the fact that locally convergent series are Henselian [Nag62, Theorem 45.5] to define a morphism $\mathcal{I}_2 \to \mathrm{An}(k_0)$ to the ring of meromorphic functions around k_0 which extends the natural inclusion of $\mathcal{I}_1$. Once this morphism is defined, we can derive elements of $\mathcal{I}_2$ as if they were locally analytic functions.

There are some cases in which this morphism is known not to be étale; for example, for certain weight one forms [Dim13, DG12] and some critical CM forms [Bel12, Proposition 1]. (Note that in these cases no regular submodule D exists.)

We would like to explain what we can do in the case when $\mathcal{I}_1 \to \mathcal{I}_2$ is not étale. We also hope that what we say will clarify the situation in the case where the morphism is étale.

We have that $\Omega_{\Lambda/K}$ is a free rank 1 Λ-module. Using the universal property of differentials

$$\mathrm{Hom}_{\Lambda}(\Omega_{\Lambda/K}, \Lambda) = \mathrm{Der}_K(\Lambda, \Lambda),$$

we shall say, by slight abuse of notation, that $\Omega_{\Lambda/K}$ is generated as Λ-module by the derivation $\frac{\mathrm{d}}{\mathrm{d}T}$. Similarly, we identify $\Omega_{\mathcal{I}_1/K}$ with the free $\mathcal{I}_1$-module generated by $\frac{\mathrm{d}}{\mathrm{d}T}$. As the first arrow in the first fundamental sequence is injective, there exists an element $d \in \Omega_{\mathcal{I}_2/K}$ (which we see as K-linear derivation from $\mathcal{I}_2$ to $\mathcal{I}_2$) which extends $\frac{\mathrm{d}}{\mathrm{d}T}$. If $\mathcal{I}_1 \to \mathcal{I}_2$ is étale, then $\Omega_{\mathcal{I}_2/\mathcal{I}_1} = 0$ and the choice of d is unique.

Under the above hypotheses, we can then define a new analytic $\mathcal{L}$-invariant (which a priori depends on the deformation $V(\kappa)$ and d) by

$$\begin{aligned}\mathcal{L}_d^{\mathrm{an}}(V) &:= -\log_p(u)\, d(s(\kappa))^{-1} d(E(\kappa))\Big|_{\kappa=\kappa_0} \\ &= -\, d(s(\kappa))^{-1} \frac{d(L_p(0,\kappa))}{L_p^*(\kappa)}\Bigg|_{\kappa=\kappa_0}.\end{aligned}$$

We remark that with the notation above we have $\log_p(u)\frac{\mathrm{d}}{\mathrm{d}T} = \frac{\mathrm{d}}{\mathrm{d}s}$. We apply d to $L_p(\kappa, s(\kappa)) = 0$ to obtain

$$\begin{aligned}0 &= d(L_p(\kappa, s(\kappa))) \\ &= \log_p(u)^{-1} \frac{\mathrm{d}}{\mathrm{d}s} L_p(\kappa, s)\Big|_{s=s(\kappa)} d(s(\kappa)) + d(L_p(\kappa, s))\big|_{s=s(\kappa)}.\end{aligned}$$

Evaluating at $\kappa = \kappa_0$ we deduce

$$\frac{\mathrm{d}}{\mathrm{d}s} L_p(V, s)\Big|_{s=0} = \mathcal{L}_d^{\mathrm{an}}(V) L_p^*(\kappa_0)$$

and consequently

$$\mathcal{L}^{\mathrm{an}}(V) = \mathcal{L}^{\mathrm{an}}_{d}(V).$$

In the cases in which $\mathcal{L}^{\mathrm{al}}$ has been calculated, namely symmetric powers of Hilbert modular forms, it is expressed in terms of the logarithmic derivative of Hecke eigenvalues at p of certain finite slope families [Ben10, HJ13, Hid06, Mok12]. Consequently, the above formula should allow us to prove the equality $\mathcal{L}^{\mathrm{an}} = \mathcal{L}^{\mathrm{al}}$.

Moreover, the fact that the $\mathcal{L}$-invariant is a derivative of a non-constant function shows that $\mathcal{L}^{\mathrm{an}} \neq 0$ outside a codimension 1 subspace of the weight space. In this direction, positive results for a given V have been obtained only in the cases of a Hecke character of a quadratic imaginary field and of an elliptic curve with p-adic uniformization, using a deep theorem in transcendent number theory [BSDGP96].

15.3 Eisenstein measures

In this section we first fix the notation concerning genus two Siegel forms. We then recall a normalization of certain Eisenstein series for $\mathrm{GL}_2 \times \mathrm{GL}_2$ and develop a theory of p-adic families of modular forms (of parallel weight) on $\mathrm{GL}_2 \times \mathrm{GL}_2$. Finally we construct two Eisenstein measures which will be used in the next section to construct two p-adic L-functions.

15.3.1 Siegel modular forms

We now recall the basic theory of Siegel modular forms. We follow closely the notation of [BS00] and we refer to the first section of *loc. cit.* for more details. Let us denote by $\mathbb{H}_1$ the complex upper half-plane and by $\mathbb{H}_2$ the Siegel space for GSp_4. We have explicitly

$$\mathbb{H}_2 = \left\{ Z = \left(\begin{array}{cc} z_1 & z_2 \\ z_3 & z_4 \end{array} \right) \middle| Z^t = Z \text{ and } \mathrm{Im}(Z) > 0 \right\}.$$

It has a natural action of $\mathrm{GSp}_4(\mathbb{R})$ via fractional linear transformation; for any $M = \left(\begin{array}{cc} A & B \\ C & D \end{array} \right)$ in $\mathrm{GSp}_4^+(\mathbb{R})$ and Z in $\mathbb{H}_2$ we define

$$M(Z) = (AZ + B)(CZ + D)^{-1}.$$

Let $\Gamma = \Gamma_0^{(2)}(N)$ be the congruence subgroup of $\mathrm{Sp}_4(\mathbb{Z})$ of matrices whose lower block C is congruent to 0 modulo N. We consider the space $M_k^{(2)}(N, \phi)$ of scalar Siegel forms of weight k and Nebentypus ϕ:

$$\Big\{F : \mathbb{H}_2 \to \mathbb{C} \Big| F(M(Z))(CZ + D)^{-k} \\ = \phi(M)F(Z) \ \forall M \in \Gamma, f \text{ holomorphic}\Big\}.$$

Each F in $M_k(\Gamma, \phi)$ admits a Fourier expansion

$$F(Z) = \sum_T a(T)e^{2\pi i \mathrm{tr}(TZ)},$$

where $T = \begin{pmatrix} T_1 & T_2 \\ T_2 & T_4 \end{pmatrix}$ ranges over all matrices T positive and semi-defined, with T_1, T_4 integer and T_2 half-integer.

We have two embeddings (of algebraic groups) of SL_2 in Sp_4:

$$\mathrm{SL}_2^{\uparrow}(R) = \left\{ \begin{pmatrix} a & 0 & b & 0 \\ 0 & 1 & 0 & 0 \\ c & 0 & d & 0 \\ 0 & 0 & 0 & 1 \end{pmatrix} \middle| \begin{pmatrix} a & b \\ c & d \end{pmatrix} \in \mathrm{SL}_2(R) \right\},$$

$$\mathrm{SL}_2^{\downarrow}(R) = \left\{ \begin{pmatrix} 1 & 0 & 0 & 0 \\ 0 & a & 0 & b \\ 0 & 0 & 1 & 0 \\ 0 & c & 0 & d \end{pmatrix} \middle| \begin{pmatrix} a & b \\ c & d \end{pmatrix} \in \mathrm{SL}_2(R) \right\}.$$

We can embed $\mathbb{H}_1 \times \mathbb{H}_1$ in $\mathbb{H}_2$ in the following way

$$(z_1, z_4) \mapsto \begin{pmatrix} z_1 & 0 \\ 0 & z_4 \end{pmatrix}.$$

If γ belongs to $\mathrm{SL}_2(\mathbb{R})$, we have

$$\gamma^{\uparrow} \begin{pmatrix} z_1 & 0 \\ 0 & z_4 \end{pmatrix} = \begin{pmatrix} \gamma(z_1) & 0 \\ 0 & z_4 \end{pmatrix},$$

$$\gamma^{\downarrow} \begin{pmatrix} z_1 & 0 \\ 0 & z_4 \end{pmatrix} = \begin{pmatrix} z_1 & 0 \\ 0 & \gamma(z_4) \end{pmatrix}.$$

Consequently, evaluation at $z_2 = 0$ gives us a map

$$M_k^{(2)}(N, \phi) \hookrightarrow M_k(N, \phi) \otimes_{\mathbb{C}} M_k(N, \phi),$$

where $M_k(N, \phi)$ denotes the space of elliptic modular forms of weight k, level N and Nebentypus ϕ.

This also induces a closed embedding of two copies of the modular curve in the Siegel threefold. We shall call its image the Igusa divisor. On points, it corresponds to abelian surfaces which decompose as the product of two elliptic curves.

We consider the following differential operators on $\mathbb{H}_2$:

$$\partial_1 = \frac{\partial}{\partial z_1}, \quad \partial_2 = \frac{1}{2}\frac{\partial}{\partial z_2}, \quad \partial_4 = \frac{\partial}{\partial z_4}.$$

We define

$$\begin{aligned}\mathfrak{D}_l =& z_2\left(\partial_1\partial_4 - \partial_2^2\right) - \left(l - \frac{1}{2}\right)\partial_2,\\ \mathfrak{D}_l^s =& \mathfrak{D}_{l+s-1} \circ \ldots \circ \mathfrak{D}_l,\\ \mathring{\mathfrak{D}}_l^s =& \mathfrak{D}_l^s|_{z_2=0}.\end{aligned}$$

The importance of $\mathring{\mathfrak{D}}_l^s$ is that it preserves holomorphicity.

Let $I = \begin{pmatrix} T_1 & T_2 \\ T_2 & T_4 \end{pmatrix}$. We define $\mathfrak{b}_l^s(I)$ to be the only homogeneous polynomial in the indeterminates T_1, T_2, T_4 of degree s such that

$$\mathring{\mathfrak{D}}_l^s e^{T_1z_1+2T_2z_2+T_4z_4} = \mathfrak{b}_l^s(I)e^{T_1z_1+T_4z_4}.$$

We need to know a little bit more about the polynomial $\mathfrak{b}_l^s(I)$. Let us write

$$\mathfrak{D}_l^s e^{T_1z_1+2T_2z_2+T_4z_4} = P_l^s(z_2, I)e^{T_1z_1+2T_2z_2+T_4z_4},$$

we have

$$\begin{aligned}\mathfrak{D}_l^{s+1}e^{\mathrm{tr}(ZI)} =& \Bigg[P_{l+s}^1(z_2, I)P_l^s(z_2, I)\\ &-\left(z_2\frac{1}{4}\frac{\partial^2}{\partial z_2^2} + \left(l+s-\frac{1}{2}\right)\frac{1}{2}\frac{\partial}{\partial z_2}\right)P_l^s(z_2, I)\Bigg]e^{tr(ZI)}.\end{aligned} \tag{15.2}$$

We obtain easily

$$\begin{aligned}P_l^1(z_2, I) =& \left(z_2\det(I) - \left(l - \frac{1}{2}\right)T_2\right),\\ \mathfrak{b}_l^s(I) =& P_l^s(0, I).\end{aligned} \tag{15.3}$$

We have

$$\begin{aligned}\mathfrak{b}_l^{s+1}(I) =& \mathfrak{b}_l^s(I)\mathfrak{b}_{l+s}^1(I) + (l+s-1/2)\partial_2 P_l^s(z_2, I)\\ =& (l+s-1/2)(-T_2\mathfrak{b}_l^s(I) + \partial_2 P_l^s(z_2, I)|_{z_2=0}).\end{aligned}$$

Let $J = \{j_1, j_2, j_4\}$. We shall write ∂^J resp. z^J for $\partial_1^{j_1}\partial_2^{j_2}\partial_4^{j_4}$ resp. $z_1^{j_1}z_2^{j_2}z_4^{j_4}$. We can write [BS00, page 1381]

$$\mathring{\mathfrak{D}}_l^s = \sum_{j_1+j_2+j_4=s} c_l^J \partial^J, \tag{15.4}$$

where $c_l^J = \frac{\mathring{\mathfrak{D}}_l^s(z^J)}{\partial^J(z^J)}$. We have easily:

$$\begin{aligned}\partial^J(z^J) &= j_1!j_2!j_4!2^{-j_2},\\ \partial^J(e^{\mathrm{tr}(ZI)}) &= T_1^{j_1}T_2^{j_2}T_4^{j_4}(e^{\mathrm{tr}(ZI)}),\end{aligned}$$

We pose

$$\begin{aligned}c_l^s &:= c_l^{0,s,0},\\ (-1)^s c_l^s &= \prod_{i=1}^{s}\left(l-1+s-\frac{i}{2}\right) = 2^{-s}\frac{(2l-2+2s-1)!}{(2l-2+s-1)!}.\end{aligned}$$

Consequently, for $L \mid T_1, T_4$ and for any positive integer d, we obtain

$$4^s\mathfrak{b}_{t+1}^s(I) \equiv (-1)^s 4^s \sum_{j_1+j_4<d} c_{t+1}^J T_1^{j_1}T_2^{s-j_1-j_4}T_4^{j_4} \pmod{L}^d.$$

15.3.2 Eisenstein series

The aim of this section is to recall certain Eisenstein series which can be used to construct the p-adic L-functions, as in [BS00]. In *loc. cit.* the authors consider certain Eisenstein series for GSp_{4g} whose pullback to the Igusa divisor is a **holomorphic** Siegel modular form.

We now fix a (parallel weight) Siegel modular form f for GSp_{2g}. We write the standard L-function of f as a double Petersson product between f and these Eisenstein series (see Proposition 15.4). When $g = 1$, the standard L-function of f coincides, up to a twist, with the symmetric square L-function of f we are interested in.

In general, for an algebraic group bigger than GL_2 it is quite hard to find the normalization of the Eisenstein series which maximizes, in a suitable sense, the p-adic behavior of its Fourier coefficients. In [BS00, §2] the authors develop a *twisting method* which allow them to define Eisenstein series whose Fourier coefficients satisfy Kummer's congruences when the character associated with the Eisenstein series varies p-adically. This is the key for their construction of the one variable (cyclotomic) p-adic L-function and of our two-variable p-adic L-function.

When the character is trivial modulo p there exists a simple relation between the twisted and the not-twisted Eisenstein series [BS00, §6 Appendix]. To construct the improved p-adic L-function, we shall simply interpolate the not-twisted Eisenstein series.

Let us now recall these Fourier developments.

We fix a weight k, an integer N prime to p and a Nebentypus ϕ. Let f be an eigenform in $M_k(Np, \phi)$, of finite slope for the Hecke operator U_p. We write $N = N_{\mathrm{ss}} N_{\mathrm{nss}}$, where N_{ss} (resp. N_{nss}) is divisible by all primes $q \mid N$ such that $U_q f = 0$ (resp. $U_q f \neq 0$). Let R be an integer coprime with N and p and N_1 a positive integer such that $N_{\mathrm{ss}} \mid N_1 \mid N$. We fix a Dirichlet character χ modulo $N_1 R p$ which we write as $\chi_1 \chi' \varepsilon_1$, with χ_1 defined modulo N_1, χ' primitive modulo R and ε_1 defined modulo p. We shall explain after Proposition 15.4 why we introduce χ_1.

Let $t \geq 1$ be an integer and $\mathbb{F}^{t+1}\left(w, z, R^2N^2p^{2n}, \phi, u\right)^{(\chi)}$ be the twisted Eisenstein series of [BS00, (5.3)]. We define

$$\mathcal{H}'_{L,\chi}(z, w) := L(t+1+2s, \phi\chi)\mathring{\mathfrak{D}}^s_{t+1}\left(\mathbb{F}^{t+1}\left(w, z, R^2N^2p^{2n}, \phi, u\right)^{(\chi)}\right) |^z U_{L^2} |^w U_{L^2}$$

for s a non-negative integer and $p^n \mid L$, with L a p-power. It is a form for $\Gamma_0(N^2R^2p) \times \Gamma_0(N^2R^2p)$ of weight $t+1+s$.

We shall sometimes choose $L = 1$ and in this case the level is $N^2R^2p^{2n}$.

For any prime number q and matrices I as in the previous section, let $B_q(X, I)$ be the polynomial of degree at most 1 of [BS00, Proposition 5.1]. We pose

$$B(t) = (-1)^{t+1}\frac{2^{1+2t}}{\Gamma(3/2)}\pi^{\frac{5}{2}}.$$

We deduce easily from [BS00, Theorem 7.1] the following theorem

Theorem 15.3. *The Eisenstein series defined above has the following Fourier development;*

$$\mathcal{H}'_{L,\chi}(z, w)|_{u=\frac{1}{2}-t} = B(t)(2\pi i)^s G(\chi) \sum_{T_1 \geq 0} \sum_{T_4 \geq 0} \left(\sum_I \mathfrak{b}^s_{t+1}(I)(\chi)^{-1}(2T_2) \sum_{G \in \mathrm{GL}_2(\mathbb{Z}) \backslash \mathbf{D}(I)} (\phi\chi)^2(\det(G))|\det(G)|^{2t-1} L(1-t, \sigma_{-\det(2I)}\phi\chi) \prod_{q \mid \det(2G^{-t}IG^{-1})} B_q\left(\chi\phi(q)q^{t-2}, G^{-t}IG^{-1}\right)\right) e^{2\pi i(T_1 z + T_2 w)},$$

where the sum over I runs along the matrices $\begin{pmatrix} L^2T_1 & T_2 \\ T_2 & L^2T_4 \end{pmatrix}$ *positive definite and with* $2T_2 \in \mathbb{Z}$, *and*

$$\mathbf{D}(I) = \left\{ G \in M_2(\mathbb{Z}) | G^{-t} I G^{-1} \text{ is a half-integral symmetric matrix} \right\}.$$

Proof. The only difference from *loc. cit.* is that we do not apply $\left| \begin{pmatrix} 1 & 0 \\ 0 & N^2 S \end{pmatrix} \right.$. □

In fact, contrary to [BS00], we prefer to work with $\Gamma_0(N^2S)$ and not with the opposite congruence subgroup.

In particular each sum over I is finite because I must have positive determinant. Moreover, we can rewrite it as a sum over T_2, with $(2T_2, p) = 1$ and $T_1T_4 - T_2^2 > 0$.

It is proved in [BS00, Theorem 8.5] that (small modifications of) these functions $\mathcal{H}'_{L,\chi}(z, w)$ satisfy Kummer's congruences. The key fact is what they call the twisting method [BS00, (2.18)]; the Eisenstein series $\mathbb{F}^{t+1}\left(w, z, R^2N^2p^{2n}, \phi, u\right)^{(\chi)}$ are obtained weighting $\mathbb{F}^{t+1}\left(w, z, R^2N^2p^{2n}, \phi, u\right)$ with respect to χ over integral matrices modulo NRp^n. To ensure these Kummer's congruences, even when p does not divide the conductor of χ, the authors are forced to consider χ of level divisible by p. Using nothing more than Tamagawa's rationality theorem for GL_2, they find the relation [BS00, (7.13')]:

$$\begin{aligned} &\mathbb{F}^{t+1}\left(w, z, R^2N^2p, \phi, u\right)^{(\chi)} \\ &\quad = \mathbb{F}^{t+1}\left(w, z, R^2N^2p, \phi, u\right)^{(\chi_1\chi')} \left| \left(\mathrm{id} - p \begin{pmatrix} 1 & p \\ 0 & 1 \end{pmatrix} \right) \right. . \end{aligned}$$

So the Eisenstein series we want to interpolate to construct the improved p-adic L-function is

$$\mathcal{H}'^{*}_{L,\chi'}(z, w) := L(t + 1 + 2s, \phi\chi)\mathring{\mathfrak{D}}^s_{t+1}\left(\mathbb{F}^{t+1}\left(w, z, R^2N^2p, \phi, u\right)^{(\chi_1\chi')}\right).$$

In what follows, we shall specialize $t = k - k_0 + 1$ (for k_0 the weight of the form in the theorem of the introduction) to construct the *improved* one variable p-adic L-function.

For each prime q, let us denote by α_q and β_q the roots of the Hecke polynomial at q associated to f. We define

$$D_q(X) := (1 - \alpha_q^2 X)(1 - \alpha_q\beta_q X)(1 - \beta_q^2 X).$$

For each Dirichlet character χ we define

$$\mathcal{L}(s, \mathrm{Sym}^2(f), \chi) := \prod_q D_q(\chi(q)q^{-s})^{-1}.$$

This L-function differs from the motivic L-function $L(s, \mathrm{Sym}^2(\rho_f) \otimes \chi)$ by a finite number of Euler factors at prime dividing N. We conclude with the integral formulation of $\mathcal{L}(s, \mathrm{Sym}^2(f), \chi)$ [BS00, Theorem 3.1, Proposition 7.1 (7.13)].

Proposition 15.4. *Let f be a form of weight k, Nebentypus ϕ. We put $t + s = k - 1$, $s_1 = \frac{1}{2} - t$ and $\mathcal{H}' = \mathcal{H}'_{1,\chi}(z, w)|_{u=s_1}$; we have*

$$\begin{aligned}
&\left\langle f(w) \middle| \begin{pmatrix} 0 & -1 \\ N^2 p^{2n} R^2 & 0 \end{pmatrix}, \mathcal{H}' \right\rangle_{N^2 p^{2n} R^2} \\
&\quad = \frac{\Omega_{k,s}(s_1) p_{s_1}(t+1)}{\chi(-1) d_{s_1}(t+1)} (R N_1 p^n)^{s+3-k} \left(\frac{N}{N_1}\right)^{2-k} \times \\
&\quad \times \mathcal{L}(s+1, \mathrm{Sym}^2(f), \chi^{-1}) f(z) | U_{N^2/N_1^2}
\end{aligned}$$

for

$$\frac{p_{s_1}(t+1)}{d_{s_1}(t+1)} = \frac{c^s_{t+1}}{c^s_{\frac{3}{2}}} = \frac{\prod_{i=1}^s \left(s + t - \frac{i}{2}\right)}{\prod_{i=1}^s \left(s - \frac{i-1}{2}\right)},$$

$$\Omega_{k,s}\left(\frac{1}{2} - t\right) = 2^{2t}(-1)^{\frac{k}{2}} \pi \frac{\Gamma(k-t)\Gamma\left(k - t - \frac{1}{2}\right)}{\Gamma\left(\frac{3}{2}\right)}.$$

Proof. With the notation of [BS00, Theorem 3.1] we have $M = R^2 N^2 p^{2n}$ and $N = N_1 R p^n$. We have that $\frac{d_{s_1}(t+1)}{p_{s_1}(t+1)} \mathcal{H}'$ is the holomorphic projection of the Eisenstein series of [BS00, Theorem 3.1] (see [BS00, (1.30),(2.1),(2.25)]). The final remark to make is the following relation between the standard (or adjoint) L-function of f and the symmetric square one:

$$\mathcal{L}(1 - t, \mathrm{Ad}(f) \otimes \phi, \chi) = \mathcal{L}(k - t, \mathrm{Sym}^2(f), \chi). \qquad \square$$

The authors of [BS00] prefer to work with an auxiliary character modulo N to remove all the Euler factors at bad primes of f, but we do not want to do this. Still, we have to make some assumptions on the level χ. Suppose that there is a prime q dividing N such that $q \nmid N_1$ and $U_q f = 0$, then the above formula would give us zero. That is why we introduce the character χ_1 defined modulo a multiple of N_{ss}.

At the level of L-function this does not change anything as for $q \mid N_{ss}$ we have $D_q(X) = 1$.

For f as in Theorem 15.2 we can take $N_1 = 1$.

15.3.3 Families for $\mathrm{GL}_2 \times \mathrm{GL}_2$

The aim of this section is the construction of families of modular forms on two copies of the modular curves. Let us fix a tame level N, and let us denote by X the compactified modular curve of level $\Gamma_1(N) \cap \Gamma_0(p)$. For $n \geq 2$, we shall denote by $X(p^n)$ the modular curve of level $\Gamma_1(N) \cap \Gamma_0(p^n)$.

We denote by $X(v)$ the tube of ray p^{-v} of the ordinary locus. We fix a p-adic field K. We recall from [AIS12, Pil13] that, for v and r suitable, there exists an invertible sheaf ω^κ on $X(v) \times_K \mathcal{W}(r)$. This allows us to define families of overconvergent forms as

$$M_\kappa(N) := \varinjlim_{v} H^0(X(v) \times_K \mathcal{W}(r), \omega^\kappa).$$

We denote by $\omega^{\kappa,(2)}$ the sheaf on $X(v) \times_K X(v) \times_K \mathcal{W}(r)$ obtained by base change over $\mathcal{W}(r)$.

We define

$$\begin{aligned} M^{(2)}_{\kappa,v}(N) &:= H^0(X(v) \times_K X(v) \times_K \mathcal{W}(r), \omega^{\kappa,(2)}) \\ &= H^0(X(v) \times \mathcal{W}(r), \omega^\kappa) \hat{\otimes}_{\mathcal{O}(\mathcal{W}(r))} H^0(X(v) \times \mathcal{W}(r), \omega^\kappa); \\ M^{(2)}_\kappa(N) &:= \varinjlim_{v} M^{(2)}_{\kappa,v}(N). \end{aligned}$$

We believe that this space should correspond to families of Siegel modular forms [AIP12] of parallel weight restricted on the Igusa divisor, but for shortness of exposition we do not examine this now.

We have a correspondence C_p above $X(v)$ defined as in [Pil13, §4.2]. We define by fiber product the correspondence C_p^2 on $X(v) \times_K X(v)$ which we extend to $X(v) \times_K X(v) \times_K \mathcal{W}(r)$. This correspondence induces a Hecke operator $U_p^{\otimes^2}$ on $H^0(X(v) \times_K X(v) \times_K \mathcal{W}(r), \omega^{\kappa,(2)})$ which corresponds to $U_p \otimes U_p$. These are potentially orthonormalizable $\mathcal{O}(\mathcal{W}(r))$-modules [Pil13, §5.2] and $U_p^{\otimes^2}$ acts on these spaces as a completely continuous operator (or compact, in the terminology of [Buz07, §1]) and this allows us to write

$$M^{(2)}_\kappa(N)^{\leq \alpha} = \bigoplus_{\alpha_1 + \alpha_2 \leq \alpha} M_\kappa(N)^{\leq \alpha_1} \hat{\otimes}_{\mathcal{O}(\mathcal{W}(r))} M_\kappa(N)^{\leq \alpha_2}. \tag{15.5}$$

Here and in what follows, for A a Banach ring, M a Banach A-module and U a completely continuous operator on M, we write $M^{\leq\alpha}$ for the finite dimensional submodule of generalized eigenspaces associated to the eigenvalues of U of valuation smaller or equal than α. We write $\mathrm{Pr}^{\leq\alpha}$ for the corresponding projection.

We remark [Urb11, Lemma 2.3.13] that there exists $v > 0$ such that

$$M_{\kappa}^{(2)}(N)^{\leq\alpha} = M_{\kappa,v'}^{(2)}(N)^{\leq\alpha}$$

for all $0 < v' < v$. We define similarly $M_{\kappa}^{(2)}(Np^n)$.

We now use the above Eisenstein series to give examples of families. More precisely, we shall construct a two-variable measure (which will be used for the two-variable p-adic L-function in the ordinary case) and a one-variable measure (which will be used to construct the improved one variable L-function) without the ordinary assumption.

Let us fix $\chi = \chi_1\chi'\varepsilon_1$ as before. We suppose χ even. We recall the Kubota-Leopoldt p-adic L-function;

Theorem 15.5. *Let η be a even Dirichlet character. There exists a p-adic L-function $L_p(\kappa,\eta)$ satisfying for any integer $t \geq 1$ and finite-order character ε of $1+p\mathbb{Z}_p$*

$$L_p(\varepsilon(u)[t],\eta) = (1-(\varepsilon\omega^{-t}\eta)_0(p))L(1-t,\varepsilon\omega^{-t}\eta),$$

where η_0 stands for the primitive character associated to η. If η is not trivial then $L_p([t],\eta)$ is holomorphic. Otherwise, it has a simple pole at $[0]$.

We can consequently define a p-adic analytic function interpolating the Fourier coefficients of the Eisenstein series defined in the previous section; for any z in $\mathbb{Z}_p^*$, we define $l_z = \frac{\log_p(z)}{\log_p(u)}$. We define also

$$\begin{aligned} &a_{T_1,T_4,L}(\kappa,\kappa') \\ &= \Bigg(\sum_I \kappa(u_{2T_2})\chi^{-1}(2T_2) \sum_{G\in \mathrm{GL}_2(\mathbb{Z})\backslash \mathbf{D}(I)} (\phi\chi)^2(\det(G))|\det(G)|^{-1}\kappa'^2(u^{l_{|\det(G)|}}) \\ &\quad L_p(\kappa',\sigma_{-\det(2I)}\phi\chi) \prod_{q|\det(2G^{-t}IG^{-1})} B_q\left(\phi(q)\kappa'(u^{l_q})q^{-2}, G^{-t}IG^{-1}\right)\Bigg). \end{aligned}$$

If $p=2$, then $\chi^{-1}(2T_2)$ vanishes when T_2 is an integer, so the above above sum is only on half-integral T_2 and $\kappa(u_{2T_2})$ is a well-defined two-analytic function. Moreover, it is two-integral.

We recall that if $p^j \mid T_1, T_4$ we have [BS00, (1.21, 1.34)]

$$4^s \mathfrak{b}^s_{t+1}(I) \equiv (-1)^s 4^s c^s_{t+1} T_2^s \pmod{p}^d,$$

for $s = k - t - 1$. Consequently, if we define

$$\mathcal{H}_L(\kappa, \kappa') = \sum_{T_1 \geq 0} \sum_{T_4 \geq 0} a_{T_1,T_4,L}(\kappa[-1]\kappa'^{-1}, \kappa') q_1^{T_1} q_2^{T_4}$$

we have,

$$(-1)^s 2^s c^s_{t+1} A \mathcal{H}_L([k], \varepsilon[t]) \equiv 2^s \mathcal{H}'_{L,\chi\varepsilon\omega^{-s}}(z, w) \pmod{L}^r,$$

with

$$A = A(t, k, \varepsilon) = B(t)(2\pi i)^s G\left(\chi\varepsilon\omega^{-s}\right).$$

We have exactly as in [Pan03, Definition 1.7] the following lemma;

Lemma 15.6. *There exists a projector*

$$\mathrm{Pr}^{\leq\alpha}_{\infty} : \bigcup_n M^{(2)}_{\kappa}(Np^n) \to M^{(2)}_{\kappa}(N)^{\leq\alpha}$$

which on $M^{(2)}_{\kappa}(Np^n)$ *is* $(U_p^{\otimes 2})^{-i} \mathrm{Pr}^{\leq\alpha} (U_p^{\otimes 2})^i$, *independent of* $i \geq n$.

When $\alpha = 0$, we shall write $\mathrm{Pr}^{\mathrm{ord}}_{\infty}$.

We shall now construct the improved Eisenstein family. Fix $k_0 \geq 2$ and $s_0 = k_0 - 2$. It is easy to see from (15.3) and (15.2) that when k varies p-adically the value $\mathfrak{b}^{s_0}_k(I)$ varies p-adically analytic too. We define $\mathfrak{b}^{s_0}_{\kappa}(I)$ to be the only polynomial in $\mathcal{O}(\mathcal{W}(r))[T_i]$, homogeneous of degree s_0 such that $\mathfrak{b}^{s_0}_{[t]}(I) = \mathfrak{b}^{s_0}_{t+1}(I)$. Its coefficients are products of $\mathrm{Log}_p(\kappa[i])$. We let $\chi = \chi_1 \chi' \omega^{k_0-2}$ and we define

$$\begin{aligned} & a^*_{T_1,T_4}(\kappa) \\ & = \Bigg(\sum_I \mathfrak{b}^{s_0}_{\kappa}(I) (\chi'\chi_1)^{-1}(2T_2) \sum_{G \in \mathrm{GL}_2(\mathbb{Z}) \backslash \mathbf{D}(I)} (\phi\chi'\chi_1)^2(\det(G)) |\det(G)|^{-1} \kappa(u^{l_{2|\det(G)|}}) \times \\ & \times L_p(\kappa, \sigma_{-\det(2I)}\phi\chi'\chi_1) \prod_{q | \det(2G^{-t}IG^{-1})} B_q\left(\phi\chi'\chi_1(q)\kappa(u^{l_q})q^{-2}, G^{-t}IG^{-1}\right) \Bigg). \end{aligned}$$

We now construct another p-adic family of Eisenstein series:

$$\mathcal{H}^*(\kappa) = \mathrm{Pr}^{\leq\alpha}(-1)^{k_0} \sum_{T_1 \geq 0} \sum_{T_4 \geq 0} a^*_{T_1,T_4}(\kappa[1 - k_0]) q_1^{T_1} q_2^{T_2}.$$

Proposition 15.7. *Suppose $\alpha = 0$. We have a p-adic family $\mathcal{H}(\kappa, \kappa') \in M_{\kappa}^{(2)}(N^2R^2)^{\mathrm{ord}}$ such that*

$$\mathcal{H}([k], \varepsilon[t]) = \frac{1}{(-1)^s c_{t+1}^s A} (U_p^{\otimes^2})^{-2i} \mathrm{Pr}^{\mathrm{ord}} \mathcal{H}'_{p^i, \chi\varepsilon\omega^{-s}}(z, w).$$

For any α, we have $\mathcal{H}^*(\kappa) \in M_{\kappa}^{(2)}(N^2R^2)^{\leq \alpha}$ such that

$$\mathcal{H}^*([k]) = A^{*-1}(-1)^{k_0} \mathrm{Pr}^{\leq \alpha} \mathcal{H}'^{*}_{1, \chi_1\chi'}(z, w)|_{u=\frac{1}{2}-k+k_0-1},$$

for

$$A^* = B(k - k_0 + 1)(2\pi i)^{k_0 - 2} G(\chi'\chi_1).$$

Proof. From its own definition we have $(U_p^{\otimes^2})^{2j} \mathcal{H}'_{1,\chi}(z, w) = \mathcal{H}'_{p^j,\chi}(z, w)$. We define

$$\mathcal{H}(\kappa, \kappa') = \varinjlim_j (U_p^{\otimes^2})^{-2j} \mathrm{Pr}^{\mathrm{ord}} \mathcal{H}_{p^j,\chi}(\kappa, \kappa').$$

With Lemma 15.6 and the previous remark we obtain

$$\begin{aligned} 4^s (U_p^{\otimes^2})^{-2i} \mathrm{Pr}^{\mathrm{ord}} \mathcal{H}'_{p^i, \chi\varepsilon\omega^{-s}}(z, w) &= 4^s (U_p^{\otimes^2})^{-2j} \mathrm{Pr}^{\mathrm{ord}} \mathcal{H}'_{p^j, \chi\varepsilon\omega^{-s}}(z, w) \\ &\equiv A(-1)^s 4^s c_t^s (U_p^{\otimes^2})^{-2j} \\ &\quad \times \mathrm{Pr}^{\mathrm{ord}} \mathcal{H}_{p^j}([k], \varepsilon[t]) \pmod{p}^{2j}, \end{aligned}$$

as U_p^{-1} acts on the ordinary part with norm 1.

The punctual limit is then a well-defined classical finite slope form. By the same method of proof of [Urb13, Corollary 3.4.7] or [Ros13b, Proposition 2.17] (in particular, recall that when the slope is bounded, the ray of overconvergence v can be fixed), we see that this q-expansion defines a family of finite slope forms.

For the second family, we remark that $\mathrm{Log}_p(\kappa)$ is bounded on $\mathcal{W}(r)$, for any r and we reason as above. □

We want to explain briefly why the construction above works in the ordinary setting and not in the finite slope one.

It is slightly complicated to explicitly calculate the polynomial $\mathfrak{b}_{t+1}^s(I)$ and in particulat to show that they vary p-adically when varying s. But we know that $\mathfrak{D}_l^s$ is an homogeneous polynomial in ∂_i of degree s. Suppose for now $\alpha = 0$, i.e. we are in the ordinary case. We have a single monomial of $\mathfrak{D}_l^s$ which does not involve ∂_1 and ∂_4, namely $c_{t+1}^s \partial_2^s$. Consequently, in $\mathfrak{b}_{t+1}^s(I)$ there is a single monomial without T_1 and T_4.

When the entries on the diagonal of I are divisible by p^i, $4^s\mathfrak{b}^s_{t+1}(I)$ reduces to $(-1)^s 4^s c^s_{t+1}$ modulo p^i. Applying $U_p^{\otimes^2}$ many times ensures us that T_1 and T_4 are very divisible by p. Speaking p-adically, we approximate $\mathfrak{D}^s_l$ by ∂_2 (multiplied by a constant). The more times we apply $U_p^{\otimes^2}$, the better we can approximate p-adically $\mathfrak{D}^s_l$ by ∂^s_2. At the limit, we obtain equality.

For $\alpha > 0$, when we apply $U_p^{\otimes^{2}-1}$ we introduce denominators of the order of p^α. In order to construct a two-variable family, we should approximate $\mathfrak{D}^s_l$ with higher precision. For example, it would be enough to consider the monomials ∂^J for $j_1 + j_4 \leq \alpha$. In fact, ∂_1 and ∂_4 increase the slope of U_p; we have the relation

$$\left(U_p^{\otimes^2}\partial_1\partial_4\right)_{|z_2=0} = p^2\left(\partial_1\partial_4 U_p^{\otimes^2}\right)_{|z_2=0}.$$

Unluckily, it seems quite hard to determine explicitly the coefficients c^J_{t+1} of (15.4) or even show that they satisfy some p-adic congruences as done in [CP04, Gor06, Ros13b]. We guess that it could be easier to interpolate p-adically the projection to the ordinary locus of the Eisenstein series of [BS00, Theorem 3.1] rather than the holomorphic projection as we are doing here (see also [Urb13, §3.4.5] for the case of nearly holomorphic forms for GL_2).

This should remind the reader of the fact that on p-adic forms the Maaß-Shimura operator and Dwork Θ-operator coincide [Urb13, §3.2].

15.4 p-adic L-functions

We now construct two p-adic L-functions using the above Eisenstein measure: the two-variable one in the ordinary case and the improved one for any finite slope. Necessary for the construction is a p-adic Petersson product [Pan03, §6] which we now recall. We fix a family $F = F(\kappa)$ of finite slope modular forms which we suppose primitive, i.e. all its classical specializations are primitive forms, of prime-to-p conductor N. We consider characters χ, χ' and χ_1 as in Section 15.3.2. We keep the same decomposition for N (because the local behavior at q is constant along the family). We shall write N_0 for the conductor of χ_1.

(**notCM**) We suppose that F has not complex multiplication by χ.

Let $\mathcal{C}_F$ be the corresponding irreducible component of the Coleman-Mazur eigencurve. It is finite flat over $\mathcal{W}(r)$, for a certain r. Let $\mathcal{I}$ be the coefficients ring of F and $\mathcal{K}$ its field of fraction. For a classical form f, let us denote by f^c

the complex conjugated form. We denote by τ_N the Atkin-Lehner involution of level N normalized as in [Hid90, h4]. When the level is clear from the context, we shall simply write τ.

Standard linear algebra allows us to define a $\mathcal{K}$-linear form l_F on $M_\kappa(N)^{\leq\alpha} \otimes_{\mathcal{I}} \mathcal{K}$ with the following property [Pan03, Proposition 6.7];

Proposition 15.8. *For all $G(\kappa)$ in $M_\kappa(N)^{\leq\alpha} \otimes_{\mathcal{I}} \mathcal{K}$ and any κ_0 classical point, we have*

$$l_F(G(\kappa))_{|\kappa=\kappa_0} = \frac{\langle F(\kappa_0)^c|\tau_{Np}, G(\kappa_0)\rangle_{Np}}{\langle F(\kappa_0)^c|\tau_{Np}, F(\kappa_0)\rangle_{Np}}.$$

We can find $H_F(\kappa)$ in $\mathcal{I}$ such that $H_F(\kappa)l_F$ is defined over $\mathcal{I}$. We shall refer sometimes to $H_F(\kappa)$ as the *denominator* of l_F.

We define consequently a $\mathcal{K}$-linear form on $M_\kappa^{(2)}(N)^{\leq\alpha} \otimes_{\mathcal{I}} \mathcal{K}$ by

$$l_{F\times F} := l_F \otimes l_F$$

under the decomposition in (15.5).

Before defining the p-adic L-functions, we need an operator to lower the level of the Eisenstein series constructed before. We follow [Hid88, §1 VI]. Fix a prime-to-p integer L, with $N|L$. We define for classical weights k:

$$\begin{array}{rcl} T_{L/N,k}: M_k(Lp, A) & \rightarrow & M_k(Np, A) \\ f & \mapsto & (L/N)^{k/2} \sum_{[\gamma]\in\Gamma(N)/\Gamma(N,L/N)} f|_k \begin{pmatrix} 1 & 0 \\ 0 & L/N \end{pmatrix} |_k\gamma \end{array}.$$

As L is prime to p, it is clear that $T_{L/N,k}$ commutes with U_p. It extends uniquely to a linear map

$$T_{L/N}: \quad M_\kappa(L) \quad \rightarrow \quad M_\kappa(N)$$

which in weight k specializes to $T_{L/N,k}$.

We have a map $M_\kappa(N)^{\leq\alpha} \hookrightarrow M_\kappa(N^2R^2)^{\leq\alpha}$. We define $1_{N^2R^2/N}$ to be one left inverse. We define

$$L_p(\kappa,\kappa') = \frac{N_0}{N^2R^2N_1} l_F \otimes l_F \left((U_{N^2/N_1^2}^{-1} \circ 1_{N^2R^2/N}) \otimes T_{N^2R^2/N}(\mathcal{H}(\kappa,\kappa')) \right)$$

(for $\alpha = 0$),

$$L_p^*(\kappa) = \frac{N_0}{N^2R^2N_1} l_F \otimes l_F \left((U_{N^2/N_1^2}^{-1} \circ 1_{N^2R^2/N}) \otimes T_{N^2R^2/N}(\mathcal{H}^*(\kappa) \right)$$

(for $\alpha < +\infty$).

We will see in the proof of the following theorem that it is independent of the left inverse $1_{N^2R^2/N}$ which we have chosen.

We fix some notations. For a Dirichlet character η, we denote by η_0 the associated primitive character. Let $\lambda_p(\kappa) \in \mathcal{I}$ be the U_p-eigenvalue of F. We say that $(\kappa, \kappa') \in \mathcal{C}_F \times \mathcal{W}$ is of type $(k; t, \varepsilon)$ if:

- $\kappa_{|\mathcal{W}(r)} = [k]$ with $k \geq 2$,
- $\kappa' = \varepsilon[t]$ with $1 \leq t \leq k-1$ and ε finite order character defined modulo p^n, $n \geq 1$.

Let as before $s = k - t - 1$. We define

$$E_1(\kappa, \kappa') = \lambda_p(\kappa)^{-2n_0}(1 - (\chi\varepsilon\omega^{-s})_0(p)\lambda_p(\kappa)^{-2}p^s))$$

where $n_0 = 0$ (resp. $n_0 = n$) if $\chi\varepsilon\omega^{-s}$ is (resp. is not) trivial at p.

If $F(\kappa)$ is primitive at p we define $E_2(\kappa, \kappa') = 1$, otherwise

$$\begin{aligned} E_2(\kappa, \kappa') = &(1 - (\chi^{-1}\varepsilon^{-1}\omega^s\phi)_0(p)p^{k-2-s})\times \\ &(1 - (\chi^{-1}\varepsilon^{-1}\omega^s\phi^2)_0(p)\lambda_p(\kappa)^{-2}p^{2k-3-s}). \end{aligned}$$

We denote by $F^\circ(\kappa)$ the primitive form associated to $F(\kappa)$. We shall write $W'(F(\kappa))$ for the prime-to-p part of the root number of $F^\circ(\kappa)$. If $F(\kappa)$ is not p-primitive we pose

$$S(F(\kappa)) = (-1)^k\left(1 - \frac{\phi_0(p)p^{k-1}}{\lambda_p(\kappa)^2}\right)\left(1 - \frac{\phi_0(p)p^{k-2}}{\lambda_p(\kappa)^2}\right),$$

and $S(F(\kappa)) = (-1)^k$ otherwise. We pose

$$\begin{aligned} C_{\kappa,\kappa'} &= i^{1-k}s!G\left((\chi\varepsilon\omega^{-s})^{-1}\right)(\chi\varepsilon\omega^{-s})_0(p^{n-n_0})(N_1Rp^{n_0})^sN^{-k/2}2^{-s}, \\ C_\kappa &= C_{\kappa,[k-k_0+1]}, \\ \Omega(F(\kappa), s) &= W'(F(\kappa))(2\pi i)^{s+1}\left\langle F^\circ(\kappa), F^\circ(\kappa)\right\rangle. \end{aligned}$$

We have the following theorem, which will be proven at the end of the section.

Theorem 15.9. *i) The function $L_p(\kappa, \kappa')$ is defined on $\mathcal{C}_F \times \mathcal{W}$, it is meromorphic in the first variable and bounded in the second variable. For all classical points (κ, κ') of type $(k; t, \varepsilon)$ with $k \geq 2$, $1 \leq t \leq k-1$, we have the following interpolation formula*

$$L_p(\kappa, \kappa') = C_{\kappa,\kappa'}E_1(\kappa, \kappa')E_2(\kappa, \kappa')\frac{\mathcal{L}(s+1, \mathrm{Sym}^2(F(\kappa)), \chi^{-1}\varepsilon^{-1}\omega^s)}{S(F(\kappa))\Omega(F(\kappa), s)}.$$

ii) The function $L^*_p(\kappa)$ is meromorphic on $\mathcal{C}_F$. For κ of type k with $k \geq k_0$, we have the following interpolation formula

$$L^*_p(\kappa) = C_\kappa E_2(\kappa, [k - k_0 + 1]) \frac{\mathcal{L}(k_0 - 1, \mathrm{Sym}^2(F(\kappa)), \chi'^{-1}\chi_1^{-1})}{S(F(\kappa))\Omega(F(\kappa), k_0 - 2)}.$$

Let us denote by $\tilde{L}_p(\kappa, \kappa')$ the two-variable p-adic L-function of [Ros13b, Theorem 4.14], which is constructed for any slope but NOT for $p = 2$. We can deduce the fundamental corollary which allows us to apply the method of Greenberg and Stevens;

Corollary 15.10. *For $\alpha = 0$ (resp. $\alpha > 0$ and $p \neq 2$) we have the following factorization of locally analytic functions around κ_0 in $\mathcal{C}_F$:*

$$L_p(\kappa, [k - k_0 + 1]) = (1 - \chi'\chi_1(p)\lambda_p(\kappa)^{-2} p^{k_0 - 2}) L^*_p(\kappa)$$

$$(\textit{resp. } C \frac{\kappa^{-1}(2)\tilde{L}_p(\kappa, [k - k_0 + 1])}{1 - \phi^{-2}\chi^2\kappa(4)2^{-2k_0}} = (1 - \chi'\chi_1(p)\lambda_p(\kappa)^{-2} p^{k_0 - 2}) L^*_p(\kappa)),$$

where C is a constant independent of κ, explicitly determined by the comparison of C_κ here and in [Ros13b, Theorem 4.14].

We can now prove the main theorem of the paper;

Proof of Theorem 15.2. Let f be as in the statement of the theorem; we take $\chi' = \chi_1 = \mathbf{1}$. For $\alpha = 0$ resp. $\alpha > 0$ we define

$$L_p(\mathrm{Sym}^2(f), s) = C^{-1}_{\kappa_0, [1]} L_p(\kappa_0, [k_0 - s]),$$

$$L_p(\mathrm{Sym}^2(f), s) = C \frac{\kappa(2)\tilde{L}_p(\kappa, [k - k_0 + 1])}{C_{\kappa_0, [1]}(1 - \phi^{-2}\chi^2\kappa(4)2^{-2k_0})}.$$

The two variables p-adic L-function vanishes on $\kappa' = [1]$. As f is Steinberg at p, we have $\lambda_p(\kappa_0)^2 = p^{k_0 - 2}$.

Consequently, the following formula is a straightforward consequence of Section 15.2 and Corollary 15.10;

$$\lim_{s \to 0} \frac{L_p(\mathrm{Sym}^2(f), s)}{s} = -2 \frac{\mathrm{d} \log \lambda_p(\kappa)}{\mathrm{d}\kappa}|_{\kappa = \kappa_0} \frac{\mathcal{L}(\mathrm{Sym}^2(f), k_0 - 1)}{S(F(\kappa))\Omega(f, k_0 - 2)}.$$

From [Ben10, Mok12] we obtain

$$\mathcal{L}^{\mathrm{al}}(\mathrm{Sym}^2(f)) = -2 \frac{\mathrm{d} \log \lambda_p(\kappa)}{\mathrm{d}\kappa}|_{\kappa = \kappa_0}.$$

Under the hypotheses of the theorem, f is Steinberg at all primes of bad reduction and we see from [Ros13b, §3.3] that

$$\mathcal{L}(\mathrm{Sym}^2(f), k_0 - 1) = L(\mathrm{Sym}^2(f), k_0 - 1)$$

and we are done. □

Proof of Theorem 15.9. We point out that most of the calculations we need in this proof and have not already been quoted can be found in [Hid90, Pan03].

If ε is not trivial at p, we shall write p^n for the conductor of ε. If ε is trivial, then we let $n = 1$.

We recall that $s = k - t - 1$; we have

$$L_p(\kappa, \kappa') = \frac{N_0 \left\langle F(\kappa)^c|\tau_{Np}, U_{N^2/N_1^2}^{-1}\left\langle F(\kappa)^c|\tau_{Np}, T_{N^2R^2/N,k}\mathcal{H}([k], \varepsilon[t])\right\rangle\right\rangle}{N^2R^2N_1\langle F(\kappa)^c|\tau, F(\kappa)\rangle^2}.$$

We have as in [Pan03, (7.11)]

$$\begin{aligned}\left\langle F(k)^c|\tau_{Np}, U_p^{-2n+1}\mathrm{Pr}^{\mathrm{ord}}U_p^{2n-1}g\right\rangle \\ = \lambda_p(\kappa)^{1-2n}p^{(2n-1)(k-1)}\left\langle F(\kappa)^c|\tau_{Np}|[p^{2n-1}], g\right\rangle,\end{aligned} \tag{15.6}$$

where $f|[p^{2n-1}](z) = f(p^{2n-1}z)$. We recall the well-known formulae [Hid88, page 79]:

$$\begin{aligned}\left\langle f|[p^{2n-1}], T_{N^2R^2/N,k}g\right\rangle &= (NR^2)^k\left\langle f|[p^{2n}NR^2], g\right\rangle,\\ \tau_{Np}|[p^{2n-1}N^2R^2] &= \left(NR^2p^{2n-1}\right)^{-k/2}\tau_{N^2R^2p^{2n}},\\ \frac{\left\langle F(\kappa)^c|\tau_{Np}, F(\kappa)\right\rangle}{\langle F(\kappa)^\circ, F(\kappa)^\circ\rangle} &= (-1)^kW'(F(\kappa))p^{(2-k)/2}\lambda_p(\kappa)\times\\ &\left(1-\frac{\phi_0(p)p^{k-1}}{\lambda_p(\kappa)^2}\right)\left(1-\frac{\phi_0(p)p^{k-2}}{\lambda_p(\kappa)^2}\right).\end{aligned}$$

So we are left to calculate

$$\begin{aligned}&\frac{N_0(NR^2)^{k/2}\lambda_p(\kappa)^{1-2n}p^{(2n-1)\left(\frac{k}{2}-1\right)}}{A(-1)^sc_{t+1}^sN_1N^2R^2}\\ &\times\left\langle F(\kappa)^c|\tau_{Np}, U_{N^2/N_1^2}^{-1}\left\langle F(\kappa)^c|\tau_{N^2R^2p^n}, \mathcal{H}'_{1,\chi\varepsilon\omega^{-t}}(z,w)|_{s=\frac{1}{2}-t}\right\rangle\right\rangle.\end{aligned}$$

We use Proposition 15.4 and [BS00, (3.29)] (with the notation of *loc. cit.* $\beta_1 = \frac{\lambda_p^2(\kappa)}{p^{k-1}}$) to obtain that the interior Petersson product is a scalar multiple of

$$\frac{\mathcal{L}(\mathrm{Sym}^2(f), s+1)\tilde{E}(\kappa,\kappa')E_2(\kappa,\kappa')F(\kappa)|U_{N^2/N_1^2}}{\langle F(\kappa)^\circ, F(\kappa)^\circ\rangle\, W'(F(\kappa))p^{(2-k)/2}\lambda_p(\kappa)S(F(\kappa))}. \tag{15.7}$$

Here

$$\tilde{E}(\kappa,\kappa') = -p^{-1}(1-\chi^{-1}\varepsilon^{-1}\omega^s(p)\lambda_p(\kappa)^2 p^{t+1-k})$$

if $\chi\varepsilon\omega^{-s}$ is trivial modulo p, and 1 otherwise. The factor $E_2(\kappa,\kappa')$ appears because $F(\kappa)$ could not be primitive. Clearly it is independent of $1_{N^2R^2/N}$ and we have $l_F(F(\kappa)) = 1$. We explicit the constant which multiplies (15.7):

$$\begin{aligned}
&(-1)^s \frac{N_0(NR^2)^{k/2}\lambda_p(\kappa)^{1-2n}p^{(2n-1)\left(\frac{k}{2}-1\right)}N_1^{s+1}(Rp^n)^{s+3-k}N^{2-k}}{B(t)(2\pi i)^s c_{t+1}^s G\left((\chi\varepsilon\omega^{-s})\right)N^2R^2N_1}\frac{\omega^{k-t-1}(-1)\Omega_{k,s}(s_1)p_{s_1}(t+1)}{d_{s_1}(t+1)}\\
&\quad = (-1)^{-\frac{k}{2}}\frac{N^{-\frac{k}{2}}N_0N_1^sR^{s+1}\lambda_p(\kappa)^{1-2n}p^{(2-k)/2}p^{n(s+1)}(2s)!2^{-2s}2^{2t}\pi^{3/2}}{2^{1+2t}\pi^{\frac{5}{2}}(2\pi i)^s G\left((\chi\varepsilon\omega^{-s})\right)2^{-s}\frac{(2s)!}{s!}}.
\end{aligned}$$

If $\varepsilon_1\varepsilon\omega^{-s}$ is not trivial we obtain

$$= i^{1-k}\frac{N^{-\frac{k}{2}}N_1^sR^s\lambda_p(\kappa)^{1-2n}p^{(2-k)/2}p^{ns}s!G\left((\chi\varepsilon\omega^{-s})^{-1}\right)}{(2\pi i)^{s+1}2^s},$$

otherwise

$$= i^{1-k}\frac{N^{-\frac{k}{2}}N_1^sR^s\lambda_p(\kappa)^{-1}p^{(2-k)/2}p^{s+1}s!G\left((\chi_1\chi')^{-1}\right)}{(2\pi i)^{s+1}2^s\chi_1\chi'(p)}.$$

The calculations for $L_p^*(\kappa)$ are similar. We have to calculate

$$\begin{aligned}
&\frac{N_0(NR^2)^{k/2}}{A^*N^2R^2N_1}\\
&\quad\times\left\langle F(\kappa)^c|\tau_{Np}, U_{N^2/N_1^2}^{-1}\left\langle F(\kappa)^c|\tau_{N^2R^2p}, (-1)^{k_0}\mathcal{H}'^*_{1,\chi_1\chi'}(z,w)|_{u=s_1}\right\rangle\right\rangle,
\end{aligned}$$

where $s_1 = \frac{1}{2} - k + k_0 - 1$. The interior Petersson product equals (see [BS00, Theorem 3.1] with M, N of *loc. cit.* as follows: $M = R^2N^2p$, $N = N_1R$)

$$\frac{R^{k_0+1-k}(N^2p)^{\frac{2-k}{2}}(-1)^{k_0}\Omega_{k,k_0-2}(s_1)p_{s_1}(k-k_0+2)}{d_{s_1}(k-k_0+2)\langle F(\kappa)^\circ, F(\kappa)^\circ\rangle\, W'(F(\kappa))p^{(2-k)/2}\lambda_p(\kappa)S(F(\kappa))}F(\kappa)|U_p|U_{N^2/N_1^2},$$

so we have

$$L_p^*(\kappa) = i^{1-k} \frac{N_1^{k_0-2} R^{k_0-2} N^{k/2} (k_0-2)!}{(2\pi i)^{k_0-1} G(\chi_1\chi')2^{k_0-2}} \times \frac{E_2(\kappa, [k-k_0+1])\mathcal{L}(k_0-1, \mathrm{Sym}^2(F(\kappa)), (\chi_1\chi')^{-1})}{S(F(\kappa))W'(F(\kappa))\langle F(\kappa)^\circ, F(\kappa)^\circ \rangle}. \qquad \square$$

We give here some concluding remarks. As $F(\kappa)$ has not complex multiplication by χ, we can see exactly as in [Hid90, Proposition 5.2] that $H_F(\kappa)L_p(\kappa, \kappa')$ is holomorphic along $\kappa' = [0]$ (which is the pole of the Kubota Leopoldt p-adic L-function).

We point out that the analytic $\mathcal{L}$-invariant for CM forms has already been studied in literature [DD97, Har12, HL13].

Note also that our choice of periods is not optimal [Ros13b, §6].

References

[AIP12] Fabrizio Andreatta, Adrian Iovita, and Vincent Pilloni. p-adic families of Siegel modular cuspforms. *to appear Ann. of Math.*, 2012.

[AIS12] Fabrizio Andreatta, Adrian Iovita, and Glenn Stevens. Overconvergent modular sheaves and modular forms for $\mathrm{GL}_{2/F}$. *to appear Israel J. Math.*, 2012.

[BDR14] Massimo Bertolini, Henri Darmon, and Victor Rotger. Beilinson-Flach elements and Euler systems II: the Birch and Swinnerton-Dyer conjecture for Hasse-Weil-Artin L-functions. *in preparation*, 2014.

[Bel12] Joël Bellaïche. p-adic L-functions of critical CM forms. *preprint available at http://people.brandeis.edu/ jbellaic/preprint/CML-functions4.pdf*, 2012.

[Ben10] Denis Benois. Infinitesimal deformations and the ℓ-invariant. *Doc. Math.*, (Extra volume: Andrei A. Suslin sixtieth birthday):5–31, 2010.

[Ben11] Denis Benois. A generalization of Greenberg's $\mathcal{L}$-invariant. *Amer. J. Math.*, 133(6):1573–1632, 2011.

[Ben12] Denis Benois. Trivial zeros of p-adic L-functions at near central points. *to appear J. Inst. Math. Jussieu http://dx.doi.org/10.1017/ S1474748013000261*, 2012.

[Bra13] Riccardo Brasca. Eigenvarieties for cuspforms over PEL type Shimura varieties with dense ordinary locus. *in preparation*, 2013.

[BS00] S. Böcherer and C.-G. Schmidt. p-adic measures attached to Siegel modular forms. *Ann. Inst. Fourier (Grenoble)*, 50(5):1375–1443, 2000.

[BSDGP96] Katia Barré-Sirieix, Guy Diaz, François Gramain, and Georges Philibert. Une preuve de la conjecture de Mahler-Manin. *Invent. Math.*, 124(1-3):1–9, 1996.

[Buz07] Kevin Buzzard. Eigenvarieties. In *L-functions and Galois representations*, volume 320 of *London Math. Soc. Lecture Note Ser.*, pages 59–120. Cambridge University Press, Cambridge, 2007.

[Col05] Pierre Colmez. Zéros supplémentaires de fonctions L p-adiques de formes modulaires. In *Algebra and number theory*, pages 193–210. Hindustan Book Agency, Delhi, 2005.

[CP04] Michel Courtieu and Alexei Panchishkin. *Non-Archimedean L-functions and arithmetical Siegel modular forms*, volume 1471 of *Lecture Notes in Mathematics*. Springer-Verlag, Berlin, second edition, 2004.

[DD97] Andrzej Dabrowski and Daniel Delbourgo. S-adic L-functions attached to the symmetric square of a newform. *Proc. London Math. Soc. (3)*, 74(3):559–611, 1997.

[DG12] Mladen Dimitrov and Eknath Ghate. On classical weight one forms in Hida families. *J. Théor. Nombres Bordeaux*, 24(3):669–690, 2012.

[Dim13] Mladen Dimitrov. On the local structure of ordinary hecke algebras at classical weight one points. *to appear in Automorphic Forms and Galois Representations, proceedings of the LMS Symposium, 2011*, to appear, 2013.

[Gor06] B. Gorsse. *Mesures p-adiques associées aux carrés symétriques*. 2006. Thesis (Ph.D.)–Université Grenoble 1.

[Gre94] Ralph Greenberg. Trivial zeros of p-adic L-functions. In *p-adic monodromy and the Birch and Swinnerton-Dyer conjecture (Boston, MA, 1991)*, volume 165 of *Contemp. Math.*, pages 149–174. Amer. Math. Soc., Providence, RI, 1994.

[GS93] Ralph Greenberg and Glenn Stevens. p-adic L-functions and p-adic periods of modular forms. *Invent. Math.*, 111(2):407–447, 1993.

[Har12] Robert Harron. The exceptional zero conjecture for symmetric powers of CM modular forms: the ordinary case. *to appear Int. Math. Res. Notices doi: 10.1093/imrn/rns161*, 2012.

[Hid88] Haruzo Hida. A p-adic measure attached to the zeta functions associated with two elliptic modular forms. II. *Ann. Inst. Fourier (Grenoble)*, 38(3):1–83, 1988.

[Hid89] Haruzo Hida. On nearly ordinary Hecke algebras for GL(2) over totally real fields. In *Algebraic number theory*, volume 17 of *Adv. Stud. Pure Math.*, pages 139–169. Academic Press, Boston, MA, 1989.

[Hid90] Haruzo Hida. p-adic L-functions for base change lifts of GL_2 to GL_3. In *Automorphic forms, Shimura varieties, and L-functions, Vol. II (Ann Arbor, MI, 1988)*, volume 11 of *Perspect. Math.*, pages 93–142. Academic Press, Boston, MA, 1990.

[Hid06] Haruzo Hida. *Hilbert modular forms and Iwasawa theory*. Oxford Mathematical Monographs. The Clarendon Press Oxford University Press, Oxford, 2006.

[HJ13] Robert Harron and Andrei Jorza. On symmetric power $\mathcal{L}$-invariants of Iwahori level Hilbert modular forms. *preprint available at www.its.caltech.edu/ ajorza/papers/harron-jorza-non-cm.pdf*, 2013.

[HL13] Robert Harron and Antonio Lei. Iwasawa theory for symmetric powers of CM modular forms at non-ordinary primes. *preprint available at http://arxiv.org/abs/1208.1278*, 2013.

[HTU97] Haruzo Hida, Jacques Tilouine, and Eric Urban. Adjoint modular Galois representations and their Selmer groups. *Proc. Nat. Acad. Sci.*

U.S.A., 94(21):11121–11124, 1997. Elliptic curves and modular forms (Washington, DC, 1996).

[Liu13] Ruochuan Liu. Triangulation of refined families. *preprint available at http://arxiv.org/abs/1202.2188*, 2013.

[LLZ12] Antonio Lei, David Loeffler, and Sarah Livia Zerber. Euler systems for Rankin-Selberg convolutions of modular forms. *to appear Ann. of Math.*, 2012.

[Mat80] Hideyuki Matsumura. *Commutative algebra*, volume 56 of *Mathematics Lecture Note Series*. Benjamin/Cummings Publishing Co., Inc., Reading, Mass., second edition, 1980.

[Mok09] Chung Pang Mok. The exceptional zero conjecture for Hilbert modular forms. *Compos. Math.*, 145(1):1–55, 2009.

[Mok12] Chung Pang Mok. $\mathcal{L}$-invariant of the adjoint Galois representation of modular forms of finite slope. *J. Lond. Math. Soc. (2)*, 86(2):626–640, 2012.

[MTT86] B. Mazur, J. Tate, and J. Teitelbaum. On p-adic analogues of the conjectures of Birch and Swinnerton-Dyer. *Invent. Math.*, 84(1):1–48, 1986.

[Nag62] Masayoshi Nagata. *Local rings*. Interscience Tracts in Pure and Applied Mathematics, No. 13. Interscience Publishers a division of John Wiley & Sons New York-London, 1962.

[Pan94] Alexei A. Panchishkin. Motives over totally real fields and p-adic L-functions. *Ann. Inst. Fourier (Grenoble)*, 44(4):989–1023, 1994.

[Pan03] A. A. Panchishkin. Two variable p-adic L functions attached to eigenfamilies of positive slope. *Invent. Math.*, 154(3):551–615, 2003.

[Pil13] Vincent Pilloni. Overconvergent modular forms. *Ann. Inst. Fourier (Grenoble)*, 63(1):219–239, 2013.

[Pot13] Jonathan Pottharst. Cyclotomic Iwasawa theory of motives. *preprint available at http://vbrt.org/writings/cyc.pdf*, 2013.

[PR95] Bernadette Perrin-Riou. Fonctions L p-adiques des représentations p-adiques. *Astérisque*, (229):198, 1995.

[PR98] Bernadette Perrin-Riou. Zéros triviaux des fonctions L p-adiques, un cas particulier. *Compositio Math.*, 114(1):37–76, 1998.

[Ros13a] Giovanni Rosso. Derivative at $s = 1$ of the p-adic L-function of the symmetric square of a Hilbert modular form. *preprint available at http://arxiv.org/abs/1306.4935*, 2013.

[Ros13b] Giovanni Rosso. A formula for the derivative of the p-adic L-function of the symmetric square of a finite slope modular form. *preprint at http://arxiv.org/abs/1310.6583*, 2013.

[Urb06] Éric Urban. Groupes de Selmer et fonctions L p-adiques pour les représentations modulaires adjointes. *preprint available at http://www.math.jussieu.fr/~urban/eurp/ADJMC.pdf*, 2006.

[Urb11] Eric Urban. Eigenvarieties for reductive groups. *Ann. of Math. (2)*, 174(3):1685–1784, 2011.

[Urb13] Éric Urban. Nearly overconvergent modular forms. *to appear in the Proceedings of conference IWASAWA 2012*, 2013.

16

Uniform bounds for rational points on cubic hypersurfaces

Per Salberger

Mathematics, Chalmers University of Technology,
SE-412 96 Göteborg, Sweden
E-mail address: salberg@chalmers.se

ABSTRACT. We use a global version of Heath-Brown's p-adic determinant method to show that there are $O_{N,\varepsilon}(B^{\dim X+1/7+\varepsilon})$ rational points of height at most B on a geometrically integral variety $X \subset \mathbf{P}^N$ of degree three defined over $\mathbf{Q}$. By the same method we also show that there are $O_\varepsilon(B^{12/7+\varepsilon})$ rational points of height at most B outside the lines on any cubic surface in $\mathbf{P}^3$.

16.0 Introduction

We shall in this paper study the number $N(W; B)$ of rational points of height at most $B \geq 1$ on quasi-projective subvarieties W of $\mathbf{P}^N$ defined over $\mathbf{Q}$. The height $H(x)$ of a rational point x on W will be given by $\max(|x_0|, \ldots, |x_N|)$ for a primitive integral $N+1$-tuple $(x_0, \ldots, x_N)$ representing x. We will use the O notation for functions defined for $B \geq 1$. If g is a function from $[1, \infty)$ to $[0, \infty)$, then we write $N(W; B) = O(g(B))$ when there is a constant $C > 0$ such that $N(W; B) \leq Cg(B)$ for all $B \geq 1$. If the constant C depends on parameters, then we include these as indices.

One of the goals of the paper is to establish the following uniform bound, in which the implicit constant is independent of X.

Theorem 16.1. *Let $X \subset \mathbf{P}^3$ be an integral cubic surface defined over $\mathbf{Q}$ and U be the complement of the union of all lines on X. Then,*

$$N(U; B) = O_\varepsilon(B^{12/7+\varepsilon}).$$

The first general bound where the growth order on U is less than two is due to Heath-Brown. He proved in [HB02] that $N(U; B) = O_\varepsilon(B^{52/27+\varepsilon})$.

Arithmetic and Geometry, ed. Luis Dieulefait *et al.* Published by Cambridge University Press.

This was then improved in [Sal08], where it was shown that $N(U; B) = O_\varepsilon(B^{2/3+73/36\sqrt{3}+\varepsilon})$. Finally, it was shown that $N(U; B) = O_\varepsilon(B^{\sqrt{3}+\varepsilon})$ in [Sal]. The bound in 16.1 is superior to these bounds.

It has been conjectured by Manin that $N(U; B) = O_{X,\varepsilon}(B^{1+\varepsilon})$, but this has not been proved for any non-singular cubic surface with a rational point.

To prove theorem 16.1, we first establish the following lemma (cf. 16.12).

Lemma 16.2. *Let $X \subset \mathbf{P}^3$ be an integral cubic surface over $\mathbf{Q}$ and $B \geq 1$. Let $\varepsilon > 0$ and $p_1, \ldots, p_t$ be distinct primes with $q = p_1 \ldots p_t \geq B^{6/7+\varepsilon}$. For $i = 1, \ldots, t$, let P_i be a non-singular $\mathbf{F}_{p_i}$-point on X_{p_i} and $X(\mathbf{Q}; B; P_1, \ldots, P_t)$ be the set of rational points of height at most B on X, which specialize to P_i on X_{p_i} for $i = 1, \ldots, t$. Then there exists a surface $Y \subset \mathbf{P}^3$ not containing X of degree bounded solely in terms of ε with $X(\mathbf{Q}; B; P_1, \ldots, P_t) \subset Y$.*

Here X_p is the reduction (mod p) of $\operatorname{Proj} \mathbf{Z}[x_0, \ldots, x_3]/(F(x_0, \ldots, x_3))$ for a primitive cubic form $F(x_0, \ldots, x_3)$ defining $X \subset \mathbf{P}^3$. This lemma is inspired by theorem 14 in [HB02], from which one gets the same conclusion for primes $q \geq B^{\sqrt{3}/2+\varepsilon}$. To establish 16.2, we use a refined version of the p-adic determinant method in [HB02], where the equations for the tangent planes play a special role.

The basic idea to deduce 16.1 from 16.2 is to reduce to a counting problem for curves for each residue class (mod q). But to obtain the estimate in 16.1, it is important to apply 16.2 to many smooth integers q even if the number of such q will be bounded in terms of ε. We use thereby a dichotomy (see (16.3.2) and (16.3.3)) similar to the one used for the global determinant method in [Sal]. It seems at first sight that the method in this paper is cruder than the method in [Sal], where one makes use of conguences modulo almost all primes. But we have not been able to deduce 16.1 from 16.2 with the global method in [Sal]. The present method which might be called "semiglobal" is easier to use as the degrees of the auxiliary hypersurfaces do not depend on B. But there are varieties where the method in [Sal] gives better results. It is thus likely that both methods will be used in the future.

Another goal of this paper is to prove the following result, which improves upon the previous estimates in [BHBS06] and [Sal].

Theorem 16.3. *Let $N \geq 3$ and $f(y_1, \ldots, y_N) \in \mathbf{Q}[y_1, \ldots, y_N]$ be a cubic polynomial such that its leading cubic form is absolutely irreducible. Let $n(f; B)$ be the number of N-tuples $(y_1, \ldots, y_N)$ of integers in $[-B, B]$ such that $f(y_1, \ldots, y_N) = 0$. Then*

$$n(f; B) = O_{N,\varepsilon}(B^{N-2+1/7+\varepsilon}).$$

To prove this, we proceed as in [BHBS06] and reduce to the case $N = 3$ by means of hyperplane sections. The proof is then similar to the proof of theorem 16.1. On applying theorem 1 in [BHBS06], we obtain the following corollary of theorem 16.3.

Corollary 16.4. *Let $X \subset \mathbf{P}^N$ be a geometrically integral closed subvariety over $\mathbf{Q}$ of degree three. Then $N(X; B) = O_{N,\varepsilon}(B^{\dim X + 1/7+\varepsilon})$.*

It is known (see [Sal]) that $N(X; B) = O_{X,\varepsilon}(B^{\dim X+\varepsilon})$ for such varieties. But the bound in 16.4 is better than previous *uniform* bounds for varieties of degree three (cf. [BHBS06] and [Sal]).

16.1 Some consequences of Siegel's lemma

We present in this section some consequences of Siegel's lemma on bounds for solutions to linear equations. These will be needed for the p-adic determinant method to work well. In the sequel a hypersurface X in $\mathbf{P}^N_{\mathbf{Q}}$ will always mean an equidimensional closed subscheme of codimension one. We shall in this paper use the following notation.

Notation 16.5. Let $X \subset \mathbf{P}^N$ be a hypersurface over $\mathbf{Q}$ and $F(x_0, \ldots, x_N) \in \mathbf{Q}[x_0, \ldots, x_N]$ be a form which defines X. Let $\mathbf{B} = (B_0, \ldots, B_N) \in \mathbf{R}^{N+1}_{\geq 0}$. Then,

(i) $X(\mathbf{Q}; \mathbf{B})$ is the set of rational points on X which may be represented by an integral $(N+1)$-tuple $(x_0, \ldots, x_N)$ with $|x_j| \leq B_j$ for $0 \leq j \leq N$. If $B_0 = \cdots = B_N = B$, then we also denote this set by $X(\mathbf{Q}; B)$.
(ii) $N(X; \mathbf{B}) = \#X(\mathbf{Q}; \mathbf{B})$ and $N(X; B) = \#X(\mathbf{Q}; B)$.
(iii) $V = B_0 \ldots B_N$.
(iv) $T = \max \left\{ B_0^{m_0} \ldots B_N^{m_N} \right\}$ with the maximum taken over all $(N+1)$-tuples $(m_0, \ldots, m_N)$ for which the corresponding monomial $x_0^{m_0} \ldots x_N^{m_N}$ occurs in $F(x_0, \ldots, x_N)$ with non-zero coefficient.
(v) If $\underline{m} = (m_0, \ldots, m_N)$ is an $(N+1)$-tuple of non-negative integers, then $x^{\underline{m}} = x_0^{m_0} \ldots x_N^{m_N}$.
(vi) If $X \subset \mathbf{P}^N_{\mathbf{Q}}$ is a hypersurface over $\mathbf{Q}$ defined by a primitive form $F(x_0, \ldots, x_N) = \sum a_{\underline{m}} x^{\underline{m}}$ in $\mathbf{Z}[x_0, \ldots, x_N]$, then $X^{cl} = \operatorname{Proj} \mathbf{Z}[x_0, \ldots, x_N]/(F(x_0, \ldots, x_N))$ and $H(X) = \max |a_{\underline{m}}|$.

X^{cl} is thus the scheme-theoretic closure of X in $\mathbf{P}^N_{\mathbf{Z}}$. We shall also write $X_p = X^{cl} \times \mathbf{F}_p$ for primes p and make extensive use of the specialization map from $X^{cl}(\mathbf{Z}) = X(\mathbf{Q})$ to $X_p(\mathbf{F}_p)$.

Lemma 16.6. *Let $X \subset \mathbf{P}^N_{\mathbf{Q}}$ be an integral hypersurface of degree D and $\mathbf{B} = (B_0, \ldots, B_N) \in \mathbf{R}^{N+1}_{\geq 1}$. Then one of the following two statements holds.*

(a) There exists a projective $\mathbf{Q}$-hypersurface $Y \subset \mathbf{P}^N$ of degree D disinct from X which contains $X(\mathbf{Q}; \mathbf{B})$.
(b) $H(X) = O_{D,N}(V^{(D+N)!/(D-1)!N!})$.

Proof. This is proved in [HB02, th.4] for $N = 2$ and the same argument can be used to prove 16.6 for $N > 2$.

We shall use lemma 16.6 together with the following elementary result.

Lemma 16.7. *Let $\mathbf{B} = (B_0, \ldots, B_N) \in \mathbf{R}^{N+1}_{\geq 1}$ and $X \subset \mathbf{P}^N_{\mathbf{Q}}$ be a hypersurface of degree D and of height $H(X) = O_{D,N}(V^{(D+N)!/(D-1)!N!})$. Let x be a non-singular point in $X(\mathbf{Q}; \mathbf{B})$ and π_x be the product of all primes p where x specializes to a singular $\mathbf{F}_p$-point on X_p. Then,*

$$\log \pi_x = O_{D,N}(1 + \log V).$$

Proof. Let $F(x_0, \ldots, x_N) = \sum a_{\underline{m}} x^{\underline{m}}$ be a primitive form defining $X \subset \mathbf{P}^N$ and ξ be an integral primitive $(N+1)$-tuple representing $x \in X(\mathbf{Q}; \mathbf{B})$. Suppose that $(\delta F/\delta x_j)(\xi) \neq 0$ for some j. Then, $\pi_x \mid (\delta F/\delta x_j)(\xi)$ and $\log \pi_x \leq \log |(\delta F/\delta x_j)(\xi)| = O_{D,N}(1 + \log V)$.

We shall also need a generalization of 16.6 and 16.7 to arbitrary integral closed subschemes $Z \subset \mathbf{P}^N_{\mathbf{Q}}$. If $\mathbf{B} = (B_0, \ldots, B_N) \in \mathbf{R}^{N+1}_{\geq 0}$, then $Z(\mathbf{Q}; \mathbf{B})$ will denote the set of rational points on X, which may be represented by an integral $(N+1)$-tuple $(x_0, \ldots, x_N)$ with $|x_j| \leq B_j$ for $1 \leq j \leq N$ and Z^{cl} will denote the scheme-theoretic closure of Z in $\mathbf{P}^N_{\mathbf{Z}}$.

Lemma 16.8. *Let $Z \subset \mathbf{P}^N_{\mathbf{Q}}$ be an integral subscheme of degree $\leq D$ and $\mathbf{B} = (B_0, \ldots, B_N) \in \mathbf{R}^{N+1}_{\geq 1}$. Let $V = B_0 \ldots B_N$. Then one of the following two statements holds.*

(a) There exists a hypersurface $Y \subset \mathbf{P}^N_{\mathbf{Q}}$ of degree $O_{D,N}(1)$ which contains $Z(\mathbf{Q}; \mathbf{B})$ but not Z.
(b) If z is a non-singular point in $Z(\mathbf{Q}; \mathbf{B})$, then the product $\underline{\pi}_z$ of all primes p where z specializes to a singular $\mathbf{F}_p$-point on $Z_p = Z^{cl} \times \mathbf{F}_p$ satisfies $\log \underline{\pi}_z = O_{D,N}(1 + \log V)$.

Proof. It follows from lemma 16.7 and lemma 16.8 in [Sal07] that $Z \subset \mathbf{P}^N_{\mathbf{Q}}$ is defined as a subscheme by homogeneous polynomials of degree $\delta = O_{D,N}(1)$.

We may therefore apply lemma 5 in [Bro04] and conclude that either (a) holds or that we may find a set of $t = O_{D,N}(1)$ homogeneous polynomials $F_1, \dots, F_t$ of degree at most δ with integer coefficients, which define $Z \subset \mathbf{P}^N_{\mathbf{Q}}$ as a subscheme and with $\prod_{i=1}^t \|F_i\| = O_{D,N}(V^g)$ for $g = \delta^2\binom{N+\delta}{\delta}$. It is now easy to see from the Jacobian criterion that (b) holds for schemes defined by such polynomials $F_1, \dots, F_t$.

16.2 The p-adic determinant method for cubic surfaces

We shall in this section describe an improvement of Heath-Brown's p-adic determinant method for the special case of cubic surfaces. We will keep the notation of 16.5, but only consider the case where X is an integral cubic surface over $\mathbf{Q}$. If I is a prime ideal in a commutative ring R, then we let R_I be the localization of R at I. In particular, if $R = \mathbf{Z}$ and $I = (p)$, then $\mathbf{Z}_{(p)} = \{m/n \in \mathbf{Q} : m, n \in \mathbf{Z} \text{ such that } n/p \notin \mathbf{Z}\}$.

The following lemma should be viewed as a refinement of lemma 16.8 in [Sal07].

Lemma 16.9. *Let R be a commutative noetherian regular local ring containing $\mathbf{Z}_{(p)}$ as subring. Let $\phi_1, \dots, \phi_s$ be ring homomorphisms from R to $\mathbf{Z}_{(p)}$, I be the kernel of ϕ_1 and $\mathfrak{m} = IR_I$ be the maximal ideal of $A = R_I$. Let $W \subset A$ be the vector subspace over $\mathbf{Q}$ generated by elements $r_1, \dots, r_s$ of R and $f = \sum_{k \geq 1} \dim_{\mathbf{Q}}(W \cap \mathfrak{m}^k)$. Then the determinant of the $s \times s$-matrix $(\phi_i(r_j))$ is divisible by p^f.*

Proof. Let I be the kernel of ϕ_1 and $I^0 = R$. Then I is generated by $\dim R - 1$ regular parameters (see [AK70, p.56]). Since these form a regular sequence on R (see [Eis95, p.243]), we conclude that $\mathrm{Sym}^k(I/I^2) = I^k/I^{k+1}$ is a free R/I-module for each $k \geq 1$ (see [FL85, p.75]). So if $ur \in I^{k+1}$ for $u \in S$ and $r \in I^k$, then $r \in I^{k+1}$ as $(u + I)(r + I^{k+1}) = I^{k+1}$. Therefore, $\mathfrak{m}^{k+1} \cap I^k = I^{k+1}R_I \cap I^k = I^{k+1}$ for $k \geq 0$.

Now suppose there are more than $\dim_{\mathbf{Q}}(W \cap \mathfrak{m}^k)/(W \cap \mathfrak{m}^{k+1})$ elements among $r_1, \dots, r_s$ in $I^k \setminus I^{k+1}$ for some $k \geq 0$. If we denote these by $r_1, \dots, r_q$, then we may find $\beta_1, \dots, \beta_q \in \mathbf{Q}$, not all 0, such that $\beta_1 r_1 + \dots + \beta_{q-1} r_{q-1} + \beta_q r_q \in \mathfrak{m}^{k+1}$. After scaling and reordering we may assume that $\beta_q = 1$ and $\beta_1, \dots, \beta_{q-1} \in \mathbf{Z}_{(p)}$. Then $\rho_q = \beta_1 r_1 + \dots \beta_{q-1} r_{q-1} + r_q \in \mathfrak{m}^{k+1} \cap I^k = I^{k+1}$. We may therefore by performing elementary tranformations of this kind replace $r_1, \dots, r_s$ by new elements $\rho_1, \dots, \rho_s$ in R such that $\det(\phi_i(\rho_j)) = \det(\phi_i(r_j))$ and such that we have at most $\dim_{\mathbf{Q}}(W \cap \mathfrak{m}^k) - \dim_{\mathbf{Q}}(W \cap \mathfrak{m}^{k+1})$ elements ρ_j in $I^k \setminus I^{k+1}$ for each $k \geq 0$. But then we have at least $\dim_{\mathbf{Q}}(W \cap m^k)$ elements ρ_j in I^k for each $k \geq 1$. As $p^k \mid \phi_i(\rho_j)$ for $i = 1, \ldots, s$, for each such ρ_j, we find that $p^f \mid \det(\phi_i(\rho_j))$, as was to be proved.

We now apply this result to a cubic surface $X \subset \mathbf{P}^3_{\mathbf{Q}}$ and let $Q(n)$ be its Hilbert polynomial. Then, $Q(n) = \binom{n+3}{3} - \binom{n}{3} = (3n^2+3n+2)/2$. We shall also consider the integer $e(n)$ given by

$$e(n) = \sum_{1 \leq k \leq 3n/2} \big(Q(n) - (k^2+k)/2\big) + \sum_{3n/2 < k \leq 2n} Q(2n-k). \tag{16.1}$$

Lemma 16.10. *Let $X \subset \mathbf{P}^3$ be an integral cubic surface over $\mathbf{Q}$ and $F_1, \ldots, F_{Q(n)} \subset \mathbf{Z}[x_0, \ldots, x_3]$ be homogeneous polynomials of degree $n \geq 0$. Let $\xi_i = (\xi_{i,0}, \ldots, \xi_{i,3})$, $1 \leq i \leq Q(n)$ be primitive quadruples of integers such that all $\xi_i \pmod p$ represent the same non-singular $\mathbf{F}_p$-point P on X_p. Then the determinant of the $Q(n) \times Q(n)$-matrix $(F_j(\xi_i))$ is divisible by $p^{e(n)}$.*

Proof. We may suppose that $F_1, \ldots, F_{Q(n)}$ are linearly independent on X, since otherwise $\det(F_j(\xi_i)) = 0$. Let R be the stalk of the structure sheaf of X^{cl} at P and assume without loss of generality that $x_0(P) \neq 0$. Then $r_j = F_j/x_0^n \in R$ for $j = 1, \ldots, Q(n)$. Let $\phi_i : R \to \mathbf{Z}_{(p)}$, $1 \leq i \leq Q(n)$ be the local ring homomorphisms at P corresponding to the $\mathbf{Z}_{(p)}$-points on X^{cl} given by ξ_i for $1 \leq i \leq Q(n)$. Then, $\phi_i(r_j) = F_j(\xi_i)/\xi_{i,0}^n$ and

$$\det(F_j(\xi_i)) = \det(\phi_i(r_j)) \prod_{i=1}^{Q(n)} \xi_{i,0}^n. \tag{16.2}$$

Let I be the kernel of ϕ_1. Then $A = R_I$ is the stalk of the structure sheaf of X at the non-singular rational point x defined by ξ_1. It is thus a two-dimensional regular local ring with $m = IR_I$ as maximal ideal and $\dim_{\mathbf{Q}}(A/m^k) = 1 + \cdots + k = (k^2+k)/2$ for $k \geq 1$.

Let W be the $\mathbf{Q}$-subspace of A generated by $r_1, \ldots, r_{Q(n)}$. Then $\dim_{\mathbf{Q}} W = Q(n)$ as $F_1, \ldots, F_{Q(n)}$ are linearly independent on X. Further, $\dim_{\mathbf{Q}}(W/(W \cap m^k)) \leq \dim_{\mathbf{Q}}(A/m^k)$. Hence,

$$\dim_{\mathbf{Q}}(W \cap m^k) = \dim_{\mathbf{Q}}(W) - \dim_{\mathbf{Q}}(W/(W \cap m^k)) \geq Q(n) - (k^2+k)/2. \tag{16.3}$$

We shall use (16.3) for $1 \leq k \leq 3n/2$. For $3n/2 < k \leq 2n$, we will use another lower estimate for $\dim_{\mathbf{Q}}(W \cap m^k)$. Let $G_1, \ldots, G_{Q(4n-2k)}$ be homogeneous polynomials in $\mathbf{Q}[x_0, \ldots, x_3]$ of degree $4n - 2k \geq 0$, which are linearly independent on X and W_0 be the $\mathbf{Q}$-subspace of A generated by G_j/x_0^{4n-2k}, $1 \le j \le Q(4n-2k)$. Then, $\dim_{\mathbf{Q}}(W_0 \cap m^{6n-3k}) \ge \dim_{\mathbf{Q}}(W_0) - \dim_{\mathbf{Q}}(A/m^{6n-3k})$ and

$$\dim_{\mathbf{Q}}(W_0 \cap m^{6n-3k}) \ge Q(4n-2k) - ((6n-3k)^2 + (6n-3k))/2 = Q(2n-k). \tag{16.4}$$

There are thus linearly independent $\mathbf{Q}$-linear combinations $H_1, \dots, H_{Q(2n-k)}$ of $G_1, \dots, G_{Q(4n-2k)}$ such that $H_j/x_0^{4n-2k} \in W_0 \cap m^{6n-3k}$ for $1 \le j \le Q(2n-k)$. If now $L(x_0, \dots, x_3) = 0$ is a linear equation of the tangent plane of X of ξ_1, we deduce that $L^{2k-3n} H_j/x_0^n \in m^k$ for $1 \le j \le Q(2n-k)$ as $L/x_0 \in m^2$.

Let $\Phi_1, \dots, \Phi_{Q(2n-k)}$ be linear combinations of $F_1, \dots, F_{Q(n)}$ such that $\Phi_j - L^{2k-3n} H_j$ vanish on X for $1 \le j \le Q(2n-k)$. Then, $\Phi_1/x_0^n, \dots, \Phi_{Q(2n-k)}/x_0^n$ are linearly independent elements of $W \cap m^k$ and

$$\dim_{\mathbf{Q}}(W \cap m^k) \ge Q(2n-k) \quad \text{for } k \in (3n/2, 2n]. \tag{16.5}$$

The assertion now follows from lemma 16.9, (16.2), (16.3) and (16.5).

Lemma 16.11. $e(n) = 7n^3/4 + O(n^2)$.

Proof. $Q(n) - (k^2+k)/2 = \int_{k-1}^{k} \frac{1}{2}(3n^2 - x^2)dx + O(n)$ for integers $k \in [1, 3n/2]$, while $Q(2n-k) = \int_{(3n-1)/2}^{3n/2} \frac{1}{2}(3n^2-x^2)dx + \int_{3n/2}^{(3n+1)/2} \frac{3}{2}(2n-x)^2dx + O(n)$ for $k = (3n+1)/2$ if n is odd and $Q(2n-k) = \int_{k-1}^{k} \frac{3}{2}(2n-x)^2dx + O(n)$ for integers $k \in [3n/2+1, 2n]$.

Hence, $e(n) = I(n) + O(n^2)$ for

$$I(n) = \int_0^{3n/2} \tfrac{1}{2}(3n^2-x^2)dx + \int_{3n/2}^{2n} \tfrac{3}{2}(2n-x)^2dx = 7n^3/4.$$

Remark. The arguments in (16.5) and 16.11 can be generalized to integral hypersurfaces $X \subset \mathbf{P}^N$ of degree $2 \le D < 2^{N-1}$. Let $Q(n) = \binom{n+N}{N} - \binom{n+N-D}{N}$ be the Hilbert polynomial of $X \subset \mathbf{P}^N$, $F_1, \dots, F_{Q(n)}$ be homogeneous polynomials in $\mathbf{Z}[x_0, \dots, x_N]$ of degree $n \ge 0$ and $\xi_i = (\xi_{i,0}, \dots, \xi_{i,N})$, $1 \le i \le Q(n)$ be primitive $(N+1)$-tuples of integers such that all $\xi_i \pmod p$ represent the same non-singular $\mathbf{F}_p$-point P on X_p. Then the determinant of the $Q(n) \times Q(n)$-matrix $(F_j(\xi_i))$ is divisible by $p^{e(n)}$ for an integer $e(n) = I(n) + O_{D,N}(n^{N-1})$ where $g(D,N) = (D/2)^{1/(N-2)}$ and

$$\begin{aligned} I(n) &= \int_0^{g(D,N)n} \frac{Dn^{N-1} - x^{N-1}}{(N-1)!} dx \\ &+ \int_{g(D,N)n}^{2n} \frac{\left(Dn^{N-1} - (g(D,N)n)^{N-1}\right)}{(N-1)!} \left(\frac{2n-x}{2n - g(D,N)n}\right)^{N-1} dx \\ &= \left(\frac{2D}{N!} + \frac{D(D/2)^{1/(N-2)}}{(N-1)!}\left(1 - \frac{2}{N}\right)\right) n^N \\ &> D^{N/(N-1)} n^N/(N-2)!N \qquad \text{for } 2 \le D < 2^{N-1}. \end{aligned}$$

We have here used the inequality of arithmetic and geometric means in the last line. The assertion that $p^{e(n)} \mid \det(F_j(\xi_i))$ is stronger than (3.11) in [HB02], which says $p^{f(n)} \mid \det(F_j(\xi_i))$ for $f(n) = D^{N/(N-1)} n^N/(N-2)!N + O_{D,N}(n^{N-1})$. If X is a quadric, then $e(n) = 2n^N/(N-1)! + O_N(n^{N-1})$.

We now introduce some new notation, which will be needed in the sequel.

Notation. Let $X \subset \mathbf{P}^N$ be a closed subscheme over $\mathbf{Q}$ and $\mathbf{B} = (B_0, \ldots, B_N) \in \mathbf{R}_{\ge 1}^{N+1}$. Let $p_1, \ldots, p_t$ be distinct primes and P_i be an $\mathbf{F}_{p_i}$-point on X_{p_i} for $1 \le i \le t$. Then $X(\mathbf{Q}; \mathbf{B}; P_1, \ldots, P_t)$ is the set of points in $X(\mathbf{Q}; \mathbf{B})$, which specialize to P_i on X_{p_i} for $1 \le i \le t$. If $B_0 = \cdots = B_N = B$, then we also write $X(\mathbf{Q}; B; P_1, \ldots, P_t)$ for this set.

Theorem 16.12. *Let $X \subset \mathbf{P}^3$ be an integral cubic surface over $\mathbf{Q}$ and $\mathbf{B} = (B_0, \ldots, B_3) \in \mathbf{R}_{\ge 1}^4$. Let $\varepsilon > 0$ and $p_1, \ldots, p_t$ be distinct primes with $q := p_1 \ldots p_t \ge (V/T^{1/3})^{2/7} V^{\varepsilon}$ (cf. 16.5). For $i = 1, \ldots, t$, let P_i be a non-singular $\mathbf{F}_{p_i}$-point on X_{p_i}. Then there exists a surface $Y(\mathbf{B}; P_1, \ldots, P_t) \subset \mathbf{P}^3$ not containing X of degree bounded solely in terms of ε with $X(\mathbf{Q}; \mathbf{B}; P_1, \ldots, P_t) \subset Y(\mathbf{B}; P_1, \ldots, P_t)$.*

Proof. Suppose first that $q > 2V^{1/4}$. We will then show that $X(\mathbf{Q}; \mathbf{B}; P_1, \ldots, P_t)$ is contained in a plane. To see this, let $\xi_k = (\xi_{k,0}, \ldots, \xi_{k,3})$, $1 \le k \le 4$ be primitive quadruples of integers representing points in $X(\mathbf{Q}; \mathbf{B}; P_1, \ldots, P_t)$. Then, by [HB02, lemma 6] we have that $p_i^4 \mid \det(\xi_{k,j})$ for $i = 1, \ldots, t$. As $|\det(\xi_{k,j}))| \le 16V$ by Hadamard's inequality, we conclude that $\det(\xi_{k,j}) = 0$ for $q > 2V^{1/4}$. Hence, any four points in $X(\mathbf{Q}; \mathbf{B}; P_1, \ldots, P_t)$ are coplanar and $X(\mathbf{Q}; \mathbf{B}; P_1, \ldots, P_t)$ contained in a plane when $q > 2V^{1/4}$. We may and shall thus in the sequel suppose that $q \leq 2V^{1/4}$. We will also assume that $V^{1/4} \geq 2$ since the assertion is trivial when $V^{1/4} < 2$.

Suppose that $X \subset \mathbf{P}^3$ is defined by $F(x_0, \dots, x_3)$. There is then by the remark after (5) in [Bro04] a graded monomial ordering of the monomials in $(x_0, \dots, x_3)$ such that $\Phi(\mathbf{B}) = T$ for the leading monomial $\Phi(x_0, \dots, x_3) = x_0^{m_0} \dots x_3^{m_3}$ occurring in $F(x_0, \dots, x_3)$ with non-zero coefficient. We also note that the $Q(n)$ monomials $F_1, \dots, F_{Q(n)} \in \mathbf{Z}[x_0, \dots, x_3]$ of degree n not divisible by Φ form a basis for the summand of degree n of the homogeneous coordinate ring of X (see [Bro04, p.163]).

Now let $I_n(x_0, \dots, x_3)$ be the product of all $\binom{n+3}{3}$ monomials of degree n in $(x_0, \dots, x_3)$ and $J_n(x_0, \dots, x_3)$ be the product of all $\binom{n}{3}$ monomials of degree n in $(x_0, \dots, x_3)$ divisible by Φ. Then $F_1 \dots F_{Q(n)} = I_n/J_n$, $J_n = I_{n-3}\Phi^{\binom{n}{3}}$ and $I_k^4 = (x_0x_1x_2x_3)^{\binom{k+3}{3}}$ for $k = n-3$ and n. Hence

$$\log(F_1(\mathbf{B}) \dots F_{Q(n)}(\mathbf{B})) \tag{16.6}$$
$$= \binom{n+3}{3} \log\left(V^{n/4}\right) - \binom{n}{3} \log\left(V^{(n-3)/4}\right) - \binom{n}{3} \log(T),$$

where the right-hand side is equal to $\frac{1}{2}n^3 \log(V/T^{1/3}) + \frac{1}{2}n \log V + \frac{1}{6}(3n^2 - 2n) \log T$. Hence as $1 \leq T \leq V^3$ for $\mathbf{B} \in \mathbf{R}^4_{\geq 1}$, we deduce from (16.6) that

$$\log(F_1(\mathbf{B}) \dots F_{Q(n)}(\mathbf{B})) \leq (n^3/2) \log(V/T^{1/3}) + \tfrac{3}{2}n^2 \log V + \tfrac{1}{2}n \log V. \tag{16.7}$$

Now fix $\varepsilon > 0$ and assume that $\xi_i = (\xi_{i,0}, \dots, \xi_{i,3})$, $1 \leq i \leq Q(n)$ are primitive quadruples of integers representing points in $X(\mathbf{Q}; \mathbf{B}; P_1, \dots, P_t)$. Then,

$$|\det(F_j(\xi_i))| \leq Q(n)^{Q(n)/2} F_1(\mathbf{B}) \dots F_{Q(n)}(\mathbf{B})$$

by the inequality of Hadamard. Hence, if $\det(F_j(\xi_i)) \neq 0$, we obtain from (16.7) that:

$$\log|\det(F_j(\xi_i))| \leq (n^3/2) \log(V/T^{1/3}) + (Q(n)/2) \log Q(n) + 2n^2 \log V. \tag{16.8}$$

We now give a lower estimate for $\log|\det(F_j(\xi_i))|$. By the previous two lemmas $\det(F_j(\xi_i))$ is divisible by $q^{e(n)}$ for a non-negative integer $e(n) =$

$7n^3/4 + O(n^2)$. We have thus if $\det(F_j(\xi_i)) \neq 0$ and $(V/T^{1/3})^{2/7}V^\varepsilon \leq q \leq 2V^{1/4}$ that

$$\log|\det(F_j(\xi_i))| \geq e(n)\log q \geq (n^3/2)\log(V/T^{1/3}) + (7n^3/4)\log V^\varepsilon - C_1(n^2\log V). \tag{16.9}$$

for some absolute constant $C_1 > 0$.

Therefore, by (16.8) and (16.9) we will have that $\det(F_j(\xi_i)) = 0$ if

$$(7n^3/4)\log V^\varepsilon > (Q(n)/2)\log Q(n) + (C_1+2)n^2\log V. \tag{16.10}$$

But if $V^{1/4} \geq 2$, then it is easy to see that there exists an integer n_0 depending only on $\varepsilon > 0$ such that (16.10) holds for all $n \geq n_0$. There is thus for such n a non-trivial linear combination $\lambda_1 F_1 + \cdots + \lambda_{Q(n)} F_{Q(n)}$ which vanishes at $X(\mathbf{Q}; \mathbf{B}; P_1, \ldots, P_t)$. This completes the proof.

16.3 The semiglobal determinant method for cubic surfaces

We now come to the main lemma of this paper in which we apply theorem 16.12 to several smooth square-free integers q simultaneously.

Main lemma 16.3.1. *Let $X \subset \mathbf{P}^3$ be an integral cubic surface over $\mathbf{Q}$. Let $\mathbf{B} = (B_0, \ldots, B_3) \in \mathbf{R}^4_{\geq 1}$ be a quadruple with $V^{1/4} \geq 2$ for which X is the only cubic surface containing $X(\mathbf{Q}, \mathbf{B})$. Then there exist for each $\varepsilon > 0$ a set Ω of primes depending only on $\mathbf{B}$ and ε, and a set of surfaces $Y(\mathbf{B}; P_1, \ldots, P_t) \subset \mathbf{P}^3$ indexed by sequences $(P_1, \ldots, P_t)$ of non-singular $\mathbf{F}_{p_i}$-points P_i on X_{p_i} for sequences of primes $p_1 < \cdots < p_t$ in Ω with $p_1 \ldots p_t \geq (V/T^{1/3})^{2/7}V^\varepsilon$ such that the following properties hold.*

(a) If $p \in \Omega$, then $V^\varepsilon \leq p \leq C_0 V^\varepsilon$ for some constant C_0 depending only on ε.

(b) $\#\Omega$ is bounded solely in terms of ε.

(c) $X(\mathbf{Q}; \mathbf{B}; P_1, \ldots, P_t) \subset Y(\mathbf{B}; P_1, \ldots, P_t)$ for any sequence $(P_1, \ldots, P_t)$ as above and all surfaces $Y(\mathbf{B}; P_1, \ldots, P_t)$ have the same degree d bounded solely in terms of ε.

(d) Let x be a non-singular point in $X(\mathbf{Q}; \mathbf{B})$. Then one of the following alternatives holds.

(16.3.2) Case (i): There exist primes $p_0 < \cdots < p_t$ in Ω with $p_0 \ldots p_t = O_\varepsilon((V/T^{1/3})^{2/7}V^{3\varepsilon})$ and $p_0 \ldots p_{t-1} \geq (V/T^{1/3})^{2/7}V^\varepsilon$ such that x specializes to a non-singular $\mathbf{F}_{p_i}$-point P_i on X for $0 \leq i \leq t$ and

where x lies on a component of $X \cap Y(\mathbf{B}; P_0, \ldots, P_{t-1})$ not contained in $Y(\mathbf{B}; P_1, \ldots, P_t)$.

(16.3.2) *Case (ii): There exist primes $p_0 < \cdots < p_t$ in Ω with $p_0 \ldots p_t = O_\varepsilon((V/T^{1/3})^{2/7} V^{2\varepsilon})$ such that x specializes to a non-singular $\mathbf{F}_{p_i}$-point P_i on X for $0 \le i \le t$ and where for some $i \in \{0, \ldots, t\}$ with $p_0 \ldots \hat{p}_i \ldots p_t \ge (V/T^{1/3})^{2/7} V^{\varepsilon}$ x lies on a component of $X \cap Y(\mathbf{B}; P_0, \ldots, P_t)$ not contained in $Y(\mathbf{B}; P_0, \ldots, \hat{P}_i, \ldots P_t)$. (Here $\hat{p}_i$ means that p_i is omitted and $\hat{P}_i$ that P_i is omitted.)*

(16.3.3) *There are primes $p_1 < \cdots < p_t$ in Ω with $p_1 \ldots p_t = O_\varepsilon((V/T^{1/3})^{2/7} V^{2\varepsilon})$ and $p_1 \ldots p_t \ge (V/T^{1/3})^{2/7} V^{\varepsilon}$ such that x specializes to a non-singular $\mathbf{F}_{p_i}$-point P_i on X for $1 \le i \le t$ and a component Z of $X \cap Y(\mathbf{B}; P_1, \ldots, P_t)$ containing x where $\#Z(\mathbf{Q}; \mathbf{B}) = O_\varepsilon(1)$, x is singular on Z or x specializes to a non-singular $\mathbf{F}_{p_i}$-point P_i on Z_{p_i} for each $i \in \{1, \ldots, t\}$.*

Proof. As X is the only cubic surface containing $X(\mathbf{Q}; \mathbf{B})$ and $V^{1/4} \ge 2$, we have by 16.6 and 16.7 that $\log \pi_x = O(\log V)$ for any non-singular point x in $X(\mathbf{Q}; \mathbf{B})$. There is thus in this case an absolute constant k such that $\pi_x \le V^k$ for any non-singular point x in $X(\mathbf{Q}; \mathbf{B})$.

Now consider sequences $p_1 < \cdots < p_t$ of primes with $p_1 \ldots p_t \ge (V/T^{1/3})^{2/7} V^{\varepsilon}$ and choose a surface $Y(\mathbf{B}; P_1, \ldots, P_t) \subset \mathbf{P}^3$ not containing X as in theorem 16.12 for each sequence $(P_1, \ldots, P_t)$ of non-singular $\mathbf{F}_{p_i}$-points on X_{p_i}. Then all $Y(\mathbf{B}; P_1, \ldots, P_t)$ are of the same degree $d = O_\varepsilon(1)$ and the components Z of $X \cap Y(\mathbf{B}; P_1, \ldots, P_t)$ of degree at most $D = 3d$ by the theorem of Bezout in [Ful84, 8.4.6]. By lemma 16.8 there are two alternatives for these curves. If 16.8(a) holds, then $\#Z(\mathbf{Q}; \mathbf{B}) = O_d(1)$ by [Ful84, 8.4.6] and hence $\#Z(\mathbf{Q}; \mathbf{B})$ bounded in terms of ε. If 16.8(b) holds, then $\log \underline{\pi}_z = O_d(\log V)$ for any non-singular point z in $Z(\mathbf{Q}; \mathbf{B})$. There is thus for such curves Z a constant l depending only on d such that $\underline{\pi}_z \le V^l$ for any non-singular point z in $Z(\mathbf{Q}; \mathbf{B})$.

Let m be the smallest positive integer such that $p_1 \ldots p_m \ge V^{k+l} (V/T^{1/3})^{2/7} V^{\varepsilon}$ for the m smallest primes with $V^{\varepsilon} \le p_1 < \cdots < p_m$. Let $\Omega = \{p_1, \ldots, p_m\}$. Then,

$$p \ge V^{\varepsilon} \text{ for all } p \in \Omega. \tag{16.11}$$

$$\prod_{p \in \Omega} p \ge V^{k+l} (V/T^{1/3})^{2/7} V^{\varepsilon}. \tag{16.12}$$

We next note that $p_1 p_2 \ldots p_{m-1} < V^{k+l} (V/T^{1/3})^{2/7} V^{\varepsilon} \le V^{k+l+2/7} V^{\varepsilon}$ by the minimality of m. This is also true for $m = 1$ if we put $p_1 p_2 \ldots p_{m-1} = 1$.

Hence by (16.11) we get that $V^{(m-2)\varepsilon} < V^{k+l+2/7}$ and

$$\#\Omega < 2 + (k + l + 2/7)/\varepsilon. \tag{16.13}$$

It also follows from (16.11) and Bertrand's postulate that

$$p \leq 2^{\#\Omega} V^{\varepsilon} \text{ for all } p \in \Omega. \tag{16.14}$$

Hence from (16.11), (16.13) and (16.14) we obtain that (a) and (b) hold, while (c) follows from the definition of $Y(\mathbf{B}; P_1, \ldots, P_t)$ for sequences $p_1 < \cdots < p_t$ in Ω with $p_1 \ldots p_t \geq (V/T^{1/3})^{2/7} V^{\varepsilon}$.

To prove (d), let x be a non-singular point in $X(\mathbf{Q}; \mathbf{B})$ and $\Omega(x) \subset \Omega$ be the subset of all primes p where x specializes to a non-singular $\mathbf{F}_p$-point on X_p. We shall also for square-free products $q = p_1 \ldots p_t \geq (V/T^{1/3})^{2/7} V^{\varepsilon}$ of primes in $\Omega(x)$, write $Y(x, q) = Y(\mathbf{B}; P_1, \ldots, P_t)$ where P_i is the specialization of x in X_{p_i}.

From (16.12) and the bound $\pi_x \leq V^k$, we obtain :

$$\prod_{p \in \Omega(x)} p \geq V^l (V/T^{1/3})^{2/7} V^{\varepsilon}. \tag{16.15}$$

Now list the primes in $\Omega(x)$ in increasing order and stop as soon as $p_1 \ldots p_t \geq (V/T^{1/3})^{2/7} V^{\varepsilon}$. Then $p_1 \ldots p_{t-1} < (V/T^{1/3})^{2/7} V^{\varepsilon}$, with $p_1 \ldots p_{t-1} = 1$ for $t = 1$. Let q_0 be the product $p_1 \ldots p_t$ of these primes and Z be a component of $X \cap Y(x, q_0)$ containing x. Further, let $\Omega(x; Z)$ be the subset of primes p in $\Omega(x)$ where x specializes to a non-singular $\mathbf{F}_p$-point on Z_p. On applying (16.15) and the bound $\underline{\pi}_x \leq V^l$, we obtain the following inequality:

$$\prod_{p \in \Omega(x; Z)} p \geq (V/T^{1/3})^{2/7} V^{\varepsilon}. \tag{16.16}$$

We now define a finite sequence $(q_0, \ldots, q_m)$ of square-free integers recursively as follows. If Z is not a component of $X \cap Y(x, q_i)$, then we stop. If all prime factors of q_i are in $\Omega(x; Z)$ or if all primes in $\Omega(x; Z)$ divide q_i, then we also stop. Otherwise, we replace the largest prime factor p of q_i not in $\Omega(x; Z)$ by the smallest prime $\pi \in \Omega(x; Z)$, which is not a prime factor of q_i and let $q_{i+1} = q_i \pi / p$. Then $q_i < q_{i+1}$ for all i as the prime factors of q_0 are smaller than the other primes in $\Omega(x)$. Further, $q_{i+1} \leq C_0 q_i$ by (a). Hence, as $q_0 < (V/T^{1/3})^{2/7} V^{\varepsilon} p_t \leq C_0 (V/T^{1/3})^{2/7} V^{2\varepsilon}$, we get that $\operatorname{lcm}(q_{i-1}, q_i) < C_0^{i+1} (V/T^{1/3})^{2/7} V^{3\varepsilon}$ and $q_i < C_0^{i+1} (V/T^{1/3})^{2/7} V^{2\varepsilon}$ for all $i \geq 1$.

This process will finish after at most $t \leq \#\Omega$ steps and produce an integer q_m with $(V/T^{1/3})^{2/7} V^{2\varepsilon} \leq q_m < C_0^{t+1} (V/T^{1/3})^{2/7} V^{2\varepsilon}$ satisfying one of the following alternatives.

(16.3.4) $m \geq 1$ and Z is a component of $X \cap Y(x, q_{m-1})$, but not of $X \cap Y(x, q_m)$.

(16.3.5) Z is a component of $X \cap Y(x, q_m)$ and all factors of q_m are in $\Omega(x; Z)$.

(16.3.6) Z is a component of $X \cap Y(x, q_m)$ and q_m is divisible by all primes in $\Omega(x; Z)$.

Then, as $\mathrm{lcm}(q_{m-1}, q_m) < C_0^{t+1}(V/T^{1/3})^{2/7}V^{3\varepsilon}$ all the conditions in the first case of (16.3.2) will hold if (16.3.4) holds. We also obtain that the conditions in (16.3.3) hold if (16.3.5) holds.

If (16.3.6) holds, then we define a new sequence $(q_m, \ldots, q_{m+n})$ of square-free integers starting with q_m. If Z is not a component of $X \cap Y(x, q_{m+j})$ or if all factors of q_{m+j} are in $\Omega(x; Z)$, then we stop. Otherwise, we let $q_{m+j+1} = q_{m+j}/p$ for the largest prime factor p of q_{m+j}, which is not in $\Omega(x; Z)$. This process will finish after less than t steps and all integers in the sequence will be divisible by all primes in $\Omega(x; Z)$. By (16.16) we have thus that $(V/T^{1/3})^{2/7}V^{\varepsilon} \leq q_{m+n}$. The last integer q_{m+n} will also satisfy $q_{m+n} \leq q_m < C_0^{t+1}(V/T^{1/3})^{2/7}V^{2\varepsilon}$ and one of the following conditions:

(16.3.7) $n \geq 1$ and Z is a component of $X \cap Y(x, q_{m+n-1})$, but not of $X \cap Y(x, q_{m+n})$.

(16.3.8) Z is a component of $X \cap Y(x, q_{m+n})$ and all factors of q_{m+n} are in $\Omega(x; Z)$.

If (16.3.7) holds, then all the conditions in the second case of (16.3.2) hold. If instead (16.3.8) holds, then the conditions in (16.3.3) will hold. This completes the proof.

Remark. Let $(P_0, \ldots, P_t)$ be a sequence where P_i is a non-singular $\mathbf{F}_{p_i}$-point on X_{p_i} for $0 \leq i \leq t$. It will be important to count the number of non-singular points $x \in X(\mathbf{Q}; \mathbf{B})$, which specialize to $(P_0, P_1, \ldots, P_t)$ and satisfy (16.3.2). In case (i), any such point x belongs to $Y(\mathbf{B}; P_1, \ldots, P_t)$ (see 16.3.1(c)) and to a component of $X \cap Y(\mathbf{B}; P_0, \ldots, P_{t-1})$ not contained in $Y(\mathbf{B}; P_1, \ldots, P_t)$. There are thus by the theorem of Bezout [Ful84, 8.4.6] at most $3d^2$ non-singular points in $X(\mathbf{Q}; \mathbf{B})$ which specialize to $(P_0, P_1, \ldots, P_t)$ and satisfy the condition in case (i). Similarily, we get that there are at most $3d^2$ non-singular points in $X(\mathbf{Q}; \mathbf{B})$, which specialize to $(P_0, P_1, \ldots, P_t)$ and satisfy the condition in case (ii) with $x \in Y(\mathbf{B}; P_0, \ldots, \hat{P}_i, \ldots, P_t)$ for a given $i \in \{0, 1, \ldots, t\}$. There are thus altogether at most $3(t+2)d^2$ points $x \in X(\mathbf{Q}; \mathbf{B})$, which specialize to $(P_0, P_1, \ldots, P_t)$ and satisfy (16.3.2).

Lemma 16.13. *Let $Z \subset \mathbf{P}^3$ be an integral curve over $\mathbf{Q}$ of degree δ. Let $\varepsilon > 0$ and $p_1, \ldots, p_t$ be distinct primes with $p_1 \ldots p_t \geq B^{2/\delta} V^{\varepsilon}$. Let $(P_1, \ldots, P_t)$ be a sequence where P_i is a non-singular $\mathbf{F}_{p_i}$-point on X_{p_i} for $1 \leq i \leq t$. Then there exists a surface $W \subset \mathbf{P}^3$ of degree $O_{\delta,\varepsilon}(1)$ not containing Z with $Z(\mathbf{Q}; B; P_1, \ldots, P_t) \subset W$. In particular, $N(Z; B; P_1, \ldots, P_t) = O_{\delta,\varepsilon}(1)$.*

Proof. This is a special case of theorem 3.2 in [Sal07]. To obtain the second assertion from the first, use the theorem of Bezout in [Ful84, 8.4.6].

Lemma 16.14. *Let $Z \subset \mathbf{P}^3 = \operatorname{Proj} \mathbf{Q}[x_0, \ldots, x_3]$ be an integral curve over $\mathbf{Q}$ of degree δ, which is not contained in the hyperplane $H_0 \subset \mathbf{P}^3$ defined by $x_0 = 0$. Let $\varepsilon > 0$, $\mathbf{B} = (1, B, B, B)$ and $p_1, \ldots, p_t$ be distinct primes with $p_1 \ldots p_t \geq B^{1/\delta} V^{\varepsilon}$. Let $(P_1, \ldots, P_t)$ be a sequence where P_i is a non-singular $\mathbf{F}_{p_i}$-point on X_{p_i} for $1 \leq i \leq t$. Then there exists a surface $W \subset \mathbf{P}^3$ of degree $O_{\delta,\varepsilon}(1)$ not containing Z with $Z(\mathbf{Q}; \mathbf{B}; P_1, \ldots, P_t) \subset W$. In particular, $N(Z; \mathbf{B}; P_1, \ldots, P_t) = O_{\delta,\varepsilon}(1)$.*

Proof. This is a special case of corollary 3.7 in [Sal07]. To obtain the second assertion from the first, use the theorem of Bezout in [Ful84, 8.4.6].

16.4 Integral points on affine cubic hypersurfaces

We shall in this section prove theorem 16.3 for affine hypersurfaces.

Lemma 16.15. *Let $X \subset \mathbf{P}^3$ be a cubic surface over some field. Then the sum of the degrees of the components of the singular locus of X is at most* 16.

Proof. This follows from the theorem of Bezout [Ful84, 8.4.6] applied to the system of quadratic equations given by the gradient of a cubic homogeneous polynomial defining $X \subset \mathbf{P}^3$.

We may now derive the following corollary of 16.3.1 and 16.14. We shall thereby use the trivial uniform estimate $\#X_p(\mathbf{F}_p) = O(p^2)$ for cubic surfaces X over $\mathbf{Q}$, when we sum over tuples $(P_1, \ldots, P_t)$ of non-singular $\mathbf{F}_{p_i}$-points P_i on X_{p_i}. □

Lemma 16.16. *Let $X \subset \mathbf{P}^3 = \operatorname{Proj} \mathbf{Q}[x_0, \ldots, x_3]$ be an integral cubic surface over $\mathbf{Q}$ such that the plane defined by $x_0 = 0$ intersects X properly. Let $B \geq 1$ and $\mathbf{B} = (1, B, B, B)$. Then there exists a set of $O_\varepsilon(B^{8/7+\varepsilon})$ rational lines*

$l \subset X$ such that all but $O_\varepsilon(B^{8/7+\varepsilon})$ points in $X(\mathbf{Q}; \mathbf{B})$ lie on one of these lines.

Proof. Suppose first that there is more than one cubic surface containing $X(\mathbf{Q}; \mathbf{B})$. Then $X(\mathbf{Q}; \mathbf{B})$ is contained in the union of at most nine integral curves C of degree at most nine by [Ful84, 8.4.6]. As $\#C(\mathbf{Q}; B) = O_{\delta,\varepsilon}(B^{2/\delta+\varepsilon})$ for any integral curve $C \subset \mathbf{P}^3$ of degree δ defined over $\mathbf{Q}$ (see [HB02, th.5]), there is thus a set of at most nine rational lines $l \subset X$ such that all but $O_\varepsilon(B^{1+\varepsilon})$ points in $X(\mathbf{Q}; \mathbf{B})$ lie on one of these lines in this case. We may hence assume that X is the only cubic surface containing $X(\mathbf{Q}; \mathbf{B})$ and we may also suppose that $B \geq 2$ since the assertion is trivial when $B < 2$. We have then by lemma 16.3.1 that any non-singular point in $X(\mathbf{Q}; \mathbf{B})$ will satisfy (16.3.2) or (16.3.3).

It is clear from lemma 16.15 and [HB02, th.5] that there are $O(1)$ lines in the singular locus and that there are at most $O_\varepsilon(B^{1+\varepsilon})$ singular points in $X(\mathbf{Q}; B)$ outside these lines. It is thus enough to consider the points in $X(\mathbf{Q}; \mathbf{B})$ in the non-singular locus X_{ns}. We also observe that $V = B^3$ and that $V/T^{1/3} = B^2$ by lemma 1.12 in [Sal07].

Let us first estimate the number of points in $X_{ns}(\mathbf{Q}; \mathbf{B})$ satisfying (16.3.2). We may then fix the $t+1$ primes $p_0, p_1, \ldots, p_t$ in (16.3.2) as the number of such sets of primes is bounded in terms of ε (cf. 16.3.1(b)). There are $O_\varepsilon((p_0 p_1 \ldots p_t)^2)$ tuples $(P_0, P_1, \ldots, P_t)$ of non-singular $\mathbf{F}_{p_i}$-points P_i on X_{p_i} as $t+1 \leq \#\Omega$ is bounded in terms of ε (see 16.3.1(b)). There are, thus, by (16.3.2) $O_\varepsilon(B^{8/7}B^{18\varepsilon})$ such $(t+1)$-tuples. There are further by the remark after (16.3.8) at most $3(t+2)d^2 = O_\varepsilon(1)$ points in $X_{ns}(\mathbf{Q}; \mathbf{B})$ which specialize to a given $(t+1)$-tuple $(P_0, P_1, \ldots, P_t)$ and satisfy (16.3.2). There are thus altogether $O_\varepsilon(B^{8/7+18\varepsilon})$ points in $X_{ns}(\mathbf{Q}; \mathbf{B})$ satisfying (16.3.2).

We now estimate the number of points in $X_{ns}(\mathbf{Q}; \mathbf{B})$ satisfying (16.3.3). We may then fix the primes $p_1, \ldots, p_t$ in (16.3.3) as the number of such sets of primes is bounded in terms of ε. There are $O_\varepsilon((p_1 \ldots p_t)^2)$ and hence $O_\varepsilon(B^{8/7}B^{12\varepsilon})$ t-tuples $(P_1, \ldots, P_t)$ of non-singular $\mathbf{F}_{p_i}$-points P_i on X_{p_i} for $1 \leq i \leq t$. By [Ful84, 8.4.6], there are at most $3d$ components Z in $X \cap Y(P_1, \ldots, P_t)$ and each component is of degree at most $3d$. The points in $X_{ns}(\mathbf{Q}; \mathbf{B})$ satisfying (16.3.3) will thus lie on $O_\varepsilon(B^{8/7+12\varepsilon})$ curves Z of degree $\leq 3d = O_\varepsilon(1)$ (see 16.3.1(c)). As there are $O_d(1)$ singular points on each curve of degree $\leq 3d$, it only remains to estimate the total contribution from all $Z(\mathbf{Q}; \mathbf{B}; P_1, \ldots, P_t)$ such that P_i is a non-singular $\mathbf{F}_{p_i}$-point on Z_{p_i} for $1 \leq i \leq t$. If $\deg Z \geq 2$, then $p_1 \ldots p_t \geq B^{4/7}V^\varepsilon \geq B^{1/\deg Z}V^\varepsilon$ such that $\#Z(\mathbf{Q}; \mathbf{B}; P_1, \ldots, P_t) = O_\varepsilon(1)$ by 16.14. There are thus altogether $O_\varepsilon(B^{8/7+12\varepsilon})$ points in $X_{ns}(\mathbf{Q}; \mathbf{B})$ satisfying (16.3.3) for some curve Z of degree at least two. The other points satisfying (16.3.3) will lie on one of $O_\varepsilon(B^{8/7+12\varepsilon})$ rational lines Z on X, thereby completing the proof.

We shall need the following notation for counting problems for affine varieties.

Notation 16.17. Let $X \subset \mathbf{P}^N$ be a quasi-projective variety over $\mathbf{Q}$.

(a) $S_1(X; B)$ is the set of rational points on X which may be represented by an integral $(N+1)$-tuple $(1, x_1, \ldots, x_N)$ with $|x_i| \leq B$ for $i \in \{1, \ldots, N\}$.
(b) $N_1(X; B) = \#S_1(X; B)$.

Lemma 16.18. *Let $X \subset \mathbf{P}^3$ be an integral surface of degree $D > 1$ over some field k and $H \subset \mathbf{P}^3$ be a k-rational plane such that X is not a cone with vertex in H. Then, the number of lines on X passing through a given point y on $H \cap X$ is bounded in terms of D.*

Proof. The assertion is proved in [BHBS06, lemma 9] under the stronger hypothesis that $H \cap X$ is integral. This hypothesis is only used to ensure that the schemes $Z_{X,y}$ parametrizing lines $l \subset X$ through points y on $H \cap X$ are empty or zero-dimensional. Hence as $\dim(Z_{X,y}) \geq 1$ would imply that X is a cone with vertex in y, we obtain the desired result from the proof in (op. cit.).

Theorem 16.19. *Let $X \subset \mathbf{P}^3 = \operatorname{Proj} \mathbf{Q}[x_0, \ldots, x_3]$ be an integral cubic surface such that X is not a cone with vertex in the plane H defined by $x_0 = 0$. Then $N_1(X; B) = O_\varepsilon(B^{8/7+\varepsilon})$.*

Proof. Let $X_0 \subset X$ be the open subset where $x_0 \neq 0$. Then $2N_1(X; B) = N(X_0; 1, B, B, B)$. By 16.16 it thus suffices to show that there are $O_\varepsilon(B^{8/7+2\varepsilon})$ points on a set of $O_\varepsilon(B^{8/7+\varepsilon})$ lines not contained in H. This follows from the previous lemma and the proof of proposition 1 in [BHBS06].

Theorem 16.20. *Let $f(y_1, y_2, y_3) \in \mathbf{Q}[y_1, y_2, y_3]$ be a cubic irreducible polynomial which is not a polynomial of two linear forms in $\mathbf{Q}[y_1, y_2, y_3]$. Let $n(f; B)$ be the number of triples (y_1, y_2, y_3) of integers in $[-B, B]$ such that $f(y_1, y_2, y_3) = 0$. Then*

$$n(f; B) = O_\varepsilon(B^{8/7+\varepsilon}).$$

Proof. Let $F(x_0, x_1, x_2, x_3) = x_0^3 f(x_1/x_0, x_2/x_0, x_3/x_0)$. Then the cubic surface $X \subset \mathbf{P}^3$ over $\mathbf{Q}$ defined by F will satisfy the hypothesis in 16.19. Hence $n(f; B) = N_1(X; B) = O_\varepsilon(B^{8/7+\varepsilon})$.

Theorem 16.21. *Let $N \geq 3$ and $f(y_1, \ldots, y_N) \in \mathbf{Q}[y_1, \ldots, y_N]$ be a cubic polynomial such that its leading cubic form is absolutely irreducible. Let $n(f; B)$ be the number of n-tuples $(y_1, \ldots, y_N)$ of integers in $[-B, B]$ such that $f(y_1, \ldots, y_N) = 0$. Then*

$$n(f; B) = O_\varepsilon(B^{n-2+1/7+\varepsilon}).$$

Proof. This is a special case of theorem 16.20 when $N = 3$. The general case is deduced from this case by means of an argument with repeated hyperplane sections (see [BHBS06, lemma 8]).

Remark. One can use a finer hyperplane section argument than in (op.cit.) to deduce a more general result than in 16.21 from 16.20. It suffices that $f(y_1, \ldots, y_N) \in \mathbf{Q}[y_1, \ldots, y_N]$ is absolutely irreducible and not a polynomial of two linear forms in $(y_1, \ldots, y_N)$ to have the estimate in 16.21.

16.5 Counting points on families of conics in $\mathbf{P}^3$

We shall in this section use the word conic for a closed subscheme of $\mathbf{P}^3$ with Hilbert polynomial $2t+1$. A conic may thus be a union of two intersecting lines or a double line. It is well known (cf. [Har82, 1.b]) that the Hilbert scheme $\mathbf{H}$ of conics in $\mathbf{P}^3$ is a $\mathbf{P}^5$-bundle $\pi : \mathbf{H} \to \mathbf{P}^{3\vee}$ over the dual projective space $\mathbf{P}^{3\vee}$. This morphism π sends the point on $\mathbf{H}$ parametrizing a conic $C \subset \mathbf{P}^3$ to the point in the dual plane $\mathbf{P}^{3\vee}$ parametrizing the plane $\Pi = \langle C \rangle \subset \mathbf{P}^3$ spanned by C. We shall also let $\mathbf{H}_X \subset \mathbf{H}$ be the subscheme of all conics $C \subset X$ on a cubic surface $X \subset \mathbf{P}^3$.

Lemma 16.22. *Let $X \subset \mathbf{P}^3$ be a cubic surface over some field, which is not covered by its lines. Then the number of lines on X is uniformly bounded.*

Proof. This is well known and easy to prove using elimination theory. □

Lemma 16.23. *Let $X \subset \mathbf{P}^3$ be an integral cubic surface over some field, which is not covered by its lines. There is then for each component Y of $\mathbf{H}_X$ a line $l \subset X$ such that Y parametrizes the residual conics of l in $\Pi \cap X$ for the planes $\Pi \subset \mathbf{P}^3$ passing through l.*

Proof. The residual lines l in $\langle C\rangle \cap X$ of the conics C parametrized by Y form a family of lines over Y. As X and Y are irreducible and X is not covered by its lines, theses lines will coincide. Hence Y is the family of residual conics of l in $\Pi \cap X$ for the planes Π passing through l. □

Lemma 16.24. *Let $C \subset \mathbf{P}^3$ be a non-singular conic over $\mathbf{Q}$, $q = p_1 \dots p_t$ be a square-free integer and $(P_1, \dots, P_t)$ be a t-tuple where each P_i is a non-singular $\mathbf{F}_{p_i}$-point P_i on C_{p_i}. Let $H(\Pi)$ be the height of the point in the dual plane $\mathbf{P}^{3\vee}$ parametrizing $\Pi = \langle C\rangle \subset \mathbf{P}^3$. Then the following holds.*

(a) $\#C(\mathbf{Q}; B; P_1, \dots, P_t) \leq 2$ in case $H(\Pi) > 6B^3/q^3$.
(b) $\#C(\mathbf{Q}; B) = O_\varepsilon(B/H(\Pi)^{1/3} + B^\varepsilon)$.

Proof. To prove (a), suppose that we had $\#C(\mathbf{Q}; B; P_1, \dots, P_t) > 2$. We may then find primitive quadruples $\xi_k = (\xi_{k,0}, \dots, \xi_{k,3})$, $k = 1, 2, 3$ of integers in $[-B, B]$ representing different points in $\#C(B; P_1, \dots, P_t)$. These cannot be collinear. So they will span the plane $\Pi = \langle C\rangle$. Its height $H(\Pi)$ is given by $\max_{0\leq j\leq 3} |\Delta_j| / \gcd_{0\leq j\leq 3}(\Delta_j)$ for the determinants Δ_j, $0 \leq j \leq 3$ of the four 3×3-submatrices of the 3×4-submatrix with rows ξ_k, $k = 1, 2, 3$. We now apply 16.9 to the stalk R of the structure sheaf of C^{cl} at P_i. If say $x_0(P_i) \neq 0$, then $r_j = x_j/x_0 \in R$ for $0 \leq j \leq 3$ and we conclude as in (16.2) that $p_i^3 \mid \Delta_j$ for $0 \leq j \leq 3$. Further, $\max_{0\leq j\leq 3} |\Delta_j| \leq 6B^3$ as all $\xi_{k,j} \in [-B, B]$. This proves that $H(\Pi) \leq 6B^3/q^3$ when $\#C(\mathbf{Q}; B; P_1, \dots, P_t) > 2$, as desired.

To prove (b), we first note that (a) also holds when $q = 1$. Hence $\#C(\mathbf{Q}; B) \leq 2$ when $H(\Pi) > 6B^3$. It remains to treat the case where $H(\Pi) \leq 6B^3$. We may then assume that $\#C(\mathbf{Q}; B; P_1, \dots, P_t) \geq 5$ and $B \geq 2$. Let $\xi_k = (\xi_{k,0}, \dots, \xi_{k,3})$, $k = 1, 2, 3, 4, 5$ be primitive quadruples representing five such points and $\pi \leq 60B^3$ be the product of all primes p such that three of these quadruples become collinear in C_p after reduction (mod p). There is then by Bertrand's postulate a prime p not dividing π such that $p > (6B^3/H(\Pi))^{1/3}$ and $p = O_\varepsilon(B/H(\Pi)^{1/3} + B^\varepsilon)$. As C_p is non-singular and $H(\Pi) > 6B^3/p^3$, we may now apply (a) for $q = p$. We then get that $\#C(\mathbf{Q}; B; P) \leq 2$ for all $\mathbf{F}_p$-points P on C_p and that $\#C(\mathbf{Q}; B) \leq 2(p + 1) = O_\varepsilon(B/H(\Pi)^{1/3} + B^\varepsilon)$, as was to be shown. □

Lemma 16.25. *Let $X \subset \mathbf{P}^3$ be an integral cubic surface over $\mathbf{Q}$, which is not covered by its lines. Let $q = p_1 \dots p_t \geq B^{6/7}$ be a square-free integer with all prime factors $p_i \geq B^\varepsilon$ and Λ be a set of $O_\varepsilon(B^{12/7+\varepsilon})$ tuples $(P_1, \dots, P_t)$ where P_i is a non-singular $\mathbf{F}_{p_i}$-point on X_{p_i} for each i. Suppose that for each* *such* $\lambda = (P_1, \dots, P_t) \in \Lambda$ *we are given a non-singular conic* $C_\lambda \subset X$ *where each* P_i *is a non-singular* $\mathbf{F}_{p_i}$*-point on* $(C_\lambda)_{p_i}$*. Then,*

$$\sum_{\lambda \in \Lambda} \#C_\lambda(\mathbf{Q}; B; P_1, \dots, P_t) = O_\varepsilon(B^{12/7+\varepsilon}).$$

Proof. If $H(\langle C_\lambda \rangle) > 6B^{3/7}$ for $\lambda = (P_1, \dots, P_t) \in \Lambda$, then $H(\langle C_\lambda \rangle) > 6B^3/q^3$ and $\#C_\lambda(B; P_1, \dots, P_t) \leq 2$ by 16.24(a). The total contribution to the sum from all t-tuples λ with $H(\langle C_\lambda \rangle) > 6B^{3/7}$ is thus $O_\varepsilon(B^{12/7+\varepsilon})$. For the remaining C_λ, we first consider the contribution from conics C with $H(\langle C \rangle) \in [R, 2R)$ for a fixed $R \geq 1$. There are $O(R^2)$ such conics C on X by 16.22 and 16.23, and $O_\varepsilon(B/R^{1/3} + B^\varepsilon)$ rational points of height $\leq B$ on each such conic by 16.24. The sum of $\#C(\mathbf{Q}; B)$ for all such C will thus be of order $O_\varepsilon(BR^{5/3} + R^2B^\varepsilon)$ and of order $O_\varepsilon(B^{12/7})$ for $R \leq 6B^{3/7}$. After summing over $O(\log B)$ dyadic intervals $[R, 2R)$, we thus obtain the bound $O_\varepsilon(B^{12/7+\varepsilon})$ for the total contribution of all C_λ with $H(\langle C_\lambda \rangle) \leq 6B^{3/7}$. This completes the proof. □

16.6 Rational points on projective cubic surfaces

The following result improves upon theorem 0.1 in [Sal] and theorem 7 in [HB02].

Theorem 16.26. *Let* $X \subset \mathbf{P}^3$ *be an integral cubic surface over* $\mathbf{Q}$ *and* U *be the complement of the union of the lines on* X*. Then,*

$$N(U; B) = O_\varepsilon(B^{12/7+\varepsilon}).$$

Proof. We may just as in the proof of lemma 16.16 assume that X is the only cubic surface containing $X(\mathbf{Q}; B)$ and that $B \geq 2$. Then, by 16.3.1 we get that any non-singular point in $X(\mathbf{Q}; B)$ will satisfy (16.3.2) or (16.3.3). We have also seen in the proof of lemma 16.16 that there are $O_\varepsilon(B^{1+\varepsilon})$ singular points of height $\leq B$ on U. It is thus enough to consider the contribution to $N(U; B)$ from the non-singular locus X_{ns} of X. Finally, we note that $V = B^4$ and $V/T^{1/3} = B^3$.

Let us first estimate the number of points in $X_{ns}(\mathbf{Q}; \mathbf{B})$ satisfying (16.3.2).We may then fix the $t+1$ primes $p_0, p_1, \dots, p_t$ in (16.3.2) as the number of such sets of primes is bounded in terms of ε (cf. 16.3.1(b)). There are $O_\varepsilon((p_0p_1 \dots p_t)^2)$ $(t+1)$-tuples $(P_0, P_1, \dots, P_t)$ of non-singular $\mathbf{F}_{p_i}$-points P_i on X_{p_i} as $t+1 \leq \#\Omega$ is bounded in terms of ε (see 16.3.1(b)). There are thus by (16.3.2) $O_\varepsilon(B^{12/7}B^{24\varepsilon})$ such $(t+1)$-tuples. There are also by the remark after (16.3.8) at most $3(t+2)d^2 = O_\varepsilon(1)$ points in $X_{ns}(\mathbf{Q}; \mathbf{B})$ which specialize to a given $(t+1)$-tuple $(P_0, P_1, \ldots, P_t)$ and satisify (16.3.2). There are hence altogether $O_\varepsilon(B^{12/7+24\varepsilon})$ points in $X_{ns}(\mathbf{Q}; \mathbf{B})$ satisfying (16.3.2).

We now estimate the contribution to $N(U; B)$ from the points in $X_{ns}(\mathbf{Q}; B)$ satisfying (16.3.3). We may then fix the primes $p_1, \ldots, p_t$ in (16.3.3) as the number of such sets of primes is bounded in terms of ε. There are $O_\varepsilon((p_0p_1 \ldots p_t)^2)$ and hence $O_\varepsilon(B^{12/7}B^{16\varepsilon})$ tuples $(P_1, \ldots, P_t)$ of non-singular $\mathbf{F}_{p_i}$-points P_i for $1 \le i \le t$. By [Ful84, 8.4.6], there are at most $3d$ components Z in $X \cap Y(P_1, \ldots, P_t)$ and each such component is of degree at most $3d$. The points in $X_{ns}(\mathbf{Q}; B)$ satisfying (16.3.3) will thus lie on $O_\varepsilon(B^{12/7+16\varepsilon})$ curves Z of degree $\le 3d = O_\varepsilon(1)$ (see 16.3.1(c)). As there are $O_d(1)$ singular points on each curve of degree $\le 3d$, it only remains to estimate the total contribution to $N(U; B)$ from all $Z(\mathbf{Q}; B; P_1, \ldots, P_t)$ such that P_i is a non-singular $\mathbf{F}_{p_i}$-point on Z_{p_i} for $1 \le i \le t$. If $\deg Z \ge 3$, then $p_1 \ldots p_t \ge B^{6/7}V^\varepsilon \ge B^{2/\deg Z}V^\varepsilon$ such that $\#Z(\mathbf{Q}; B; P_1, \ldots, P_t) = O_\varepsilon(1)$ by 16.13. There are thus altogether $O_\varepsilon(B^{12/7+16\varepsilon})$ points in $X_{ns}(\mathbf{Q}; B)$ satisfying (16.3.3) for some curve Z of degree at least three. By 16.25 we get the same total bound for the number of points in $X_{ns}(\mathbf{Q}; B)$ satisfying (16.3.3) for some non-singular conic Z. This finishes the proof as the components Z of degree one and the singular components Z of degree two are disjoint from U. □

Remark. The method of this paper applies to other del Pezzo surfaces. If $X \subset \mathbf{P}^3$ is an integral quadric, one gets thereby another proof of the bound $N(X; B) = O_\varepsilon(B^{2+\varepsilon})$ in [HB02, theorem 2]. If U is the complement of the lines on X on an anticanonically embedded non-singular del Pezzo surface over $\mathbf{Q}$ of degree $3 \le d \le 8$, then one obtains that $N(U; B) = O_\varepsilon(B^{12/(d+4)+\varepsilon})$.

One can combine the method of this paper with conic bundle techniques to get non-uniform upper bounds for classes of cubic surface with rational lines. If X contains one rational line then $N(U; B) = O_{X,\varepsilon}(B^{8/5+\varepsilon})$ while $N(U; B) = O_{X,\varepsilon}(B^{3/2+\varepsilon})$ for a cubic surface with two rational lines.

References

[AK70] Allen Altman and Steven Kleiman. *Introduction to Grothendieck duality theory*. Lecture Notes in Mathematics, Vol. 146. Springer-Verlag, Berlin-New York, 1970.

[BHBS06] T. D. Browning, D. R. Heath-Brown, and P. Salberger. Counting rational points on algebraic varieties. *Duke Math. J.*, 132(3):545–578, 2006.

[Bro04] Niklas Broberg. A note on a paper by R. Heath-Brown: "The density of rational points on curves and surfaces" [Ann. of Math. (2) **155** (2002), no. 2, 553–595; mr1906595]. *J. Reine Angew. Math.*, 571:159–178, 2004.

[Eis95] David Eisenbud. *Commutative Algebra with a View toward Algebraic Geometry*, volume 150 of *Graduate Texts in Mathematics*. Springer-Verlag, New York, 1995.

[FL85] William Fulton and Serge Lang. *Riemann-Roch algebra*, volume 277 of *Grundlehren der Mathematischen Wissenschaften [Fundamental Principles of Mathematical Sciences]*. Springer-Verlag, New York, 1985.

[Ful84] William Fulton. *Intersection theory*, volume 2 of *Ergebnisse der Mathematik und ihrer Grenzgebiete (3) [Results in Mathematics and Related Areas (3)]*. Springer-Verlag, Berlin, 1984.

[Har82] Joe Harris. *Curves in projective space*, volume 85 of *Séminaire de Mathématiques Supérieures [Seminar on Higher Mathematics]*. Presses de l'Université de Montréal, Montreal, Que., 1982. With the collaboration of David Eisenbud.

[HB02] D. R. Heath-Brown. The density of rational points on curves and surfaces. *Ann. of Math. (2)*, 155(2):553–595, 2002.

[Sal] Per Salberger. Counting rational points on projective varieties. submitted.

[Sal07] Per Salberger. On the density of rational and integral points on algebraic varieties. *J. Reine Angew. Math.*, 606:123–147, 2007.

[Sal08] Per Salberger. Rational points of bounded height on projective surfaces. *Math. Z.*, 258(4):805–826, 2008.

17

Descent on toric fibrations

Alexei N. Skorobogatov

Department of Mathematics, South Kensington Campus, Imperial College London, SW7 2BZ England, U.K. – and – Institute for the Information Transmission Problems, Russian Academy of Sciences, 19 Bolshoi Karetnyi, Moscow, 127994 Russia

E-mail address: a.skorobogatov@imperial.ac.uk

ABSTRACT. We describe descent on families of torsors for a constant torus. A recent result of Browning and Matthiesen then implies that the Brauer–Manin obstruction controls the Hasse principle and weak approximation when the ground field is $\mathbb{Q}$ and the singular fibres are all defined over $\mathbb{Q}$.

17.1 Introduction

Let T be a torus over a number field k. Let X be a smooth, proper, geometrically integral variety with a surjective morphism $f : X \to \mathbb{P}^1_k$ whose generic fibre $X_{k(t)}$ is geometrically integral and is birationally equivalent to a $k(t)$-torsor for T. The main result of this note says that the set $X(k)$ is dense in $X(\mathbf{A}_k)^{\mathrm{Br}}$ if certain auxiliary varieties satisfy the Hasse principle and weak approximation.

These varieties are given by explicit equations. Choose $\mathbb{A}^1_k \subset \mathbb{P}^1_k$ so that the fibre of f at $\infty = \mathbb{P}^1_k \setminus \mathbb{A}^1_k$ is smooth. Let $P_1, \ldots, P_r$ be closed points of $\mathbb{P}^1_k$ such that each fibre of the restriction of f to $\mathbb{P}^1_k \setminus (P_1 \cup \ldots \cup P_r)$ is *split*, i.e. contains an irreducible component of multiplicity 1 that is geometrically irreducible. By a well-known result (Lemma 17.8) the fibre X_{P_i}, for $i = 1, \ldots, r$, has an irreducible component of multiplicity 1. We fix such a component in each X_{P_i} and define k_i as the algebraic closure of the residue field $k(P_i)$ in the function field of this component. For $i = 1, \ldots, r$ let $p_i(t) \in k[t]$ be the monic irreducible polynomial such that P_i is the zero set of $p_i(t)$ in $\mathbb{A}^1_k$, and let a_i be the image of t in $k(P_i) = k[t]/(p_i(t))$. Let u, v be independent variables, and let z_i be a k_i-variable, for $i = 1, \ldots, r$. For $\alpha = \{\alpha_i\}$, where $\alpha_i \in k(P_i)^*$, we define the quasi-affine variety $W_\alpha \subset \mathbb{A}^2_k \times \prod_{i=1}^r R_{k_i/k}(\mathbb{A}^1_{k_i})$ by

$$\alpha_i(u - a_i v) = N_{k_i/k(P_i)}(z_i), \quad (u, v) \neq (0, 0), \quad i = 1, \ldots, r. \tag{17.1}$$

Arithmetic and Geometry, ed. Luis Dieulefait *et al.* Published by Cambridge University Press.

The varieties W_α are smooth and geometrically irreducible. Over an algebraic closure of k such a variety is given by $\sum_{i=1}^r [k(P_i) : k]$ equations in $2 + \sum_{i=1}^r [k_i : k]$ variables. We can now state our main result.

Theorem 17.1. *Suppose that for each* $\alpha \in \prod_{i=1}^r k(P_i)^*$ *the variety* W_α *satisfies the Hasse principle and weak approximation. Then* $X(k)$ *is dense in* $X(\mathbb{A}_k)^{\mathrm{Br}}$.

The results of this kind are obtained by the descent method of Colliot-Thélène and Sansuc, and have a long history. When the relative dimension and the number of singular geometric fibres of f are small, geometric proofs of the Hasse principle and weak approximation for W_α were obtained by Colliot-Thélène, Sansuc, Swinnerton-Dyer and others, see [8, 5, 22, 27, 10] and [24, Ch. 7]. The analytic tool in these proofs is Dirichet's theorem on primes in an arithmetic progression. Over $k = \mathbb{Q}$ one can do more if one uses more advanced analytic methods: the circle method [18, 9], sieve methods [1] (see also [11] where the results of [1] are used) and recent powerful results from additive combinatorics [12, 13, 14, 3, 17, 25]. Note that the circle method can sometimes be applied over arbitrary number fields, see [26, 21].

When $k = k(P_1) = \ldots = k(P_r) = \mathbb{Q}$, a recent result of Browning and Matthiesen obtained by methods of additive combinatorics [2, Thm 1.3] establishes the Hasse principle and weak approximation for W_α. Hence we deduce the following corollary of Theorem 17.1.

Corollary 17.2. *Let* X *be a smooth, proper, geometrically integral variety over* $\mathbb{Q}$*, and let* $f : X \to \mathbb{P}^1_{\mathbb{Q}}$ *be a surjective morphism satisfying the following properties.*

(a) *There is a torus* T *over* $\mathbb{Q}$ *such that the generic fibre* $X_{\mathbb{Q}(t)}$ *of* f *is birationally equivalent to a* $\mathbb{Q}(t)$*-torsor for* $T \times_{\mathbb{Q}} \mathbb{Q}(t)$.
(b) *There exists a finite subset* $E \subset \mathbb{P}^1_{\mathbb{Q}}(\mathbb{Q})$ *such that* $X \setminus f^{-1}(E) \to \mathbb{P}^1_{\mathbb{Q}} \setminus E$ *has split fibres.*

Then $X(\mathbb{Q})$ *is dense in* $X(\mathbb{A}_{\mathbb{Q}})^{\mathrm{Br}}$.

This generalizes a recent result due to A. Smeets, namely the unconditional counterpart of [25, Thm. 1.1, Rem. 1.3]. For a version of this statement over $\mathbb{P}^n_{\mathbb{Q}}$ see Proposition 17.11.

For a number field k of degree $n = [k : \mathbb{Q}]$ we write $\mathbf{x} = (x_1, \ldots, x_n)$ and denote by $N_{k/\mathbb{Q}}(\mathbf{x})$ the norm form $\mathrm{Norm}_{k/\mathbb{Q}}(x_1\omega_1 + \ldots + x_n\omega_n)$, where $\omega_1, \ldots, \omega_n$ is a basis of k as a vector space over $\mathbb{Q}$. The following corollary extends [2, Thm. 1.1] to the case of a product of norm forms. It generalizes the unconditional version of [25, Cor. 1.6], a number of statements from [17, Section 4] and the main result of [21] in the case of the ground field $\mathbb{Q}$.

Corollary 17.3. *Let $k_1, \dots, k_n$ be number fields and let $m_1, \dots, m_n$ be positive integers with $gcd(m_1, \dots, m_n) = 1$. Let $L_i \in \mathbb{Q}[t_1, \dots, t_s]$ be polynomials of degree 1, for $i = 1, \dots, r$. Let X be a smooth and proper variety over $\mathbb{Q}$ that is birationally equivalent to the affine hypersurface*

$$\prod_{i=1}^{r} L_i(t_1, \dots, t_s) = \prod_{j=1}^{n} N_{k_j/\mathbb{Q}}(\boldsymbol{x}_j)^{m_j}. \tag{17.2}$$

Then $X(\mathbb{Q})$ is dense in $X(\mathbf{A}_\mathbb{Q})^{\mathrm{Br}}$.

Note that repetitions among $L_1, \dots, L_r$ are allowed here. Corollary 17.3 is a particular case of Proposition 17.12 which deals with several equations like (17.2) and extends [2, Thm. 1.3].

This note consists of two sections. In §2 we make preliminary remarks, some of which, like Corollary 17.7, are not needed in the proof of our main results but could possibly be of independent interest. In §3 we prove Theorem 17.1, Corollary 17.3 and their generalizations.

The density of $X(\mathbb{Q})$ in $X(\mathbf{A}_\mathbb{Q})^{\mathrm{Br}}$ can sometimes be proved by the method of fibration, when rational points are found in specific subvarieties of X, or by the method of descent, when rational points are found in certain X-torsors which are varieties with a surjective morphism to X. Our proof of Theorem 17.1 uses descent like [3], [11] or [21], and not the fibration method like [25] or [17]. It was inspired by the approach of Colliot-Thélène and Sansuc [6] to degeneration of torsors for tori, and by their computation of universal torsors on conic bundles [7, Section 2.6]. We apply open descent based on Harari's formal lemma [15, Cor. 2.6.1] as in [10], with an improvement found in [9] (see the proof of Thm. 3.1, pages 84–85).

The author is grateful to Jean-Louis Colliot-Thélène for useful discussions over the past many years. I would like to thank Daniel Loughran for his question that led me to Proposition 17.6, and Olivier Wittenberg for useful comments. This work was carried out at the Institute for the Information Transmission Problems of the Russian Academy of Sciences at the expense of the Russian Foundation for Sciences (project number 14-50-00150).

17.2 Reduction of varieties over discretely valued fields

Let R be a discrete valuation ring with the field of fractions K, the maximal ideal $\mathfrak{m}$ and the residue field $\kappa = R/\mathfrak{m}$. The following lemma is well known, cf. [6, p. 163] or [24, p. 70].

Lemma 17.4. *Let T be a torus over R, and let $T_\kappa = T \times_R \kappa$. Let T_κ° be the torus dual to T_κ, that is, the κ-torus whose module of characters $\widehat{T}_\kappa^\circ$ is the Galois module $Hom(\widehat{T}_\kappa, \mathbb{Z})$.*

(i) *The restriction map from R to K fits into an exact sequence*

$$H^1(R,T) \to H^1(K,T) \to H^1(\kappa, \widehat{T}_\kappa^\circ). \tag{17.3}$$

(ii) *If R is a subring of a discrete valuation ring R' with the residue field κ', the fraction field K' and the maximal ideal $\mathfrak{m}'$ such that $\mathfrak{m}' = \mathfrak{m}R'$, then the restriction maps give rise to a commutative diagram*

$$\begin{array}{ccccc} H^1(R',T) & \to & H^1(K',T) & \to & H^1(\kappa', \widehat{T}_\kappa^\circ) \\ \uparrow & & \uparrow & & \uparrow \\ H^1(R,T) & \to & H^1(K,T) & \to & H^1(\kappa, \widehat{T}_\kappa^\circ) \end{array} \tag{17.4}$$

Proof. (i) Let $\mathrm{Spec}(K) \xrightarrow{j} \mathrm{Spec}(R) \xleftarrow{i} \mathrm{Spec}(\kappa)$ be the embeddings of the generic and the special points, respectively. By [19, Example II.3.9] we have an exact sequence of étale sheaves on $\mathrm{Spec}(R)$

$$0 \to \mathbb{G}_{m,R} \to j_*\mathbb{G}_{m,K} \to i_*\mathbb{Z}_\kappa \to 0. \tag{17.5}$$

For any scheme X and any X-torus T the local-to-global spectral sequence for Ext gives a canonical isomorphism $\mathrm{H}^1(X,T) = \mathrm{Ext}^1_X(\widehat{T}, \mathbb{G}_m)$, see [24, Lemma 2.3.7] (although in *loc. cit.* X is a variety over a field, the same proof works in general). Thus an application of $\mathrm{Ext}^1_R(\widehat{T}, \cdot)$ to (17.5) gives an exact sequence

$$\mathrm{H}^1(R,T) \to \mathrm{Ext}^1_R(\widehat{T}, j_*\mathbb{G}_{m,K}) \to \mathrm{Ext}^1_R(\widehat{T}, i_*\mathbb{Z}_\kappa).$$

The spectral sequence $\mathrm{Ext}^p_R(\widehat{T}, R^q i_*\mathbb{Z}_\kappa) \Rightarrow \mathrm{Ext}^{p+q}_\kappa(\widehat{T}_\kappa, \mathbb{Z})$, in view of the vanishing of $R^1 i_*\mathbb{Z}_\kappa = 0$ [19, Example III.2.22], gives rise to the first canonical isomorphism in

$$\mathrm{Ext}^1_R(\widehat{T}, i_*\mathbb{Z}_\kappa) = \mathrm{Ext}^1_\kappa(\widehat{T}_\kappa, \mathbb{Z}) = \mathrm{H}^1(\kappa, \widehat{T}_\kappa^\circ).$$

The second canonical isomorphism in this line is provided by the spectral sequence $\mathrm{H}^p(\kappa, \mathrm{Ext}^q_{\mathbb{Z}}(\widehat{T}_\kappa, \mathbb{Z})) \Rightarrow \mathrm{Ext}^{p+q}_\kappa(\widehat{T}_\kappa, \mathbb{Z})$. Hilbert's theorem 90 gives $R^1 j_*\mathbb{G}_{m,K} = 0$, which produces canonical isomorphisms

$$\mathrm{Ext}^1_R(\widehat{T}, j_*\mathbb{G}_{m,K}) = \mathrm{Ext}^1_K(\widehat{T}, \mathbb{Z}) = \mathrm{H}^1(K,T).$$

Putting all these together gives the exact sequence (17.3).

(ii) The left-hand square in (17.4) is obviously commutative. The commutativity of the right-hand square follows from the fact that the restriction of the discrete valuation of R' to R is the given discrete valuation on R.

□

For an R-scheme X we denote by X_K and X_κ the generic and special fibres of X, respectively.

Lemma 17.5. *Let Y be a K-torsor for T_K. Suppose that there is an integral normal scheme X and a surjective morphism $X \to \mathrm{Spec}(R)$ of finite type with integral fibres such that the generic fibre X_K is K-birationally equivalent to Y.*

(i) *Let κ' be the separable closure of κ in the function field of the special fibre $\kappa(X_\kappa)$. Then the image of $[Y]$ in $H^1(\kappa, \widehat{T}^\circ)$ is in the kernel of the restriction map $H^1(\kappa, \widehat{T}^\circ) \to H^1(\kappa', \widehat{T}^\circ)$.*

(ii) *If X_κ is geometrically integral, there is an R-torsor Z for T such that $Y \cong Z_K$.*

Proof. (i) The morphism $X \to \mathrm{Spec}(R)$ is surjective, so the (integral) special fibre X_κ has codimension 1. Let R' be the local ring of X_κ. Since X is normal, R' is an integrally closed domain of dimension 1. Since $X \to \mathrm{Spec}(R)$ is of finite type, R' is Noetherian, and hence a discrete valuation ring. As X is integral, the field of fractions of R' is $K(X_K)$. A local parameter of R is also a local parameter of R', because X_κ is integral. Thus $\mathfrak{m}' = \mathfrak{m}R'$ is the maximal ideal of R', and the residue field $R'/\mathfrak{m}'$ is the field of functions on the special fibre X_κ. In our case the commutative diagram (17.4) takes the form

$$\begin{array}{ccccc} \mathrm{H}^1(R',T) & \to & \mathrm{H}^1(K(X_K),T) & \to & \mathrm{H}^1(\kappa(X_\kappa),\widehat{T}^\circ_\kappa) \\ \uparrow & & \uparrow & & \uparrow \\ \mathrm{H}^1(R,T) & \to & \mathrm{H}^1(K,T) & \to & \mathrm{H}^1(\kappa,\widehat{T}^\circ_\kappa) \end{array}$$

The restriction of the diagonal $Y \to Y \times_K Y$ to the generic point of Y is a $K(Y)$-point of Y. Hence the torsor Y is split by the field extension $K(X_K) = K(Y)$, so that the class $[Y] \in \mathrm{H}^1(K,T)$ goes to zero in $\mathrm{H}^1(K(X_K),T)$.

The right-hand vertical map in the diagram factorises as follows:

$$\mathrm{H}^1(\kappa,\widehat{T}^\circ_\kappa) \longrightarrow \mathrm{H}^1(\kappa',\widehat{T}^\circ_\kappa) \xrightarrow{\sim} \mathrm{H}^1(\kappa(X_\kappa),\widehat{T}^\circ_\kappa).$$

The second map here is an isomorphism because $\widehat{T}^\circ_\kappa$ is a finitely generated free abelian group and κ' is separably closed in $\kappa(X_\kappa)$. This implies (i).

(ii) If the κ-scheme X_κ is geometrically integral, the field κ is separably closed in $\kappa(X_\kappa)$, that is, $\kappa' = \kappa$. The exact sequence (17.3) then shows that $[Y]$ is in the image of the map $H^1(R, T) \to H^1(K, T)$, and this proves (ii).

□

The statement of Lemma 17.5 (i) leaves open the question to what extent κ' is determined by the field $K(X_K)$. We treat this as a question about integral, regular, proper schemes over a discretely valued field, see Corollary 17.7 below.

For an integral variety V over a field k, we write k_V for the separable closure of k in the field of functions $k(V)$.

Proposition 17.6. *Let Y and Y' be integral regular schemes that are proper and flat over R. If there is a dominant rational map from Y_K to Y'_K, then for any irreducible component $C \subset Y_\kappa$ of multiplicity* 1 *there exists an irreducible component $C' \subset Y'_\kappa$ of multiplicity* 1 *such that $\kappa_{C'} \subset \kappa_C$.*

Proof. Write $F = K(Y_K)$ and $F' = K(Y'_K)$. We are given an inclusion $F' \subset F$, or, equivalently, a morphism $\mathrm{Spec}(F) \to \mathrm{Spec}(F')$. Let $\mathcal{O}_C$ be the local ring of the generic point of C. Since Y is regular and C has codimension 1 in Y, $\mathcal{O}_C$ is a discrete valuation ring. The multiplicity of C is 1, that is, the maximal ideal of $\mathcal{O}_C$ is $\mathfrak{m}\mathcal{O}_C = \mathfrak{m} \otimes_{\mathcal{R}} \mathcal{O}_C$. Since Y is integral, the field of fractions of $\mathcal{O}_C$ is F. The residue field of $\mathcal{O}_C$ is $\kappa(C)$.

After the base change from R to $\mathcal{O}_C$ we obtain a morphism $Y' \times_R \mathcal{O}_C \to \mathcal{O}_C$. Its generic fibre $Y'_F = Y'_K \times_K F$ has an F-point $\mathrm{Spec}(F) \to \mathrm{Spec}(F') \to Y'_F$ coming from the F'-point defined by the diagonal morphism $Y'_K \to Y'_K \times_K Y'_K$. The morphism $Y' \times_R \mathcal{O}_C \to \mathcal{O}_C$ is proper, and by the valuative criterion of properness any F-point of its generic fibre extends to a section of the morphism.

Since C has multiplicity 1, the special fibre of $Y' \times_R \mathcal{O}_C \to \mathcal{O}_C$ is $Y'_\kappa \times_\kappa \kappa(C)$. A section of $Y' \times_R \mathcal{O}_C \to \mathcal{O}_C$ thus gives rise to a $\kappa(C)$-point of $Y'_\kappa \times_\kappa \kappa(C)$, which can be viewed also as a $\kappa(C)$-point of Y'_κ. Since $Y' \times_R \mathcal{O}_C$ is regular, any section of $Y' \times_R \mathcal{O}_C \to \mathcal{O}_C$ meets the special fibre at a smooth point. In particular, this point belongs to exactly one geometric irreducible component, moreover, this component has multiplicity 1 (see the calculation in the proof of [23, Lemma 1.1 (b)], or [28, Lemme 3.8]).

Let $U \subset Y'_\kappa$ be the smooth locus of Y'_κ. Since U contains a $\kappa(C)$-point, it is non-empty. By Stein factorization, the structure morphism $U \to \mathrm{Spec}(\kappa)$ factors through the surjective morphism $U \to \mathrm{Spec}(L)$ with geometrically connected fibres, where L is an étale κ-algebra. Explicitly, L is the direct sum of finite field extensions of κ, each of which is the separable closure of κ in the function field of an irreducible component of Y'_κ of multiplicity 1. The image of the composed map $\mathrm{Spec}(\kappa(C)) \to \mathrm{Spec}(L)$ is connected, hence this is $\mathrm{Spec}(\kappa_{C'})$ for some irreducible component $C' \subset Y'_\kappa$ of multiplicity 1. Thus $\kappa_{C'} \subset \kappa(C)$ and hence $\kappa_{C'} \subset \kappa_C$. □

For a discussion of some related results see Section 17.3 of [4].

Corollary 17.7. *Let X be an integral regular scheme that is proper and flat over R. Let Σ_X be the partially ordered set of irreducible components of multiplicity 1 of X_κ, where C dominates D if $\kappa_D \subset \kappa_C$. The set of finite field extensions $\kappa \subset \kappa_C$, where C is a minimal element of Σ_X, is a birational invariant of the generic fibre X_K.*

Proof. Suppose that proper R-schemes X and Y are integral and regular with birationally equivalent generic fibres, that is, $K(X_K) \cong K(Y_K)$. Let C be a minimal element of Σ_X. By Proposition 19.19 there exists $C' \in \Sigma_Y$ such that $\kappa_{C'} \subset \kappa_C$. By the same proposition, there is $C'' \in \Sigma_X$ such that $\kappa_{C''} \subset \kappa_{C'}$. By minimality of C we have $\kappa_{C''} = \kappa_C$, hence $\kappa_C = \kappa_{C'}$. If C' is not minimal in Σ_Y, then, by Proposition 19.19, C is not minimal in Σ_X. □

This set of finite extensions of the residue field can be explicitly determined when the generic fibre is a conic, and, more generally, a Severi–Brauer variety, or a quadric of dimension 2.

Remark. Olivier Wittenberg suggested a somewhat different approach to Proposition 17.6 and Corollary 17.7. Consider a discrete valuation $v : K(X_K)^* \to \mathbb{Z}$ such that the restriction of v to K^* is the given discrete valuation of K. Let R' be the valuation ring of v, and let κ' be the separable closure of κ in the residue field of R'. Let us call Θ_X the resulting set of finite field extensions of κ partially ordered by inclusion. It is clear that Θ_X is a birational invariant of X_K. The discrete valuation associated to an irreducible component of X_κ of multiplicity 1 is an example of such a valuation, so we have an inclusion of partially ordered sets $\Sigma_X \subset \Theta_X$. Since X is proper over R and regular, it can be shown that any extension of the given discrete valuation of K to $K(X_K)$ gives rise to a morphism of R-schemes $\mathrm{Spec}(R') \to X$ that factors through the smooth locus of X/R. Hence Σ_X and Θ_X have the same set of minimal elements, which is thus a birational invariant of X_K.

17.3 Torsors over toric fibrations

Let k be a field of characteristic zero. Let $\bar{k}$ be an algebraic closure of k, and let $\Gamma_k = \mathrm{Gal}(\bar{k}/k)$. For a variety X over k we write $\overline{X} = X \times_k \bar{k}$. We use the standard notation $\mathrm{Br}_0(X)$ for the image of the natural map $\mathrm{Br}(k) \to \mathrm{Br}(X)$, and $\mathrm{Br}_1(X)$ for the kernel of the natural map $\mathrm{Br}(X) \to \mathrm{Br}(\overline{X})$.

Let T be a k-torus. We write $\widehat{T}$ for the Γ_k-module of characters of T.

Let X be a smooth, proper, geometrically integral variety with a surjective morphism $f : X \to \mathbb{P}^1_k$ and the geometrically integral generic fibre $X_{k(t)}$ which is birationally equivalent to a $k(t)$-torsor for T.

Lemma 17.8. *Each fibre of* $X \to \mathbb{P}^1_k$ *has an irreducible component of multiplicity* 1.

Proof. The generic fibre of $\overline{X} \to \mathbb{P}^1_{\bar{k}}$ is birationally equivalent to a torsor for $\overline{T} \simeq \mathbb{G}^d_{m,\bar{k}}$, where $d = \dim(T)$. By Hilbert's theorem 90 we have $\mathrm{H}^1(\bar{k}(t), T) = \mathrm{H}^1(\bar{k}(t), \mathbb{G}_m)^d = 0$, so this torsor has a $\bar{k}(t)$-point. By the lemma of Lang and Nishimura, $X_{\bar{k}(t)}$ has a $\bar{k}(t)$-point too. By the valuative criterion of properness, this point extends to a section of the proper morphism $\overline{X} \to \mathbb{P}^1_{\bar{k}}$. Since X is smooth, by a standard argument (see the proof of [23, Lemma 1.1 (b)], or [28, Lemme 3.8]) any section intersects each fibre of $\overline{X} \to \mathbb{P}^1_{\bar{k}}$ in an irreducible component of multiplicity 1. □

The same conclusion holds in far greater generality: by a theorem of Graber, Harris and Starr it is enough to assume the generic fibre to be rationally connected. We refer the reader to Section 7 of [4] for a survey of known results, in particular, see [4, Thm. 7.7].

Proof of Theorem 17.1. We keep the notation of §1. By Lemma 17.8 there is an irreducible component of multiplicity 1 in each X_{P_i}, for $i = 1, \ldots, r$. We mark these components. We also mark a geometrically integral irreducible component of multiplicity 1 in each of the remaining fibres of f. Define $Y \subset X$ as the complement to the union of all the unmarked irreducible components of the fibres of $f : X \to \mathbb{P}^1_k$. It is clear that Y is a dense open subset of X. The restriction of f to Y is a surjective morphism $f : Y \to \mathbb{P}^1_k$ with integral fibres, and with proper and geometrically integral generic fibre $Y_{k(t)} = X_{k(t)}$. An standard easy argument then shows that $\bar{k}[Y]^* = \bar{k}^*$ and $\mathrm{Pic}(\overline{Y})$ is torsion-free, cf. the proof of Thm. A in [10, p. 391].

Let $\mathrm{Pic}(\overline{Y}) \to \mathrm{Pic}(Y_{\bar{k}(t)})$ be the homomorphism of Γ_k-modules induced by the inclusion of the generic fibre $Y_{\bar{k}(t)}$ into $\overline{Y}$. This homomorphism is surjective since Y is smooth. Let S be the k-torus defined by the exact sequence of Γ_k-modules

$$0 \to \widehat{S} \to \mathrm{Pic}(\overline{Y}) \to \mathrm{Pic}(Y_{\bar{k}(t)}) \to 0. \tag{17.6}$$

Thus the abelian group $\widehat{S}$ is generated by the geometric irreducible components of the fibres $Y_{P_1}, \dots, Y_{P_r}$ (this holds when $r \geq 1$; for $r = 0$ the group $\widehat{S} \cong \mathbb{Z}$ is generated by the class of a geometric fibre). Recall that a *vertical torsor* $\mathcal{T} \to Y$ is a torsor of S whose type is the injective map $\widehat{S} \to \mathrm{Pic}(\overline{Y})$ from (17.6). According to [24, Prop. 4.4.1] such torsors exist. For any vertical torsor $\mathcal{T}$ the injectivity of type $\widehat{S} \to \mathrm{Pic}(\overline{Y})$ and $\bar{k}[Y]^* = \bar{k}^*$ imply $\bar{k}[\mathcal{T}]^* = \bar{k}^*$, see [7, Prop. 2.1.1] or formula (2.6) in [24]. By (17.6) and [7, Prop. 2.1.1] we have isomorphisms of Galois modules

$$\mathrm{Pic}(\overline{\mathcal{T}}) \cong \mathrm{Pic}(Y_{\bar{k}(t)}) \cong \mathrm{Pic}(X_{\bar{k}(t)}).$$

The Picard group of the proper rational variety $X_{\bar{k}(t)}$ is torsion-free. It is well known that the absence of torsion in $\mathrm{Pic}(\overline{\mathcal{T}})$ implies the finiteness of $\mathrm{H}^1(k, \mathrm{Pic}(\overline{\mathcal{T}}))$, and then $\bar{k}[\mathcal{T}]^* = \bar{k}^*$ implies the finiteness of $\mathrm{Br}_1(\mathcal{T})/\mathrm{Br}_0(\mathcal{T})$.

Recall that for $i = 1, \dots, r$ we write $k(P_i)$ for the residue field of P_i, and k_i for the algebraic closure of $k(P_i)$ in the function field $k(Y_{P_i})$. The variety W_α is defined by (17.1), which is also equation (4.34) of [24].

Lemma 17.9. *The variety W_α is smooth and geometrically irreducible, and the morphism $\pi : W_\alpha \to \mathbb{P}^1_k$ given by $(u : v)$ is faithfully flat. We have $\bar{k}[W_\alpha]^* = \bar{k}^*$ and Pic$(\overline{W}_\alpha) = 0$.*

Proof. The first statement is [24, Lemma 4.4.5]. The second statement is a straightforward adaptation of [7, Lemme 2.6.1]. □

The reader may want to compare the following statement with [7, Thm. 2.6.4].

Proposition 17.10. *Let $f : Y \to \mathbb{P}^1_k$ be as above. Then any vertical torsor over Y is birationally equivalent to the product $W_\alpha \times_k Z$, where $\alpha \in \prod_{i=1}^r k(P_i)^*$ and Z is a k-torsor for T.*

Proof. The local description of torsors due to Colliot-Thélène and Sansuc, see [24, Thm. 4.3.1, Cor. 4.4.6], can be stated as follows. (Recall that the fibre of $f : Y \to \mathbb{P}^1_k$ at infinity is assumed to be smooth.) Let

$$U' = \mathbb{A}^1_k \setminus \{P_1, \dots, P_r\}, \quad U = Y \cap f^{-1}(U'), \quad V_\alpha = \pi^{-1}(U').$$

Then for any vertical torsor $\mathcal{T}/Y$ there exists an $\alpha \in \prod_{i=1}^{r} k(P_i)^*$ such that the restriction $\mathcal{T}_U = \mathcal{T} \times_Y U$ is isomorphic to the fibred product $U \times_{U'} V_\alpha$. Let us write $W = W_\alpha$, $V = V_\alpha$.

Let $j' : \mathrm{Spec}(k(t)) \to U'$ be the embedding of the generic point of $\mathbb{A}^1_k$ into the open set U'. Let $i_P : P \to U'$ be the embedding of a closed point $P \subset U'$. Since $\mathbb{A}^1_k$ is smooth, there is the following exact sequence of étale sheaves on U':

$$0 \to \mathbb{G}_{m,U'} \to j'_*\mathbb{G}_{m,k(t)} \to \oplus_{P \in U'} i_{P*}\mathbb{Z}_{k(P)} \to 0,$$

see [19, Examples II.3.9, III.2.22]. Repeating the proof of Lemma 17.5, we see that an application of $\mathrm{Ext}^1_{U'}(\widehat{T}, \cdot)$ to this sequence produces an exact sequence

$$\mathrm{H}^1(U', T) \to \mathrm{H}^1(k(t), T) \to \oplus_{P \in U'} \mathrm{H}^1(k(P), \widehat{T}^{\circ}_{k(P)}). \tag{17.7}$$

Let $\xi \in \mathrm{H}^1(k(t), T)$ be the class of a $k(t)$-torsor for T birationally equivalent to $X_{k(t)}$. The map $\mathrm{H}^1(k(t), T) \to \mathrm{H}^1(k(P), \widehat{T}^{\circ}_{k(P)})$ in (17.7) can be computed in the local ring $R = O_P$ of P in $\mathbb{A}^1_k$. The fibres of $f : U \to U'$ are geometrically integral, hence Lemma 17.5 implies that this map is trivial. Thus we see from (17.7) that ξ comes from some $\xi' \in \mathrm{H}^1(U', T)$. By Lemma 17.5 (i) applied to $f : Y \to \mathbb{P}^1_k$ the image of ξ' in $\mathrm{H}^1(k(P_i), \widehat{T}^{\circ}_{k(P)})$ goes to zero in $\mathrm{H}^1(k_i, \widehat{T}^{\circ}_{k(P)})$.

The equations (17.1) show that the fibre $W_i = \pi^{-1}(P_i)$ is the product of a torsor for a $k(P_i)$-torus and the affine $k(P_i)$-variety defined by $N_{k_i/k(P_i)}(z_i) = 0$. Thus W_i is integral over $k(P_i)$ and the field k_i is the algebraic closure of $k(P_i)$ in the function field of W_i. If we write $P_0 = \infty$, then by setting $v = 0$ in (17.1) we see that $W_0 = \pi^{-1}(P_0)$ is geometrically integral, and thus the algebraic closure of $k(P_0) = k$ in $k(W_0)$ is $k_0 = k$. We have a commutative diagram with exact rows similar to (17.4):

$$\begin{array}{ccccc}
\mathrm{H}^1(W, T) & \to & \mathrm{H}^1(V, T) & \to & \oplus_{i=0}^{r} \mathrm{H}^1(k(W_i), \widehat{T}^{\circ}_{k(P_i)}) \\
\uparrow & & \uparrow & & \uparrow \\
\mathrm{H}^1(k, T) & \to & \mathrm{H}^1(U', T) & \to & \oplus_{i=0}^{r} \mathrm{H}^1(k(P_i), \widehat{T}^{\circ}_{k(P_i)})
\end{array}$$

By the structure of degenerate fibres of $\pi : W \to \mathbb{P}^1_k$ described above, the image of ξ' in $\mathrm{H}^1(k(W_i), \widehat{T}^{\circ}_{k(P_i)})$ is zero, for $i = 0, \ldots, r$. Now the upper sequence in the diagram shows that $\pi^*(\xi') \in \mathrm{H}^1(V, T)$ comes from $\mathrm{H}^1(W, T)$.

By Lemma 17.9 we have $\bar{k}[W]^* = \bar{k}^*$ and $\mathrm{Pic}(\overline{W}) = 0$, and thus the fundamental exact sequence of Colliot-Thélène and Sansuc (see [7, Thm. 1.5.1] or [24, Cor. 2.3.9]) shows that the natural map $\mathrm{H}^1(k, T) \to \mathrm{H}^1(W, T)$ is an isomorphism. It follows that any W-torsor for T is the product of W and a k-torsor for T. □

End of proof of Theorem 17.1. Let $(M_v) \in X(\mathbf{A}_k)^{\mathrm{Br}}$. By a theorem of Grothendieck, $\mathrm{Br}_1(X)$ is naturally a subgroup of $\mathrm{Br}_1(Y)$. We have $\bar{k}[Y]^* = \bar{k}^*$, and this implies that $\mathrm{Br}_1(Y)/\mathrm{Br}_0(Y)$ is a subgroup of $\mathrm{H}^1(k, \mathrm{Pic}(\overline{Y}))$, which is finite because $\mathrm{Pic}(\overline{Y})$ is torsion-free. Thus we can use [10, Prop. 1.1] (a consequence of Harari's formal lemma) which says that the natural injective map of topological spaces $Y(\mathbf{A}_k)^{\mathrm{Br}_1(Y)} \to X(\mathbf{A}_k)^{\mathrm{Br}_1(X)}$ has a dense image. Thus we can assume without loss of generality that $(M_v) \in Y(\mathbf{A}_k)^{\mathrm{Br}_1(Y)}$.

The main theorem of the descent theory of Colliot-Thélène and Sansuc states that for any adelic point in $Y(\mathbf{A}_k)^{\mathrm{Br}_1(Y)}$ there exists a universal torsor $\mathcal{T}_0 \to Y$ such that our adelic point is in the image of the map $\mathcal{T}_0(\mathbf{A}_k) \to Y(\mathbf{A}_k)$, see [7, Section 3] or [24, Thm. 6.1.2(a)]. Thus we can find a point $(N_v) \in \mathcal{T}_0(\mathbf{A}_k)$ such that the image of N_v in Y is M_v for all v.

The structure group of $\mathcal{T}_0 \to Y$ is the Néron–Severi torus T_0 defined by the property $\widehat{T_0} = \mathrm{Pic}(\overline{Y})$. The dual of the injective map in (17.6) is a surjective morphism of tori $T_0 \to S$. Let S_0 be its kernel. Then $\mathcal{T} = \mathcal{T}_0/S_0$ is a Y-torsor with the structure group S whose type is the natural map $\widehat{S} \to \mathrm{Pic}(\overline{Y})$, so $\mathcal{T}$ is a vertical torsor. Since $\mathcal{T}_0$ is a universal torsor, [7, Prop. 2.1.1] tells us that $\bar{k}[\mathcal{T}_0]^* = \bar{k}^*$ and $\mathrm{Pic}(\overline{\mathcal{T}}_0) = 0$, hence $\mathrm{Br}_1(\mathcal{T}_0) = \mathrm{Br}_0(\mathcal{T}_0)$. Let $(P_v) \in \mathcal{T}(\mathbf{A}_k)$ be the image of (N_v). By the functoriality of the Brauer–Manin pairing we see that $(P_v) \in \mathcal{T}(\mathbf{A}_k)^{\mathrm{Br}_1(\mathcal{T})}$.

By Proposition 17.10 the variety $\mathcal{T}$ is birationally equivalent to $E \times W_\alpha$ for some $\alpha \in \prod_{i=1}^r k(P_i)^*$. Let E^c be a smooth compactification of E. Then we have a rational map g from the smooth variety $\mathcal{T}$ to the proper variety E^c, and a rational map h from $\mathcal{T}$ to W_α. By the valuative criterion of properness there is an open subset $\Omega \subset \mathcal{T}$ with complement of codimension at least 2 in $\mathcal{T}$ such that g is a morphism $\Omega \to E^c$. Let $\Omega' \subset \Omega$ be a dense open subset such that (g, h) defines an open embedding $\Omega' \subset E^c \times W_\alpha$.

By Grothendieck's purity theorem the natural restriction map gives an isomorphism $\mathrm{Br}(\mathcal{T}) \tilde{\longrightarrow} \mathrm{Br}(\Omega)$. Thus $g^*\mathrm{Br}_1(E^c) \subset \mathrm{Br}_1(\Omega)$ is naturally a subgroup of $\mathrm{Br}_1(\Omega) = \mathrm{Br}_1(\mathcal{T})$. (The argument given in the last three paragraphs is inspired by the proof of Thm. 3.1 in [9].)

Evaluating Brauer classes at local points is continuous and hence locally constant. Since $\mathrm{Br}_1(\mathcal{T})/\mathrm{Br}_0(\mathcal{T})$ is finite, by a small deformation we can assume $P_v \in \Omega'(k_v)$ for each v. Thus $(g(P_v))$ is a well-defined element of $E^c(\mathbf{A}_k)^{\mathrm{Br}_1(E^c)}$. By Sansuc's theorem [20], $E(k)$ is then non-empty and, moreover, $E(k)$ is dense in $E^c(\mathbf{A}_k)^{\mathrm{Br}_1(E^c)}$.

Let Σ be a finite set of places of k containing all the places where we need to approximate. By the Hasse principle and weak approximation for W_α we can find a k-point in Ω' that is arbitrarily close to $(g(P_v), h(P_v))$ for $v \in \Sigma$. We conclude that there is a k-point in $\mathcal{T}$ that is arbitrarily close to P_v for $v \in \Sigma$.

The image of this point in Y approximates (M_v). This finishes the proof of Theorem 17.1. □

Proposition 17.11. *Let X be a smooth, projective, geometrically integral variety over $\mathbb{Q}$, and let $f : X \to \mathbb{P}^n_{\mathbb{Q}}$ be a surjective morphism satisfying the following properties.*

(a) *There is a torus T over $\mathbb{Q}$ such that the generic fibre $X_{\mathbb{Q}(\eta)}$ of f is birationally equivalent to a $\mathbb{Q}(\eta)$-torsor for $T \times_{\mathbb{Q}} \mathbb{Q}(\eta)$.*

(b) *There exist hyperplanes $H_1, \ldots, H_r \subset \mathbb{P}^n_{\mathbb{Q}}$ such that f has split fibres at all points of codimension 1 of $\mathbb{P}^n_{\mathbb{Q}}$ other than $H_1, \ldots, H_r$.*

Then $X(\mathbb{Q})$ is dense in $X(\mathbf{A}_{\mathbb{Q}})^{\mathrm{Br}}$.

Proof. We deduce this from a fibration theorem due to D. Harari [16, Thm. 3.2.1]. There is an open subset $U \subset \mathbb{P}^n_{\mathbb{Q}}$ such that each fibre of $f^{-1}(U) \to U$ is geometrically integral and birationally equivalent to a torsor for T. Choose a point $M \in U(\mathbb{Q})$ and a hyperplane $L \subset \mathbb{P}^n_{\mathbb{Q}}$ through M, and let $V = f^{-1}(\mathbb{P}^n_{\mathbb{Q}} \setminus L)$. Obviously, V is a quasi-projective dense open subset of X. The projective lines through M that are not contained in L are in a natural bijection with the points of $\mathbb{A}^{n-1}_{\mathbb{Q}}$, so we have a surjective morphism $p : V \to \mathbb{A}^{n-1}_{\mathbb{Q}}$. Each fibre of p is split, because the fibre of f at the generic point of the corresponding line through M is geometrically integral and hence its Zariski closure is a geometrically integral irreducible component of multiplicity 1. Any point of $X_M(\overline{\mathbb{Q}})$ defines a section of p over $\overline{\mathbb{Q}}$. The generic fibre of p is birationally equivalent to a family of torsors for T parameterized by an open subset of the projective line, hence it is geometrically integral and geometrically rational. In particular, a smooth and proper model of the generic fibre has trivial geometric Brauer group and torsion-free geometric Picard group. The fibres of p over $\mathbb{Q}$-points of a non-empty open subset of $\mathbb{A}^{n-1}_{\mathbb{Q}}$ satisfy the assumptions of Corollary 17.2, hence all the conditions of [16, Thm. 3.2.1] are satisfied. An application of this result proves the proposition. □

Proposition 17.12. *Let $k_1, \ldots, k_n$ be number fields, and let $L_1, \ldots, L_r \in \mathbb{Q}[t_1, \ldots, t_s]$ be polynomials of degree 1. Let $m_{ij} \geq 0$, for $i = 1, \ldots, \ell$ and $j = 1, \ldots, n$, be integers such that the sublattice of $\mathbb{Z}^n$ generated by the rows of the matrix (m_{ij}) is primitive. Finally, let $d_{ip} \geq 0$ be integers, where $i = 1, \ldots, \ell$ and $p = 1, \ldots, r$. Let X be a smooth and proper variety over $\mathbb{Q}$ that is birationally equivalent to the affine variety given by the system of equations*

$$c_i \prod_{p=1}^{r} L_p(t_1, \ldots, t_s)^{d_{ip}} = \prod_{j=1}^{n} N_{k_j/\mathbb{Q}}(\mathbf{x}_j)^{m_{ij}}, \quad i = 1, \ldots, \ell, \tag{17.8}$$

where $c_i \in \mathbb{Q}^*$ *for* $i = 1, \ldots, \ell$. *Then* $X(\mathbb{Q})$ *is dense in* $X(\mathbb{A}_\mathbb{Q})^{\mathrm{Br}}$.

Proof. The condition on the matrix (m_{ij}) implies that the affine variety given by

$$\prod_{j=1}^{n} N_{k_j/\mathbb{Q}}(\mathbf{x}_j)^{m_{ij}} = 1, \quad i = 1, \ldots, \ell,$$

is a torus. Let us call it T. The affine variety Y given by (17.8) has a morphism $g : Y \to \mathbb{A}^s_\mathbb{Q}$ given by the coordinates $t_1, \ldots, t_s$. Let H_i be the hyperplane $L_i = 0$. The restriction of g to $\mathbb{A}^s_\mathbb{Q} \setminus (H_1 \cup \ldots \cup H_r)$ is a torsor for T. We can choose a smooth compactification X of the smooth locus Y_{sm} in such a way that there is a surjective morphism $f : X \to \mathbb{P}^s_\mathbb{Q}$ extending $g : Y_{\mathrm{sm}} \to \mathbb{A}^s_\mathbb{Q}$. Now we apply Proposition 17.11. □

For $\ell = 1$ this statement is Corollary 17.3.

References

[1] T. Browning and D.R. Heath-Brown. Quadratic polynomials represented by norm forms. *GAFA* **22** (2012) 1124–1190.

[2] T. Browning and L. Matthiesen. Norm forms for arbitrary number fields as products of linear polynomials. arXiv:1307.7641

[3] T. Browning, L. Matthiesen and A.N. Skorobogatov. Rational points on pencils of conics and quadrics with many degenerate fibres. *Ann. of Math.* **180** (2014) 381–402.

[4] J.-L. Colliot-Thélène. Variétés presque rationnelles, leurs points rationnels et leurs dégénérescences. *Arithmetic Geometry*. P. Corvaja, C. Gasbarri, eds. Lecture Notes Math. **2009**, Springer-Verlag, 2011, pp. 1–44.

[5] J.-L. Colliot-Thélène and P. Salberger. Arithmetic on some singular cubic hypersurfaces. *Proc. London Math. Soc.* **58** (1989) 519–549.

[6] J.-L. Colliot-Thélène and J.-J. Sansuc. Principal homogeneous spaces under flasque tori: applications. *J. Algebra* **106** (1987) 148–205.

[7] J.-L. Colliot-Thélène and J.-J. Sansuc. La descente sur les variétés rationnelles, II. *Duke Math. J.* **54** (1987) 375–492.

[8] J.-L. Colliot-Thélène, J.-J. Sansuc and Sir Peter Swinnerton-Dyer. Intersections of two quadrics and Châtelet surfaces, I, II. *J. reine angew. Math.* **373** (1987) 37–168.

[9] J.-L. Colliot-Thélène, D. Harari and A.N. Skorobogatov. Valeurs d'un polynôme à une variable représentées par une norme, *Number theory and algebraic geometry*,

London Math. Soc. Lecture Note Ser. **303**. Cambridge University Press, 2003, pp. 69–89.

[10] J.-L. Colliot-Thélène and A.N. Skorobogatov. Descent on fibrations over $\mathbb{P}^1_k$ revisited. *Math. Proc. Camb. Phil. Soc.* **128** (2000) 383–393.

[11] U. Derenthal, A. Smeets and D. Wei. Universal torsors and values of quadratic polynomials represented by norms. *Math. Annalen* **361** (2015), 1021–1042.

[12] B. Green and T. Tao. Linear equations in primes. *Ann. of Math.* **171** (2010) 1753–1850.

[13] B. Green and T. Tao. The Möbius function is strongly orthogonal to nilsequences. *Ann. of Math.* **175** (2012) 541–566.

[14] B. Green, T. Tao and T. Ziegler. An inverse theorem for the Gowers $U^{s+1}[N]$-norm. *Ann. of Math.* **176** (2012) 1231–1372.

[15] D. Harari, Méthode des fibrations et obstruction de Manin. *Duke Math. J.* **75** (1994), 221–260.

[16] D. Harari. Flèches de spécialisation en cohomologie étale et applications arithmétiques. *Bull. Soc. Math. France* **125** (1997) 143–166.

[17] Y. Harpaz, A.N. Skorobogatov and O. Wittenberg. The Hardy–Littlewood conjecture and rational points. *Compositio Math.* **150** (2014), 2095–2111.

[18] R. Heath-Brown and A.N. Skorobogatov. Rational solutions of certain equations involving norms. *Acta Math.* **189** (2002) 161–177.

[19] J.S. Milne. *Étale cohomology*. Princeton University Press, 1980.

[20] J.-J. Sansuc. Groupe de Brauer et arithmétique des groupes algébriques linéaires sur un corps de nombres. *J. reine angew. Math.* **327** (1981) 12–80.

[21] D. Schindler and A.N. Skorobogatov. Norms as products of linear polynomials. *J. London Math. Soc.* **89** (2014) 559–580.

[22] A.N. Skorobogatov. Arithmetic on certain quadric bundles of relative dimension 2, I. *J. reine angew. Math.* **407** (1990) 57–74.

[23] A.N. Skorobogatov. Descent on fibrations over the projective line. *Amer. J. Math.* **118** (1996) 905–923.

[24] A. Skorobogatov. *Torsors and rational points*. Cambridge University Press, 2001.

[25] A. Smeets. Principes locaux-globaux pour certaines fibrations en torseurs sous un tore. *Math. Proc. Camb. Phil. Soc.* **158** (2015), 131–145.

[26] M. Swarbrick Jones. A note on a theorem of Heath-Brown and Skorobogatov. *Quat. J. Math.* **64** (2013) 1239–1251.

[27] Sir Peter Swinnerton-Dyer. Rational points on some pencils of conics with 6 singular fibres. *Ann. Fac. Sci. Toulouse* **8** (1999) 331–341.

[28] O. Wittenberg. *Intersections de deux quadriques et pinceaux de courbes de genre 1*. Lecture Notes Math. **1901**, Springer-Verlag, 2007.

18

On filtrations of vector bundles over $\mathbb{P}^1_{\mathbb{Z}}$

*A. Smirnov**

St. Petersburg Department of Steklov Math. Institute

E-mail address: smirnov@pdmi.ras.ru

Introduction

A positive definite lattice Γ may be regarded as a vector bundle on a compactification of Spec $\mathbb{Z}$, that is on an arithmetic curve. From this point of view, the number $[-1/2 \log \operatorname{disc} \Gamma]$ can be thought of as the first Chern class, and the short vectors of the lattice correspond to the global sections.

On the other hand, there is a well-developed theory of vector bundles on algebraic surfaces. One could study, for example, the discrete invariants and the moduli spaces of such bundles. Some very interesting questions relate to the exceptional collections of bundles on algebraic surfaces.

The study of vector bundles on arithmetic surfaces, for instance on $\mathbb{P}^1_{\mathbb{Z}}$, seems to be a rather promising enterprise and may be regarded as a synthesis of two directions above. The "right" problem is, of course, a study of bundles on some compactifications of arithmetic surfaces, for instance, on Arakelov's compactifications. As an approach to this problem, one may study vector bundles on the classical scheme $\mathbb{P}^1_{\mathbb{Z}}$. We shall make a few steps in this direction here.

Theorems 18.16, 18.18, and 18.24 are the main results of this paper. The first theorems give a complete classification of the rank 2 bundles with the trivial generic fiber and simple jumps. The third of these theorems gives a criterion, allowing to determine whether the global restrictions for a minimal filtration of a bundle with linear quotients can be reduced to local ones. We give examples, showing that it is insufficient to know the local restrictions.

Parts of this paper were written during the program "Arithmetic and Geometry 2013" at the Hausdorff Institute for Mathematics in Bonn. I thank

*Partially supported by RFFI (grant 13-01-00429 ГЪ).

Arithmetic and Geometry, ed. Luis Dieulefait *et al.* Published by Cambridge University Press.

the organizers of this program for their invitation and hospitality. It is a pleasure for me to thank Boris Moroz for useful discussions.

18.1 General considerations

We shall study vector bundles over $\mathbb{P}^1_A$ for a Dedekind domain A, in particular for $A = \mathbb{Z}$. It is instructive and more natural, however, to state a few results in greater generality. Thus let A be a Noetherian commutative ring.

We write usually $\mathcal{O}$ and $\mathcal{O}(d)$ instead of $\mathcal{O}_X$ and $\mathcal{O}_X(d)$ for a suitable scheme X, especially if X can be easily specified. As usual, let

$$\mathbb{P}^1_A = \operatorname{Proj} A[t_0, t_1], \quad (\deg t_0 = \deg t_1 = 1).$$

In addition, $\mathcal{O}(U_0) = \mathbb{Z}[x]$, $\mathcal{O}(U_1) = \mathbb{Z}[x^{-1}]$, and $\mathcal{O}(U_{01}) = \mathbb{Z}[x, x^{-1}]$, where $x = t_1/t_0$, U_i denotes the complement to the zero locus of t_i, and $U_{01} = U_0 \cap U_1$.

To construct vector bundles on $X = \mathbb{P}^1_A$, let us define the following gluing operation.

18.1.1 Gluing

Let $\sigma \in \mathrm{GL}_n(A[x, x^{-1}])$. To σ one associates a vector bundle E of rank n over $\mathbb{P}^1_A$, as follows: $E|U_0 = \mathcal{O}e_1 + \cdots + \mathcal{O}e_n$, $E|U_1 = \mathcal{O}f_1 + \cdots + \mathcal{O}f_n$, and

$$[e_1, \ldots, e_n]\sigma = [f_1, \ldots, f_n]$$

over U_{01}, so that $f_j = \sum_{i=1}^n \sigma_{i,j} e_i$.

If every projective A-module is free, by the Quillen–Suslin theorem [6], [8], any vector bundle of rank n over $\mathbb{P}^1_A$ can be obtained in that way. In this case, the isomorphism class of such a bundle is an element of the double quotient

$$\mathrm{Vect}_n(\mathbb{P}^1_A) = \mathrm{GL}_n A[x] \backslash \mathrm{GL}_n A[x, x^{-1}] / \mathrm{GL}_n A[x^{-1}]. \qquad (18.1)$$

Example 18.1. The line bundle $\mathcal{O}(d)$ can be given by the matrix $\sigma = [x^d] \in \mathrm{GL}_1(\mathbb{Z}[x, x^{-1}])$.

Example 18.2. Let $A = \mathbb{Z}_p$ and E be given by the matrix

$$\sigma = \begin{bmatrix} x^{-1} & p \\ 0 & x \end{bmatrix}.$$

This bundle is not isomorphic to a sum of line bundles (see 18.1.3).

18.1.2 Jumps

When studying vector bundles on $\mathbb{P}^1_A$ it is important to understand the phenomenon of jumps. Let us start by stating a few general results of independent interest.

Theorem 18.3 (Grothendieck, [5]). *Let F be a field. Any vector bundle on* $\mathbb{P}^1_{\mathrm{F}}$ *is isomorphic to a sum of line bundles with uniquely defined summands.*

Theorem 18.4 (Serre, [7]). *Any line bundle on* $\mathbb{P}^n_A$ *is isomorphic to a bundle of the form* $p^*L \otimes \mathcal{O}(d)$*, where L is a line bundle on* Spec A *and* $p : \mathbb{P}^n_A \to$ Spec A *is the structure projection.*

In particular, any line bundle on $\mathbb{P}^1_{\mathrm{F}}$ is isomorphic to $\mathcal{O}(d)$ for some $d \in \mathbb{Z}$.

18.1.2.1 Splitting types

Let E be a vector bundle over $\mathbb{P}^1_A$, $r = \operatorname{rk} E$, and $y \in \operatorname{Spec} A$. Let E_y be the pullback of E to $\mathbb{P}^1_F$, where F is the residue field at y. By the theorems 18.3 and 18.4, the pullback E_y is isomorphic to the sum

$$E_y = \mathcal{O}(d_1) + \cdots + \mathcal{O}(d_r),$$

where the collection $d_1 \leq \cdots \leq d_r$ is uniquely defined. This collection is called the splitting type of E at y.

Example 18.5. To illustrate the notion of a splitting type, let us use the bundle E of (18.2). Let y be the closed point and η the generic point Spec $\mathbb{Z}_p$. In this example, E_y is obtained by gluing with the matrix

$$\sigma_y = \begin{bmatrix} x^{-1} & p \\ 0 & x \end{bmatrix} = \begin{bmatrix} x^{-1} & 0 \\ 0 & x \end{bmatrix} \in \mathrm{GL}_2(\mathbb{F}_p[x, x^{-1}]).$$

Since σ_y is diagonal it is clear that $E_y \cong \mathcal{O}(-1) + \mathcal{O}(1)$. On the other hand, the residue field at the point η coincides with $\mathbb{Q}_p$ and p is invertible in this field. Therefore we have

$$\begin{bmatrix} 1 & 0 \\ -x/p & 1 \end{bmatrix} \begin{bmatrix} x^{-1} & p \\ 0 & x \end{bmatrix} \begin{bmatrix} 0 & -p \\ 1/p & 1/x \end{bmatrix} = \begin{bmatrix} 1 & 0 \\ 0 & 1 \end{bmatrix}.$$

Thus (18.1) shows that $E_\eta \cong \mathcal{O}^2$.

18.1.2.2 Variations of splitting types

To describe the character of these variations we need a function and an order on its value domain. Fix a bundle E on $\mathbb{P}^1_A$ and consider the following function

$s(E, y)$ with values in $\mathbb{Z}^r$, where $r = \operatorname{rk} E$. It maps $y \in \operatorname{Spec} A$ to the splitting type of the bundle E_y. Let

$$s(E, y) = (d_1(E, y), \dots, d_r(E, y)).$$

Consider the lexicographic order on the set of the nondecreasing collections $(d_1, \dots, d_r)$, so that $(d_1, \dots, d_r) < (e_1, \dots, e_r)$ if $d_r < e_r$, or $d_r = e_r$ and $d_{r-1} < e_{r-1}$, etc. For $(d_1, \dots, d_r) \in \mathbb{Z}^r$, the set

$$\{y \in \operatorname{Spec} A | s(E, y) \geq (d_1, \dots, d_r)\}$$

is Zariski closed, i.e. the splitting type can not decrease under specializations. Moreover, the sum $d_1(E, y) + \cdots + d_r(E, y)$ doesn't depend on y.

These assertions follow easily from the base change theorem [3, Ch. III, §12].

18.1.2.3 Consequences of the base change theorem

Let us state the necessary results. Let $f : X \to Y$ be a proper morphism of Noetherian schemes and $\mathcal{F}$ be a coherent $\mathcal{O}$-module, flat over Y. As usually, X_y is the fiber of f over $y \in Y$ and $\mathcal{F}_y = \mathcal{F}|_{X_y}$. Then $y \mapsto \chi(\mathcal{F}_y)$ is a locally constant function on Y and $y \mapsto \dim_{k(y)} H^p(X_y, \mathcal{F}_y)$ is an upper semicontinuous function. Moreover, if Y is reduced and connected, then

(1) the function $y \mapsto \dim_{k(y)} H^p(X_y, \mathcal{F}_y)$ is constant on Y iff $R^p f_* \mathcal{F}$ is locally free and the natural arrow $R^p f_* \mathcal{F} \otimes_{\mathcal{O}_Y} k(y) \to H^p(X_y, \mathcal{F}_y)$ is an isomorphism for all $y \in Y$;
(2) if the equivalent conditions of the previous item hold, then the natural morphism $R^{p-1} f_* \mathcal{F} \otimes_{\mathcal{O}_Y} k(y) \to H^{p-1}(X_y, \mathcal{F}_y)$ is an isomorphism for all $y \in Y$.

Example 18.6. In the example 18.5 the underlying topological space of $\operatorname{Spec} A$ is the set $\{y, \eta\}$ with

$$s(E, y) = (-1, 1), \text{ and } s(E, \eta) = (0, 0).$$

Thus the splitting type of E at y "jumps" relatively to the splitting type at η.

18.1.3 Linear filtrability of bundles

Any vector bundle on the projective line over a field is the sum of linear bundles. Moreover, the splitting type of the sum of linear bundles on the projective line over a Dedekind ring is obviously constant. However, the splitting type of the bundles in the examples 18.2, 18.5, and 18.6 is not constant and therefore those vector bundles are not isomorphic to sums of linear bundles.

It is an open question whether any vector bundle on $\mathbb{P}^1_A$ for a Dedekind ring A admits a filtration with linear bundles as the quotients. Let us cite a few known results in this direction.

Theorem 18.7 (Horrocks, [4]). *If A is a local Dedekind ring, then any vector bundle on $\mathbb{P}^1_A$ admits a filtration with linear bundles as the quotients.*

The following result is valid for arbitrary Dedekind ring A.

Theorem 18.8 (Hanna, [2]). *Any vector bundle on $\mathbb{P}^1_A$ admits a filtration with linear and 2-dimensional bundles as the quotients.*

The question is answered completely for any Euclidian ring.

Theorem 18.9 (Hanna, [2]). *Any vector bundle over $\mathbb{P}^1_A$, where A is an Euclidian ring, admits a filtration with linear bundles as the quotients.*

In particular, any bundle on $\mathbb{P}^1_{\mathbb{Z}}$ admits a filtration with linear bundles as the quotients.

18.1.4 Beilinson's spectral sequence

Some of the methods of constructing bundles on $\mathbb{P}^n_{\mathbb{C}}$ (see [5]) can be applied to $\mathbb{P}^1_A$. For example, it concerns two Beilinson's spectral sequences.

Theorem 18.10 (Beilinson, [5]). *Let F be a vector bundle on $\mathbb{P}^1_A$ and let $p : \mathbb{P}^1_A \to \operatorname{Spec} A$ be the structure projection. There exists a spectral sequence with $E_1^{pq} = Rp^q_*(F(p)) \otimes \Omega^{-p}(p)$, which converges to*

$$F^i = \begin{cases} F, & \text{if } i = 0; \\ 0, & \text{if } i \neq 0. \end{cases}$$

In particular, the E_1-term of this sequence is concentrated in the second quadrant. Moreover, its nontrivial part is concentrated in the first two rows:

$$\mathrm{H}^1(F(-1)) \otimes \mathcal{O}(-1) \xrightarrow{d^1} \mathrm{H}^1(F) \otimes \mathcal{O}\,.$$

$$\mathrm{H}^0(F(-1)) \otimes \mathcal{O}(-1) \xrightarrow{d^1} \mathrm{H}^0(F) \otimes \mathcal{O}$$

18.1.4.1 Normalizations

The theorem (18.10) is especially useful when $\mathrm{H}^1(F(-1)) = \mathrm{H}^1(F) = 0$ and both $\mathrm{H}^0(F(-1))$ and $\mathrm{H}^0(F)$ are not too large.

A bundle F is called normalized if $H^1(F(-1)) = 0$ and $H^1(F(-2)) \neq 0$. The base change theorem (see 18.1.2.3) implies that for a Dedekind ring A there exists one and only one normalized bundle of the form $F(d)$. Let $\delta(F) = d$ in that case; for example, $\delta(\mathcal{O}) = 0$.

18.2 Bundles of rank 2 with simple jumps

Let A be a Dedekind ring. We shall study bundles E with $\mathrm{rk}\, E = 2$, with the trivial generic fiber, and with simple jumps. In such a case $E_\eta \cong \mathcal{O}^2$ for the generic point $\eta \in \operatorname{Spec} A$ and $E_y \cong \mathcal{O}^2$ or $E_y \cong \mathcal{O}(-1) + \mathcal{O}(1)$ for each $y \in \operatorname{Spec} A$. It follows from these conditions that the jumping points, i.e. the points y with $E_y \cong \mathcal{O}(-1) + \mathcal{O}(1)$, form a finite set (see 18.1.2.2).

18.2.1 Classification of the bundles with simple jumps

First of all we construct some bundles of rank 2 with simple jumps. Such a bundle will be denoted by

$$F^{\nu,\varepsilon_1,\varepsilon_2}$$

with $(\nu, \varepsilon_1, \varepsilon_2) \in A^3$. The triples $(\nu, \varepsilon_1, \varepsilon_2)$ are subjected to the only condition that ν and ε_1 are coprime. In other words, the row $[\nu, \varepsilon_1]$ is supposed to be unimodular.

We shall see that any 2-bundle with simple jumps can be obtained in this way. If the ring A is factorial, we can restrict ourselves to the triples with $\varepsilon_2 = 0$ (see 18.16). In that case, theorem 18.18 solves the classification problem.

18.2.1.1

We construct the bundle $F^{\nu,\varepsilon_1,\varepsilon_2}$ as the cokernel of an arrow of the form

$$\phi : \mathcal{O}^2(-2) \longrightarrow \mathcal{O}^4(-1). \tag{18.2}$$

Let e_1, e_2 and $f_1, \dots, f_4$ be the standard bases of $\mathcal{O}^2$ and $\mathcal{O}^4$. The choice of bases fixes an identification $\mathrm{Hom}\,(\mathcal{O}^2(-2), \mathcal{O}^4(-1)) \simeq M_{4,2}(\mathrm{Hom}\,(\mathcal{O}, \mathcal{O}(1)))$. Since $\mathrm{Hom}\,(\mathcal{O}, \mathcal{O}(1)) = At_0 + At_1$, we have $\phi = t_0\phi_0 + t_1\phi_1$ with $\phi_0, \phi_1 \in M_{4,2}(A)$.

18.2.1.2

The following particular case of ϕ (see (18.2)) is of special interest.

$$\phi_0 = \begin{bmatrix} M \\ N(\nu) \end{bmatrix}, \quad \phi_1 = \begin{bmatrix} 1_2 \\ 0_2 \end{bmatrix}, \quad N(\nu) = \begin{bmatrix} \nu & 0 \\ 0 & 1 \end{bmatrix}, \tag{18.3}$$

where $M \in M_{2,2}(A)$, $\nu \in A$. In this case Coker ϕ does not depend on the 2nd column of M. This can be easily proved by means of the elementary transformations. They adjust the first two rows of ϕ by means of the 4th row. Thus without loss of generality we may assume that

$$M = M(\varepsilon_1, \varepsilon_2) = \begin{bmatrix} \varepsilon_2 & 0 \\ \varepsilon_1 & 0 \end{bmatrix}. \tag{18.4}$$

18.2.1.3

The arrow ϕ in (18.2) is called nondegenerate if F = Coker ϕ is a bundle of rank 2. This is equivalent to the assertion that ϕ can be split locally. For a nondegenerate ϕ we have an exact sequence

$$0 \longrightarrow \mathcal{O}^2(-2) \xrightarrow{\phi} \mathcal{O}^4(-1) \longrightarrow F \longrightarrow 0\,. \tag{18.5}$$

In particular, for the nondegenerate ϕ there is a canonical isomorphism

$$H^0(X, F) \to H^1(X, \mathcal{O}(-2))^2 \simeq A^2.$$

It arises from the cohomology sequence for (18.5) and the standard identification of $\mathcal{O}(-2)$ with the dualizing sheaf.

Proposition 18.11. *The map ϕ of (18.2.1.2) is nondegenerate if and only if the row $[\nu, \varepsilon_1]$ is unimodular.*

Proof. The map ϕ is nondegenerate iff the pullbacks of ϕ to U_0 and U_1 are nondegenerate. It is easy to see that $\phi|_{U_0}$ is nondegenerate iff the map $A[x] \to A[x]^3$, $1 \mapsto (\nu, \varepsilon_1, \varepsilon_2 + x)$ is an injection with the projective cokernel. This happens if and only if the inclusion $A[x] \to A[x]^3$ splits, that is if and only if the row $[\nu, \varepsilon_1, \varepsilon_2 + x]$ is unimodular. On taking $x = -\varepsilon_2$ we see that this is equivalent to the unimodularity of $[\nu, \varepsilon_1]$.

Conversely, suppose that $\alpha\nu + \beta\varepsilon_1 = 1$. It is easy to see that the projectivity of $F|_{U_1}$ is equivalent to the unimodularity of the row $[\nu y, \varepsilon_1 y^2, 1 + \varepsilon_2 y]$ in $A[y]$, where $y = t_0/t_1$. We show step by step, that $y^2 = (\alpha y)\cdot(\gamma y) + \beta\cdot(\varepsilon_1 y^2)$, $y = y\cdot(1+\varepsilon_2 y^2) - \varepsilon_2 \cdot y^2$, and $1 = 1\cdot(1+\varepsilon_2 y) - \varepsilon_2 \cdot y$ are linear combinations of νy, $\varepsilon_1 y^2$ and $1 + \varepsilon_2 y$. □

Thus we have associated to any unimodular row $[\nu, \varepsilon_1]$ and any $\varepsilon_2 \in A$ a vector 2-bundle

$$F^{\nu,\varepsilon_1,\varepsilon_2} = \operatorname{Coker}\phi, \tag{18.6}$$

where the arrow ϕ in (18.5) is given by (18.2.1.2).

Proposition 18.12. *If A is a field, then*

$$F^{\nu,\varepsilon_1,\varepsilon_2} \simeq \begin{cases} \mathcal{O}^2, & \text{if } \nu \neq 0; \\ \mathcal{O}(-1) + \mathcal{O}(1), & \text{if } \nu = 0. \end{cases}$$

Proof. The exactness of (18.5) shows that $\operatorname{Det} F \simeq \mathcal{O}$ and $F \simeq \mathcal{O}(-d)+\mathcal{O}(d)$ for some $d \geq 0$. The cohomology sequence for (18.5) shows that $h^0(X, F) = 2$ and $d \leq 1$. To distinguish between the cases $d = 0$ and $d = 1$ let us consider the cohomology sequence for (18.5) twisted by $\mathcal{O}(-1)$, that is the sequence

$$0 \to H^0(F(-1)) \to H^1(\mathcal{O}(-3))^2 \to H^1(\mathcal{O}(-2))^4 \to H^1(F(-1)) \to 0,$$

where the scheme X is omitted in the notation for cohomology.

We have to calculate the middle arrow $H^1(\phi(-1))$. It is more convenient to calculate the adjoint arrow $H^0([\phi(-1)]^\vee \otimes K_X) : H^0(X, \mathcal{O})^4 \to H^0(X, \mathcal{O}_X(1))^2$, where $K_X \simeq \mathcal{O}(-2)$ is the canonical class. It suffices to note that $H^1(\psi)^\vee = H^0(\psi^\vee)$ and, consequently, the arrow $H^0([\phi(-1)]^\vee \otimes K_X)$ is given by multiplication with the adjoint matrix $\phi^* \in M_{2,4}(\operatorname{Hom}(\mathcal{O}, \mathcal{O}(1)))$. A direct calculation shows that the matrix of ϕ^* with respect to the bases $f_1^*, \ldots, f_4^*$ and $t_1e_1^*, t_1e_2^*, t_0e_1^*, t_0e_2^*$ has the form:

$$\begin{bmatrix} 1 & 0 & 0 & 0 \\ 0 & 1 & 0 & 0 \\ \varepsilon_2 & \varepsilon_1 & \nu & 0 \\ 0 & 0 & 0 & 1 \end{bmatrix}.$$

In particular, $H^0(X, F(-1)) = H^1(X, F(-1)) = A/(\nu)$. □

Corollary 18.13. *If A is a domain and $\nu \neq 0$, then the generic fiber of $F^{\nu,\varepsilon_1,\varepsilon_2}$ is isomorphic to $\mathcal{O}^2$, all the jumps have the form $\mathcal{O}(-1)+\mathcal{O}(1)$ and lie exactly over the divisors of ν.*

Corollary 18.14. *Let $F = F^{\nu,\varepsilon_1,\varepsilon_2}$. Then $H^0(X, F(-1)) \simeq H^1(X, F(-1)) \simeq A/(\nu)$. In particular, the ideal, generated by ν, depends on the isomorphism class of F only.*

Proof. This statement is proved in the course of the proof of Proposition 18.12. □

18.2.1.4

Further, let us assume that A is a factorial Dedekind ring. This assumption allows us to describe explicitly the orbits of the action of $\mathrm{GL}_n(A)$ on A^n and the orbits of the action of certain congruence subgroups on $\mathbb{P}^1(A)$. Let K be the fraction field of A. For $a, b \in A$ we write $c = (a, b)$ for any c with $Aa + Ab = (c)$.

To $\nu \in A$ one associates the following congruence subgroup

$$\tilde{\Gamma}_0(\nu) = \{\begin{bmatrix} \alpha & \beta \\ \gamma & \delta \end{bmatrix} \in \mathrm{GL}_2(A) | \quad \gamma = 0 \pmod{\nu}\}.$$

For $v \in K^2 \setminus \{0\}$, let $\bar{v} = Kv$, $\bar{v} \in \mathbb{P}^1(K)$. Since A is a Dedekind ring, we have $\mathbb{P}^1(K) = \mathbb{P}^1(A)$, so that $\bar{v} \in \mathbb{P}^1(A)$. We denote the line $0 \times K$ by ∞.

Lemma 18.15. *Let $v = \begin{bmatrix} v_0 \\ v_1 \end{bmatrix} \in A^2$ be a nontrivial vector. Then $\bar{v}$ and ∞ belong to the same orbit for the standard (left) action of $\tilde{\Gamma}_0(\nu)$ on $\mathbb{P}^1(A)$ if and only if $v_1/(v_0, v_1)$ and ν are coprime.*

Proof. Replacing v by $v/(v_0, v_1)$ we can assume that $(v_0, v_1) = 1$ (here we use the assumption 18.2.1.4). The orbit of ∞ consists of the projectivizations of the vectors $\begin{bmatrix} \beta \\ \delta \end{bmatrix}$, where $\begin{bmatrix} \alpha & \beta \\ \gamma & \delta \end{bmatrix} \in \tilde{\Gamma}_0(\nu)$ and, in particular, $(\gamma, \delta) = 1$. Since $\gamma = 0 \pmod{\nu}$, it follows that $(\nu, \delta) = 1$.

Conversely, let $(v_1, \nu) = 1$. The condition $(v_0, v_1) = 1$ implies the condition $(\nu v_0, v_1) = 1$. Let $\alpha v_1 - \gamma_0 \nu v_0 = 1$. Then

$$\begin{bmatrix} v_0 \\ v_1 \end{bmatrix} = \sigma \begin{bmatrix} 0 \\ 1 \end{bmatrix} \text{ with } \sigma = \begin{bmatrix} \alpha & v_0 \\ \nu\gamma_0 & v_1 \end{bmatrix}.$$

It remains to note that $\sigma \in \tilde{\Gamma}_0(\nu)$. □

Theorem 18.16. *Let F be a vector 2-bundle on $\mathbb{P}^1_A$. Assume that its generic fiber F_K on $\mathbb{P}^1_K$ is isomorphic to $\mathcal{O}^2$ and its jumps have the form $\mathcal{O}(-1)+\mathcal{O}(1)$. Then F is isomorphic to a bundle of the form $F^{\nu,\varepsilon,0}$ with coprime ν and ε.*

Proof. Taking into account the base change theorem (see 18.1.2.3), the form of the generic fiber and the form of the jumps it is easy to see that $H^1(\mathbb{P}^1, F) \simeq H^1(\mathbb{P}^1, F(-1)) \simeq 0$, $H^0(\mathbb{P}^1, F) \simeq A^2$, $H^0(\mathbb{P}^1, F(1)) \simeq A^4$. Applying to

$F(1)$ the Beilinson theorem (see 18.10) and a twist, we get F as a quotient in the exact sequence

$$0 \to \mathcal{O}^2(-2) \xrightarrow{\phi} \mathcal{O}^4(-1) \to F \to 0, \tag{18.7}$$

where $\phi = t_0\phi_0 + t_1\phi_1$ and $\phi_0, \phi_1 \in M_{4,2}(A)$. We can identify the bundles, constructed by means of equivalent matrices, where $\phi \sim \phi'$, iff $\phi' = \rho\phi\sigma^{-1}$, $\rho \in \mathrm{GL}_4(A), \sigma \in \mathrm{GL}_2(A)$.

Let us show first that any nondegenerate ϕ (see 18.2.1.3) is given by a matrix of the form (18.2.1.2). Indeed, the pullback of (18.7) to $t_0 = 0$ shows that $\mathrm{rk}\,\phi_1 = 2$. The theory of elementary divisors (here we use the assumption 18.2.1.4) allows to suppose that $\phi \sim \phi^{(1)}$ with

$$\phi_1^{(1)} = \begin{bmatrix} 1_2 \\ 0_2 \end{bmatrix} \in M_{4,2}(A).$$

The stabilizer of $\phi_1^{(1)}$ in $\mathrm{GL}_4(A) \times \mathrm{GL}_2(A)$ consists of the pairs (ρ, α^{-1}) with $\rho = \begin{bmatrix} \alpha & \beta \\ 0 & \delta \end{bmatrix}$. The theory of elementary divisors implies that $\phi^{(1)} \sim \phi^{(2)}$, where

$$\phi_1^{(2)} = \phi_1^{(1)}, \quad \phi_0^{(2)} = \begin{bmatrix} M \\ N \end{bmatrix}, \text{ and } N = \begin{bmatrix} \nu_1\nu & 0 \\ 0 & \nu_1 \end{bmatrix}.$$

Since ϕ is nondegenerate, it follows that $\nu_1 \in A^*$. Indeed, otherwise we could find a prime π dividing ν_1. But $\phi_0^{(2)}$ has only one 2-minor, which can be nontrivial mod π. This minor, being a homogeneous polynomial of degree 2, has zeros on $\mathbb{P}^1 \otimes A/(\pi)$. This contradicts to the assumption that ϕ is nondegenerate.

Without loss of generality we can assume that

$$\phi^{(1)} \sim \phi^{(2)} = t_1 \begin{bmatrix} 1 \\ 0 \end{bmatrix} + t_0 \begin{bmatrix} M \\ N(\nu) \end{bmatrix} \text{ with } N(\nu) = \begin{bmatrix} \nu & 0 \\ 0 & 1 \end{bmatrix},$$

that $\nu \neq 0$ and that $M = M(\varepsilon_1, \varepsilon_2)$; this follows from the invertibility of ν_1, (18.2.1.2) and the conditions on the generic fiber of F.

The stabilizer of $\phi^{(2)}$ in $\mathrm{GL}_4(A) \times \mathrm{GL}_2(A)$ consists of the pairs (ρ, α^{-1}) with $\delta N(\nu)\alpha^{-1} = N(\nu)$. The integrality of δ is equivalent to the condition that $\alpha \in \tilde{\Gamma}_0(\nu)$. It remains to show that for any matrix $M = M(\varepsilon_1, \varepsilon_2)$ with $(\nu, \varepsilon_1) = 1$, we can solve the equation $\alpha M\alpha^{-1} + \beta N(\nu)\alpha^{-1} = M(\varepsilon, 0)$ or, in other words, the equation

$$\alpha \begin{bmatrix} \varepsilon_2 & 0 \\ \varepsilon_1 & 0 \end{bmatrix} + \beta \begin{bmatrix} \nu & 0 \\ 0 & 1 \end{bmatrix} = \begin{bmatrix} 0 & 0 \\ \varepsilon & 0 \end{bmatrix} \alpha. \tag{18.8}$$

Here $\alpha = (\alpha_{i,j}) \in \tilde{\Gamma}_0(\nu)$, $\beta = (\beta_{i,j}) \in M_{2,2}(A)$ and $\varepsilon \in A$ are the unknowns. First of all, we use Lemma 18.15 and choose $\alpha \in \tilde{\Gamma}_0(\nu)$ such that $\alpha \begin{bmatrix} \varepsilon_2 \\ \varepsilon_1 \end{bmatrix} = \lambda \begin{bmatrix} 0 \\ 1 \end{bmatrix}$ for some $\lambda \in A$. This is possible because ν and ε_1 are coprime (see 18.11). The equation (18.8) takes the form:

$$\begin{bmatrix} 0 & 0 \\ \lambda & 0 \end{bmatrix} + \begin{bmatrix} \nu\beta_{1,1} & \beta_{1,2} \\ \nu\beta_{2,1} & \beta_{2,2} \end{bmatrix} = \begin{bmatrix} 0 & 0 \\ \varepsilon\alpha_{1,1} & \varepsilon\alpha_{1,2} \end{bmatrix}.$$

It remains to find β and ε. Let $\beta_{1,1} = \beta_{1,2} = 0$ and choose $\beta_{2,1}$ and ε to satisfy the equation $\lambda = \varepsilon\alpha_{1,1} - \xi_{2,1}\nu$ (it is possible since $(\alpha_{1,1}, \nu) = 1$). Then take $\beta_{2,2} = \varepsilon\alpha_{1,2}$. □

We shall describe morphisms between the bundles in (18.16). In the following two results it sufficies to assume that A is a domain.

Proposition 18.17. *Let $F = F^{\nu,\varepsilon,0}$, $G = F^{\mu,\zeta,0}$ with $(\nu, \varepsilon) = 1$, $(\mu, \zeta) = 1$, $\nu \neq 0$, $\mu \neq 0$. The functor H^0 and the canonical isomorphisms $H^0(X, F) \simeq A^2$ and $H^0(X, G) \simeq A^2$ (see 18.2.1.3) identify $\mathrm{Hom}_{\mathcal{O}}(F, G)$ with the set of those θ in $M_{2,2}(A)$, which satisfy the following conditions: $N(\mu)\theta N(\nu)^{-1} \in M_{2,2}(A)$, $(M(\zeta, 0)\theta - \theta M(\varepsilon, 0))N(\nu)^{-1} \in M_{2,2}(A)$. Here $N(\nu)^{-1} \in M_{2,2}(K)$. That identification turns the composition into the product.*

Proof. The functoriality of the Beilinson spectral sequence (see 18.10) reduces the computation of $\mathrm{Hom}_{\mathcal{O}}(F, G)$ to a description of the commutative diagrams of the form

$$\begin{array}{ccccccc} \mathcal{O}^2(-2) & \xrightarrow{\phi} & \mathcal{O}^4(-1) & \longrightarrow & F & \longrightarrow & 0, \\ \downarrow{\scriptstyle\theta} & & \downarrow{\scriptstyle\lambda} & & & & \\ \mathcal{O}^2(-2) & \xrightarrow{\psi} & \mathcal{O}^4(-1) & \longrightarrow & G & \longrightarrow & 0 \end{array}$$

where ϕ and ψ are the arrows in the definition of F and G. We have the following commutativity equation

$$\begin{bmatrix} t_1 & 0 \\ \zeta t_0 & t_1 \\ \mu t_0 & 0 \\ 0 & t_0 \end{bmatrix} \theta = \begin{bmatrix} \lambda_{1,1} & \lambda_{1,2} \\ \lambda_{2,1} & \lambda_{2,2} \end{bmatrix} \begin{bmatrix} t_1 & 0 \\ \varepsilon t_0 & t_1 \\ \nu t_0 & 0 \\ 0 & t_0 \end{bmatrix}.$$

For $t_0 = 0$ it is reduced to the equalities $\theta = \lambda_{1,1}$, $\lambda_{2,1} = 0$. Taking them into account, we see that for $t_1 = 0$ the commutativity is equivalent to the relations $\lambda_{2,2} = N(\mu)\theta N(\nu)^{-1}$, $\lambda_{1,2} = (M(\zeta,0)\theta - \theta M(\varepsilon,0))N(\nu)^{-1}$. Thus λ is uniquely determined by θ; moreover, $\lambda_{1,2} \in A$ and $\lambda_{2,2} \in A$. □

Theorem 18.18. *Let* $F = F^{\nu,\varepsilon,0}$, $G = F^{\mu,\zeta,0}$ *with* $(\nu,\varepsilon) = 1$, $(\mu,\zeta) = 1$, $\nu \neq 0$, $\mu \neq 0$. *The vector bundles F and G are isomorphic iff* $(\nu) = (\mu)$ *and there are* $\eta \in A^*$ *and* $\lambda \in A$ *such that*

$$\zeta\lambda^2 \equiv \varepsilon\eta \pmod{\nu}.$$

Proof. From the relation $F \simeq G$ it follows that $(\mu) = (\nu)$ (see 18.14). If the isomorphism $F \to G$ is given by $\theta \in \mathrm{GL}_2(A)$, as defined in Prop. 18.17, then θ and θ^{-1} satisfy the integrality relations in that Proposition. It can be proved by a calculation that $\theta = (\theta_{i,j})$ defines an isomorphism iff $\theta_{2,1} \equiv 0$, $\varepsilon\theta_{2,2} \equiv \zeta\theta_{1,1}$, $\varepsilon\theta_{1,2} \equiv 0$, $\zeta\theta_{1,2} \equiv 0 \mod \nu$. Since ε and ζ are invertible modulo (ν), it follows that

$$\theta_{2,1} \equiv \theta_{1,2} \equiv 0 \pmod{\nu} \quad \text{and } \varepsilon\theta_{2,2} \equiv \zeta\theta_{1,1} \pmod{\nu}. \tag{18.9}$$

Let $\eta = \det\theta \in A^*$. Then $\eta \equiv \theta_{1,1}\theta_{2,2} \equiv \zeta\varepsilon^{-1}\theta_{1,1}^2 \pmod{\nu}$. This proves that if $F \simeq G$, then $(\mu) = (\nu)$ and there are $\eta \in A^*$ and $\lambda \in A$ with $\zeta\lambda^2 \equiv \varepsilon\eta \pmod{\nu}$. Conversely, suppose μ, ν, λ, and ε satisfy those conditions and let $\chi \in A$ such that

$$\chi\varepsilon \equiv 1 \pmod{\nu}. \tag{18.10}$$

Then it follows from the relation $\zeta\lambda^2 \equiv \varepsilon\eta \pmod{\nu}$ that $\eta = \zeta\chi\lambda^2 + \kappa\nu$ with $\kappa \in A$. Let $\omega, \tau \in A$ so that $\omega\lambda = \kappa + \tau\nu$. Let $\theta_{1,1} = \lambda$, $\theta_{2,2} = \zeta\chi\lambda + \omega\nu$, $\theta_{2,1} = \tau\nu$, $\theta_{1,2} = \nu$. Then θ induces an isomorphism $F \to G$. The condition $\theta \in \mathrm{GL}_2(A)$ follows from the definitions of θ, τ, ω, κ and the relation $\det\theta \in A^*$. It remains to prove (18.9). The relations $\theta_{2,1} \equiv \theta_{1,2} \equiv 0 \pmod{\nu}$ and $\varepsilon\theta_{2,2} \equiv \zeta\theta_{1,1}$ follow from the definition of θ and (18.10). □

Example 18.19. Theorems 18.16 and 18.18 classify completely the 2-bundles with the trivial generic fiber and simple jumps in the case of a factorial Dedekind ring A.

For $A = \mathbb{C}[t]$ the isomorphism classes of vector bundles in question can be identified with the effective divisors on $\mathbb{A}^1$.

Let $A = \mathbb{R}[t]$. If ν is an irreducible polynomial then there is only one isomorphism class of bundles related to ν, independently on $\deg\nu$. This is related to the fact that for degree 1 the negative residues are represented by global units. However, if $\nu = p_1^{n_1}\cdots p_e^{n_d} q_1^{m_1}\cdots q_d^{m_d}$ with irreducible p_i of degree 1 and irreducible q_i of degree 2, then for $e \geq 1$ there are 2^{e-1} isomorphism classes of bundles with this ν.

We have a similar, but slightly more interesting picture for $A = \mathbb{F}_q[t]$ and $A = \mathbb{Z}$. For example, in the case $A = \mathbb{Z}$ there are two bundles with $\nu = 5$, because the only nontrivial global unit (-1) is a square $\mod \nu$ (the same is true for any prime $\nu \equiv 1 \pmod 4$), and only one bundle with $\nu = 7$ (the same is true for any prime $\nu \equiv 3 \pmod 4$).

18.3 Minimal linear filtrations

Let E be a vector bundle on $\mathbb{P}^1_A$ such that there exists at least one filtration of E with linear quotients. We are interested in filtrations with simplest quotients. We shall study this problem in the special case when $\operatorname{rk} E = 2$ and A is a factorial Dedekind ring. By Serre's theorem (see 18.4) each linear quotient is isomorphic to $\mathcal{O}(d)$ for some d in this case.

18.3.1 Amplitudes of bundles

Assume additionally that the generic fiber of E is trivial, i.e. that $E_\eta \cong \mathcal{O}^2$. Then for each filtration with linear quotients

$$0 = E_0 \subset E_1 \subset E_2 = E \tag{18.11}$$

the line bundles E_1 and E/E_1 are dual each to other.

Indeed, the linear bundle $\operatorname{Det} E \cong \mathcal{O}(d)$ for some integer d, by the factoriality of A and Serre's theorem (see 18.4). The pullback of this isomorphism to the generic fiber and the triviality of the generic fiber E_η show that $d = 0$ and therefore $\operatorname{Det} E \cong \mathcal{O}$. But then $E_1 \otimes E/E_1 \cong \operatorname{Det} E \cong \mathcal{O}$. This isomorphism gives the claimed duality.

Thus, for any filtration of E of the form (18.11), there is an integer d such that

$$E/E_1 \cong \mathcal{O}(d) \text{ and } E_1 \cong \mathcal{O}(-d).$$

Moreover, $d \geq 0$. Indeed, the filtration (18.11) induces an element in

$$\operatorname{Ext}^1_{\mathcal{O}}(E/E_1, E_1) = \operatorname{Ext}^1_{\mathcal{O}}(\mathcal{O}(d), \mathcal{O}(-d));$$

for $d < 0$ this group is trivial and consequently $E \cong \mathcal{O}(-d) + \mathcal{O}(d)$. This contradicts the triviality of the generic fiber, however.

Definition 18.20. Let E be a vector bundle on $\mathbb{P}^1_A$ with $\operatorname{rk} E = 2$ and $E_\eta \cong \mathcal{O}^2$. The minimal non-negative integer d, for which there exists a filtration on E of the form (18.11) with $E/E_1 \cong \mathcal{O}(d)$, is called the amplitude of E. The amplitude of E is denoted by $a(E)$.

18.3.2 Local amplitudes

Let A be a local ring. In this case, the amplitude can be easily computed and is determined by the splitting type at the generic point. This follows from the following complement to the Horrocks result (see 18.7).

Proposition 18.21. *Let A be a local Dedekind ring, E be a vector bundle on $\mathbb{P}^1_A$, y be the closed point of* Spec A, *and $E_y = \mathcal{O}(d_1) + \cdots + \mathcal{O}(d_n)$ with $d_1 \leq \cdots \leq d_n$. Then there exists a filtration $0 = E_0 \subset E_1 \subset \cdots \subset E_r = E$, where E_i is a vector bundle and $E_i/E_{i-1} \cong \mathcal{O}(d_i)$ for $i = 1, \ldots, r$.*

Proof. By induction, it suffices to find an inclusion $\mathcal{O}(d_1) \to E$ such that $E/\mathcal{O}(d_1)$ is a bundle. We can twist E and, without loss of generality, assume that $d_1 = 0$. Let $H^n(E) = R^n p_* E$, where $p : \mathbb{P}^1_A \to \text{Spec}\, A$ is the structure projection. Let F be the residue field at y. Making use of the base change theorem (see 18.1.2.3), one can prove that $H^1(E) = 0$ and the pullback $H^0(E) \otimes_A \mathrm{F} \to H^0(E_y)$ is an isomorphism. Thus we can choose $s \in H^0(E)$, identifying $\mathcal{O}$ with a summand of E_y. Then the quotient of $s : \mathcal{O} \to E$ is locally free. □

18.3.3 Local restrictions for amplitudes

Let

$$a_{loc}(E) = \max_y a(E_y),$$

where y runs over the closed points of Spec A. It follows from the definition of the amplitude that

$$a(E) \geq a_{loc}(E). \tag{18.12}$$

18.3.4 Bundles without jumps

Proposition 18.22. *Let A be a Dedekind ring and let E be a vector bundle on $\mathbb{P}^1_A$. If the function $y \mapsto d_1(E, y)$ (see 18.1.2.1) is constant on the set of the closed points of* Spec A, *then there is an inclusion $\mathcal{O}(d) \to E$ such that $E/\mathcal{O}(d)$ is a vector bundle, where $d = d_1(E, y)$ for each closed point y.*

Proof. Essentially this is proved in the course of the proof of Proposition 18.21. □

It follows from Proposition 18.22 that E is a trivial bundle if $a_{loc}(E) = 0$.

18.3.5 Bundles with simple jumps

Let E be a vector bundle with $a_{loc}(E) = 1$. As we shall see, the local data, i.e. the local amplitudes, do not determine the global amplitude. Let

$$E = F^{\beta,\alpha,0} = \mathrm{Coker}\,[\mathcal{O}(-2)^2 \to \mathcal{O}(-1)^4],$$

where the arrow is given by the matrix

$$\phi = \begin{bmatrix} t_1 & 0 \\ \alpha & t_1 \\ \beta t_0 & 0 \\ 0 & t_0 \end{bmatrix}, \quad (\beta, \alpha) = 1, \beta \neq 0, \beta \notin A^*. \tag{18.13}$$

Since $(\alpha, \beta) = 1$, the sheaf E is a bundle of rank 2 (see 18.11). It follows from the condition $\beta \neq 0$ that the generic fiber is trivial and the jumps are simple. The condition $\beta \notin A^*$ ensures the existence of jumps (see 18.13).

18.3.5.1 Gluing matrices

To compute $\mathrm{Im}\,\phi$ and the corresponding quotient let us introduce the following notation. Denote by e_1, e_2 the standard $\mathcal{O}$-basis for $\mathcal{O}^2$ and by f_1, f_2, f_3, f_4 the standard $\mathcal{O}$-basis for $\mathcal{O}^4$. Let $\gamma, \delta \in A$ and

$$\alpha\delta - \beta\gamma = 1. \tag{18.14}$$

In the standard bases on U_0, the arrow ϕ is given by

$$\phi|U_0 = t_0 \begin{bmatrix} x & 0 \\ \alpha & x \\ \beta & 0 \\ 0 & 1 \end{bmatrix}.$$

Let the basis of $\mathrm{Im}\,\phi$ on U_0 consist of $\phi(t_0^{-2}e_1)$ and $\phi(t_0^{-2}e_2)$, that is of the columns of $\phi|_{U_0}$. Extend $\phi|_{U_0}$ to an invertible matrix by letting

$$\widetilde{\phi|U_0} = t_0 \begin{bmatrix} x & 0 & 0 & 1 \\ \alpha & x & \gamma & 0 \\ \beta & 0 & \delta & 0 \\ 0 & 1 & 0 & 0 \end{bmatrix}, \quad \det \widetilde{\phi|U_0} = t_0^4.$$

The $\mathcal{O}_{U_0}$-module $\mathcal{O}(-1)^4|U_0$ is thereby equipped with the basis

$$t_0^{-1}[g_1, \ldots, g_4] = t_0^{-2}[f_1, \ldots, f_4]\widetilde{\phi|U_0}, \tag{18.15}$$

where $t_0^{-1}[\bar{g}_3, \bar{g}_4]$ is a basis of the quotient, i.e. a basis of $E|U_0$.

In the standard bases on U_1 the arrow ϕ is given by

$$\phi|_{U_1} = t_1 \begin{bmatrix} 1 & 0 \\ \alpha y & 1 \\ \beta y & 0 \\ 0 & y \end{bmatrix}, \text{ where } y = t_0/t_1 = x^{-1}.$$

Extend $\phi|_{U_1}$ to an invertible matrix, by letting

$$\widetilde{\phi|_{U_1}} = t_1 \begin{bmatrix} 1 & 0 & 0 & 0 \\ \alpha y & 1 & 0 & 0 \\ \beta y & 0 & 1 & 0 \\ 0 & y & 0 & 1 \end{bmatrix}, \quad \det \widetilde{\phi|_{U_1}} = t_1^4.$$

The $\mathcal{O}_{U_1}$-module $\mathcal{O}(-1)^4|U_1$ is thereby equipped with the basis

$$t_1^{-1}[h_1, \ldots, h_4] = t_1^{-2}[f_1, \ldots, f_4]\widetilde{\phi|_{U_1}}, \tag{18.16}$$

whereby $t_1^{-1}[\bar{h}_3, \bar{h}_4]$ is a basis of the quotient, i.e. a basis of $E|U_1$.

Let us compute the gluing matrix on U_{01}. In order to do it, let us express $t_1^{-1}h_3$ and $t_1^{-1}h_4$ in terms of $t_0^{-1}g_1, t_0^{-1}g_2, t_0^{-1}g_3, t_0^{-1}g_4$. Let us start by expressing $t_0^{-1}g_i$, $1 \le i \le 4$, in terms of $t_0^{-2}f_j$, $1 \le j \le 4$, (see (18.15)). It follows then that

$$\widetilde{\phi|_{U_0}}^{-1} = t_0^{-1} \begin{bmatrix} 0 & \delta & -\gamma & -\delta x \\ 0 & 0 & 0 & 1 \\ 0 & -\beta & \alpha & \beta x \\ 1 & -\delta x & \gamma x & \delta x^2 \end{bmatrix}.$$

Relations (18.15) and (18.16) give:

$$t_1^{-1}[h_1, h_2, h_3, h_4] = x^{-2} t_0^{-1}[g_1, g_2, g_3, g_4]\widetilde{\phi|_{U_0}}^{-1}\widetilde{\phi|_{U_1}}.$$

It follows that

$$\widetilde{\phi|_{U_0}}^{-1}\widetilde{\phi|_{U_1}} = x \begin{bmatrix} y & -\delta & -\gamma & -\delta x \\ 0 & y & 0 & 1 \\ 0 & 0 & \alpha & \beta x \\ 0 & 0 & \gamma x & \delta x^2 \end{bmatrix}.$$

Thus

$$t_1^{-1}\begin{bmatrix} h_3 & h_4 \end{bmatrix} = x^{-2} x \begin{bmatrix} \alpha & \beta x \\ \gamma x & \delta x^2 \end{bmatrix} t_0^{-1}\begin{bmatrix} g_3 & g_4 \end{bmatrix} \quad \text{mod } (g_1, g_2).$$

Therefore the bundle E can be given by the gluing matrix

$$\sigma = \begin{bmatrix} \alpha y & \beta \\ \gamma & \delta x \end{bmatrix}, \quad \det \sigma = 1. \tag{18.17}$$

In other words, there are a basis $[e_1, e_2]$ of the pullback of E to U_0 and a basis $[f_1, f_2]$ of the pullback of E to U_1 with

$$[e_1, e_2]\begin{bmatrix}\alpha x^{-1} & \beta \\ \gamma & \delta x\end{bmatrix} = [f_1, f_2] \text{ or } [e_1, e_2] = [f_1, f_2]\begin{bmatrix}\delta x & -\beta \\ -\gamma & \alpha x^{-1}\end{bmatrix} \text{ on } U_{01}.$$

Although E has the trivial generic fiber and simple jumps at the divisors of β (see 18.12), we need an explicit identification of the corresponding fibers with $\mathcal{O}^2$ and $\mathcal{O}(-1) + \mathcal{O}(1)$.

18.3.5.2 Generic fibers

Let us fix an isomorphism $E \otimes \mathbb{Z}[1/\beta] \simeq \mathcal{O}^2$. Let $[g_1, g_2]$ be the standard basis for $\mathcal{O}^2$ on U_0 and $[h_1, h_2]$ be the standard basis for $\mathcal{O}^2$ on U_1, so that

$$[g_1, g_2]\begin{bmatrix}1 & 0 \\ 0 & 1\end{bmatrix} = [h_1, h_2] \text{ or } [g_1, g_2] = [h_1, h_2]\begin{bmatrix}1 & 0 \\ 0 & 1\end{bmatrix} \text{ on } U_{01}.$$

The isomorphism $E \otimes \mathbb{Z}[1/\beta] \simeq \mathcal{O}^2$ is given as follows:

$$[g_1, g_2]\begin{bmatrix}\delta x & -\beta \\ 1 & 0\end{bmatrix} = [e_1, e_2] \quad \text{and} \quad [f_1, f_2]\begin{bmatrix}1 & 0 \\ -\alpha\beta^{-1}x^{-1} & \beta^{-1}\end{bmatrix} = [h_1, h_2].$$

18.3.5.3 Special fibers

Let us fix an isomorphism $E \otimes A/\beta \simeq \mathcal{O}(-1) + \mathcal{O}(1)$. Choose $[t_0^{-1}g_1, t_0 g_2]$ and $[t_1^{-1}h_1, t_1 h_2]$ as the bases of $\mathcal{O}(-1) + \mathcal{O}(1)$ on U_0 and U_1. Here $[g_1, g_2]$ and $[h_1, h_2]$ are the standard bases for $\mathcal{O}^2$ on U_0 and U_1. Then

$$[t_0^{-1}g_1, t_0 g_2]\begin{bmatrix}x^{-1} & 0 \\ 0 & x\end{bmatrix} = [t_1^{-1}h_1, t_1 h_2] \text{ or } [t_0^{-1}g_1, t_0 g_2]$$
$$= [t_0^{-1}h_1, t_0 h_2]\begin{bmatrix}x & 0 \\ 0 & x^{-1}\end{bmatrix} \text{ on } U_{01}.$$

The isomorphism $E \otimes A/(\beta) \simeq \mathcal{O}(-1) + \mathcal{O}(1)$ is given as follows:

$$[t_0^{-1}g_1, t_0 g_2]\begin{bmatrix}\delta & 0 \\ -\gamma x & \alpha\end{bmatrix} = [e_1, e_2] \quad \text{and} \quad [f_1, f_2]\begin{bmatrix}1 & 0 \\ 0 & 1\end{bmatrix} = [t_1^{-1}h_1, t_1 h_2].$$

Let us check whether $a(E) = 1$, that is whether there is a filtration of the form $E_1 \subset E$ with $E_1 \cong \mathcal{O}(-1)$ and $E/E_1 \cong \mathcal{O}(1)$. The existence of such a filtration of E is equivalent to the existence of a global nonvanishing section of $V = E(1)$.

In further calculations, take $t_0[e_1, e_2]$ and $t_1[f_1, f_2]$ as bases of V on U_0 and U_1, where $[e_1, e_2]$ is the basis of E on U_0 and $[f_1, f_2]$ is the basis of E on U_1. Then

$$t_0[e_1, e_2]\begin{bmatrix} \alpha & \beta x \\ \gamma x & \delta x^2 \end{bmatrix} = t_1[f_1, f_2]$$

$$\text{or} \quad t_0[e_1, e_2] = t_1[f_1, f_2]\begin{bmatrix} \delta & -\beta x^{-1} \\ -\gamma x^{-1} & \alpha x^{-2} \end{bmatrix} \text{ on } U_{01}.$$

18.3.5.4 Calculating $H^0(V)$

The generic U_0-section $s = (s_1e_1 + s_2e_2)t_0$, where $s_i \in A[x]$, can be written in the U_1-basis as $s = (\delta s_1 - \beta x^{-1}s_2)t_1 f_1 + (-\gamma x^{-1}s_1 + \alpha x^{-2}s_2)t_1 f_2$. Therefore the conditions for s to be a global section are as follows:

$$\begin{cases} c_1: & \delta s_1 - \beta x^{-1}s_2 \in A[x^{-1}]; \\ c_2: & -\gamma x^{-1}s_1 + \alpha x^{-2}s_2 \in A[x^{-1}]. \end{cases}$$

It can be easily seen that $\{x^{-1}s_1, x^{-2}s_2\} \subset A[x^{-1}]$: it suffices to note that $\gamma x^{-1}c_1 + \delta c_2 = x^{-2}s_2$. Therefore $s_1 = u_0 + u_1x$, $s_2 = v_0 + v_1x + v_2x^2$. We obtain an equation for u_i and v_j: $\delta u_1 = \beta v_2$. Since α and γ are coprime, then the generic global section of V has the form:

$$s = (u_0 + \beta wx)t_0e_1 + (v_0 + v_1x + \delta wx^2)t_0e_2, \text{ where } u_0, v_0, v_1, w \in A. \tag{18.18}$$

18.3.5.5 Nonvanishing over $A[1/\beta]$

Let us determine when the section s, defined by (18.18), vanishes nowhere on $\mathbb{P}^1 \otimes A[1/\beta]$. To this end, we shall use the isomorphism in 18.3.5.2 and regard s as a section of $\mathcal{O}(1)^2$. Choose $t_0[g_1, g_2]$ as a basis of $\mathcal{O}^2(1)$ on U_0, where $[g_1, g_2]$ is the standard basis for $\mathcal{O}^2$ on U_0. The isomorphism $V \otimes \mathbb{Z}[1/\beta] \simeq \mathcal{O}^2(1)$ (see 18.3.5.2) on U_0 is given as follows:

$$t_0[g_1, g_2]\begin{bmatrix} \delta x & -\beta \\ 1 & 0 \end{bmatrix} = t_0[e_1, e_2].$$

The generic section of $\mathcal{O}(1)^2$ is of the form: $(a_{10}t_0 + a_{01}t_1)g_1 + (b_{10}t_0 + b_{01}t_1)g_2$. Let us express the generic section of V in these terms: $s = (u_0 + \beta wx)t_0e_1 + (v_0 + v_1x + \delta wx^2)t_0e_2 = (-\beta v_0 + [\delta u_0 - \beta v_1]x)t_0g_1 + (u_0 + \beta wx)t_0g_2$. Therefore

$$a_{10} = -\beta v_0, a_{01} = \delta u_0 - \beta v_1, b_{10} = u_0, b_{01} = \beta w. \tag{18.19}$$

Thus we have constructed a section of $\mathcal{O}(1)^2$ over $A[\beta^{-1}]$, i.e. a pair of polynomials of degree 1, by means of $s \in H^0(V)$. The existence of their common zero can be determined by means of the resultant

$$R_{11}(a_{10}, a_{01}; b_{10}, b_{01}) = \det\begin{bmatrix} a_{10} & a_{01} \\ b_{10} & b_{01} \end{bmatrix} = a_{10}b_{01} - a_{01}b_{10}.$$

Using (18.19) we obtain the following condition

$$-\delta u_0^2 + \beta u_0 v_1 - \beta^2 v_0 w \in A[\beta^{-1}]^*. \tag{18.20}$$

The rows $(u_0, v_0, v_1, w) \in A^4$, satisfying this condition, correspond to the global sections of V without zeros outside of the divisors of β.

18.3.5.6 Nonvanishing over divisors of β

Let us determine when the section s, defined by (18.18), vanishes nowhere on $\mathbb{P}^1 \otimes A/(\pi)$, where π is a prime divisor of β. To this end, we shall use the isomorphism of 18.3.5.3 and regard s as a section of $\mathcal{O} + \mathcal{O}(2)$. Choose $[g_1, t_0^2 g_2]$ as a basis for $\mathcal{O} + \mathcal{O}(2)$ on U_0, where $[g_1, g_2]$ is the standard basis for $\mathcal{O}^2$ on U_0. The isomorphism $V \otimes A/\pi \simeq \mathcal{O} + \mathcal{O}(2)$ (see 18.3.5.3) on U_0 is given as follows:

$$[g_1, t_0^2 g_2]\begin{bmatrix} \delta & 0 \\ -\gamma x & \alpha \end{bmatrix} = t_0[e_1, e_2].$$

The generic section of $\mathcal{O} + \mathcal{O}(2)$ is of the form: $a_{00}g_1 + (b_{20}t_0^2 + b_{11}t_0t_1 + b_{02}t_1^2)g_2$. Let us express the generic section of V in these terms: $s \pmod{\pi} = (u_0 + \beta w x)t_0 e_1 + (v_0 + v_1 x + \delta w x^2)t_0 e_2 = (\delta u_0 + \beta\delta w x)g_1 + (\alpha v_0 + [\alpha v_1 - \gamma u_0]x + w x^2)t_0^2 g_2$. In other words,

$$a_{00} = \delta u_0,\ b_{20} = \alpha v_0,\ b_{11} = \alpha v_1 - \gamma u_0,\ b_{02} = w. \tag{18.21}$$

Thus, by $s \in H^0(V)$ we have constructed a section of $\mathcal{O} + \mathcal{O}(2)$ over A/π, i.e. a pair of polynomials of degree 0 and 2. The existence of their common zero can be determined by means of the resultant

$$R_{02}(a_{00}; b_{20}, b_{11}, b_{02}) = \det\begin{bmatrix} a_{00} & 0 \\ 0 & a_{00} \end{bmatrix} = a_{00}^2.$$

Using (18.21) we obtain the condition $\delta^2 u_0^2 \neq 0 \pmod{\pi}$ or

$$u_0 \neq 0 \pmod{\pi}. \tag{18.22}$$

The rows $(u_0, v_0, v_1, w) \in A^4$, satisfying this condition, correspond to global sections of V without zeros over $A/(\pi)$.

18.3.5.7 Conclusion

The existence of $E_1 \subset E$ with $E_1 \cong \mathcal{O}(-1)$ and $E/E_1 \cong \mathcal{O}(1)$ is equivalent to the existence of a row $(u_0, v_0, v_1, w) \in A^4$, satisfying the following conditions:

$$\begin{cases} \delta u_0^2 - \beta u_0 v_1 + \beta^2 v_0 w \in A[\beta^{-1}]^* \\ u_0 \pmod{\beta} \in (A/(\beta))^*. \end{cases} \tag{18.23}$$

Of course, the solvability of these conditions can be verified by giving such a row. However, if we fail to find a solution it is desirable to prove the insolvability. We have already seen that there are no local obstructions to the solvability. However, we can use some global information, namely the knowledge of units in $A[\beta^{-1}]$, to modify slightly the system (18.23). After such a modification local obstructions can arise.

Namely, let $\pi_1, \dots, \pi_r$ be all the prime divisors of β. The solvability of (18.23) is equivalent to the existence of nonnegative integers $n_1, \dots, n_r$ and of $\theta \in A^*$ such that the system

$$\begin{cases} \delta u_0^2 - \beta u_0 v_1 + \beta^2 v_0 w = \theta \pi_1^{n_1} \cdots \pi_r^{n_r}. \\ v_0 \pmod{\beta} \in (A/(\beta))^* \end{cases} \tag{18.24}$$

is solvable. Moreover, it is clear that $\mathrm{ord}_\pi(\delta) = 0$ for any p, dividing β. Therefore the second condition implies that $n_i = 0$ for every i, so that the solvability of the system (18.24) is equivalent to the existence of a global unit $\theta \in A^*$ such that the equation

$$\delta u_0^2 - \beta u_0 v_1 + \beta^2 v_0 w = \theta. \tag{18.25}$$

is solvable.

Thus let the bundle E be given by the gluing matrix

$$\sigma = \begin{bmatrix} \alpha x^{-1} & \beta \\ \gamma & \delta x \end{bmatrix}, \text{ where } \begin{bmatrix} \alpha & \beta \\ \gamma & \delta \end{bmatrix} \in \mathrm{SL}_2(A).$$

The existence of $E_1 \subset E$ with $E_1 \cong \mathcal{O}(-1)$ and $E/E_1 \cong \mathcal{O}(1)$ is equivalent to the existence of $\theta \in A^*$ and of a row $(u_0, v_0, v_1, w) \in A^4$, satisfying (18.25). In other words, it is equivalent to the existence of $\theta \in A^*$ such that the equation (18.25) can be solved in A.

Example 18.23. Let $A = \mathbb{Z}$ and the bundle E be given by the gluing matrix

$$\sigma = \begin{bmatrix} 2x^{-1} & 5 \\ 1 & 2x \end{bmatrix}.$$

The equality $a(E) = 1$ is equivalent to the existence of a row $(u_0, v_0, v_1, w) \in \mathbb{Z}^4$ with $2u_0^2 - 5u_0 v_1 + 25 v_0 w = \pm 1$. These equations are inconsistent because ± 2 is not a square in $\mathbb{Z}/(5)$. It can be shown, however, that there exists a filtration $E_1 \subset E$ with $E_1 \cong \mathcal{O}(-2)$ and $E/E_1 \cong \mathcal{O}(2)$. In other words, $a(E) = 2$.

Theorem 18.24. *Let $A = \mathbb{Z}$. Then $a(E) = 1$, iff either $\alpha \pmod{\beta}$ is a square and the equation $\delta u_0^2 - \beta u_0 v_1 + \beta^2 v_0 w = 1$ can be solved in $\mathbb{Z}_2$, or $(-\alpha$*

(mod β)) *is a square and the equation* $\delta u_0^2 - \beta u_0 v_1 + \beta^2 v_0 w = -1$ *can be solved in* $\mathbb{Z}_2$.

Proof. It follows from the equation $\alpha\delta - \beta\gamma = 1$ (see (18.17)) that $\varepsilon\alpha \pmod{\beta}$ is a square iff $\varepsilon\delta \pmod{\beta}$ is a square, where $\varepsilon \in \{\pm 1\}$. Therefore it is enough to prove the theorem with δ instead of α. Further, $a(E) = 1$ iff (see 18.3.5.7) at least one of the equations $\delta u_0^2 - \beta u_0 v_1 + \beta^2 v_0 w = \pm 1$ has a solution in $\mathbb{Z}$. If the equation $\delta u_0^2 - \beta u_0 v_1 + \beta^2 v_0 w = 1$ has a solution in $\mathbb{Z}$, then $\delta \pmod{\beta}$ is a square. And the same is true for the second equation. This proves the "only if" part.

Conversely, suppose that $\delta \pmod{\beta}$ is a square and the equation $\delta u_0^2 - \beta u_0 v_1 + \beta^2 v_0 w = 1$ has a solution in $\mathbb{Z}_2$. We have to show the solvability of the following equation

$$\delta u_0^2 - \beta u_0 v_1 + \beta^2 v_0 w = 1. \tag{18.26}$$

We shall apply results of [1]. The definition of a quadratic form in [1, I, §2] requires the evenness of the coefficients for the products. So consider the equation

$$q(u_0, v_0, v_1, w) = 2, \qquad \text{where } q(u_0, v_0, v_1, w) = 2\delta u_0^2 - 2\beta u_0 v_1 + 2\beta^2 v_0 w. \tag{18.27}$$

It is obtained by multiplication (18.26) with 2. This does not change the solvability. The form q is indefinite and $\mathbb{Z}$-regular in the sense of [1], i.e. it is nondegenerate over $\mathbb{Q}$. There is an important result (the strong approximation theorem is applied) concerning such forms [1, Ch.9, Th.1.5]. It says that if $\operatorname{rk} q \geq 4$ (in our case $\operatorname{rk} q = 4$), then an integer is represented by q over $\mathbb{Z}$ iff it is represented by q over $\mathbb{Z}_p$ for every p .

It remains to study the local solvability of (18.27). Since $\operatorname{disc}(q) = 16\beta^6$, we have the local solvability outside the jumps and 2. It follows from the smoothness, the existence of rational points on quadrics over finite fields for dim > 1 and Hensel's lemma.

Let us study the local solvability at an odd prime p dividing β. If $\delta \pmod{p}$ is a square, we can take $v_0 = v_1 = w = 0$ and use Hensel's lemma. The solvability in $\mathbb{Z}_2$ is assumed in the statement. The case when $(-\delta) \pmod{\beta}$ is a square is similar. □

Corollary 18.25. *Let* $A = \mathbb{Z}$ *and* E *has no jump at 2. Then* $a(E) = 1$ *iff either* $\alpha \pmod{\beta}$ *is a square, or* $(-\alpha) \pmod{\beta}$ *is a square.*

Proof. By Theorem 18.24, it is enough to show that the equations $\delta u_0^2 - \beta u_0 v_1 + \beta^2 v_0 w = \pm 1$ are solvable in $\mathbb{Z}_2$ if there is no jump at 2. The last condition is equivalent (see 18.13) to the condition of β being odd. Let $u_0 = 1$ and $v_0 = w = 0$. The equations in the question take the form $\delta - \beta v_1 = \pm 1$. Their solution $v_1 = -(\pm 1 - \delta)/\beta$ lies in $\mathbb{Z}_2$ since β is invertible in $\mathbb{Z}_2$. □

Theorem 18.24 and Corollary 18.25 show when $a(E) = 1$ for $A = \mathbb{Z}$ and $E = F^{\beta,\alpha,0}$. Theorem 18.16 asserts that any bundle with the trivial generic fiber and simple jumps is isomorphic to a bundle of the form $F^{\beta,\alpha,0}$. Therefore Theorem 18.24 and Corollary 18.25 allow to check whether $a(E) = 1$ for any bundle with the trivial generic fiber and simple jumps.

References

[1] *J.W.S. Cassels.* Rational Quadratic Forms. Academic Press, 1978.

[2] *Ch.C. Hanna.* Subbundles of vector bundles on the projective line. J. Algebra, 52, no. 2, 322–327, 1978.

[3] *R. Hartshorne.* Algebraic Geometry. Graduate Texts in Math. 52, Springer Verlag, 1977.

[4] *G. Horrocks.* Projective modules over an extension of a local ring, Proc. London Math. Soc. (3) 14 (1964), 714–718.

[5] *Ch. Okonek, M. Shneider, H. Spindler.* Vector Bundles On Complex Projective Spaces. Progress in Math 3, Birkhäuser, 1980.

[6] *D. Quillen.* Projective modules over polynomial rings, Invent. Math., 1976, vol. 36, p. 167–171.

[7] *J.-P. Serre.* Faisceaux algébriques cohérents. Ann. of Math., 61 (1955), 197–278.

[8] *A. Suslin.* Projective modules over polynomial rings are free, Doklady Akademii Nauk SSSR (in Russian), 1976, 229 (5), p. 1063–1066. Translated in Soviet Mathematics, 1976, 17 (4), p. 1160–1164.

19

On the dihedral Euler characteristics of Selmer groups of Abelian varieties

Jeanine Van Order
Einstein Institute of Mathematics, The Hebrew University of Jerusalem
E-mail address: jeaninevanorder@gmail.com

ABSTRACT. This note shows how to use the framework of Euler characteristic formulae to study Selmer groups of abelian varieties in certain dihedral or anticyclotomic extensions of CM fields via Iwasawa main conjectures, and in particular how to verify the p-part of the refined Birch and Swinnerton-Dyer conjecture in this setting. When the Selmer group is cotorsion with respect to the associated Iwasawa algebra, we obtain the p-part of formula predicted by the refined Birch and Swinnerton-Dyer conjecture. When the Selmer group is not cotorsion with respect to the associated Iwasawa algebra, we give a conjectural description of the Euler characteristic of the cotorsion submodule, and explain how to deduce inequalities from the associated main conjecture divisibilities of Perrin-Riou and Howard.

Contents

1991 *Mathematics Subject Classification.* Primary 11, Secondary 11G05, 11G10, 11G40, 11R23
Keywords and phrases: Iwasawa theory, abelian varieties, elliptic curves
The author acknowledges partial support from the Swiss National Science Foundation (FNS) grant 200021-125291.

Arithmetic and Geometry, ed. Luis Dieulefait *et al.* Published by Cambridge University Press.

19.1 Introduction

Fix a prime number p. Let G be a compact p-adic Lie group without p-torsion. Writing $\mathcal{O}$ to denote the ring of integers of some fixed finite extension of $\mathbf{Q}_p$, let

$$\Lambda(G) = \varprojlim_{U} \mathcal{O}[G/U]$$

denote the $\mathcal{O}$-Iwasawa algebra of G. Here, the projective limit runs over all open normal subgroups U of G, and $\mathcal{O}[G/U]$ denotes the usual group ring of G/U with coefficients in $\mathcal{O}$. Let M be any discrete, cofinitely generated $\Lambda(G)$-module. If the homology groups $H_i(G, M)$ are finite for all integers $i \geq 0$, then we say that the G-Euler characteristic $\chi(G, M)$ of M is *well-defined*, and given by the (finite) value

$$\chi(G, M) = \prod_{i \geq 0} |H^i(G, M)|^{(-1)^i}.$$

If G is topologically isomorphic to $\mathbf{Z}_p^\delta$ for some integer $\delta \geq 1$, then a clearer picture of this Euler characteristic $\chi(G, M)$ emerges. In particular, a classical result shows that $\chi(G, M)$ is well defined if and only if $H_0(G, M)$ is finite. Moreover, if M in this setting is a pseudonull $\Lambda(G)$-module, then $\chi(G, M) = 1$. Now, the structure theory of finitely generated torsion $\Lambda(G)$-modules shown in Bourbaki [4] also gives the following Iwasawa theoretic picture of the G-Euler characteristic $\chi(G, M)$. That is, suppose M is a discrete, cofinitely generated $\Lambda(G)$-cotorsion module. Let $M^\vee = \mathrm{Hom}(M, \mathbf{Q}_p/\mathbf{Z}_p)$ denote the Pontragin dual of M, which has the structure of a compact $\Lambda(G)$-module. Then, there exists a finite collection of non-zero elements $f_1, \ldots, f_r \in \Lambda(G)$, along with a pseudonull $\Lambda(G)$-module $D^\vee$, such that the following sequence is exact:

$$0 \longrightarrow \bigoplus_{i=1}^{r} \Lambda(G)/f_i\Lambda(G) \longrightarrow M^\vee \longrightarrow D^\vee \longrightarrow 0.$$

The product

$$\mathrm{char}(M^\vee) = \mathrm{char}_{\Lambda(G)}(M^\vee) = \prod_{i=1}^{r} f_i,$$

is unique up to unit in $\Lambda(G)$, and defines the $\Lambda(G)$-*characteristic power series of* $M^\vee$. If the homology group $H_0(G, M)$ is finite, then the G-Euler characteristics $\chi(G, M)$ and $\chi(G, D)$ are both well-defined, and given by the formulae

$$\chi(G, M) = |\operatorname{char}(M^\vee)(0)|_p^{-1}$$
$$\chi(G, D) = 1.$$

Here, $\operatorname{char}(M^\vee)(0)$ denotes the image of the characteristic power series $\operatorname{char}(M^\vee)$ under the natural map $\Lambda(G) \longrightarrow \mathbf{Z}_p$, and D denotes the discrete dual of $D^\vee$.

In this note, we show how to use the framework of Euler characteristics to study the Selmer groups of abelian varieties in certain abelian extensions of CM fields, and in particular to verify the p-part of the refined Birch and Swinnerton-Dyer conjecture via the associated Iwasawa main conjectures. To be more precise, let A be a principally polarized abelian variety defined over a totally real number field F. We shall assume for simplicity that A is principally polarized, though all of the arguments given below can be carried through to the more general case with some extra care. Fix a prime $\mathfrak{p}$ above p in F. Let K be a totally imaginary quadratic extension of F. Assume that A has good ordinary reduction at each prime above p in K. Let $K_{\mathfrak{p}^\infty}$ denote the dihedral or anticyclotomic $\mathbf{Z}_p^\delta$-extension of K, where $\delta = [F_\mathfrak{p} : \mathbf{Q}_p]$ is the residue degree. This p-adic Lie extension, whose construction we recall below, factors through the tower of ring class fields of $\mathfrak{p}$-power conductor over K. Let G denote the Galois group $\operatorname{Gal}(K_{\mathfrak{p}^\infty}/K)$, which is topologically isomorphic to $\mathbf{Z}_p^\delta$. Let $\operatorname{Sel}(A/K_{\mathfrak{p}^\infty})$ denote the p^∞-Selmer group of A over $K_{\mathfrak{p}^\infty}$. Hence by definition,

$$\operatorname{Sel}(A/K_{\mathfrak{p}^\infty}) = \varinjlim_{L} \operatorname{Sel}(A/L),$$

where the inductive limit ranges over all finite extensions L of K contained in $K_{\mathfrak{p}^\infty}$ of the classically defined Selmer groups

$$\operatorname{Sel}(A/L) = \ker\left(H^1(G_S(L), A_{p^\infty}) \longrightarrow \bigoplus_{v \in S} J_v(L)\right).$$

Here, S is any (fixed) finite set of primes of K that includes both the primes above p and the primes where A has bad reduction. We write K^S to denote the maximal Galois extension of K unramified outside of S and the archimedean places of K, and $G_S(L)$ to denote the Galois group $\operatorname{Gal}(K^S/L)$. We write A_{p^∞} to denote the collection of all p-power torsion of A over the relevant extension of K in K^S, i.e. $A_{p^\infty} = \bigcup_{n \geq 0} A_{p^n}$ with $A_{p^n} = \ker([p^n] : A \longrightarrow A)$, which is standard notation. In general, however, given an abelian group B, we shall write $B(p)$ to denote its p-primary component. We define

$$J_v(L) = \bigoplus_{w|v} H^1(L_w, A(\overline{L}_w))(p),$$

where the sum runs over primes w above v in L. Note that each Selmer group $\mathrm{Sel}(A/L)$ fits into the short exact descent sequence

$$0 \longrightarrow A(L) \otimes \mathbf{Q}_p/\mathbf{Z}_p \longrightarrow \mathrm{Sel}(A/L) \longrightarrow Ш(A/L)(p) \longrightarrow 0,$$

where $A(L)$ denotes the Mordell-Weil group of A over L, and $Ш(A/L)(p)$ the p-primary part of the Tate-Shafarevich group $Ш(A/L)$ of A over L. Now, the p-primary of p^∞-Selmer group $\mathrm{Sel}(A/K_{\mathfrak{p}^\infty})$ has the structure of a discrete, cofinitely generated $\Lambda(G)$-module. It is often (but not always) the case that $\mathrm{Sel}(A/K_{\mathfrak{p}^\infty})$ is $\Lambda(G)$-cotorsion. For instance, this is known by a standard argument to be the case when the p-primary Selmer group $\mathrm{Sel}(A/K)$ is finite. Let us first consider this case where $\mathrm{Sel}(A/K_{\mathfrak{p}^\infty})$ is $\Lambda(G)$-cotorsion. This setting is generally known as the *definite case* in the literature, for reasons that we do not dwell on here.[1] In any case, the conjecture of Birch and Swinnerton-Dyer predicts that the Hasse-Weil L-function $L(A/K, s)$ has an analytic continuation to $s = 1$, with order of vanishing at $s = 1$ given by the rank of the Mordell-Weil group $A(K)$. The refined conjecture predicts that Tate-Shafarevich group $Ш(A/K)$ is finite, and moreover that the leading coefficient of $L(A/K, s)$ at $s = 1$ is given by the formula

$$\frac{|Ш(A/K)| \cdot \mathrm{Reg}(A/K) \cdot \Phi(A/K)}{|A(K)_{\mathrm{tors}}| \cdot |A^t(K)_{\mathrm{tors}}| \cdot |D|^{\frac{\dim(A)}{2}}} \cdot \prod_{\substack{v|\infty \\ v:K\to\mathbf{R}}} \int_{A(K_v)} |\omega| \cdot \prod_{\substack{v|\infty \\ v:K\to\mathbf{C}}} 2\int_{A(K_v)} \omega \wedge \overline{\omega}. \tag{19.1}$$

Here, $\mathrm{Reg}(A/K)$ denotes the regulator of A over K, i.e. the determinant of the canonical height pairing on a basis of $A(K)/A(K)_{\mathrm{tors}}$, $\Phi(A/K)$ denotes the product

$$\Phi(A/K) = \prod_{v\nmid\infty} c_v \left|\frac{\omega}{\omega_v^*}\right|_v,$$

where $c_v = c_v(A) = [A(K_v) : A_0(K_v)]$ is the local Tamagawa factor of A at a finite prime v of K, $\omega = \omega_A$ a fixed nonzero global exterior form (an invariant differential if A is an elliptic curve), and $\omega_v^* = \omega_{A,v}^*$ the Néron differential at v. Moreover, A^t denotes the dual abelian variety associated to A, and $D = D_K$ the absolute discriminant of K. We refer the reader to the article of Tate [36] for details.[2] We consider a p-adic analogue of this formula (19.1). Namely,

[1] Roughly, when A/F is a modular abelian variety, the associated Hasse-Weil L-function $L(A/K, s)$ has an analytic continuation a product of Rankin-Selberg L-functions of cuspidal eigenforms on some totally definite quaternion algebra over F, and this in particular can be used to deduce that the root number of $L(A/K, s)$ is equal to 1 (as opposed to -1). We refer the reader to the e.g. works [3], [29], [24], or [38] for some of the many known Iwasawa theoretic results in this direction.

[2] The reader will note that we have included extraneous real places in the formula (19.1) above, i.e. as K here is totally imaginary.

let $\widetilde{A}(\kappa_v)$ denote the reduction of A at a finite prime v of K. Write $A(K)(p)$ to denote the p-primary component of $A(K)$, and $\widetilde{A}(\kappa_v)(p)$ that of $\widetilde{A}(\kappa_v)$. We first establish the following result, using standard techniques.

Theorem 19.1 (Theorem 19.30). *Assume that A is a principally polarized abelian variety having good ordinary reduction at each prime above p in K. Assume additionally that the p^{∞}-Selmer group* $\mathrm{Sel}(A/K)$ *is finite. If the p^{∞}-Selmer group* $\mathrm{Sel}(A/K_{\mathfrak{p}^\infty})$ *is $\Lambda(G)$-cotorsion, then the G-Euler characteristic $\chi(G, \mathrm{Sel}(A/K_{\mathfrak{p}^\infty}))$ is well defined, and given by the formula*

$$\chi(G, \mathrm{Sel}(A/K_{\mathfrak{p}^\infty})) = \frac{|\text{Ш}(A/K)(p)|}{|A(K)(p)|^2} \cdot \prod_{v|p} |\widetilde{A}(\kappa_v)(p)|^2 \cdot \prod_{\substack{v\nmid p \\ \in S_{\mathrm{ns}}}} \mathrm{ord}_p \left(\frac{c_v(A)}{L_v(A/K,1)} \right). \tag{19.2}$$

Here, S_{ns} denotes the subset of S of primes which do not split completely in $K_{\mathfrak{p}^\infty}$, and $L_v(A/K,1)$ denotes the Euler factor at v of the global L-value $L(A/K,1)$.

This result has various antecedents in the literature (see e.g. [6], [8], or [42]), though the setting of dihedral extensions considered here differs greatly from the cyclotomic or single-variable settings considered in these works. Anyhow, let us now examine the complementary setting where the p^{∞}-Selmer group $\mathrm{Sel}(A/K_{\mathfrak{p}^\infty})$ is *not* expected to be $\Lambda(G)$-cotorsion. This setting is generally known as the *indefinite case* in the literature, for reasons that we likewise do not dwell on here.[3] To fix ideas, let us assume the so-called *weak Heegner hypothesis* that $\chi(N) = (-1)^{d-1}$, where χ denotes the quadratic character associated to K/F, N the arithmetic conductor of A, and $d = [F : \mathbf{Q}]$ the degree of F. This condition ensures that the abelian variety A comes equipped with families of Heegner or CM points defined over ring class extensions of K. We refer the reader to [19] or the discussion below for details. Let us first consider the issue of describing the G-Euler characteristic of the $\Lambda(G)$-cotorsion submodule $\mathrm{Sel}(A/K_{\mathfrak{p}^\infty})_{\mathrm{cotors}}$ of $\mathrm{Sel}(A/K_{\mathfrak{p}^\infty})$, following the conjecture of Perrin-Riou [26] and Howard [19]. To state this version of

[3] Roughly (as before), in the setting where A is a modular abelian variety defined over F, the Hasse-Weil L-function $L(A/K,s)$ has an analytic continuation given by the product of Rankin-Selberg L-functions of cuspidal eigenforms on some totally indefinite quaternion algebra over F (where indefinite here means ramified at all but one real place of F), and this in particular can be used to deduce that the root number of $L(A/K,s)$ is equal to -1 (as opposed to 1). We refer the reader to the excellent paper of Howard [19] for an account of the Iwasawa theory in this setting.

the Iwasawa main conjecture, let us write $\mathfrak{S}_\infty = \mathfrak{S}(A/K_{\mathfrak{p}^\infty})$ denote the compactified p^∞-Selmer group of A in $K_{\mathfrak{p}^\infty}$, which has the structure of a compact $\Lambda(G)$-module. Let $\mathfrak{H}_\infty$ denote the $\Lambda(G)$-submodule of $\mathfrak{S}_\infty$ generated by the images of Heegner points of $\mathfrak{p}$-power conductor under the appropriate norm homomorphisms. We refer the reader to [18] or to the passage below for more details of this construction. It is often known and generally expected to be the case that the $\Lambda(G)$-modules $\mathfrak{S}_\infty$ and $\mathfrak{H}_\infty$ are torsionfree of $\Lambda(G)$-rank one, in which case the quotient $\mathfrak{S}_\infty/\mathfrak{H}_\infty$ has $\Lambda(G)$-rank zero. Hence in this setting, the quotient $\mathfrak{S}_\infty/\mathfrak{H}_\infty$ is a torsion $\Lambda(G)$-module, so has a well defined $\Lambda(G)$-characteristic power series $\mathrm{char}(\mathfrak{S}_\infty/\mathfrak{H}_\infty)$. Let $\mathrm{char}(\mathfrak{S}_\infty/\mathfrak{H}_\infty)^*$ denote the image of $\mathrm{char}(\mathfrak{S}_\infty/\mathfrak{H}_\infty)$ under the involution of $\Lambda(G)$ induced by inversion in G. Given an integer $n \geq 0$, let $K[\mathfrak{p}^n]$ denote the ring class field of conductor $\mathfrak{p}^n$ over K, with Galois group $G[\mathfrak{p}^n]$. Write $K[\mathfrak{p}^\infty] = \bigcup_{n\geq 0} K[\mathfrak{p}^n]$ to denote the union of all ring class extensions of $\mathfrak{p}$-power conductor, with Galois group $G[\mathfrak{p}^\infty] = \mathrm{Gal}(K[\mathfrak{p}^\infty]/K) = \varprojlim_n G[\mathfrak{p}^n]$. Recall that we let $K_{\mathfrak{p}^\infty}$ denote the dihedral or anticyclotomic $\mathbf{Z}_p^\delta$-extension of K. We write $G = \varprojlim_n G_{\mathfrak{p}^n}$ to denote the associated profinite Galois group $G_{\mathfrak{p}^n} = \mathrm{Gal}(K_{\mathfrak{p}^n}/K)$. Let $\mathfrak{K}_n$ denote the Artin symbol of $\mathfrak{d}_n = (\sqrt{D}\mathcal{O}_K) \cap \mathcal{O}_{p^n}$. Here, $D = D_K$ denotes the absolute discriminant of K, and $\mathcal{O}_{p^n}$ the $\mathcal{O}_F$-order of conductor $\mathfrak{p}^n$ in K, i.e. $\mathcal{O}_{p^n} = \mathcal{O}_F + \mathfrak{p}^n\mathcal{O}_K$. Let $\mathfrak{K} = \varprojlim_n \mathfrak{K}_n$. Let $\langle\, , \,\rangle_{A,K[p^n]}$ denote the p-adic height pairing

$$\langle\, , \,\rangle_{A,K[p^n]} : A^t(K[p^n]) \times A(K[p^n]) \longrightarrow \mathbf{Q}_p,$$

defined in Howard [18, (9), §3.3] and Perrin-Riou [28]. Let us again assume for simplicity that A is principally polarized. There exists by a construction of Perrin-Riou [26] (see also [28]) a p-adic height pairing

$$\mathfrak{h}_n : \mathfrak{S}(A/K_{p^n}) \times \mathfrak{S}(A/K_{p^n}) \longrightarrow c^{-1}\mathbf{Z}_p$$

whose restriction to the image of the Kummer map $A(K[p^n]) \otimes \mathbf{Z}_p \rightarrow \mathfrak{S}(E/K[p^n])$ coincides with the pairing $\langle\, , \,\rangle_{A,K[p^n]}$ after identifying $A^t \cong A$ in the canonical way (cf. [18, Proposition 0.0.4]). Here, $c \in \mathbf{Z}_p$ is some integer that does not depend on the choice of n. Following Perrin-Riou [26] and Howard [18], these pairings can be used to construct a pairing

$$\begin{aligned} \mathfrak{h}_\infty : \mathfrak{S}_\infty \times \mathfrak{S}_\infty &\longrightarrow c^{-1}\mathbf{Z}_p[[G]] \\ (\varprojlim_n a_n, \varprojlim_n b_n) &\longmapsto \varprojlim_n \sum_{\sigma \in G_{p^n}} \mathfrak{h}_n(a_n, b_n^\sigma) \cdot \sigma. \end{aligned}$$

We then define the *p-adic regulator* $\mathcal{R} = \mathcal{R}(A/K_{p^\infty})$ to be the image of $\mathfrak{S}_\infty$ in $c^{-1}\mathbf{Z}_p[[G]]$ under this pairing $\mathfrak{h}_\infty$. Let $\mathbf{e}$ denote the natural projection

$$\mathbf{e} : \mathbf{Z}_p[[G[p^\infty]]] \longrightarrow \mathbf{Z}_p[[G]].$$

We then make the following conjecture, reformulating those of [26], [18], and [19].

Conjecture 19.2 (Conjecture 19.32). *Let A be a principally polarized abelian variety defined over F having good ordinary reduction at each prime above p in K. Assume that A satisfies the weak-Heegner hypothesis with respect to N and K. Then, the G-Euler characteristic $\chi(G, \mathrm{Sel}(A/K_{\mathfrak{p}^\infty})_{\mathrm{cotors}})$ is well-defined, and given by the formula*

$$\chi(G, \mathrm{Sel}(A/K_{\mathfrak{p}^\infty})_{\mathrm{cotors}}) = |\mathrm{char}(\mathfrak{S}_\infty/\mathfrak{H}_\infty)(0) \cdot \mathrm{char}(\mathfrak{S}_\infty/\mathfrak{H}_\infty)^*(0) \cdot \mathfrak{N}(0)|_p^{-1}.$$

Here, $\mathfrak{R}$ denotes the ideal $(\mathbf{e}(\mathfrak{K}))^{-1}\mathcal{R}$, which lies in the $\mathbf{Z}_p$-Iwasawa algebra $\mathbf{Z}_p[[G]]$.

We refer the reader to the divisibility proved in Howard [19, Theorem B] for results towards this conjecture. We can also deduce the following result toward this conjecture in the setting where A is an elliptic curve defined over the totally real field $\mathbf{Q}$. Hence, A is modular by the fundamental work of Wiles [41], Taylor-Wiles [37], and Breuil-Conrad-Diamond-Taylor [5]. Let $f \in S_2(\Gamma_0(N))$ denote the cuspidal newform of level N associated to A by modularity. Let K be an imaginary quadratic field in which the fixed prime p os not ramified. Let $K(\mu_{p^\infty})$ denote the extension obtained from K by adjoining all primitive p-power roots of unity, with $\Gamma = \mathrm{Gal}(K(\mu_{p^\infty})/K)$ its Galois group. Thus, Γ is topologically isomorphic to $\mathbf{Z}_p^\times$. Let $\mathcal{L}_f$ denote the two-variable p-adic L-function constructed by Hida [16] and Perrin-Riou [26], which lies in the $\mathbf{Z}_p$-Iwasawa algebra $\mathbf{Z}_p[[G[p^\infty] \times \Gamma]]$, as explained in the proof of [39, Theorem 2.9]. Fixing a topological generator γ of Γ, we write the expansion of $\mathcal{L}_f$ in the cyclotomic variable $\gamma - 1$ as

$$\mathcal{L}_f = \mathcal{L}_{f,0} + \mathcal{L}_{f,1}(\gamma - 1) + \mathcal{L}_{f,2}(\gamma - 1)^2 + \ldots, \tag{19.3}$$

with $\mathcal{L}_{f,i} \in \mathbf{Z}_p[[G[p^\infty]]]$ for each $i \geq 0$. If we assume the so-called *Heegner hypothesis* that each prime dividing N splits in K, then root number of the Hasse-Weil L-function $L(A/K, s)$ equals -1, which forces the leading term $\mathcal{L}_{f,0}$ in (19.3) to vanish. Let $\Lambda(G)$ denote the $\mathbf{Z}_p$-Iwasawa algebra $\mathbf{Z}_p[[G]]$. We deduce the following result from the $\Lambda(G)$-adic Gross-Zagier theorem of Howard [18, Theorem B] with the two-variable main conjecture shown in Skinner-Urban [35, Theorem 3] and the integrality of the two-variable p-adic L-function $\mathcal{L}_f$ shown in [39].

Corollary 19.3 (Corollary 19.37). *Let A be an elliptic curve defined over $\mathbf{Q}$, associated by modularity to a newform $f \in S_2(\Gamma_0(N))$. Let K be an imaginary quadratic extension of $\mathbf{Q}$ of discriminant D in which the fixed prime p is not ramified. Assume that the elliptic curve A has good ordinary reduction at p, that each prime dividing N splits in K, and that D is odd and not equal to -3. Assume in addition that the absolute Galois group $G_K = \mathrm{Gal}(\overline{K}/K)$ surjects onto the $\mathbf{Z}_p$-automorphisms of $T_p(A)$, that $\chi(p) = 1$, that p does not divide the class number of K, and that the compactified Selmer group $\mathfrak{S}_\infty$ has $\Lambda(G)$-rank one. Then the G-Euler characteristic of $\chi(G, \mathrm{Sel}(A/K_{p^\infty})_{\mathrm{cotors}})$ is well-defined, and given by the formulae*

$$\begin{aligned}\chi(G, \mathrm{Sel}_{p^\infty}(A/K_{p^\infty})_{\mathrm{cotors}}) &\geq |\mathbf{e}(\mathcal{L}_{f,1}(0))|_p^{-1} \\ &= |\mathbf{e}(\mathfrak{K} \cdot \log_p(\gamma))^{-1} \mathfrak{h}_\infty(\widetilde{x}_\infty, \widetilde{x}_\infty)(0)|_p^{-1} \\ &= \left|\mathrm{char}(\mathfrak{S}_\infty/\mathfrak{H}_\infty)(0) \cdot \mathrm{char}(\mathfrak{S}_\infty/\mathfrak{H}_\infty)^*(0) \cdot \mathfrak{R}(0)\right|_p^{-1}.\end{aligned}$$

Here, $\log_p$ is the p-adic logarithm (composed with a fixed isomorphism $\Gamma \cong \mathbf{Z}_p^\times$), $\widetilde{x}_\infty$ is a generator of $\mathfrak{H}_\infty$ constructed from a compatible sequence of regularized Heegner points (described below), and $\mathfrak{R}$ is the ideal $(\mathbf{e}(\mathfrak{K} \cdot \log_p(\gamma))^{-1}\mathcal{R} \in \Lambda(G)$. Moreover, these formulae for $\chi(G, \mathrm{Sel}(A/K_{p^\infty})_{\mathrm{cotors}})$ do not depend on the choice of topological generator $\gamma \in \Gamma$.

Though relatively simple to deduce, this result is revealing in that it links the nonvanishing of the coefficient $\mathcal{L}_{f,1}$ to the nondegeneracy of the height pairing $\mathfrak{h}_\infty$ and the p-adic regulator $\mathcal{R}$. Anyhow, we have shown here we can verify the p-part of the refined Birch and Swinnerton-Dyer formula (and variations) by relatively straightforward computations of Euler characteristics after knowing partial results towards the associated Iwasawa main conjectures. It would be interesting to relate these formula more precisely to the modular setting outlined in Pollack-Weston [29], as well as perhaps to the conjectures of Shimura/Prasanna [30] via Ribet-Takahashi [31]. It would also be interesting to relate the conjectural formula described above in the indefinite setting to the conjecture of Kolyvagin [22] on indivisibility of Heegner points (and natural extensions to CM points on quaternionic Shimura curves). However, such problems lie beyond the scope of the present note.

19.2 Ring class towers

The dihedral or anticyclotomic $\mathbf{Z}_p^\delta$-extension $K_{\mathfrak{p}^\infty}$ of K. Suppose that F is any totally real number field, and K any totally imaginary quadratic extension of F. Recall that we fix throughout a rational prime p, and let $\mathfrak{p}$ denote a fixed prime above p in F. Given an integral ideal $\mathfrak{c} \subseteq \mathcal{O}_F$, we write $\mathcal{O}_\mathfrak{c} = \mathcal{O}_F + \mathfrak{c}\mathcal{O}_K$ to denote the $\mathcal{O}_F$-order of conductor $\mathfrak{c}$ in K. The *ring class field of conductor* $\mathfrak{c}$ *of* K is the Galois extension $K[\mathfrak{c}]$ of K characterized via class field theory by the identification

$$\widehat{K}^\times / \widehat{F}^\times \widehat{\mathcal{O}}_\mathfrak{c}^\times K^\times \xrightarrow{\mathrm{rec}_K} \mathrm{Gal}(K[\mathfrak{c}]/K).$$

Here, rec_K denotes the Artin reciprocity map, normalized to send uniformizers to geometric Frobenius automorphisms. Let us write $G[\mathfrak{c}]$ to denote the Galois group $\mathrm{Gal}(K[\mathfrak{c}]/K)$. We consider the union of all ring class extensions of $\mathfrak{p}$-power conductor

$$K[\mathfrak{p}^\infty] = \bigcup_{n \geq 0} K[\mathfrak{p}^n],$$

along with its Galois group $G[\mathfrak{p}^\infty] = \mathrm{Gal}(K[\mathfrak{p}^\infty]/K)$, whose profinite structure we write as

$$G[\mathfrak{p}^\infty] = \varprojlim_n G[\mathfrak{p}^n].$$

A standard argument in the theory of profinite groups shows that the torsion subgroup $G[\mathfrak{p}^\infty]_{\mathrm{tors}} \subseteq G[\mathfrak{p}^\infty]$ is finite, and moreover that the quotient $G[\mathfrak{p}^\infty]/G[\mathfrak{p}^\infty]_{\mathrm{tors}}$ is topologically isomorphic to $\mathbf{Z}_p^\delta$, where $\delta = [F_\mathfrak{p} : \mathbf{Q}_p]$ is the residue degree of $\mathfrak{p}$. We refer the reader e.g. to [12, Corollary 2.2] for details on how to deduce this fact via a basic computation with adelic quotient groups. Let $G_{\mathfrak{p}^\infty}$ denote the Galois group $G[\mathfrak{p}^\infty]/G[\mathfrak{p}^\infty]_{\mathrm{tors}} \cong \mathbf{Z}_p^\delta$, which has the structure of a profinite group

$$G_{\mathfrak{p}^\infty} = \varprojlim_n G_{\mathfrak{p}^n}.$$

Let us then write $K_{\mathfrak{p}^n}$ to denote the abelian extension of K contained in $K[\mathfrak{p}^n]$ such that $G_{\mathfrak{p}^n} = \mathrm{Gal}(K_{\mathfrak{p}^n}/K)$. When there is no risk of confusion, we shall simply write G to denote the Galois group $G_{\mathfrak{p}^\infty}$.

Lemma 19.4. *The extension $K_{\mathfrak{p}^\infty}$ over K is unramified outside of p.*

Proof. The result is a standard deduction in local class field theory, as shown for instance in Iwasawa [21, Theorem 1 §2.2]. In general, given any p-adic field L, the abelianization of $G_L = \mathrm{Gal}(\overline{L}/L)$ is a completion of $L^\times$. The subgroup corresponding to ramified extensions is $\mathcal{O}_L^\times$, which is a profinite abelian group whose open subgroups are pro-p. The result is then easy to deduce. □

Finally, let us recall that the Galois extension $K[\mathfrak{p}^{\infty}]$ is of generalized dihedral type. Equivalently, the complex conjugation automorphism of $\mathrm{Gal}(K/F)$ acts by inversion on $G[\mathfrak{p}^{\infty}]$. It follows that $K[\mathfrak{p}^{\infty}]$ is linearly disjoint over K to the cyclotomic extension $K(\mu_{p^{\infty}})$, where $\mu_{p^{\infty}}$ denotes the set of all p-power roots of unity, because the complex conjugation automorphism acts trivially on $\mathrm{Gal}(K(\mu_{p^{\infty}})/K)$. For this reason, the extension $K_{\mathfrak{p}^{\infty}}/K$ is often called the *anticyclotomic* $\mathbf{Z}_p^{\delta}$-extension of K.

Decomposition of primes in $K_{\mathfrak{p}^{\infty}}$. Let us now write G to denote the Galois group $G_{\mathfrak{p}^{\infty}} = \mathrm{Gal}(K_{\mathfrak{p}^{\infty}}/K)$. Given a finite prime v of K, let $D_v \subseteq G$ denote the decomposition subgroup at a fixed prime above v in $K_{\mathfrak{p}^{\infty}}$. We shall often make the standard identification D_v with the Galois group $G_w = \mathrm{Gal}(K_{\mathfrak{p}^{\infty},w}/K_v)$, where w is a fixed prime above v in $K_{\mathfrak{p}^{\infty}}$.

Proposition 19.5. *Let v be a finite prime of K.*

(i) *If v does not divide p and splits completely in $K_{\mathfrak{p}^{\infty}}$, then D_v is trivial.*
(ii) *If v does not divide p and does not split completely in $K_{\mathfrak{p}^{\infty}}$, then D_v can be identified with a finite-index subgroup of $G \approx \mathbf{Z}_p^{\delta}$.*
(iii) *If v divides p, then D_v is topologically isomorphic to $\mathbf{Z}_p^{\delta}$.*

Proof. The result is easy to deduce from standard properties of local fields, see e.g. [13, Proposition 5.10]. □

19.3 Galois cohomology

We now record for later use some basic facts from Galois cohomology.

p-cohomological dimension. Recall that the p-cohomological dimension $\mathrm{cd}_p(G)$ of a profinite group G is the smallest integer n satisfying the condition that for any discrete torsion G-module M and any integer $q > n$, the p-primary component of $H^q(G, M)$ is zero ([33, §3.1]). We have the following crucial characterization of cd_p.

Theorem 19.6 (Serre-Lazard)**.** *Let G be any p-adic Lie group. Let M be any discrete G-module. If G does not contain an element of order p, then $\mathrm{cd}_p(G)$ is equal to the dimension of G as a p-adic Lie group.*

Proof. See [34] and [25]. □

Since $G_{\mathfrak{p}^\infty} \cong \mathbf{Z}_p^\delta$ has no point of order p, we obtain the following consequence.

Corollary 19.7. *Let G denote the Galois group $G_{\mathfrak{p}^\infty} \cong \mathbf{Z}_p^\delta$ defined above. Then, $\mathrm{cd}_p(G) = \delta$. Moreover, given a finite prime v of K,*

(i) $\mathrm{cd}_p(D_v) = 0$ *if v does not divide p and splits completely in $K_{\mathfrak{p}^\infty}$;*
(ii) $\mathrm{cd}_p(D_v) = \delta$ *if v does divide p.*

Proof. The claim follows from Theorem 19.6, using Proposition 19.5 for (i) and (ii). □

The Hochschild-Serre spectral sequence. Fix S any finite set of primes of K containing the primes above p in K, and the primes where A has bad reduction. Let K_S denote the maximal extension of K that is unramified outside of S and the archimedean places of K. Let $G_S(K) = \mathrm{Gal}(K_S/K)$ denote the corresponding Galois group. Observe that there is an inclusion of fields $K_{\mathfrak{p}^\infty} \subset K_S$. Given any intermediate field L with $K \subset L \subset K_{\mathfrak{p}^\infty}$, let

$$G_S(L) = \mathrm{Gal}(K_S/L).$$

Recall again that we write G to denote the Galois group $G_{\mathfrak{p}^\infty} = \mathrm{Gal}(K_{\mathfrak{p}^\infty}/K)$.

Lemma 19.8. *For all integers $i \geq 1$, we have bijections*

$$H^i(G, H^i(G_S(K_{\mathfrak{p}^\infty}), A_{p^\infty}) \cong H^{i+2}(G, A_{p^\infty}). \tag{19.4}$$

Proof. See [33, § 2.2.6] or [34]. We claim that for all $i \geq 0$, a direct application of the Hochschild-Serre spectral sequence gives the exact sequence

$$H^{i+1}(G_S(K), A_{p^\infty}) \longrightarrow H^i(G, H^1(G_S(K_{\mathfrak{p}^\infty}), A_{p^\infty})) \longrightarrow H^{i+2}(G, A_{p^\infty}).$$

On the other hand, it is well known that $H^i(G_S(K), A_{p^\infty}) = 0$ for all $i \geq 2$. Hence, the claim holds for all $i \geq 1$. □

Some identifications in local cohomology. Let L be a finite extension of K. Recall that for each finite prime v of K, we define

$$J_v(L) = \bigoplus_{w|v} H^1(L_w, A(\overline{L}_w))(p), \tag{19.5}$$

where the sum runs over all primes w above v in L. If L_∞ is an infinite extension of K, then we define the associated group $J_v(L_\infty)$ by taking the inductive limit over finite extensions L of K contained in L_∞ with respect to restriction maps,

$$J_v(L_\infty) = \varinjlim_{L} J_v(L). \tag{19.6}$$

Given a finite prime v of K, let us fix a prime w above v in $K_{\mathfrak{p}^\infty}$. Recall that we then write G_w to denote the Galois group $\mathrm{Gal}(K_{\mathfrak{p}^\infty,w}/K_v)$, which we identify with the decomposition subgroup $D_v \subseteq G$.

Lemma 19.9. *Let v be a finite prime of K, and w any prime above v in $K_{\mathfrak{p}^\infty}$. Then, for each integer $i \geq 0$, there is a canonical bijection*

$$H^i(G, J_v(K_{\mathfrak{p}^\infty})) \cong H^i(G_w, H^1(K_{\mathfrak{p}^\infty,w}, A)(p)). \tag{19.7}$$

Proof. Recall that we let $G_{\mathfrak{p}^n}$ denote the Galois group $\mathrm{Gal}(K_{\mathfrak{p}^n}/K)$, which is isomorphic to $(\mathbf{Z}/p^n\mathbf{Z})^\delta$. Let $G_{\mathfrak{p}^n,w} \subseteq G$ denote the decomposition subgroup of the restriction of w to $K_{\mathfrak{p}^n}$. Shapiro's lemma implies that for all integers $n \geq 1$ and $i \geq 0$, we have canonical bijections

$$H^i(G_{\mathfrak{p}^n}, J_v(K_{\mathfrak{p}^n})) \cong H^i(G_{\mathfrak{p}^n,w}, H^1(K_{\mathfrak{p}^n,w}, A)(p)).$$

Passing to the inductive limit with n then proves the claim. □

Corollary 19.10. *Let v be any finite prime of K not dividing p which splits completely in $K_{\mathfrak{p}^\infty}$, and let w be any prime above v in $K_{\mathfrak{p}^\infty}$. Then, for each integer $i \geq 1$, we have that $H^i(G_w, J_v(K_{\mathfrak{p}^\infty})) = 0$.*

Proof. The result follows from (19.7), since G_w is trivial by Proposition 19.5. □

Lemma 19.11. *Let v be any finite prime of K that divides p, and write w to denote the prime above v in $K_{\mathfrak{p}^\infty}$. Then, for each $i \geq \delta - 1$, we have that*

$$H^i(G_w, H^1(K_{\mathfrak{p}^\infty,w}, A)(p)) = 0.$$

Moreover, for each $1 \leq i \leq \delta - 2$, we have bijections

$$H^i(G_w, H^1(K_{\mathfrak{p}^\infty,w}, A)(p)) \cong H^{i+2}(G_w, \widetilde{A}_{v,p^\infty}). \tag{19.8}$$

Proof. See [6, Lemma 3.5]. Let G_{K_v} denote the Galois group $\mathrm{Gal}(\overline{K}_v/K_v)$, where $\overline{K}_v$ is a fixed algebraic closure of K_v. Let $I_{K_v} \subseteq G_{K_v}$ denote the inertia subgroup at a fixed prime above v in $\overline{K}_v$. Consider the short exact sequence of G_{K_v}-modules

$$0 \longrightarrow C \longrightarrow A_{p^\infty} \longrightarrow D \longrightarrow 0. \tag{19.9}$$

Here, C denotes the canonical subgroup associated to the formal group of the Néron model of A over $\mathcal{O}_{K_v}$ (see [7, § 4]). Moreover, (19.9) is characterized by the fact that C is divisible, with D being the maximal quotient of A_{p^∞} by a divisible subgroup such that the inertia subgroup I_{K_v} acts via a finite quotient. Since $K_{\mathfrak{p}^\infty,w}$ is deeply ramified in the sense of Coates-Greenberg [7], we have by [7, Propositions 4.3 and 4.8] a canonical G_w-isomorphism

$$H^1(K_{\mathfrak{p}^\infty,w}, A)(p) \cong H^1(K_{\mathfrak{p}^\infty,w}, D). \tag{19.10}$$

Moreover, the G_{K_v}-module D vanishes if and only if A has potentially supersingular reduction at v, as explained in [7, § 4].

Observe now that $\mathrm{Gal}(\overline{K}_v/K_{\mathfrak{p}^\infty,w})$ has p-cohomological dimension equal to 1, as the profinite degree of $K_{\mathfrak{p}^\infty,w}$ over K_v is divisible by p^∞ (see [33]). Hence, for all $i \geq 2$, we have identifications

$$H^i(K_{\mathfrak{p}^\infty,w}, D) = 0. \tag{19.11}$$

Taking (19.11) along with the Hochschild-Serre spectral sequence then gives us for each $i \geq 1$ exact sequences

$$H^{i+1}(K_v, D) \longrightarrow H^i(G_w, H^1(K_{\mathfrak{p}^\infty,w}, D)) \longrightarrow H^{i+2}(G_w, D). \tag{19.12}$$

While we would like to follow the argument of [6, Lemma 3.5] in showing that the terms on either side of (19.12) must vanish, we have not been able to find an argument to deal with the fact that δ could be greater than 4. Anyhow, we can deduce that

$$H^i(K_v, D) = 0 \tag{19.13}$$

for all $i \geq 2$. That is, we know that $\mathrm{cd}_p\left(G_{K_v}\right) = 2$, whence (19.13) holds trivially for all $i \geq 3$. To deduce the vanishing for $i = 2$, we take cohomology of the exact sequence (19.9) to obtain a surjection from $H^2(K_v, A_{p^\infty})$ onto $H^2(K_v, D)$. We know by local Tate duality that $H^2(K_v, A_{p^\infty})$ vanishes (see e.g. [8, 1.12 Lemma]). Hence, we deduce that (19.13) holds for all $i \geq 2$. It then follows from the exact sequence (19.12) that we have bijections

$$H^i(G_w, H^1(K_{\mathfrak{p}^\infty,w}, D)) \longrightarrow H^{i+2}(G_w, D)$$

for all $i \geq 1$. Now, observe that both of these groups vanish for all $i \geq \delta - 1$, as $\mathrm{cd}_p(G_w) = \delta$. The result now follows from the natural identification of D with the G_w-module defined by $\widetilde{A}_{p^\infty,v} = A(K_{\mathfrak{p}^\infty,w})_{p^\infty}$. □

19.4 The torsion subgroup $A(K_{\mathfrak{p}^\infty})_{\mathrm{tors}}$

We start with the following basic result, whose proof relies on the fact that there exist primes $v \subset \mathcal{O}_K$ which split completely in the anticyclotomic extension $K_{\mathfrak{p}^\infty}$.

Lemma 19.12. *The torsion subgroup $A(K_{\mathfrak{p}^\infty})_{\mathrm{tors}}$ of $A(K_{\mathfrak{p}^\infty})$ is finite.*

Proof. Fix a finite prime v of F that remains inert in K. Let us also write v to denote the prime above v in K. By class field theory, $v \subseteq \mathcal{O}_K$ splits completely in $K_{\mathfrak{p}^\infty}$. Hence, fixing a prime w above v in $K_{\mathfrak{p}^\infty}$, we can identify the union of all completions $K_{\mathfrak{p}^\infty,w}$ with K_v itself. In particular, we find that $A(K_{\mathfrak{p}^\infty,w}) = A(K_v)$, which gives us an injection $A(K_{\mathfrak{p}^\infty})_{\mathrm{tors}} \longrightarrow A(K_v)_{\mathrm{tors}}$. On the other hand, consider the short exact sequence

$$0 \longrightarrow A_1(K_v) \longrightarrow A_0(K_v) \longrightarrow \widetilde{A}(\kappa_v) \longrightarrow 0. \tag{19.14}$$

Here, $A_1(K_v)$ denotes the kernel of reduction modulo v, which can be identified with the formal group $\widehat{A}(\mathfrak{m}_v)$, and κ_v denotes the residue field of K at v. Let $n \geq 1$ be any integer prime to the residue characteristic q_v. It is well-known that $\widehat{A}(\mathfrak{m}_v)[n] = 0$. Hence, taking n-torsion in (19.14), we find for any such prime $v \subset \mathcal{O}_K$ an injection

$$A_0(K_v)[n] \longrightarrow \widetilde{A}(\kappa_v). \tag{19.15}$$

The group $\widetilde{A}(\kappa_v)$ is of course finite. Since we have this injection (19.15) for any such prime $v \subset \mathcal{O}_K$, it follows that $A(K_{\mathfrak{p}^\infty})_{\mathrm{tors}}$ must be finite. □

We now consider the G-Euler characteristic of the p-primary subgroup A_{p^∞} of $A(K_{\mathfrak{p}^\infty})$, denoted here by $\chi(G, A_{p^\infty})$. Recall that this is defined by the product

$$\chi(G, A_{p^\infty}) = \prod_{i \geq 0} |H^i(G, A_{p^\infty})|^{(-1)^i},$$

granted that it is well-defined.

Corollary 19.13. *We have that $\chi(G, A_{p^\infty})$ is well-defined and equal to* 1.

Proof. This is a standard result, since G is a pro-p group with $A_{p^\infty}(K_{\mathfrak{p}^\infty})$ a finite group of order equal to some power of p. See [33, Exercise (a) §4.1]. □

Recall that given a finite prime v if K, we let G_w denote the Galois group $\mathrm{Gal}(K_{\mathfrak{p}^\infty,w}/K_v)$, where w is a fixed prime above v in $K_{\mathfrak{p}^\infty}$. Let $\widetilde{A}_{v,p^\infty}$ denote the G_w-module defined by $A_{p^\infty}(K_{\mathfrak{p}^\infty,w})$. Consider the G_w-Euler characteristic of $\widetilde{A}_{p^\infty,v}$, which we denote by $\chi(G_w, \widetilde{A}_{v,p^\infty})$. Recall that this is defined by the product

$$\chi(G_w, \widetilde{A}_{p^\infty,v}) = \prod_{i\geq 0} |H^i(G_w, \widetilde{A}_{v,p^\infty})|^{(-1)^i},$$

granted that it is well-defined.

Corollary 19.14. *Let v be any finite prime of K, with w a fixed prime above v in $K_{\mathfrak{p}^\infty}$. The G_w-Euler characteristic $\chi(G_w, A_{\mathfrak{p}^\infty,v})$ is well-defined and equal to* 1.

Proof. In any case on the decomposition of v in $K_{\mathfrak{p}^\infty}$, G_w is pro-p, with $\widetilde{A}_{p^\infty,v}$ a finite group of order equal to some power of p, whence the result is standard. Note as well that since $G_w \cong D_v$ is in any case a closed subgroup of G, the result can also be deduced from Corollary 19.13 above, since Shapiro's lemma gives canonical isomorphisms $H^i(G, A_{p^\infty}) \cong H^i(G_w, \widetilde{A}_{v,p^\infty})$ for each $i \geq 1$. □

Putting this all together, we have shown the following result.

Proposition 19.15. *Let v be any finite prime of K, with w a fixed prime above K in $K_{\mathfrak{p}^\infty}$. Then, $\chi(G, A_{p^\infty}) = \chi(G_w, \widetilde{A}_{v,p^\infty}) = 1$.*

19.5 The definite case

Let us keep all of the notations of the section above. Here, we shall assume that the p^∞-Selmer group $\mathrm{Sel}(A/K_{\mathfrak{p}^\infty})$ is $\Lambda(G)$-cotorsion, where $\Lambda(G)$ denotes the $\mathcal{O}$-Iwasawa algebra $\Lambda(G_{\mathfrak{p}^\infty}) = \mathcal{O}[[G_{\mathfrak{p}^\infty}]]$ for $\mathcal{O}$ the ring of integers of some fixed finite extension of $\mathbf{Q}_p$. We shall also use here the standard notations for abelian varieties and local fields, following Coates-Greenberg [7].

Strategy. We consider the snake lemma in the following commutative diagram

$$\begin{array}{ccccc}
\mathrm{Sel}(A/K_{\mathfrak{p}^\infty})^G & \longrightarrow & H^1(G_S(K_{\mathfrak{p}^\infty}), A_{p^\infty})^G & \xrightarrow{\psi_S(K_{\mathfrak{p}^\infty})} & \bigoplus_{v\in S} J_v(K_{\mathfrak{p}^\infty})^G \\
\uparrow \alpha & & \uparrow \beta & & \uparrow \gamma=\bigoplus_{v\in S}\gamma_v \\
\mathrm{Sel}(A/K) & \longrightarrow & H^1(G_S(K), A_{p^\infty}) & \xrightarrow{\lambda_S(K)} & \bigoplus_{v\in S} J_v(K).
\end{array} \tag{19.16}$$

Here, recall that S denotes any finite set of places of K containing the primes above p, as well as all places where A has bad reduction. The map $\psi_S(K_{\mathfrak{p}^\infty})$ is induced from the localization map

$$\lambda_S(K_{\mathfrak{p}^\infty}) : H^1(G_S(K_{\mathfrak{p}^\infty}), A_{p^\infty}) \longrightarrow \bigoplus_{s \in S} J_v(K_{\mathfrak{p}^\infty})$$

defining the p^∞-Selmer group $\mathrm{Sel}(A/K_{\mathfrak{p}^\infty})$. The vertical arrows are induced by restriction on cohomology. We also consider the associated diagram

$$\begin{array}{ccccc} \mathrm{Im}\left(\psi_S(K_{\mathfrak{p}^\infty})\right) & \longrightarrow & \bigoplus_{v \in S} J_v(K_{\mathfrak{p}^\infty}) & \longrightarrow & \mathrm{coker}\left(\psi_S(K_{\mathfrak{p}^\infty})\right) \\ \uparrow{\scriptstyle \epsilon_1} & & \uparrow{\scriptstyle \gamma = \bigoplus_{v \in S} \gamma_v} & & \uparrow{\scriptstyle \epsilon_2} \\ \mathrm{Im}\,(\lambda_S(K)) & \longrightarrow & \bigoplus_{v \in S} J_v(K) & \longrightarrow & \mathrm{coker}\,(\lambda_S(K)) \,. \end{array} \tag{19.17}$$

Local restrictions maps. We now compute the cardinalities of the kernel and cokernel of the restriction map $\gamma = \bigoplus_{v \in S} \gamma_v$ in (19.16), in a series of lemmas leading to Theorem 19.23 below. Throughout this section, we shall assume for simplicity that the abelian variety is principally polarized.

Lemma 19.16. *Let v be a finite prime of K which does not divide p. Assume that A is principally polarized. Then, the group $J_v(K) = H^1(K_v, A)(p)$ is finite, and its order is given by the exact power of p dividing the quotient $c_v(A)/L_v(A, 1)$. Here, $c_v(A) = [A(K_v) : A_0(K_v)]$ denotes the local Tamagawa factor at v, and $L_v(A, 1)$ the Euler factor at v of the Hasse-Weil L-function $L(A, s)$ at $s = 1$.*

Proof. The result is standard, see e.g. [6, Lemma 3.6]. □

Lemma 19.17. *Let v be a finite prime of K which does not divide p. Let L be an arbitrary Galois extension of K_v with Galois group $\Omega = \mathrm{Gal}(L/K_v)$. Let*

$$\tau_v : H^1(K_v, A)(p) \longrightarrow H^1(L, A)(p)^\Omega$$

denote the natural map induced by restriction. Then, we have bijections

$$\ker(\tau_v) \cong H^1(\Omega, A_{p^\infty}(L))$$
$$\mathrm{coker}(\tau_v) \cong H^2(\Omega, A_{p^\infty}(L)).$$

Proof. See [6, Lemma 3.7]. That is, consider the commutative diagram

$$\begin{array}{ccc} H^1(K_v, A_{p^\infty}) & \xrightarrow{s_v} & H^1(L, A_{p^\infty})^\Omega \\ \downarrow & & \downarrow \\ H^1(K_v, A)(p) & \xrightarrow{\tau_v} & H^1(L, A)(p)^\Omega. \end{array} \tag{19.18}$$

Here, s_v denotes the restriction map on cohomology. The vertical arrows denote the surjective maps derived from Kummer theory on A. Now, observe that since v does not divide p, we have identifications

$$A(K_v) \otimes \mathbf{Q}_p/\mathbf{Z}_p = A(L) \otimes \mathbf{Q}_p/\mathbf{Z}_p = 0.$$

Hence, the vertical arrows in (19.18) are isomorphisms, and we have bijections

$$\ker(\tau_v) \cong \ker(s_v)$$
$$\operatorname{coker}(\tau_v) \cong \operatorname{coker}(s_v).$$

The result now follows from the Hochschild-Serre spectral sequence associated to s_v, using the fact that $H^2(K_v, A_{p^\infty}) = 0$ by local Tate duality. □

Corollary 19.18. *Let v be a prime of S not dividing p. If v splits completely $K_{\mathfrak{p}^\infty}$, then the local restriction map γ_v is an isomorphism, i.e.* $\ker(\gamma_v) = \operatorname{coker}(\gamma_v) = 0$. *If v does not split completely in $K_{\mathfrak{p}^\infty}$, then γ_v is the zero map, and the order of its kernel $J_v(K) = H^1(K_v, A)(p)$ is the exact power of p dividing $c_v(A)/L_v(A, 1)$.*

Proof. The first assertion follows from the result of Corollary 19.7, which implies that $\mathrm{cd}_p(G_w) = 0$ for any such prime v of K not dividing p. To see the second assertion, it can be argued in the same manner as [6, Lemma 3.3] that the $J_v(K_{\mathfrak{p}^\infty})$ vanishes in this setting (i.e. since the Galois group of the extension $K_{\mathfrak{p}^\infty,w}$ over K_v is isomorphic to a finite number of copies of the Galois group of the maximal unramified pro-p extension of K_v). The result then follows from that of Lemma 19.16. Note as well that if we take $L = K_{\mathfrak{p}^\infty,w}$ with $\Omega = G_w$ in Lemma 19.17, we obtain identifications

$$\ker(\gamma_v) = H^1(G_w, A_{p^\infty})$$
$$\operatorname{coker}(\gamma_v) = H^2(G_w, A_{p^\infty}),$$

whence we can also deduce that $|H^1(G_w, A_{p^\infty})| = \mathrm{ord}_p(c_v(A)/L_v(A, 1))$ and that $H^2(G_w, A_{p^\infty}) = 0$. □

To study the localization maps γ_v with v dividing p in the manner of [6], we start with the following basic result.

Lemma 19.19. *Let v be a prime of K dividing p. Then, we have identifications*

$$\ker(\gamma_v) \cong H^1(G_w, A(K_{\mathfrak{p}^\infty,w}))(p)$$
$$\operatorname{coker}(\gamma_v) \cong H^2(G_w, A(K_{\mathfrak{p}^\infty,w}))(p).$$

Proof. Taking the Hochschild-Serre spectral sequence associated to $K_{\mathfrak{p}^\infty,w}$ over K_v with the fact that $H^2(K_v, A_{p^\infty}) = 0$, the result follows from the definition of γ_v. □

We now consider the primes $v \mid p$ of K where A has good ordinary reduction.

Lemma 19.20. *Let v be a prime of K dividing p where A has good ordinary reduction. Assume that A is principally polarized. Then, we have identifications*

$$|H^i(K_v, \widehat{A}_v(\mathfrak{m}))| = \begin{cases} 0 & \text{if } i \geq 2 \\ |\widetilde{A}(\kappa_v)(p)| & \text{if } i = 1. \end{cases}$$

Proof. See [6, Lemma 3.13], the proof carries over without changes to the setting of principally polarized abelian varieties. □

Lemma 19.21. *Let v be a prime of K above p where A has good ordinary reduction. Assume additionally that A is principally polarized. Then, for all $i \geq 2$, we have isomorphisms*

$$H^i(G_w, A(K_{\mathfrak{p}^\infty,w}))(p) \cong H^i(G_w, \widetilde{A}_{v,p^\infty}).$$

We also have for $i = 1$ the short exact sequence

$$H^1(K_v, \widehat{A}_v(\mathfrak{m})) \longrightarrow H^1(G_w, A(K_{\mathfrak{p}^\infty,w}))(p) \longrightarrow H^1(G_w, \widetilde{A}_{v,p^\infty}). \tag{19.19}$$

Proof. The results are deduced from the lemmas above, following [6, Lemmas 3.12 and 3.14], which work in the same way for principally polarized abelian varieties using the results of [7]. □

Putting these identifications together, we obtain the following result. Let us write $h_{v,i}$ to denote the cardinality of $H^i(G_w, \widetilde{A}_{v,p^\infty})$ for any integer $i \geq 0$.

Proposition 19.22. *Let v be a prime of S above p where A has good ordinary reduction. Assume additionally that A is principally polarized. Then,* $\ker(\gamma_v)$ *and* $\operatorname{coker}(\gamma_v)$ *are finite, and we have the identification*

$$\frac{|\ker(\gamma_v)|}{|\operatorname{coker}(\gamma_v)|} = \frac{h_{v,0} \cdot h_{v,1}}{h_{v,2}}.$$

Proof. Lemmas 19.19 and 19.21 imply that $|\mathrm{coker}(\gamma_v)| = h_{v,2}$. Using the exact sequence (19.19) along with Lemma 19.20 for $i = 0$, we then deduce that $|\mathrm{coker}(\gamma_v)| = h_{v,0} \cdot h_{v,1}$, as required. □

To summarize what we have shown in this section, let us for a prime $v \in S$ not dividing p write $\mathfrak{m}_v(A)$ to denote the exponent $\mathrm{ord}_p(c_v(A)/L_v(A, 1))$.

Theorem 19.23. *Let S_{ns} denote the subset of primes of the fixed set S which (i) do not dividing p and (ii) do not split completely in $K_{\mathfrak{p}^\infty}$. We have the following description of the global restriction map γ in* (19.16)*:*

$$\frac{|\ker(\gamma)|}{|\mathrm{coker}(\gamma)|} = \prod_{\substack{v|p \\ v\in S}} \frac{h_{v,0} \cdot h_{v,1}}{h_{v,2}} \times \prod_{\substack{v\nmid p \\ v\in S_{\mathrm{ns}}}} \mathfrak{m}_v(A).$$

Global calculation. We start with the easy part of the calculation, which in some sense is just diagram chasing in (19.16). Let us now assume for simplicity that A has good ordinary reduction at all primes above p in K.

Lemma 19.24. *We have the following bijections for the restriction map β:*

$$\ker(\beta) \cong H^1(G, A_{p^\infty})$$
$$\mathrm{coker}(\beta) \cong H^2(G, A_{p^\infty}).$$

Proof. This follows immediately from the inflation–restriction exact sequence, using the fact that $H^2(G_S(K), A_{p^\infty}) = 0$. □

Recall that given a prime $v \in S$ and an integer $i \geq 0$, we write $h_{v,i}$ to denote the cardinality of $H^i(G_w, \widetilde{A}_{v,p^\infty})$. Let us also write h_i to denote the cardinality of $H^i(G, A_{p^\infty})$. We now give the first part of the global calculation:

Lemma 19.25. *Assume that $\mathrm{Sel}(A/K)$ and $\mathrm{coker}(\lambda_S(K))$ are finite. Assume that A has good ordinary reduction at each prime above p in K. Then, the cardinality $|H^0(G, \mathrm{Sel}(A/K_{\mathfrak{p}^\infty}))|$ is given by the formula*

$$|\mathrm{Sel}(A/K)| \cdot |\mathrm{coker}(\psi_S(K_{\mathfrak{p}^\infty}))| \cdot \frac{h_2}{h_1 \cdot h_0} \cdot \prod_{v|p} \frac{h_{v,0} \cdot h_{v,1}}{h_{v,2}} \cdot \prod_{\substack{v\nmid p \\ v\in S_{\mathrm{ns}}}} \mathfrak{m}_v(A). \tag{19.20}$$

Proof. Theorem 19.23 implies that

$$\frac{|\ker(\gamma)|}{|\mathrm{coker}(\gamma)|} = \prod_{v|p} \frac{h_{v,0} \cdot h_{v,1}}{h_{v,2}} \cdot \prod_{\substack{v \nmid p \\ v \in S_{\mathrm{ns}}}} \mathfrak{m}_v(A). \tag{19.21}$$

Observe that the numerator and denominator of (19.21) are finite by our assumption that A has good ordinary reduction at each prime above p in K. Using this result in the snake lemma associated to (19.17), we deduce that $\ker(\epsilon_1)$ and $\mathrm{coker}(\epsilon_1)$ must be finite. Moreover, we deduce that

$$\frac{|\ker(\epsilon_1)|}{|\mathrm{coker}(\epsilon_1)|} = \frac{|\ker(\gamma)|}{|\mathrm{coker}(\gamma)|} \cdot \frac{|\mathrm{coker}\left(\psi_S(K_{\mathfrak{p}^\infty})\right)|}{|\mathrm{coker}\left(\lambda_S(K)\right)|}. \tag{19.22}$$

Now, it is well-known that if $\mathrm{Sel}(A/K)$ and $\mathrm{coker}(\lambda_S(K))$ are finite, then

$$|\mathrm{coker}\left(\lambda_S(K)\right)| = A(K)_{p^\infty} = h_0. \tag{19.23}$$

On the other hand, taking the snake lemma in (19.16), we find that

$$\frac{|\mathrm{Sel}(A/K_{\mathfrak{p}^\infty})^G|}{|\mathrm{Sel}(A/K)|} = \frac{|\mathrm{coker}(\beta)|}{|\ker(\beta)|} \cdot \frac{|\ker(\epsilon_1)|}{|\mathrm{coker}(\epsilon_1)|}. \tag{19.24}$$

The result now follows from (19.24) by (19.21), (19.22) and (19.23). □

Corollary 19.26. *If the localization map $\lambda_S(K_{\mathfrak{p}^\infty})$ is surjective, then*

$$\chi(G, \mathrm{Sel}(A/K_{\mathfrak{p}^\infty})) = \frac{|\mathrm{Sel}(A/K)|}{h_0^2} \cdot \prod_{v|p} h_{0,v}^2 \cdot \prod_{\substack{v \nmid p \\ v \in S_{\mathrm{ns}}}} \mathfrak{m}_v(A). \tag{19.25}$$

Proof. If the localization map $\lambda_S(K_{\mathfrak{p}^\infty})$ is surjective, then we have the short exact sequence

$$0 \longrightarrow \mathrm{Sel}(A/K_{\mathfrak{p}^\infty}) \longrightarrow B_\infty \longrightarrow C_\infty \longrightarrow 0, \tag{19.26}$$

where

$$B_\infty = H^1(G_S(K_{\mathfrak{p}^\infty}), A_{p^\infty})$$
$$C_\infty = \bigoplus_{v \in S} J_v(K_{\mathfrak{p}^\infty}).$$

Taking G-cohomology of (19.26) then gives the long exact sequence

$$\begin{array}{llllll} 0 & \longrightarrow H^0(G, \mathrm{Sel}(A/K_{\mathfrak{p}^\infty})) & \longrightarrow H^0(G, B_\infty) & \xrightarrow{\psi_S(K_{\mathfrak{p}^\infty})} H^0(G, C_\infty) \\ & \longrightarrow H^1(G, \mathrm{Sel}(A/K_{\mathfrak{p}^\infty})) & \longrightarrow H^1(G, B_\infty) & \longrightarrow H^1(G, C_\infty) \\ & \longrightarrow \cdots & & \\ & \longrightarrow H^\delta(G, \mathrm{Sel}(A/K_{\mathfrak{p}^\infty})) & \longrightarrow H^\delta(G, B_\infty) & \longrightarrow H^\delta(G, C_\infty) \\ & \longrightarrow 0. & & \end{array} \tag{19.27}$$

Let us now write

$$W_{i,v} = H^i(G_w, H^1(K_{\mathfrak{p}^\infty,v}, A_{p^\infty})(p)).$$

We can then extract from (19.27) the long exact sequence

$$\begin{array}{llll} 0 & \longrightarrow H^0(G, \mathrm{Sel}(A/K_{\mathfrak{p}^\infty})) & \longrightarrow H^0(G, B_\infty) & \xrightarrow{\psi_S(K_{\mathfrak{p}^\infty})} \bigoplus_{v \in S} W_{0,v} \\ & \longrightarrow H^1(G, \mathrm{Sel}(A/K_{\mathfrak{p}^\infty})) & \longrightarrow H^3(G, A_{p^\infty}) & \longrightarrow \bigoplus_{v \in S} H^3(G_w, \widetilde{A}_{v,p^\infty}) \\ & \longrightarrow \cdots & & \\ & \longrightarrow H^\delta(G, \mathrm{Sel}(A/K_{\mathfrak{p}^\infty})) & \longrightarrow 0. & \end{array} \tag{19.28}$$

Here, the identifications for the central terms come from the Hochschild-Serre spectral sequence, (19.4). The identifications of the terms on the right come from Corollary 19.10 for the first row, then from the version of Shapiro's lemma given in (19.7) along with the bijections (19.8) for the subsequent rows. Note that we have ignored the vanishing of higher cohomology groups here. The reason for this will me made clear below. That is, using the definition of the map $\psi_S(K_{\mathfrak{p}^\infty})$ in (19.16), we can then extract from (19.28) the exact sequence

$$\begin{array}{llll} 0 & \longrightarrow \mathrm{coker}\left(\psi_S(K_{\mathfrak{p}^\infty})\right) & \longrightarrow H^1(G, \mathrm{Sel}(A/K_{\mathfrak{p}^\infty})) & \longrightarrow H^3(G, A_{p^\infty}) \\ & \longrightarrow \cdots & \longrightarrow H^\delta(G, \mathrm{Sel}(A/K_{\mathfrak{p}^\infty})) & \longrightarrow 0. \end{array} \tag{19.29}$$

Using (19.29), we deduce that

$$\chi(G, \mathrm{Sel}(A/K_{\mathfrak{p}^\infty})) = \frac{|H^1(G, \mathrm{Sel}(A/K_{\mathfrak{p}^\infty}))|}{|\mathrm{coker}\left(\psi_S(K_{\mathfrak{p}^\infty})\right)|} \cdot \chi^{(3)}(G, A_{p^\infty}) \cdot \prod_{v \in S} \chi^{(3)}(G_w, \widetilde{A}_{v,p^\infty})^{-1}. \tag{19.30}$$

Here, we have written

$$\chi^{(3)}(G, A_{p^\infty}) = \prod_{i\geq 3} |H^i(G, A_{p^\infty})|^{(-1)^i},$$

and

$$\chi^{(3)}(G_w, A_{p^\infty}) = \prod_{i\geq 3} |H^i(G_w, \widetilde{A}_{v,p^\infty})|^{(-1)^i}.$$

Substituting (19.20) into (19.30), we then find that $\chi(G, \mathrm{Sel}(A/K_{\mathfrak{p}^\infty}))$ is equal to

$$|\mathrm{Sel}(A/K)| \cdot \frac{h_2}{h_0 h_1} \cdot \prod_{v|p} \frac{h_{v,0} h_{v,1}}{h_{v,2}} \cdot \prod_{\substack{v\nmid p \\ v\in S_{\mathrm{ns}}}} \mathfrak{m}_v(A) \cdot \chi^{(3)}(G, A_{p^\infty}) \cdot \prod_{v|p} \chi^{(3)}(G_w, \widetilde{A}_{v,p^\infty})^{-1}$$

$$= |\mathrm{Sel}(A/K)| \cdot \frac{\chi(G, A_{p^\infty})}{h_0^2} \cdot \prod_{\substack{v\nmid p \\ v\in S_{\mathrm{ns}}}} \mathfrak{m}_v(A) \cdot \prod_{v|p} h_{0,v}^2 \cdot \chi(G_w, \widetilde{A}_{v,p^\infty})^{-1}.$$

The result then follows from Proposition 19.15, which gives for each prime $v \mid p$ the identifications $\chi(G, A_{p^\infty}) = \chi(G_w, \widetilde{A}_{v,p^\infty}) = 1$. □

Surjectivity of the localization map. Recall that we write Λ to denote the $\mathcal{O}$-Iwasawa algebra $\Lambda(G) = \mathcal{O}[[G_{\mathfrak{p}^\infty}]]$. We now show that the localization map $\lambda_S(K_{\mathfrak{p}^\infty})$ is surjective if $\mathrm{Sel}_{p^\infty}(A/K_{\mathfrak{p}^\infty})$ is Λ-cotorsion, at least when A is known to be prinicipally polarized. Given L a Galois extension of K, let $\mathfrak{S}(A/L)$ denote the compactified Selmer group of A over L,

$$\mathfrak{S}(A/L) = \varprojlim_n \ker\left(H^1(G_S(L), A_{p^n}) \longrightarrow \bigoplus_{v\in S} J_v(L)\right).$$

Proposition 19.27. *Let $\Omega = \mathrm{Gal}(L/K)$ be an infinite, pro-p group. If $A(L)_{p^\infty}$ is finite, then there is a $\Lambda(\Omega)$-module injection*

$$\mathfrak{S}(A/L) \longrightarrow \mathrm{Hom}_{\Lambda(\Omega)}(X(A/L), \Lambda(\Omega)).$$

Proof. The result is well-known, see for instance [15, Theorem 7.1], the proof of which carries over without change. □

Theorem 19.28. (Cassels-Poitou-Tate exact sequence for abelian varieties). *Let A^t denote the dual abelian variety associated to A. Let $\mathfrak{S}(A/K_{\mathfrak{p}^\infty})^\vee$ denote the Pontryagin dual of the compactified Selmer group $\mathfrak{S}(A/K_{\mathfrak{p}^\infty})$. There is a canonical exact sequence*

$$\begin{array}{ccccccc} 0 & \longrightarrow & \mathrm{Sel}(A/K_{\mathfrak{p}^\infty}) & \longrightarrow & H^1(G_S(K_{\mathfrak{p}^\infty}), A_{p^\infty}) & \longrightarrow & \bigoplus_{v\in S} H^1(K_{\mathfrak{p}^\infty,w}, A)(p) \\ & \longrightarrow & \mathfrak{S}^\vee(A^t/K_{\mathfrak{p}^\infty}) & \longrightarrow & H^2(G_S(K_{\mathfrak{p}^\infty}), A_{p^\infty}) & \longrightarrow & 0. \end{array} \tag{19.31}$$

Proof. See Coates-Sujatha [8, § 1.7]. The same deduction using the generalized Cassels-Poitou-Tate exact sequence [8, Theorem 1.5] works here, keeping track of the dual abelian variety A^t. □

Corollary 19.29. *Assume that A is principally polarized. If* $\mathrm{Sel}(A/K_{\mathfrak{p}^\infty})$ *is Λ-cotorsion, then the localization map $\lambda_S(K_{\mathfrak{p}^\infty})$ is surjective, i.e. there is a short exact sequence*

$$0 \longrightarrow \mathrm{Sel}(A/K_{\mathfrak{p}^\infty}) \longrightarrow H^1(G_S(K_{\mathfrak{p}^\infty}), A_{p^\infty}) \longrightarrow \bigoplus_{v\in S} J_v(K_{\mathfrak{p}^\infty}) \longrightarrow 0.$$

Proof. If $\mathrm{Sel}(A/K_{\mathfrak{p}^\infty})$ is Λ-cotorsion, then $X(A/K_{\mathfrak{p}^\infty})$ is Λ-torsion by duality. It follows that $\mathrm{Hom}_\Lambda(X(A/K_{\mathfrak{p}^\infty}), \Lambda) = 0$. Hence by Proposition 19.27, we find that $\mathfrak{S}(A/K_{\mathfrak{p}^\infty}) = 0$. It follows that $\mathfrak{S}^\vee(A/K_{\mathfrak{p}^\infty}) = 0$. Since we assume that A is principally polarized, we can assume that there is an isomorphism $A \cong A^t$, in which case it also follows that $\mathfrak{S}^\vee(A^t/K_{\mathfrak{p}^\infty}) = 0$. The claim then follows from (19.31). Observe that we also obtain from this the vanishing of $H^2(G_S(K_{\mathfrak{p}^\infty}), A_{p^\infty})$. □

Hence, we obtain from Corollary 19.26 the following main result of this section.

Theorem 19.30. *Assume that A is a principally polarized, and that A has good ordinary reduction at each prime above p in K. Assume additionally that* $\mathrm{Sel}(A/K)$ *is finite. If* $\mathrm{Sel}_{p^\infty}(A/K_{\mathfrak{p}^\infty})$ *is $\Lambda(G)$-cotorsion, then*

$$\chi(G, \mathrm{Sel}(A/K_{\mathfrak{p}^\infty})) = \frac{|\mathrm{Ш}(A/K)(p)|}{|A(K)(p)|^2} \cdot \prod_{v|p} |\widetilde{A}(\kappa_v)(p)|^2 \cdot \prod_{\substack{v\nmid p \\ v\in S_{\mathrm{ns}}}} \mathrm{ord}_p\left(\frac{c_v(A)}{L_v(A,1)}\right).$$

Proof. The computations above show that the Euler characteristic is well-defined, and given by the formula

$$\chi(G, \mathrm{Sel}(A/K_{\mathfrak{p}^\infty})) = \frac{|\mathrm{Sel}(A/K)|}{h_0^2} \cdot \prod_{v|p} h_{0,v}^2 \cdot \prod_{\substack{v\nmid p \\ v\in S_{\mathrm{ns}}}} \mathfrak{m}_v(A).$$

The result then follows from the definitions, as well as the standard identification in this setting of $\mathrm{Sel}(A/K)$ with $\mathrm{Ш}(A/K)[p^\infty]$. □

19.6 The indefinite case

Let us keep all of the setup above, but with the following crucial condition. Let $N \subset \mathcal{O}_F$ denote the arithmetic conductor of A. Recall that we write η to denote the quadratic character associated to K/F. We assume here that the so-called *weak Heegner hypothesis* holds, which is that $\eta(N) = (-1)^{d-1}$ (cf. [19, 0.]). In this setting, the Selmer group $\mathrm{Sel}(A/K_{\mathfrak{p}^\infty})$ is generally not Λ-cotorsion, but rather free of rank one over Λ. For results in this direction, see Howard [19, Theorem B], which generalizes earlier results of Bertolini [1] and Nekovar (cf. [?] or [?]).

The conjectures of Perrin-Riou and Howard. We refer the reader to [19] or [26] for background. The version of the Iwasawa main conjectures posed in these works has the following interpretation in terms of Euler characteristic formulae. Recall that we fix a prime $\mathfrak{p}$ above p in F. For each finite extension L of K, we consider the two $\mathfrak{p}$-primary Selmer groups $\mathrm{Sel}_{\mathfrak{p}^\infty}(A/L)$ and $\mathfrak{S}(A/L)$, which can be defined implicitly via their inclusion in the descent exact sequences

$$0 \to A(L) \otimes_{\mathcal{O}_F} \mathcal{O}_{F_\mathfrak{p}} \to \mathfrak{S}(A/L) \to \varprojlim_n Ш(A/L)[\mathfrak{p}^n]$$

$$0 \to A(L) \otimes_{\mathcal{O}_F} \left(F_\mathfrak{p}/\mathcal{O}_{F_\mathfrak{p}}\right) \to \mathrm{Sel}_{\mathfrak{p}^\infty}(A/L) \to Ш(A/L)[\mathfrak{p}^\infty] \to 0.$$

The abelian variety A comes equipped with a family of Heegner (or CM) points defined over the ring class extensions $K[\mathfrak{p}^n]$, as described e.g. in [19, §1.2], which gives rise to the following submodule. Given an integer $n \geq 0$, let h_n denote the image under the norm from $K[\mathfrak{p}^{n+1}]$ to $K_{\mathfrak{p}^n}$ of the Heegner/CM point of conductor $\mathfrak{p}^{n+1}$. Let $\mathfrak{H}_n$ denote the Λ-module generated by all the images h_m with $m \leq n$. Let $\mathfrak{H}_\infty = \varprojlim_n \mathfrak{H}_n$ denote the associated *Heegner module.* Taking the usual limits, let us also define the finitely generated Λ-modules

$$\mathfrak{S}_\infty = \mathfrak{S}(A/K_{\mathfrak{p}^\infty}) = \varprojlim_n \mathfrak{S}(A/K_{\mathfrak{p}^n})$$

$$\mathrm{Sel}_{\mathfrak{p}^\infty}(A/K_{\mathfrak{p}^\infty}) = \varinjlim_n \mathrm{Sel}_{\mathfrak{p}^\infty}(A/K_{\mathfrak{p}^n}).$$

Let $X(A/K_{\mathfrak{p}^\infty})$ denote the Pontryagin dual of $\mathrm{Sel}_{\mathfrak{p}^\infty}(A/K_{\mathfrak{p}^\infty})$, with $X(A/K_{\mathfrak{p}^\infty})$ its Λ-cotorsion submodule. Given an element $\lambda \in \Lambda$, let λ^* denote the image of λ under the involution of Λ induced by inversion in $G = G_{\mathfrak{p}^\infty}$. The main conjecture posed by Howard [19, Theorem B], following Perrin-Riou [26], can then be summarized as in the following way.

Conjecture 19.31 (Iwasawa main conjecture). *If A has ordinary reduction at $\mathfrak{p}$, then*

(i) *The Λ-modules $\mathfrak{H}_\infty$ and $\mathfrak{S}_\infty$ are torsionfree of rank one.*
(ii) *The dual Selmer group $X(A/K_{\mathfrak{p}^\infty})$ has rank one as a Λ-module.*
(iii) *There is a pseudoisomorphism of Λ-modules*

$$X(A/K_{\mathfrak{p}^\infty})_{\mathrm{tors}} \longrightarrow M \oplus M \oplus M_{\mathfrak{p}}.$$

Here, the Λ-characteristic power series $\mathrm{char}(M)$ *is prime to* $\mathfrak{p}\Lambda$ *with the property that* $\mathrm{char}(M) = \mathrm{char}(M)^*$, *and* $\mathrm{char}(M_{\mathfrak{p}})$ *is a power of* $\mathfrak{p}\Lambda$.
(iv) *The following equality of ideals holds in Λ, at least up to powers of $\mathfrak{p}\Lambda$:*

$$(\mathrm{char}(M)) = (\mathrm{char}\,(\mathfrak{S}_\infty/\mathfrak{H}_\infty))\,.$$

Remark. Note that many cases of this conjecture (with one divisibility in (iv)) are established by Howard [19, Theorem B], using the relevant nonvanishing theorem of Cornut-Vatsal [12] to ensure nontriviality.

Let us now re-state this conjecture in terms of Euler characteristic formulae. Let $\mathrm{Sel}_{\mathfrak{p}^\infty}(A/K_{\mathfrak{p}^\infty})_{\mathrm{cotors}}$ denote the Λ-cotorsion submodule of $\mathrm{Sel}_{\mathfrak{p}^\infty}(A/K_{\mathfrak{p}^\infty})$. Recall that given an element $\lambda \in \Lambda$, we write $\lambda(0)$ to denote the image of λ under the natural map $\Lambda \longrightarrow \mathbf{Z}_p$. Recall as well that we let $\mathfrak{K}_n$ denote the Artin symbol of $\mathfrak{d}_n = (\sqrt{D}\mathcal{O}_K) \cap \mathcal{O}_{p^n}$. Here, D denotes the absolute discriminant of K, and $\mathcal{O}_{p^n}$ the $\mathcal{O}_F$-order of conductor $\mathfrak{p}^n$ in K, i.e. $\mathcal{O}_{p^n} = \mathcal{O}_F + \mathfrak{p}^n\mathcal{O}_K$. Let $\mathfrak{K} = \varprojlim_n \mathfrak{K}_n$. We let $\langle\ ,\ \rangle_{A,K[p^n]}$ denote the p-adic height pairing

$$\langle\ ,\ \rangle_{A,K[p^n]} : A^t(K[p^n]) \times A(K[p^n]) \longrightarrow \mathbf{Q}_p,$$

defined e.g. in [18, (9), §3.3] or [28]. Assume for simplicity that A is principally polarized. There exists by a construction of Perrin-Riou [26] (see also [28]) a p-adic height pairing

$$\mathfrak{h}_n : \mathfrak{S}(A/K_{p^n}) \times \mathfrak{S}(A/K_{p^n}) \longrightarrow c^{-1}\mathbf{Z}_p$$

whose restriction to the image of the Kummer map $A(K[p^n]) \otimes \mathbf{Z}_p \to \mathfrak{S}(E/K[p^n])$ coincides with the pairing $\langle\ ,\ \rangle_{A,K[p^n]}$ after identifying $A^t \cong A$ in the canonical way (cf. [18, Proposition 0.0.4]). Here, $c \in \mathbf{Z}_p$ is some integer that does not depend on the choice of n. Following [26] and [18], these pairings can be used to construct a pairing

$$\begin{aligned}\mathfrak{h}_\infty : \mathfrak{S}_\infty \times \mathfrak{S}_\infty &\longrightarrow c^{-1}\mathbf{Z}_p[[G]] \\ (\varprojlim_n a_n, \varprojlim_n b_n) &\longmapsto \varprojlim_n \sum_{\sigma \in G_{p^n}} \mathfrak{h}_n(a_n, b_n^\sigma)\cdot \sigma.\end{aligned}$$

We then define the *p-adic regulator* $\mathcal{R} = \mathcal{R}(A/K_{p^\infty})$ to be the image of $\mathfrak{S}_\infty$ in $c^{-1}\mathbf{Z}_p[[G]]$ under this pairing $\mathfrak{h}_\infty$. Let $\mathbf{e}$ denote the natural projection

$$\mathbf{e} : \mathbf{Z}_p[[G[p^\infty]]] \longrightarrow \mathbf{Z}_p[[G]].$$

We then make the following conjecture in this situation.

Conjecture 19.32. *Let A be a principally polarized abelian variety of arithmetic conductor N defined over F having good ordinary reduction at each prime above p in K. Assume as well that A satisfies the weak-Heegner hypothesis with respect to N and K, i.e. that $\eta(N) = (-1)^{d-1}$. Then, the G-Euler characteristic $\chi(G, \mathrm{Sel}(A/K_{\mathfrak{p}^\infty})_{\mathrm{cotors}})$ is well-defined, and given by the expression*

$$\chi(G, \mathrm{Sel}(A/K_{\mathfrak{p}^\infty})_{\mathrm{cotors}}) = |\mathrm{char}(\mathfrak{S}_\infty/\mathfrak{H}_\infty)(0)\cdot \mathrm{char}(\mathfrak{S}_\infty/\mathfrak{H}_\infty)^*(0)\cdot \mathfrak{R}(0)|_p^{-1}.$$

Here, $\mathfrak{R}$ denotes the ideal $(\mathbf{e}(\mathfrak{K}))^{-1}\mathcal{R}$, which lies in the $\mathbf{Z}_p$-Iwasawa algebra $\mathbf{Z}_p[[G]]$. Thus, we also conjecture that the p-adic regulator $\mathcal{R} = \mathcal{R}(A/K_{p^\infty})$ is not identically zero, and hence that the p-adic height pairing $\mathfrak{h}_\infty$ is nondegenerate.

Modular elliptic curves over imaginary quadratic fields. Let us now give a more precise description of the conjectural formula for elliptic curves defined over the totally real field $F = \mathbf{Q}$. We first recall the Λ-adic Gross-Zagier theorem of Howard [18, Theorem B]. Let p be an odd prime. Fix embeddings $\overline{\mathbf{Q}} \to \overline{\mathbf{Q}}_p$ and $\overline{\mathbf{Q}} \to \mathbf{C}$. Let E be an elliptic curve of conductor N defined over $\mathbf{Q}$. Hence, E is modular by fundamental work of Wiles, Taylor-Wiles, and Breuil-Conrad-Diamond-Taylor. Let $f \in S_2(\Gamma_0(N))$ denote the cuspidal newform of weight 2 and level N associated to E by modularity. Let K be an imaginary quadratic field of discriminant D, and associated quadratic character η. We impose the following

Hypothesis 19.33. *Assume (i) that f is p-ordinary in the sense that the image of its T_p-eigenvalue under the fixed embedding $\overline{\mathbf{Q}} \to \overline{\mathbf{Q}}_p$ is a p-adic unit, (ii) that the Heegner hypothesis holds with respect to N and K, i.e. that each prime divisor of N split in K, and (iii) that $(p, ND) = 1$.*

Let us for clarity also fix the following notations here. We write $K[p^{\infty}]$ to denote the p^{∞}-ring class tower over K, with Galois group $\Omega = \mathrm{Gal}(K[p^{\infty}]/K)$. We then write $K_{p^{\infty}}$ to denote the dihedral or anticyclotomic $\mathbf{Z}_p$-extension of K, with Galois group $\Omega = \mathrm{Gal}(K_{p^{\infty}}/K) \approx \mathbf{Z}_p$. We also write $K(\mu_{p^{\infty}})$ to denote the extension of K obtained by adjoining all primitive, p-th power roots of unity, with Galois group $\Gamma = \mathrm{Gal}(K(\mu_{p^{\infty}})/K)$. We then write K^{cyc} to denote the cyclotomic $\mathbf{Z}_p$-extension of K, with Galois group $\Gamma = \mathrm{Gal}(K^{\mathrm{cyc}}/K) \approx \mathbf{Z}_p$. We shall consider the compositum extension $R_{\infty} = K[p^{\infty}] \cdot K(\mu_{p^{\infty}})$ with Galois group $\mathcal{G} = \mathrm{Gal}(R_{\infty}/K) = \Omega \times \Gamma$. We shall also also consider the $\mathbf{Z}_p^2$-extension K_{∞} of K contained in R_{∞}, with Galois group $G = \mathrm{Gal}(K_{\infty}/K) = \Omega \times \Gamma \approx \mathbf{Z}_p^2$. The construction of Hida [16] and Perrin-Riou [27] (whose integrality is shown in [39, Theorem 2.9]) gives a two-variable p-adic L-function

$$\mathcal{L}_f \in \mathbf{Z}_p[[\mathcal{G}]] = \mathbf{Z}_p[[\Omega \times \Gamma]].$$

This element interpolates the algebraic values

$$L(E/K, \mathcal{W}, 1)/8\pi^2\langle f, f\rangle_N = L(f \times g_{\mathcal{W}}, 1)/8\pi^2\langle f, f\rangle,$$

which we view as elements of $\overline{\mathbf{Q}}_p$ under our fixed embedding $\overline{\mathbf{Q}} \longrightarrow \overline{\mathbf{Q}}_p$. Here, $\mathcal{W}$ is any finite order character of $\mathcal{G}$, and $\langle f, f\rangle_N$ the Petersson inner product of f with itself. Let us now commit a minor abuse of notation in also writing $\mathcal{L}_f$ to denote the image of this p-adic L-function in the Iwasawa algebra $\mathbf{Z}_p[[\Omega \times \Gamma]]$. Fixing a topological generator γ of Γ, we can then write this two-variable p-adic L-function $\mathcal{L}_f$ as a power series in the cyclotomic variable $(\gamma - 1)$,

$$\mathcal{L}_f = \mathcal{L}_{f,0} + \mathcal{L}_{f,1}(\gamma - 1) + \mathcal{L}_{f,2}(\gamma - 1)^2 + \ldots \qquad (19.32)$$

Here, the coefficients $\mathcal{L}_{f,i}$ are elements of the completed group ring $\mathbf{Z}_p[[\Omega]]$.

Now, since we assume the Heegner hypothesis in Hypothesis 19.33 (ii), it is a well known consequence that the functional equation for $L(E/K, s) = L(f \times g_K, s)$ forces the central value $L(E/K, 1) = L(f \times g_K, 1)$ to vanish. In fact, it is also well known that $L(E/K, \mathcal{W}, 1) = L(f \times g_{\mathcal{W}}, 1)$ vanishes for $\mathcal{W}$ any finite order character of Ω in this situation (see e.g. [12, §1]). Hence, in this situation, it follows from the interpolation property of $\mathcal{L}_f$ that we must have $\mathcal{L}_{f,0} = 0$. The main result of Howard [18] shows that the linear term $\mathcal{L}_{f,1}$ in this setting is related to height pairings of Heegner points in the Jacobian $J_0(N)$ of the modular curve $X_0(N)$. To be more precise, the Heegner hypothesis implies for each integer $n \geq 0$ the existence of a family of Heegner points $h_n \in X_0(N)(\mathbf{C})$ of conductor p^n. These Heegner points h_n are defined in the usual way: each h_n is a cyclic N-isogeny $h_n : \mathcal{E}_n \rightarrow \mathcal{E}_n'$ of elliptic curves $\mathcal{E}_n$ and $\mathcal{E}_n'$ having exact CM by the $\mathbf{Z}$-order $\mathcal{O}_{p^n} = \mathbf{Z} + p^n\mathcal{O}_K$. As explained in

[18], the family $\{h_n\}_n$ can be chosen in such a way that the following diagram commutes, where the vertical arrows are p-isogenies:

$$\begin{array}{ccc} \mathcal{E}_n & \xrightarrow{h_n} & \mathcal{E}'_n \\ \downarrow & & \downarrow \\ \mathcal{E}_{n-1} & \xrightarrow{h_{n-1}} & \mathcal{E}'_{n-1}. \end{array}$$

This in particular makes $\mathcal{E}_{n-1}$ a quotient of $\mathcal{E}_n$ by its $p\mathcal{O}_{p^{n-1}}$-torsion, and $\mathcal{E}'_{n-1}$ a quotient of $\mathcal{E}'_n$ by its $p\mathcal{O}_{p^{n-1}}$-torsion. By the theory of complex multiplication, the CM elliptic curves $\mathcal{E}_n$ and $\mathcal{E}'_n$, as well as the isogeny h_n between them, can all be defined over the ring class field $K[p^n]$ of conductor p^n over K. Hence, we have for each integer $n \geq 0$ a family of Heegner points $h_n \in X_0(N)(K[p^n])$. Let us commit an abuse of notation in also writing h_n to denote the image of any such point under the usual embedding $X_0(N) \to J_0(N)$ that sends the cusp at infinity to the origin.

Let $\mathbf{T}$ denote the $\mathbf{Q}$-algebra generated by the Hecke operators T_l at primes l not dividing N acting on $J_0(N)$. The semisimplicity of $\mathbf{T}$ induces $\mathbf{T}$-module isomorphism

$$J_0(N)(K[p^n]) \longrightarrow \bigoplus_{\beta} J(K[p^n])_{\beta}.$$

Here, the direct sum runs over all $\mathrm{Gal}(\overline{\mathbf{Q}}_p/\mathbf{Q}_p)$-orbits of algebra homomorphisms $\beta : \mathbf{T} \to \overline{\mathbf{Q}}_p$, and each summand is fixed by the natural action of $\mathrm{Gal}(K[p^n]/\mathbf{Q})$. If $\beta(\mathbf{T})$ is contained in $\mathbf{Q}_p$, then the Hecke algebra $\mathbf{T}$ acts on the summand $J(K[p^n])_{\beta}$ via the homomorphism β. Our fixed newform $f \in S_2(\Gamma_0(N))$ determines such a homomorphism, which we shall also denote by f in an abuse of notation. We shall then write $h_{n,f}$ to denote the image of a Heegner point $h_n \in J_0(N)(K[p^n])$ in the associated summand $J(K[p^n])_f$. Let α_p denote the p-adic unit root of the Hecke polynomial $X^2 - a_p(f)X + p$, where $a_p(f)$ denotes the T_p-eigenvalue of f. Using the language of [2], we define for each integer $n \geq 0$ a sequence of *regularized Heegner points*:

$$z_n = \frac{1}{\alpha_p^n} h_{n,f} - \frac{1}{\alpha_p^{n-1}} h_{n-1,f}$$

if $n \geq 1$, otherwise

$$z_0 = \frac{1}{u} \cdot \begin{cases} \left(1 - \frac{\sigma_p}{\sigma_p^*}\right)\left(1 - \frac{\sigma_p^*}{\sigma_p}\right) h_{0,f} & \text{if } \eta(p) = 1 \\ \left(1 - \frac{1}{\alpha_p^2}\right) h_{0,f} & \text{if } \eta(p) = -1. \end{cases}$$

Here, $u = \frac{1}{2}|\mathcal{O}_K^\times|$, and σ_p and σ_p^* denote the Frobenius automorphisms in the Galois group $\mathrm{Gal}(K[1]/K) \approx \mathrm{Pic}(\mathcal{O}_K)$ of the Hilbert class field $K[1]$ over K at the two primes above p in the case where $\eta(p) = 1$. It can be deduced from the Euler system relations shown in Howard [18] that these regularized Heegner points are compatible under norm and trace maps on the summand $J(K[p^n])_f$. We refer the reader to [18, § 1.2] for details.

The following main result of Howard [18] was first proved by Perrin-Riou [26] for the case of $n = 0$. To state this result, let $\mathfrak{K}_n$ denote the Artin symbol of $\mathfrak{d}_n = (\sqrt{D}\mathcal{O}_K) \cap \mathcal{O}_{p^n}$. Let $\langle\ ,\ \rangle_{E,K[p^n]}$ denote the canonical p-adic height pairing

$$\langle\ ,\ \rangle_{E,K[p^n]} : E^t(K[p^n]) \times E(K[p^n]) \longrightarrow \mathbf{Q}_p,$$

as defined in [18, (9), § 3.3]. Note that this pairing is canonical as a consequence of the fact that E is ordinary at p, along with the uniqueness claims of [18, Proposition 3.2.1]. The reader will also note that we do not extend this pairing $\mathbf{Q}_p$-linearly as done throughout [18], to obtain the results stated here after taking tensor products $\otimes \mathbf{Q}_p$. Indeed, such extensions are not necessary as $\mathcal{L}_f$ is integral by [39, Theorem 2.9], and hence belongs to the Iwasawa algebra $\mathbf{Z}_p[[\mathcal{G}]]$ (as opposed to just $\mathbf{Z}_p[[\mathcal{G}]] \otimes_{\mathbf{Z}_p} \mathbf{Q}_p$). Let $\log_p$ denote the p-adic logarithm, composed with a fixed isomorphism $\Gamma \cong \mathbf{Z}_p^\times$. Let us also write z_n^t denote the image of the regularized Heegner point z_n under the canonical principal polarization $J_0(N) \cong J_0(N)^t$. We deduce from [18, Theorem A] the following result.

Theorem 19.34 (Howard). *Assume that D is odd and not equal to -3. Assume as well that $\eta(p) = 1$. Then, for any dihedral character $\rho : \mathrm{Gal}(K[p^n]/K) \longrightarrow \overline{\mathbf{Q}}_p^\times$, we have the identity*

$$\rho(\mathfrak{K}_n) \cdot \log_p(\gamma) \cdot \rho(\mathcal{L}_{f,1}) = \sum_{\sigma \in \mathrm{Gal}(K[p^n]/K)} \rho(\sigma) \langle z_n^t, z_n^\sigma \rangle_{E,K[p^n]}. \qquad (19.33)$$

This identity (19.33) *is independent of the choice of topological generator $\gamma \in \Gamma$.*

This result has the following implications for our modular elliptic curve E, which belongs the isogeny class of ordinary elliptic curves associated to the newform f. Fix a modular parametrization $\varphi : X_0(N) \to E$. Let $\varphi_* : J_0(N) \to E$ denote the induced Albanese map, and $\varphi^* : E \to J_0(N)$ the induced Picard map. Let

$$x_n = \varphi_*(z_n) \in E(K[p^n])$$

denote the image of the regularized Heegner point z_n under the Albanese map φ_*, and let x_n^t denote the unique point of $E(K[p^n])$ for which $\varphi^*(x_n^t) = z_n^t$. Hence, we can identify x_n with $\deg(\varphi)x_n^t$. Note that the points x_n and x_n^t are norm compatible as n varies, thanks to the norm compatibility of the regularized Heegner points z_n and z_n^t. Granted this property, we define the *Heegner p-adic L-function* $\mathcal{L}_{\mathrm{Heeg}}$ in $\mathbf{Z}_p[[\Omega]]$ by

$$\mathcal{L}_{\mathrm{Heeg}} = \varprojlim_n \sum_{\sigma \in \mathrm{Gal}(K[p^n]/K)} \langle x_n^t, x_n^\sigma \rangle_{E,K[p^n]} \cdot \sigma.$$

We deduce from Theorem 19.34 the following integral version of [18, Theorem B].

Theorem 19.35. *Keep the setup and hypotheses of Theorem 19.34 above. Let us write* $\mathfrak{K} = \varprojlim_n \mathfrak{K}_n \in \Omega$. *Then, as ideals in the Iwasawa algebra* $\mathbf{Z}_p[[\Omega]]$, *we have*

$$\mathfrak{K} \cdot \log_p(\gamma) \cdot \mathcal{L}_{f,1} = \mathcal{L}_{\mathrm{Heeg}}.$$

This result in turn has applications to the Λ-adic Gross-Zagier theorem of Mazur-Rubin [23, Conjecture 9], where Λ denotes the Iwasawa algebra $\mathbf{Z}_p[[\Omega]]$. Following [18, § 0], we define a pairing

$$\langle\ ,\ \rangle^{\Gamma}_{E,K[p^n]} : E(K[p^n]) \times E(K[p^n]) \longrightarrow \Gamma$$

by the implicit relation $\langle\ ,\ \rangle_{E,K[p^n]} = \log_p \circ \langle\ ,\ \rangle^{\Gamma}_{E,K[p^n]}$. Here, we should mention that the construction of $\langle\ ,\ \rangle_{E,K[p^n]}$ given in [18, § 3.3] requires as input an auxiliary (idele class) character

$$\rho_{K[p^n]} : \mathbf{A}^{\times}_{K[p^n]}/K[p^n]^{\times} \longrightarrow \Gamma \xrightarrow{\log_p} \mathbf{Z}_p.$$

Anyhow, we then define from this a completed group ring element

$$\mathcal{L}^{\Gamma}_{\mathrm{Heeg}} = \varprojlim_n \sum_{\sigma \in \mathrm{Gal}(K[p^n]/K)} \langle x_n, x_n^\sigma \rangle^{\Gamma}_{E,K[p^n]} \cdot \sigma \in \mathbf{Z}_p[[\Omega]] \otimes \Gamma.$$

Let I denote the kernel of the natural projection

$$\mathbf{Z}_p[[\Omega \times \Gamma]] \longrightarrow \mathbf{Z}_p[[\Omega]].$$

Let ϑ denote the isomorphism

$$\begin{aligned} \vartheta : \mathbf{Z}_p[[\Omega]] \otimes \Gamma &\longrightarrow I/I^2 \\ (\lambda \otimes \gamma) &\longmapsto \lambda(\gamma - 1). \end{aligned}$$

As explained in [18], we can derive from this the formula

$$\vartheta\left(\mathcal{L}_{\mathrm{Heeg}}^{\Gamma}\right) = \frac{\deg(\varphi)}{\log_p(\gamma)} \cdot \mathcal{L}_{\mathrm{Heeg}}(\gamma - 1).$$

Now, observe that since $\mathcal{L}_{f,0} = 0$ with our hypothesis (i.e. Hypothesis 19.33 (ii)), the two-variable p-adic L-function $\mathcal{L}_f$ is contained in the kernel I. Hence, Theorem 19.35 gives us the following identifications in the quotient I/I^2:

$$\mathcal{L}_f = \mathcal{L}_{f,1}(\gamma - 1) = \frac{\mathfrak{L}_{\mathrm{Heeg}}}{\mathfrak{K} \cdot \log_p(\gamma)} \cdot (\gamma - 1) = \frac{\vartheta\left(\mathcal{L}_{\mathrm{Heeg}}^{\Gamma}\right)}{\mathfrak{K} \cdot \deg(\varphi)}. \tag{19.34}$$

Let us now keep the hypotheses of Theorem 19.34 in force. We assume additionally that the absolute Galois group $G_K = \mathrm{Gal}(\overline{K}/K)$ of K surjects onto the $\mathbf{Z}_p$-automorphisms of the p-adic Tate module $T_p(E)$ of E, and that p does not divide the class number of K. Let us now write $\widetilde{x}_\infty \in \mathfrak{S}(E/K_{p^\infty})$ to denote the projective limit of the sequence of norm compatible elements

$$\widetilde{x}_n = \mathrm{Norm}_{K[p^{n+1}]/K_{p^n}}(x_{n+1}) \in \mathfrak{S}(E/K_{p^n}).$$

We again write $\mathfrak{H}_\infty = \mathfrak{H}_\infty(E/K_{p^\infty})$ to denote the Heegner or CM submodule of the compactified Selmer group $\mathfrak{S}_\infty = \mathfrak{S}(E/K_{p^\infty})$ generated by this $\widetilde{x}_\infty$. By the main theorem of Cornut [11] (see also [12]), $\mathfrak{H}_\infty$ has rank one as a Λ-module. By work of Howard [17] (see also [18]) and Bertolini [1], we also know (i) that $X(E/K_{p^\infty})$ is a rank one Λ-module, (ii) that $\mathfrak{S}_\infty = \mathfrak{S}(E/K_{p^\infty})$ is free of rank one over Λ (whence the quotient $\mathfrak{S}_\infty/\mathfrak{H}_\infty$ is a torsion Λ-module), and (iii) that the following divisibility of ideals holds in Λ:

$$\mathrm{char}\, X(E/K_{p^\infty})_{\mathrm{tors}} \mid \mathrm{char}(\mathfrak{S}_\infty/\mathfrak{H}_\infty) \cdot \mathrm{char}(\mathfrak{S}_\infty/\mathfrak{H}_\infty)^*. \tag{19.35}$$

Here again, we have written $X(E/K_{p^\infty})_{\mathrm{tors}}$ to denote the Λ-torsion submodule of $X(E/K_{p^\infty})$. Let us now consider the following result of Perrin-Riou [26], as described in [18, Proposition 0.0.4].

Theorem 19.36 (Perrin-Riou). *There exists a p-adic height pairing*

$$\mathfrak{h}_n : \mathfrak{S}(E/K_{p^n}) \times \mathfrak{S}(E/K_{p^n}) \longrightarrow c^{-1}\mathbf{Z}_p \tag{19.36}$$

whose restriction to the image of the Kummer map $E(K[p^n]) \otimes \mathbf{Z}_p \to \mathfrak{S}(E/K[p^n])$ coincides with the pairing $\langle\ ,\ \rangle_{E,K[p^n]}$. Here, $c \in \mathbf{Z}_p$ is some p-adic integer that does not depend on choice of n.

Using this pairing (19.36), we can then define a pairing

$$\mathfrak{h}_\infty : \mathfrak{S}_\infty \times \mathfrak{S}_\infty \longrightarrow c^{-1}\mathbf{Z}_p[[\Omega]]$$
$$(\varprojlim_n a_n, \varprojlim_n b_n) \longmapsto \varprojlim_n \sum_{\sigma \in \mathrm{Gal}(K_{p^n}/K)} \mathfrak{h}_n(a_n, b_n^\sigma) \cdot \sigma.$$

We again define the p-adic regulator $\mathcal{R} = \mathcal{R}(E/K_{p^\infty})$ of E over K_{p^∞} to be the image in $c^{-1}\mathbf{Z}_p[[\Omega]]]$ of this pairing $\mathfrak{h}_\infty$. Let $\mathbf{e}$ denote the natural projection

$$\mathbf{e} : \mathbf{Z}_p[[\Omega]] \longrightarrow \mathbf{Z}_p[[\Omega]].$$

Following Howard [18] (using [18, Remark 3.2.2] with the uniqueness claims of [18, Proposition 3.2.1] and the fact that E is ordinary at p), we deduce that the height pairing $\mathfrak{h}_\infty$ is norm compatible. This allows us to deduce that

$$\mathbf{e}\left(\mathcal{L}_{\mathrm{Heeg}}\right) = \mathfrak{h}_\infty(\widetilde{x}_\infty, \widetilde{x}_\infty) = \mathrm{char}\,(\mathfrak{S}_\infty/H_\infty) \cdot \mathrm{char}\,(\mathfrak{S}_\infty/H_\infty)^* \,\mathcal{R} \quad (19.37)$$

as ideals in $c^{-1}\mathbf{Z}_p[[\Omega]]$. On the other hand, we obtain from (19.34) the identifications

$$\mathbf{e}(\mathcal{L}_{f,1}) = \frac{\mathfrak{h}_\infty(\widetilde{x}_\infty, \widetilde{x}_\infty)}{\mathbf{e}(\mathfrak{K} \cdot \log_p(\gamma))} = \frac{\mathrm{char}\,(\mathfrak{S}_\infty/H_\infty) \cdot \mathrm{char}\,(\mathfrak{S}_\infty/H_\infty)^* \,\mathcal{R}}{\mathbf{e}(\mathfrak{K} \cdot \log_p(\gamma))}$$

as ideals in $\Lambda = \mathbf{Z}_p[[\Omega]]$. Putting this all together, we obtain the following result.

Corollary 19.37. *Assume that D is odd and not equal to -3. Assume additionally that the absolute Galois group $G_K = \mathrm{Gal}(\overline{K}/K)$ surjects onto the $\mathbf{Z}_p$-automorphisms of $T_p(E)$, that $\eta(p) = 1$, and that p does not divide the class number of K. Then, we have the following formal relations:*

$$\begin{aligned}
&\chi(G, \mathrm{Sel}_{p^\infty}(A/K_{p^\infty})_{\mathrm{cotors}}) \\
&\quad \geq |\,\mathrm{char}(\mathfrak{S}_\infty/\mathfrak{H}_\infty)(0) \cdot \mathrm{char}(\mathfrak{S}_\infty/\mathfrak{H}_\infty)^*(0)|_p^{-1} \\
&\quad \geq |\,\mathrm{char}(\mathfrak{S}_\infty/\mathfrak{H}_\infty)(0) \cdot \mathrm{char}(\mathfrak{S}_\infty/\mathfrak{H}_\infty)^*(0)\mathcal{R}(0)|_p^{-1} \\
&\quad = |\mathbf{e}(\mathcal{L}_{f,1}(0))|_p^{-1} \\
&\quad = |\mathfrak{h}_\infty(\widetilde{x}_\infty, \widetilde{x}_\infty)(0)|_p^{-1}
\end{aligned}$$

Proof. Note that $\mathbf{e}(\mathfrak{K} \cdot \log_p(\gamma))$ is a unit, and so we can ignore contributions from the obvious $\mathbf{e}(\mathfrak{K} \cdot \log_p(\gamma))^{-1}$ terms. The first inequality follows from the main conjecture divisibility shown in [19]. The second inequality is trivial, at least granted the conjectural nontriviality of $\mathcal{R}$. The third and fourth equalities follow from the Iwasawa theoretic Gross-Zagier theorem shown in [18], as described above. □

Note as well that this result is formal, as we have not assumed the G-Euler characteristic to be well-defined, i.e. as we have not assumed that the p-adic regulator $\mathcal{R}$ is nondegenerate. We therefore end this section with the following speculation.

Conjecture 19.38. *Assume the conditions of Hypothesis 19.33 are true. Recall that we let Λ denote the Iwasawa algebra $\Lambda(G) = \mathbf{Z}_p[[G]]$. Then, (i) the G-Euler characteristic $\chi(G, \mathrm{Sel}_{p^\infty}(A/K_{p^\infty}))$ is well-defined, and given by the formula*

$$\begin{aligned}\chi(G, \mathrm{Sel}_{p^\infty}(A/K_{p^\infty})_{\mathrm{cotors}}) &= |\mathrm{char}(\mathfrak{S}_\infty/\mathfrak{H}_\infty)(0) \cdot \mathrm{char}(\mathfrak{S}_\infty/\mathfrak{H}_\infty)^*(0)\mathcal{R}(0)|_p^{-1} \\ &= |\mathbf{e}(\mathcal{L}_{f,1}(0))|_p^{-1} \\ &= |\mathfrak{h}_\infty(\widetilde{\mathfrak{x}}_\infty, \widetilde{\mathfrak{x}}_\infty)(0)|_p^{-1}\end{aligned}$$

Moreover, (ii) the image in the Iwasawa algebra Λ of the two-variable p-adic L-function $\mathcal{L}_f = \mathcal{L}_{f,1}(\gamma - 1)$ is not identically zero, (iii) the Λ-adic height pairing $\mathfrak{h}_\infty$ of Perrin-Riou is nondegenerate, and (iv) the p-adic regulator $\mathcal{R} = \mathcal{R}(E/K_{p^\infty})$ of E over K_{p^∞} is nondegenerate, and satisfies the relation $\mathcal{R} = \mathbf{e}(\mathfrak{K} \cdot \log_p(\gamma))^{-1}\Lambda$.

Remark. Note that the nonvanishing condition for $\mathcal{L}_f$ presented in part (i) of this conjecture does not follow a priori from the nonvanishing theorems of Cornut [11] or Cornut-Vatsal [12]. Those results show here that the complex values $L'(E/K, \rho, 1)$ do not vanish, where ρ a ring class character of $G[p^\infty]$ of sufficiently large conductor. In particular, what is required is a strengthening of their result to include twists by characters of the cyclotomic Galois group $\Gamma = \mathrm{Gal}(K(\mu_{p^\infty})/K)$.

Acknowledgements. The author is grateful to John Coates, Christophe Cornut, Olivier Fouquet, Ben Howard, and Dimitar Jetchev for useful discussions, as well as to an anonymous referee for pointing out a bad error in a earlier version of this work (which has since been corrected).

References

[1] M. Bertolini, *Selmer groups and Heegner points in anticyclotomic $\mathbf{Z}_p$-extensions*, Compositio Math. **99** (1995), 153-182.

[2] M. Bertolini and H. Darmon, *Heegner points on Mumford-Tate curves*, Invent. math. **126** (1996), 413-456.

[3] M. Bertolini and H. Darmon, *Iwasawa's Main Conjecture for elliptic curves over anticyclotomic $\mathbf{Z}_p$-extensions*, Annals of Math. **162** (2005), 1-64.

[4] N. Bourbaki, *Élements de mathématique, Fasc. XXXI, Algèbre commutative, Chapitre 7: Diviseurs*, Actualiés Scientifiques et Industrielles, **1314**, Paris, Hermann, (1965).

[5] C. Breuil, B. Conrad, F. Diamond, and R. Taylor, *On the modularity of elliptic curves over* $\mathbf{Q}$*: wild* 3*-adic exercises*, J. Amer. Math. Soc. **14** (2001), 843-939.

[6] J. Coates, *Fragments of* GL_2 *Iwasawa Theory of Elliptic Curves without Complex Multiplication,* Arithmetic Theory of Elliptic Curves (C.I.M.E.) Ed. Viola, Lecture Notes in Mathematics, Springer (1999), 1-50.

[7] J. Coates and R. Greenberg, *Kummer theory for abelian varieties over local fields,* Invent. math. **124** (1996), 129-174.

[8] J. Coates and R. Sujatha, *Galois Cohomology of Elliptic Curves,* T.I.F.R. Lectures in Mathematics **88**, Narosa Publishing House (1996).

[9] J. Coates, P. Schneider, and R. Sujatha, *Links between cyclotomic and* GL_2 *Iwasawa theory*, Doc. Math. Extra Volume: Kazuya Kato's 50th birthday (2003), 187-215.

[10] J. Coates and A. Wiles, *On the Conjecture of Birch and Swinerton-Dyer*, Inventiones math. **39** (1977), 223-251.

[11] C. Cornut, *Mazur's conjecture on higher Heegner points*, Inventiones math. **148** (2002), 495-523.

[12] C. Cornut and V. Vatsal, *Nontriviality of Rankin-Selberg L-functions and CM points*, L-functions and Galois Representations, Ed. Burns, Buzzard and Nekovář, Cambridge University Press (2007), 121-186

[13] D.A. Cox, *Primes of the form* x^2+ny^2*: Fermat, Class Field Theory, and Complex Multiplication*, John Wiley & Sons (1989).

[14] R. Greenberg, *Iwasawa theory for Elliptic Curves,* Arithmetic Theory of Elliptic Curves (C.I.M.E.) Ed. Viola, Lecture Notes in Mathematics, Springer (1999), 50-142.

[15] Y. Hachimori and O. Venjakob, *Completely faithful Selmer groups over Kummer extensions*, Doc. Math. Extra Volume: Kazuya Kato's 50th birthday (2003), 2-36.

[16] H. Hida, *A p-adic measure attached to the zeta functions associated to two elliptic modular forms I*, Inventiones math. **79**, 159-195 (1985).

[17] B. Howard, *The Heegner point Kolyvagin system*, Compositio Math. **140** No. 6 (2004), 1439-1472.

[18] B. Howard, *The Iwasawa theoretic Gross-Zagier theorem*, Compositio Math. **141** No. 4 (2005), 811-846.

[19] B. Howard, *Iwasawa theory of Heegner points on abelian varieties of* GL_2*-type*, Duke Math. Jour. **124** No. 1 (2004), 1-45.

[20] H. Imai, *A remark on the rational points of abelian varieties with values in cyclotomic* $\mathbb{Z}_l$*-extensions*, Proc. Japan. Acad. Math. Sci. **51** (1975), 12-16.

[21] K. Iwasawa, *On the* $\mathbf{Z}_l$*-extensions of algebraic number fields,* Ann. of Math., **98** No. 2 (1973), 246-326.

[22] V.A. Kolyvagin *On the structure of Selmer groups*, Math. Ann **291** (1991), no. 2, 253-259.

[23] B. Mazur and K. Rubin, *Elliptic curves and class field theory*, Proceedings of the ICM, Beijing 2002, **2** (2002), 185-196.

[24] J. Nekovar, *Level raising and anticyclotomic Selmer groups for Hilbert modular forms of weight* 2, to appear in Canadian J. Math.

[25] M. Lazard, *Groupes analytiques p-adiques,* I.H.E.S. Publ. Math. (1965), 389-603.

[26] B. Perrin-Riou, *Fonctions L p-adiques, théorie d'Iwasawa et points de Heegner*, Bull. Soc. Math. France **115** (1987), 399-456.

[27] B. Perrin-Riou, *Fonctions L p-adiques associées à une forme modulaire et à un corps quadratique imaginaire*, Jour. London Math. Soc. (2) **38** (1988), 1-32.

[28] B. Perrin-Riou, *Théorie d'Iwasawa et hauteurs p-adiques (cas des variétés abéliennes)*, Séminaire de Théorie des Nombres, Paris, 1990-1991, Birkh auser, 203-220.

[29] R. Pollack and T. Weston, *On anticyclotomic μ-invariants of modular forms,* to appear in Compositio Math.

[30] K. Prasanna, *Arithmetic properties of the theta correspondence and periods of modular forms*, Eisenstein series and Applications, Progress in Math. **258**, Birkhauser (2008), 251-269.

[31] K. Ribet and S. Takahashi, *Parametrizations of elliptic curves by Shimura curves and by classical modular curves*, Elliptic curves and modular forms (Washington, DC, 1996), Proc. Nat. Acad. Sci. U.S.A. **94** (1997), 11110-11114.

[32] D. Rohrlich, *On L-functions of elliptic curves and cyclotomic towers*, Inventiones math. **75** (1984), 409-423.

[33] J.-P. Serre, *Cohomologie Galoisienne,* Lecture Notes in Mathematics, 5th Ed., Springer, Berlin (1994).

[34] J.-P. Serre, *Sur la dimension des groupes profinis,* Topology **3** (1965), 413-420.

[35] C. Skinner and E. Urban, *The Main Conjecture for $GL(2)$*, Inventiones math. (2013).

[36] J. Tate, *On the conjecture of Birch and Swinnerton-Dyer and a geometric analogue*, Seminaire Bourbaki 1965/66, exposé 306.

[37] R. Taylor and A. Wiles, *Ring theoretic properties of certain Hecke algebras*, Ann. of Math., **141** (1995), 553-572.

[38] J. Van Order, *On the dihedral main conjectures of Iwasawa theory for Hilbert modular eigenforms*, Can. J. Math. (2013), 403-466, available at `http://arxiv.org/abs/1112.3821`

[39] J. Van Order, *Some remarks on the two-variable main conjecture of Iwasawa theory for elliptic curves without complex multiplication*, J. of Algebra **350**, Issue 1 (2012), 273-299.

[40] V. Vatsal, *Uniform distribution of Heegner points*, Inventiones math. **148** 1-46 (2002).

[41] A. Wiles, *Modular elliptic curves and Fermat's last theorem*, Ann. of Math. **141** (1995), 443-551.

[42] S.L. Zerbes, *Euler characteristics of Selmer groups I*, J. London Math. Soc. **70** no. 3 (2004), 586-608.

20

CM values of higher Green's functions and regularized Petersson products

Maryna Viazovska

Department of Mathematical Analysis, Taras Shevchenko National University of Kyiv, Volodymyrska 64, 01033 Kyiv, Ukraine

E-mail address: viazovska@gmail.com

ABSTRACT. This paper is a written version of a talk given in the Trimester Seminar "Arithmetic and Geometry" on 21.02.2012.

Higher Green functions are real-valued functions of two variables on the upper half-plane, which are bi-invariant under the action of a congruence subgroup, have a logarithmic singularity along the diagonal, and satisfy the equation $\Delta f = k(1-k)f$; here Δ is a hyperbolic Laplace operator and k is a positive integer. The significant arithmetic properties of these functions were disclosed in the paper of B. Gross and D. Zagier "Heegner points and derivatives of L-series"(1986). In particular, it was conjectured that higher Green's functions have "algebraic" values at the CM points. In the case when $k = 2$ and one of the CM points is equal to $\sqrt{-1}$, the conjecture was proved by A. Mellit in his doctoral dissertation. In this lecture we prove this conjecture for arbitrary k, assuming that both CM points under consideration lie in the same quadratic imaginary field.

The two main parts of the proof are as follows. We first show that the regularized Petersson scalar product of a binary theta series and a weight one weakly holomorphic cusp form is equal to the logarithm of the absolute value of an algebraic number and then prove that the special values of weight k Green's function, occurring at the conjecture of Gross and Zagier, can be written as the Petersson product of that type, where the form of weight one is the $(k-1)$-st Rankin-Cohen bracket of an explicitly given weakly holomorphic modular form of weight $2-2k$ and a binary theta series. Algebraicity of regularized Petersson products was also proved at about the same time by W. Duke and Y. Li by a different method; however, our result is stronger than theirs since we also give a formula for the factorization of the algebraic number in question.

Arithmetic and Geometry, ed. Luis Dieulefait *et al.* Published by Cambridge University Press.

Notations

We shall use the following conventions and notations:

Let $\mathbb{N} := \{n | n \in \mathbb{Z}, n \geq 1\}$ be the set of natural numbers;

We denote the real and the imaginary part of a complex number z by $\Re(z)$ and $\Im(z)$, respectively;

We shall regard $\overline{\mathbb{Q}}$ as a subfield of $\mathbb{C}$.

20.1 CM values of elliptic j-invariant

Let $\mathfrak{H} := \{\tau | \tau \in \mathbb{C},\ \Im(\tau) > 0\}$ be the upper half-plane; the group

$$\mathrm{SL}_2(\mathbb{Z}) = \left\{ \left(\begin{smallmatrix} a & b \\ c & d \end{smallmatrix}\right) \middle| a, b, c, d \in \mathbb{Z},\ ad - bc = 1 \right\}$$

acts on $\mathfrak{H}$ by linear fractional transformations $\tau \to \frac{a\tau+b}{c\tau+d}$. The elliptic j-invariant is the unique holomorphic function on $\mathfrak{H}$ such that

1. $j\left(\frac{a\tau+b}{c\tau+d}\right) = j(\tau)$ for any $\tau \in \mathfrak{H}$ and $\left(\begin{smallmatrix} a & b \\ c & d \end{smallmatrix}\right) \in \mathrm{SL}_2(\mathbb{Z})$;
2. $j(\tau) = q^{-1} + 744 + \sum_{n=1}^{\infty} a_n q^n$, where $q = e^{2\pi i \tau}$ and $a_n \in \mathbb{Z}$ for $n \in \mathbb{N}$.

A point $\mathfrak{z} \in \mathfrak{H}$ that satisfies a quadratic equation

$$A\mathfrak{z}^2 + B\mathfrak{z} + C = 0,$$

where

$$A, B, C \in \mathbb{Z},\ (A, B, C) = 1,\ B^2 - 4AC = -D < 0,$$

is called a *point of complex multiplication (a CM point)* of discriminant $-D$. Let $\mathcal{Z}_D$ be the set of the CM-points of discriminant $-D$ on $\mathrm{SL}_2(\mathbb{Z})\backslash\mathfrak{H}$.

Let $K := \mathbb{Q}(\sqrt{-D})$ and let $D = f^2\tilde{D}$ with $f \in \mathbb{N}$ and $-\tilde{D}$ a fundamental discriminant.

The values of j-invariant at the CM points are known as *singular moduli*. The following classical result is often called "The first main theorem of complex multiplication".

Theorem 20.1. *For each $\mathfrak{z} \in \mathcal{Z}_D$ the value $j(\mathfrak{z})$ is an algebraic integer. Moreover, the field $H_D = K(j(\mathfrak{z}))$ is abelian over K and dihedral over $\mathbb{Q}$; it is the ring class field of conductor f over K.*

Another remarkable property of the singular moduli is that their differences are highly divisible numbers. An explicit factorization formula for these numbers was discovered by B. Gross and D. Zagier [4]. In particular, they computed the prime factorization of the integer numbers

$$A(D_1, D_2) := \prod_{\mathfrak{z}_1 \in \mathcal{Z}_{D_1}} \prod_{\mathfrak{z}_2 \in \mathcal{Z}_{D_2}} (j(\mathfrak{z}_1) - j(\mathfrak{z}_2)) \in \mathbb{Z}, \tag{20.1}$$

where $-D_1$ and $-D_2$ are coprime fundamental discriminants, and for the algebraic numbers

$$j(\mathfrak{z}_1) - j(\mathfrak{z}_2), \tag{20.2}$$

where $\mathfrak{z}_1, \mathfrak{z}_2 \in \mathcal{Z}_D$ and $-D$ is an odd prime discriminant. Now we will state the factorization formula for the prime ideal generated by $j(\mathfrak{z}_1) - j(\mathfrak{z}_2)$.

We denote by K the imaginary quadratic field $\mathbb{Q}(\sqrt{-D})$, by h the class number, and by H the Hilbert class field of K. For a rational prime p with $(\frac{p}{D}) = -1$ let $\mathcal{P}_p = \{\mathfrak{P}_i\}_{i=1}^h$ be the set of prime ideals of H lying above p. Recall that we consider H as a subset of $\mathbb{C}$. The complex conjugation acts on H and also on the set $\mathcal{P}_p$. There exists the unique prime ideal in $\mathcal{P}_p$, say $\mathfrak{P}_1$, such that $\mathfrak{P}_1 = \overline{\mathfrak{P}}_1$ (see [4]). For each prime ideal $\mathfrak{P} \in \mathcal{P}_p$ there exists a unique element $\sigma \in \mathrm{Gal}(H/K)$ such that

$$\mathfrak{P}^\sigma = \mathfrak{P}_1. \tag{20.3}$$

Let $\mathcal{A} = \mathcal{A}(\mathfrak{P})$ be the ideal class of K that corresponds to σ under the Artin isomorphism.

Theorem 20.2. *([4, Proposition 3.8]) Let $\mathfrak{z}_1, \mathfrak{z}_2$ be two CM points of prime odd discriminant $-D$. Denote by $\mathcal{B}$ and $\mathcal{C}$, respectively, the ideal classes of $\mathbb{Q}(\sqrt{-D})$ containing the fractional ideals $\mathbb{Z}\mathfrak{z}_1 + \mathbb{Z}$ and $\mathbb{Z}\mathfrak{z}_2 + \mathbb{Z}$, respectively. Then for a rational prime p and a prime $\mathfrak{P}$ lying above p in the Hilbert class field H we have*

$$\mathrm{ord}_{\mathfrak{P}}(j(\mathfrak{z}_1) - j(\mathfrak{z}_2)) = \sum_{n=0}^{D} r_{\mathcal{B}\overline{\mathcal{C}}}(D-n)\, r_{\mathcal{B}\mathcal{C}\mathcal{A}^2}\left(\frac{n}{p}\right) \mathrm{ord}_p(n) \quad \text{if } \left(\frac{p}{D}\right) = -1, \tag{20.4}$$

$$\mathrm{ord}_{\mathfrak{P}}(j(\mathfrak{z}_1) - j(\mathfrak{z}_2)) = 0 \quad \text{if } \left(\frac{p}{D}\right) = 1. \tag{20.5}$$

Here $r_{\mathcal{C}}(n)$ stands for the number of integral ideals of norm n in the ideal class $\mathcal{C}$.

The key technique used in the proof of Theorem 20.2 is the computation of local height pairings between images of CM points on the modular curve X_1. These technique was further developed in the well-known paper [5].

20.2 CM values of higher Green's functions

For any integer $k > 1$ and subgroup $\Gamma \subset \mathrm{PSL}_2(\mathbb{Z})$ of finite index there is a unique function $G_k^{\Gamma\backslash\mathfrak{H}}$ on the product of two upper half planes $\mathfrak{H} \times \mathfrak{H}$ that satisfies the following conditions:

1. $G_k^{\Gamma\backslash\mathfrak{H}}$ is a smooth function on $\mathfrak{H} \times \mathfrak{H} \setminus \{(\tau, \gamma\tau), \tau \in \mathfrak{H}, \gamma \in \Gamma\}$ with values in $\mathbb{R}$.
2. $G_k^{\Gamma\backslash\mathfrak{H}}(\tau_1, \tau_2) = G_k^{\Gamma\backslash\mathfrak{H}}(\gamma_1\tau_1, \gamma_2\tau_2)$ for all $\gamma_1, \gamma_2 \in \Gamma$.
3. $\Delta_i G_k^{\Gamma\backslash\mathfrak{H}} = k(1-k)G_k^{\Gamma\backslash\mathfrak{H}}$, where Δ_i is the hyperbolic Laplacian with respect to the i-th variable, $i = 1, 2$.
4. $G_k^{\Gamma\backslash\mathfrak{H}}(\tau_1, \tau_2) = m \log|\tau_1 - \tau_2| + O(1)$ when τ_1 tends to τ_2 (m is the order of the stabilizer of τ_2, which is almost always 1).
5. $G_k^{\Gamma\backslash\mathfrak{H}}(\tau_1, \tau_2)$ tends to 0 when τ_1 tends to a cusp.

These functions have been introduces in [5] and are called *higher Green's functions*. Existence of a function satisfying 1–5 is shown in [5] by an explicit construction and the uniqueness follows from the maximum principle for subharmonic functions.

In the case $k = 1$, there is a unique function $G_1^{\Gamma\backslash\mathfrak{H}}(\tau_1, \tau_2)$ satisfying conditions 1-4 and condition 5 should be slightly modified in that case.

Let $S_{2k}(\Gamma)$ be the space of cusp forms of weight $2k$ for a group $\Gamma \subset \mathrm{PSL}_2(\mathbb{Z})$. A sequence of integer numbers $\boldsymbol{\lambda} = \{\lambda_m\}_{m=1}^{\infty}$, which belongs to $\oplus_{m=1}^{\infty}\mathbb{Z}$, is called a *relation* for $S_{2k}(\Gamma)$ if $\sum_{m=1}^{\infty} \lambda_m a_m = 0$ for any cusp form $f = \sum_{m=1}^{\infty} a_m q^m \in S_{2k}(\Gamma)$.

We shall study the CM values of the functions

$$G_{k,\boldsymbol{\lambda}}^{\mathrm{PSL}_2(\mathbb{Z})\backslash\mathfrak{H}} := \sum_{m=1}^{\infty} \lambda_m \, m^{k-1} \, G_k^{\mathrm{PSL}_2(\mathbb{Z})\backslash\mathfrak{H}}(\tau_1, \tau_2)|T_m$$

where T_m is a Hecke operator and $\boldsymbol{\lambda}$ is a relation for $S_{2k}(\mathrm{SL}_2(\mathbb{Z}))$. We shall write for simplicity $G_{k,\boldsymbol{\lambda}} := G_{k,\boldsymbol{\lambda}}^{\mathrm{PSL}_2(\mathbb{Z})\backslash\mathfrak{H}}$.

Conjecture 20.3. *Suppose $\boldsymbol{\lambda}$ is a relation for $S_{2k}(\mathrm{SL}_2(\mathbb{Z}))$. Then for any pair of CM points $\mathfrak{z}_1 \in \mathcal{Z}_{D_1}$ and $\mathfrak{z}_2 \in \mathcal{Z}_{D_2}$ there is an algebraic number α such that*

$$G_{k,\boldsymbol{\lambda}}(\mathfrak{z}_1, \mathfrak{z}_2) = (D_1 D_2)^{\frac{1-k}{2}} \log|\alpha|.$$

This conjecture was formulated in [5] for $D_1 = D_2$ and in [6] for any pair of CM points. Moreover, D. Zagier has made a more precise conjecture about the field of definition and prime factorization of this number α [12].

A. Mellit has proved Conjecture 20.1 in the case $k = 2$, $D_1 = -4$, and arbitrary D_2 in his doctoral dissertation [8]. In the thesis [11] this conjecture is proved for arbitrary k and any pair of CM points under the additional assumption that both CM points under consideration lie in the same imaginary quadratic field. Furthermore, in this case we have found the prime factorization of the algebraic number α.

20.3 Regularized theta lifts

The main technique used in [11] is the theory of regularized theta lifts, which was developed by Borcherds [1], Bruinier [2], Kudla [7], and others. Let us give a brief review of this theory.

Let V be a finite dimensional rational vector space and let $Q : V \to \mathbb{Q}$ be a quadratic form. The corresponding bilinear form on $V \times V$ is defined as

$$(x, y) := \frac{1}{2}(Q(x+y) - Q(x-y)).$$

Let $L \subset V$ be a lattice. The *dual lattice* of L is defined as

$$L' := \{x \in V | (x, L) \subset \mathbb{Z}\}.$$

The lattice L is said to be *even* if $(l, l) \in 2\mathbb{Z}$ for all $l \in L$. In that case $L' \subset L$ and L'/L is a finite abelian group. We denote by $\mathrm{Aut}(L)$ the group of those isometries of $L \otimes \mathbb{R}$ that fix each coset of L in L'.

Suppose now that L is an even lattice of signature (b^+, b^-). Denote by $\mathrm{Gr}^+(L)$ the positive Grassmanian, i.e. the set of positive-definite b^+-dimensional subspaces of $L \otimes \mathbb{R}$. For lattices L of signature $(2, b)$ the Grassmanian $\mathrm{Gr}^+(L)$ has the structure of a Hermitian symmetric domain.

Example 20.4. Consider the lattice $\mathrm{S}_2(\mathbb{Z})$ of 2×2 symmetric integral matrices equipped with the quadratic form $Q(x) := -\det(x)$. This is an even lattice of signature $(2, 1)$. The Grassmanian $\mathrm{Gr}^+(\mathrm{S}_2(\mathbb{Z}))$ is isomorphic to the upper half-plane $\mathfrak{H}$. There is an isomorphism $\mathfrak{H} \to \mathrm{Gr}^+(\mathrm{S}_2(\mathbb{Z}))$ given by

$$\tau \to v^+(\tau), \quad v^+(\tau) := \Re\begin{pmatrix} \tau^2 & \tau \\ \tau & 1 \end{pmatrix}\mathbb{R} + \Im\begin{pmatrix} \tau^2 & \tau \\ \tau & 1 \end{pmatrix}\mathbb{R},$$
$$v^+(\tau) \subset \mathrm{S}_2(\mathbb{Z}) \otimes \mathbb{R}. \tag{20.6}$$

The group $\mathrm{SL}_2(\mathbb{Z})$ acts on the lattice $\mathrm{S}_2(\mathbb{Z})$ by $\gamma(x) = \gamma x \gamma^t$, $x \in \mathrm{S}_2(\mathbb{Z})$, $\gamma \in \mathrm{SL}_2(\mathbb{Z})$. This action preserves the quadratic form Q.

Example 20.5. The lattice $M_2(\mathbb{Z})$ of 2×2 integral matrices equipped with the quadratic form $Q(x) := -\det(x)$ is an even unimodular lattice of signature $(2, 2)$. The Grassmanian $\mathrm{Gr}^+(M_2(\mathbb{Z}))$ is isomorphic to $\mathfrak{H} \times \mathfrak{H}$. An isomorphism $\mathfrak{H} \times \mathfrak{H} \to \mathrm{Gr}^+(M_2(\mathbb{Z}))$ is given by

$$(\tau_1, \tau_2) \to v^+(\tau_1, \tau_2), \quad v^+(\tau_1, \tau_2) := \Re\begin{pmatrix} \tau_1\tau_2 & \tau_1 \\ \tau_2 & 1 \end{pmatrix}\mathbb{R} + \Im\begin{pmatrix} \tau_1\tau_2 & \tau_1 \\ \tau_2 & 1 \end{pmatrix}\mathbb{R}, \tag{20.7}$$

$$v^+(\tau_1, \tau_2) \subset M_2(\mathbb{Z}) \otimes \mathbb{R}.$$

The group $\mathrm{SL}_2(\mathbb{Z}) \times \mathrm{SL}_2(\mathbb{Z})$ acts on the lattice $M_2(\mathbb{Z})$ by $(\gamma_1, \gamma_2)(x) = \gamma_1 x \gamma_2^t$, $x \in M_2(\mathbb{Z})$, $(\gamma_1, \gamma_2) \in \mathrm{SL}_2(\mathbb{Z}) \times \mathrm{SL}_2(\mathbb{Z})$.

Recall that the group $\mathrm{SL}_2(\mathbb{Z})$ has a double cover $\mathrm{Mp}_2(\mathbb{Z})$, called the *metaplectic group*. Elements of the metaplectic group can be written in the form

$$\left(\begin{pmatrix} a & b \\ c & d \end{pmatrix}, \sqrt{c\tau + d}\right)$$

where $\begin{pmatrix} a & b \\ c & d \end{pmatrix} \in \mathrm{SL}_2(\mathbb{Z})$ and $\sqrt{c\tau + d}$ is one of the two holomorphic functions of τ in the upper half-plane whose square is $c\tau + d$. The multiplication is defined so that the usual formulas for the transformation of modular forms of half integral weight work, i.e.

$$(A, f(\tau))(B, g(\tau)) = (AB, f(B(\tau))g(\tau))$$

for $A, B \in \mathrm{SL}_2(\mathbb{Z})$ and f, g are square roots of $c_A\tau + d_A$ and $c_B\tau + d_B$, respectively. The *Weil representation* ρ_L is a unitary representation of $\mathrm{Mp}_2(\mathbb{Z})$ on the vector space $\mathbb{C}^{L'/L}$ (see [1] for a definition). A *vector-valued modular form* of half-integral weight k and representation ρ_L is a function $f : \mathfrak{H} \to \mathbb{C}^{L'/L}$ that satisfies the following transformation law

$$f\left(\frac{a\tau + b}{c\tau + d}\right) = \sqrt{c\tau + d}^{\,2k} \rho_L\left(\begin{pmatrix} a & b \\ c & d \end{pmatrix}, \sqrt{c\tau + d}\right) f(\tau). \tag{20.8}$$

For an even lattice L of signature $(2, b)$ the vector-valued *Siegel theta function* $\Theta_L : \mathfrak{H} \times \mathrm{Gr}^+(L) \to \mathbb{C}^{L'/L}$ is defined as follows

$$\Theta_L(\tau, v^+) := y^{b/2} \sum_{\lambda \in L'/L} \mathbf{e}_\lambda \sum_{\ell \in L + \lambda} \mathrm{e}\big(Q(\ell_{v^+})\tau + Q(\ell_{v^-})\bar{\tau}\big). \tag{20.9}$$

Here $\{\mathbf{e}_\lambda\}_{\lambda \in L'/L}$ is the canonical basis of $\mathbb{C}^{L'/L}$, $\mathrm{e}(x) := e^{2\pi i x}$, and ℓ_{v^+} and ℓ_{v_-} denote the orthogonal projections of a vector $\ell \in V$ to the subspaces v^+ and $(v^+)^\perp$, respectively.

Theorem 4.1 in [1] asserts that $\Theta_L(\tau, v^+)$ satisfies the following properties

1. it is real-analytic in both variables;
2. transforms as a vector-valued modular form of weight $1 - b/2$ and representation ρ_L with respect to the variable τ;
3. is invariant in the variable v^+ under the group $\mathrm{Aut}(L) \subset \mathrm{O}(L \otimes \mathbb{R})$.

Let $\mathfrak{M}_k(\rho_L)$ be the space of real-analytic (vector-valued) modular forms of weight k and representation ρ_L. For $f \in \mathfrak{M}_{1-b/2}(\rho_L)$ the *theta lift* Φ_L is defined as follows:

$$\Phi_L(v^+, f) := \int\limits_{\mathrm{SL}_2(\mathbb{Z})\backslash\mathfrak{H}} \left(f(\tau), \overline{\Theta_L(\tau, v^+)}\right)_{L'/L} y^{-1-b/2}\, dx\, dy \tag{20.10}$$

(here the product $(\cdot,\cdot)_{L'/L}$ is defined by $\left(\mathbf{e}_\mu, \mathbf{e}_\nu\right)_{L'/L} = \delta_{\mu,\nu}$).

The integral (20.10) is often divergent and has to be regularized. For $f, g \in \mathfrak{M}_k(\rho_L)$ we define the *regularized Petersson product* as follows

$$(f, g)^{\mathrm{reg}} := \lim_{T\to\infty} \int_{\mathcal{F}_T} \left(f(\tau), \overline{\Theta_L(\tau, v^+)}\right)_{L'/L} y^{-1-b/2}\, dx\, dy,$$

where

$$\mathcal{F}_T := \left\{\tau \in \mathfrak{H} \middle| -1/2 < \Re(\tau) < 1/2,\ |\tau| > 1,\ \text{and } \Im(\tau) < T\right\}$$

is a truncated fundamental domain of $\mathrm{SL}_2(\mathbb{Z})$.

The *regularized theta lift* of $f \in \mathfrak{M}_k(\rho_L)$ is defined as

$$\Phi_L(v^+, f) := (f(\tau), \Theta_L(v^+, \tau))^{\mathrm{reg}}.$$

It is clear that the regularized integral $\Phi_L(v^+, f)$ is invariant under the action of the group $\mathrm{Aut}(L)$ on the Grassmannian $\mathrm{Gr}^+(L)$.

The regularized theta lift is well-defined for a much larger class of modular forms than the classical theta lift. For instance, regularized theta lift can be computed for *weakly holomorphic cusp forms*. A weakly holomorphic (vector-valued) modular form is called a *weakly holomorphic cusp form* if its constant term in the Fourier expansion is zero. Let $S^!_k(\rho)$ be the space of weakly holomorphic cusp forms of weight k and representation ρ.

Theorem 20.6. *([1, Theorem 13.3]) Let* $f \in S^!_{1-b/2}(\rho_L)$, *let*

$$f(\tau) = \sum_{\lambda\in L'/L} \mathbf{e}_\lambda \sum_{n\gg-\infty} c_\lambda(n)\, \mathrm{e}(n\tau)$$

and suppose that the Fourier coefficients $c_\lambda(n) \in \mathbb{Z}$ *for* $n \leq 0$. *Then there is a meromorphic function* $\Psi_L(Z, f)$ *on* $\mathrm{Gr}^+(L)$, *satisfying the following conditions:*

1. Ψ_L *is an automorphic function for the group* $\mathrm{Aut}(L)$ *with respect to a unitary character of* $\mathrm{Aut}(L)$;
2. *The zeros and the poles of* Ψ_L *lie on the rational quadratic divisors* $\ell^\perp$ *for* $\ell \in L$ *with* $Q(\ell) < 0$. *The order of* Ψ_L *at* $l^\perp$ *is equal to*
$$-4 \sum_{\substack{x \in \mathbb{R}^+:\\ xl \in L'}} c_{xl}\big(Q(xl)\big);$$
3.
$$\Phi_L(v^+, f) = \log|\Psi_L(v^+, f)|;$$
4. *One can write an explicit infinite product expansion converging in a neighborhood of each cusp of* $\mathrm{Gr}^+(L)$.

Let us consider several examples of regularized theta lifts associated to the lattice $\mathrm{M}_2(\mathbb{Z})$. Recall that the Grassmanian $\mathrm{Gr}^+(\mathrm{M}_2(\mathbb{Z}))$ is isomorphic to $\mathfrak{H} \times \mathfrak{H}$. The theta function $\Theta(\tau, \tau_1, \tau_2) := \Theta_{\mathrm{M}_2(\mathbb{Z})}(\tau, v^+(\tau_1, \tau_2))$ is a real analytic modular form of weight 0 in each variable.

Example 20.7. Set $J_1 := j(\tau) - 744 = q^{-1} + \sum_{n=1}^\infty a_n q^n$. By Theorem 20.6
$$\Phi_{\mathrm{M}_2(\mathbb{Z})}(v^+(\tau_1, \tau_2), J_1) = \log|j(\tau_1) - j(\tau_2)|.$$

Example 20.8. Let λ be a relation for $S_{2k}(\mathrm{SL}_2(\mathbb{Z}))$. Then by Serre duality there exists the unique $g_\lambda \in M^!_{2-2k}$ with the Fourier expansion $g_\lambda = \sum_m \lambda_m q^{-m} + \sum_{n \geq 0} b_n q^n$. It has been proved in [2] that
$$\Phi_{\mathrm{M}_2(\mathbb{Z})}(v^+(\tau_1, \tau_2), R^{k-1}(g_\lambda)) = G_{k,\lambda}(\tau_1, \tau_2).$$

This result provides the connection between regularized theta lifts and higher Green's functions. Here $R_k = \frac{1}{2\pi i}(\frac{\partial}{\partial \tau} + \frac{k}{\tau - \bar{\tau}})$ is the raising Maass operator and $R^{k-1} = R_{-2} \circ \cdots \circ R_{2-2k}$.

20.4 CM-values of the theta lifts

In this section we shall study the CM values of the theta lifts.

Let L be an even lattice of signature $(2, b_1)$ and let $N \subset L$ be an even lattice of signature $(2, b_2)$. Then there is an embedding $i : \mathrm{Gr}^+(N)/\mathrm{Aut}(N) \hookrightarrow \mathrm{Gr}^+(L)/\mathrm{Aut}(L)$.

Example 20.9. The pairs of the CM points $\mathfrak{z}_1, \mathfrak{z}_2$ such that $\mathbb{Q}(\mathfrak{z}_1) = \mathbb{Q}(\mathfrak{z}_2)$ are in one-to-one correspondence with signature $(2, 0)$ rational subspaces of $M_2(\mathbb{Q})$.

For simplicity, assume that $L = M \oplus N$. Then for $f \in \mathfrak{M}_{1-b_1/2}(\rho_L)$

$$\Phi_L(i(v^+), f) = \Phi_N(v^+, (f, \overline{\Theta}_M)_{M'/M}).$$

Therefore, the CM-values of theta lifts can be expressed as regularized Petersson products between (real-analytic) weight one modular forms and binary theta series.

In [11], we have obtained the following result for regularized Petersson products of such type.

Theorem 20.10. *([11, Theorem 5.6]) Let N be an even lattice of signature $(2, 0)$ and let $f \in S_1^!(\rho_N)_{\mathbb{Z}}$. Then*

$$(f, \Theta_N)_{\mathrm{reg}} = \log|\alpha| \tag{20.11}$$

for some $\alpha \in \overline{\mathbb{Q}}$.

After this result had been announced [10], the author learned that a similar result was obtained independently in a slightly different context by W. Duke and Y. Li [3].

Furthermore, we obtained the prime factorization of the number α.

Theorem 20.11. *([11, Theorem 5.8]) Suppose that $-D$ is an odd prime discriminant and $K := \mathbb{Q}(\sqrt{-D})$. Given a fractional ideal $\mathcal{B} \subset K$, consider the following even lattice*

$$(N, Q) = \left(\mathcal{B}, \frac{\mathrm{N}_{K/\mathbb{Q}}(\cdot)}{\mathrm{N}_{K/\mathbb{Q}}(\mathcal{B})}\right).$$

Let $f \in S_1^!(\rho_N)$ be a weakly holomorphic cusp form with the Fourier expansion

$$f = \sum_{\nu \in \mathbb{Z}/D\mathbb{Z}} \mathbf{e}_\nu \sum_{\substack{t \in \mathbb{Z} \\ t \gg -\infty}} c_\nu(t)\, \mathrm{e}\Big(\frac{t}{D}\tau\Big).$$

Suppose that the Fourier coefficients $c_\nu(n) \in \mathbb{Z}$ for $n \leq 0$. Then there exists a number α in the Hilbert class field H of K such that

$$(f, \Theta_{\mathcal{B}})^{\mathrm{reg}} = \log|\alpha|.$$

Moreover, for a rational prime p and a prime $\mathfrak{P}$ lying above p in the Hilbert class field H we have

$$\mathrm{ord}_{\mathfrak{P}}(\alpha) = \sum_{t<0} \sum_{\nu \in \mathbb{Z}/D\mathbb{Z}} c_\nu(t)\, r_{\mathcal{B}\mathcal{A}^2}\left(\frac{-t}{p}\right) \mathrm{ord}_p(t) \quad \text{in the case } \left(\frac{p}{D}\right) = -1, \tag{20.12}$$

$$\mathrm{ord}_{\mathfrak{P}}(\alpha) = 0 \quad \text{in the case } \left(\frac{p}{D}\right) = 1. \tag{20.13}$$

Here $r_{\mathcal{C}}(n)$ denotes the number of integral ideals on norm n in the ideal class $\mathcal{C}$.

Finally, the following result relates CM values of higher Green's functions and regularized Petersson products (20.11).

Theorem 20.12. *([11, Theorem 5.4]) Let $\mathfrak{z}_1, \mathfrak{z}_2$ be two CM points of negative fundamental discriminant $-D$. Denote by $\mathcal{B}$ and $\mathcal{C}$, respectively, the ideal classes of $\mathbb{Q}(\sqrt{-D})$ containing the fractional ideals $\mathbb{Z}\mathfrak{z}_1 + \mathbb{Z}$ and $\mathbb{Z}\mathfrak{z}_2 + \mathbb{Z}$, respectively. Let $\boldsymbol{\lambda}$ be a relation for S_{2k}. Then*

$$G_{k,\boldsymbol{\lambda}}(\mathfrak{z}_1, \mathfrak{z}_2) = \left([g_{\boldsymbol{\lambda}}, \Theta_{\mathcal{B}\overline{\mathcal{C}}}]_{k-1}, \Theta_{\mathcal{B}\mathcal{C}}\right)^{\mathrm{reg}}, \tag{20.14}$$

where $g_{\boldsymbol{\lambda}}$ is defined in Example 20.8 and $[\cdot,\cdot]_{k-1}$ denotes the $(k-1)$-st Rankin-Cohen brackets.

This result combined with Theorem 20.10 proves Conjecture 20.1 for each pair of CM points lying in the same imaginary quadratic field. Finally, Theorems 20.11 and 20.12 imply the factorization formula for $\exp(G_{k,\boldsymbol{\lambda}}(\mathfrak{z}_1, \mathfrak{z}_2))$.

References

[1] R. E. Borcherds, *Automorphic forms with singularities on Grassmannians*, Invent. Math. 132, 491–562 (1998).

[2] J. H. Bruinier, *Borcherds products on $O(2,l)$ and Chern classes of Heegner divisors*, Springer Lecture Notes in Mathematics 1780, Springer-Verlag (2002).

[3] W. Duke, Y. Li, *Mock-modular forms of weight one*, preprint, `http://www.math.ucla.edu/~wdduke/preprints/mock%20weight%20one.pdf`, (2012).

[4] B. Gross and D. Zagier, *On singular moduli*, J. Reine Angew. Math. 355, 191–220 (1985).

[5] B. Gross, D. Zagier, *Heegner points and derivatives of L-series*, Invent. Math. 84, 225–320 (1986).
[6] B. Gross, W. Kohnen, D. Zagier, *Heegner points and derivatives of L-series. II* Math. Ann. 278, 497–562 (1987).
[7] S. Kudla, *Integrals of Borcherds forms*, Compositio Math. 137 (2003), 293–349.
[8] A. Mellit, *Higher Greens functions for modular forms*, PhD dissertation, University of Bonn (2008).
[9] A. Néron, *Quasi-fonctions et hauteurs sur les variétés abéliennes,* Ann. of Math. 82, 249–331 (1965).
[10] M. Viazovska, *CM Values of Higher Green's Functions*, arXiv:1110.4654 [math.NT] (2011).
[11] M. Viazovska, *Modular functions and special cycles*, PhD dissertation, University of Bonn (2013).
[12] D. Zagier, Private communication.

Printed in the United States
By Bookmasters